Universe: Feature Papers 2023—Cosmology

Universe: Feature Papers 2023—Cosmology

Kazuharu Bamba

Basel • Beijing • Wuhan • Barcelona • Belgrade • Novi Sad • Cluj • Manchester

Editor
Kazuharu Bamba
Faculty of Symbiotic Systems Science
Fukushima University
Fukushima
Japan

Editorial Office
MDPI AG
Grosspeteranlage 5
4052 Basel, Switzerland

This is a reprint of articles from the Special Issue published online in the open access journal *Universe* (ISSN 2218-1997) (available at: www.mdpi.com/journal/universe/special_issues/W5S81790I9).

For citation purposes, cite each article independently as indicated on the article page online and as indicated below:

Lastname, A.A.; Lastname, B.B. Article Title. *Journal Name* **Year**, *Volume Number*, Page Range.

ISBN 978-3-7258-2322-2 (Hbk)
ISBN 978-3-7258-2321-5 (PDF)
doi.org/10.3390/books978-3-7258-2321-5

Contents

About the Editor

Kazuharu Bamba

Kazuharu Bamba is a Professor at Fukushima University in Japan. He has been working on inflationary cosmology, dark energy problems, and modified gravity theories. He has been an editor for several MDPI journals, i.e., *Universe*, *Symmetry*, and *Entropy*.

Preface

Various recent precise cosmological observations, such as type Ia supernovae (NaIa), the cosmic microwave background (CMB) radiation, the large-scale structure of the universe, the baryon acoustic oscillations (BAO), and the effect of weak lensing, have suggested the accelerated expansion of the current universe, in addition to the early universe at the inflationary stage. For the spatially flat universe at the present time, it is estimated that the energy density of the present universe consists of dark energy (about 70%), dark matter (about 25%), and baryons (about 5%).

Two main approaches for studying the late-time cosmic acceleration have been proposed. The first is introducing an unknown energy component called dark energy with negative pressure in general relativity. The second is extending a gravity theory from general relativity on large scales. This second method is a type of geometrical interpretation of dark energy components.

Furthermore, there are a number of important subjects in modern cosmology, including inflationary cosmology, bounce cosmology, cosmic string/superstring theory, quantum cosmology, big bang nucleosynthesis (BBN), observational cosmology, standard LCDM cosmology, large-scale structure of the universe, dark matter, dark energy, modified gravity, axion cosmology, cosmic acceleration, cosmological constants, cosmological perturbation theory, cosmic microwave background, Hubble's law/constants, machine learning, and cosmology.

The main aim of this Special Issue is to understand the various recent subjects in modern cosmology and related foundations of physics. The organization of this book (twenty research articles and one review) is as follows: The first part (three articles) details inflationary cosmology and physics in the early universe. The second part (three articles) concerns observational cosmology. The third part (eleven articles) covers dark energy problems and extended theories of gravitation from general relativity. The fourth part (three articles) discusses gravitational waves and black holes. The final part (one review) presents an overview of the Hubble tension.

Finally, I would like to sincerely acknowledge MDPI and am greatly appreciative of the Managing Editor, Ms. Athena Li, for her very kind support and warm assistance during this project. Moreover, I am highly grateful to the Editor-in-Chief, Professor Dr. Lorenzo Iorio, for giving me the chance to serve as Guest Editor of this Special Issue. I would also like to express my sincere gratitude to all the authors for their article submissions to this Special Issue of *Universe*.

Kazuharu Bamba
Editor

Editorial

Editorial to the Special Issue "Universe: Feature Papers 2023—Cosmology"

Kazuharu Bamba 

Faculty of Symbiotic Systems Science, Fukushima University, Fukushima 960-1296, Japan; bamba@sss.fukushima-u.ac.jp

Citation: Bamba, K. Editorial to the Special Issue "Universe: Feature Papers 2023—Cosmology". *Universe* **2024**, *10*, 380. https://doi.org/10.3390/universe10100380

Received: 18 September 2024
Accepted: 23 September 2024
Published: 27 September 2024

According to recent observational data, including Supernovae Ia (SNe Ia) [1,2], the cosmic microwave background (CMB) radiation [3–8], the large-scale structure (LSS) of the universe [9–11], baryon acoustic oscillations (BAOs) [12,13], and the weak lensing effect [14–18], not only in the early universe of the inflationary stage [19–22] but also in the late-time universe, the cosmic expansion is accelerating. Moreover, the recent Planck observations [7,8] suggest that if the universe is spatially flat, the energy density of the current universe consists of dark energy (about 70%) with negative pressure and dark matter (about 25%) with only a gravitational interaction and baryon acoustic oscillations (about 5%). The identities of dark energy have not yet been understood so clearly, although a large number of studies have been carried out very actively.

Through the future observations by the Euclid satellite [23] of the European Space Agency (ESA) [24–30], the Roman Space Telescope [31], the Simons Observatory [32,33], and the James Webb Space Telescope [34,35], a further understanding of modern cosmology will be developed in more precise and detail.

In addition, by the direct detection of gravitational waves [36,37], the era of gravitational wave cosmology has kicked off. It is expected that not only gravitational waves originated from astrophysical objects but also that primordial gravitational waves originated from the early universe, including the inflationary stage and cosmological first-order phase transitions, such as the electroweak phase transition (EWPT) and the QCD phase transition (QCDPT) [38–44].

Two main methods to explain the late-time accelerated expansion of the spatially flat universe have been proposed in the literature. One approach is to explore the unknown energy component with negative pressure, the so-called dark energy, in the framework of general relativity. The other is to investigate the extension of a theory of gravitation from general relativity at cosmological scales. Such a way can be regarded as geometrical dark energy. Regarding the latter approach, many extended theories of gravitation have been proposed (for reviews in terms of extended theories of gravitation, including their applications to astrophysics and the properties of gravitational theories, as well as the dark energy problem (see, for instance, refs. [45–85]).

Furthermore, there are various important subjects in modern cosmology, such as physics in the early universe [86–97], cosmological perturbation theory [98–101], the Swampland picture [102–104], cosmological constant [105], neutrino cosmology [106–110], axion cosmology [111], primordial black holes [112–116], dark matter [117–128], primordial magnetic fields [129–133], Big Bang nucleosynthesis (BBN) [134,135], the Hubble tension [136–142], 21 cm cosmology [143,144], gravitational lensing effects [145–147], and other related topics, including applications to astrophysical aspects [148–155].

In the present Special Issue, titled "Universe: Feature Papers 2023—Cosmology" of Universe, twenty-one original research manuscripts in terms of modern cosmology are collected. It was organized as follows. In the first part, the subjects of physics in the early universe [156,157] including inflationary cosmology [158] are explored. In the second part, the generic topics of observational cosmology [159–161] are studied. In the third part, the issues of dark energy [162–168] and extended theories of gravity from general relativity [169–172] are investigated. In the fourth part, the themes of gravitational waves [173] and physics of black holes [174,175] are given. As a kind of summary of this Special Issue, in the final part, a review on the cosmological recent problem of the Hubble tension [176] is presented. In the following, twenty research articles and one review are overviewed briefly.

In ref. [156], through a metric procedure, the so-called big bounce dynamics of an isotropic spacetime are studied by introducing a scalar field with its self-interaction. For the case that there is no potential, the cosmic expansion and collapse can occur along with the mode of positive and negative frequencies for the Wheeler–DeWitt equation. On the other hand, if there exists a potential, a transition from a state with a positive frequency to that with negative one can happen. It is demonstrated that the probability of a transition from the collapsing phase to the expanding one can be analyzed.

In ref. [157], a scenario to realize the de Sitter expansion of the early universe is proposed, which originated from stochastic gravitational waves based on the arguments of the Lorentzian and Euclidean spacetimes and the Wick rotation.

In ref. [158], through the Bayesian analysis, the cosmological evolution of the extrinsic energy density is numerically examined for the inflationary universe in the context of the five-dimensional spacetime. The effective potential leading to inflation is derived, originating from the extrinsic geometry. The possibility for a unified scenario of inflation with the late-time cosmic acceleration is also argued.

In ref. [159], the propagation of electromagnetic waves is studied for homogeneous and isotropic spacetimes. For plane and spherical electromagnetic waves, exact solutions are derived. Moreover, the redshift, the change in amplitude, and the dispersion of the electromagnetic waves are examined. Furthermore, the relation between the equation of electromagnetic waves and the Proca equation is argued. The importance of the relation is discussed for physics in the early universe.

In ref. [160], the methods of the number count are investigated in the context of observational cosmology, such as the number of galaxies. In particular, with the method of the so-called geodesic-light-cone gauge, the procedures to count the number are investigated.

In ref. [161], the basis of genetic algorithms and the way of parameter estimations for cosmological models are explained as a complementary technique with a conventional method of the Markov chain Monte Carlo. It is shown that the space of the parameter can be explored effectively by using models of dark energy.

In ref. [162], a possible origin of the cosmological constant is proposed from a boundary condition. The relation between the issue of the cosmological constant and the Hubble tension is investigated. It is argued that the cosmological constant represents a covariant integration constant arising from a spatial boundary condition, which is only applicable to a three-dimensional bounded subspace representing our universe.

In ref. [163], a fractional differential equation in a quantum K-essence scalar field is studied in homogeneous and isotropic spacetime. For the Wheeler–DeWitt equation of the scalar field in quantum cosmology, a fractional differential equation is found, and the solutions of the equation are examined.

In ref. [164], cosmology is studied in the Einstein–Newcomb–de Sitter space, where antipodal points are topologically identified. In particular, with the type Ia supernovae data of Pantheon+ and the sample data of gamma-ray bursts, the fitting analysis to this cosmological model is performed. As a result, it is reported that the minimum value of the χ^2 analysis for this model could be smaller than that for the Λ CDM model.

In ref. [165], with the autonomous system analysis, the phase-space diagram of the cosmological evolutions for a scalar field model is investigated. For two types of potentials, the stationary points of the phase-space are examined. It is demonstrated that for the case in which the potential is close to asymptotically constant, the de Sitter expansion of the universe can be realized, while for that with the exponential form of the potential, the de Sitter cosmic expansion can occur only in the infinity limit.

In ref. [166], a cosmological transition from the anti-de Sitter phase to the de Sitter one is analyzed based on a motivation to explain the so-called Hubble tension. It is assumed that there exist two kinds of fluids playing a role of the dark energy component. One of these two fluids can lead to the anti-de Sitter phase in the early universe. The self-interaction of these fluids may influence the energy density of the effective dark energy to realize late-time cosmic acceleration.

In ref. [167], a relation between a scalar field theory of K-essence and the Vaidya metric is discussed. Especially, a non-canonical action classified into a type of Dirac–Born–Infeld one is considered. The geodesics within a comoving plane are examined for certain forms of the mass function. In addition, for a kind of Vaidya spacetime, a solution of wormholes, the tunneling effect, the existence of (quasi-)circular orbits of the event horizon, and the central singularity of the spacetime are analyzed.

In ref. [168], the (im)possibility of the discrimination of the de Sitter phase of the cosmic expansion from the following stationary phase is examined. In relation with this issue, a signal (a kind of non-standard photons) related to dark energy is argued.

In ref. [169], as a possibility to avoid the initial singularity in the early universe, a context of Loop Quantum Cosmology is considered. In particular, the Friedmann equations for $f(R)$ gravity in metric formalism are analyzed. As a result, the effective actions with covariance, in which a bounce can be realized, are derived through a reduction method.

In ref. [170], cosmological fluctuations for cosmic microwave background radiations are studied in delta gravity, which is proposed as an alternative theory of gravity to general relativity to explain the type Ia SN data. With a semi-analytic method, the temperature–temperature power spectrum of the scalar mode of the cosmological fluctuations is analyzed by using the gauge transformations of the spacetime metric as well as perfect fluid, so that observational constraints on delta gravity can be obtained.

In ref. [171], a homogeneous and anisotropic Bianchi type I universe is explored for the local limit in nonlocal gravity. In such a theory of gravity, there exists a function representing susceptibility, which vanishes in the limit of general relativity. The influence of the function of susceptibility is investigated on the accelerated anisotropic expansion of the universe.

In ref. [172], late-time cosmology is studied in dilaton gravity, which consists of the scalar curvature and the trace of the energy–momentum tensor of the dilatonic field. In particular, the issue of the Hubble tension is discussed by using cosmological observational data.

In ref. [173], through the Bayesian statistical analysis with the pulsar timing of the NANOGrav 12.5-year data, the curvature perturbations from inflation and the mass of primordial black holes are investigated. The possibility that primordial black holes play a role of a fraction of dark matter is also studied. Moreover, detectability with future experiments of gravitational waves is discussed.

In ref. [174], the existence of white holes is argued. In general relativity, a white hole is recognized as a reversed process of the solution of a collapsing black hole. This phenomenon leads to the crossing of the event horizon of the black hole, at which causality is broken. It is indicated that the Big Bang of our universe is a kind of black hole, and not a white hole. It is explained why the crossing of the event horizon can be realized from the inside to the outside if an event decelerates when the cosmic expansion accelerates by discussing that an apparent solution of a white hole has an exterior of a regular solution of a Schwarzschild black hole.

In ref. [175], for the Schwarzschild–Finsler–Randers spacetime, the equation of the geodesic deviation and the Raychaudhuri equation are explored. In this spacetime, there exist multiple curvature tensors, enabling the Raychaudhuri equation to have extra degrees of freedom. The astrophysical results of the limit of the weak field and the limit of Newtonian mechanics are also presented.

In ref. [176], the issue of the "Hubble tension" is reviewed in detail, including the history of the Hubble constant. The significance of the discovery of our universe's expansion and the pioneering measurement of the cosmic expansion rate are explained. For the measurement, high-level technology and small uncertainties in terms of parameters are important. A review of various method to measure the Hubble constant is also necessary at the present time.

In conclusion, we believe that the twenty-one articles introduced above in the present Special Issue are useful for relevant future studies for various aspects of modern physics and cosmology.

Funding: This work was supported by the JSPS KAKENHI Grant Number JP21K03547 and 24KF0100.

Acknowledgments: Contributions of all the authors are highly appreciated by the Guest Editor (Kazuharu Bamba) of this Special Issue.

Conflicts of Interest: The author declares no conflicts of interest.

References

1. Perlmutter, S. et al. [SNCP Collaboration]. Measurements of Omega and Lambda from 42 High-Redshift Supernovae. *Astrophys. J.* **1999**, *517*, 565. [CrossRef]
2. Riess, A.G. et al. [Supernova Search Team Collaboration]. Observational Evidence from Supernovae for an Accelerating Universe and a Cosmological Constant. *Astron. J.* **1998**, *116*, 1009. [CrossRef]
3. Spergel, D.N. et al. [WMAP Collaboration]. First Year Wilkinson Microwave Anisotropy Probe (WMAP) Observations: Determination of Cosmological Parameters. *Astrophys. J. Suppl.* **2003**, *148*, 175. [CrossRef]
4. Spergel, D.N. et al. [WMAP Collaboration]. Wilkinson Microwave Anisotropy Probe (WMAP) three year results: Implications for cosmology. *Astrophys. J. Suppl.* **2007**, *170*, 377. [CrossRef]
5. Komatsu, E. et al. [WMAP Collaboration]. Five-Year Wilkinson Microwave Anisotropy Probe (WMAP) Observations: Cosmological Interpretation. *Astrophys. J. Suppl.* **2009**, *180*, 330. [CrossRef]
6. Komatsu, E. et al. [WMAP Collaboration]. Seven-Year Wilkinson Microwave Anisotropy Probe (WMAP) Observations: Cosmological Interpretation. *Astrophys. J. Suppl.* **2011**, *192*, 18. [CrossRef]
7. Aghanim, N. et al. [Planck]. Planck 2018 results. VI. Cosmological parameters,. *Astron. Astrophys.* **2020**, *641*, A6; Erratum in *Astron. Astrophys.* **2021**, *652*, C4.
8. Akrami, Y. et al. [Planck]. Planck 2018 results. X. Constraints on inflation. *Astron. Astrophys.* **2020**, *641*, A10. [CrossRef]
9. Tegmark, M. et al. [SDSS Collaboration]. Cosmological parameters from SDSS and WMAP. *Phys. Rev. D* **2004**, *69*, 103501. [CrossRef]
10. Seljak, U. et al. [SDSS Collaboration]. Cosmological parameter analysis including SDSS Ly-alpha forest and galaxy bias: Constraints on the primordial spectrum of fluctuations, neutrino mass, and dark energy. *Phys. Rev. D* **2005**, *71*, 103515. [CrossRef]
11. Tsagas, C.G.; Challinor, A.; Maartens, R. Relativistic cosmology and large-scale structure. *Phys. Rept.* **2008**, *465*, 61–147. [CrossRef]
12. Eisenstein, D.J. et al. [SDSS Collaboration]. Detection of the Baryon Acoustic Peak in the Large-Scale Correlation Function of SDSS Luminous Red Galaxies. *Astrophys. J.* **2005**, *633*, 560. [CrossRef]
13. Alam, S. et al. [eBOSS]. Completed SDSS-IV extended Baryon Oscillation Spectroscopic Survey: Cosmological implications from two decades of spectroscopic surveys at the Apache Point Observatory. *Phys. Rev. D* **2021**, *103*, 083533. [CrossRef]
14. Jain, B.; Taylor, A. Cross-correlation Tomography: Measuring Dark Energy Evolution with Weak Lensing. *Phys. Rev. Lett.* **2003**, *91*, 141302. [CrossRef]
15. Munshi, D.; Valageas, P.; Waerbeke, L.V.; Heavens, A. Cosmology with Weak Lensing Surveys. *Phys. Rept.* **2008**, *462*, 67–121. [CrossRef]
16. Troxel, M.A.; Ishak, M. The Intrinsic Alignment of Galaxies and its Impact on Weak Gravitational Lensing in an Era of Precision Cosmology. *Phys. Rept.* **2014**, *558*, 1–59. [CrossRef]
17. Abbott, T.M.C. et al. [DES]. Dark Energy Survey year 1 results: Cosmological constraints from galaxy clustering and weak lensing. *Phys. Rev. D* **2018**, *98*, 043526. [CrossRef]
18. Abbott, T.M.C. et al. [DES]. Dark Energy Survey Year 3 results: Cosmological constraints from galaxy clustering and weak lensing. *Phys. Rev. D* **2022**, *105*, 023520. [CrossRef]
19. Guth, A.H. The Inflationary Universe: A Possible Solution to the Horizon and Flatness Problems. *Phys. Rev. D* **1981**, *23*, 347. [CrossRef]
20. Sato, K. First Order Phase Transition of a Vacuum and Expansion of the Universe. *Mon. Not. Roy. Astron. Soc.* **1981**, *195*, 467–479. [CrossRef]
21. Starobinsky, A.A. A New Type of Isotropic Cosmological Models Without Singularity. *Phys. Lett.* **1980**, *91B*, 99. [CrossRef]
22. Linde, A.D. A New Inflationary Universe Scenario: A Possible Solution of the Horizon, Flatness, Homogeneity, Isotropy and Primordial Monopole Problems. *Phys. Lett.* **1982**, *108B*, 389. [CrossRef]
23. Available online: https://www.esa.int/Science_Exploration/Space_Science/Euclid (accessed on 16 September 2024).
24. Laureijs, R. et al. [EUCLID]. Euclid Definition Study Report. *arXiv* **2011**, arXiv:1110.3193.

25. Amendola, L. et al. [Euclid Theory Working Group]. Cosmology and fundamental physics with the Euclid satellite. *Living Rev. Rel.* **2013**, *16*, 6. [CrossRef] [PubMed]
26. Amendola, L.; Appleby, S.; Avgoustidis, A.; Bacon, D.; Baker, T.; Baldi, M.; Bartolo, N.; Blanchard, A.; Bonvin, C.; Borgani, S.; et al. Cosmology and fundamental physics with the Euclid satellite. *Living Rev. Rel.* **2018**, *21*, 2.
27. Blanchard, A. et al. [Euclid]. Euclid preparation. VII. Forecast validation for Euclid cosmological probes. *Astron. Astrophys.* **2020**, *642*, A191. [CrossRef]
28. Nesseris, S. et al. [Euclid]. Euclid: Forecast constraints on consistency tests of the ΛCDM model. *Astron. Astrophys.* **2022**, *660*, A67. [CrossRef]
29. Casas, S. et al. [Euclid]. Euclid: Constraints on f(R) cosmologies from the spectroscopic and photometric primary probes. *arXiv* **2023**, arXiv:2306.11053.
30. Ballardini, M. et al. [Euclid]. *Euclid*: The search for primordial features. *arXiv* **2024**, arXiv:2309.17287.
31. Eifler, T.; Miyatake, H.; Krause, E.; Heinrich, C.; Miranda, V.; Hirata, C.; Xu, J.; Hemmati, S.; Simet, M.; Capak, P.; et al. Cosmology with the Roman Space Telescope–multiprobe strategies. *Mon. Not. Roy. Astron. Soc.* **2021**, *507*, 1746–1761. [CrossRef]
32. Available online: https://simonsobservatory.org/ (accessed on 16 September 2024).
33. Ade, P. et al. [Simons Observatory]. The Simons Observatory: Science goals and forecasts. *JCAP* **2019**, *2*, 056.
34. Available online: https://webbtelescope.org/home (accessed on 16 September 2024).
35. Gardner, J.P.; Mather, J.C.; Clampin, M.; Doyon, R.; Greenhouse, M.A.; Hammel, H.B.; Hutchings, J.B.; Jakobsen, P.; Lilly, S.J.; Long, K.S.; et al. The James Webb Space Telescope. *Space Sci. Rev.* **2006**, *123*, 485. [CrossRef]
36. Abbott, B.P. et al. [LIGO Scientific and Virgo]. Observation of Gravitational Waves from a Binary Black Hole Merger. *Phys. Rev. Lett.* **2016**, *116*, 061102. [CrossRef]
37. Abbott, B.P. et al. [LIGO Scientific and Virgo]. GW170817: Observation of Gravitational Waves from a Binary Neutron Star Inspiral. *Phys. Rev. Lett.* **2017**, *119*, 161101. [CrossRef]
38. Abbott, B.P. et al. [LIGO Scientific, Virgo, Fermi GBM, INTEGRAL, IceCube, AstroSat Cadmium Zinc Telluride Imager Team, IPN, Insight-Hxmt, ANTARES, Swift, AGILE Team, 1M2H Team, Dark Energy Camera GW-EM, DES, DLT40, GRAWITA, Fermi-LAT, ATCA, ASKAP, Las Cumbres Observatory Group, OzGrav, DWF (Deeper Wider Faster Program), AST3, CAASTRO, VINROUGE, MASTER, J-GEM, GROWTH, JAGWAR, CaltechNRAO, TTU-NRAO, NuSTAR, Pan-STARRS, MAXI Team, TZAC Consortium, KU, Nordic Optical Telescope, ePESSTO, GROND, Texas Tech University, SALT Group, TOROS, BOOTES, MWA, CALET, IKI-GW Follow-up, H.E.S.S., LOFAR, LWA, HAWC, Pierre Auger, ALMA, Euro VLBI Team, Pi of Sky, Chandra Team at McGill University, DFN, ATLAS Telescopes, High Time Resolution Universe Survey, RIMAS, RATIR and SKA South Africa/MeerKAT]. Multi-messenger Observations of a Binary Neutron Star Merger. *Astrophys. J. Lett.* **2017**, *848*, L12. [CrossRef]
39. Barack, L.; Cardoso, V.; Nissanke, S.; Sotiriou, T.P.; Askar, A.; Belczynski, C.; Bertone, G.; Bon, E.; Blas, D.; Brito, R.; et al. Black holes, gravitational waves and fundamental physics: A roadmap. *Class. Quant. Grav.* **2019**, *36*, 143001. [CrossRef]
40. Nakar, E. The electromagnetic counterparts of compact binary mergers. *Phys. Rept.* **2020**, *886*, 1–84. [CrossRef]
41. Yokoyama, J. Implication of pulsar timing array experiments on cosmological gravitational wave detection. *AAPPS Bull.* **2021**, *31*, 17. [CrossRef]
42. Domènech, G. Scalar Induced Gravitational Waves Review. *Universe* **2021**, *7*, 398. [CrossRef]
43. Mandel, I.; Farmer, A. Merging stellar-mass binary black holes. *Phys. Rept.* **2022**, *955*, 1–24. [CrossRef]
44. Agazie, G. et al. [NANOGrav]. The NANOGrav 15 yr Data Set: Evidence for a Gravitational-wave Background. *Astrophys. J. Lett.* **2023**, *951*, L8. [CrossRef]
45. Nojiri, S.; Odintsov, S.D. Introduction to modified gravity and gravitational alternative for dark energy. *Int. J. Geom. Meth. Mod. Phys.* **2007**, *4*, 115. [CrossRef]
46. Lue, A. The phenomenology of dvali-gabadadze-porrati cosmologies. *Phys. Rept.* **2006**, *423*, 1–48. [CrossRef]
47. Copeland, E.J.; Sami, M.; Tsujikawa, S. Dynamics of dark energy. *Int. J. Mod. Phys. D* **2006**, *15*, 1753–1936. [CrossRef]
48. Fujii, Y.; Maeda, K. *The Scalar-Tensor Theory of Gravitation*; Cambridge University Press: Cambridge, UK, 2007.
49. Padmanabhan, T. Dark energy and gravity. *Gen. Rel. Grav.* **2008**, *40*, 529–564. [CrossRef]
50. Durrer, R.; Maartens, R. Dark energy and dark gravity: theory overview. *Gen. Rel. Grav.* **2008**, *40*, 301–328. [CrossRef]
51. Alexander, S.; Yunes, N. Chern-Simons Modified General Relativity. *Phys. Rept.* **2009**, *480*, 1–55. [CrossRef]
52. Sotiriou, T.P.; Faraoni, V. $f(R)$ Theories of Gravity. *Rev. Mod. Phys.* **2010**, *82*, 451–497. [CrossRef]
53. Cai, Y.F.; Saridakis, E.N.; Setare, M.R.; Xia, J.Q. Quintom Cosmology: Theoretical implications and observations. *Phys. Rept.* **2010**, *493*, 1–60. [CrossRef]
54. Felice, A.D.; Tsujikawa, S. $f(R)$ theories. *Living Rev. Rel.* **2010**, *13*, 3. [CrossRef]
55. Amendola, L.; Tsujikawa, S. *Dark Energy*; Cambridge University Press: Cambridge, UK, 2010.
56. Faraoni, V.; Capozziello, S. *Beyond Einstein Gravity*; Springer: Dordrecht, The Netherlands, 2010.
57. Nojiri, S.; Odintsov, S.D. Unified cosmic history in modified gravity: From $F(R)$ theory to Lorentz non-invariant models. *Phys. Rept.* **2011**, *505*, 59–144. [CrossRef]
58. Capozziello, S.; Laurentis, M.D. Extended Theories of Gravity. *Phys. Rept.* **2011**, *509*, 167–321. [CrossRef]
59. Clifton, T.; Ferreira, P.G.; Padilla, A.; Skordis, C. Modified Gravity and Cosmology. *Phys. Rept.* **2012**, *513*, 1–189.
60. Bamba, K.; Capozziello, S.; Nojiri, S.; Odintsov, S.D. Dark energy cosmology: The equivalent description via different theoretical models and cosmography tests. *Astrophys. Space Sci.* **2012**, *342*, 155–228. [CrossRef]

61. Padmanabhan, T.; Kothawala, D. Lanczos-Lovelock models of gravity. *Phys. Rept.* **2013**, *531*, 115–171. [CrossRef]
62. Weinberg, D.H.; Mortonson, M.J.; Eisenstein, D.J.; Hirata, C.; Riess, A.G.; Rozo, E. Observational Probes of Cosmic Acceleration. *Phys. Rept.* **2013**, *530*, 87–255.
63. Will, C.M. The Confrontation between General Relativity and Experiment. *Living Rev. Rel.* **2014**, *17*, 4. [CrossRef]
64. Joyce, A.; Jain, B.; Khoury, J.; Trodden, M. Beyond the Cosmological Standard Model. *Phys. Rept.* **2015**, *568*, 1–98. [CrossRef]
65. Bamba, K.; Odintsov, S.D. Inflationary cosmology in modified gravity theories. *Symmetry* **2015**, *7*, 220–240. [CrossRef]
66. Cai, Y.F.; Capozziello, S.; Laurentis, M.D.; Saridakis, E.N. $f(T)$ teleparallel gravity and cosmology. *Rept. Prog. Phys.* **2016**, *79*, 106901. [CrossRef]
67. Wang, S.; Wang, Y.; Li, M. Holographic Dark Energy. *Phys. Rept.* **2017**, *696*, 1–57. [CrossRef]
68. Nojiri, S.; Odintsov, S.D.; Oikonomou, V.K. Modified Gravity Theories on a Nutshell: Inflation, Bounce and Late-time Evolution. *Phys. Rept.* **2017**, *692*, 1–104. [CrossRef]
69. Jimenez, J.B.; Heisenberg, L.; Olmo, G.J.; Rubiera-Garcia, D. Born–Infeld inspired modifications of gravity. *Phys. Rept.* **2018**, *727*, 1–129. [CrossRef]
70. Bahamonde, S.; Böhmer, C.G.; Carloni, S.; Copeland, E.J.; Fang, W.; Tamanini, N. Dynamical systems applied to cosmology: Dark energy and modified gravity. *Phys. Rept.* **2018**, *775–777*, 1–122. [CrossRef]
71. Adami, H.; Setare, M.R.; Sisman, T.C.; Tekin, B. Conserved Charges in Extended Theories of Gravity. *Phys. Rept.* **2019**, *834*, 1. [CrossRef]
72. Heisenberg, L. A systematic approach to generalisations of General Relativity and their cosmological implications. *Phys. Rept.* **2019**, *796*, 1–113. [CrossRef]
73. Langlois, D. Dark energy and modified gravity in degenerate higher-order scalar–tensor (DHOST) theories: A review. *Int. J. Mod. Phys. D* **2019**, *28*, 1942006. [CrossRef]
74. Frusciante, N.; Perenon, L. Effective field theory of dark energy: A review. *Phys. Rept.* **2020**, *857*, 1–63. [CrossRef]
75. Olmo, G.J.; Rubiera-Garcia, D.; Wojnar, A. Stellar structure models in modified theories of gravity: Lessons and challenges. *Phys. Rept.* **2020**, *876*, 1–75. [CrossRef]
76. Saridakis, E.N.; Lazkoz, R.; Salzano, V.; Moniz, P.V.; Capozziello, S.; Jimenez, J.B.; Laurentis, M.D.; Olmo, G.J. *Modified Gravity and Cosmology*; Springer: Cham, Switzerland, 2021.
77. Faraoni, V.; Giusti, A.; Fahim, B.H. Spherical inhomogeneous solutions of Einstein and scalar–tensor gravity: A map of the land. *Phys. Rept.* **2021**, *925*, 1–58. [CrossRef]
78. Bamba, K. Review on Dark Energy Problem and Modified Gravity Theories. *LHEP* **2022**, *2022*, 352. [CrossRef]
79. Bahamonde, S.; Dialektopoulos, K.F.; Escamilla-Rivera, C.; Farrugia, G.; Gakis, V.; Hendry, M.; Hohmann, M.; Said, J.L.; Mifsud, J.; Valentino, E.D. Teleparallel gravity: From theory to cosmology. *Rept. Prog. Phys.* **2023**, *86*, 026901. [CrossRef] [PubMed]
80. Arai, S.; Aoki, K.; Chinone, Y.; Kimura, R.; Kobayashi, T.; Miyatake, H.; Yamauchi, D.; Yokoyama, S.; Akitsu, K.; Hiramatsu, T.; et al. Cosmological gravity probes: Connecting recent theoretical developments to forthcoming observations. *PTEP* **2023**, *2023*, 072E01. [CrossRef]
81. Odintsov, S.D.; Oikonomou, V.K.; Giannakoudi, I.; Fronimos, F.P.; Lymperiadou, E.C. Recent Advances in Inflation. *Symmetry* **2023**, *15*, 1701. [CrossRef]
82. de Haro, J.; Nojiri, S.; Odintsov, S.D.; Oikonomou, V.K.; Pan, S. Finite-time cosmological singularities and the possible fate of the Universe. *Phys. Rept.* **2023**, *1034*, 1–114. [CrossRef]
83. Heisenberg, L. Review on f(Q) gravity. *Phys. Rept.* **2024**, *1066*, 1–78. [CrossRef]
84. Avsajanishvili, O.; Chitov, G.Y.; Kahniashvili, T.; Mandal, S.; Samushia, L. Observational Constraints on Dynamical Dark Energy Models. *Universe* **2024**, *10*, 122. [CrossRef]
85. Yousaf, Z.; Bamba, K.; Bhatti, M.Z.; Farwa, U. Quasi-static evolution of axially and reflection symmetric large-scale configuration. *Int. J. Geom. Meth. Mod. Phys.* **2024**, *21*, 2430005. [CrossRef]
86. Gasperini, M.; Veneziano, G. The Pre-big bang scenario in string cosmology. *Phys. Rept.* **2003**, *373*, 1–212. [CrossRef]
87. Kiritsis, E. D-branes in standard model building, gravity and cosmology. *Phys. Rept.* **2005**, *421*, 105–190; Erratum in *Phys. Rept.* **2006**, *429*, 121–122. [CrossRef]
88. Davidson, S.; Nardi, E.; Nir, Y. Leptogenesis. *Phys. Rept.* **2008**, *466*, 105–177. [CrossRef]
89. Novello, M.; Bergliaffa, S.E.P. Bouncing Cosmologies. *Phys. Rept.* **2008**, *463*, 127–213. [CrossRef]
90. Lehners, J.L. Ekpyrotic and Cyclic Cosmology. *Phys. Rept.* **2008**, *465*, 223–263. [CrossRef]
91. Mazumdar, A.; Rocher, J. Particle physics models of inflation and curvaton scenarios. *Phys. Rept.* **2011**, *497*, 85–215. [CrossRef]
92. Maleknejad, A.; Sheikh-Jabbari, M.M.; Soda, J. Gauge Fields and Inflation. *Phys. Rept.* **2013**, *528*, 161–261. [CrossRef]
93. Battefeld, D.; Peter, P. A Critical Review of Classical Bouncing Cosmologies. *Phys. Rept.* **2015**, *571*, 1–66. [CrossRef]
94. Sato, K.; Yokoyama, J. Inflationary cosmology: First 30+ years. *Int. J. Mod. Phys. D* **2015**, *24*, 1530025. [CrossRef]
95. Asadi, P.; Bansal, S.; Berlin, A.; Co, R.T.; Croon, D.; Cui, Y.; Curtin, D.; Cyr-Racine, F.Y.; Davoudiasl, H.; Rose, L.D.; et al. Early-Universe Model Building. *arXiv* **2022**, arXiv:2203.06680.
96. Cicoli, M.; Conlon, J.P.; Maharana, A.; Parameswaran, S.; Quevedo, F.; Zavala, I. String cosmology: From the early universe to today. *Phys. Rept.* **2024**, *1059*, 1–155.
97. Donnay, L. Celestial holography: An asymptotic symmetry perspective. *Phys. Rept.* **2024**, *1073*, 1–41. [CrossRef]
98. Kodama, H.; Sasaki, M. Cosmological Perturbation Theory. *Prog. Theor. Phys. Suppl.* **1984**, *78*, 1–166. [CrossRef]

99. Mukhanov, V.F.; Feldman, H.A.; Brandenberger, R.H. Theory of cosmological perturbations. Part 1. Classical perturbations. Part 2. Quantum theory of perturbations. Part 3. Extensions. *Phys. Rept.* **1992**, *215*, 203–333. [CrossRef]

100. Bernardeau, F.; Colombi, S.; Gaztanaga, E.; Scoccimarro, R. Large scale structure of the universe and cosmological perturbation theory. *Phys. Rept.* **2002**, *367*, 1–248. [CrossRef]

101. Malik, K.A.; Wands, D. Cosmological perturbations. *Phys. Rept.* **2009**, *475*, 1–51. [CrossRef]

102. Palti, E. The Swampland: Introduction and Review. *Fortsch. Phys.* **2019**, *67*, 1900037. [CrossRef]

103. van Beest, M.; Calderón-Infante, J.; Mirfendereski, D.; Valenzuela, I. Lectures on the Swampland Program in String Compactifications. *Phys. Rept.* **2022**, *989*, 1–50. [CrossRef]

104. Van Riet, T.; Zoccarato, G. Beginners lectures on flux compactifications and related Swampland topics. *Phys. Rept.* **2024**, *1049*, 1–51. [CrossRef]

105. Padmanabhan, T. Cosmological constant: The Weight of the vacuum. *Phys. Rept.* **2003**, *380*, 235–320. [CrossRef]

106. Dolgov, A.D. Neutrinos in cosmology. *Phys. Rept.* **2002**, *370*, 333–535. [CrossRef]

107. Lesgourgues, J.; Pastor, S. Massive neutrinos and cosmology. *Phys. Rept.* **2006**, *429*, 307–379. [CrossRef]

108. Kusenko, A. Sterile neutrinos: The Dark side of the light fermions. *Phys. Rept.* **2009**, *481*, 1–28. [CrossRef]

109. Abazajian, K.N. Sterile neutrinos in cosmology. *Phys. Rept.* **2017**, *711-712*, 1–28. [CrossRef]

110. Dasgupta, B.; Kopp, J. Sterile Neutrinos. *Phys. Rept.* **2021**, *928*, 1–63. [CrossRef]

111. Marsh, D.J.E. Axion Cosmology. *Phys. Rept.* **2016**, *643*, 1–79. [CrossRef]

112. Yokoyama, J. Formation of primordial black holes in the inflationary universe. *Phys. Rept.* **1998**, *307*, 133–139. [CrossRef]

113. Carr, B.; Kohri, K.; Sendouda, Y.; Yokoyama, J. Constraints on primordial black holes. *Rept. Prog. Phys.* **2021**, *84*, 116902. [CrossRef]

114. Özsoy, O.; Tasinato, G. Inflation and Primordial Black Holes. *Universe* **2023**, *9*, 203. [CrossRef]

115. Carr, B.; Clesse, S.; Garcia-Bellido, J.; Hawkins, M.; Kuhnel, F. Observational evidence for primordial black holes: A positivist perspective. *Phys. Rept.* **2024**, *1054*, 1–68. [CrossRef]

116. Domènech, G. Lectures on Gravitational Wave Signatures of Primordial Black Holes. *arXiv* **2023**, arXiv:2307.06964.

117. Bertone, G.; Hooper, D.; Silk, J. Particle dark matter: Evidence, candidates and constraints. *Phys. Rept.* **2005**, *405*, 279–390. [CrossRef]

118. Hooper, D.; Profumo, S. Dark Matter and Collider Phenomenology of Universal Extra Dimensions. *Phys. Rept.* **2007**, *453*, 29–115. [CrossRef]

119. Zurek, K.M. Asymmetric Dark Matter: Theories, Signatures, and Constraints. *Phys. Rept.* **2014**, *537*, 91–121. [CrossRef]

120. Baer, H.; Choi, K.Y.; Kim, J.E.; Roszkowski, L. Dark matter production in the early Universe: Beyond the thermal WIMP paradigm. *Phys. Rept.* **2015**, *555*, 1–60. [CrossRef]

121. Aramaki, T.; Boggs, S.; Bufalino, S.; Dal, L.; von Doetinchem, P.; Donato, F.; Fornengo, N.; Fuke, H.; Grefe, M.; Hailey, C.; et al. Review of the theoretical and experimental status of dark matter identification with cosmic-ray antideuterons. *Phys. Rept.* **2016**, *618*, 1–37. [CrossRef]

122. Mayet, F.; Green, A.M.; Battat, J.B.R.; Billard, J.; Bozorgnia, N.; Gelmini, G.B.; Gondolo, P.; Kavanagh, B.J.; Lee, S.K.; Loomba, D.; et al. A review of the discovery reach of directional Dark Matter detection. *Phys. Rept.* **2016**, *627*, 1–49. [CrossRef]

123. Tulin, S.; Yu, H.B. Dark Matter Self-interactions and Small Scale Structure. *Phys. Rept.* **2018**, *730*, 1–57. [CrossRef]

124. Buckley, M.R.; Peter, A.H.G. Gravitational probes of dark matter physics. *Phys. Rept.* **2018**, *761*, 1–60. [CrossRef]

125. Arcadi, G.; Djouadi, A.; Raidal, M. Dark Matter through the Higgs portal. *Phys. Rept.* **2020**, *842*, 1–180. [CrossRef]

126. Buen-Abad, M.A.; Essig, R.; McKeen, D.; Zhong, Y.M. Cosmological constraints on dark matter interactions with ordinary matter. *Phys. Rept.* **2022**, *961*, 1–35. [CrossRef]

127. Ahluwalia, D.V.; da Silva, J.M.H.; Lee, C.Y.; Liu, Y.X.; Pereira, S.H.; Sorkhi, M.M. Mass dimension one fermions: Constructing darkness. *Phys. Rept.* **2022**, *967*, 1–43. [CrossRef]

128. Bramante, J.; Raj, N. Dark matter in compact stars. *Phys. Rept.* **2024**, *1052*, 1–48. [CrossRef]

129. Grasso, D.; Rubinstein, H.R. Magnetic fields in the early universe. *Phys. Rept.* **2001**, *348*, 163–266. [CrossRef]

130. Barrow, J.D.; Maartens, R.; Tsagas, C.G. Cosmology with inhomogeneous magnetic fields. *Phys. Rept.* **2007**, *449*, 131–171. [CrossRef]

131. Subramanian, K. Magnetic fields in the early universe. *Astron. Nachr.* **2010**, *331*, 110–120. [CrossRef]

132. Kandus, A.; Kunze, K.E.; Tsagas, C.G. Primordial magnetogenesis. *Phys. Rept.* **2011**, *505*, 1–58. [CrossRef]

133. Yamazaki, D.G.; Kajino, T.; Mathew, G.J.; Ichiki, K. The Search for a Primordial Magnetic Field. *Phys. Rept.* **2012**, *517*, 141–167. [CrossRef]

134. Iocco, F.; Mangano, G.; Miele, G.; Pisanti, O.; Serpico, P.D. Primordial Nucleosynthesis: From precision cosmology to fundamental physics. *Phys. Rept.* **2009**, *472*, 1–76. [CrossRef]

135. Pitrou, C.; Coc, A.; Uzan, J.P.; Vangioni, E. Precision big bang nucleosynthesis with improved Helium-4 predictions. *Phys. Rept.* **2018**, *754*, 1–66. [CrossRef]

136. Knox, L.; Millea, M. Hubble constant hunter's guide. *Phys. Rev. D* **2020**, *101*, 043533. [CrossRef]

137. Asgari, M. et al. [KiDS]. KiDS-1000 Cosmology: Cosmic shear constraints and comparison between two point statistics. *Astron. Astrophys.* **2021**, *645*, A104. [CrossRef]

138. Valentino, E.D.; Mena, O.; Pan, S.; Visinelli, L.; Yang, W.; Melchiorri, A.; Mota, D.F.; Riess, A.G.; Silk, J. In the realm of the Hubble tension—A review of solutions. *Class. Quant. Grav.* **2021**, *38*, 153001. [CrossRef]

139. Perivolaropoulos, L.; Skara, F. Challenges for ΛCDM: An update. *New Astron. Rev.* **2022**, *95*, 101659.
140. Schöneberg, N.; Abellán, G.F.; Sánchez, A.P.; Witte, S.J.; Poulin, V.; Lesgourgues, J. The H0 Olympics: A fair ranking of proposed models. *Phys. Rept.* **2022**, *984*, 1–55. [CrossRef]
141. Abdalla, E.; Abellán, G.F.; Aboubrahim, A.; Agnello, A.; Akarsu, O.; Akrami, Y.; Alestas, G.; Aloni, D.; Amendola, L.; Anchordoqui, L.A.; et al. Cosmology intertwined: A review of the particle physics, astrophysics, and cosmology associated with the cosmological tensions and anomalies. *J. High Energy Astrophys.* **2022**, *34*, 49–211.
142. Poulin, V.; Smith, T.L.; Karwal, T. The Ups and Downs of Early Dark Energy solutions to the Hubble tension: A review of models, hints and constraints circa 2023. *Phys. Dark Univ.* **2023**, *42*, 101348. [CrossRef]
143. Furlanetto, S.; Oh, S.P.; Briggs, F. Cosmology at Low Frequencies: The 21 cm Transition and the High-Redshift Universe. *Phys. Rept.* **2006**, *433*, 181–301. [CrossRef]
144. Barkana, R. The Rise of the First Stars: Supersonic Streaming, Radiative Feedback, and 21-cm Cosmology. *Phys. Rept.* **2016**, *645*, 1–59. [CrossRef]
145. Bartelmann, M.; Schneider, P. Weak gravitational lensing. *Phys. Rept.* **2001**, *340*, 291–472. [CrossRef]
146. Lewis, A.; Challinor, A. Weak gravitational lensing of the CMB. *Phys. Rept.* **2006**, *429*, 1–65. [CrossRef]
147. Barnacka, A. Gravitational Lenses as High-Resolution Telescopes. *Phys. Rept.* **2018**, *778–779*, 1–46. [CrossRef]
148. Durrer, R.; Kunz, M.; Melchiorri, A. Cosmic structure formation with topological defects. *Phys. Rept.* **2002**, *364*, 1–81. [CrossRef]
149. Padmanabhan, T. Gravity and the thermodynamics of horizons. *Phys. Rept.* **2005**, *406*, 49–125. [CrossRef]
150. Hollands, S.; Wald, R.M. Quantum fields in curved spacetime. *Phys. Rept.* **2015**, *574*, 1–35. [CrossRef]
151. Porto, R.A. The effective field theorist's approach to gravitational dynamics. *Phys. Rept.* **2016**, *633*, 1–104. [CrossRef]
152. Dayal, P.; Ferrara, A. Early galaxy formation and its large-scale effects. *Phys. Rept.* **2018**, *780–782*, 1–64. [CrossRef]
153. Adams, F.C. The degree of fine-tuning in our universe—And others. *Phys. Rept.* **2019**, *807*, 1–111. [CrossRef]
154. Perlick, V.; Tsupko, O.Y. Calculating black hole shadows: Review of analytical studies. *Phys. Rept.* **2022**, *947*, 1–39. [CrossRef]
155. Brout, D.; Scolnic, D.; Popovic, B.; Riess, A.G.; Zuntz, J.; Kessler, R.; Carr, A.; Davis, T.M.; Hinton, S.; Jones, D.; et al. The Pantheon+ Analysis: Cosmological Constraints. *Astrophys. J.* **2022**, *938*, 110. [CrossRef]
156. Giovannetti, E.; Maione, F.; Montani, G. Quantum Big Bounce of the Isotropic Universe Using Relational Time. *Universe* **2023**, *9*, 373. [CrossRef]
157. Marochnik, L. Nothing into Something and Vice Versa: A Cosmological Scenario. *Universe* **2023**, *9*, 445. [CrossRef]
158. Capistrano, A.J.S.; Cabral, L.A. Effective Potential for Quintessential Inflation Driven by Extrinsic Gravity. *Universe* **2023**, *9*, 497. [CrossRef]
159. Staicova, D.; Stoilov, M. Electromagnetic Waves in Cosmological Spacetime. *Universe* **2023**, *9*, 292. [CrossRef]
160. Fanizza, G.; Gasperini, M.; Marozzi, G. A Simple, Exact Formulation of Number Counts in the Geodesic-Light-Cone Gauge. *Universe* **2023**, *9*, 327. [CrossRef]
161. Medel-Esquivel, R.; Gómez-Vargas, I.; Sánchez, A.A.M.; García-Salcedo, R.; Vázquez, J.A. Cosmological Parameter Estimation with Genetic Algorithms. *Universe* **2024**, *10*, 11. [CrossRef]
162. Stenflo, J.O. Cosmological Constant from Boundary Condition and Its Implications beyond the Standard Model. *Universe* **2023**, *9*, 103. [CrossRef]
163. Socorro, J.; Rosales, J.J. Quantum Fractionary Cosmology: K-Essence Theory. *Universe* **2023**, *9*, 185. [CrossRef]
164. Yershov, V.N. Fitting Type Ia Supernova Data to a Cosmological Model Based on Einstein-Newcomb-De Sitter Space. *Universe* **2023**, *9*, 204. [CrossRef]
165. Paliathanasis, A. Revise the Phase-Space Analysis of the Dynamical Spacetime Unified Dark Energy Cosmology. *Universe* **2023**, *9*, 406. [CrossRef]
166. Ong, Y.C. An Effective Sign Switching Dark Energy: Lotka–Volterra Model of Two Interacting Fluids. *Universe* **2023**, *9*, 437. [CrossRef]
167. Majumder, B.; Khlopov, M.; Ray, S.; Manna, G. Geodesic Structure of Generalized Vaidya Spacetime through the K-Essence. *Universe* **2023**, *9*, 510. [CrossRef]
168. Alcántara-Pérez, Y.B.; García-Aspeitia, M.A.; Martínez-Huerta, H.; Hernández-Almada, A. MeV Dark Energy Emission from a De Sitter Universe. *Universe* **2023**, *9*, 513. [CrossRef]
169. Ribeiro, A.R.; Vernieri, D.; Lobo, F.S.N. Effective f(R) Actions for Modified Loop Quantum Cosmologies via Order Reduction. *Universe* **2023**, *9*, 181. [CrossRef]
170. Alfaro, J.; Rubio, C.; Martín, M.S. Cosmological Fluctuations in Delta Gravity. *Universe* **2023**, *9*, 315. [CrossRef]
171. Tabatabaei, J.; Banihashemi, A.; Baghram, S.; Mashhoon, B. Anisotropic Cosmology in the Local Limit of Nonlocal Gravity. *Universe* **2023**, *9*, 377. [CrossRef]
172. Brito, F.A.; Borges, C.H.A.B.; Campos, J.A.V.; Costa, F.G. Weak Coupling Regime in Dilatonic Cosmology. *Universe* **2024**, *10*, 134. [CrossRef]
173. Zhao, Z.C.; Wang, S. Bayesian Implications for the Primordial Black Holes from NANOGrav's Pulsar-Timing Data Using the Scalar-Induced Gravitational Waves. *Universe* **2023**, *9*, 157. [CrossRef]
174. Gaztanaga, E. Do White Holes Exist? *Universe* **2023**, *9*, 194. [CrossRef]

175. Triantafyllopoulos, A.; Kapsabelis, E.; Stavrinos, P.C. Raychaudhuri Equations, Tidal Forces, and the Weak-Field Limit in Schwarzshild–Finsler–Randers Spacetime. *Universe* **2024**, *10*, 26. [CrossRef]
176. Cervantes-Cota, J.L.; Galindo-Uribarri, S.; Smoot, G.F. The Unsettled Number: Hubble's Tension. *Universe* **2023**, *9*, 501. [CrossRef]

Article

Quantum Big Bounce of the Isotropic Universe Using Relational Time

Eleonora Giovannetti [1,2,*], Fabio Maione [2] and Giovanni Montani [2,3]

1   Aix Marseille Université, CNRS Luminy—Case 907, Centre de Physique Théorique—UMR 7332, 13288 Marseille, France
2   Department of Physics, "La Sapienza" University of Rome, P.le Aldo Moro 5, 00185 Roma, Italy; fabiomaione93@gmail.com (F.M.); giovanni.montani@enea.it (G.M.)
3   ENEA, Fusion and Nuclear Safety Department, C.R. Frascati, Via E. Fermi 45, 00044 Frascati, Italy
*   Correspondence: eleonora.giovannetti@uniroma1.it

Abstract: We analyze the canonical quantum dynamics of the isotropic Universe with a metric approach by adopting a self-interacting scalar field as relational time. When the potential term is absent, we are able to associate the expanding and collapsing dynamics of the Universe with the positive- and negative-frequency modes that emerge in the Wheeler–DeWitt equation. On the other side, when the potential term is present, a non-zero transition amplitude from positive- to negative-frequency states arises, as in standard relativistic scattering theory below the particle creation threshold. In particular, we are able to compute the transition probability for an expanding Universe that emerges from a collapsing regime both in the standard quantization procedure and in the polymer formulation. The probability distribution results similar in the two cases, and its maximum takes place when the mean values of the momentum essentially coincide in the in-going and out-going wave packets, as it would take place in a semiclassical Big Bounce dynamics.

Keywords: quantum cosmology; isotropic Universe; quantum Big Bounce; polymer quantum mechanics

Citation: Giovannetti, E.; Maione, F.; Montani, G. Quantum Big Bounce of the Isotropic Universe Using Relational Time. *Universe* **2023**, *9*, 373. https://doi.org/10.3390/universe9080373

Academic Editors: Kazuharu Bamba and Yi-Fu Cai

Received: 14 June 2023
Revised: 26 July 2023
Accepted: 9 August 2023
Published: 16 August 2023

1. Introduction

The most relevant and long-standing question of relativistic cosmology is surely the presence of an initial singularity in the dynamics of the isotropic Universe as a general feature of the Einstein equations under cosmological hypotheses [1,2]. Thought as a shortcoming of the underlying theory, the initial singularity has been the subject of a wide number of attempts devoted to its removal and originally focused on modifications of the Einstein–Hilbert action, able to alter the Friedmann dynamics of the Universe [3–5]. Since the canonical quantization of gravity was initially formulated [2,6], the dominant proposal has been the possibility that quantum effects in the Planck age could prevent the zero-volume limit characterizing the singularity. In this respect, understanding that in loop quantum gravity (LQG) [2,7–10], the volume operator has a discrete spectrum [11] and then showing the emergence of the Big Bounce semiclassical dynamics in so-called loop quantum cosmology (LQC) theory [12–19] were the most relevant successes. Polymer quantum cosmology (PQC) is also related to this scenario [20,21], due to its several morphological features in common with the quasi-classical limit of LQC. However, all these approaches are based on a semiclassical representation of the Universe dynamics, in which the Big Bounce is the consequence of a regularization of the standard Friedmann evolution of the Universe near the Planck era. In this regard, in [22], it is underlined that a quasi-classical representation of the primordial phase is not always allowed, since the Universe is in a fully quantum state. So, the concept of a transition amplitude from collapsing to expanding Bianchi I Universes in the Wheeler–DeWitt (WDW) formulation is introduced, where the isotropic Misner variable plays the role of time. The associated transition probability results well defined and with a Gaussian-like distribution.

In this work, we follow the same spirit of [22], limiting our attention to the Friedmann–Lemaître–Robertson–Walker (FLRW) Universe but generalizing the quantum formulation towards the implementation of a real matter field as relational time, here, a self-interacting scalar field [23]. Firstly, we construct a parallelism between the WDW equation and a Klein—Gordon (KG) one by identifying the expanding Universe and the collapsing Universe as positive- and negative-frequency solutions, respectively. Then, following the relativistic scattering procedure (below the particle creation threshold) [24], we calculate the transition amplitude from negative- to positive-frequency states, which results in a "Quantum Big Bounce" picture. The probability associated to this transition is maximum when the momentum mean value of the in-going Universe wave packet is nearly that of the out-going one. Finally, we consider the same model in polymer quantum mechanics (PQM), which is not able to remove the singularity in the associated WDW equation when written in Misner variables (see [25,26], where the singular behavior of the Bianchi I model, and thus of the Bianchi IX model, is outlined). However, the polymer paradigm singles out a spreading feature that prevents a semiclassical description, making the concept of a Quantum Big Bounce more robust. In consideration of the study developed in [27], the surprising result is that the profile of the probability density is essentially unchanged in the polymer formulation with respect to the standard quantum scheme. Basically, the present analysis suggests that a Quantum Big Bounce is allowed at a probabilistic level in the WDW formalism for the isotropic Universe, thanks to a proper interpretation of the WDW solutions as negative- and positive-frequency states defined in terms of matter relational time. We remark that the possibility to have a transition from a collapsing Universe to an expanding Universe is due to the breaking of the frequency separation because of the presence of potential energy density in the dynamics of the matter clock and that the position of the peak in the probability density reflects the same symmetrical reconnection of the singular branches typical of a semiclassical bouncing picture [2].

2. Quantum FLRW Dynamics

The isotropic Universe is described by the FLRW model, and its classical dynamics is affected by past and future singularities, i.e., the Big Bang and the Big Crunch, respectively. The Hamiltonian of the system in the presence of a free scalar field ϕ is

$$H = N\mathcal{H} = N\left[-\frac{2}{3}\frac{\pi G}{c^3 V}p_\alpha^2 + \frac{c}{2V}p_\phi^2\right] = 0, \tag{1}$$

where N is the lapse function, $\alpha = \ln a$ is the isotropic Misner variable, G is the Einstein constant, c is the speed of light in vacuum and V is the fiducial volume. In order to preserve the covariant structure of the theory, we use the so-called Dirac procedure and quantize the theory before having explicitly solved the constraints, now promoted to quantum operators. So, by implementing the super-Hamiltonian constraint as a quantum operator $\hat{\mathcal{H}}$ that annihilates the Universe wave function, $\Psi(\alpha, \phi)$, we obtain the WDW equation

$$\hat{\mathcal{H}}\Psi(\alpha, \phi) = \frac{2}{3}\frac{\pi G\hbar^2}{c^3 V}\frac{\partial^2\Psi(\alpha, \phi)}{\partial\alpha^2} - \frac{c\hbar^2}{2V}\frac{\partial^2\Psi(\alpha, \phi)}{\partial\phi^2} = 0, \tag{2}$$

where we use the coordinate representation and the normal ordering convention ($\hbar = h/2\pi$ is the Planck constant). We note that Equation (2) is formally analogous to a two-dimensional massless KG equation for a relativistic particle. Thanks to this analogy, we can identify ϕ as a relational-time variable and α as the spatial degree of freedom. Hence, according to the superposition principle, the most general solution is a linear combination of plane waves, i.e., a wave packet:

$$\psi(\alpha, \phi) = \int_{-\infty}^{+\infty}[A_+(k)\varphi_+(\alpha, \phi) + A_-(k)\varphi_-(\alpha, \phi)]\,dk, \tag{3}$$

$$\varphi_\pm(\alpha, \phi) = e^{i(k\alpha \mp \omega_k\phi)}, \tag{4}$$

where k is the wave number, $\omega_k = v|k|$ is the dispersion relation (i.e., the frequency), $A_\pm(k)$ are arbitrary functions and $v = \sqrt{4\pi G\hbar^2/3c^4}$.

Actually, when dealing with highly localized wave packets, it is possible to relate the classical values of the momenta (i.e., p_α and p_ϕ) and the corresponding quantum eigenvalues (i.e., k and ω_k, respectively), in agreement with the Ehrenfest theorem. In this respect, we note that by deriving the Hamilton equations

$$\dot{\phi} = \frac{\partial H}{\partial p_\phi} \frac{Ne^{-3\alpha}cp_\phi}{V}, \qquad \dot{p}_\phi = -\frac{\partial H}{\partial \phi} = 0, \tag{5a}$$

$$\dot{\alpha} = -\frac{Ne^{-3\alpha}4\pi G p_\alpha}{3c^3 V}, \qquad \dot{p}_\alpha = 3H = 0 \tag{5b}$$

from (1) and by combining the equations for $\dot{\alpha}$ and $\dot{\phi}$, we obtain the relation

$$\frac{d\alpha}{d\phi} = -\frac{4\pi G}{3c^4} \frac{p_\alpha}{p_\phi} \tag{6}$$

which has a covariant character, i.e., it is valid for any temporal gauge. From (6), we obtain that the distinction between the classical branches depends on the relative sign between the constants of motion p_α and p_ϕ: the Universe collapses if the sign is concordant; otherwise, it expands. In particular, once the sign of p_α is fixed, the expanding or collapsing feature depends on the sign of p_ϕ. Therefore, if we consider sufficiently localized (i.e., semiclassical) wave packets far from the fully quantum region of the initial singularity, we can only select the states with $k > 0$ from the full Hilbert space and infer that the positive-frequency solutions of the type

$$\psi_+(\alpha, \phi) = \int_0^{+\infty} A_+(k)e^{i(k\alpha - \omega_k\phi)} \, dk \tag{7}$$

describe an expanding Universe, whereas the negative-frequency ones of the type

$$\psi_-(\alpha, \phi) = \int_0^{+\infty} A_-(k)e^{i(k\alpha + \omega_k\phi)} \, dk \tag{8}$$

describe a collapsing one. Thus, we can associate positive energy states (i.e., "particles states") with the expanding branch and negative energy ones (i.e., "antiparticles states") with the collapsing phase. We remark that considering negative values of p_α leads to the opposite identification. Actually, retaining both signs of p_α would provide redundant information. So, in (7) and (8), we choose the portion of the spectrum of operator $\hat{p}_\alpha$ associated to $k > 0$, and we impose $A_\pm(k) = e^{-\frac{(k-\bar{k})^2}{2\sigma^2}}/\sqrt{2\pi\sigma^2}$, in which $\bar{k}$ and σ are fixed by the initial condition on the wave function at a given $\phi = \phi_0$.

We remark that frequency separation can only be performed in the absence of a self-interacting potential of the scalar field (here, a time-dependent term) and that the choice of the semi-axis of k on which to perform the integration is conventional. We also stress that the pure classical dynamics in function of time ϕ would require a fixed sign for p_ϕ; then, the two branches would depend on the Cauchy condition on the sign of p_α. Actually, in this picture, we use a "time after quantization" approach thanks to the analogy with the KG equation. Hence, in this fully quantum scenario, we have a pure relational dynamics in which states can propagate forward and backward in time as in relativistic quantum particle interactions. Furthermore, approaching the classical limit, the emergent state is that of an expanding Universe with fixed (here, negative) p_ϕ; therefore, the consistency of the classical dynamics is preserved. In addition, from (5a), we can justify the viability of ϕ as a relational-time variable a posteriori. In fact, p_ϕ is a constant of motion; so, ϕ is a monotonic function of synchronous time. In this respect, the advantage of using matter relational time is that the monotonicity requirement would also be fulfilled in a bouncing picture.

Klein–Gordon-like Formulation for Homogeneous Cosmology

In the original works of DeWitt [6,28,29], the different signature of the three-metric determinant (actually, its power to $1/4$) with respect to the other five configurational variables was first stressed. In other words, the Supermetric has a Lorentzian signature point by point in space, and the three-metric determinant plays the role of the time-like degree of freedom. As a result of this feature, the Wheeler–DeWitt equation seems to have the morphology of a Klein–Gordon-like equation in the presence of a potential term, due to the spatial curvature [30]. Thus, we can introduce a scalar product for the wavefunctional of the Universe, which corresponds to the natural extension of the Klein–Gordon one, and approach an interpretation of the Wheeler–DeWitt equation as being associated to a Hilbert space, having features in common with relativistic quantum field theory [31]. In [32,33], the idea of a Wheler–DeWitt equation as a Klein–Gordon-like setting is implemented into the Minisuperspace of Bianchi homogeneous Universes [2]. In fact, the homogeneity constraint reduces the problem to a finite number of degrees of freedom, i.e., the three Misner variables [34,35]. In particular, isotropic variable α plays the role of time, while the two anisotropies correspond to the space coordinates. This study allowed us to clarify that a scalar product can be defined when the potential term vanishes in one time direction. In the recent investigation pursued in [22], this approach was generalized in the spirit of a scattering scenario, resulting in a non-zero amplitude for the transition between a collapsing solution and an expanding solution. It is important to stress that both in [22,32,33], the Wheeler–DeWitt equation is interpreted as a Klein–Gordon formalism below the threshold of particle creation. Actually, we are not considering the third quantization scenario, where creation and annihilation operators are introduced; see, for instance [36]. In fact, while the possibility of creating Universes with different quantum numbers appears to be essentially a mathematical procedure whose physical meaning is unclear, the parallelism with a simpler scattering procedure turns out to be a much more solid prescription. In constructing such a parallelism, the driving idea consists of two main points: (i) when the potential is absent, the free states with positive frequencies can be interpreted as expanding Universe states, while those ones with negative frequencies, as collapsing configurations; (ii) the presence of the potential term (to be thought as relevant only in the Planck era) introduces an avoidably superposition of expanding and collapsing Universe states, whose transition amplitude can be calculated using the standard equipment of relativistic quantum scattering.

In the present study, we deal with a Wheeler–DeWitt equation associated to an isotropic Universe in the presence of a self-interacting scalar field ϕ, i.e.,

$$\partial_\alpha^2 \psi - \partial_\phi^2 \psi + V_{Pl}(\alpha, \phi)\psi = 0, \tag{9}$$

where $\psi = \psi(\alpha, \phi)$ and V_{Pl} is a Planck scale potential. This equation is clearly isomorphic to a $1 + 1$ Klein–Gordon equation in which we choose to identify the clock with the scalar field, in agreement with the relational-time approach [3,23,31]. Furthermore, this choice would be viable even throughout a Big Bounce configuration, while logarithmic scale factor α would not be monotonic time in that scenario. It is immediate to check that (9) is associated to the Klein–Gordon-like probability density:

$$\mathcal{P}_\psi = i(\psi^* \partial_\alpha \psi - \psi \partial_\alpha \psi^*). \tag{10}$$

Clearly, as in relativistic quantum mechanics, the probability density above becomes positive-defined only when a frequency separation procedure can be performed. However, as pointed out in [33], when the potential term is vanishing in one time direction (and not spatially diverging), it is possible to asymptotically recover positive energy states and define a norm to construct a proper Hilbert space. In particular, the model we take in consideration meets this requirement. A key point of our theoretical proposal is the identification of the frequency separation with the expanding and collapsing branches of the Universe as non-overlapped solutions. We finally remark that it is probability density (10) that induces the scalar product adopted to project the final state onto the initial one, as discussed in [24].

In particular, the explicit expression of the transition amplitude is then derived by using the propagator formalism (for more details, see Section 4).

3. Quantum FLRW Dynamics with a Self-Interacting Scalar Field

In this section, we introduce a potential term $U(\phi)$ at a quantum level. We note that the WDW approach is not able to regularize the dynamics of the present model. However, the proposed approach has the purpose to demonstrate the presence of the Big Bounce in the WDW formulation by treating it as quantum scattering. Hence, we consider the quantum potential $U(\phi) = \frac{\lambda}{2}e^{-n\phi}$, where $\lambda > 0$ and $n \in \mathbb{R}$ (for cosmological implementations of such a potential, see [37–39]). So, the WDW equation turns out to be

$$\left[\frac{\partial^2}{\partial\alpha^2} - \frac{1}{v^2}\frac{\partial^2}{\partial\phi^2} + Ce^{6\alpha - n\phi}\right]\psi(\alpha, \phi) = 0,\tag{11}$$

where $C = 3\lambda(Vc)^2/4\hbar^2\pi G$ (v can be taken out from the equation by redefining $\bar{\phi} = v\phi$ and $\bar{n} = n/v$ and then renaming the new variables as the old ones). As above, (11) can be interpreted as a KG equation where the potential term depends on both the spatial and time variables. To solve the equation, we use the transformation

$$\begin{cases} \alpha - \sqrt{2}\phi = -\eta \\ \sqrt{2}\alpha - \phi = \xi \end{cases}\tag{12}$$

which has the properties of a proper Lorentz transformation in the Minisuperspace (note that n is set equal to $6\sqrt{2}$ in order to satisfy the orthochronous criterion). Thus, the kinetic term remains diagonal, and the WDW equation in the new variables becomes

$$\left[\frac{\partial^2}{\partial\xi^2} - \frac{\partial^2}{\partial\eta^2} + Ce^{-6\eta}\right]\Psi(\xi, \eta) = 0.\tag{13}$$

For a detailed discussion on the issue of selecting the unitarily equivalent physical solutions at a quantum level in constrained systems, see [40]. We notice that (13) is isomorphic to the equation analyzed in [33], in which the possibility of recovering a proper Hilbert space in Wheeler–DeWitt theory in those cases in which "asymptotically positive energy solutions" can be recovered is widely discussed.

It is worth noting that the change in variables in (12) mixes the spatial and time coordinates, making their immediate interpretation difficult. Recalling the hypothesis that the introduced potential is time-dependent, η can be interpreted as the time variable, and ξ, as the spatial one. It is worth stressing that such a potential term is only relevant near the singularity and thus guarantees the existence of the free expanding branch of the Universe dynamics. Actually, besides the cosmological justification mentioned above, this potential allows for the choice of purely positive-frequency solutions. Moreover, in this approach, it plays the role of a quantum scattering source; therefore, it has to be considered significantly different from zero only in the Planck era.

By analyzing the Hamilton equations for the new variables, we obtain $\frac{d\xi}{d\eta} = -\frac{p_\xi}{p_\eta}$, so the classical relation between the new spatial and time coordinates is the same as (6). This is due to the fact that proper Lorentz transformations do not invert the arrow of time. Therefore, in the new variables, it remains valid that the positive-frequency solutions can be associated to an expanding Universe, and the negative-frequency ones, to a collapsing Universe.

Now, we propose the solution

$$\Psi^k(\xi, \eta) = \Phi^k(\eta)\psi^k(\xi)\tag{14}$$

to (13), where $\psi_k(\zeta) = e^{ik\zeta}$ and

$$\begin{aligned}
\Phi^k(\eta) &= \Phi^k_+(\eta) + \Phi^k_-(\eta) \\
&= 6^{\frac{-ik}{3}} \mathcal{B}_I\big(-ik/3, e^{-3\eta}\sqrt{C}/3\big)\Gamma(1 - ik/3) \\
&+ 6^{\frac{ik}{3}} \mathcal{B}_I\big(ik/3, e^{-3\eta}\sqrt{C}/3\big)\Gamma(1 + ik/3),
\end{aligned} \tag{15}$$

in which $\mathcal{B}_I$ are the modified Bessel functions of the first kind and Γ is the Euler gamma function. As before, we write the general solution $\Psi(\zeta, \eta)$ using Gaussian packets as

$$\Psi(\zeta, \eta) = \frac{1}{\sqrt{2\pi\sigma^2}} \int_{-\infty}^{+\infty} e^{-\frac{(k-\bar{k})^2}{2\sigma^2}} \Phi^k(\eta) e^{ik\zeta}\, dk. \tag{16}$$

Differently from the previous case, here, the potential term has mixed the branches, making it impossible to distinguish them in terms of frequencies. Moreover, it is no longer possible to find an explicit expression for the eigenvalue of operator $\hat{p}_\eta$ in terms of the eigenvalue of $\hat{p}_\zeta$, since separation constant k appears to be a complex index of the Bessel functions. We finally note that the Bessel functions tend to the free solutions in the limit $\eta \to +\infty$, i.e.,

$$\lim_{\eta \to +\infty} \Phi^k_+(\eta) = e^{ik\eta} \quad \text{and} \quad \lim_{\eta \to +\infty} \Phi^k_-(\eta) = e^{-ik\eta}. \tag{17}$$

In other words, when the time-dependent potential becomes negligible, $\Psi^k(\zeta, \eta)$ is that of the free case; therefore, we recover the picture described in the previous section. In particular, in (15), the two different asymptotic behaviors (i.e., expanding and collapsing) are equally weighed.

4. Quantum Big Bounce

Here, we propose a probabilistic approach to the Big Bounce within WDW theory. At the basis of the formalism used below, there is the analogy with relativistic quantum scattering theory. In the same spirit, we assume that the Bounce is a quantum interaction between single particle states, i.e., they can be described by a wave function. In this framework, we associate a probability to the transition of the Universe from a collapsing state to an expanding one, in analogy with the fundamental interaction processes between particles below the particle creation threshold [24]. In the presence of a potential, the general solution of the modified KG equation

$$(\Box_x + m^2 + U(x))\Psi(x, t) = 0 \tag{18}$$

can be found in terms of the propagator for the free scalar field and calculated with arbitrary precision as

$$\Psi(x, t) = \Phi(x, t) + \int \Delta_F(x - y) U(y)\Psi(y, t)\, d^4y, \tag{19}$$

where $\Phi(x, t)$ is the solution of the homogeneous problem and $\Delta_F(x - y)$ is the propagator, which has the property of making the positive-frequency solutions evolve forwards in time, and the negative-frequency ones, backwards. In this formalism, the scattering amplitude between two states is simply the Klein–Gordon projection of the interacting solution onto the initial free one, by using expression (19) for the former (for the complete derivation, see [24]). In this case, it can be demonstrated that the probability associated to the transition amplitude can be written as

$$|S_{\bar{k}', \bar{k}}|^2 = \left| -i \iint_{-\infty}^{+\infty} \psi^*_+(\zeta, \eta) U(\eta)\Psi(\zeta, \eta)\, d\eta\, d\zeta \right|^2, \tag{20}$$

where $\Psi(\xi, \eta)$ coincides with the solution resulting from the interaction of a collapsing Universe with potential $U(\eta) = Ce^{-6\eta}$ (see (16)) and $\psi_+^*(\xi, \eta)$ is an emerging free wave associated only to positive frequencies:

$$\psi_+^*(\xi, \eta) = \frac{1}{\sqrt{2\pi\sigma'^2}} \int_0^{+\infty} \frac{e^{-\frac{(k'-\bar{k}')^2}{2\sigma'^2}}}{\sqrt{2w_{k'}}} e^{i(k'\xi - w_k'\eta)} \, dk' , \tag{21}$$

i.e., the branch of the expanding Universe. We remark that the probability amplitude in (20) refers to a transition from a negative (collapsing) frequency state to a positive (expanding) frequency state by definition. Actually, the scattering amplitude between two "particles" or "antiparticles" would have contained a Dirac delta term added to the integral in (20) (see [24]). After both analytical and numerical integrations, transition probability $|S(\bar{k}, \bar{k}')|^2$ results to be only dependent on values $(\bar{k}, \bar{k}')$ associated to the in-going (collapsing) and out-going (expanding) wave packets, respectively. So, once the value of $\bar{k}$ is fixed, the result is a function of $\bar{k}'$, as shown in Figure 1.

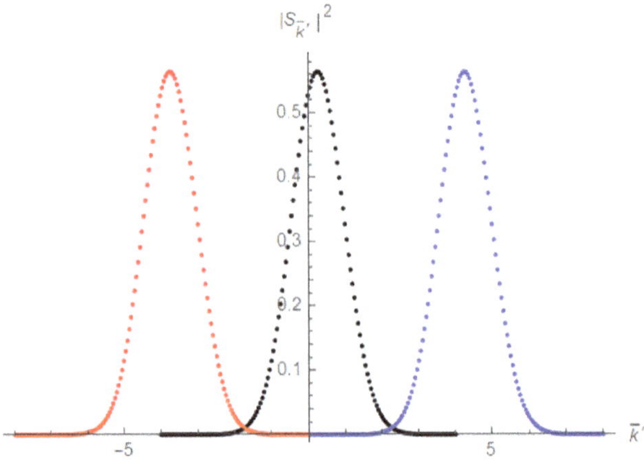

Figure 1. Plot of $|S(\bar{k}, \bar{k}')|^2$ as a function of $\bar{k}'$. Starting from the left, the transition probability is calculated for $\bar{k} = -4, 0, 4$, respectively. In all the three cases, $\sigma = \sigma' = 1$, $C = 1$, and an integration step equal to 0.05 is used.

We notice that the transition probability between a collapsing Universe and an expanding Universe is well defined. In particular, its peak position indicates that the wave packet with $\bar{k}' \approx \bar{k}$ maximizes the probability of the Bounce. It is worth noting that the phase shift observed in Figure 1 between $\bar{k}'$ and $\bar{k}$ is due to a mathematical feature of the Bessel functions, and it has no physical meaning. Finally, we stress that the KG probability density becomes positive-defined only when frequency separation is performed (i.e., when the potential term is negligible). Surprisingly, this feature holds also when we overlap frequency modes of the same sign by using Gaussian weights (for a detailed discussion of this question, see [22]). However, when a time-dependent potential is turned on, the frequency separation is forbidden, and the KG norm is not globally conserved anymore, so its interpretation as a probability completely loses its meaning. Actually, the approach we adopt here goes beyond the troublesome probabilistic interpretation of the KG wave function by resorting to the relativistic quantum formalism of scattering processes described above. Here, the probability associated to the Quantum Big Bounce is simply the square modulus of the transition amplitude constructed by projecting the in-going state onto the out-going one, according to the KG scalar product. Hence, the positive nature of the Quantum Big Bounce probability density as defined in (20) is automatically guaranteed. In this respect, the present construction is a natural implementation of the relativistic quantum theory procedures into the Minisuperspace of the considered cosmological model.

5. Quantum Big Bounce in the Polymer Paradigm

In the last part of the work, we study the system by implementing the PQM formalism [21]. The ultimate goal is to make a comparison with the transition amplitude obtained in the previous formalism and analyze if there are significant variations in the two cases. What follows will briefly recall the basic features of PQM.

5.1. The Polymer Representation of Quantum Mechanics

PQM is an alternative representation of quantum mechanics introduced by Corichi in [21] that results non-equivalent to the standard Schrödinger one. In particular, it is based on the assumption that the configurational variables are discrete, and for this reason, its main applications regard the investigation of cut-off physics effects in quantum cosmology theories.

Let us consider abstract kets $|\mu\rangle$ labeled by real parameter $\mu \in \mathbb{R}$, so that a generic state in Hilbert space $\mathcal{H}_{poly}$ can be defined with a finite linear combination of them, i.e.,

$$|\psi\rangle = \sum_{i=1}^{N} a_i |\mu_i\rangle, \tag{22}$$

where $\mu_i \in \mathbb{R}$, $i = 1, \ldots, N \in \mathbb{N}$. Also, let us fix the inner product

$$\langle \mu | \nu \rangle = \delta_{\mu\nu} \tag{23}$$

in order to guarantee the orthonormality between the basis kets. It can be demonstrated that such Hilbert space $\mathcal{H}_{poly}$ is non-separable. Two fundamental operators on $\mathcal{H}_{poly}$ can be defined as follows: symmetric operator $\hat{\epsilon}$, which labels the kets, and unitary operator $\hat{s}(\lambda)$, with $\lambda \in \mathbb{R}$, which shifts them. Their action is

$$\hat{\epsilon}|\mu\rangle := \mu|\mu\rangle \tag{24}$$

and

$$\hat{s}(\lambda)|\mu\rangle := |\mu + \lambda\rangle \tag{25}$$

respectively. In particular, since kets $|\mu\rangle$ are orthogonal, the eigenvalues of the label operator constitute a discrete set, and the shift operator results to be discontinuous in λ.

In order to provide an explicit representation of these two operators, let us consider a one-dimensional system (q, p), in which configurational coordinate q has a discrete character. It is easy to see that in p-polarization, operator $\hat{q}$ acts as a differential operator as

$$\hat{q} \cdot \psi_\mu(p) = -i\frac{\partial}{\partial p}\psi_\mu(p) = \mu\psi_\mu(p) \tag{26}$$

and corresponds to label operator $\hat{\epsilon}$. We remark that the eigenvalues of $\hat{q}$ can be considered a discrete set, since they label kets that are all orthonormal. On the other hand, the shift operator acts as

$$\hat{s}(\lambda) \cdot \psi_\mu(p) = e^{\frac{i\lambda p}{\hbar}} e^{\frac{i\mu p}{\hbar}} = e^{\frac{i(\mu+\lambda)p}{\hbar}} = \psi_{\mu+\lambda}(p) \tag{27}$$

and is discontinuous, since the final state is always orthonormal to the initial one, no matter how small λ is. This feature results in the consequence that $\hat{p}$ cannot be defined rigorously, since no Hermitian operator can generate a discontinuous operator by exponentiation. Therefore, a proper definition for physical operators $\hat{p}$ and $\hat{q}$ is needed in order to deal with a well-defined dynamics. Actually, in PQM, if one of the two is discrete, then the other one must be regularized. The standard procedure consists in the introduction of a lattice with constant spacing μ

$$\gamma_\mu = \{q \in \mathbb{R} : q = n\mu, \; \forall n \in \mathbb{Z}\} \tag{28}$$

on which to restrict the action of operator $e^{\frac{i\lambda\hat{p}}{\hbar}}$, so that it is possible to use it to define an approximate version of $\hat{p}$ as

$$
\begin{aligned}
\hat{p}_\mu |\mu_n\rangle \quad &:= \frac{\hbar}{2i\mu}[e^{\frac{i\mu\hat{p}}{\hbar}} - e^{-\frac{i\mu\hat{p}}{\hbar}}]|\mu_n\rangle \\
&= \frac{\hbar}{2i\mu}(|\mu_{n+1}\rangle - |\mu_{n-1}\rangle),
\end{aligned}
\tag{29}
$$

i.e., the discretization of the derivative. Actually, for $\mu p \ll \hbar$, one obtains $p \curvearrowright \sin(\mu p)/\mu = \hbar(e^{\frac{i\mu p}{\hbar}} - e^{-\frac{i\mu p}{\hbar}})/2i\mu$. Accordingly, for $\hat{p}^2$, we obtain

$$
\begin{aligned}
\hat{p}^2_\mu |\mu_n\rangle \quad &:= \hat{p}_\mu \cdot \hat{p}_\mu |\mu_n\rangle \\
&= \frac{\hbar^2}{4\mu^2}[-|\mu_{n-2}\rangle + 2|\mu_n\rangle - |\mu_{n+2}\rangle] \\
&= \frac{\hbar^2}{\mu^2}\sin^2(\mu p)|\mu_n\rangle.
\end{aligned}
\tag{30}
$$

We remind that $\hat{q}$ is well defined, so the regularized version of the Hamiltonian is

$$\hat{H}_\mu := \frac{1}{2m}\hat{p}^2_\mu + \hat{V}(q) \tag{31}$$

and represents a symmetric and well-defined operator on $\mathcal{H}_{\gamma_\mu}$.

5.2. Transition Amplitude

Let us now compute the Quantum Bounce transition amplitude in the polymer formulation. First of all, in accordance with the theoretical structure of PQM, it is necessary to select which Minisuperspace variables are discrete and thus which ones should be regularized. When PQM is applied in a cosmological context, functions of the scale factor are usually chosen as discrete with the aim of trying to solve the singularity. Accordingly, in this model, α is the discrete variable; so, operator $\hat{p}_\alpha$ is formally replaced with

$$\hat{p}_\alpha \rightarrow \frac{\hbar}{\mu_\alpha}\sin\left(\frac{\mu_\alpha p_\alpha}{\hbar}\right) \tag{32}$$

in the momentum representation. From the semiclassical study of the Hamilton equations in the absence of the potential term, we obtain that p_α is still a constant of motion and that

$$\dot{\alpha} = -\frac{4\pi GNe^{-3\alpha}\hbar}{3c^3 V\mu_\alpha}\sin\left(\frac{2\mu_\alpha p_\alpha}{\hbar}\right). \tag{33}$$

Hence, it is clear that the same previous considerations regarding the possibility of discerning the expanding and collapsing branches of the Universe remain valid, since $\sin(2\mu_\alpha p_\alpha/\hbar)$ is odd in its argument. As usual, the polymer version of Hamiltonian (1) is obtained using (32) as a semiclassical approximation. Then, by promoting it to a quantum operator, we obtain the polymer-modified WDW equation, which corresponds to (2), in which (32) is rigorously implemented as a regularized version of operator $\hat{p}_\alpha$ in the momentum representation. The expression of the Universe wave packet as a superposition of functions weighed with Gaussian coefficients is analogous to the previous one (see (16)).

$$\psi_\pm(\alpha, \phi) = \frac{1}{\sqrt{2\pi\sigma^2}}\int_0^{+\infty} e^{-\frac{(k-\bar{k})^2}{2\sigma^2}} e^{i(\bar{p}_\alpha\alpha \mp k\phi)}\, dk, \tag{34}$$

where $\bar{p}_\alpha = \arcsin(\mu_\alpha k)/\mu_\alpha$. The main difference is represented by the dispersion relation (i.e., the relation between the quantum eigenvalues of the momenta), which stops being linear. It can be shown that this fact is responsible for the spreading of the wave packet (see Figure 2). Therefore, the comparison between the semiclassical trajectory and the evolution of the quantum wave packet loses its meaning, since the variance sooner or later becomes of the same order as the expectation value, making it necessary to resort to a fully quantum treatment. This behavior of the FLRW Universe wave packet in the polymer representation further consolidates the study of the Big Bounce as quantum scattering. We remark that PQM is not able to solve the singularity of the FLRW model in the Misner variable (see [25,26], where this argument is used to demonstrate the presence of the initial singularity in the Bianchi IX model); so, no semiclassical bouncing dynamics is present in this model.

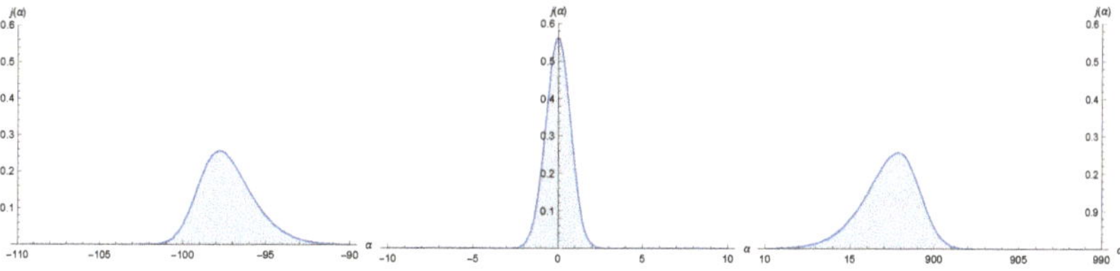

Figure 2. Probability density $j(\alpha) = i(\psi_+^*(\alpha,\bar{\phi})\partial_\phi\psi_+(\alpha,\bar{\phi}) - \psi_+(\alpha,\bar{\phi})\partial_\phi\psi_+^*(\alpha,\bar{\phi}))$ for the values of time $\bar{\phi} = -100, 0, 100$, respectively. As it can be seen, the variance in the probability density becomes larger as time goes by, showing the spreading of the wave packet during evolution. Both the integration step and μ_α are set equal to 0.1.

Since PQM does not commute with other coordinate transformations, we introduce it after the Lorentz transformation in (12). Hence, we start by considering the polymer version of the WDW equation in the (ξ, η) variables, i.e.,

$$\left[\frac{1}{\mu_\xi^2}\sin^2(\mu_\xi p_\xi) - \frac{\partial^2}{\partial\eta^2} + Ce^{-6\eta}\right]\Psi(p_\xi, \eta) = 0, \tag{35}$$

where substitution (32) is used. Then, the polymer transition probability can be computed as in (20), in which

$$\psi_+^*(\xi, \eta) = \frac{1}{\sqrt{2\pi\sigma'^2}}\int_0^{+\infty}\frac{e^{-\frac{(k'-\bar{k}')^2}{2\sigma'^2}}}{\sqrt{2w_{k'}}}e^{i(\bar{p}_\xi'\xi - w_k'\eta)}\,dk', \tag{36}$$

$U(\eta) = Ce^{-6\eta}$ and

$$\Psi(\xi, \eta) = \frac{1}{\sqrt{2\pi\sigma^2}}\int_{-\infty}^{+\infty}e^{-\frac{(k-\bar{k})^2}{2\sigma^2}}\Phi^k(\eta)e^{i\bar{p}_\xi\xi}\,dk. \tag{37}$$

After analytically integrating over variables ξ and k', we obtain

$$|S^{\mu_\xi}(\bar{k}', \bar{k})|^2 = \tag{38}$$

$$\left|-i\int_0^{+\infty}\frac{Ce^{-\frac{(k-\bar{k})^2}{2\sigma^2}}e^{-\frac{(k-\bar{k}')^2}{2\sigma'^2}}}{\sqrt{2w_k\sigma^2\sigma'^2}}\sqrt{1-\mu_\xi^2k^2}I(k)\,dk\right|^2,$$

where

$$I(k) = \int_{-\infty}^{+\infty}e^{\eta(ik-6)}\Phi^k(\eta)\,d\eta. \tag{39}$$

Then, polymer transition probability $|S^{\mu_\xi}(\bar{k}', \bar{k})|^2$ has been computed by using both analytical and numerical integrations. We notice that the polymer modification induced on (38) consists in the presence of global factor $\sqrt{1 - \mu_\zeta^2 k^2}$ and that in the limit $\mu_\zeta \to 0$, the standard scattering amplitude is recovered. Therefore, as we can see from Figure 3, the introduction of PQM does not significantly change neither the shape of transition probability $|S^{\mu_\zeta}(\bar{k}, \bar{k}')|^2$ nor the position of its peak (for a comparison with the previous case, see Figure 1). In particular, as in the standard case analyzed in Section 4, the maximum of the probability density occurs in correspondence of $\bar{k}' \approx \bar{k}$, where $\bar{k}'$ is associated to the expanding Universe wave packet emerging after the interaction, and $\bar{k}$, to the collapsing one. The absence of relevant effects is due to the fact that in (38), the exponential factors of the Gaussian coefficients dominate over polymer global factor $\sqrt{1 - \mu_\zeta^2 k^2}$ in the integral.

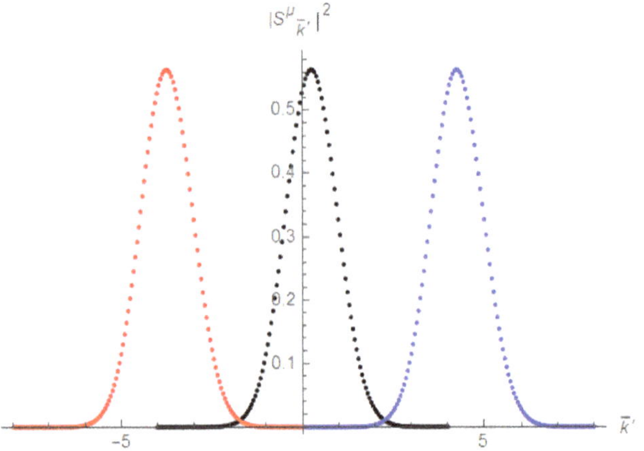

Figure 3. Plot of $|S^{\mu_\zeta}(\bar{k}', \bar{k})|^2$ as a function of $\bar{k}'$ in the polymer paradigm. Starting from the left, the polymer transition probability is calculated for $\bar{k} = -4, 0, 4$, respectively. In all the three cases, $\sigma = \sigma' = 1$; $C = 1, \mu_\zeta = 0.1$; and an integration step equal to 0.05 is used.

6. Discussion of the Results

Here, we develop some basic considerations that are useful to provide a physical interpretation of the analysis above. We start by observing that the collapsing and expanding branches are separated at a classical level and that no semiclassical dynamics is present, both in the Einsteinian and in the polymer picture addressed here. The situation is significantly different at a quantum level, where the expanding and collapsing branches co-exist as positive- and negative-frequency solutions. The crucial point is that the two frequency solutions, i.e., the collapsing and expanding branches, can no longer be separated when a self-interaction potential term (to be considered relevant only in the Planck regime) is introduced. As a result, the quantum nature of the singularity acquires a very different morphology compared with the standard expectancy, according to which the trajectories are localized near the two singular branches. Actually, in the polymer picture, a quasi-classical localization of the dynamics is only possible for a finite time interval, and the solution describing the two branches should be regarded as a non-localized state near the singularity, even in the absence of a quantum potential term. Thus, here, we state that it is possible to obtain an emerging classical expanding Universe for $\eta \to +\infty$ from the Planck quantum state in which the frequency solutions are not separable. For the reasons explained above, a quantum transition between a mixed Planck state and the expanding (but, in principle, also collapsing) Universe emerges as a new scenario that can be interpreted as a Quantum Big Bounce. We note that the precise morphology of a quantum scattering process would require the quantum potential to be a time transient, i.e., only relevant in a finite interval of ϕ. Nevertheless, we stress that dealing with the exact solution of the interaction regime makes it possible to use the relativistic quantum treatment, even if such a potential term is not perturbative in this region. We conclude by observing that the present study must be regarded as an improvement to the original proposal in [22], in which the Universe volume is adopted as time. Here, the choice of relational time would intrinsically fulfill the requirement of a monotonic behavior even when considering a semiclassical/quantum bouncing dynamics.

7. Conclusions

In this work, we analyze the quantum dynamics of the isotropic Universe in the presence of a self-interacting scalar field, which plays the role of relational time [23]. When the scalar field is free of its potential, the WDW equation has the morphology of a two-dimensional massless KG equation, and we are able to identify the positive- and negative-frequency modes as the expanding and collapsing phases of the Universe, respectively. Such an identification is found by comparing the classical dynamics with the behavior of localized quantum wave packets. Then, when the potential term of the matter clock is introduced, the frequency separation is broken; so, we apply the standard techniques of relativistic quantum scattering below the particle creation threshold [24], in order to obtain a transition probability from a collapsing Universe to an expanding Universe. It is worth noting that we use an orthochronous Lorentz-like transformation in the Minisuperspace when performing such a calculation, in order to make it possible to analytically solve the WDW equation. In both cases, i.e., the standard WDW equation and the polymer-modified scheme, the transition probability results similar in morphology. Furthermore, in the latter scenario, the Universe wave packet unavoidably spreads towards the (still existing) initial singularity, and the Big Bounce description in terms of a probabilistic phenomenon becomes mandatory. Actually, in both cases, the idea of dealing with a Quantum Big Bounce is made precise by observing that the maximum value of the transition probability is taken when the expectation value of the in-going wave packet momentum is close to the corresponding one in the out-going state, as it takes place before and after a semiclassical Big Bounce. Since the existence of a self-interacting scalar field in the early Universe appears natural in many fundamental approaches [2], the possibility of dealing with a Quantum Big Bounce (in the sense defined here) can be considered a rather general feature of the canonical quantum dynamics in the metric formalism, even in more general cosmological models like the Bianchi Universes or the generic inhomogeneous cosmological solution [41,42].

Author Contributions: Conceptualization, G.M., E.G. and F.M.; methodology, G.M., E.G. and F.M.; formal analysis, E.G., F.M. and G.M.; writing-original draft preparation, F.M., E.G. and G.M.; writingreview and editing, E.G., G.M. and F.M. All authors have read and agreed to the published version of the manuscript.

Funding: The work of E.G. is supported by the Della Riccia Foundation grant for the year 2023.

Institutional Review Board Statement: Not applicable.

Informed Consent Statement: Not applicable.

Data Availability Statement: Not applicable.

Conflicts of Interest: The authors declare no conflict of interest.

References

1. Landau, L.D.; Lifshitz, E.M. *The Classical Theory of Fields*; Butterworth-Heinemann: Oxford, UK, 1980.
2. Montani, G. *Primordial Cosmology*; World Scientific: Singapore, 2011.
3. Bombacigno, F.; Cianfrani, F.; Montani, G. Big-bounce cosmology in the presence of Immirzi field. *Phys. Rev. D* **2016**, *94*, 064021. [CrossRef]
4. Bombacigno, F.; Montani, G. Big bounce cosmology for Palatini R^2 gravity with a Nieh–Yan term. *Eur. Phys. J. C* **2019**, *79*, 405. [CrossRef]
5. Olmo, G.J. Palatini Approach Beyond Einstein's Gravity. 2011. Available online: http://xxx.lanl.gov/abs/1112.1572 (accessed on 8 August 2023).
6. DeWitt, B.S. Quantum theory of gravity. 1. The canonical theory. *Phys. Rev.* **1967**, *160*, 1113–1148. [CrossRef]

7. Cianfrani, F.; Lecian, O.; Lulli, M.; Montani, G. *Canonical Quantum Gravity: Fundamentals and Recent Developments*; World Scientific Publishing Company: Singapore, 2014.

8. Rovelli, C.; Smolin, L. Knot Theory and Quantum Gravity. *Phys. Rev. Lett.* **1988**, *61*, 1155–1158. [CrossRef] [PubMed]

9. Ashtekar, A.; Rovelli, C.; Smolin, L. Weaving a classical metric with quantum threads. *Phys. Rev. Lett.* **1992**, *69*, 237–240. [CrossRef]

10. Rovelli, C.; Smolin, L. The physical Hamiltonian in nonperturbative quantum gravity. *Phys. Rev. Lett.* **1994**, *72*, 446–449. [CrossRef]

11. Rovelli, C.; Smolin, L. Discreteness of area and volume in quantum gravity. *Nucl. Phys. B* **1995**, *442*, 593–622. [CrossRef]

12. Ashtekar, A.; Bojowald, M.; Lewandowski, J. Mathematical Structure of Loop Quantum Cosmology. *Adv. Theor. Math. Phys.* **2003**, *7*, 233–268. [CrossRef]

13. Ashtekar, A.; Pawlowski, T.; Singh, P. Quantum nature of the big bang: An analytical and numerical investigation. *Phys. Rev. D* **2006**, *73*, 124038. [CrossRef]

14. Ashtekar, A.; Pawlowski, T.; Singh, P. Quantum nature of the big bang: Improved dynamics. *Phys. Rev. D* **2006**, *74*, 084003. [CrossRef]

15. Ashtekar, A.; Corichi, A.; Singh, P. Robustness of key features of loop quantum cosmology. *Phys. Rev. D* **2008**, *77*, 024046. [CrossRef]

16. Ashtekar, A. Singularity resolution in loop quantum cosmology: A brief overview. *J. Phys. Conf. Ser.* **2009**, *189*, 012003. [CrossRef]

17. Ashtekar, A.; Singh, P. Loop Quantum Cosmology: A Status Report. *Class. Quantum Gravity* **2011**, *28*, 213001. [CrossRef]

18. Bojowald, M. Isotropic loop quantum cosmology. *Class. Quantum Gravity* **2002**, *19*, 2717–2741. [CrossRef]

19. Bojowald, M. Loop quantum cosmology: Recent progress. *Pramana* **2004**, *63*, 765–776. [CrossRef]

20. Barca, G.; Giovannetti, E.; Montani, G. An Overview on the Nature of the Bounce in LQC and PQM. *Universe* **2021**, *7*, 327. [CrossRef]

21. Corichi, A.; Vukašinac, T.; Zapata, J.A. Polymer Quantum Mechanics and its continuum limit. *Phys. Rev. D* **2007**, *76*, 044016. [CrossRef]

22. Giovannetti, E.; Montani, G. Is Bianchi I a Bouncing Cosmology in the Wheeler-DeWitt picture? *Phys. Rev. D* **2022**, *106*, 044053. [CrossRef]

23. Rovelli, C. Time in quantum gravity: An hypothesis. *Phys. Rev. D* **1991**, *43*, 442–456. [CrossRef]

24. Bjorken, J.D.; Drell, S.D. *Relativistic Quantum Mechanics*; International Series in Pure and Applied Physics; McGraw-Hill: New York, NY, USA, 1965.

25. Crinò, C.; Montani, G.; Pintaudi, G. Semiclassical and quantum behavior of the Mixmaster model in the polymer approach for the isotropic Misner variable. *Eur. Phys. J. C* **2018**, *78*, 886. [CrossRef]

26. Giovannetti, E.; Montani, G. Polymer representation of the Bianchi IX cosmology in the Misner variables. *Phys. Rev. D* **2019**, *100*, 104058. [CrossRef]

27. Ziprick, J.; Gegenberg, J.; Kunstatter, G. Polymer quantization of a self-gravitating thin shell. *Phys. Rev. D* **2016**, *94*, 104076. [CrossRef]

28. DeWitt, B.S. Quantum theory of gravity. 2. The manifestly covariant theory. *Phys. Rev.* **1967**, *162*, 1195–1239. [CrossRef]

29. DeWitt, B.S. Quantum theory of gravity. 3. Applications of the covariant theory. *Phys. Rev.* **1967**, *162*, 1239–1256. [CrossRef]

30. Kuchar, K. Canonical Methods of Quantization. In Proceedings of the Oxford Conference on Quantum Gravity, Oxford, UK, 15–19 April 1980.

31. Isham, C.J. Canonical quantum gravity and the problem of time. *NATO Sci. Ser. C* **1993**, *409*, 157–287.

32. Wald, R.M. Proposal for solving the "problem of time" in canonical quantum gravity. *Phys. Rev. D* **1993**, *48*, R2377–R2381. [CrossRef]

33. Higuchi, A.; Wald, R.M. Applications of a new proposal for solving the "problem of time" to some simple quantum cosmological models. *Phys. Rev. D* **1995**, *51*, 544–561. [CrossRef]

34. Misner, C.W. Mixmaster Universe. *Phys. Rev. Lett.* **1969**, *22*, 1071–1074. [CrossRef]

35. Misner, C.W. Quantum Cosmology. I. *Phys. Rev.* **1969**, *186*, 1319–1327. [CrossRef]

36. Kan, N.; Aoyama, T.; Hasegawa, T.; Shiraishi, K. Third quantization for scalar and spinor wave functions of the Universe in an extended minisuperspace. *Class. Quant. Grav.* **2022**, *39*, 165010. [CrossRef]

37. Geng, C.Q.; Lee, C.C.; Sami, M.; Saridakis, E.N.; Starobinsky, A.A. Observational constraints on successful model of quintessential Inflation. *J. Cosmol. Astropart. Phys.* **2017**, *2017*, 011. [CrossRef]

38. Dimopoulos, K.; Karam, A.; López, S.S.; Tomberg, E. Palatini R^2 Quintessential Inflation. *J. Cosmol. Astropart. Phys.* **2022**, *2022*, 076. [CrossRef]

39. Dimopoulos, K.; Karam, A.; López, S.; Tomberg, E. Modelling Quintessential Inflation in Palatini-Modified Gravity. *Galaxies* **2022**, *10*, 57. [CrossRef]

40. Barvinsky, A.O.; Kamenshchik, A.Y. Selection rules for the Wheeler-DeWitt equation in quantum cosmology. *Phys. Rev. D* **2014**, *89*, 043526. [CrossRef]

41. Benini, R.; Montani, G. Frame independence of the inhomogeneous mixmaster chaos via Misner-Chitré -like variables. *Phys. Rev. D* **2004**, *70*, 103527. [CrossRef]
42. Benini, R.; Montani, G. Inhomogeneous quantum Mixmaster: From classical towards quantum mechanics. *Class. Quantum Gravity* **2006**, *24*, 387–404. [CrossRef]

Article

Nothing into Something and Vice Versa: A Cosmological Scenario

Leonid Marochnik

Physics Department, University of Maryland, College Park, MD 20742, USA; lmarochnik@gmail.com

Abstract: In the almost empty universe (with almost no matter in it), stochastic gravitational waves (SGW) of finite amplitude produce a de Sitter regime as a solution, which is invariant with respect to the Wick rotation. Asymptotically, super horizon SGWs do not "feel" difference between Lorentzian and Euclidean spacetime and belong simultaneously to both of them. The universe is finishing its evolution in Euclidean spacetime, i.e., it disappears into nothing. Quantum fluctuations of the gravitational field (gravitons) produce a de Sitter regime again in Euclidean spacetime where the current universe finished its existence, and due to the invariance of the de Sitter regime with respect to Wick rotation, the next universe starts its life with de Sitter inflation in Lorentzian spacetime. Such a scenario assumes that a permanent process of birth, death and rebirth of an infinite sequence of universes takes place on an infinite time axis.

Keywords: universes; birth; death; rebirth; permanently

check for **updates**

Citation: Marochnik, L. Nothing into Something and Vice Versa: A Cosmological Scenario. *Universe* **2023**, *9*, 445. https://doi.org/10.3390/universe9100445

Academic Editor: Kazuharu Bamba

Received: 29 August 2023
Revised: 24 September 2023
Accepted: 27 September 2023
Published: 9 October 2023

1. Introduction

The contribution of visible and invisible matter to the dynamics of the modern universe is approximately 30% [1,2], i.e., very small. In the process of expansion of the universe, the energy density of matter in it decreases as $a(t)^{-3}$, where $a(t)$ is a scale factor. Therefore, when the universe expands, for example, three times, then the energy density of matter in it will be 1%, i.e., the influence of matter on the dynamics of the universe can be neglected. The universe will turn out to be composed of gravitational waves of finite amplitude, which will have to determine its asymptotic behavior. Of course, some contribution to the overall energy balance will be made by the already-discovered gravitational waves caused by the collisions of black holes and other similar processes [3–6]. However, these facts only confirm the reality of the existence of gravitational waves, which were hypothetical until confirmed by the LIGO and Virgo detectors. When we are dealing with small amplitude fluctuations of a metric tensor, as Lifshitz once showed [7], they can be divided into three types: scalar, vector and tensor (gravitational waves). Such a separation is not possible for fluctuations of finite amplitude due to the non-linearity of the Einstein equations. However, in the absence of matter, any fluctuations of finite amplitude are gravitational waves of finite amplitude, and since only they remain asymptotically, they must also determine the evolution of the Universe. The present work is devoted precisely to this issue.

The exact equations for the stochastic gravitational waves (SGW) in spacetime with fluctuations of the metrical tensor g_{ik} of arbitrary form were obtained by the late Grigory Vereshkov [[8], Appendix A.1.][1]. The general theory of such waves with arbitrary tensor g_{ik} covers three full journal pages. Thus, we are unable to reproduce them here for the reader's convenience. However, in Appendix A.2., the exact equations for SGWs in the FLRW metric are also presented. For the convenience of the reader, we present these equations again in Section 2. In Section 3, we present the exact solution to the set of these equations. The discussion of the obtained solution is the content of Section 4. In this work we assumed from the very beginning that the universe is empty in the sense that it is already at that stage of expansion when the contribution of matter to the overall dynamics is already negligibly small.

2. Grigori Vereshkov's Equations for Stochastic Gravitational Waves of Finite Amplitude

The general approach to the problem is described in Appendix A.1. of [8]. The situation was considered when the geometric characteristics of spacetime metric $\hat{g}_{ik}$, the connection $\hat{\Gamma}^l_{ik}$ and the curvature $\hat{R}_{ik}$ are fluctuating functions for some physical reasons. It is assumed, however, that there exist regularly determined components of these functions g_{ik}, $\Gamma^l{}_{ik}$ and R_{ik}. It was assumed also that the standard relations of Riemannian geometry are satisfied. Extracting the background geometry from the fluctuating geometry is a non-trivial problem because of the nonlinearity of Einstein's equations. However, the functional integration method allows us to do that. Also, it is considered that the mean value of the random function $< \psi_i{}^k >$ in the statistical ensemble is zero, i.e., $< \psi_i{}^k >= 0$ by definition. Referring the reader for physical and mathematical details to Section A.1. of Appendix A of work [8], I will proceed to Section A.2. "Stochastic Nonlinear Gravitational Waves over the FLRW Background", where the equations of stochastic gravitational waves in the FLRW metric are written explicitly.

We take the FLRW metric of the flat universe as the background metric. In this case, we have

$$ds^2 = dt^2 - a(t)^2 \gamma_{\alpha\beta} dx^\alpha dx^\beta = dt^2 - a(t)^2 \left(dx^2 + dy^2 + dz^2\right)$$
$$R_0{}^0 = -3\frac{\ddot{a}}{a} \quad R_\alpha{}^\beta = -\delta_\alpha{}^\beta \left(\frac{\ddot{a}}{a} + 2\frac{\dot{a}^2}{a^2}\right) \tag{1}$$

In the synchronous gauge we have

$$\psi_0{}^0 = 0 \quad \psi_0{}^\alpha = 0 \tag{2}$$

Einstein's equations in the explicit form read (before averaging, [4], Appendix A)

$$\sqrt{\frac{\hat{g}}{g}}\hat{g}^{0l}\hat{R}_{0l} \equiv -3\frac{\ddot{a}}{a} - \frac{1}{4}\left\{\ddot{\psi} - \frac{\dot{a}}{a}\dot{\psi} - \frac{1}{a^2}\left(X_\mu{}^\nu \psi'^\mu\right)_{,\nu}\right\} - \frac{1}{4}\dot{\psi}_\mu{}^\nu \dot{\psi}_\nu{}^\mu + \frac{1}{8}\dot{\psi}^2 = 0 \tag{3}$$

$$\sqrt{\frac{\hat{g}}{g}}\hat{g}^{\beta l}\hat{R}_{\alpha l} \equiv \ -\delta_\alpha{}^\beta \left(\frac{\ddot{a}}{a} + 2\frac{\dot{a}^2}{a^2}\right) + \frac{1}{2}\left\{-\frac{1}{2}\delta_\alpha{}^\beta \left[\ddot{\psi} + 3\frac{\dot{a}}{a}\dot{\psi} - \frac{1}{a^2}\left(X_\mu{}^\nu \psi'^\mu\right)_{,\nu}\right] \right.$$
$$-\frac{1}{a^2}\left(X_\mu{}^\nu \dot{\psi}_\alpha{}^{\beta,\mu} - X_\mu{}^\beta \psi_\alpha{}^{\nu,\mu} - X_\mu{}^\nu \psi_\alpha{}^{\mu,\beta}\right)_{,\nu}\bigg\} \tag{4}$$
$$+\frac{1}{4a^2}\left[X_\lambda{}^\beta \psi_{\mu,\alpha}{}^\nu \psi_\nu{}^{\mu,\lambda} - \frac{1}{2}X_\lambda{}^\beta \psi_{,\alpha}\psi'^\lambda - 2X_\alpha{}^\nu \psi_{\nu,\mu}{}^\lambda \psi^{\mu,\beta}{}_\lambda\right]$$

$$\sqrt{\frac{\hat{g}}{g}}\hat{g}^{0l}\hat{R}_{\alpha l} \equiv \frac{1}{2}\left(-\dot{\psi}_{\alpha,\mu}{}^\mu + \frac{\dot{a}}{a}\psi_{,\alpha}\right) - \frac{1}{4}\left(\psi_{\nu,\alpha}{}^\mu \dot{\psi}_\mu{}^\nu\right) - \frac{1}{2}\psi_{,\alpha}\dot{\psi} - 2X_\alpha \tag{5}$$

$$X_l^k \equiv (\exp\psi)_l^k = \delta_l^k + \hat{\psi}_l^k + \frac{1}{2!}\hat{\psi}_l^m\hat{\psi}_m^k + \frac{1}{3!}\hat{\psi}_l^m\hat{\psi}_m^n\hat{\psi}_n^k + \dots \tag{6}$$

In this approach, the exponential parameterization Equation (6) has been used, which automatically provides conservation of the energy–momentum tensor (EMT) of gravitational waves [9][2]. In these equations, dots are derivatives over physical time t. All operations with the spatial indexes are conducted with Euclidean metric. Averaging of Equations (3) and (4)[3] leads to the fact that all terms in the curly brackets are zeroed (because the mean of the random variable ψ is zero) and we get

$$-3\frac{\ddot{a}}{a} = \frac{1}{4} < \dot{\psi}_\mu{}^\nu \dot{\psi}_\nu{}^\mu - \frac{1}{2}\dot{\psi}^2 > \tag{7}$$

$$3\frac{\ddot{a}}{a} + 6\frac{\dot{a}^2}{a^2} = \frac{1}{4a^2} < X_\lambda{}^\alpha \psi_{\mu,\alpha}{}^\nu \psi_\nu{}^{\mu,\lambda} - \frac{1}{2}X_\lambda{}^\alpha \psi_{,\alpha}\psi'^\lambda - 2X_\alpha{}^\nu \psi_{\nu,\mu}{}^\lambda \psi_\lambda{}^{\mu,\alpha} > \tag{8}$$

We can define the energy density and pressure of nonlinear gravitational wave medium as follows

$$3\frac{\dot{a}^2}{a^2} = \kappa\rho_{GW} \tag{9}$$

$$\frac{\ddot{a}}{a} = -\frac{\kappa}{6}(\rho_{GW} + 3p_{GW}) \tag{10}$$

where $\kappa = 8\pi G$, speed of light $c = 1$ and G is the gravitational constant. Thus, the energy density of such a nonlinear gravitational wave medium reads

$$\kappa\rho_{GW} = \frac{1}{8} < \dot{\psi}_\mu{}^\nu \dot{\psi}_\nu{}^\mu - \frac{1}{2}\dot{\psi}^2 + \frac{1}{a^2}\left(X_\mu{}^\nu \psi_\lambda{}^{\sigma,\mu}\psi_{\sigma,\nu}{}^\lambda - 2X_\mu{}^\nu\psi_\lambda{}^{\sigma,\mu}\psi_{\nu,\sigma}{}^\lambda - \frac{1}{2}X_\mu{}^\nu\psi_{,\nu}\psi^{,\mu}\right) > \tag{11}$$

The pressure of such a nonlinear gravitational wave medium can be found from Equations (7), (10) and (11). Finally, we turn to the equations for gravitational waves. They need to be divided into equations of constraints and equations of proper dynamics. The constraint equations can be obtained from (3)–(8). The equations of proper dynamics follow from (4). They read

$$\begin{aligned}
&\ddot{\psi}_\alpha{}^\beta + 3\frac{\dot{a}}{a}\dot{\psi}_\alpha{}^\beta - \frac{1}{a^2}\left(X_\nu{}^\mu\psi_\alpha{}^{\beta,\mu} - X_\mu{}^\beta\psi_\alpha{}^{\nu,\mu} - X_\mu{}^\nu\psi_\alpha{}^{\mu,\beta}\right)_{,\nu} \\
&-\frac{1}{2}\delta_\alpha{}^\beta\left[\ddot{\psi} + 3\frac{\dot{a}}{a}\dot{\psi} - \frac{1}{a^2}\left(X_\mu{}^\nu\psi^{,\mu}\right)_{,\nu}\right] = \\
&+\frac{1}{2a^2}\left[X_\lambda{}^\beta\psi^\nu{}_{\mu,\alpha}\psi_\nu{}^{\mu,\lambda} - \frac{1}{2}X_\lambda{}^\beta\psi_{,\alpha}\psi^{,\lambda} - 2X_\alpha{}^\nu\psi^\lambda{}_{\nu,\mu}\psi_\lambda{}^{\mu,\beta}\right] - \\
&\frac{1}{2a^2} < \left[X_\lambda{}^\beta\psi^\nu{}_{\mu,\alpha}\psi_\nu{}^{\mu,\lambda} - \frac{1}{2}X_\lambda{}^\beta\psi_{,\alpha}\psi^{,\lambda} - 2X_\alpha{}^\nu\psi^\lambda{}_{\nu,\mu}\psi_\lambda{}^{\mu,\beta}\right] >
\end{aligned} \tag{12}$$

Thus, Equations (7)–(12) describe the backreaction of non-linear SGWs of the arbitrary amplitude $\psi_\alpha{}^\beta$ on the background $a(t), p(t), \rho(t)$, which after averaging depends on t only. Equations (8), (11) and (12) can be rewritten in the following convenient form

$$3\frac{\ddot{a}}{a} + 6\frac{\dot{a}^2}{a^2} = \frac{1}{4} < EMT > \tag{13}$$

$$\kappa\rho_{GW} = \frac{1}{8} < \dot{\psi}_\mu{}^\nu \dot{\psi}_\nu{}^\mu - \frac{1}{2}\dot{\psi}^2 + EMT\right) > \tag{14}$$

$$\kappa p_{GW} = \frac{1}{8} < \dot{\psi}_\mu{}^\nu \dot{\psi}_\nu{}^\mu - \frac{1}{2}\dot{\psi}^2 - \frac{1}{3}EMT > \tag{15}$$

$$\begin{aligned}
&\ddot{\psi}_\alpha{}^\beta + 3\frac{\dot{a}}{a}\dot{\psi}_\alpha{}^\beta - \frac{1}{a^2}\left(X_\nu{}^\mu\psi_\alpha{}^{\beta,\mu} - X_\mu{}^\beta\psi_\alpha{}^{\nu,\mu} - X_\mu{}^\nu\psi_\alpha{}^{\mu,\beta}\right)_{,\nu} \\
&-\frac{1}{2}\delta_\alpha{}^\beta\left[\ddot{\psi} + 3\frac{\dot{a}}{a}\dot{\psi} - \frac{1}{a^2}\left(X_\mu{}^\nu\psi^{,\mu}\right)_{,\nu}\right] = \frac{1}{2}(EMT - < EMT >)
\end{aligned} \tag{16}$$

where the energy–momentum tensor (EMT) is

$$EMT = \frac{1}{a^2}\left[X_\lambda{}^\beta\psi^\nu{}_{\mu,\alpha}\psi_\nu{}^{\mu,\lambda} - \frac{1}{2}X_\lambda{}^\beta\psi_{,\alpha}\psi^{,\lambda} - 2X_\alpha{}^\nu\psi^\lambda{}_{\nu,\mu}\psi_\lambda{}^{\mu,\beta}\right] \tag{17}$$

3. Solution for (13)–(16) Is the de Sitter Regime

In this section, we show that the de Sitter regime is a solution to Equations (13)–(16). One can check this statement by substitution of $a(t) = exp(Ht)$ into the LHSs of (13)–(16). From the LHS of (13) one can find the average value of EMT, which is

$$< EMT >= 36H^2 \tag{18}$$

After that, we get from the LHSs of (14) and (15)

$$\kappa \rho_{GW} = 3H^2 \quad \kappa p_{GW} = -3H^2 \tag{19}$$

Thus, (19) confirms our hypothesis that SGWs of finite amplitude produce de Sitter expansion (20)

$$a(t) = exp(Ht) \tag{20}$$

To figure out the behavior of SGW themselves, we have to solve (12) over the background (20), which looks impossible even numerically. As a result of averaging, all background functions $a(t), \rho(t), p(t)$ are functions of time only; meanwhile, the SGW tensor $\psi_\alpha{}^\beta$ is a function of coordinates and time, which can be found using some approximations. In paper [8], it was shown that gravitational waves of small amplitude are unable to produce a de Sitter expansion; meanwhile, Equations (19) and (20) tell us that SGW of arbitrary amplitude can do that. There is no contradiction here. The gravitational waves of small amplitude are unable to build a de Sitter expansion if their wavelengths are sub horizon (shorter than the distance to the horizon) because formally speaking integrals (6) and (7) in work [8] are divergent. Meanwhile, super horizon wavelengths can produce a de Sitter regime because of changing the limits of integrations from $[0, \infty\}$ for sub horizon waves to interval $[0, 1]$ for super horizon waves in Equations (6) and (7) of work [8]. Thus, one can think that the solutions to Equations (19) and (20) describe waves of finite amplitude with the super horizon wavelengths. This fact testifies in favor of super horizon wavelengths as solutions to Equations (19) and (20).

4. Invariance of de Sitter Regime with Respect to Wick Rotation

In a Euclidean spacetime interval between events, (1) reads

$$ds^2 = d\tau^2 + a\left(\tau\right)^2 \gamma_{\alpha\beta} dx^\alpha dx^\beta = d\tau^2 + a\left(\tau\right)^2 \left(dx^2 + dy^2 + dz^2\right) \tag{21}$$

where $t = i\tau$. Recall that $H = \frac{1}{a}\frac{da}{dt}$ by definition. Thus, the Hubble constant in Euclidean spacetime τ $H_E = \frac{1}{a}\frac{da}{d\tau}$ is

$$H_E = iH \tag{22}$$

Thus, the de Sitter regime is invariant with respect to the transfer from Lorentzian spacetime to Euclidean spacetime and vice versa, which follows from (24)

$$exp(Ht) = exp(H_E \cdot \tau) \tag{23}$$

5. What Kind of SGWs are Capable to Provide the Invariance of the de Sitter Regime with Respect to Lorentzian and Euclidean Spacetime?

To answer this question, we have to go back to (16) and (17). Note that (17) does not contain time derivatives and only the two first terms contain the time derivatives in (16) for $\alpha \neq \beta$. This means that such invariance can be provided by super horizon wavelengths $\lambda >> ct$. In such a case, spatial derivatives can be neglected compared to time derivatives, and (16) takes the following form ($\alpha \neq \beta$)

$$\ddot{\psi}_\alpha{}^\beta + 3\frac{\dot{a}}{a}\dot{\psi}_\alpha{}^\beta = 0 \tag{24}$$

Obviously, (24) is invariant with respect to change of variables $t = i\tau$. This fact means that such waves belong to both Lorentzian and Euclidean spacetimes simultaneously. This means, in turn, there is no topologically impenetrable barrier between Lorentzian and Euclidean spacetimes for such waves and the de Sitter regime. Note also that such waves provide an overwhelming input to the effect because of the infinite length of the interval which they occupied $ct < \lambda \leq \infty$.

One more argument in favor of the idea of this work follows from the consideration of the back reaction of SGWs of small amplitude but of arbitrary wavelengths on the background metrics. Such waves are described by Equations (A26)–(A31) of Appendix A.3 of work [8]. As it is shown in work [8], Section 1.2, they are unable to form a de Sitter regime in Lorentzian spacetime because of divergency of integrals. However, they can easily do so in Euclidean spacetime ([8], Section 1.2)[4]. Thus, the invariance of the de Sitter regime with respect to the transfer from Lorentzian to Euclidean spacetime and vice versa, the absence of a topologically impenetrable barrier between Lorentzian and Euclidean spacetimes for super horizon wavelengths and the ability of waves of arbitrary wavelengths but small amplitude to form a de Sitter regime only in Euclidean spacetime allows us to suppose that the final stage of the evolution of the universe is "tunneling" (using instanton terminology) [12] from our Lorentzian spacetime to Euclidean spacetime. In other words, our Lorentzian universe is going to disappear into "Nothing"[5]. In the frame of such a scenario, the modern de Sitter expansion of the universe (dark energy effect) looks like the last stage of evolution of our universe before its "dives" into nothing.

6. Cosmological Scenario

One can suppose that a new universe should start from quantum fluctuations in Euclidean spacetime, i.e., from the place where the previous universe had completed its existence. In one-loop approximation, the equation of state of gravitons in the Euclidean spacetime reads [13]

$$\rho_g = \frac{3\hbar N_g \cdot H^4}{8\pi^2} = -p_g \tag{25}$$

where $\hbar$ is Planck constant, N_g is number of gravitons in the universe and ρ_g and p_g are the energy density and pressure of gravitons, respectively. Note the important fact that (25) is invariant with respect to Wick rotation $t = i\tau$ because (25) contains H^4 (recalling that $i^4 = 1$). Thus, this de Sitter expansion is invariant with respect to the transfer from Euclidean to Lorentzian spacetime and vice versa. As it follows from (25), in one loop approximation, gravitons produce de Sitter expansion in Lorentzian space-time. Thus, we can consider (25) as an initial condition for the formation of a new universe that starts with a de Sitter expansion, which eventually should lead to a new Big Bang in a new universe. This process probably should start near the Planck time $t_p \approx 5.4 \cdot 10^{-44}$ s. This number should be close to the initial conditions used by inflation theories [14–18] for the present universe. The modern inflation theories applicable to the existing universe usually start approximately at $10^8 t_p$ and last about $\left(10^{11} - 10^{12}\right) t_p$. However, we need to remember the words of Steven Weinberg [19]: "So far, the details of inflation are unknown, and the whole idea of inflation remains a speculation, though one that is increasingly plausible". The scenario proposed assumes that a permanent process of birth, death and rebirth of an infinite sequence of universes takes place on an infinite time axis.

7. Conclusions

I am deeply grateful to Zeev Dashevsky, who drew my attention to a book written 100 years ago by Alexander Friedmann [20], the man who predicted the expansion of the universe [21] a few years before its discovery by Edwin Hubble, making him the founding father of modern cosmology. In his book, Friedmann quotes the famous phrase from the book "Ecclesiastes": "What was, will be, and what has been done is what will be done- and there is nothing new under the Sun," meaning that even with all the progress in our ideas about the world, our approach fundamentally remains the same as it was millennia ago for Aristotle.

Funding: This research received no external funding.

Data Availability Statement: Data sharing is not applicable.

Acknowledgments: I am deeply grateful to John Payne for stimulating discussion that allowed me to state the content of this work more clearly.

Conflicts of Interest: The author declares no conflict of interest.

Notes

1   The material constituting the content of the Appendix A of work [8] was prepared by the late Grigory Vereshkov. It was to be included in our joint paper [9]. We did not include it in the final version of [9], since [9] was entirely devoted to quantum effects. After Grigory passed away, I decided to publish his work, including it as Appendix A to my paper [8] with a reference to its author.

2   There are other types of parameterizations as well, e.g., Fierz–Pauli parameterization [10] and a linear parametrization used in most of the works known to us (see [9] and references therein). The linear parameterization does not provide conservation of EMT, which can be conducted "by hand" [11].

3   In this paper, averaging is conducted over a three-space (see, e.g., [11]).

4   Note that in Section 1.2 of work [8] the incorrect normalization was used. In the right hand sides (RHS) of Formulas (16)–(20), (23) and (25) of [8], the Hubble constant H must be changed for 1. This correction does not change the main result of this section, which proves that the back reaction of SGWs of small amplitude can produce a de Sitter expansion in Euclidean spacetime only.

5   To the best of my knowledge, the first time such a scenario was proposed in paper [12], although the problem of tunneling gravitational instantons from Euclidean spacetime into Lorentzian spacetime of the universe has a large literature (see, e.g., [12] and references therein].

References

1.   Bennett, C.L.; Larson, D.; Weiland, J.L.; Jarosik, N.; Hinshaw, G.; Odegard, N.; Smith, K.M.; Hill, R.S.; Gold, B.; Halpern, M.; et al. Nine-year Wilkinson Microwave Anisotropy Probe (WMAP) observations: Final maps and results. *Astrophys J.* **2012**, *208*, 20. [CrossRef]

2.   Abbott, T.M.; Aguena, M.; Alarcon, A.; Allam, S.; Alves, O.; Amon, A.; Andrade-Oliveira, F.; Annis, J.; Avila, S.; Bacon, D.; et al. Dark Energy Survey Year 3 results: Cosmological constraints from galaxy clustering and weak lensing. *Phys. Rev. D* **2022**, *105*, 023520. [CrossRef]

3.   Demorest, P.B.; Ferdman, R.D.; Gonzalez, M.E.; Nice, D.; Ransom, S.; Stairs, I.H.; Arzoumanian, Z.; Brazier, A.; Burke-Spolaor, S.; Chamberlin, S.J.; et al. Limits on the stochastic gravitational wave background from the North American nanohertz observatory for gravitational waves. *Astrophys. J.* **2013**, *762*, 94. [CrossRef]

4.   Nakamal, T.; Silk, J.; Kamionkowski, M. Stochastic gravitational waves associated with the formation of primordial black holes. *Phys. Rev. D* **2017**, *95*, 043511. [CrossRef]

5.   Christensen, N. Stochastic gravitational wave backgrounds. *Rep. Prog. Phys.* **2019**, *82*, 016903. [CrossRef] [PubMed]

6.   Reardon, D.J.; Zic, A.; Shannon, R.M.; Hobbs, G.B.; Bailes, M.; Di Marco, V.; Kapur, A.; Rogers, A.F.; Thrane, E.; Askew, J.; et al. Search for an isotropic gravitational-wave background with the Parkes Pulsar Timing Array. *Astrophys. J. Lett.* **2023**, *951*, L6. [CrossRef]

7.   Lifshitz, E.M. On the gravitational stability of the expanding universe. *JETP* **1946**, *16*, 587.

8.   Marochnik, L. Dark Energy and Inflation from Gravitational Waves. *Universe* **2017**, *3*, 72. [CrossRef]

9.   Vereshkov, G.; Marochnik, L. Quantum gravity in Heisenberg representation and self-consistent theory of gravitons in macroscopic spacetime. *J. Mod. Phys.* **2013**, *4*, 2. [CrossRef]

10.   Fierz, M.; Pauli, W.E. On relativistic wave equations for particles of arbitrary spin in an electromagnetic field. *Proc. R. Soc. Lond.* **1939**, *173*, 211–232.

11.   Abramo, L.R.; Brandenberger, R.H.; Mukhanov, V.F. Energy-momentum tensor for cosmological perturbations. *Phys. Rev. D* **1997**, *56*, 3248. [CrossRef]

12.   Marochnik, L. Dark energy from instantons. *Gravit. Cosmol.* **2013**, *19*, 178–187. [CrossRef]

13.   Marochnik, L.; Usikov, D.; Vereshkov, G. Cosmological acceleration from virtual gravitons. *Found. Phys.* **2008**, *38*, 546–555. [CrossRef]

14.   Starobinsky, A.A. A new type of isotropic cosmological models without singularity. *Phys. Lett. B* **1980**, *91*, 99–102. [CrossRef]

15.   Guth, A.H. Inflationary universe: A possible solution to the horizon and flatness problems. *Phys. Rev. D* **1981**, *23*, 347. [CrossRef]

16.   Linde, A.D. A new inflationary universe scenario: A possible solution of the horizon, flatness, homogeneity, isotropy and primordial monopole problems. *Phys. Lett. B* **1982**, *108*, 389. [CrossRef]

17.   Mukhanov, V.; Chibisov, G. Vacuum energy and large-scale structure of the Universe. *Sov. Phys. JETP* **1982**, *56*, 258–265.

18.   Albrecht, A.; Steinhardt, P.J. Cosmology for grand unified theories with radiatively induced symmetry breaking. *Phys. Rev. Lett.* **1982**, *48*, 1220. [CrossRef]

19.   Weinberg, S. *Cosmology*; Oxford University Press: Oxford, UK, 2008.

20.   Friedmann, A. *The World as Space and Time*, 1st ed.; Мир а Простраco Время (English translation: The World as Space and Time); First Russian Edition, 1923; Second Russian Edition, 1965; Minkowski Institute Press: Montreal, QC, Canada, 2014; p. 62.

21. Friedmann, A. Über die krümmung des raumes. *Z. Für Physik.* **1922**, *10*, 377–386. [CrossRef]

Article

Effective Potential for Quintessential Inflation Driven by Extrinsic Gravity

Abraão J. S. Capistrano [1,2,*] and Luís Antonio Cabral [3]

[1] Departamento de Engenharias e Ciências Exatas, Universidade Federal do Paraná, Palotina 85950-000, PR, Brazil
[2] Applied Physics Graduation Program, Federal University of Latin-American Integration, Foz do Iguassu 85867-670, PR, Brazil
[3] Curso de Física, Setor Cimba, Universidade Federal do Norte do Tocantins, Araguaína 77824-838, TO, Brazil; cabral@mail.uft.edu.br
* Correspondence: capistrano@ufpr.br

Abstract: We numerically study the evolution of the extrinsic energy density in the context of an inflationary regime at the background level in a five-dimensional model using a Bayesian analysis from a dynamic nested sampler (DYNESTY) code. By means of the Nash–Greene embedding theorem, we show that the corresponding model provides an effective potential driven by the influence of extrinsic geometry. We obtain a quintessential inflation that defines a model with a potential $V(\phi) = e^{-\alpha_1 \phi}(1 - \alpha_2 \phi^2)$, where α_1 and α_2 are dimensionless parameters. Using some known phenomenological parameterizations, such as Chevallier–Polarski–Linder (CPL) and Barboza–Alcaniz (BA) parameterizations, we show that the model reflects a slow-varying inflation preferring a thawing behavior, suggesting an optimistic scenario for further research on the unification of inflation with late cosmic acceleration.

Keywords: inflation; Nash–Greene theorem; gravitational field

Citation: Capistrano, A.J.S.; Cabral, L.A. Effective Potential for Quintessential Inflation Driven by Extrinsic Gravity. *Universe* **2023**, *9*, 497. https://doi.org/10.3390/universe9120497

Academic Editor: Kazuharu Bamba

Received: 9 October 2023
Revised: 20 November 2023
Accepted: 25 November 2023
Published: 28 November 2023

1. Introduction

Cosmic inflation is one of the most fantastic mechanisms for understanding the evolution of the early universe, which underwent a rapid and exponential increase in the cosmic scale factor shortly after the Big Bang. It proposes a simple solution to the horizon problem by suggesting that far-reaching regions of the universe were in close proximity before inflation occurred. Then, it explains why the universe appears homogeneous and isotropic on large scales. Moreover, the so-called *flatness problem* that states, *Why does the universe seem to have a nearly flat geometry?* is solved in the context of inflation since it naturally leads to a flat universe [1]. It also provides an explanation of the formation of a large-scale structure in the universe, including galaxies and galaxy clusters, by quantum fluctuations in a peculiar scalar field dubbed as *inflaton field* during that cosmic inflationary period [2–9].

The nature of the inflaton field motivated the elaboration of a plethora of competing cosmological models [10–19] due to the fact that the form of the inflation field is not determined and the standard phenomenological model Λ cold dark matter (ΛCDM) with the fluid parameter $w = -1$ rapidly decays at the end of the inflationary period [20]. Thus, the inflaton field cannot be directly regarded as a cosmological constant Λ for the late acceleration of the universe. An immediate issue arises whether the inflaton field should be some form of a dark energy (DE) fluid: *how does the equation of the state parameter w significantly deviate from $w = -1$ during the inflationary era?* In this paper, we make the first steps to answer this question in the context of a modification of gravity by extra dimensions. Apart from traditional models, such as the Arkani-Hamed–Dvali–Dimopoulos (ADD) model [21], the Randall–Sundrum model [22,23], and/or the Dvali–Gabadadze–Porrati model (DPG) [24], we explore the dynamical embedding of space-times as the main mathematical structure [25–40]. In *Encyclopaedia Inflationaris* [41], the reader can find various types and classifications of inflationary models, such as supergravity brane inflation, supersymmetry (SUSY), and brane inflation models.

The paper is organized into sections. The second and third sections summarize the essentials of previously published results on the construction of the four-dimensional embedded model and cosmological applications [26–28,31,39,40]. The fourth section presents the fluid analogy necessary to make a comparison with phenomenological models. Moreover, we construct a model wherein the extrinsic curvature is thought as an inflaton field as shown in the fifth section. By means of numerical analysis generating synthetic random points using the DYNESTY [42,43] Python code, we constrain the parameters of our toy model. To analyze how the related equation of state (EoS) evolves, we use Chevallier–Polarski–Linder (CPL) [44,45] and Barboza–Alcaniz (BA) [46] parameterizations to distinguish the present model as a *thawing* [47–49] or *freezing* [50–53] pattern. Thawing models are conceived to be those that the fluid parameter as a function of the redshift $w(z)$ in EoS is moving away from -1, which means that DE density decreases over time, gradually allowing the universe to accelerate more rapidly. On the other hand, in freezing models, DE density remains approximately constant as the universe expands, with $w(z)$ approaching the value -1. Finally, the conclusion and prospects are presented in the *Remarks* section.

2. Essentials on Embeddings

We define a model endowed with a gravitational action S in the presence of confined matter fields wherein a four-dimensional embedded space-time is embedded in five dimensions as

$$S = -\frac{1}{2\kappa_5^2} \int \sqrt{|\mathcal{G}|} \left({}^5\mathcal{R} + \mathcal{L}_m^* \right) d^5x \, , \tag{1}$$

where κ_5^2 is a fundamental energy scale on the embedded space, and the curly ${}^5\mathcal{R}$ denotes the five-dimensional Ricci scalar of the bulk given by the summation of the four-dimensional Ricci scalar R and the deformation scalar Q and has the simple form ${}^5\mathcal{R} = R + Q$. Moreover, $\mathcal{L}_m^*$ denotes the bulk source Lagrangian in five dimensions. Since gravity should only propagate in the bulk, $\mathcal{L}_m^*$ is localized in the four-dimensional embedded space and obeys the confinement condition

$$\kappa_5^2 T_{\mu\nu}^* = -8\pi G T_{\mu\nu} \, , \tag{2}$$

$$\kappa_5^2 T_{\mu a}^* = 0 \, ,$$

$$\kappa_5^2 T_{ab}^* = 0 \, ,$$

from the components of $\mathcal{T}_{AB}$ of the energy–momentum tensor of the bulk. The terms of $T_{\mu\nu}^*$, $T_{\mu a}^*$, and T_{ab}^* denote the energy–momentum of the tangent (tensor), vector, and scalar components of $\mathcal{T}_{AB}$. Concerning notation, capital Latin indices are fixed to 5 to reinforce the notation of a five-dimensional space. Small-case Latin indices refer to the only one extra dimension fixed to 1. All Greek indices refer to the embedded space-time running from 1 to 4. Ordinary matter and gauge fields are represented by $T_{\mu\nu}$, and G is the gravitational Newtonian constant. From the set of equations in Equation (2), $\mathcal{L}_m^*$ reduces to the confined four-dimensional Lagrangian $\mathcal{L}_m$ related to $T_{\mu\nu}$ as

$$T_{\mu\nu} = -\frac{2}{\sqrt{|g|}} \frac{\delta\left(\sqrt{|g|}\mathcal{L}_m \right)}{\delta\, g^{\mu\nu}} \, , \tag{3}$$

where $g_{\mu\nu}$ is the metric and $|g|$ is the absolute value of $g_{\mu\nu}$. Detailed demonstrations of the fundamental field equations, from a bulk of five or in arbitrary dimensions, can be found in Refs. [26–28,31,40]. In this work, in search of constructing a viable physical model, we consider the extrinsic curvature as a main character of cosmic dynamics. Mathematically speaking, besides the metric $g_{\mu\nu}$, such new curvature is a pivotal element in the embedding of geometries. As traditionally defined in [54], the nonperturbed extrinsic curvature $k_{\mu\nu}$ reads,

$$k_{\mu\nu} = -\mathcal{X}^A_{,\mu}\, \eta^B_{,\nu}\mathcal{G}_{AB}\,.\tag{4}$$

The definition in Equation (4) essentially means a measure of *bending* of the embedded geometry; i.e., it reflects how a normal unitary vector η^A in the embedded space deviates from the tangent plane. Let the coordinate $\mathcal{X}$ define a regular map $\mathcal{X}: V_4 \to V_5$. Such a local map states that a Riemannian manifold V_4 is isometrically embedded in a larger five-dimensional Riemannian bulk space V_5. The term $\mathcal{G}_{AB}$ is the bulk metric

$$\mathcal{G}_{AB} \;=\; \begin{pmatrix} g_{\mu\nu} & 0 \\ 0 & 1 \end{pmatrix}.\tag{5}$$

Another pivotal relation is Nash's deformation formula given by

$$k_{\mu\nu} = -\frac{1}{2}\frac{\partial g_{\mu\nu}}{\partial y}\,,\tag{6}$$

where y is an arbitrary spatial coordinate. Nash's deformation formula is the pivot result of Nash's embedding theorem of 1954 applied to non-Euclidean metrics [55]. In 1956, Nash generalized his former results to Riemannian metrics [56]. Decades later, Nash's theorem was expanded to non-Riemannian metrics by Greene [57]. From a physical point of view, Equation (6) localizes gravity closer to the four-dimensional embedded space, imposing a strong bending (warping) geometrical constraint.

Nash's results in Equation (6) is the capital element to obtain a solution to well-known Gauss and Codazzi equations given by

$$^5\mathcal{R}_{ABCD}\mathcal{Z}^A_{,\alpha}\mathcal{Z}^B_{,\beta}\mathcal{Z}^C_{,\gamma}\mathcal{Z}^D_{,\delta} = R_{\alpha\beta\gamma\delta}+ \left(k_{\alpha\gamma}\,k_{\beta\delta} - k_{\alpha\delta}\,k_{\beta\gamma}\right),\tag{7}$$

$$^5\mathcal{R}_{ABCD}\mathcal{Z}^A_{,\alpha}\mathcal{Z}^B_{,\beta}\mathcal{Z}^C_{,\gamma}\eta^D = k_{\alpha[\beta;\gamma]}\,,\tag{8}$$

where $^5\mathcal{R}_{ABCD}$ denotes the five-dimensional Riemann tensor, and $R_{\alpha\beta\gamma\delta}$ is the four-dimensional Riemann tensor. The coordinate $\mathcal{Z}^A_{,\mu}$ is a perturbed coordinate in the sense that $\mathcal{Z}^A_{,\mu} = \mathcal{X}^A_{,\mu} + \delta y\,\eta^A_{,\mu}$. The semicolon sign in Equation (8) represents the ordinary covariant derivative with respect to the metric $g_{\mu\nu}$ and $k_{\alpha[\beta;\gamma]} \equiv k_{\alpha\beta;\gamma} - k_{\alpha\gamma;\beta}$. The importance of Equations (7) and (8) is that they reflect the integrability conditions for the embedding explicitly relating the bulk and embedded space, and they are the starting point to obtain the induced field equations [26–28,31,40].

3. Friedmann–Lemaître–Robertson–Walker (FLRW) Embedded Cosmology

From integration of Equations (7) and (8), one obtains the induced four-dimensional nonperturbed field equations as

$$G_{\mu\nu} - Q_{\mu\nu} \;=\; -8\pi G T_{\mu\nu}\,,\tag{9}$$

$$k_{\mu[\nu;\rho]} \;=\; 0\,,\tag{10}$$

where $T_{\mu\nu}$ is the energy–momentum tensor of the confined sources, G is the gravitational Newtonian constant, and $G_{\mu\nu}$ is the four-dimensional Einstein tensor. The deformation tensor $Q_{\mu\nu}$ is given by

$$Q_{\mu\nu} = k^\rho_\mu k_{\rho\nu} - k_{\mu\nu}h - \frac{1}{2}\left(K^2 - h^2\right)g_{\mu\nu}\,,\tag{11}$$

where the mean curvature is given by $h^2 = h\!\cdot\! h$ and $h = g^{\mu\nu}\,k_{\mu\nu}$. The quantity $K^2 = k^{\mu\nu}k_{\mu\nu}$ is the Gaussian curvature. By direct derivation, Equation (11) is conserved in the sense that

$$Q_{\mu\nu;\mu} = 0\,.\tag{12}$$

Once the field equations of Equations (9) and (10) are set, we obtain the background cosmic evolution from the usual four-dimensional line element of the FLRW metric given by

$$ds^2 = a^2 \left(dr^2 + r^2 d\theta^2 + r^2 \sin^2 \theta d\phi^2 \right) - dt^2 \,, \tag{13}$$

where $a \equiv a(t)$ is the scale factor and t is the physical time. As shown in previous works [27,31], the solutions of Equations (9) and (10) are summarized as

$$k_{ij} = \frac{b}{a^2} g_{ij} \ (i,j = 1,2,3) \,, \tag{14}$$

$$k_{44} = \frac{-1}{\dot{a}} \frac{d}{dt} \frac{b}{a} \,, \tag{15}$$

$$k_{44} = -\frac{b}{a^2} \left(\frac{B}{H} - 1 \right) \,, \tag{16}$$

$$K^2 = \frac{b^2}{a^4} \left(\frac{B^2}{H^2} - 2\frac{B}{H} + 4 \right), \quad h = \frac{b}{a^2} \left(\frac{B}{H} + 2 \right), \tag{17}$$

$$Q_{ij} = \frac{b^2}{a^4} \left(2\frac{B}{H} - 1 \right) g_{ij}, \quad Q_{44} = -\frac{3b^2}{a^4}, \tag{18}$$

$$Q = -(K^2 - h^2) = \frac{6b^2}{a^4} \frac{B}{H} \,, \tag{19}$$

where Q is the deformation scalar given by the contraction $g^{\mu\nu} Q_{\mu\nu} = Q$, and $H \equiv H(t) = \frac{\dot{a}}{a}$ is the Hubble parameter. Defined as an "extrinsic" copy of the Hubble parameter H, we have the cosmic bending parameter $B = B(t) \equiv \frac{\dot{b}}{b}$. The obtainment of these results is shown in detail in Refs. [27,31]. The *warping* or *bending* function $b(t)$ is given by

$$b(t) = \frac{b_0}{a_0^{\beta_0}} a(t)^{\beta_0} \,, \tag{20}$$

where the parameter b_0 denotes the current value of $b(t)$ and a_0 is the current value of the scale factor.

4. The Fluid Analogy

We obtain the hydrodynamical equations for a perfect fluid in comoving coordinates as the same way as in the usual general relativity (GR) framework. Thus, the stress energy tensor $T_{\mu\nu}$ has the standard form

$$T_{\mu\nu} = (\rho + p)u_\mu u_\nu + p g_{\mu\nu} \,; \ u_\mu = \delta_\mu^4 \,, \tag{21}$$

and the conservation equation

$$\dot{\rho} + 3H(\rho + p) = 0 \,, \tag{22}$$

from $T_{\mu\nu;\mu} = 0$. Hence, one obtains the resulting Friedmann equation as

$$H^2 = \frac{8}{3}\pi G\rho + \frac{b^2}{a^4} \,, \tag{23}$$

where $\rho \equiv \rho_m(t)$ is the nonperturbed matter density. Using Equations (20) and (23), we have

$$H^2 = \frac{8}{3}\pi G\rho_{m(0)} a^{-3} + \frac{b_0^2}{a_0^{2\beta_0}} a^{2\beta_0 - 4} \,, \tag{24}$$

where $\rho_{m(0)}$ is the current value of matter energy density. With respect to the cosmological parameters $\Omega_i = \frac{8\pi G}{3H_0^2}\rho_{i(0)}$, i identifies the density species, and considering only *matter* and *extrinsic* species, we write the Hubble evolution $H(z)$ simply as

$$H(z) = H_0\sqrt{\Omega_m(z) + \Omega_{ext}(z)} \,, \tag{25}$$

where $H(z)$ is the Hubble parameter in terms of the redshift $z = \frac{1-a}{a}$. The matter density parameter is denoted by $\Omega_m(z) = \Omega_{m(0)}(1+z)^3$. The term $\Omega_{ext}(z) = \Omega_{ext(0)}(1+z)^{4-2\beta_0}$ stands for the density parameter associated with the extrinsic curvature. The cosmological parameters $\Omega_{m(0)}$ and $\Omega_{ext(0)} = b_0^2/H_0^2 a_0^{2\beta_0}$ denote the present value of the matter and extrinsic contributions, respectively. The current warp (bending) of the universe denoted by b_0 has the same dimension as H_0 that is the current value of a Hubble constant in units of $\text{km}\cdot\text{s}^{-1}\,\text{Mpc}^{-1}$. If $\Omega_{ext}(z)$ vanishes with $b_0 \to 0$, one obtains GR limit as a background with the recovery of Einstein equations. For a flat universe, we have $\Omega_{ext(0)} = 1 - \Omega_{m(0)}$.

To obtain an effective cosmological model in order to make tests with real data, an effective "extrinsic fluid parameter" w_{ext} must be set. We define a "fluid analogy" by means of an effective equation of state (EoS) as

$$w_{ext} = -1 + \frac{1}{3}(4 - 2\beta_0) \,, \tag{26}$$

in which the equality $w_{ext} = w$ happens if $\beta_0 = \frac{1}{2}(1 - 3w)$, where the dimensionless fluid parameter w defines an EoS related to pressure p and energy density ρ as $w = \frac{p}{\rho}$. Thus, one defines the dimensionless Hubble parameter $E(z) \equiv \frac{H}{H_0}$ that reads

$$E^2(z) = \Omega_{m(0)}(1+z)^3 + \Omega_{ext(0)}(1+z)^{3(1+w)} \,, \tag{27}$$

which resembles an XCDM fluid [58].

5. Extrinsic Curvature as an Effective Inflaton Field

Besides Equations (26) and (27), we can explore the effective fluid approach rewriting Equation (9) in the form

$$R_{\mu\nu} - \frac{1}{2}R\bar{g}_{\mu\nu} = -8\pi G\bar{T}_{total} \,, \tag{28}$$

where $\bar{T}^{total}_{\mu\nu} = \bar{T}_{\mu\nu} + \frac{1}{8\pi G}\bar{T}^{ext}_{\mu\nu}$. The quantity $\bar{T}^{ext}_{\mu\nu} \doteq Q_{\mu\nu}$ denotes the extrinsic contribution. Mimicking Equation (21), we write $\bar{T}^{ext}_{\mu\nu}$ as

$$\bar{T}^{ext}_{\mu\nu} = (\bar{p}_{ext} + \bar{\rho}_{ext})u_\mu u_\nu + \bar{p}_{ext}\,\bar{g}_{\mu\nu}, \quad u_\mu = \delta^4_\mu \,, \tag{29}$$

where u_μ is the comoving four velocities. It is worth noting that the deformation tensor $Q_{\mu\nu}$ is independently conserved as shown in Equation (12), so is $\bar{T}^{ext}_{\mu\nu}$. We stress that the dynamical embedding naturally warrants the influence of the bulk over the embedded space, such as quantities $Q_{\mu\nu}$ with a mixture of intrinsic and extrinsic terms, as shown in Equation (11). As a consequence, $\bar{T}^{ext}$ is conserved unless an exotic matter-energy source is added. This result differs from a DE fluid in standard rigid brane-world models [22–24], where $\bar{T}^{ext}$ is not conserved due to the necessity of the bulk–brane energy exchange.

From the conservation of Equation (29), we obtain

$$\frac{d\bar{\rho}_{ext}}{dt} + 3H(\bar{\rho}_{ext} + \bar{p}_{ext}) = 0 \,. \tag{30}$$

In a perfect fluid analogy with Equation (22), $\bar{\rho}_{ext}$ denotes the nonperturbed "extrinsic" energy density given by

$$\bar{\rho}_{ext}(a) = \rho_{ext(0)}a^{-3(1+w)} \,, \tag{31}$$

where the current "extrinsic" energy density is $\rho_{ext(0)} = \frac{3b_0^2}{8\pi G a_0^{1-3w}}$. Therefore, the "extrinsic" pressure $\bar{p}_{ext}$ is straightforwardly calculated by Equation (30) to obtain

$$\bar{p}_{ext}(a) = w\rho_{ext(0)} a^{-3(1+w)} . \tag{32}$$

When $b_0 \to 0$, the extrinsic curvature ceases, and $\bar{T}_{\mu\nu}^{ext}$, $\bar{\rho}_{ext}(a)$, and $\bar{p}_{ext}(a)$ vanish, and the GR limit is reached. Alternatively, one writes the related Friedman equation in the form

$$H^2 = \frac{8}{3}\pi G(\bar{\rho}_m + \bar{\rho}_{ext}) . \tag{33}$$

5.1. The Extrinsic Inflaton Field

Since $\bar{\rho}_{ext}$ plays a role in driving the late accelerated expansion as explored in [27,31], the same energy density $\bar{\rho}_{ext}$ should have the dynamics of a scalar field potential energy density $V(\phi)$ during early inflationary periods generated by the "extrinsic scalar" field ϕ. Hence, the related Lagrangian $\mathcal{L}_\phi$ is defined as

$$\mathcal{L}_\phi = \frac{1}{2}\dot{\phi}^2 - V(\phi) . \tag{34}$$

The form written in Equation (34) shows that the field ϕ is spatially homogeneous; then it only depends on the cosmic time t. In addition, the related energy momentum of such a field is written as

$$T_{\mu\nu}^\phi = \partial_\mu\phi \, \partial_\nu\phi + g_{\mu\nu}\left(\frac{1}{2}\dot{\phi}^2 - V(\phi)\right) . \tag{35}$$

From the conservation of Equation (35), i.e., $T_{\mu\nu;\mu}^\phi = 0$, one obtains the correspondence relations

$$\bar{\rho}_{ext} = \bar{\rho}_\phi = \frac{\dot{\phi}^2}{2} + V(\phi) , \tag{36}$$

$$\bar{p}_{ext} = \bar{p}_\phi = \frac{\dot{\phi}^2}{2} - V(\phi) , \tag{37}$$

where the dot symbol denotes an ordinary derivative with respect to the cosmic time t. The forms of Equations (36) and (37) couple with the inflation dynamics given by $V(\phi)$ to the background evolution. Then, the quantities $\bar{\rho}_{ext}$ and $\bar{p}_{ext}$ can be written as a function of the extrinsic scalar field ϕ.

In terms of fluid analogy given by Equation (26) and the direct sum of Equations (36) and (37), and using Equations (31) and (32), one obtains

$$\dot{\phi}(a) = \sqrt{\rho_{ext(0)}(1+w)} \, a^{-\frac{3}{2}(1+w)} . \tag{38}$$

Using the relation $\dot{\phi} = aH\frac{\phi}{da}$ and the inflation condition $H^2(a) \sim \frac{8\pi G}{3}\rho_{ext}(a)$, we integrate Equation (38) to obtain the potential $\phi(a)$ as

$$\phi(a) = \sqrt{\left|\frac{3(1+w)}{8\pi}\right|} \, M_{pl} \ln a , \tag{39}$$

where we denote the Planck mass as $M_{pl} = \frac{1}{\sqrt{G}}$. We consider the initial conditions $\phi_{ini} = 0$ and $a_{ini} = 1$. Putting the solution of Equation (39) in Equation (36), one obtains a potential $V(\phi)$ in the form

$$V(\phi) = V_0 e^{-\alpha_0\phi/M_{pl}} . \tag{40}$$

For the sake of notation, we denote the quantity α_0 in Equation (40) as $\alpha_0 = \sqrt{|24\pi(1+w)|}$. The models of such exponential form produce a *power law inflation* (PLI), first introduced in Refs. [59–61]. For the time being, this type of potential has strong constraints imposed by Planck data [62] and is ruled out at more than 2-σ confidence regions. Moreover, PLI inflation requires a *graceful exit* in the slow-roll approximation. In such a model, the first slow-roll Hubble parameter is constant that leads to an eternal inflation [41]. Such potential was also studied in the context of M-theory [63] and RSII scenarios [64]. Due to the strong constraints on PLI, based on Equation (40), our numerical study will take an effective quintessential dimensionless potential $V(\phi)_{eff}$ to generalize Equation (40), such as

$$V(\phi)_{eff} = V_{PLI}(\phi) + V_{LMI}(\phi) ,$$ (41)

where $V_{PLI}(\phi)$ denotes our PLI potential given by Equation (40). The quantity $V_{LMI}(\phi)$ denotes a phenomenological potential that defines the so-called *logamediate inflation* (LMI) [65,66] in the form

$$V_{LMI}(\phi) = V_0 \, \gamma_3 \left(\frac{\phi}{M_{pl}} \right)^{\gamma_2} e^{-\gamma_1 \left(\frac{\phi}{M_{pl}} \right)^{\gamma_4}} ,$$ (42)

in which $(\gamma_1, \gamma_2, \gamma_3, \gamma_4)$ are free dimensionless parameters. For our purposes, we need to impose on Equation (40) $V_0 = \tilde{V}_0 = 1$, $\alpha_0 = \alpha_1$, and $\phi = \tilde{\phi} M_{pl}$. A similar adaptation is applied to Equation (42) with $V_0 = \tilde{V}_0 = 1$, $\gamma_1 = \alpha_1$, $\gamma_2 = 2$, $\gamma_3 = -|\alpha_2|$ $(\alpha_2 > 0)$, $\gamma_4 = 1$, and $\phi = \tilde{\phi} M_{pl}$. Hence, the resulting potential $V(\phi)_{eff}$ is given by

$$V(\phi)_{eff} = \tilde{V}_0 \, e^{-\alpha_1 \tilde{\phi}}(1 - \alpha_2 \tilde{\phi}^2) .$$ (43)

We point out that (α_1, α_2) are dimensionless parameters. Assuming that the field should roll in a direction wherein $\frac{\partial V(\phi)_{eff}}{\partial \phi} = 0$, starting at $\phi = \phi_{\min} = 1$, one obtains the condition $\alpha_2 = |\frac{\alpha_1}{2 - \alpha_1}| > 0$. For our purposes, the positivity of α_2 warrants a logistic equation set in the domain of real numbers. Therefore, the model turns again as being a one-parameter model from an extrinsic origin. This guarantees a reheating phase without relying on any approximation in a future work. It is worth noting that when an extrinsic curvature vanishes $\alpha_1 = \alpha_0 = 0$, then $\alpha_2 = 0$, and inflation ceases.

To study the evolution of the effective fluid parameter w and the characteristics of the potential $V(\phi)_{eff}$, we adopt a dynamical system approach as in the works of Clemson and Liddle [67] and Pantazis, Nesseris, and Perivolaropoulos [68]. In the latter, they use a quintessence potential of the form

$$V(\phi) = e^{-\beta_1 \phi}(1 + \beta_2 \phi) ,$$ (44)

where β_1 and β_2 are free dimensionless parameters with $\beta_1 > \beta_2$. For comparison purposes, such potential will be also taken into account as a reference in contrast with Equation (43).

5.2. Numerical Parameter Estimation

In order to obtain the parameter estimation of our toy model by the potential in Equation (43), we define a "true" model that produces a close curve pattern with the adopted priors to the reference model in Equation (44).

As shown in Figure 1, we have a comparison between the potentials from Equation (44) (blue line) and our model defined by Equation (43) (black line). They present a similar pattern with the same peak height, and the latter model presents a narrower basis of its curve (black line). Both cases positively respond to a remote future passing through an accelerated phase.

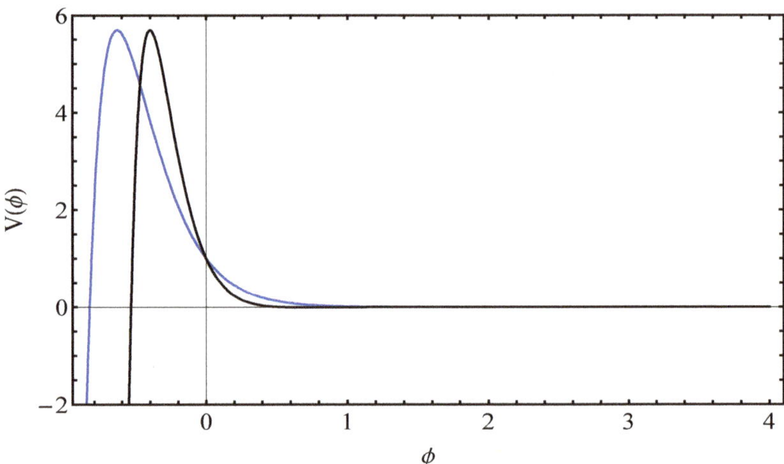

Figure 1. Comparison between the potentials from Equation (44) (blue line) and Equation (43) (black line) that indicates a close overall pattern of such potentials. The blue line was produced by adopted parameters $\{(\beta_1, \beta_2) = (5, 1.2)\}$, as shown in Ref. [68]. The black line was produced by the adopted priors that define our "true" model with the values $\{(\alpha_1, \alpha_2) = (4, 2)\}$.

In order to use the DYNESTY [42,43] Python code for estimating Bayesian posteriors and evidences, we adopt priors for the "true model" set as $\{(\alpha_{1(true)}, \alpha_{2(true)}, f_{(true)}) = (4, 2, 1.7)\}$, where the quantity f is defined as a mechanism to control the variance in the resulting Gaussian distributions. To estimate the values of the parameters α_1 and α_2 on how they can approximate to the "true" values, we define a prior transform $u_\theta = (u_{af1}, u_{af2}, u_f)$ as

$$\alpha_1 = 5u_{af1} + 0.5; \tag{45}$$
$$\alpha_2 = 2.5u_{af2} + 0.5; \tag{46}$$
$$\ln f = 2.8u_f - 1. \tag{47}$$

We start generating $n = 30000$ synthetic random points and 8000 live points for better efficiency. On the other hand, under the circumstances we have defined, during the runs, we noted that generating larger random points does not change the results. To estimate both evidence and posteriors, DYNESTY dynamically allocates points in the runs. Each point also carries information of their covariances. It allows for creating a variable number of K_i live points at each iteration i to realize a change in a prior volume X_i. From Numpy 5000 random seeds and the `celerite` [69] library for a scalable Gaussian process, we obtained a total of 144659 interactions in the final of the runs. We have that the prior volume and evidence are controlled when the variation $\Delta \ln X_i \approx K_i$, as shown in Figure 2.

In Figure 3, we show the posteriors in the contour plots with the values of the parameters (α_1, α_2, f). In the lower panels, we have confidence regions for marginalized posteriors. In the upper panels, the estimated values of the parameters and their Gaussian distributions are shown. In both panels, the vertical lines indicate the values of the "true" model, defined by the priors, and horizontal lines indicate the estimated values for the model parameters.

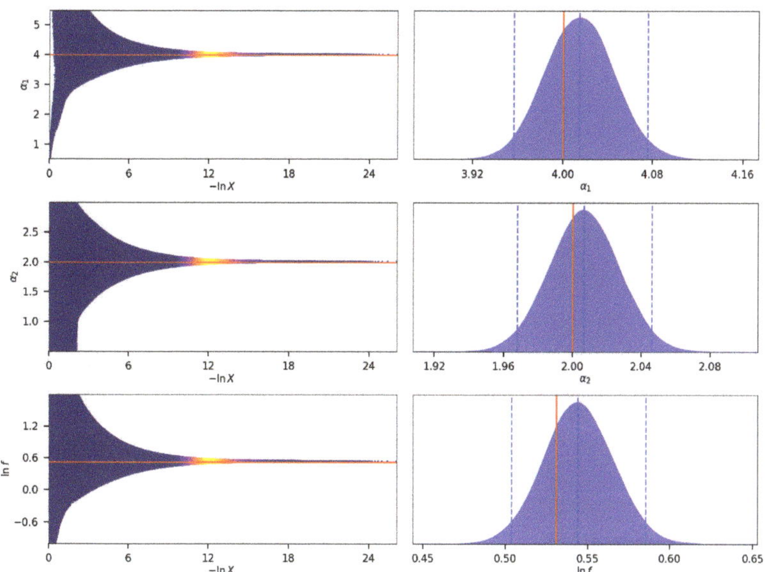

Figure 2. In the left panels, we have the control of the prior volume $\ln X_i$ of random points that shrinks exponentially over time to determine the evidence. The right panels show the posteriors (Gaussian distribution of the parameters). In both sets of panels, the red lines denote the priors ("true" values) adopted in the code.

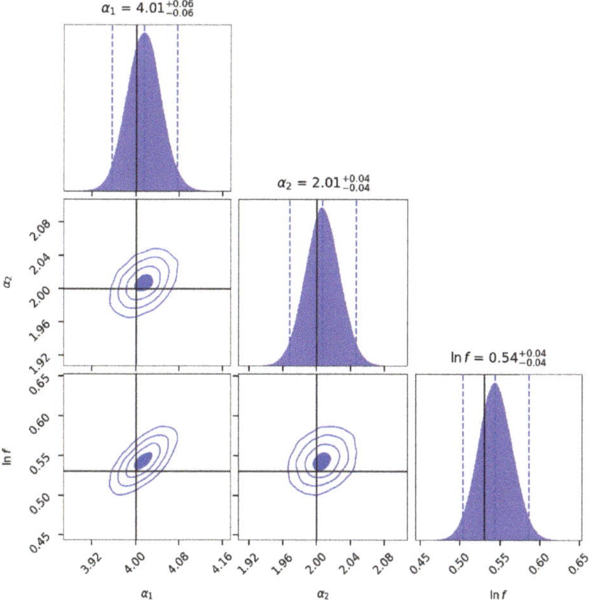

Figure 3. Contour plot for the marginalized posteriors for $(\alpha_1, \alpha_2, \ln f)$ parameters at 10%, 40%, 65%, and 85% confidence levels. Vertical dashed lines mark the 2σ region, while horizontal lines indicate the mean values of the marginalized parameters.

Using the logarithmic Jeffrey's scale [70] to input a qualitative classification of the evidence, our results indicate *barely worth-mentioning* evidence against the model as compared with the "true" model with roughly $X \sim 10^{-7}$ for the Bayes factor that is well accommodated in the range $0 \leq X < 1.1$. As traditionally known, Jeffrey's scale classifies it as weak evidence against the model with the interval $1.1 \leq X < 3$. The range $3 \leq X < 5$ indicates strong evidence disfavoring the competing model. Finally, very strong evidence against the competing model is achieved at $X \geq 5$.

5.3. Dynamical System and Comparison with the Potentials

Following Refs. [67,68], one constructs a dynamical system with the coordinates

$$x = \frac{\dot{\phi}}{\sqrt{6}H} , \tag{48}$$

$$y = \sqrt{\frac{V}{3H^2}} , \tag{49}$$

$$\lambda = \frac{-V'}{V} , \tag{50}$$

where the prime sign $(')$ indicates an ordinary derivative with respect to the scalar field ϕ. The variables (x, y, λ) can be written in the form of an autonomous system as

$$\frac{dx}{dN} = -3x + \sqrt{\frac{3}{2}}\lambda y^2 + \frac{3}{2}x(1 + x^2 - y^2) , \tag{51}$$

$$\frac{dy}{dN} = -\sqrt{\frac{3}{2}}\lambda xy + \frac{3}{2}y(1 + x^2 - y^2) , \tag{52}$$

$$\frac{d\lambda}{dN} = -\sqrt{6}\lambda^2(\Gamma - 1)x . \tag{53}$$

The quantity N is the number of e-foldings of inflation that is commonly defined as the logarithm of the scale factor $a(t)$ such as $N = N_0 + \ln a$. The term N_0 denotes the number of the e-folds set by the current time. The quantity Γ is defined in terms of the related potential as

$$\Gamma = \frac{VV''}{V'^2} , \tag{54}$$

in which the double prime symbol $('')$ denotes an ordinary second derivative with respect to the scalar field ϕ. From Equation (36), and using Equations (48) and (49), one obtains

$$\Omega_{ext}(x, y) = x^2 + y^2 , \tag{55}$$

for a locally universe flat space curvature. Defining a corresponding effective equation of the state γ_{ext}, one writes

$$\gamma_{ext} = 1 + w = \frac{2x^2}{x^2 + y^2} . \tag{56}$$

By direct calculations of Equations (48)–(50), with the initial value of $\lambda_{ini} = \alpha_1$, we obtain the logistic equation

$$\frac{d\lambda}{dN} = 2\sqrt{6}(\lambda - (\sqrt{\alpha_2} + \alpha_1))^2 x . \tag{57}$$

In Figure 4, we show the resulting curves from the models of Equation (44) (black lines) and our model defined by Equation (43) (red lines), showing the evolution for the dynamical variable λ_i running from 0.1 to 1, denoted by the outer and inner curves, respectively. The curves were obtained, setting the values of $\Omega_{ext} = \Omega_\phi = 0.68$ and the thawing initial conditions to both models as $N_0 = -15, x_i = 10^{-5}, y_i = 10^{-3}$. As a result, the evolution of the curves of our model begins earlier than the reference model in Equation (44) shown by the red lines. In both cases, the semicircles reach the x-axis at ∼0.82. The curves close to this value are compatible with the models of late cosmic acceleration of the universe. The origin represents an initial point in an early matter domination universe.

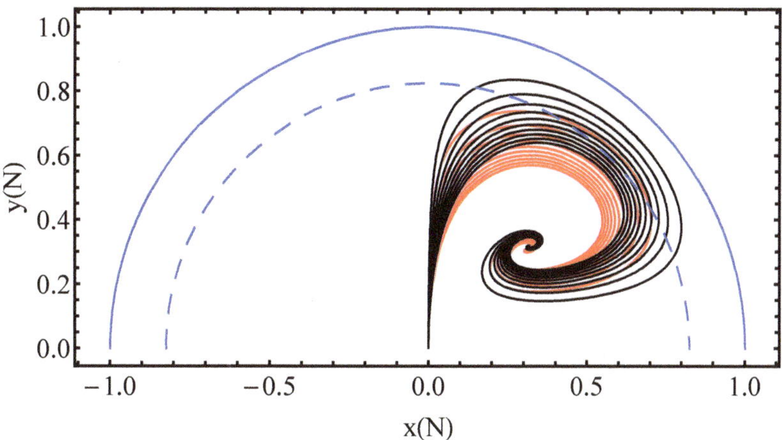

Figure 4. Comparison between the potentials from Equation (44) (black lines) and Equation (43) (red lines) that shows the evolution of the related autonomous systems. The outer continuous blue line and the dashed blue line semicircles represent the x-axis with $x \sim \sqrt{\Omega_{ext}} \sim 0.82$ and $\Omega_{ext} = \Omega_\phi = 0.68$, respectively.

Henceforth, we use the estimated values of the parameters of out model to evaluate how a resulting EoS of an effective $w(z)$ evolves. For this matter, we adapt the publicly code provided in [68]. The set of equations of Equations (48)–(50) has the initial conditions $N_0 = -15, x_i = 10^{-5}; y_i = 10^{-3}, \lambda[N_0] = \alpha_1$ for thawing models. For freezing models, one defines $N_0 = Log[10^{-6}], x_i = -0.5; y_i = 0.1, \lambda[N_0] = 0.1$. We use CPL parameterization as a function of the redshift written as

$$w(z) = w_0 + w_a \frac{z}{1+z}, \tag{58}$$

where (w_0, w_a) are parameters used to evaluate how $w(z)$ tends to depart away or converge to ΛCDM values. In conjunction with CPL, we use BA parameterization, which is given by

$$w(z) = w_0 + w_a \frac{z(1+z)}{1+z^2}. \tag{59}$$

We adopt the maximum value for the redshift as $z_{max} = 2$ and the maximum number of iterations i as $i_{max} = 50$.

In Figure 5, we have comparisons between CPL and BA parameterizations of the effective model in Equation (43) (with the values of Figure 5). Thus, we have a test adopting generic thawing and freezing models by means of Equations (48)–(50), and we obtain similar results as in Ref. [68]. The model prefers a thawing behavior from CPL (blue line) and BA (red line) parameterizations that are closer to numerical points (dotted line). When plotting the freezing pattern for the model, we obtained curves out of range with a large difference from the numerical curve, and they were omitted.

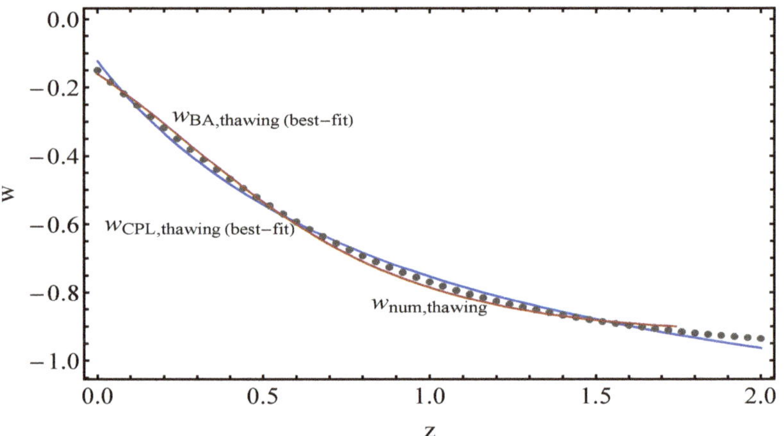

Figure 5. Behavior of numerical curves (dotted line) produced from the related autonomous system of the model preferring a thawing pattern of CPL (blue line) and BA (red line) parameterizations.

6. Remarks

We investigate the extrinsic curvature as an inflaton scalar field. The starting point lies in the correlation of the model parameter that is constrained for both accelerated regimes. For instance, the dark energy accelerated regime relies on $w < -1/3$; then we expect $w > -1/3$ for inflationary regimes in accordance with Figure 5. We also present a fluid approach in order to provide a better understanding of the related dynamics with a relation between the model parameter β_0 and the fluid parameter w, as shown in Equation (26). Our intent is to construct the basis of a geometrical model that fully relies on gravitational interactions. In this paper, we focus our attention on the inflationary period obtaining a logarithm inflaton field that generates an exponential potential in Equation (40). This form of potential creates a PLI mechanism that has strong constraints imposed by Planck data [62]. Then, we have proposed an effective form for a new potential defined in Equation (43) to generalize Equation (40). We have numerically restrained our model parameters by means of a Bayesian analysis obtained from random points using the DYNESTY code. Establishing a quintessential potential from a reference model of Equation (44) used in Ref. [68], we have obtained the related autonomous system of equations for forecasting models compatible with the universe evolution. Moreover, we have shown a comparison with CPL and BA parameterizations of our model that prefers a thawing cosmic scenario.

In this paper, we have considered all the cosmic components to be immutable. The numerical analysis gave us an optimistic scenario for further studies on the unification of inflation with late cosmic acceleration. The stability of the present model in this numerical evaluation allows us to use more realistic contexts to produce red spectra with a scalar spectral tilt and a small tensor-to-scalar ratio. As a subject of a future work, we will consider the full perturbation equations of the model in a scenario with a matter-sourced anisotropic stress and a series of latest astronomical data from cosmic microwave background radiation (CMB) experiments, such as Planck collaboration.

Author Contributions: Conceptualization, A.J.S.C.; formal analysis, L.A.C.; methodology, A.J.S.C. and L.A.C.; software, A.J.S.C.; supervision, L.A.C.; writing—original draft, A.J.S.C.; writing—review and editing, A.J.S.C. and L.A.C. All authors have read and agreed to the published version of the manuscript.

Funding: A.J.S.C. acknowledges Conselho Nacional de Desenvolvimento Científico e Tecnologico (CNPq, National Council for Scientific and Technological Development) for the partial financial support for this work (Grant No. 305881/2022-1) and Fundação da Universidade Federal do Paraná (FUN-PAR, Paraná Federal University Foundation) through public notice 04/2023-Pesquisa/PRPPG/UFPR for the partial financial support (Process No. 23075.019406/2023-92).

Data Availability Statement: All data/codes used in this work are publicly available, and correctly cited in the references.

Acknowledgments: The authors thank the referees for their useful comments and criticisms, which substantially improved this paper.

Conflicts of Interest: The authors declare no conflict of interest.

References

1. Guth, A.H. Inflationary universe: A possible solution to the horizon and flatness problems. *Phys. Rev. D* **1981**, *23*, 347–356. [CrossRef]
2. Starobinsky, A.A. Spectrum of relict gravitational radiation and the early state of the universe. *JETP Lett.* **1979**, *30*, 682–685.
3. Mukhanov, V.F.; Chibisov, G.V. Quantum Fluctuations and a Nonsingular Universe. *JETP Lett.* **1981**, *33*, 532–535.
4. Linde, A.D. Chaotic Inflation. *Phys. Lett. B* **1983**, *129*, 177–181. [CrossRef]
5. Hawking, S.W. The Development of Irregularities in a Single Bubble Inflationary Universe. *Phys. Lett. B* **1982**, *115*, 295. [CrossRef]
6. Hawking, S.W.; Moss, I.G. Fluctuations in the Inflationary Universe. *Nucl. Phys. B* **1983**, *224*, 180. [CrossRef]
7. Starobinsky, A.A. Dynamics of Phase Transition in the New Inflationary Universe Scenario and Generation of Perturbations. *Phys. Lett. B* **1982**, *117*, 175–178. [CrossRef]
8. Guth, A.H.; Pi, S.Y. Fluctuations in the New Inflationary Universe. *Phys. Rev. Lett.* **1982**, *49*, 1110–1113. [CrossRef]
9. Bardeen, J.M.; Steinhardt, P.J.; Turner, M.S. Spontaneous Creation of Almost Scale - Free Density Perturbations in an Inflationary Universe. *Phys. Rev. D* **1983**, *28*, 679. [CrossRef]
10. Giovannini, M. Spikes in the relic graviton background from quintessential inflation. *Class. Quant. Grav.* **1999**, *16*, 2905–2913. [CrossRef]
11. Giovannini, M. Production and detection of relic gravitons in quintessential inflationary models. *Phys. Rev. D* **1999**, *60*, 123511. [CrossRef]
12. Giovannini, M. Low scale quintessential inflation. *Phys. Rev. D* **2003**, *67*, 123512. [CrossRef]
13. Cognola, G.; Elizalde, E.; Nojiri, S.; Odintsov, S.D.; Sebastiani, L.; Zerbini, S. Class of viable modified $f(R)$ gravities describing inflation and the onset of accelerated expansion. *Phys. Rev. D* **2008**, *77*, 046009. [CrossRef]
14. Myrzakulov, R.; Sebastiani, L.; Vagnozzi, S. Inflation in $f(R, \phi)$ -theories and mimetic gravity scenario. *Eur. Phys. J. C* **2015**, *75*, 444. [CrossRef]
15. Salvio, A. Inflationary Perturbations in No-Scale Theories. *Eur. Phys. J. C* **2017**, *77*, 267. [CrossRef]
16. Salvio, A. Quasi-Conformal Models and the Early Universe. *Eur. Phys. J. C* **2019**, *79*, 750. [CrossRef]
17. Oikonomou, V.K. Viability of the intermediate inflation scenario with F(T) gravity. *Phys. Rev. D* **2017**, *95*, 084023. [CrossRef]
18. Agarwal, A.; Myrzakulov, R.; Sami, M.; Singh, N.K. Quintessential inflation in a thawing realization. *Phys. Lett. B* **2017**, *770*, 200–208. [CrossRef]
19. Keskin, A.I. Viable super inflation scenario from F(T) modified teleparallel gravity. *Eur. Phys. J. C* **2018**, *78*, 705. [CrossRef]
20. Castello, S.; Ilić, S.; Kunz, M. Updated dark energy view of inflation. *Phys. Rev. D* **2021**, *104*, 023522. [CrossRef]
21. Arkani–Hamed, N.; Dimopoulos, S.; Dvali, G. The hierarchy problem and new dimensions at a millimeter. *Phys. Lett. B* **1998**, *429*, 263–272. CrossRef]
22. Randall, L.; Sundrum, R. Large Mass Hierarchy from a Small Extra Dimension. *Phys. Rev. Lett.* **1999**, *83*, 3370–3373. [CrossRef]
23. Randall, L.; Sundrum, R. An Alternative to Compactification. *Phys. Rev. Lett.* **1999**, *83*, 4690–4693. [CrossRef]
24. Dvali, G.; Gabadadze, G.; Porrati, M. 4D gravity on a brane in 5D Minkowski space. *Phys. Lett. B* **2000**, *485*, 208–214. CrossRef]
25. Battye, R.A.; Carter, B. Generic junction conditions in brane-world scenarios. *Phys. Lett. B* **2001**, *509*, 331–336. CrossRef]
26. Maia, M.; Monte, E.M. Geometry of brane-worlds. *Phys. Lett. A* **2002**, *297*, 9–19. CrossRef]
27. Maia, M.D.; Monte, E.M.; Maia, J.M.F.; Alcaniz, J.S. On the geometry of dark energy. *Class. Quant. Grav.* **2005**, *22*, 1623. [CrossRef]
28. Maia, M.D.; Silva, N.; Fernandes, M.C.B. Brane-world quantum gravity. *J. High Energy Phys.* **2007**, *2007*, 47. [CrossRef]
29. Heydari-Fard, M.; Sepangi, H. Anisotropic brane gravity with a confining potential. *Phys. Lett. B* **2007**, *649*, 1–11. CrossRef]
30. Jalalzadeh, S.; Mehrnia, M.; Sepangi, H.R. Classical tests in brane gravity. *Class. Quant. Grav.* **2009**, *26*, 155007. [CrossRef]
31. Maia, M.D.; Capistrano, A.J.S.; Alcaniz, J.S.; Monte, E.M. The Deformable Universe. *Gen. Rel. Grav.* **2011**, *43*, 2685–2700. [CrossRef]
32. Ranjbar, A.; Sepangi, H.R.; Shahidi, S. Asymptotically Lifshitz Brane-World Black Holes. *Ann. Phys.* **2012**, *327*, 3170. [CrossRef]
33. Capistrano, A.J.S.; Cabral, L.A. Geometrical aspects on the dark matter problem. *Ann. Phys.* **2014**, *348*, 64–83. [CrossRef]
34. Capistrano, A.J.S.; Cabral, L.A. Implications on the cosmic coincidence by a dynamical extrinsic curvature. *Class. Quant. Grav.* **2016**, *33*, 245006. [CrossRef]

35. Capistrano, A.J.S. Constraints on cosmokinetics of smooth deformations. *Mon. Not. Roy. Astron. Soc.* **2015**, *448*, 1232–1239. [CrossRef]
36. Capistrano, A.J.S.; Gutiérrez-Piñeres, A.C.; Ulhoa, S.C.; Amorim, R.G. On classical thermal stability of black holes with a dynamical extrinsic curvature. *Ann. Phys.* **2017**, *380*, 106–120. [CrossRef]
37. Capistrano, A.J.S. Evolution of Density Parameters on a Smooth Embedded Universe. *Annalen der Physik.* **2018**, *530*, 1700232. [CrossRef]
38. Capistrano, A.J.S. Lukewarm black holes in the Nash-Greene framework. *Phys. Rev. D* **2019**, *100*, 064049. [CrossRef]
39. Capistrano, A.J.S.; Seidel, P.T.Z.; Duarte, H.R. Subhorizon linear Nash–Greene perturbations with constraints on H(z) and the deceleration parameter q(z). *Phys. Dark Univ.* **2021**, *31*, 100760. [CrossRef]
40. Capistrano, A.J.S.; Cabral, L.A.; Marão, J.A.P.F.; Coimbra-Araújo, C.H. Linear Nash-Greene fluctuations on the evolution of S8 and H0 tensions. *Eur. Phys. J.* **2022**, *82*, 1434–6052. [CrossRef]
41. Martin, J.; Ringeval, C.; Vennin, V. Encyclopædia Inflationaris. *Phys. Dark Univ.* **2014**, *5-6*, 75–235. [CrossRef]
42. Higson, E.; Handley, W.; Hobson, M.; Lasenby, A. Dynamic nested sampling: an improved algorithm for parameter estimation and evidence calculation. *Stat. Comput.* **2019**, *29*, 891–913. [CrossRef]
43. Speagle, J.S. dynesty: a dynamic nested sampling package for estimating Bayesian posteriors and evidences. *Mon. Not. Roy. Astron. Soc.* **2020**, *493*, 3132–3158. [CrossRef]
44. Chevallier, M.; Polarski, D. Accelerating universes with scaling dark matter. *Int. J. Mod. Phys. D* **2001**, *10*, 213–224. [CrossRef]
45. Linder, E.V. Exploring the expansion history of the universe. *Phys. Rev. Lett.* **2003**, *90*, 091301. [CrossRef]
46. Jr., E.M.B.; Alcaniz, J. Probing the time dependence of dark energy. *J. Cosmol. Astropart. Phys.* **2012**. *2012*, 42. [CrossRef]
47. Gupta, G.; Rangarajan, R.; Sen, A.A. Thawing quintessence from the inflationary epoch to today. *Phys. Rev. D* **2015**, *92*, 123003. [CrossRef]
48. Chiba, T. Slow-roll thawing quintessence. *Phys. Rev. D* **2009**, *79*, 083517. [CrossRef]
49. Scherrer, R.J.; Sen, A.A. Thawing quintessence with a nearly flat potential. *Phys. Rev. D* **2008**, *77*, 083515. [CrossRef]
50. Sahlen, M.; Liddle, A.R.; Parkinson, D. Quintessence reconstructed: New constraints and tracker viability. *Phys. Rev. D* **2007**, *75*, 023502. [CrossRef]
51. Schimd, C.; Tereno, I.; Uzan, J.P.; Mellier, Y.; van Waerbeke, L.; Semboloni, E.; Hoekstra, H.; Fu, L.; Riazuelo, A. Tracking quintessence by cosmic shear - constraints from virmos-descart and cfhtls and future prospects. *Astron. Astrophys.* **2007**, *463*, 405–421. :20065154 [CrossRef]
52. Chiba, T. w and w' of scalar field models of dark energy. *Phys. Rev. D* **2006**, *73*, 063501. [CrossRef]
53. Scherrer, R.J. Dark energy models in the w-w' plane. *Phys. Rev. D* **2006**, *73*, 043502. [CrossRef]
54. Eisenhart, L.P. *On Riemannian Geometry*; Dover Publications: New York, 2005.
55. Nash, J. C1 Isometric Imbeddings. *Ann. Math.* **1954**, *60*, 383–396. [CrossRef]
56. Nash, J. The Imbedding Problem for Riemannian Manifolds. *Ann. Math.* **1956**, *63*, 20–63. [CrossRef]
57. Greene, R.E. Isometric embeddings of Riemannian and pseudo-Riemannian manifolds. *Memoirs of the American Mathematical Society*; American Mathematical Society: Providence, RI, USA, 1970; Volume 63. [CrossRef]
58. Turner, M.S.; White, M.J. CDM models with a smooth component. *Phys. Rev. D* **1997**, *56*, R4439. [CrossRef]
59. Abbott, L.F.; Wise, M.B. Constraints on Generalized Inflationary Cosmologies. *Nucl. Phys. B* **1984**, *244*, 541–548. [CrossRef]
60. Lucchin, F.; Matarrese, S. Power Law Inflation. *Phys. Rev. D* **1985**, *32*, 1316. [CrossRef]
61. Davies, P.C.W.; Sahni, V. Quantum gravitational effects near cosmic strings. *Class. Quant. Grav.* **1988**, *5*, 1. [CrossRef]
62. Tristram, M.; Banday, A.J.; Górski, K.M.; Keskitalo, R.; Lawrence, C.R.; Andersen, K.J.; Barreiro, R.B.; Borrill, J.; Eriksen, H.K.; Fernandez-Cobos, R.; et al. Planck constraints on the tensor-to-scalar ratio. *Astron. Astrophys.* **2021**, *647*, A128. [CrossRef]
63. Becker, K.; Becker, M.; Krause, A. M-theory inflation from multi M5-brane dynamics. *Nucl. Phys. B* **2005**, *715*, 349–371. [CrossRef]
64. Bennai, M.; Chakir, H.; Sakhi, Z. On Inflation Potentials in Randall-Sundrum Braneworld Model. *Eur. J. Phys.* **2006**, *9*, 84–93.
65. Parsons, P.; Barrow, J.D. Generalized scalar field potentials and inflation. *Phys. Rev. D* **1995**, *51*, 6757–6763. [CrossRef] [PubMed]
66. Barrow, J.D.; Nunes, N.J. Dynamics of Logamediate Inflation. *Phys. Rev. D* **2007**, *76*, 043501. [CrossRef]
67. Clemson, T.G.; Liddle, A.R. Observational constraints on thawing quintessence models. *Mon. Not. Roy. Astron. Soc.* **2009**, *395*, 1585–1590. [CrossRef]
68. Pantazis, G.; Nesseris, S.; Perivolaropoulos, L. Comparison of thawing and freezing dark energy parametrizations. *Phys. Rev. D* **2016**, *93*, 103503. [CrossRef]
69. Foreman-Mackey, D.; Agol, E.; Ambikasaran, S.; Angus, R. Fast and Scalable Gaussian Process Modeling with Applications to Astronomical Time Series. *Astron. J.* **2017**, *154*, 220. [CrossRef]
70. Jeffreys, H. *Theory of Probability*, 3rd ed.; Oxford University Press: Oxford, UK, 1961.

Article

Electromagnetic Waves in Cosmological Spacetime

Denitsa Staicova and **Michail Stoilov** *

Institute for Nuclear Research and Nuclear Energy, Bulgarian Academy of Sciences, 1784 Sofia, Bulgaria; dstaicova@inrne.bas.bg
* Correspondence: mstoilov@inrne.bas.bg

Abstract: We consider the propagation of electromagnetic waves in the Friedmann–Lemaître–Robertson–Walker metric. The exact solutions for plane and spherical wave are written down. The corresponding redshift, amplitude change, and dispersion are discussed. We also speculate about the connection of the electromagnetic wave equation to the Proca equation and its significance for the early Universe.

Keywords: Friedmann metric; electromagnetic waves; redshift; amplitude change; dispersion; Klein–Gordon equation; tachyon inflation

---

1. Introduction

Currently, most of the known information about the Universe outside the Earth is obtained observing electromagnetic waves. This is especially true for the deep cosmos, even though gravitational astronomy [1] and neutrino telescopes [2,3] are on the path of opening the doors of the multimessenger observations [4]. So far, however, the experiments that have shaped our vision about the Universe and its evolution, such as Planck [5], Wilkinson Microwave Anisotropy Probe [6], WiggleZ Dark Energy Survey [7], Background Imaging of Cosmic Extragalactic Polarization [8], All Sky Automated Survey for Super Novae [9], Sloan Digital Sky Survey [10], Hubble Space Telescope [11,12], James Webb Space Telescope [13], etc., all detect electromagnetic signals emitted somewhere faraway that reach us after a long journey through the spacetime. From them we know that the Universe is currently in the epoch of accelerated expansion [14] and that the flat Friedmann–Lemaître–Robertson–Walker (FLRW) metric [15] is, so far, the best candidate for large-scale metrics (and is considered as a standard model for cosmological observations).

In this work we consider the problem of how the electromagnetic waves propagate in the FLRW metric, which is quite important to observational cosmology. This is not a new problem and it has attracted constant attention during the past few decades [16–21]. A wide variety of methods has been used to solve it. Some of them (such as the usage of the Debye potential or the reformulation of the problem as propagation of the electromagnetic field in Minkowski space time in medium for which susceptibilities are determined by the metric tensor) are quite general and are suitable for different gravitational backgrounds. Other ones (such as using the conformal coordinates) are tailored specifically to the FLRW metric. Some of the authors work directly with the electric and magnetic fields; other ones prefer to work with the electromagnetic potential in different gauges (usual—the Coulomb one). Here our approach is to consider the problem using electromagnetic potential in the standard co-moving frame and in the generalized Lorenz gauge. We have three reasons which give grounds for our choice: the first two explain why we use the electromagnetic potential, while the third one answers the question of why we prefer the Lorenz gauge: First, the electromagnetic potential is the canonical coordinate in the electrodynamics, considered the Hamiltonian dynamical system, while the electric field is the canonical momentum and the magnetic field is derived from the potential. The eventual (canonical) quantization will rely deeply on the potential modes. Second, the Aharonov–Bohm effect demonstrates

Citation: Staicova, D.; Stoilov, M. Electromagnetic Waves in Cosmological Spacetime. *Universe* **2023**, *9*, 292. https://doi.org/10.3390/universe9060292

Academic Editor: Kazuharu Bamba and Øyvind Grøn

Received: 30 March 2023
Revised: 12 June 2023
Accepted: 14 June 2023
Published: 16 June 2023

the fundamental role of the interaction of the potential with charged particles. Third, the usage of a generally covariant gauge condition exhibits a very interesting peculiarity of the electrodynamics in curved spacetime: The strong equivalence principle states that the laws of physics are one and the same in all inertial frames. However, there is a result [22] that on a world line one can place any metric into Minkowski form (the inertial coordinate system) and to nullify its first derivatives but not the second and higher ones. As a result, any equation in which metric's second and higher derivatives appear is not the same in Minkowski spacetime and in local inertial coordinate system in a more general spacetime. The electrodynamics in curved spacetime (more specifically—in non-Ricci flat spacetime) is an example of such theory because of the presence of the Ricci tensor in its equation of motion.

We want to emphasize that, here, we treat the electromagnetic waves as test particles in a predefined metric (flat FLRW one in our case). In other words, we consider the electromagnetic field as a perturbation, i.e., we do not take into account the gravitation produced by the wave itself. In addition, we suppose the existence of a medium in the spacetime which, first, produces the FLRW metric, and second, is inert to the electromagnetic interaction.

The paper is organized as follows: We start with some basic facts about the free electromagnetic field in curved spacetime. We give the definitions of the electromagnetic field tensor, electromagnetic Lagrangian, equation of motion, and gauge fixing. In addition, we remind the reader of how the gauge fixing can be incorporated into the Lagrangian and write down the effective Lagrangian, corresponding to our setup.

Next, we consider two simple solutions of the free electromagnetic equation of motion—a plane wave and a spherical one. We show that in both cases under consideration, each physical (transverse) component of the potential decouples from other components, while the time and longitudinal components remain entangled. The equations for the transverse components are quite simple and can be solved exactly. Their solutions demonstrate relativistic redshift, amplitude decrease (fading), and dispersion.

We also speculate about the connection between electromagnetic wave equation in FLRW metric in the generalized Lorenz gauge and the Proca equation with mass determined by the Ricci tensor. It turns out that for a special case of equation of state, the electromagnetic wave describes a tachyon. Finally, we briefly discuss the possibility to use the metric-induced transition of photons and other $U(1)$ gauge field particles from tachyons to normal particles as a source of the inflation during the early Universe.

2. Theory

The electromagnetic potential A^{ν} defines the electromagnetic field tensor $F^{\mu\nu}$ which, due to its antisymmetry and the symmetry of Christoffel symbols ($\Gamma^{\alpha}_{\mu\nu} = \Gamma^{\alpha}_{\nu\mu}$), has the same form in any frame

$$F_{\mu\nu} = \nabla_{\mu}A_{\nu} - \nabla_{\nu}A_{\mu} = \partial_{\mu}A_{\nu} - \partial_{\nu}A_{\mu}. \tag{1}$$

Here, ∇^{μ} is the covariant derivative in the spacetime. As a result, first, in any spacetime the electromagnetic field tensor is invariant under the usual gauge symmetry $A_{\mu}(x) \rightarrow A_{\mu}(x) + \partial_{\mu}f(x)$, where $f(x)$ is an arbitrary function, and second, the Lagrangian of the free electromagnetic field is exactly the same as in Minkowski spacetime:

$$\mathcal{L} = -\frac{1}{4}F_{\mu\nu}F^{\mu\nu}. \tag{2}$$

The corresponding equation of motion is

$$\partial_{\nu}(\sqrt{-g}F^{\mu\nu}) = 0, \tag{3}$$

where g is the determinant of the metric tensor $g_{\mu\nu}$. The above equation can be placed into explicitly covariant form, namely,

$$\nabla_{\nu}F^{\mu\nu} = 0. \tag{4}$$

Note that up to now we have not fixed the gauge freedom in the potential. In order to remove this arbitrariness we use the generalized Lorenz gauge

$$\nabla_\nu A^\nu = 0. \tag{5}$$

In this gauge, Equation (4) takes the form

$$\Box A^\mu - R^\mu{}_\nu A^\nu = 0. \tag{6}$$

Here, $\Box = \nabla^\nu \nabla_\nu$ is the d'Alembert operator defined for the covariant derivatives ∇^μ, and R^μ_ν is the Ricci tensor

$$R_{\mu\nu} = \partial_\alpha \Gamma^\alpha_{\mu\nu} - \partial_\mu \Gamma^\alpha_{\alpha\nu} + \Gamma^\alpha_{\mu\nu}\Gamma^\beta_{\alpha\beta} - \Gamma^\beta_{\alpha\mu}\Gamma^\alpha_{\beta\nu}. \tag{7}$$

The following relations between the Riemann tensor $R^\nu_{\alpha\mu\nu}$ and Ricci tensor are used in the derivation of Equation (6):

$$(\nabla_\nu\nabla_\mu - \nabla_\mu\nabla_\nu)A^\nu = R^\nu_{\alpha\nu\mu}A^\alpha = R_{\alpha\mu}A^\alpha = R_{\mu\alpha}A^\alpha.$$

We want to stress that Equation (6) is not the wave equation ($\Box A^\mu = 0$) and that it cannot be cast into it with the help of change of coordinates. Therefore, if the spacetime is not Ricci flat then the equation satisfied by the electromagnetic waves will differ from the one in Minkowski spacetime even in the locally inertial coordinate frame.

The gauge condition can be incorporated into the Lagrangian. The easiest way to achieve this is with the help of the Lagrangian multiplier—a new auxiliary dynamical field $\lambda(x)$, so that the gauge fixed Lagrangian is

$$\mathcal{L}_{\text{g.f.}} = -\frac{1}{4}F_{\mu\nu}F^{\mu\nu} - \lambda\nabla_\nu A^\nu. \tag{8}$$

There is a procedure, outlined, e.g., in Ref. [23], for how to integrate out the Lagrangian multiplier. Applying it, we obtain the following, fully covariant, gauge-fixed, effective Lagrangian

$$\mathcal{L}_{\text{eff}} = -\frac{1}{4}F_{\mu\nu}F^{\mu\nu} - \frac{1}{2}(\nabla_\nu A^\nu)^2. \tag{9}$$

Note that in the last term of Equation (9) there is a part $\mathcal{L}_2$ which is quadratic with respect to the potential $\mathcal{L}_2 = A_\mu M^{\mu\nu} A_\nu$, $M^{\mu\nu} = \partial_\alpha(\sqrt{-g}g^{\mu\alpha})\partial_\beta(\sqrt{-g}g^{\nu\beta})/g$, so Equation (9) resembles the Proca action (more precisely—$\mathcal{L}_4$ generalized Proca one [24]). We shall come back to this question later.

3. Results

Here, we describe some basic solutions of Equation (6) in the case of FLRW background. On the base of these solutions we give some estimates about the redshift, amplitude decrease, and dispersion which the electromagnetic wave undergoes during its propagation in spacetime.

3.1. Plane Electromagnetic Wave in the FLRW Metric

The description of a plane wave can be given most easily in the Cartesian form of the metric of a flat FLRW-universe. In this case the invariant length is

$$ds^2 = -dt^2 + a(t)^2(dx^2 + dy^2 + dz^2), \tag{10}$$

where $a(t)$ is the scale "parameter" and we work in units where the speed of light $c = 1$.

For the sake of simplicity, we consider a wave propagating in the z direction, i.e., we suppose that $A^\mu = A^\mu(t, z)$. We want to stress that in this way we can describe the most general plane wave solution because of the possibility to freely choose the spacial coordinate system orientation. For the considered functional dependence of the potential, Equation (6) simplifies, and each of the potential transverse components A^1 and A^2 (i.e., $x-$ and $y-$components of the potential) decouple from all other components. Note that, here, the transverse potential determines transverse electric and magnetic fields. Therefore, we deal with a transverse electromagnetic wave which remains as such forever. Certainly, this is valid only in the coordinate system determined by Equation (10). For instance, in the observer frame, where $x_{ob} = a(t)x$ (and analogously for the other spacial coordinates), every component of the electromagnetic potential is a mix of physical and nonphysical degrees of freedom.

The equation, which both A^1 and A^2 satisfy is

$$-\ddot{A}^i + \frac{1}{a^2}A^{i\prime\prime} - 5\frac{\dot{a}}{a}\dot{A}^i - 2\left(2\frac{\dot{a}^2}{a^2} + \frac{\ddot{a}}{a}\right)A^i = 0, \quad i = 1,2. \tag{11}$$

Here, we denote with dots the derivatives over time and with primes—derivatives with respect to z. The equations for the nonphysical components A^0 and A^3 are completely different and each of them entangles A^0 and A^3.

We look for a solution of Equation (11) with separated variables $A(t,z) = f(t) \times g(z)$. In this case, Equation (11) leads to the following two equations:

$$g'' + k^2 g = 0, \tag{12}$$
$$a^2\ddot{f} + 5a\dot{a}\dot{f} + (k^2 + 4\dot{a}^2 + 2a\ddot{a})f = 0, \tag{13}$$

where k^2 is some constant.

At this point, we want to make a small comment about Equation (12). The self-consistency of the assumption for separation of variables requires that $a(t)$ does not participate in it. In other words, the equation in question will have the same form for any $a(t)$, even for $a(t) = 1$. Thus, it will be the same as in flat Minkowski spacetime.

In view of the above comment, it is not a surprise that Equation (12) is well known. Its general real solution for $k^2 > 0$ (which we suppose) is

$$g(z) = \tilde{c}_1 \sin(kz) + \tilde{c}_2 \cos(kz) \tag{14}$$

where $\tilde{c}_i, \ i = 1,2$ are arbitrary real constants.

Now we move to Equation (13). In order to find its solution we make two consecutive ansatzes. First, let

$$f(t) = \frac{\mathfrak{f}(t)}{a(t)^2} \tag{15}$$

which leads to the following differential equation for $\mathfrak{f}(t)$:

$$a^2\ddot{\mathfrak{f}} + k^2\mathfrak{f} + a\dot{a}\dot{\mathfrak{f}} = 0. \tag{16}$$

The second ansatz is

$$\mathfrak{f}(t) = \mathbf{f}\left(\int^t \frac{d\tau}{a(\tau)}\right) \tag{17}$$

where the argument of the function $\mathbf{f}$ is the so-called conformal time and the differential equation for $\mathbf{f}$ is

$$d_\tau^2\mathbf{f}(\tau) + k^2\mathbf{f}(\tau) = 0. \tag{18}$$

Note that Equation (18) has the same form as Equation (12) and, accordingly, has the same type of solution. Therefore, the general solution of Equation (13) is

$$f(t) = \frac{1}{a(t)^2}\left(\tilde{c}_1 \sin\left(k\int^t \frac{d\tau}{a(\tau)}\right) + \tilde{c}_2 \cos\left(k\int^t \frac{d\tau}{a(\tau)}\right)\right). \tag{19}$$

3.2. Spherical Electromagnetic Wave in the FLRW Metric

We consider the spherical coordinate system as the appropriate one for description of spherical waves (provided it is positioned and oriented accordingly to the source and observer). In spherical coordinates the invariant length is

$$ds^2 = -dt^2 + a(t)^2 \left(dr^2 + r^2 d\theta^2 + r^2 \sin(\theta)^2 d\phi^2 \right) \tag{20}$$

and it determines the particular form of Equation (6) in this case.

We consider the electromagnetic wave propagation in r direction, supposing that $A^\mu = A^\mu(t, r)$, and, in addition, supposing separation of variables. Therefore, we are looking for transverse potential

$$A^\perp = (0, 0, A^\theta(t, r), A^\phi(t, r)) = f(t)(0, 0, \mathbf{A}^\theta(r), \mathbf{A}^\phi(r)) \tag{21}$$

(the time dependence of both transverse components is indeed the same—see below).

Here, we see a flaw in our considerations because we postulate the existence of a transverse constant vector field on a sphere. However, such a field does not exist. Any transverse vector field on a sphere has to have at least one zero (or singularity). Therefore, one of our assumptions, or both of them, are incorrect. Nevertheless, we continue our analysis, supposing that it is valid at most only approximately in a small vicinity on the sphere for which we suppose resides at $\theta = \pi/2$, $\phi = 0$. This should be sufficient for a cosmological observer who, anyway, can gather information only locally.

It turns out that the evolution of each of the transverse components decouples from all other components as in the plane wave case. The corresponding equations of motion are

$$-\ddot{A}^\theta + \frac{1}{a^2}(A^\theta)'' - 5\frac{\dot{a}}{a}\dot{A}^\theta + 4\frac{1}{ra^2}(A^\theta)' - 2\left(2\frac{\dot{a}^2}{a^2} + \frac{\ddot{a}}{a}\right)A^\theta + \frac{1-b^2}{r^2a^2}A^\theta = 0, \tag{22}$$

$$-\ddot{A}^\phi + \frac{1}{a^2}(A^\phi)'' - 5\frac{\dot{a}}{a}\dot{A}^\phi + 4\frac{1}{ra^2}(A^\phi)' - 2\left(2\frac{\dot{a}^2}{a^2} + \frac{\ddot{a}}{a}\right)A^\phi = 0. \tag{23}$$

Now the primes denote derivatives with respect to r coordinate and $b = \cot(\theta)$. The fact that Equations (22) and (23) are different is a surprise, taking into account that orientation of the coordinate system around the axis of observation is a matter of our choice.

Let us consider Equation (23) first. As it has been already mentioned we look for a solution with separated variables, so that $A^\phi(t, r) = f(t) \times g(r)$. It turns out that the differential equation for $f(t)$ thus defined coincides with Equation (13) considered in the plane wave case. For the equation satisfied by the function $g(r)$ we obtain

$$g'' + \frac{4}{r}g' + k^2 g = 0, \tag{24}$$

where, as in the case of plane wave, k^2 is some positive parameter. Its general solution is

$$\begin{aligned} g(r) &= -\frac{1}{kr}(\hat{c}_1 j_1(kr) + \hat{c}_2 y_1(kr)) \\ &= \frac{1}{(kr)^2}\left(\hat{c}_1\left(\cos(kr) - \frac{\sin(kr)}{kr}\right) + \hat{c}_2\left(\sin(kr) + \frac{\cos(kr)}{kr}\right)\right), \end{aligned} \tag{25}$$

where j_n and y_n are the spherical Bessel functions of first and second kind.

Next, we consider Equation (22) and we again look for a solution in the form $A^\theta(t, r) = f(t) \times g(r)$. Once again, the differential equation for $f(t)$ is exactly Equation (13). For the function $g(r)$ we obtain the following equation:

$$g'' + \frac{4}{r}g' + \left(k^2 + \frac{1-b^2}{r^2}\right)g = 0. \tag{26}$$

The general solution of Equation (26) is

$$g(r) = \frac{1}{r^{3/2}}\left(\tilde{c}_1 J_{\sqrt{5+4b^2}/2}(kr) + \tilde{c}_2 Y_{\sqrt{5+4b^2}/2}(kr)\right),\tag{27}$$

where J_α and Y_α are the Bessel functions of first and second kind. Note, however, that θ−dependence reappears in Equation (27) through the quantity b. This contradicts our initial ansatz. Therefore, we reconsider our assumption and now we are looking for a solution in the form

$$A^\theta = \mathcal{A}^\theta(t,r)\sin(\theta).\tag{28}$$

Skipping the details, it is possible to show that the equation for $\mathcal{A}^\theta$ at $\theta = \pi/2$ is exactly Equation (23). This resolves the problem with rotational symmetry around observation axis. The result also rules out the possible polarization of the light induced by its propagation through the space which is suggested by Equations (25) and (27).

3.3. Cosmological Redshift, Fading, and Dispersion

In this section we consider some quantities that are of interest for the observers of the electromagnetic waves.

First, we define the redshift z. It can be derived in number of different ways in the cosmological context. Usually, one uses the geodesic of the photon to obtain the so-called cosmological redshift [25]. One can also solve the scalar wave equation for light with certain initial data [26,27]. One can also use the Einstein–Maxwell's equations in the spacetime defined with its metric to obtain the solutions and obtain it from there. This is the approach we will use in this paper, so the redshift is defined as

$$1 + z = \frac{\omega_e}{\omega_o}\tag{29}$$

where ω_e is the wave (angular) frequency at the moment t_e and ω_o is the observed frequency at the moment t_o. Note that time dependence for spherical and plane waves is one and the same and is given by Equation (19). Therefore, in both cases the angular frequency at moment t is determined by the wave period Δt, $(\omega_t = 2\pi/\Delta t)$ and for Δt we have the following equation:

$$2\pi = k \int_t^{t+\Delta t} \frac{d\tau}{a(\tau)} - k \int^t \frac{d\tau}{a(\tau)} \approx \frac{k}{a(t)}\Delta t.\tag{30}$$

As a result, the frequency at the moment t is $\omega_t = k/a(t)$, and so the redshift is

$$z_{\text{pl, sph}} = \frac{a(t_o)}{a(t_e)} - 1,\tag{31}$$

where subscripts stand for plane and spherical wave.

Next, we define the fading $\mathcal{F}$ as the amplitude decrease due to the propagation of the wave:

$$\mathcal{F} = \frac{\mathcal{A}_o}{\mathcal{A}_e}.\tag{32}$$

Here, $\mathcal{A}_e$ is the amplitude of the wave at the moment t_e and $\mathcal{A}_o$ is the observed amplitude at the moment t_o, $t_o > t_e$.

Another interesting characteristic of the waves is their dispersion $\mathcal{D}$. Similar to the redshift, there are lot of different ways to define the dispersion; probably the simplest is an analogue of the group velocity dispersion: $\mathcal{D} = d^2\omega/d^2k$. However, in all cases under consideration here, the frequency is linear with respect to wave number, but the spatial part of the wave phase is not in the case of spherical wave (see below). Therefore, our definition for dispersion is

$$\mathcal{D} = \frac{d^2 \mathrm{arg}}{d^2 k},$$
(33)

where arg is the argument of the sine or cosine part of the wave function. The quantity is determined only at the moment t_0.

Applying the above definitions for both plane and spherical waves, we obtain the following results for the plane wave:

$$\mathcal{F}_{\mathrm{pl}} = \left(\frac{a(t_e)}{a(t_o)}\right)^2,$$
(34)

$$\mathcal{D}_{\mathrm{pl}} = 0.$$
(35)

The corresponding results for a spherical wave are different. Having in mind that $1/kr \ll 1$, we can use Equation (25) which defines the spacial-dependent component of transverse potential in the following form:

$$g(r) \approx \frac{1}{r^2}\left(\hat{c}_1 \cos\left(kr + \frac{1}{kr}\right) + \hat{c}_2 \sin\left(kr + \frac{1}{kr}\right)\right).$$
(36)

We set the origin of the coordinate system at the geometric center of the spherical wave. We consider the propagation along the $r-$axis of a fixed phase of outgoing wave, so that

$$k \int^t \frac{d\tau}{a(\tau)} - kr - \frac{1}{kr} = \mathrm{constant},$$
(37)

which can be used to determine r_o from r_e, t_e, and t_o. Therefore, the fading and dispersion are

$$\mathcal{F}_{\mathrm{sph}} = \left(\frac{r_e a(t_e)}{r_o a(t_o)}\right)^2$$

$$\approx \frac{a(t_e)^2}{a(t_o)^2\left(1 + \frac{1}{r_e}\int_{t_e}^{t_o}\frac{1}{a} + \frac{1}{k^2 r_e^2}\right)^2},$$
(38)

$$\mathcal{D}_{\mathrm{sph}} = \frac{2}{r_o k^3}.$$
(39)

3.4. The Photon Mass

The Ricci tensor in the FLRW metric is diagonal. Its space components R_i^i (no summation over i), $i = 1,2,3$ (i.e., $x-$, $y-$, and $z-$ components in Cartesian coordinates and $r-$, $\theta-$, and $\phi-$ components in spherical coordinates) have one and the same value:

$$R_0^0 = \frac{3\ddot{a}}{a},$$
(40)

$$R_i^i = \frac{2\dot{a}^2 + a\ddot{a}}{a^2} = (2 - q)H^2, \quad \forall i = 1,2,3,$$
(41)

where $q = -a\ddot{a}/\dot{a}^2$ is called, due to historical reasons, the *deceleration parameter* and $H = \dot{a}/a$ is the Hubble function. Note that q is dimensionless, so the factor $2 - q$ is well defined. Note also that only the transverse components of the potential are physical. These, in the general case, are certain linear combinations of the spatial components, so we are interested in their mass only.

Except for the very first moments of the Universe's evolution, R_i^i is a very slowly changing function of time and, therefore, the term $R_\mu^i A^\mu = R_i^i A^i$ (no summation over i) in Equation (6) behaves similar to a mass term and Equation (6) itself—similar to a variant of the Proca equation. The effective mass term for the physical degrees of freedom is zero only for $a(t) = $ constant or for $a(t) = $ constant $\times\, t^{1/3}$. These are the only two cases for which Equation (6), rewritten in the inertial coordinate system, coincides with the wave equation of Minkowski electrodynamics. Note that the power law behavior of the scale factor ($a(t) \propto t^p$, $p = 2/(3(w+1))$) is an exact solution of the Friedmann equations for the Universe full with pure fluid with constant equation of state $w = p/\rho$, where p is the fluid pressure and ρ is its density. In this more general case, the photon effective mass is

$$m_\gamma^2 \ (= R_i^i) = \frac{p(3p-1)}{t^2} = \frac{2(1-w)}{3t^2(w+1)^2} \tag{42}$$

which is positive for $w < 1$. However, in the case $w > 1$, the mass squared is negative and the photon represents a *tachyon*.

It is interesting to see the photon mass for exponentially growing scale factor $a(t) \propto \exp(ht)$. In this case we obtain

$$m_\gamma^2 = 3h^2, \tag{43}$$

i.e., it is a positive constant (exactly).

It will be instructive to have some estimate for the magnitude of the effective mass that we are speaking about. Let us note that for the present-day values of the deceleration parameter and Hubble function $q \approx -0.55$, $H \approx 70\ \mathrm{km/s/Mpc}$ the photon mass in an inertial coordinate system is $m_\gamma \approx 10^{-33}$ eV.

4. Discussion

The obtained closed expressions for the electromagnetic plane and spherical wave solutions allow us to estimate the observed redshift, fading, and dispersion. The obtained redshift (Equation (31)) coincides with the standard one known in the literature [15,28]. The amplitude decrease given in Equation (34) is referred to in [29] as "adiabatic". The extension to a nonflat Universe considered therein showed that $a(t)^{-1}$ decrease is possible. On the other hand, the amplitude decrease of a spherical wave (Equation (38)) demonstrates the dependence of the fading on the wave number. The fading is increasing when the wavelength is increasing. A result, connected to the above one, is given by Equation (39) and predicts nonzero dispersion of a spherical wave.

Finally, we want to make some comments about a possible connection between Equation (6) and the Proca equation. Here, we used the generalized Lorenz gauge to fix the gauge freedom. As a result, Equation (6) is not gauge-invariant. In the Proca case, the mass term breaks the gauge symmetry as well but it cannot be considered as a gauge-fixing term for the electromagnetic potential. Nevertheless, the equations of motion for both theories are quite similar. The main difference between them is that for the electrodynamics, in the generalized Lorenz gauge the effective mass of the photon is controlled by the metric tensor. In view of this we suppose that a solution of the Proca equation with suitably chosen mass can approximate a solution of Equation (6). We consider this fact, i.e., that the photons can be massive, as very interesting and the fact that they can be tachyons as even more interesting. Note that tachyonic models have a long history in cosmology. Some of them originate as special cases of k-essence theories with Dirac–Born–Infeld (DBI) action [30]. On the other hand, k-essence theories [31,32] are used to describe early inflation and dark energy through a minimally coupled scalar field with noncanonical kinetic term. In the tachyonic models [33,34], universe expansion (possibly accelerated) is produced while the tachyon rolls down towards its minimum. Tachyons have also been discussed in terms of the so-called tachyonic preheating [35], which may lead to explosive particle production. Ghost tachyons (i.e., models with negative sign of $\dot{\phi}^2$ in the Lagrangian) have been shown to cross the phantom line of the equation of state $w < -1$ [36]. How can our result be positioned in these studies? Note that our result is valid not only for electromagnetic field but for any $U(1)$ gauge field. For instance, it can be the fundamental $U(1)$ field of electroweak interaction which exists before spontaneous symmetry breaking. Suppose that in some early stage of the Universe the perfect fluid which determines the metric in it has $w > 1$ equation of state (which can be achieved in some k-essence models [37–42]). In this case, the fundamental $U(1)$ gauge field becomes tachyonic and, in the spirit of the articles cited above, these tachyons can drive inflation. It will continue as long as $w > 1$, and during it the tachyons will roll to their energy minimum where they are infinitely fast, thus thermalizing a significantly larger than expected part of the Universe.

Author Contributions: Conceptualization, M.S.; methodology, M.S.; software, D.S. and M.S.; validation, D.S. and M.S.; formal analysis, D.S. and M.S.; investigation, D.S. and M.S.; writing—original draft preparation, M.S.; writing—review and editing, D.S. and M.S. All authors have read and agreed to the published version of the manuscript.

Funding: This research was funded by Bulgarian National Science Fund grant KP-06-N58/5.

Data Availability Statement: No new data were created or analyzed in this study. Data sharing is not applicable to this article.

Conflicts of Interest: The authors declare no conflict of interest.

Abbreviations

The following abbreviation is used in this manuscript:

FLRW   Friedmann–Lemaître–Robertson–Walker

References

1. Abbott, B. Multi-messenger Observations of a Binary Neutron Star Merger. *Astrophys. J. Lett.* **2017**, *848*, L12. [CrossRef]
2. Aartsen, M.G.; Ackermann, M.; Adams, J.; Aguilar, J.A.; Ahlers, M.; Ahrens, M.; Altmann, D.; Anderson, T.; Arguelles, C.; Arlen, T.C.; et al. Observation of High-Energy Astrophysical Neutrinos in Three Years of IceCube Data. *Phys. Rev. Lett.* **2014**, *113*, 101101. [CrossRef] [PubMed]
3. Huang, T.; Li, Z. Neutrino observations of LHAASO sources: Present constraints and future prospects. *Mon. Not. Roy. Astron. Soc.* **2022**, *514*, 852–862. [CrossRef]
4. Addazi, A.; Alvarez-Muniz, J.; Batista, R.A.; Amelino-Camelia, G.; Antonelli, V.; Arzano, M.; Asorey, M.; Atteia, J.L.; Bahamonde, S.; Bajardi, F.; et al. Quantum gravity phenomenology at the dawn of the multi-messenger era—A review. *Prog. Part. Nucl. Phys.* **2022**, *125*, 103948. [CrossRef]
5. Aghanim, N.; Akrami, Y.; Ashdown, M.; Aumont, J.; Baccigalupi, C.; Ballardini, M.; Banday, A.J.; Barreiro, R.B.; Bartolo, N.; Basak, S.; et al. Planck 2018 results. VI. Cosmological parameters. *Astron. Astrophys.* **2020** *641*, A6; Erratum in: *Astron. Astrophys.* **2021**, *652*, C4.
6. Hinshaw, G.; Larson, D.; Komatsu, E.; Spergel, D.N.; Bennett, C.; Dunkley, J.; Nolta, M.R.; Halpern, M.; Hill, R.S.; Odegard, N.; et al. Nine-Year Wilkinson Microwave Anisotropy Probe (WMAP) Observations: Cosmological Parameter Results. *Astrophys. J. Suppl.* **2013**, *208*, 19. [CrossRef]
7. Blake, C.; Brough, S.; Colless, M.; Contreras, C.; Couch, W.; Croom, S.; Croton, D.; Davis, T.M.; Drinkwater, M.J.; Forster, K.; et al. The WiggleZ Dark Energy Survey: Joint measurements of the expansion and growth history at z < 1. *Mon. Not. Roy. Astron. Soc.* **2012**, *425*, 405–414.
8. Ade, P.A.R.; Ahmed, Z.; Amiri, M.; Barkats, D.; Thakur, R.B.; Bischoff, C.A.; Beck, D.; Bock, J.J.; Boenish, H.; Bullock, E.; et al. Improved Constraints on Primordial Gravitational Waves using Planck, WMAP, and BICEP/Keck Observations through the 2018 Observing Season. *Phys. Rev. Lett.* **2021**, *127*, 151301. [CrossRef]
9. Neumann, K.D.; Holoien, T.W.; Kochanek, C.S.; Stanek, K.Z.; Vallely, P.J.; Shappee, B.J.; Prieto, J.L.; Pessi, T.; Jayasinghe, T.; Brimacombe, J.; et al. The ASAS-SN Bright Supernova Catalog—V. 2018–2020. *Mon. Not. Roy. Astron. Soc.* **2023**, *520*, 4356. [CrossRef]
10. Zhao, C.; Variu, A.; He, M.; Forero-Sánchez, D.; Tamone, A.; Chuang, C.H.; Kitaura, F.S.; Tao, C.; Yu, J.; Kneib, J.P.; et al. The completed SDSS-IV extended Baryon Oscillation Spectroscopic Survey: Cosmological implications from multitracer BAO analysis with galaxies and voids. *Mon. Not. Roy. Astron. Soc.* **2022**, *511*, 5492–5524. [CrossRef]
11. Riess, A.G.; Strolger, L.G.; Tonry, J.; Casertano, S.; Ferguson, H.C.; Mobasher, B.; Challis, P.; Filippenko, A.V.; Jha, S.; Li, W.; et al. Type Ia supernova discoveries at z > 1 from the Hubble Space Telescope: Evidence for past deceleration and constraints on dark energy evolution. *Astrophys. J.* **2004**, *607*, 665–687. [CrossRef]
12. Freedman, W.L.; Madore, B.F.; Gibson, B.K.; Ferrarese, L.; Kelson, D.D.; Sakai, S.; Mould, J.R.; Kennicutt, R.C., Jr.; Ford, H.C.; Graham, J.A.; et al. Final results from the Hubble Space Telescope key project to measure the Hubble constant. *Astrophys. J.* **2001**, *553*, 47–72. [CrossRef]
13. Yuan, W.; Riess, A.; Casertano, S.; Macri, L. A First Look at Cepheids in a Type Ia Supernova Host with JWST. *Astrophys. J. Lett.* **2022**, *940*, L17. [CrossRef]

14. Riess, A.G.; Filippenko, A.V.; Challis, P.; Clocchiatti, A.; Diercks, A.; Garnavich, P.M.; Gilliland, R.L.; Hogan, C.J.; Jha, S.; Kirshner, R.P.; et al. Observational evidence from supernovae for an accelerating universe and a cosmological constant. *Astron. J.* **1998**, *116*, 1009–1038. [CrossRef]

15. Peacock, J.A. *Cosmological Physics*; Cambridge University Press: Cambridge, UK, 1999.

16. Mashhoon, B. Electromagnetic Waves in an Expanding Universe. *Phys. Rev. D* **1973**, *8*, 4297. [CrossRef]

17. Cohen, J.; Kegeles, L. Electromagnetic fields in curved spaces—A constructive procedure. *Phys. Rev. D* **1974**, *10*, 1070–1084. [CrossRef]

18. Cabral, F.; Lobo, F. Electrodynamics and Spacetime Geometry: Foundations. *Found. Phys.* **2017**, *47*, 208–228. [CrossRef]

19. Cabral, F.; Lobo, F. Electrodynamics and spacetime geometry: Astrophysical applications. *Eur. Phys. J. Plus* **2017**, *132*, 304. [CrossRef]

20. Asenjo, F.; Hojman, S. Birefringent light propagation on anisotropic cosmological backgrounds. *Phys. Rev. D* **2017**, *96*, 044021. [CrossRef]

21. Cotaescu, I.I. Maxwell field in spacially flat FLRW space-time. *Eur. Phys. J. C* **2021**, *81*, 908. [CrossRef]

22. Lichnerowics, A. *Elements of Tensor Calculus*; John Wiley and Sons: New York, NY, USA, 1962.

23. Faddeev, L.D.; Slavnov, A.A. *Gauge Fields: An Introduction to Quantum Theory*; CRC Press: Boca Raton, FL, USA, 1993.

24. Heisenberg, L. Generalization of the Proca Action. *J. Cosmol. Astropart. Phys.* **2014**, *2014*, 15. [CrossRef]

25. Carroll, S.M. The Cosmological constant. *Living Rev. Rel.* **2001**, *4*, 1. [CrossRef] [PubMed]

26. Abbasi, B.; Craig, W.; On the initial value problem for the wave equation in Friedmann–Robertson–Walker space–times. *Proc. R. Soc. Math. Phys. Eng. Sci.* **2014**, *470*, 20140361. [CrossRef]

27. Petersen, O.L. The mode solution of the wave equation in Kasner spacetimes and redshift. *Math. Phys. Anal. Geom.* **2016**, *19*, 26. [CrossRef]

28. Utiyama, R. *The Theory of Relativity*; Atomizdat: Moscow, Russia, 1979.

29. Tsagas, C.G. Electromagnetic fields in curved spacetimes. *Class. Quant. Grav.* **2005**, *22*, 393–408. [CrossRef]

30. Calcagni, G.; Liddle, A.R. Tachyon dark energy models: Dynamics and constraints. *Phys. Rev. D* **2006**, *74*, 043528. [CrossRef]

31. Armendariz-Picon, C.; Damour, T.; Mukhanov, V.F. k-inflation. *Phys. Lett. B* **1999**, *458*, 209–218. [CrossRef]

32. Bilic, N.; Tupper, G.B.; Viollier, R.D. Cosmological tachyon condensation. *Phys. Rev. D* **2009**, *80*, 023515. [CrossRef]

33. Gibbons, G.W. Cosmological evolution of the rolling tachyon. *Phys. Lett. B* **2002**, *537*, 1–4. [CrossRef]

34. Sen, A. Field theory of tachyon matter. *Mod. Phys. Lett. A* **2002**, *17*, 1797–1804. [CrossRef]

35. Felder, G.N.; Garcia-Bellido, J.; Greene, P.B.; Kofman, L.; Linde, A.D.; Tkachev, I. Dynamics of symmetry breaking and tachyonic preheating. *Phys. Rev. Lett.* **2001**, *87*, 011601. [CrossRef]

36. Karami, K.; Fahimi, K. Interacting viscous ghost tachyon, K-essence and dilaton scalar field models of dark energy *Class. Quant. Grav.* **2013**, *30*, 065018. [CrossRef]

37. Garriga, J.; Mukhanov, V.F. Perturbations in k-inflation. *Phys. Lett. B* **1999**, *458*, 219–225. [CrossRef]

38. Armendariz-Picon, C.; Mukhanov, V.F.; Steinhardt, P.J. Essentials of k essence. *Phys. Rev. D* **2001**, *63*, 103510. [CrossRef]

39. Bonvin, C.; Caprini, C.; Durrer, R. A no-go theorem for k-essence dark energy. *Phys. Rev. Lett.* **2006**, *97*, 081303. [CrossRef] [PubMed]

40. Bruneton, J.-P. On causality and superluminal behavior in classical field theories: Applications to k-essence theories and MOND-like theories of gravity. *Phys. Rev. D* **2007**, *75*, 085013. [CrossRef]

41. Kang, J.U.; Vanchurin, V.; Winitzki, S. Attractor scenarios and superluminal signals in k-essence cosmology *Phys. Rev. D* **2007**, *76*, 083511. [CrossRef]

42. Babichev, E.; Mukhanov, V.; Vikman, A. k-essence, superluminal propagation, causality and emergent geometry. *J. High Energy Phys.* **2008**, *2*, 101. [CrossRef]

Communication

A Simple, Exact Formulation of Number Counts in the Geodesic-Light-Cone Gauge

Giuseppe Fanizza [1,†] , Maurizio Gasperini [2,3,*,†] and Giovanni Marozzi [4,5,†]

1 Instituto de Astrofisíca e Ciências do Espaço, Faculdade de Ciências da Universidade de Lisboa, Edificio C8, Campo Grande, P-1740-016 Lisbon, Portugal; gfanizza@fc.ul.pt
2 Dipartimento di Fisica, Università di Bari, Via G. Amendola 173, 70126 Bari, Italy
3 Istituto Nazionale di Fisica Nucleare, Sezione di Bari, 70126 Bari, Italy
4 Dipartimento di Fisica, Università di Pisa, Largo B. Pontecorvo 3, 56127 Pisa, Italy; giovanni.marozzi@unipi.it
5 Istituto Nazionale di Fisica Nucleare, Sezione di Pisa, 56127 Pisa, Italy
* Correspondence: gasperini@ba.infn.it
† These authors contributed equally to this work.

Abstract: In this article, we compare different formulations of the number count prescription using the convenient formalism of the Geodesic-Light-Cone gauge. We then find a simple, exact, and very general expression of such a prescription which is suitable for generalised applications.

Keywords: observational cosmology; covariant averages of astrophysical variables; light cone gauge

Citation: Fanizza, G.; Gasperini, M.; Marozzi, G. A Simple, Exact Formulation of Number Counts in the Geodesic-Light-Cone Gauge. *Universe* **2023**, *9*, 327. https://doi.org/10.3390/universe9070327

Academic Editor: Kazuharu Bamba

Received: 13 June 2023
Revised: 3 July 2023
Accepted: 7 July 2023
Published: 10 July 2023

Galaxy number counts represent a very useful tool for understanding the Large-Scale Structure (LSS) of our universe and testing the underlying cosmological models. Indeed, the information provided by such a tool on the distribution and overall properties of galaxies can be interpreted as a probe for the associated distribution of dark matter, and thereby as a test of the early cosmological dynamics based on late time observations (see [1–6] and references therein).

One of the first studies involving galaxy number count was presented in a paper discussing the angular auto-correlation of the LSS [7] and its cross-correlation with the anisotropies of the Cosmic Microwave Background (CMB) temperature. Later, this was generalized to include linear relativistic corrections [8–10].

More recently, further developments have been achieved. Among the most relevant, we should mention the computation of the theoretically expected galaxy power spectrum [11] based on the expression of galaxy number counts containing all relativistic corrections (i.e., at the source and observer positions and along the light cone). This advance complements the evaluation of the galaxy two-point correlation function presented in [12]. In addition, ref. [13] computed the "observationally expected" galaxy power spectrum by taking into account effects such as lensing magnification which are important in principle for the evaluation of the power spectrum multipoles, with the results showing that this approach can correct the leading distortion terms due to the redshift at most at the ten percent level.

On the other hand, in a previous paper [14] we presented a new class of covariant prescriptions for averaging astrophysical variables on spatial regions typical of the chosen sources and intersecting the past light-cone of a given observer. In this short communication, we show that with the appropriate choice of ingredients, our integral prescription can exactly reproduce the number count integral introduced long ago as a useful tool in the general context of observational cosmology (see, e.g., [15,16]). In addition, it can reproduce the average integrals recently used in [17,18] as essential ingredients to obtain a reliable (i.e., observationally compatible) prescription for galaxy number counts and for their above-mentioned applications.

Let us start by considering the (possibly realistic) experimental situation in which the sources are not exactly confined on a given space such as a hypersurface (defined for instance by a scalar field $B(x)$ such that $B(x) = B_s = $ const), and are instead localized inside an (arbitrarily) extended space-time layer corresponding to the interval $B_s \leq B \leq B_s + \Delta B_s$. In thais case, the average is characterized by the following covariant integral prescription [14]:

$$I(\Delta B_s) = \int_{\mathcal{M}_4} d^4x \, \sqrt{-g} \, \rho \, \delta(V_0 - V) \Theta(B_s + \Delta B_s - B) \Theta(B - B_s) \frac{\partial^\mu A \partial_\mu V}{|\partial_\alpha A \partial^\alpha A|^{1/2}}. \tag{1}$$

Here, $\rho(x)$ is a scalar field that specifies an appropriate physical weight factor associated with the averaged sources, $V(x)$ is a scalar field (with light-like gradient) which identifies the past light-cone centered on the observer and spanned by the null momentum $k_\mu = \partial_\mu V$ of the light signals emitted by the sources, and $A(x)$ is a scalar field (with a time-like gradient) associated with the following unit vector:

$$n_\mu = \frac{\partial_\mu A}{|\partial_\alpha A \partial^\alpha A|^{1/2}}, \tag{2}$$

which possibly represents a convenient four-velocity reference, but in general depending on the particular observations we are interested in. The particular choice of ρ, A, and B obviously depends on the physical situation and on the type of observation being performed.

Suppose now that we are interested in sources localized between the constant redshift surfaces $z = z_s$ and $z = z_s + \Delta z_s$. In this case, we can choose $B = 1 + z$, where z is the standard redshift parameter defined in general by

$$1 + z = \frac{(u_\mu k^\mu)_s}{(v_\mu k^\mu)_o}, \tag{3}$$

where u_μ and v_μ are the velocities of the source and the observer, respectively, not necessarily co-moving in the given geometry, and the subscripts "s" and "o" respectively denote the source and observer positions (see, e.g., [15]). In this case, as we show below, we find that our integral (1) can exactly reproduce the standard number-count prescription of [15–17], provided the fields A and ρ are appropriately chosen, as is shown explicitly just after Equation (14).

It is convenient for this purpose to work in the so-called Geodesic Light-Cone (GLC) gauge based on the coordinates $x^\mu = (\tau, w, \theta^a)$ and $a = 1, 2$, where the most general cosmological metric is parameterized by six arbitrary functions Υ, U_a, $\gamma_{ab} = \gamma_{ba}$, and the line-element takes the following form [19]:

$$ds^2_{GLC} = -2\Upsilon dw d\tau + \Upsilon^2 dw^2 + \gamma_{ab}(d\theta^a - U^a dw)\left(d\theta^b - U^b dw\right). \tag{4}$$

Let us recall here for the reader's convenience that w is a null coordinate ($\partial_\mu w \partial^\mu w = 0$), that in this gauge the light signals travel along geodesics with constant w and θ^a, and that the time coordinate τ coincides with the time of the synchronous gauge [20]. In fact, we can easily determine $\partial_\mu \tau$ defines a geodesic flow, i.e., that $(\partial^\nu \tau)\nabla_\nu(\partial_\mu \tau) = 0$, which is in agreement with the condition $g^{\tau\tau} = -1$ following from the metric (4).

Working in the GLC gauge, we can identify $V = w$; thus, $k_\mu = \partial_\mu w$ and $k^\mu = -\Upsilon^{-1}\delta^\mu_\tau$. Moreover, we can conveniently impose the *general temporal gauge* defined by the condition[1]

$$\left(v_\mu k^\mu\right)_o = -1, \tag{5}$$

where the past light cones $w = w_0 = $ const are simply labeled by the reception time τ_0 of the corresponding light signals [21], i.e., $w_0 = \tau_0$. Hence, in this gauge we have

$$1 + z = \left(u_\tau \Upsilon^{-1}\right)_s \tag{6}$$

and we can replace the τ integration of Equation (1) with the z integration defined by

$$d\tau = dz \left(\frac{d\tau}{dz} \right) = \frac{dz}{\partial_\tau (u_\tau \Upsilon^{-1})} .$$

(7)

Note that Equation (7) has to be further integrated, as prescribed by Equation (1), on the spatial hypersurface containing the averaged source. Thus, in order to avoid any confusion between the integrated quantities and the boundaries of the integrals, we omit the explicit subscript "s" in the differential integration measure. Finally, for the metric (4) we have $\sqrt{-g} = \Upsilon \sqrt{\gamma}$, where $\gamma = \det \gamma_{ab}$. Thus, our integral prescription (1) takes the form

$$
\begin{aligned}
I(\Delta z_s) \;\; &= \;\; -\int d\tau\, dw\, d^2\theta\, \sqrt{\gamma}\, \rho\, \delta(w_0 - w)\, \Theta(z_s + \Delta z_s - z)\, \Theta(z - z_s)\, n_\tau \\
&= \;\; -\int dz\, d^2\theta\, \sqrt{\gamma}\, \rho\, \Theta(z_s + \Delta z_s - z)\, \Theta(z - z_s)\, \frac{n_\tau}{\partial_\tau (u_\tau \Upsilon^{-1})} \\
&= \;\; -\int_{z_s}^{z_s + \Delta z_s} dz\, d^2\theta\, \sqrt{\gamma}\, \rho\, \frac{n_\tau}{\partial_\tau (u_\tau \Upsilon^{-1})} ,
\end{aligned}
$$

(8)

where we have used the explicit form of n_μ of Equation (2). It should be stressed here that u_τ is the time component of the source velocity, while for the moment n_τ and ρ are both arbitrary variables to be adapted to the physical situation under consideration.[2] Finally, all the integrated functions have to be evaluated on the past light cone $w = w_0$.

We now Consider the number count integral, which evaluates the number of sources dN located inside an infinitesimal layer of thickness $d\lambda$ at a distance $d_A(z)$ and seen by a given observer within a bundle of null geodesics subtending the solid angle $d\Omega$ [15,16]:

$$dN = n\, dV \equiv n\, d\Omega\, d\lambda\, d_A^2 \left(-u_\mu k^\mu \right) .$$

(9)

Here, n is the number density of the sources per unit proper volume, d_A is the angular distance, and u_μ (as before) is the velocity field of the given sources. Finally, λ is a scalar affine parameter along the path $x^\mu(\lambda)$ of the light signals such that $k^\mu = dx^\mu/d\lambda$, and is normalized along the observer world-line by the condition

$$\left(v_\mu k^\mu \right)_0 = -1 ,$$

(10)

where v_μ is the observer velocity and the scalar product is evaluated at the observer position [16].

We now move to the GLC coordinates, where $k^\mu = g^{\mu\nu}\partial_\nu w = -\Upsilon^{-1}\delta_\tau^\mu$. The condition $k^\mu = dx^\mu/d\lambda$ then provides the following:

$$\frac{d\tau}{d\lambda} = -\Upsilon^{-1}, \qquad \frac{dw}{d\lambda} = 0, \qquad \frac{d\theta^a}{d\lambda} = 0 ,$$

(11)

and the normalization condition (10) exactly coincides with the general temporal gauge (5). In these coordinates, as anticipated in [22,23], the angular distance d_A satisfies

$$d_A^2\, d\Omega = \sqrt{\gamma}\, d^2\tilde{\theta} ;$$

(12)

see Appendix A for an explicit derivation of the above equation. Here, $\tilde{\theta}^a$ represents the angular coordinates of the so-called "observational gauge" [24,25], defined by exploiting the residual gauge freedom of the GLC gauge in such a way that the angular directions exactly coincide with those expressed in a system of Fermi Normal Coordinates (FNC).[3] Using

$$d\lambda = dz \left(\frac{dz}{d\tau} \right)^{-1} \left(\frac{d\tau}{d\lambda} \right)^{-1}$$

(13)

and applying Equations (7) and (11), we find that the number of sources of Equation (9) integrated between z_s and $z_s + \Delta z_s$ (as was the case before; see Equation (8)) finally reduces to

$$N = -\int_{z_s}^{z_s + \Delta z_s} dz\, d^2\widetilde{\theta}\,\sqrt{\gamma}\, n \frac{u_\tau}{\partial_\tau (u_\tau Y^{-1})}\,. \tag{14}$$

We are now in the position of comparing this result with our previous average integral (8). It is clear that the two are exactly the same, provided the following three conditions are satisfied: (*i*) our scalar density ρ coincides with the number density n; (*ii*) the scalar parameter A is chosen in such a way that the unit vector n_μ coincides with the source velocity u_μ (and, obviously, $n_\tau = u_\tau$); and (*iii*) the angular directions are expressed in terms of the angles fixed by the observational gauge, i.e., $\theta^a \to \widetilde{\theta}^a$.

It may be appropriate to recall at this point that the number count prescription has been recently used in (apparently) different forms by other authors. For instance, in the context of defining physically appropriate averaging prescriptions the number count has been presented in the following form [17]:

$$dN = n\, dV = \frac{n\, d_A^2}{(1+z)\, H_{||}}\, dz\, d\Omega\,, \tag{15}$$

where $H_{||}$ is a local longitudinal expansion parameter defined by

$$H_{||} \equiv (1+z)^{-2}\, k^\mu k^\nu \nabla_\mu u_\nu\,. \tag{16}$$

Moving to the GLC coordinates and using the temporal gauge (5), we can easily obtain

$$(1+z)\, H_{||} = -u_\tau^{-1} \partial_\tau \left(u_\tau Y^{-1}\right)\,. \tag{17}$$

By inserting this result in the definition in (15) and using Equation (11) in Equation (9), we can immediately determine that the number count expressions in (15) and (9) are exactly the same.

As a second example, recall the form of the number-count integral presented in [18] based on the following volume element:

$$dV = \sqrt{-g}\,\epsilon_{\mu\nu\alpha\beta}\, u^\mu dx^\nu dx^\alpha dx^\beta\,, \tag{18}$$

where, as before, u^μ is the source velocity. Moving to the GLC gauge and projecting this volume element on the light cone $w = \text{const}$, $dw = 0$, we obtain

$$dV = Y\sqrt{\gamma}\, u^w d\tau d^2\theta\,. \tag{19}$$

Recalling that $u^w = g^{w\nu} u_\nu = -u_\tau Y^{-1}$ and again using the expression $d\tau/dz$ provided in Equation (7), it is immediately clear that Equation (19) reduces to the same expression of the number count integrand in (14), provided we add the source density n and, as before, the GLC angles are identified with those of the observational gauge [24,25] $d^2\theta \to d^2\widetilde{\theta}$.

The discussion presented in this paper provides a further example of the crucial role played by the GLC coordinates in the simplification and comparison of formal non-perturbative expressions of physical observables. In addition, the simple expression obtained here for the volume element dV in terms of observable variables such as the redshift and observation angles is promising for a number of different physical applications, which will be discussed in forthcoming papers.

We would finally remark that in this brief article we have expressed the galaxy number counts in terms of the redshift. A recent interesting paper [26] has proposed studying galaxy number counts as a function of the luminosity distance of the given sources, rather than their redshift, and showed that there are already differences between the two computational methods at the first perturbative order. Thus, in the near future we plan to evaluate the exact expression of the galaxy number counts in terms of the luminosity distance by applying the covariant averaging formalism and using the GLC coordinate approach presented in this paper.

Author Contributions: Conceptualization, G.F., M.G. and G.M.; methodology, G.F., M.G. and G.M.; formal analysis, G.F., M.G. and G.M.; original draft preparation, G.F., M.G. and G.M.; review and editing, G.F., M.G. and G.M. All authors have read and agreed to the published version of the manuscript.

Funding: G. Fanizza acknowledges support by Fundação para a Ciência e a Tecnologia (FCT) under the program *"Stimulus"* with the grant no. CEECIND/04399/2017/CP1387/CT0026, and through the research project with ref. number PTDC/FIS-AST/0054/2021. M. Gasperini and G. Marozzi are supported in part by INFN under the program TAsP (*"Theoretical Astroparticle Physics"*). M. Gasperini is also supported by the research grant number 2017W4HA7S *"NAT-NET: Neutrino and Astroparticle Theory Network"*, under the program PRIN 2017 funded by the Italian Ministero dell'Università e della Ricerca (MUR). G. Fanizza and M. Gasperini wish to thank the kind hospitality and support of the TH Department of CERN, where part of this work has been carried out. Finally, we are very grateful to Gabriele Veneziano for his fundamental contribution and collaboration during the early stages of this work.

Institutional Review Board Statement: Not applicable.

Informed Consent Statement: Not applicable.

Data Availability Statement: Not applicable.

Conflicts of Interest: The authors declare no conflict of interest.

Abbreviations

The following abbreviations are used in this manuscript:

GLC   Geodesic Light-Cone
FNC   Fermi Normal Coordinates

Appendix A

In order to prove Equation (12), we can combine Equations (3.15) and (3.17) from [23] to express the angular distance in generic GLC coordinates as follows:

$$d_A^2 = 4v_\tau^2 \sqrt{\gamma_s} \left[\frac{\det\left(\partial_\tau \gamma_{ab}\right)}{\sqrt{\gamma}} \right]_o^{-1} \tag{A1}$$

where, as before, the subscripts "s" and "o" respectively denote the source and observer positions. We can now move to the observational gauge [24], where we choose a system at rest with the local observer ($v_\tau^2 = 1$) and use Equation (3.16) from [24] to obtain, after a simple calculation of γ_{ab},

$$\widetilde{d}_A^2 = \frac{\sqrt{\gamma_s}}{\sin\widetilde{\theta}}, \tag{A2}$$

from which we have

$$\widetilde{d}_A^2 \, d\widetilde{\Omega} = \sqrt{\gamma_s} \, d^2\widetilde{\theta}, \tag{A3}$$

where the tilde denotes the variables of the observational gauges. Finally, note that by using Equations (3.13)–(3.15) from [23] it can easily be shown that the left-hand side of the above equation does not depend on the particular choice of v_τ.

Notes

1.   This choice generalises the definition of the temporal gauge, already introduced in [21], such that it can be applied to the case of an arbitrary observer velocity v_μ.

2.   In our previous paper [14], we applied the above integral in the limit of the small redshift bin $\Delta z_s \to 0$.

³ The angular directions related to local observations (also used in [15,16]) are indeed those measured by a free-falling observer, and can be identified with the angles of the FNC system [24] where the metric is locally flat around all points of a given world line, with leading curvature corrections (which are quadratic) in the distance.

References

1. Jeong, D.; Schmidt, F.; Hirata, C.M. Large-scale clustering of galaxies in general relativity. *Phys. Rev. D* **2012**, *85*, 023504. [CrossRef]
2. Schmidt, F.; Jeong, D. Cosmic Rulers. *Phys. Rev. D* **2012**, *86*, 083527. [CrossRef]
3. Kehagias, A.; Riotto, A. Symmetries and Consistency Relations in the Large Scale Structure of the Universe. *Nucl. Phys. B* **2013**, *873*, 514–529. [CrossRef]
4. Bertacca, D.; Maartens, R.; Clarkson, C. Observed galaxy number counts on the lightcone up to second order: I. Main result. *J. Cosmol. Astropart. Phys.* **2014**, *9*, 037. [CrossRef]
5. Kehagias, A.; Dizgah, A.M.; Na, J.N.; Perrier, H.; Riotto, A. A Consistency Relation for the Observed Galaxy Bispectrum and the Local non-Gaussianity from Relativistic Corrections. *J. Cosmol. Astropart. Phys.* **2015**, *8*, 18. [CrossRef]
6. Ginat, Y.B.; Desjacques, V.; Jeong, D.; Schmidt, F. Covariant decomposition of the non-linear galaxy number counts and their monopole. *J. Cosmol. Astropart. Phys.* **2021**, *12*, 31. [CrossRef]
7. Yoo, J.; Fitzpatrick, A.L.; Zaldarriaga, M. Three-point correlation of the Lyman-alpha forest: An optimal redshift space distortion estimator. *Phys. Rev. D* **2009**, *80*, 083514. [CrossRef]
8. Yoo, J. A new relativistic N-body code for the clustering of cosmic neutrinos. *Phys. Rev. D* **2010**, *82*, 083508. [CrossRef]
9. Challinor, A.; Lewis, A. The linear power spectrum of observed source number counts. *Phys. Rev. D* **2011**, *84*, 043516. [CrossRef]
10. Bonvin, C.; Durrer, R. What galaxy surveys really measure. *Phys. Rev. D* **2011**, *84*, 063505. [CrossRef]
11. Grimm, N.; Scaccabarozzi, F.; Yoo, J.; Biern, S.G.; Gong, J.-O. Precision cosmology with overlapping surveys: The importance of volume and cross-correlations. *J. Cosmol. Astropart. Phys.* **2020**, *11*, 64. [CrossRef]
12. Scaccabarozzi, F.; Yoo, J.; Biern, S.G. Cross-correlation of future weak lensing surveys and Planck lensing data. *J. Cosmol. Astropart. Phys.* **2018**, *10*, 24. [CrossRef]
13. Castorina, E.; Dio, E.D. Observing the cosmic acceleration with the Kilo-Degree Survey. *J. Cosmol. Astropart. Phys.* **2022**, *1*, 61. [CrossRef]
14. Fanizza, G.; Gasperini, M.; Marozzi, G.; Veneziano, G. Generalized covariant prescriptions for averaging cosmological observables. *J. Cosmol. Astropart. Phys.* **2020**, *2*, 017. [CrossRef]
15. Ellis, G. Relativistic cosmology. *Gen. Rel. Grav.* **2009**, *41*, 581–660. [CrossRef]
16. Ellis, G.; Nell, S.D.; Maartens, R.; Stoeger, W.R.; Whitman, A.P. Ideal observational cosmology. *Phys. Rep.* **1985**, *124*, 315–417. [CrossRef]
17. Fleury, P.; Clarkson, C.; Maartens, R. How does the cosmic large-scale structure bias the Hubble diagram? *J. Cosmol. Astropart. Phys.* **2017**, *1703*, 062. [CrossRef]
18. Dio, E.D.; Durrer, R.; Marozzi, G.; Montanari, F. Galaxy number counts to second order and their bispectrum. *J. Cosmol. Astropart. Phys.* **2014**, *1412*, 17; Erratum in *J. Cosmol. Astropart. Phys.* **2015**, *1506*, E01. [CrossRef]
19. Gasperini, M.; Marozzi, G.; Nugier, F.; Veneziano, G. Light-cone averaging in cosmology: Formalism and applications. *J. Cosmol. Astropart. Phys.* **2011**, *1107*, 008. [CrossRef]
20. Ben-Dayan, I.; Gasperini, M.; Marozzi, G.; Nugier, F.; Veneziano, G. Backreaction on the luminosity-redshift relation from gauge invariant light-cone averaging. *J. Cosmol. Astropart. Phys.* **2012**, *1204*, 36. [CrossRef]
21. Fleury, P.; Nugier, F.; Fanizza, G. Geodesic-light-cone coordinates and the Bianchi I spacetime. *J. Cosmol. Astropart. Phys.* **2016**, *6*, 008. [CrossRef]
22. Ben-Dayan, I.; Gasperini, M.; Marozzi, G.; Nugier, F.; Veneziano, G. Average and dispersion of the luminosity-redshift relation in the concordance model. *J. Cosmol. Astropart. Phys.* **2013**, *1306*, 2. [CrossRef]
23. Fanizza, G.; Gasperini, M.; Marozzi, G.; Veneziano, G. An exact Jacobi map in the geodesic light-cone gauge. *J. Cosmol. Astropart. Phys.* **2013**, *11*, 019. [CrossRef]
24. Fanizza, G.; Gasperini, M.; Marozzi, G.; Veneziano, G. Observation angles, Fermi coordinates, and the Geodesic-Light-Cone gauge. *J. Cosmol. Astropart. Phys.* **2019**, *1*, 4. [CrossRef]
25. Mitsou, E.; Scaccabarozzi, F.; Fanizza, G. Observed Angles and Geodesic Light-Cone Coordinates. *Class. Quantum Grav.* **2018**, *35*, 107002. [CrossRef]
26. Fonseca, J.; Zazzera, S.; Baker, T.; Clarkson, C. The observed number counts in luminosity distance space. *arXiv* **2023**, arXiv:2304.14253.

Article

Cosmological Parameter Estimation with Genetic Algorithms

Ricardo Medel-Esquivel [1,2,3], Isidro Gómez-Vargas [2], Alejandro A. Morales Sánchez [4], Ricardo García-Salcedo [3,5] and José Alberto Vázquez [2,*]

1   Escuela Superior de Física y Matemáticas, Instituto Politécnico Nacional, Ciudad de México 07738, Mexico; rmedel@ipn.mx
2   Instituto de Ciencias Físicas, Universidad Nacional Autónoma de México, Cuernavaca 62210, Mexico; igomez@icf.unam.mx
3   CICATA-Legaria, Instituto Politécnico Nacional, Ciudad de México 11500, Mexico
4   Facultad de Ciencias, Universidad Nacional Autónoma de México, Ciudad de México 04510, Mexico
5   Escuela Superior de Ingeniería, Ciencia y Tecnología, Universidad Internacional de Valencia (VIU), 46002 Valencia, Spain
*   Correspondence: javazquez@icf.unam.mx

Abstract: Genetic algorithms are a powerful tool in optimization for single and multimodal functions. This paper provides an overview of their fundamentals with some analytical examples. In addition, we explore how they can be used as a parameter estimation tool in cosmological models to maximize the likelihood function, complementing the analysis with the traditional Markov chain Monte Carlo methods. We analyze that genetic algorithms provide fast estimates by focusing on maximizing the likelihood function, although they cannot provide confidence regions with the same statistical meaning as Bayesian approaches. Moreover, we show that implementing sharing and niching techniques ensures an effective exploration of the parameter space, even in the presence of local optima, always helping to find the global optima. This approach is invaluable in the cosmological context, where an exhaustive space exploration of parameters is essential. We use dark energy models to exemplify the use of genetic algorithms in cosmological parameter estimation, including a multimodal problem, and we also show how to use the output of a genetic algorithm to obtain derived cosmological functions. This paper concludes that genetic algorithms are a handy tool within cosmological data analysis, without replacing the traditional Bayesian methods but providing different advantages.

Keywords: parameter estimation; dark energy; machine learning; genetic algorithms

Citation: Medel-Esquivel, R.; Gómez-Vargas, I.; Morales Sánchez, A.A.; García-Salcedo, R.; Alberto Vázquez, J. Cosmological Parameter Estimation with Genetic Algorithms. *Universe* **2024**, *10*, 11. https://doi.org/10.3390/universe10010011

Academic Editor: Kazuharu Bamba

Received: 31 October 2023
Revised: 6 December 2023
Accepted: 15 December 2023
Published: 27 December 2023

1. Introduction

Genetic algorithms (GAs), established for decades, are tools from evolutionary computation [1–5] that solve many function optimization problems. Evolutionary computation is focused on algorithms exploiting randomness to solve search and optimization problems using operations inspired by natural evolution [6]. It includes several methods for stochastic or metaheuristic optimization [7,8]; notable examples are Particle Swarm Optimization (PSO) [9] based on the social behavior of organisms of the same species such as birds, the Giant Trevally Optimizer (GTO) [10–12] inspired by the hunting behavior of predatory fish, and Artificial Rabbits Optimization (ARO), drawing inspiration from social interactions among rabbits [13,14]. Within evolutionary computation, the most relevant methods are genetic algorithms [15,16], genetic programming [17], and evolutionary strategies [18]; their success is due to their ability to navigate intricate, non-linear, and high-dimensional search spaces.

In particular, genetic algorithms stand out as powerful tools for optimization problems because they mathematically always guarantee, under certain conditions, to find the best solution. Despite challenges posed by local optimum values [19], this property puts them at an advantage over other techniques. Rooted in the emulation of natural selection and evolution, the iterative process of GAs involves generating a population, subjecting it to fitness-based selection, and applying genetic operators, such as crossover and mutation. This iterative approach drives the evolution of increasingly optimal solutions over generations. GAs thrive in situations with multiple optima, irregular landscapes, or where an analytical solution is difficult to achieve. Its adaptability allows for the simultaneous exploration of numerous candidate solutions, making them effective in various optimization challenges. Unlike traditional optimization methods, GAs have the advantage of not relying on derivatives, providing excellent robustness in high-dimensional or more complex problems. Inspired by natural evolution, these algorithms efficiently explore vast and unknown search spaces [20]. Their ability to solve complex and dynamic projects makes them valuable in diverse fields, including medicine [21–23], epidemic dynamical systems [24,25], geotechnics [26], market forecasts [27], and industry [28], among others. A particularly successful application in the Deep Learning era is the optimization of neural networks, which are huge computational models in which genetic algorithms help to find optimal combinations of hyperparameters [29–31].

With the accelerated development of computational resources, genetic algorithms and other machine learning algorithms have been exploited in several scientific fields in recent years. Remarkably, they have resulted in significant advances in understanding particle physics [32–34], astronomical information [35–38], and cosmological phenomena [39–44].

Genetic programming, another method from evolutionary computation, has been widely used in astrophysics and cosmology [45–50], which allows for symbolic regression for a given dataset, treating regression as a search problem to find the best combination of mathematical operators generating an expression fitting the data. Although genetic programming and genetic algorithms solve different tasks, they use similar operators to find solutions. In this work, we focus on genetic algorithms, mentioning genetic programming for reference, assuming the astrophysical community may be more familiar with it. Moreover, genetic algorithms are the most fundamental and successful evolutionary computation technique, and understanding them is useful for studying other evolutionary computation methods, including genetic programming.

On the other hand, parameter estimation in cosmology is a very relevant task that finds a combination of values for parameters describing a cosmological model based on observational data. The goal is to refine theoretical models to align with observations for a more precise understanding of the universe. In cosmological parameter estimation, the most robust and successful algorithms are Markov chain Monte Carlo; however, these methods sometimes are computationally expensive, and recent advancements try to attack this issue with new statistical or machine learning techniques, including the iterative Gaussian emulation method [51], adaptive importance sampling, parallelizable Bayesian algorithms [52], Bayesian inference accelerated with machine learning [53–55], or likelihood-free methods [56,57].

This paper aims to achieve two primary objectives: firstly, to provide a comprehensive introduction to genetic algorithms and elucidate their application in cosmological parameter estimation, and secondly, to demonstrate the complementarity of GAs with traditional Bayesian inference methods. We include illustrative examples of optimization problems and their applications in cosmology. Particularly, we delve into using genetic algorithms to constrain the parameter space of dark energy models based on observational data. It is pertinent to mention that GAs cannot perform the same tasks as MCMC methods, and we do not try to replace them; we only perform parameter estimation with GAs by optimizing the likelihood function, whereas MCMC methods sample the posterior probability function. However, we analyze their relevance as an alternative and complementary method, as discussed in Section 4.1.

The structure of this paper is as follows: In Section 2, we present the basics of genetic algorithms and an insight into their functionality. In Section 3, we provide some examples of the optimization of analytical functions by applying genetic algorithms. Section 4.1 describes the path to perform cosmological parameter estimation using these algorithms.

Section 4.2 contains examples of multimodal problems in cosmology, and in Section 4.3, we justify how to obtain cosmological-derived parameters from a likelihood optimization. Finally, Section 5 summarizes our final remarks.

2. Fundamentals of Genetic Algorithms

2.1. Biological Fundamentals

Bioinspired computing is a field of computer science based on observing and imitating natural processes and phenomena to develop algorithms and computational systems [58]. These algorithms seek to solve complex problems. The bioinspired computation is classified into three main categories [58]: evolutionary algorithms (such as genetic algorithms), particle swarm intelligence (imitating collective behaviors) [7,59–61], and computational ecology (inspired by ecological phenomena) [8,62].

Genetic algorithms solve optimization [1–5] and search problems inspired by fundamental concepts of genetics and evolution [8,63,64]; some of its key points are as follows:

- Natural selection—This is the central principle in the theory of evolution. Just as better-adapted organisms are more likely to survive and reproduce in nature, GAs favor the fittest or most promising solutions from a population of candidate solutions. In nature, over several generations, the most promising characteristics of individuals survive to be inherited by the new generations. This is what genetic algorithms seek to do to have better solutions as more generations pass by.
- Crossing—Also called recombination, it is a process in which genes from two parents are combined to create offspring with characteristics inherited from both parents. GAs apply the idea of crossover by combining partial solutions from two individuals in the population to generate new solutions that can inherit desirable characteristics from both parents.
- Mutation—A mutation is recognized as the stochastic alterations in an organism's genetic material. In the GAs, a mutation introduces random changes in a small part of the candidate solutions, e.g., it may change the value of a bit, which increases the diversity of possible solutions and improves the exploration of the search space.
- Reproduction and inheritance—In the same sense as in nature, in genetic algorithms, these operations allow for the transmission of some characteristics of the parent solutions to the solutions of the next generation (offspring).

2.2. Genetic Algorithm Operations

John Holland was the first to introduce the genetic algorithm in 1975 in his book Adaptation in Natural and Artificial Systems [3,15]. According to the GA context, a population is a set of possible solutions to a given problem. Each individual has a genotype encoded in bits, which is then expressed as a phenotype in the problem context. The way to encode the possible solutions is fundamental to attacking a problem with GAs, and there are several options to perform it, for example, with binary, integer, or real encoding, among others [65].

Alternatively, assessing an individual's quality or a potential solution involves employing a metric or target function, which is ideally expected to approach its optimal value in the final generations. For the analogy of natural selection, this target function, or objective function, is called the fitness function. In practice, in GAs, the fitness function is directly the function to be optimized. This is unlike genetic programming, where the fitness function is a measure of the error between the algebraic expression found and the dataset used due to the regression task that genetic programming addresses.

The continuous evaluation of all the individuals (possible solutions) of a population with this fitness function and the applications of genetic operations to produce new generations allow for GAs to find the optimal value of this function. In the following list, we describe the fundamental procedures of genetic algorithms [66]:

- Selection—It is the method of choosing the best solutions to play the role of parents and improve the quality of the offspring. Several selection methods include the roulette [67], random [68], ranking [69], tournament [70], and Boltzmann entropy selections [71].

- Crossover—It is also called recombination, which generates a new possible solution given two previously selected parents. There are several crossover methods, such as one point, two points, N points, uniform, three parents, random, and order. The crossover operation has an associated probability (P_c) that determines how many individuals recombine given the population, with $P_c = 1$ indicating that all the products come from the recombination and $P_c = 0$, meaning they are exact copies of the parents.
- Mutation—After crossover, mutations make it possible to maintain diversity in the population and prevent it from stagnating at the local optima [72]. There are several types of mutation operators, such as flipping a gene if it is in the same position as in the parent; swapping values at random positions; flipping values from left to right, or in a random sequence; and shuffling random positions. Mutation also has a probability associated with it that indicates how likely it is to randomly change a gene (bit) of a possible solution. The mutation value must be low for an efficient search within the genetic algorithm[1].
- Replacement—The last step is the replacement, which keeps the population size constant by eliminating individuals after recombination. There are three methods: strong replacement (random), weak replacement (the two fittest), and replacing both parents (the children replace both parents).
- Elitism and Hall-of-Fame—The elitism method ensures that the best individuals are not discarded but transferred directly to the next generation. Hall-of-Fame is an integer that indicates how many individuals are considered under elitism to be retained in the next generation. Elitism is necessary to ensure that genetic algorithms always find the best solution [19]. Elitism and Hall-of-Fame are often considered distinct from the general replacement strategy. While the replacement strategy primarily focuses on selecting individuals for reproduction and forming the next generation, the elitism and Hall-of-Fame mechanisms specifically address preserving the best-performing individuals.
- Stopping criteria—A mechanism is needed to finalize the execution of the genetic algorithm. Some ways to perform it are to stop after a fixed number of generations, after a specific time-lapse, to finish the process if the best fitness does not change for several generations (steady fitness), or to stop it if there are no improvements in the objective function for several consecutive generations (generation stagnation).

In this way, we can summarize that genetic algorithms are a process that involves some crucial steps: the initialization of a population form of solutions, selection of parents according to their fitness, recombination of genes by crossing, introduction of variability by mutation, substitution of individuals, and running the algorithm until the stopping criterion is satisfied. The operations described above are repeated within a loop, generation after generation, until a satisfactory solution or convergence criterion is reached.

2.3. Schema Theorem

The heuristic search of genetic algorithms is based on Holland's schema theorem, which states that the chromosomes have patterns called schemas. This schema theorem deals with the decomposition of chromosomes into schemas and their influence on the evolutionary dynamics of the population.

A schema is a binary string of fixed length representing a chromosome pattern. For example, in a chromosome of length 6, the schema 001X00 defines a string that starts with 001, has an unknown bit X, and ends with 00.

The fitness of a schema refers to how many individuals in the population contain that specific schema. It can be represented as a fitness function $F(S)$ that denotes the fitness of the schema S.

The schema theorem states that high-fitness schemas are more prevalent in future generations. This is because schemas with high fitness are more likely to be selected and recombined, leading to population improvement in terms of fitness. Mathematically, we can express this as:

$$F(S_{t+1}) \geq (1 - p_m) \cdot F(S_t),\tag{1}$$

where $F(S_{t+1})$ is the fitness of the schema S at the next generation $(t+1)$, $F(S_t)$ is the fitness of the schema S in the current generation (t), and finally, p_m is the mutation probability.

This equation indicates that the fitness of the schema in the next generation is at least equal to the current fitness, modulated by the mutation probability. If p_m is low, schemas with high fitness will likely survive and propagate in future generations, contributing to population improvement.

3. Genetic Algorithm Application

In this section, we implement a genetic algorithm to optimize univariate functions and extend its application to higher-dimensional problems. The general structure of a genetic algorithm is provided in the Algorithm 1.

Algorithm 1 Simple Genetic Algorithm

Parents $\leftarrow$ {randomly generated population}
While not (termination)
Calculate the fitness of each parent in the
population
Children $\leftarrow \varnothing$
while | Children | $<$ | Parents |
Use fitness to probabilistically select a pair of
parents for mating
Mate the parents to create children c_1 and c_2
Children $\leftarrow$ Children $\cup \{c_1, c_2\}$
Loop
Randomly mutate some of the children
Parents $\leftarrow$ Children
Next generation

Several libraries incorporate genetic algorithms, such as Distributed Evolutionary Algorithms (DEAP) [73], Karoo GP [74], Tiny Genetic Programming [75], and Symbiotic Bid-Based GP [76]. These libraries simplify the implementation of genetic algorithms. In this paper, we have utilized the DEAP library, which boasts comprehensive documentation.

3.1. Single Variable Functions

Considering the following three functions, we aim to use a custom genetic algorithm to find their global maxima:

- $f_1(x) = (x^2 + x) \cos(2x) + x^2$;
- $f_2(x) = \sin^2(3x + 45) + 0.9 \sin^3(9x) - \sin(15x + 50) \cos(2x - 30)$;
- $f_3(x) = -x^6/60 - x^5/50 + x^4/2 + 2x^3/3 - 3.2x^2 - 6.4x$.

In Figure 1, it can be seen how the above functions are optimized by a genetic algorithm, using a population size of 100 individuals, with a Hall-of-Fame size equal to 1, a mutation probability of 0.2, and a crossover probability of 0.5, over 50 generations. Note that as the generations progress, the individuals are closer to the global maxima. Another interesting feature is that, despite the local optima, the genetic algorithm in all the functions can find the global optima, as it is mentioned in the Introduction and Ref. [19].

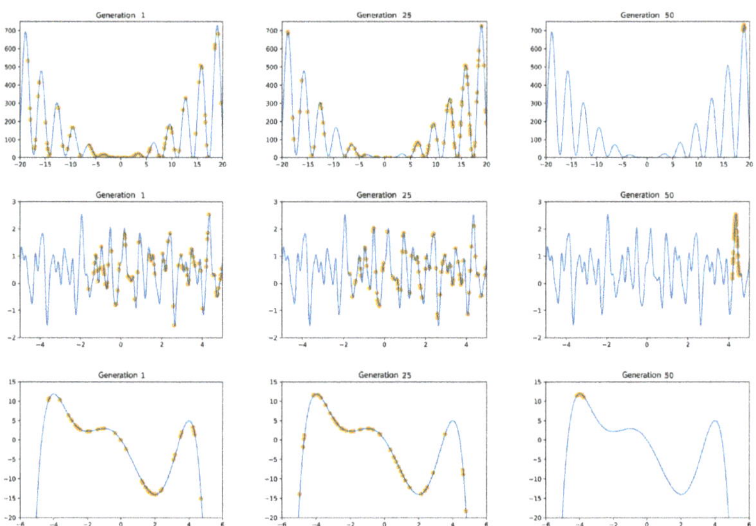

Figure 1. The search space exploration is presented for three different generations: 1, 25, and 50. As we advance through the generations, a greater concentration of individuals (yellow dots) is seen at the global maxima. In the top panels, $f_1(x)$. In the central panels, $f_2(x)$. In the bottom panels, $f_3(x)$.

3.2. Multimodal Functions

Genetic algorithms can also address problems with multiple dimensions and maxima by modifying the representation of candidate solutions and the operators used to generate new solutions. They can explore complex search spaces efficiently and identify global or local optima by appropriately designing crossover and mutation operators and analyzing different encoding techniques.

We use the Himmelblau function to demonstrate how genetic algorithms can be used to optimize these types of multimodal functions. We use the DEAP library, a robust Python framework for evolutionary computation, to achieve our goal. The following equation defines Himmelblau's function:

$$f(x,y) = (x^2 + y - 11)^2 + (x + y^2 - 7)^2. \tag{2}$$

The niching and sharing technique is employed to identify all the global optima within a single genetic algorithm run. This concept draws inspiration from nature, where regions are divided into sub-environments or niches, enhancing population efficiency and survival. Individuals compete for resources in these niches independently of those in other niches. By integrating a sharing mechanism into the genetic algorithm, individuals are incentivized to explore new niches, discovering multiple optimal solutions, each considered as a niche. Typically, this is achieved by dividing an individual's fitness value by the sum of distances from all other individuals. This approach penalizes overpopulated niches by distributing the local rewards among their individuals [77].

Niching involves dividing the population into subpopulations, each assigned to explore a specific region in the solution space. This encourages diversity by allowing genetically engineered individuals to compete for fitness locally. Conversely, sharing ensures a fair distribution of the fitness resources among individuals within the same niche. An individual's fitness is influenced not only by its performance but also by the performance of its neighbors, preventing overemphasis on a specific region and promoting a balanced exploration. This approach prevents premature convergence to a local maximum, allowing for the simultaneous exploration of different regions and ultimately facilitating the identification of the global maximum. Applying this technique effectively requires a larger population size and more generations than a simple genetic algorithm. This is essential to spread the population across the sample space, targeting different niches and, consequently, identifying multiple optimal maxima. In our experiment, we executed the algorithm with 200 individuals and 200 generations, and the outcomes are summarised in Figure 2 and Table 1.

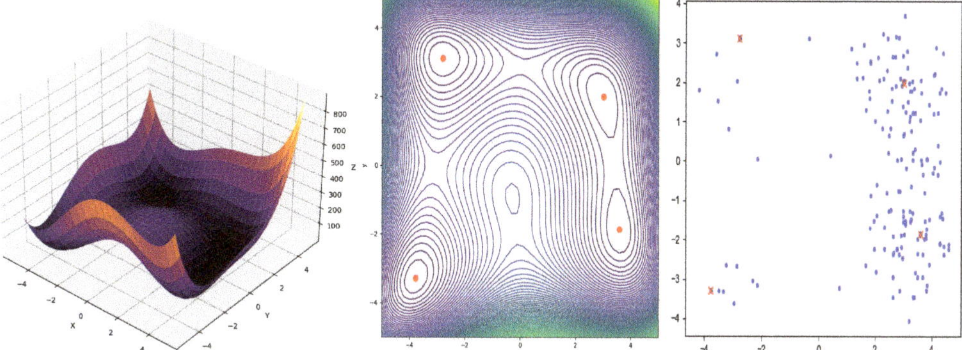

Figure 2. On the left panel, we have Himmelblau's function, while the center panel displays its contour diagram. The red points on the contours represent the global minima of the function. On the right panel, we can observe the application of the genetic algorithm with niching and sharing, specifically for Himmelblau's function, the blue dots are individuals and the red dots represent the best solutions.

Table 1. A comparison is made among the four real global optima of Himmelblau's function [77] and those found by the genetic algorithm using niching and sharing.

Real Optimum	Optimum Found by GA
$(3.000, 2.000)$	$(3.010, 1.998)$
$(-2.805, 3.131)$	$(-2.802, 3.133)$
$(-3.779, -3.283)$	$(-3.774, -3.292)$
$(3.584, -1.848)$	$(3.585, -1.847)$

As can be seen in Table 1, these results are remarkably similar to the real values. Improving these results is possible by increasing the number of individuals and generations. It should also be noted that this technique is not limited to three dimensions but can be generalized to N dimensions and can support the search for global M optima. However, it is important to remember that as the number of dimensions increases, more computational resources are required to search effectively.

3.3. Statistical Analysis

Genetic algorithms are handy tools in statistical applications for optimizing likelihood functions, thereby determining the parameters of a scientific model (which is precisely what this article aims to demonstrate). However, reporting a confidence interval for the output of a genetic algorithm can be more complex than in classical statistical methods. The most rigorous technique relies on having a mathematical model of the genetic algorithm's convergence that extends beyond Holland's schema theory for the simple genetic algorithm published in 1975.

Because the state of the population in a genetic algorithm depends solely on the previous state in a probabilistic manner, Markov chains have been studied as suitable models for specific applications, and more recently, others have been modeled as martingales [78,79].

However, it is possible to resort to less rigorous techniques. One approach is to assume a distribution for the optimized parameters. For instance, assuming the parameters follow a normal distribution, the confidence interval can be calculated based on standard deviations, and the confidence ellipses can be computed using Fisher matrices. This is the procedure employed in this article. Another procedure involves using the Bootstrap method or other re-sampling techniques [80].

4. Application in Cosmology

In observational cosmology, one of the fundamental tasks is to determine the values of the free parameters for a given theoretical model based on observational measurements. This involves creating a function that captures the discrepancies between observed data and theoretical predictions and using it to obtain a parameter estimate that fits the data well. The likelihood function is typically used to represent the data's conditional probability given the theory and its parameters. Although Bayesian inference is the most robust method for parameter estimation in cosmology, as it allows for sampling the posterior probability of parameters given the data, it can be computationally intensive (see the nomenclature under the Bayesian formalism of Bayes' theorem [81,82]); instead of sampling the posterior probability function to estimate parameter values efficiently, optimization algorithms can be used to find the maximum likelihood function. In Reference [82], there is an exciting overview of the difference between sampling and optimization, and it can be seen that they are two different tasks that can be complementary. This section presents three applications that show how genetic algorithms can be applied to analyze cosmological data. First, we offer parameter estimation in three cosmological models: ΛCDM, CPL, and PolyCDM. We then discuss how genetic algorithms can be used in a cosmological model with multiple maximum values, such as the graduated dark energy model presented in Ref. [83].

The datasets utilized in this section comprise 31 cosmic chronometers [84–91], Baryon Acoustic Oscillation measurements (BAO) [92–97], 1048 Type Ia supernovae (SNeIa) sourced from the Pantheon compilation [98], and binned data from the Joint Light Analysis SNeIa compilation [99].

Considering the datasets mentioned above, we employ the following log-likelihood functions for the Bayesian inference and optimization methods:

$$\log \mathcal{L}_i = -\frac{1}{2}(D^i_{\text{th}} - D^i_{\text{obs}})^T \cdot C_i^{-1} \cdot (D^i_{\text{th}} - D^i_{\text{obs}}), \tag{3}$$

where the index i ranges from 1 to 3, corresponding to the three datasets: cosmic chronometers [$D^{i=1} = H(z)$] and BAO [$D^{i=2} = D_A(z)$], where $D_A(z)$ represents the Hubble, volume averaged, and angular distance; and SNeIa [$D^{i=3} = \mu(z)$], where $\mu(z)$ denotes the distance modulus. In this context, D_{obs} represents the observed measurements, while D_{th} represents the theoretical values for the cosmological models. The matrices C_i encompass the covariance information, accounting for systematic and statistical errors.

We implemented a module to work with the `DEAP` genetic algorithms within the `SimpleMC`^{2} code for our cosmological parameter estimation [100]. In some of the subsequent results, we compare the genetic algorithm's outcomes with those of Bayesian inference obtained using the nested sampling algorithms, a specialized type of Markov chain Monte Carlo (MCMC) technique [81,101]. Additionally, we utilize the Fisher matrix formalism described in Refs. [102,103] to calculate the confidence intervals and generate error plots for the genetic algorithm-based parameter estimation. It is important to emphasize that genetic algorithms are not employed to generate posterior samples; instead, they are used to explore maximum likelihood estimation, which can yield similar and quicker results than parameter estimation. However, they cannot replace the robustness of MCMC methods. Furthermore, we conducted maximum likelihood estimation using a classical optimization method, specifically the L-BFGS algorithm [104], for comparison purposes and to assess the advantages of genetic algorithms.

4.1. Cosmological Parameter Estimation

As previously mentioned, we employ genetic algorithms to evaluate their effectiveness in parameter estimation. As a proof of the concept, and for simplicity, we consider three cosmological models, ΛCDM, CPL, and PolyCDM, which are described below:

- ΛCDM. The ΛCDM model serves as the standard cosmological model and comprises two primary components: cold dark matter (CDM), which plays a pivotal role in the universe's structure formation, and dark energy, which exhibits a counter-gravitational behavior, leading to the universe's accelerated expansion. The cosmological constant, denoted by Λ, is the simplest and most straightforward representation of dark energy, which exerts a pressure equal in magnitude but opposite in sign to the universe's energy density ($p = -\rho$). For a flat universe in the late stages of its evolution, the equation governing its expansion is given by $H^2 \equiv \left(\frac{\dot{a}}{a}\right)^2 = \rho_m(t) + \rho_\Lambda(t)$, where a represents the scale factor, the dot denotes the derivative with respect to time, ρ_m signifies the density of dark matter and baryons, and ρ_Λ accounts for the dark energy content in the form of a cosmological constant. These two parameters describe the evolution of the universe's content. Incorporating their initial conditions denoted with a subscript 0, this equation can be re-expressed in terms of the redshift $1 + z = 1/a$ as follows:

$$H^2 = H_0^2[\Omega_{\text{CDM},0}(1+z)^3 + \Omega_{\Lambda,0}], \tag{4}$$

  where H_0 denotes the Hubble constant, providing the present rate of expansion of the universe. The parameters $\Omega_{\text{CDM},0}$ and $\Omega_{\Lambda,0}$ are specific to the ΛCDM model. The former represents the current dimensionless density of dark matter (plus baryons), while the latter signifies the dimensionless density of dark energy. These parameters are subject to the constraint $\Omega_{\text{CDM},0} + \Omega_{\Lambda,0} = 1$; when this equality holds, we have a flat universe [105]. Consequently, for this model, we effectively have two free parameters, namely, h and $\Omega_{\text{CDM},0}$, which we simplify by denoting Ω_{CDM} as Ω_m for brevity.
- CPL model. One can discern dark energy's characteristics by investigating its state equation, denoted as $w(z)$, where p and ρ represent the pressure and dark energy density, respectively [106]. Chevallier, Polarski, and Linder introduced the following parametrization for the equation of state, $w(z) = w_0 + w_a \frac{z}{1+z}$, where w_0 signifies the current value of the equation of state. In contrast, w_a represents its rate of change over time [106]. This equation of state leads to the following derivation:

$$\begin{aligned} H(z)^2 = H_0^2[\Omega_{m,0}(1+z)^3 + \\ (1 - \Omega_{m,0})(1+z)^{3(1+w_0+w_a)}e^{-\frac{3w_a z}{1+z}}]. \end{aligned} \tag{5}$$

  Now, the parameter estimation consists of finding the free parameters H_0, $\Omega_{m,0}$, and w_0 and w_a.
- PolyCDM. We can consider an extension of dynamical dark energy by introducing spatial curvature, Ω_1, which adapts to the evolution of dark energy at low redshifts [41]. By performing a Taylor series expansion of the Equation (4) [107], we arrive at the PolyCDM model:

$$\begin{aligned} H^2 = H_0^2(\Omega_{m,0}(1+z)^3 + \\ \Omega_{1,0}(1+z)^2 + \Omega_{2,0}(1+z) \\ + (1 - \Omega_{m,0} - \Omega_{1,0} - \Omega_{2,0})), \end{aligned} \tag{6}$$

  where $\Omega_{m,0}$ represents the dark matter; and baryon, contribution, and $\Omega_{2,0}$ can be interpreted as the "lost matter" [107]. PolyCDM can be considered a parametrization of the Hubble parameter [108].

For all the models mentioned above, we use a genetic algorithm with elitism, using 50 generations, a mutation probability of 0.2, a crossover probability of 0.7, a population comprising 100 individuals, and a Hall-of-Fame size of 2 to maximize the likelihood probability function. Table 2 and Figure 3 present the parameter estimation results obtained throughout the three methods outlined earlier. It is noticeable that, in most cases, the genetic algorithm results closely align with the parameter estimations derived from the nested sampling. Consequently, although they are slower than optimization methods like the L-BFGS method, genetic algorithms offer greater precision while remaining faster than MCMC algorithms. It is important to note that genetic algorithms maximize the likelihood function rather than sampling the posterior distribution. This distinction can be computationally advantageous compared to Bayesian inference procedures in specific scenarios. However, GAs lack the assignment of weights to individuals, as found in Bayesian inference samples, and their exploration of parameter space differs from MCMC methods, which rely on Markov chains and probabilistic conditions. Genetic algorithms, instead, focus on achieving improved solutions in each generation.

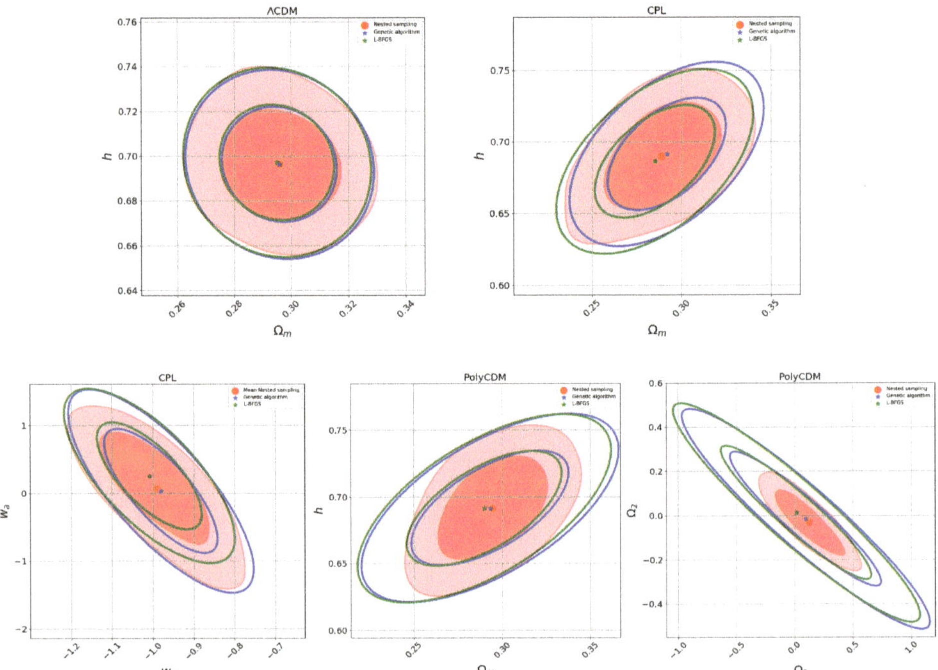

Figure 3. Two-dimensional posterior distribution plots showing the parameter mean estimates from nested sampling and the parameter values obtained through likelihood maximization using the L-BFGS and genetic algorithm methods (see color labels). Note that the confidence intervals are different due to their nature: optimization methods that maximize the likelihood function (L-BFGS and genetic algorithms) make use of the Fisher matrix formalism to approximate the errors (see Section 3.3), while the MCMC (nested sampling) method constructs its confidence intervals from sampling the posterior probability function. In the nested sampling results, the darker red regions represent 1σ, and the lighter red regions represent 2σ.

Table 2. Parameter estimation via genetic algorithms for the ΛCDM, CPL, and PolyCDM models utilizing cosmic chronometers, BAO, and SNeIa datasets. The $-2\log\mathcal{L}$ value represents the optimal fitness value.

		Data: CC + BAO + SNeIa		
Model	**Parameters**	**L-BFGS Optimizer**	**Genetic**	**Nested**
ΛCDM	h_0	0.6972 ± 0.0170	0.6964 ± 0.0170	0.6963 ± 0.0160
	Ω_m	0.2950 ± 0.0133	0.2958 ± 0.0133	0.2960 ± 0.0134
	$-2\log\mathcal{L}$	1049.2424	1049.2476	1049.2445
CPL	h_0	0.6864 ± 0.0259	0.6916 ± 0.0258	0.6901 ± 0.0240
	Ω_m	0.2853 ± 0.0221	0.2919 ± 0.0218	0.2892 ± 0.0211
	w_0	-1.0082 ± 0.0840	-0.9803 ± 0.0912	-0.9909 ± 0.0861
	w_a	0.2556 ± 0.5188	0.0330 ± 0.6035	0.0679 ± 0.5296
	$-2\log\mathcal{L}$	10483.9018	1049.0778	1048.9415
PolyCDM	h_0	0.6913 ± 0.0283	0.6916 ± 0.0283	0.6916 ± 0.0250
	Ω_m	0.2899 ± 0.0290	0.2931 ± 0.0294	0.2945 ± 0.0198
	$\Omega_{1,0}$	0.0150 ± 0.4254	0.0947 ± 0.4271	0.1232 ± 0.1795
	$\Omega_{2,0}$	0.0136 ± 0.1995	-0.0147 ± 0.2007	-0.0298 ± 0.0903
	Ω_k	-0.0013 ± 0.0703	-0.0076 ± 0.0702	-0.0004 ± 0.0117
	$-2\log\mathcal{L}$	1049.0688	1049.0660	1049.1286

4.2. Multimodal Models

Parameter inference in some models can lead to the identification of multiple optima, meaning that posterior probability functions can have multimodal distributions. To address this complexity, Bayesian nested inference algorithms, such as multinest [109], are a sampling method designed to deal with multimodal distributions, allowing for the effective sampling of the parameter space. In contrast, classical optimization algorithms are limited to finding a single maximum. Genetic algorithms, thanks to niche and sharing techniques (see Section 3.2), have the ability to exhaustively explore the parameter space, even in the presence of local maxima. An example of a model with multiple maxima in its posterior distribution is the case of graduated dark energy [83], which is governed by the following Friedmann equation:

$$H^2 = H_0^2[\Omega_{r,0}(1+z)^{-4} + \Omega_{m,0}(1+z)^{-3} + \Omega_{\mathrm{DE},0}sgn[1 - \psi \ln a]|1 - \psi \ln a|^{\frac{1}{1-\lambda}}], \tag{7}$$

where $\Omega_{\mathrm{DE},0}$ is the dimensionless density parameter of the dark energy with $\psi < 0$ and $\lambda = 0, -2, -4, \ldots$. Also, ψ is defined in terms of λ and another parameter γ in the following way: $\psi \equiv -3\gamma(\lambda - 1)$. One maximum value corresponds to the ΛCDM model, whereas the other is present to alleviate the Hubble tension. This model resembles a rapid transition of the universe from anti-de Sitter vacua to de Sitter vacua; see the details of the model in the references [83,110–113].

For the genetic algorithm with elitism used in this case, we set 20 generations, 200 individuals for the population, crossover and mutation probabilities of 0.5 and 0.2, respectively, and a Hall-of-Fame of size 2. Therefore, the free parameters for the graduated dark energy model are $\Omega_{m,0}$, h_0, λ, and γ. For this example, to appreciate the multimodality in the graduated DE model, we use the same data as that in the original work (Ref. [83]), i.e., cosmic chronometers, BAO, and SNeIa (binned data from the Joint Light Analysis compilation [99]), but for simplicity, we do not use the Planck information. We also fix $\lambda = -20$. Performing Bayesian inference on this model, the posterior distribution for the γ parameter is shown in Figure 4, in which two modes exist. In Table 3, we can analyze the outputs of the parameter estimation using nested sampling through posterior distribution sampling, the L-BFGS optimization method, and a genetic algorithm maximizing the likelihood distribution function; we can notice that the results maximizing the likelihoods are roughly consistent with the parameter estimation with Bayesian inference; however, for the γ value, the L-BFGS method is unable to find a value different to zero, and it is far from the estimation of this parameter using the same data.

As mentioned above, some algorithms for Bayesian inference, such as multinest nested sampling, could explore the regions with these two maxima; however, most MCMC methods cannot achieve this task. Using genetic algorithms with the niching and sharing techniques, we can quickly find and explore the parameter space with these two optima without performing a Bayesian inference process; we can notice them in the histograms of Figure 5, in which the GAs explore the regions of both modes of the γ parameter. Therefore, we can have more confidence in the results of a genetic algorithm than a classical optimization method.

To conclude this section, it is worth noting that there are other multimodal cosmological models, mainly involving neutrinos and spatial curvature, documented in the literature [114–118], and it is worth exploring in future works where these techniques could prove valuable for conducting efficient and rapid assessments.

Table 3. Parameter estimation with nested sampling (sampling the posterior probability distribution function), L-BFGS, and genetic algorithm. In these cases, we only consider the maximum likelihood found in the three methods and their corresponding parameter values.

	Nested Sampling	L-BFGS	Genetic
Ω_m	0.3264	0.2991	0.2959
h	0.6947	0.6760	0.6765
γ	−0.0129	0.0000	−0.0127
$-2\log\mathcal{L}$	55.8700	60.5781	61.6997

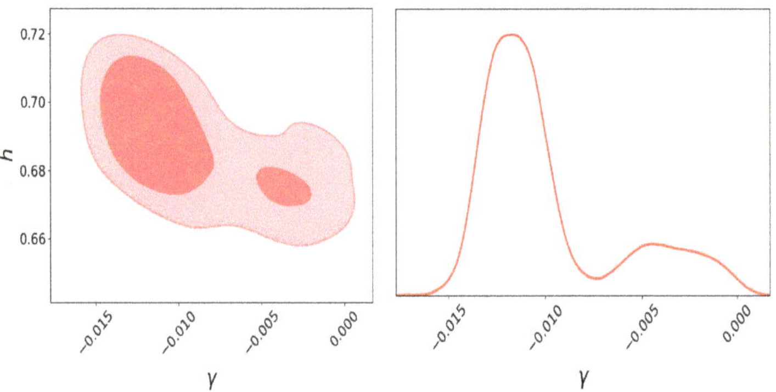

Figure 4. Posterior plots with nested sampling for h and γ parameters of the graduated DE model using HD + BAO + SN, where the bi-modality is shown. **Left**: Two-dimensional posterior plot for h vs. γ. The darker red region represents 1σ, and the lighter red region represents 2σ. **Right**: One-dimensional posterior distribution plot for γ parameter.

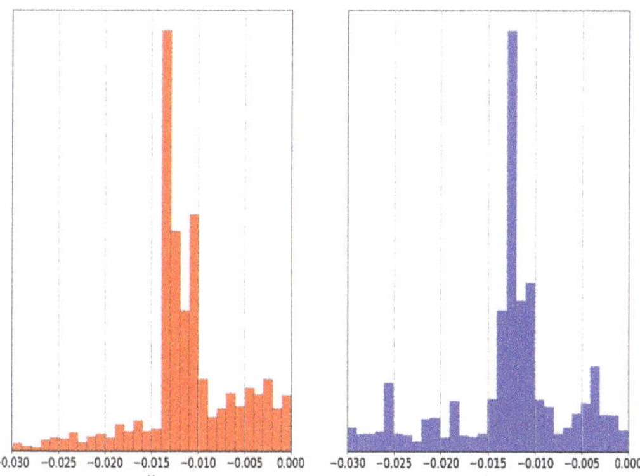

Figure 5. Comparison between the histograms of nested sampling (red) and individuals through generations of the genetic algorithm (blue) for γ parameter of graduated dark energy model.

4.3. Derived Functions

As an additional application, taking advantage of the genetic algorithms' nature, we can use the saved individuals along generations to maximize the likelihood function and calculate the derived functions to analyze their phenomenological behavior. This technique is usually used with the samples of the posterior probability with Bayesian inference algorithms, mapping the sampling of an estimated parameter to another derived one. For example, the library `fgivenx` [119] allows for this mapping. In the case of the individuals of likelihood optimization using genetic algorithms, the statistical meaning of the plots is not directly related to the posterior probability function; however, it can provide an idea of the behavior of the derived functions given the estimated parameters.

In Figure 6, we compare the equation of state reconstructed from the outputs of Section 4.1 for the CPL model, and we use the samples for the w_0 and w_a from the nested sampling and the values of these same parameters from the history of the individuals of the genetic algorithm population. We can notice that the behavior of the equation of state, analyzing the darkest regions, is similar in both cases, and it suggests that for a quick test, we can use this technique with genetic algorithms. Regarding the confidence regions, because we are only optimizing the likelihood function with the genetic algorithms, we cannot have a formal way to estimate them correctly.

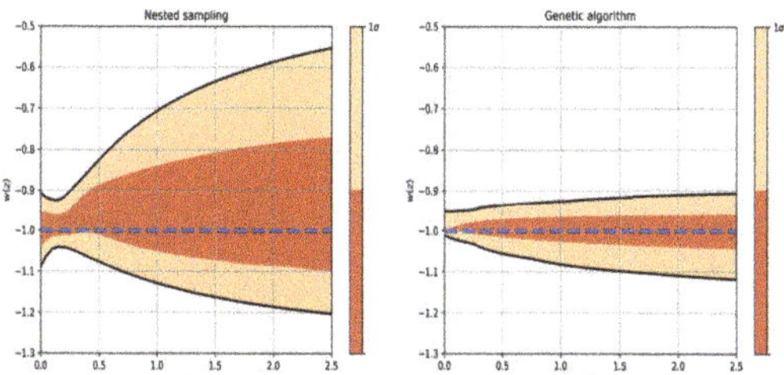

Figure 6. Equation of state for CPL model plotted with `fgivenx` from (**left**) nested sampling and (**right**) genetic algorithms. 'The darker zones represent 1σ' and the lighter 2σ. The dotted blue line represents the value of the Equation of State to ΛCDM.

5. Conclusions

In this study, we have leveraged genetic algorithms as an effective tool to estimate the free parameters of four cosmological models. Individuals generated in each genetic algorithm population have demonstrated the ability to achieve faster parameter estimates than those obtained using MCMC methods, thus reducing the number of likelihood function evaluations required. In addition, these genetic algorithms allow for a rapid computation of derived parameters, which adds flexibility and efficiency to the estimation process.

However, it is important to note that genetic algorithms differ from Bayesian approaches in their sampling process. While MCMC methods fully sample the posterior probability function, genetic algorithms focus on maximizing the likelihood function. This distinction implies that genetic algorithms cannot directly provide confidence regions with the same statistical significance as Bayesian inference procedures. However, they offer significant advantages, such as a faster speed and better results than other optimization methods, such as the L-BFGS algorithm.

Additionally, we have explored the usefulness of sharing and niche techniques in genetic algorithms, ensuring practical parameter space exploration, even in local or global optima. These features may be especially valuable in cosmology as a prior analysis to maximize the likelihood function before undertaking more computationally expensive Bayesian parameter estimation.

Throughout this paper, we can understand why genetic algorithms have been a very promising field of research over the last decades. Their flexibility allows for their application in diverse tasks, such as optimization, combinatorics, statistics, and even to speed up computational algorithms. The potential future applications of genetic algorithms in cosmological research are vast. With the presented study, we show the prospect of using them as a complement within cosmological data analysis. This is in agreement and complementary with the existing research that also focuses on the statistical applications of evolutionary computation [39,120]. In our case, we have not proposed a novel method or algorithm; however, we have analyzed how to use GAs so that they can complement a traditional analysis of cosmological data and be an alternative to optimize the likelihood function. We are convinced that genetic algorithms are a great technique with diverse cosmological and statistical applications. For example, in a parallel work, we have explored their usefulness to improve cosmological neural reconstructions [31] and to reduce the computational time of Bayesian inference routines. Therefore, we are confident that genetic algorithms are an excellent complementary element to the cosmological data analysis toolkit.

Author Contributions: All authors have contributed equally to all stages of the research: conceptualisation, methodology, formal analysis, and original writing draft. All authors have read and agreed to the published version of the manuscript.

Funding: This research was funded by FOSEC SEP-CONACYT Investigación Básica A1-S-21925, PRONACES-CONACYT/304001/2020, UNAM-DGAPA-PAPIIT IN117723, SIP20230505-IPN, FORDECYT-PRONACES-CONACYT CF-MG-2558591 and COFAA-IPN, and EDI-IPN grants.

Data Availability Statement: The datasets utilised comprise 31 cosmic chronometers [84–91], Baryon Acoustic Oscillation measurements (BAO) [92–97], 1048 Type Ia supernovae (SNeIa) sourced from the Pantheon compilation [98], and binned data from the Joint Light Analysis SNeIa compilation [99].

Conflicts of Interest: The authors declare no conflicts of interest.

Notes

[1] Let us consider a binary representation of a genetic algorithm where each individual is a sequence of binary values representing a potential solution. Suppose an individual's chromosome (binary sequence) is 101010. A mutation operation might involve flipping one of the bits, resulting in a new chromosome, like 111010 or 100010. A mutation probability determines the choice of which bit to flip. If the mutation probability is low, only a few bits are expected to change, maintaining some of the original information. This process introduces diversity in the population, allowing the algorithm to explore different regions of the search space and preventing premature convergence to suboptimal solutions. In a genetic algorithm, a bit denotes the smallest unit of information representing a decision within a solution. Unlike a bit in memory, it symbolizes binary choices in a solution space rather than directly storing data.

[2] https://igomezv.github.io/SimpleMC (accessed on 18 December 2023).

References

1. Srinivas, M.; Patnaik, L.M. Genetic algorithms: A survey. *Computer* **1994**, *27*, 17–26. [CrossRef]
2. Tomassini, M. A survey of genetic algorithms. In *Annual Reviews of Computational Physics III*; World Scientific: Singapore, 1995; pp. 87–118.
3. Mitchell, M. Genetic algorithms: An overview. *Complex* **1995**, *1*, 31–39. [CrossRef]
4. Kumar, M.; Husain, M.; Upreti, N.; Gupta, D. *Genetic Algorithm: Review and Application*; SSRN: New York, NY, USA, 2010.
5. Katoch, S.; Chauhan, S.S.; Kumar, V. A review on genetic algorithm: Past, present, and future. *Multimed. Tools Appl.* **2021**, *80*, 8091–8126. [CrossRef]
6. Dumitrescu, D.; Lazzerini, B.; Jain, L.C.; Dumitrescu, A. *Evolutionary Computation*; CRC Press: Boca Raton, FL, USA, 2000.
7. Yang, X.S. *Nature-Inspired Metaheuristic Algorithms*; Luniver Press: Bristol, UK, 2010.
8. Sadeeq, H.T.; Abdulazeez, A.M. Metaheuristics: A Review of Algorithms. *Int. J. Online Biomed. Eng.* **2023**, *19*, 142–164. [CrossRef]
9. Kennedy, J.; Eberhart, R. Particle swarm optimization. In Proceedings of the ICNN'95-International Conference on Neural Networks, Perth, Australia, 27 November–1 December 1995; IEEE: Piscataway, NJ, USA, 1995; Volume 4, pp.1942–1948.
10. Sadeeq, H.T.; Abdulazeez, A.M. Giant trevally optimizer (GTO): A novel metaheuristic algorithm for global optimization and challenging engineering problems. *IEEE Access* **2022**, *10*, 121615–121640. [CrossRef]
11. Sadeeq, H.T.; Abdulazeez, A.M. Car side impact design optimization problem using giant trevally optimizer. *Structures* **2023**, *55*, 39–45. [CrossRef]
12. Hashish, M.S.; Hasanien, H.M.; Ullah, Z.; Alkuhayli, A.; Badr, A.O. Giant Trevally Optimization Approach for Probabilistic Optimal Power Flow of Power Systems Including Renewable Energy Systems Uncertainty. *Sustainability* **2023**, *15*, 13283. [CrossRef]
13. Wang, L.; Cao, Q.; Zhang, Z.; Mirjalili, S.; Zhao, W. Artificial rabbits optimization: A new bio-inspired meta-heuristic algorithm for solving engineering optimization problems. *Eng. Appl. Artif. Intell.* **2022**, *114*, 105082. [CrossRef]
14. Alsaiari, A.O.; Moustafa, E.B.; Alhumade, H.; Abulkhair, H.; Elsheikh, A. A coupled artificial neural network with artificial rabbits optimizer for predicting water productivity of different designs of solar stills. *Adv. Eng. Softw.* **2023**, *175*, 103315. [CrossRef]
15. Holland, J.H. *Adaptation in Natural and Artificial Systems: An Introductory Analysis with Applications to Biology, Control, and Artificial Intelligence*; MIT Press: Cambridge, MA, USA, 1992.
16. Holland, J.H. Genetic algorithms and the optimal allocation of trials. *SIAM J. Comput.* **1973**, *2*, 88–105. [CrossRef]
17. Langdon, W.B. *Genetic Programming and Data Structures: Genetic Programming + Data Structures = Automatic Programming!* Kluwer Academic Publishers: Boston, MA, USA, 1998.
18. Beyer, H.G.; Schwefel, H.P. Evolution strategies—A comprehensive introduction. *Nat. Comput.* **2002**, *1*, 3–52. [CrossRef]
19. Rudolph, G. Convergence analysis of canonical genetic algorithms. *IEEE Trans. Neural Netw.* **1994**, *5*, 96–101. [CrossRef]
20. García, J.; Acosta, C.; Mesa, M. Genetic algorithms for mathematical optimization. *J. Phys. Conf. Ser.* **2020**, *1448*, 012020. [CrossRef]
21. Anastasio, M.A.; Yoshida, H.; Nagel, R.; Nishikawa, R.M.; Doi, K. A genetic algorithm-based method for optimizing the performance of a computer-aided diagnosis scheme for detection of clustered microcalcifications in mammograms. *Med. Phys.* **1998**, *25*, 1613–1620. [CrossRef]
22. Bevilacqua, A.; Campanini, R.; Lanconelli, N. A distributed genetic algorithm for parameters optimization to detect microcalcifications in digital mammograms. In Proceedings of the Workshops on Applications of Evolutionary Computation, Como, Italy, 18–20 April 2001; Springer: Berlin/Heidelberg, Germany, 2001; pp. 278–287.
23. Ghaheri, A.; Shoar, S.; Naderan, M.; Hoseini, S.S. The applications of genetic algorithms in medicine. *Oman Med. J.* **2015**, *30*, 406. [CrossRef]
24. Zelenkov, Y.; Reshettsov, I. Analysis of the COVID-19 pandemic using a compartmental model with time-varying parameters fitted by a genetic algorithm. *Expert Syst. Appl.* **2023**, *224*, 120034. [CrossRef]
25. Esquivel, R.M.; Gómez-Vargas, I.; Montalvo, T.R.; Vázquez, J.A.; García-Salcedo, R. The inverse problem of a dynamical system solved with genetic algorithms. *J. Phys. Conf. Ser.* **2021**, *1723*, 012021. [CrossRef]
26. Simpson, A.R.; Priest, S.D. The application of genetic algorithms to optimisation problems in geotechnics. *Comput. Geotech.* **1993**, *15*, 1–19. [CrossRef]
27. Drachal, K.; Pawłowski, M. A review of the applications of genetic algorithms to forecasting prices of commodities. *Economies* **2021**, *9*, 6. [CrossRef]
28. Victorino, I.R.d.S.; Maciel Filho, R. Application of Genetic Algorithms To the Optimization of an Industrial Reactor. *IFAC Proc. Vol.* **2006**, *39*, 857–862. [CrossRef]
29. Kuri-Morales, A. Closed determination of the number of neurons in the hidden layer of a multi-layered perceptron network. *Soft Comput.* **2017**, *21*, 597–609. [CrossRef]
30. Whitley, D.; Starkweather, T.; Bogart, C. Genetic algorithms and neural networks: Optimizing connections and connectivity. *Parallel Comput.* **1990**, *14*, 347–361. [CrossRef]
31. Gómez-Vargas, I.; Andrade, J.B.; Vázquez, J.A. Neural networks optimized by genetic algorithms in cosmology. *Phys. Rev. D* **2023**, *107*, 043509. [CrossRef]
32. Abel, S.; Constantin, A.; Harvey, T.R.; Lukas, A. Evolving Heterotic Gauge Backgrounds: Genetic Algorithms versus Reinforcement Learning. *Fortschritte Der Phys.* **2022**, *70*, 2200034. [CrossRef]
33. Bourilkov, D. Machine and deep learning applications in particle physics. *Int. J. Mod. Phys. A* **2019**, *34*, 1930019. [CrossRef]

34. Akrami, Y.; Scott, P.; Edsjö, J.; Conrad, J.; Bergström, L. A profile likelihood analysis of the constrained MSSM with genetic algorithms. *J. High Energy Phys.* **2010**, *2010*, 57. [CrossRef]
35. Charbonneau, P. Genetic algorithms in astronomy and astrophysics. *Astrophys. J. Suppl.* **1995**, *101*, 309. [CrossRef]
36. Fridman, P. Radio astronomy image enhancement in the presence of phase errors using genetic algorithms. In Proceedings of the 2001 International Conference on Image Processing (Cat. No. 01CH37205), Thessaloniki, Greece, 7–10 October 2001; IEEE: Piscataway, NJ, USA, 2001; Volume 3, pp. 612–615.
37. Rajpaul, V. Genetic algorithms in astronomy and astrophysics. *arXiv* **2012**, arXiv:1202.1643.
38. Holl, B.; Sozzetti, A.; Sahlmann, J.; Giacobbe, P.; Ségransan, D.; Unger, N.; Delisle, J.B.; Barbato, D.; Lattanzi, M.; Morbidelli, R.; et al. Gaia Data Release 3-Astrometric orbit determination with Markov chain Monte Carlo and genetic algorithms: Systems with stellar, sub-stellar, and planetary mass companions. *Astron. Astrophys.* **2023**, *674*, A10. [CrossRef]
39. Axiak, M.; Kitching, T.; van Hemert, J. Evolution Strategies for Cosmology: A Comparison of Nested Sampling Methods. *arXiv* **2011**, arXiv:1101.0717.
40. Luo, X.L.; Feng, J.; Zhang, H.H. A genetic algorithm for astroparticle physics studies. *Comput. Phys. Commun.* **2020**, *250*, 106818. [CrossRef]
41. Gómez-Vargas, I.; Medel-Esquivel, R.; García-Salcedo, R.; Vázquez, J.A. Neural network reconstructions for the Hubble parameter, growth rate and distance modulus. *Eur. Phys. J. C* **2023**, *83*, 304. [CrossRef]
42. Kamerkar, A.; Nesseris, S.; Pinol, L. Machine learning cosmic inflation. *Phys. Rev. D* **2023**, *108*, 043509. [CrossRef]
43. Chacón, J.; Gómez-Vargas, I.; Méndez, R.M.; Vázquez, J.A. Analysis of dark matter halo structure formation in N-body simulations with machine learning. *Phys. Rev. D* **2023**, *107*, 123515. [CrossRef]
44. de Dios Rojas Olvera, J.; Gómez-Vargas, I.; Vázquez, J.A. Observational cosmology with artificial neural networks. *Universe* **2022**, *8*, 120. [CrossRef]
45. Arjona, R.; Nesseris, S. What can Machine Learning tell us about the background expansion of the Universe? *Phys. Rev. D* **2020**, *101*, 123525. [CrossRef]
46. Nesseris, S.; Garcia-Bellido, J. A new perspective on Dark Energy modeling via Genetic Algorithms. *J. Cosmol. Astropart. Phys.* **2012**, *2012*, 033. [CrossRef]
47. Wang, K.; Guo, P.; Yu, F.; Duan, L.; Wang, Y.; Du, H. Computational intelligence in astronomy: A survey. *Int. J. Comput. Intell. Syst.* **2018**, *11*, 575. [CrossRef]
48. Bogdanos, C.; Nesseris, S. Genetic algorithms and supernovae type Ia analysis. *J. Cosmol. Astropart. Phys.* **2009**, *2009*, 6. [CrossRef]
49. Nesseris, S.; Shafieloo, A. A model-independent null test on the cosmological constant. *Mon. Not. R. Astron. Soc.* **2010**, *408*, 1879–1885. [CrossRef]
50. Alestas, G.; Kazantzidis, L.; Nesseris, S. Machine learning constraints on deviations from general relativity from the large scale structure of the Universe. *Phys. Rev. D* **2022**, *106*, 103519. [CrossRef]
51. Pellejero-Ibáñez, M.; Angulo, R.E.; Aricó, G.; Zennaro, M.; Contreras, S.; Stücker, J. Cosmological parameter estimation via iterative emulation of likelihoods. *Mon. Not. R. Astron. Soc.* **2020**, *499*, 5257–5268. [CrossRef]
52. Wraith, D.; Wraith, D.; Kilbinger, M.; Benabed, K.; Capp'e, O.; Cardoso, J.F.; Cardoso, J.F.; Fort, G.; Prunet, S.; Robert, C.P. Estimation of cosmological parameters using adaptive importance sampling. *Phys. Rev. D* **2009**, *80*, 023507. [CrossRef]
53. Graff, P.; Feroz, F.; Hobson, M.P.; Lasenby, A. BAMBI: Blind accelerated multimodal Bayesian inference. *Mon. Not. R. Astron. Soc.* **2012**, *421*, 169–180. [CrossRef]
54. Nygaard, A.; Holm, E.B.; Hannestad, S.; Tram, T. CONNECT: A neural network based framework for emulating cosmological observables and cosmological parameter inference. *J. Cosmol. Astropart. Phys.* **2023**, *2023*, 025. [CrossRef]
55. Gómez-Vargas, I.; Esquivel, R.M.; García-Salcedo, R.; Vázquez, J.A. Neural network within a bayesian inference framework. *J. Phys. Conf. Ser.* **2021**, *1723*, 012022. [CrossRef]
56. Alsing, J.; Charnock, T.; Feeney, S.; Wandelt, B. Fast likelihood-free cosmology with neural density estimators and active learning. *Mon. Not. R. Astron. Soc.* **2019**, *488*, 4440–4458. [CrossRef]
57. Leclercq, F. Bayesian optimization for likelihood-free cosmological inference. *Phys. Rev. D* **2018**, *98*, 063511. [CrossRef]
58. Bagavathi, C.; Saraniya, O. Evolutionary Mapping Techniques for Systolic Computing System. In *Deep Learning and Parallel Computing Environment for Bioengineering Systems*; Elsevier: Amsterdam, The Netherlands, 2019; pp. 207–223.
59. Passino, K.M. Biomimicry of bacterial foraging for distributed optimization and control. *IEEE Control Syst. Mag.* **2002**, *22*, 52–67.
60. Mirjalili, S.; Mirjalili, S.M.; Lewis, A. Grey wolf optimizer. *Adv. Eng. Softw.* **2014**, *69*, 46–61. [CrossRef]
61. Faris, H.; Aljarah, I.; Al-Betar, M.A.; Mirjalili, S. Grey wolf optimizer: A review of recent variants and applications. *Neural Comput. Appl.* **2018**, *30*, 413–435. [CrossRef]
62. Simon, D. Biogeography-based optimization. *IEEE Trans. Evol. Comput.* **2008**, *12*, 702–713. [CrossRef]
63. Mitchell, M. *An Introduction to Genetic Algorithms*; MIT Press: Cambridge, MA, USA, 1998.
64. Waddington, C.H. *An Introduction to Modern Genetics*; Routledge: London, UK, 2016.
65. Kumar, A. Encoding schemes in genetic algorithm. *Int. J. Adv. Res. Eng.* **2013**, *2*, 1–7.
66. Beasley, D.; Bull, D.R.; Martin, R.R. An overview of genetic algorithms: Part 1, fundamentals. *Univ. Comput.* **1993**, *15*, 56–69.
67. Mirjalili, S.; Mirjalili, S. Genetic algorithm. *Evolutionary Algorithms and Neural Networks: Theory and Applications*; Springer: Cham, Switzerland, 2019; pp. 43–55.
68. Sivanandam, S.; Deepa, S.; Sivanandam, S.; Deepa, S. *Genetic Algorithms*; Springer: Berlin/Heidelberg, Germany, 2008.

69. Goldberg, D.E.; Deb, K. A comparative analysis of selection schemes used in genetic algorithms. In *Foundations of Genetic Algorithms*; Elsevier: Amsterdam, The Netherlands, 1991; Volume 1, pp. 69–93.
70. Miller, B.L.; Goldberg, D.E. Genetic algorithms, tournament selection, and the effects of noise. *Complex Syst.* **1995**, *9*, 193–212.
71. Lee, C.Y. Entropy-Boltzmann selection in the genetic algorithms. *IEEE Trans. Syst. Man Cybern. Part B (Cybern.)* **2003**, *33*, 138–149.
72. Marsili Libelli, S.; Alba, P. Adaptive mutation in genetic algorithms. *Soft Comput.* **2000**, *4*, 76–80. [CrossRef]
73. Fortin, F.A.; De Rainville, F.M. Distributed Evolutionary Algorithms. Available online: https://github.com/deap (accessed on 18 December 2023).
74. Staats, K. Karoo_gp. Available online: http://kstaats.github.io/karoo_gp/ (accessed on 18 December 2023).
75. Sipper, M. Tiny Genetic Programming. Available online: https://github.com/moshesipper/tiny_gp (accessed on 18 December 2023).
76. Bonson, J.P.C. Symbiotic Bid-Based GP. Available online: https://github.com/jpbonson/SBBFramework (accessed on 18 December 2023).
77. Wirsansky, E. *Hands-On Genetic Algorithms with Python: Applying Genetic Algorithms to Solve Real-World Deep Learning and Artificial Intelligence Problems*; Packt Publishing: Birmingham, UK, 2020.
78. Eiben, A.E.; Rudolph, G. Theory of evolutionary algorithms: A bird's eye view. *Theor. Comput. Sci.* **1999**, *229*, 3–9. [CrossRef]
79. Oliveto, P.S.; Witt, C. On the runtime analysis of the simple genetic algorithm. *Theor. Comput. Sci.* **2014**, *545*, 2–19. [CrossRef]
80. Šílený, J. Earthquake source parameters and their confidence regions by a genetic algorithm with a 'memory'. *Geophys. J. Int.* **1998**, *134*, 228–242. [CrossRef]
81. Esquivel, R.M.; Gómez-Vargas, I.; Vázquez, J.A.; Salcedo, R.G. An introduction to Markov Chain Monte Carlo. *Boletín EstadíStica Investig. Oper.* **2021**, *1*, 47–74.
82. Hogg, D.W.; Foreman-Mackey, D. Data analysis recipes: Using markov chain monte carlo. *Astrophys. J. Suppl. Ser.* **2018**, *236*, 11. [CrossRef]
83. Akarsu, Ö.; Barrow, J.D.; Escamilla, L.A.; Vazquez, J.A. Graduated dark energy: Observational hints of a spontaneous sign switch in the cosmological constant. *Phys. Rev. D* **2020**, *101*, 063528. [CrossRef]
84. Jimenez, R.; Verde, L.; Treu, T.; Stern, D. Constraints on the equation of state of dark energy and the Hubble constant from stellar ages and the cosmic microwave background. *Astrophys. J.* **2003**, *593*, 622. [CrossRef]
85. Simon, J.; Verde, L.; Jimenez, R. Constraints on the redshift dependence of the dark energy potential. *Phys. Rev. D* **2005**, *71*, 123001. [CrossRef]
86. Stern, D.; Jimenez, R.; Verde, L.; Kamionkowski, M.; Stanford, S.A. Cosmic chronometers: Constraining the equation of state of dark energy. I: $H(z)$ measurements. *J. Cosmol. Astropart. Phys.* **2010**, *2010*, 008. [CrossRef]
87. Moresco, M.; Verde, L.; Pozzetti, L.; Jimenez, R.; Cimatti, A. New constraints on cosmological parameters and neutrino properties using the expansion rate of the Universe to $z \sim 1.75$. *J. Cosmol. Astropart. Phys.* **2012**, *2012*, 053. [CrossRef]
88. Zhang, C.; Zhang, H.; Yuan, S.; Liu, S.; Zhang, T.J.; Sun, Y.C. Four new observational $H(z)$ data from luminous red galaxies in the Sloan Digital Sky Survey data release seven. *Res. Astron. Astrophys.* **2014**, *14*, 1221. [CrossRef]
89. Moresco, M. Raising the bar: New constraints on the Hubble parameter with cosmic chronometers at $z \sim 2$. *Mon. Not. R. Astron. Soc. Lett.* **2015**, *450*, L16–L20. [CrossRef]
90. Moresco, M.; Pozzetti, L.; Cimatti, A.; Jimenez, R.; Maraston, C.; Verde, L.; Thomas, D.; Citro, A.; Tojeiro, R.; Wilkinson, D. A 6% measurement of the Hubble parameter at $z \sim 0.45$: Direct evidence of the epoch of cosmic re-acceleration. *J. Cosmol. Astropart. Phys.* **2016**, *2016*, 014. [CrossRef]
91. Ratsimbazafy, A.; Loubser, S.; Crawford, S.; Cress, C.; Bassett, B.; Nichol, R.; Väisänen, P. Age-dating luminous red galaxies observed with the Southern African Large Telescope. *Mon. Not. R. Astron. Soc.* **2017**, *467*, 3239–3254. [CrossRef]
92. Alam, S.; Ata, M.; Bailey, S.; Beutler, F.; Bizyaev, D.; Blazek, J.A.; Bolton, A.S.; Brownstein, J.R.; Burden, A.; Chuang, C.H.; et al. The clustering of galaxies in the completed SDSS-III Baryon Oscillation Spectroscopic Survey: Cosmological analysis of the DR12 galaxy sample. *Mon. Not. R. Astron. Soc.* **2017**, *470*, 2617–2652. [CrossRef]
93. Ata, M.; Baumgarten, F.; Bautista, J.; Beutler, F.; Bizyaev, D.; Blanton, M.R.; Blazek, J.A.; Bolton, A.S.; Brinkmann, J.; Brownstein, J.R.; et al. The clustering of the SDSS-IV extended Baryon Oscillation Spectroscopic Survey DR14 quasar sample: First measurement of baryon acoustic oscillations between redshift 0.8 and 2.2. *Mon. Not. R. Astron. Soc.* **2018**, *473*, 4773–4794. [CrossRef]
94. Blomqvist, M.; Des Bourboux, H.D.M.; de Sainte Agathe, V.; Rich, J.; Balland, C.; Bautista, J.E.; Dawson, K.; Font-Ribera, A.; Guy, J.; Le Goff, J.M.; et al. Baryon acoustic oscillations from the cross-correlation of Lyα absorption and quasars in eBOSS DR14. *Astron. Astrophys.* **2019**, *629*, A86. [CrossRef]
95. de Sainte Agathe, V.; Balland, C.; Des Bourboux, H.D.M.; Blomqvist, M.; Guy, J.; Rich, J.; Font-Ribera, A.; Pieri, M.M.; Bautista, J.E.; Dawson, K.; et al. Baryon acoustic oscillations at z = 2.34 from the correlations of Lyα absorption in eBOSS DR14. *Astron. Astrophys.* **2019**, *629*, A85. [CrossRef]
96. Beutler, F.; Blake, C.; Colless, M.; Jones, D.H.; Staveley-Smith, L.; Campbell, L.; Parker, Q.; Saunders, W.; Watson, F. The 6dF Galaxy Survey: Baryon acoustic oscillations and the local Hubble constant. *Mon. Not. R. Astron. Soc.* **2011**, *416*, 3017–3032. [CrossRef]
97. Ross, A.J.; Samushia, L.; Howlett, C.; Percival, W.J.; Burden, A.; Manera, M. The clustering of the SDSS DR7 main Galaxy sample–I. A 4 per cent distance measure at z = 0.15. *Mon. Not. R. Astron. Soc.* **2015**, *449*, 835–847. [CrossRef]

98. Scolnic, D.M.; Jones, D.; Rest, A.; Pan, Y.; Chornock, R.; Foley, R.; Huber, M.; Kessler, R.; Narayan, G.; Riess, A.; et al. The complete light-curve sample of spectroscopically confirmed SNe Ia from Pan-STARRS1 and cosmological constraints from the combined pantheon sample. *Astrophys. J.* **2018**, *859*, 101. [CrossRef]
99. Betoule, M.; Kessler, R.; Guy, J.; Mosher, J.; Hardin, D.; Biswas, R.; Astier, P.; El-Hage, P.; Konig, M.; Kuhlmann, S.; et al. Improved cosmological constraints from a joint analysis of the SDSS-II and SNLS supernova samples. *Astron. Astrophys.* **2014**, *568*, A22. [CrossRef]
100. Vazquez, J.; Gomez-Vargas, I.; Slosar, A. Updated Version of a Simple MCMC Code for Cosmological Parameter Estimation Where Only Expansion History Matters. 2023. Available online: https://github.com/ja-vazquez/SimpleMC (accessed on 18 December 2023).
101. Skilling, J. Nested sampling. *AIP Conf. Proc.* **2004**, *735*, 395–405.
102. Padilla, L.E.; Tellez, L.O.; Escamilla, L.A.; Vazquez, J.A. Cosmological parameter inference with Bayesian statistics. *Universe* **2021**, *7*, 213. [CrossRef]
103. Sivia, D.; Skilling, J. *Data Analysis: A Bayesian Tutorial*; OUP, Oxford University Press: Oxford, UK, 2006.
104. Zhu, C.; Byrd, R.H.; Lu, P.; Nocedal, J. Algorithm 778: L-BFGS-B: Fortran subroutines for large-scale bound-constrained optimization. *ACM Trans. Math. Softw. (TOMS)* **1997**, *23*, 550–560. [CrossRef]
105. Liddle, A. *An Introduction to Modern Cosmology*; John Wiley & Sons: Hoboken, NJ, USA, 2015.
106. Linden, S.; Virey, J.M. Test of the Chevallier-Polarski-Linder parametrization for rapid dark energy equation of state transitions. *Phys. Rev. D* **2008**, *78*, 023526. [CrossRef]
107. Vazquez, J.A.; Hee, S.; Hobson, M.; Lasenby, A.; Ibison, M.; Bridges, M. Observational constraints on conformal time symmetry, missing matter and double dark energy. *J. Cosmol. Astropart. Phys.* **2018**, *2018*, 062. [CrossRef]
108. Zhai, Z.; Blanton, M.; Slosar, A.; Tinker, J. An evaluation of cosmological models from the expansion and growth of structure measurements. *Astrophys. J.* **2017**, *850*, 183. [CrossRef]
109. Feroz, F.; Hobson, M.; Bridges, M. MultiNest: An efficient and robust Bayesian inference tool for cosmology and particle physics. *Mon. Not. R. Astron. Soc.* **2009**, *398*, 1601–1614. [CrossRef]
110. Acquaviva, G.; Akarsu, Ö.; Katırcı, N.; Vazquez, J.A. Simple-graduated dark energy and spatial curvature. *Phys. Rev. D* **2021**, *104*, 023505. [CrossRef]
111. Akarsu, Ö.; Kumar, S.; Özülker, E.; Vazquez, J.A. Relaxing cosmological tensions with a sign switching cosmological constant. *Phys. Rev. D* **2021**, *104*, 123512. [CrossRef]
112. Akarsu, O.; Di Valentino, E.; Kumar, S.; Nunes, R.C.; Vazquez, J.A.; Yadav, A. $Lambda_s$CDM model: A promising scenario for alleviation of cosmological tensions. *arXiv* **2023**, arXiv:2307.10899.
113. Akarsu, Ö.; Kumar, S.; Özülker, E.; Vazquez, J.A.; Yadav, A. Relaxing cosmological tensions with a sign switching cosmological constant: Improved results with Planck, BAO, and Pantheon data. *Phys. Rev. D* **2023**, *108*, 023513. [CrossRef]
114. Kreisch, C.D.; Park, M.; Calabrese, E.; Cyr-Racine, F.Y.; An, R.; Bond, J.R.; Dore, O.; Dunkley, J.; Gallardo, P.; Gluscevic, V.; et al. The Atacama Cosmology Telescope: The Persistence of Neutrino Self-Interaction in Cosmological Measurements. *arXiv* **2022**, arXiv:2207.03164.
115. Camarena, D.; Cyr-Racine, F.Y.; Houghteling, J. The two-mode puzzle: Confronting self-interacting neutrinos with the full shape of the galaxy power spectrum. *arXiv* **2023**, arXiv:2309.03941.
116. Cedeno, F.X.L.; Nucamendi, U. Revisiting cosmological diffusion models in Unimodular Gravity and the H0 tension. *Phys. Dark Universe* **2021**, *32*, 100807. [CrossRef]
117. Park, M.; Kreisch, C.D.; Dunkley, J.; Hadzhiyska, B.; Cyr-Racine, F.Y. Λ CDM or self-interacting neutrinos: How CMB data can tell the two models apart. *Phys. Rev. D* **2019**, *100*, 063524. [CrossRef]
118. de Cruz Pérez, J.; Park, C.G.; Ratra, B. Current data are consistent with flat spatial hypersurfaces in the Λ CDM cosmological model but favor more lensing than the model predicts. *Phys. Rev. D* **2023**, *107*, 063522. [CrossRef]
119. Handley, W. fgivenx: A Python package for functional posterior plotting. *arXiv* **2019**, arXiv:1908.01711.
120. Surendran, S.P.; Thomas, R.; Joy, M. Evolutionary optimization of cosmological parameters using metropolis acceptance criterion. *arXiv* **2022**, arXiv:2205.01752.

Article

Cosmological Constant from Boundary Condition and Its Implications beyond the Standard Model

Jan O. Stenflo [1,2]

[1] Institute for Particle Physics and Astrophysics, Eidgenössische Technische Hochschule, ETH Zurich, CH-8093 Zurich, Switzerland; stenflo@astro.phys.ethz.ch
[2] IRSOL Istituto Ricerche Solari "Aldo e Cele Daccò," Università della Svizzera Italiana, Via Patocchi 57, CH-6605 Locarno, Switzerland

Abstract: Standard cosmology has long been plagued by a number of persistent problems. The origin of the apparent acceleration of the cosmic expansion remains enigmatic. The cosmological constant has been reintroduced as a free parameter with a value in energy density units that "happens" to be of the same order as the present matter energy density. There is an internal inconsistency with regards to the Hubble constant, the so-called H_0 tension. The derived value of H_0 depends on the type of data that is used. With supernovae as standard candles, one gets a H_0 that is 4–5 σ larger than the value that one gets from CMB (Cosmic Microwave Background) data for the early universe. Here we show that these problems are related and can be solved if the cosmological constant represents a covariant integration constant that arises from a spatial boundary condition, instead of being a new type of hypothetical physical field, "dark energy", as assumed by standard cosmology. The boundary condition only applies to the bounded 3D subspace that represents the observable universe, the hypersurface of the past light cone.

Keywords: cosmological constant; cosmological models; accelerating universe; stellar ages

Citation: Stenflo, J.O. Cosmological Constant from Boundary Condition and Its Implications beyond the Standard Model. *Universe* **2023**, *9*, 103. https://doi.org/10.3390/universe9020103

Academic Editor: Kazuharu Bamba

Received: 29 January 2023
Revised: 13 February 2023
Accepted: 14 February 2023
Published: 17 February 2023

1. Introduction

While the standard model of cosmology, which is often referred to as the concordance model or the ΛCDM model, has been highly successful in the modeling of a wealth of cosmological observations, it has not been able to provide resolutions to several fundamental problems. The cosmological constant Λ had to be introduced as a free fitting parameter to allow modeling of the apparently accelerated cosmic expansion, which had been discovered unexpectedly from the observed redshift—brightness relation for standard candles in the form of supernovae type Ia [1,2]. Explanations have been sought in terms of some new field, referred to as "dark energy", a source of repulsive gravity with an equation of state, such as the vacuum energy of quantum fields [3–8]. A problem with this interpretation is that quantum field theory predicts a value of Λ that is about 54 orders of magnitude larger than the observed value (not the often quoted 122 orders of magnitude, cf. [9]). This is considered as one of the worst predictions in the history of physics. Although a variety of alternative explanations have been explored, e.g., [10,11], the cosmological constant has remained enigmatic.

In the absence of any known constraint on the value of the cosmological constant, it has been argued that any value of Λ may be physically permissible, that there may exist parallel universes where its value is different, and that only universes where the value is very small are conducive to the emergence of biological life [12]. If, however, a unique constraint condition for Λ could be identified, this "anthropic argument" would become obsolete and the multiverse option would be ruled out.

Standard cosmology has persistently stuck to the interpretation that the cosmological constant is some kind of unspecified physical field that is yet to be discovered, in spite of unsuccessful decades in search of such a hypothetical field. Internal inconsistencies also plague standard cosmology, the most prominent of which—generally referred to as the "H_0 tension"—is related to the Hubble constant, a parameter that establishes the fundamental scale for cosmological distances and for the age of the universe. While the use of supernovae type Ia as standard candles gives $H_{0,\mathrm{SN}} = 73.2 \pm 1.3\,\mathrm{km\,s^{-1}\,Mpc^{-1}}$ [13,14], the CMB analysis gives $H_{0,\mathrm{CMB}} = 67.4 \pm 0.5\,\mathrm{km\,s^{-1}\,Mpc^{-1}}$ [15]. It is a disagreement that exceeds 4σ.

There is a similar tension in another fundamental distance scale—the radius of the sound horizon r_d at the "drag epoch," when the baryons decouple from the photons. If we allow r_d to be a free fitting parameter, then BAO (Baryon Acoustic Oscillations) observations for small redshifts give $r_d \approx 136 \pm 3\,\mathrm{Mpc}$, much smaller than the CMB-derived value of $r_d = 147.2 \pm 0.3\,\mathrm{Mpc}$ [16].

The tensions in H_0 and r_d are, however, anti-correlated. There is no significant tension in the $H_0\,r_d$ product [16]. Late-time modifications of the standard model, e.g., by allowing the cosmological constant to vary, would only affect H_0 but not r_d. Early-time modifications would require new physics to be postulated beyond the standard model of particle physics, but this would still fail to resolve the problem, because a reduction of r_d would exacerbate a tension in the S_8 parameter [17]. S_8 is a measure of the matter clustering amplitude on a scale of $8/h\,\mathrm{Mpc}$ (where h is the Hubble constant in units of $100\,\mathrm{km\,s^{-1}\,Mpc^{-1}}$).

The present paper aims to show that these enigmas and inconsistencies can be resolved if one abandons the standard dark-energy interpretation of the cosmological constant and instead treats it as a covariant integration constant that arises from a spatial boundary condition that only applies to the bounded hypersurface of the past light cone, the 3D subspace that represents the observable universe. With this interpretational framework, which keeps the formalism within the realm of general relativity (in contrast to many other more or less ad hoc alternative cosmologies that have been put forward), it is possible to resolve the outstanding enigmas and to predict the numerical values of the cosmological constant and the H_0 tension, while remaining consistent with CMB data and Big Bang nucleosynthesis. Without the use of adjustable parameters, the predictions are found to agree with the observationally determined values within the observational uncertainties.

Section 2 clarifies why the Einstein field equation for gravity and cosmology must contain a covariant integration constant. Section 3 discusses the fundamental distinction between the case when the scale factor a is a function of proper time t that can be measured by comoving clocks in 4D spacetime, and when it is a function of cosmological distance r along light rays in a finite 3D subspace. Section 4 is devoted to the derivation of the expression and numerical value for the cosmological constant. Section 5 deals with two further implications beyond the standard model: the age of the universe, and a resolution of the H_0 tension. The conclusions are summarized in Section 6.

2. Why the Metric Needs to Be Constrained by a Boundary Condition

Einstein's equation for gravity is generally written with the expression for the spacetime metric on the left-hand side and the sources of matter and energy in the form of the stress-energy tensor $T_{\mu\nu}$ on the right-hand side. While mass–energy governs how spacetime should curve, the geometry determines how matter and radiation should move.

The metric is constrained by the requirement of mass–energy conservation: the covariant divergence of the stress–energy tensor must be zero. This implies that the Bianchi identities for the metric must be satisfied. The solution of such differential constraints generally requires the introduction of integration constants that need to be fixed with boundary conditions.

Integration of the Bianchi identities leads to the left-hand side of the Einstein field equation plus an integration constant that must be given in covariant form, a constant multiplied by the metric tensor $g_{\mu\nu}$, because the covariant divergence of such a term is zero. This is the form of the cosmological constant Λ term, which Einstein [18] introduced in 1917 for the mistaken purpose of obtaining a static cosmological solution.

The $g_{\mu\nu}\Lambda$ term can be placed on either the left or the right-hand side of the equation, to be considered as a degree of freedom for either the metric field or for the stress–energy tensor. Standard cosmology interprets the Λ term as a component of the stress–energy tensor $T_{\mu\nu}$ and refers to it as "dark energy." Instead of being an integration constant it is treated as a new kind of field, although there is no known physical basis for the existence of such a peculiar field (which needs to be repulsive with an equation of state $w \approx -1$).

The choice between the two interpretations of the Λ term, integration constant or physical field, is not just a matter of viewpoint but has profound consequences for both the conceptual and quantitative understanding and modeling of observational data. This makes it possible to use observations to unambiguously discriminate between the two possible choices.

3. Scale Factor as a Function of Distance

3.1. Measures of Time and Distance

Although space and time are tied together as 4D spacetime, which is the general arena where physical processes play out, there are fundamental differences between the temporal and spatial aspects in the cosmological context. Observers move in time but not in space, while light rays move in space but not in time. Cosmological observers are defined to be comoving, at rest with respect to the spatial grid. The observable universe represents the static 3D subspace along light rays that can be seen by the observer at a given proper time (which defines the tip of the past light cone).

The time scale that governs all dynamics is the proper time t that can be measured by comoving clocks. It is the time scale of stellar evolution, structure formation, and the density and temperature fluctuations of the CMB.

All information that we receive from cosmological objects comes to us in the form of radiation, electromagnetic or gravitational. While the universe may be infinite in both space and time, the part of it that is accessible to observation represents a time-frozen 3D subspace of finite volume, the 3D hypersurface of the past light cone. The distance scale r along the light cone is bounded by the Big Bang, where r equals the radius r_c of the causal or particle horizon. It is the distance where the scale factor a goes to zero and the redshift z goes to infinity. At the other end the distance scale is bounded where $r = 0$ (location of the observer), because negative distances $r < 0$ have no defined meaning.

3.2. Interpretation of Redshift Observations

A basic assumption of cosmological modeling is that the universe is spatially homogeneous and isotropic (on large scales), and that the expansion of space can be specified in terms of a scale factor $a(t)$, which only depends on the proper time t that can be measured by comoving clocks (at rest with respect to the spatial grid). The FLRW (Friedmann–Lemaître–Robertson–Walker) models represent solutions of the Einstein equation in the form of scale factor a vs. proper time t. The function $a(t)$ depends on the (spatially homogeneous) mass densities ρ_M of matter, ρ_R of radiation, and ρ_Λ (the cosmological constant when expressed in mass density units).

The function $a(t)$ is however not an observable. When we observe objects at different cosmological distances, like standard candles in the form of SN (supernovae) type Ia, we measure their redshifts z and apparent brightnesses or received energy fluxes $\mathcal{F}$, to obtain the function z vs. $\mathcal{F}$. To interpret this function in terms of the model parameters $\rho_{M,R,\Lambda}$, we need to relate the observational $z(\mathcal{F})$ function to the theoretical $a(t)$ function.

Redshift z is directly related to scale factor a via

$$a = 1/(1+z).\tag{1}$$

Because the energy flux $\mathcal{F}$ that is received from an object decreases with the square of the distance, it is related to the luminosity L of the source through

$$\mathcal{F} \equiv \frac{L}{4\,\pi\,d_L^2}, \tag{2}$$

which defines the so-called luminosity distance d_L. L can be assumed to be known for standard candles through applications of the distance ladder with its intricate calibrations, based on parallaxes for relatively nearby objects and Cepheid variable stars for intermediate distances. Equation (2) then gives us d_L from the observed brightness $\mathcal{F}$. If the universe were not expanding, d_L would equal the coordinate distance r. Because of the expansion, the relation between d_L and r depends on the scale factor a:

$$r = a(r)\,d_L. \tag{3}$$

Note that in Equation (3), the scale factor a has been written as a function of distance r instead of proper time t because t is not an observable and it does not flow along r. While time dependence with dynamical evolution can be observed for astrophysical structures on small or intermediate scales, they are unobservable on the cosmic time scale that governs the evolution of the scale factor, because the human time scale is negligible in comparison with cosmic time. Therefore, the observable universe in terms of the observed $z(\mathcal{F})$ function is a time-frozen representation of the universe at the present epoch t_0 of human observers.

Since the evolutionary cosmic time scale t is not an observable, we need a mapping of the $a(t)$ evolution (which is unbounded in the future direction) onto the bounded $(0 < r < r_c)$ distance scale to obtain the $a(r)$ function that is part of Equation (3). The problem is that the distance scale is not simply obtained with mathematical projection onto the light cone. Distances in space and time are only given a meaning in terms of a metric field, which represents a solution of the Einstein field equation for the given spacetime domain. The scale factor a is a component of the metric tensor that is obtained by solving the field equation. In the finite and static 3D subspace that represents the observable universe, the metric field can be subject to a boundary constraint that may be satisfied by a non-zero covariant integration constant, which would appear in the form of the Λ term in the equation for $a(r)$. In contrast, because 4D spacetime is unbounded, no integration constant can be induced in the equation for $a(t)$.

3.3. Meaning of the Hubble Constant

The $a(t)$ function is needed to model cosmological evolution, in particular, for numerical simulations of structure formation or for calculations of the evolution of the density and temperature fluctuations in the early universe. The most important observational signatures of this evolution are the acoustic peaks in the CMB (Cosmic Microwave Background) spectrum, which represent the anisotropies of the cosmic radiation "bath" in which we are immersed.

While the $a(t)$ function should be used for interpretations of the CMB spectrum, the $a(r)$ function is to be used for the redshift observations of supernovae. An inconsistency of standard cosmology, the so-called H_0 tension, arises when the different roles of the $a(t)$ and $a(r)$ functions are not accounted for (cf. Section 5.2).

The H_0 tension has to do with the assumption of standard cosmology that the Hubble constant H can be expressed as $\dot{a}/a$, where $\dot{a} \equiv \mathrm{d}a/\mathrm{d}t$ and t represents proper time. When the $a(r)$ function is affected by a boundary-induced cosmological constant, but the $a(t)$ function is not, the representation of the Hubble constant as $\dot{a}/a$ becomes invalid.

Historically, the Hubble constant was introduced to describe the observed linear relation between the velocity v of a galaxy that moves away from us and its distance r using the equation $v = Hr$. The measured quantity was, however, redshift z and not v. Since $v = cz$ for $v \ll c$ according to the Doppler effect, Hubble's law should be written $z = Hr/c$, or, in differential form,

$$\frac{\mathrm{d}z}{\mathrm{d}r} = H/c. \tag{4}$$

Since redshift z is directly related to scale factor a via Equation (1), one can define the Hubble constant in terms of the scale factor as

$$H \equiv -\frac{c}{a^2}\frac{\mathrm{d}a}{\mathrm{d}r} \neq \frac{\dot{a}}{a}. \tag{5}$$

The last expression, to the right of the $\neq$ sign, is the expression that standard cosmology uses for the Hubble constant. It is obtained from our definition in terms of $\mathrm{d}a/\mathrm{d}r$ if one does the substitution $\mathrm{d}r \to -c\,\mathrm{d}t/a$ (with the minus sign because t decreases when r increases). It is conceptually incorrect because $\dot{a}$ represents a time derivative with respect to proper time and depends on the $a(t)$ function, while the relevant spatial derivative depends on the $a(r)$ function. The inequality happens when Λ is different for the $a(t)$ and $a(r)$ functions.

4. Derivation of the Value for Λ

Standard FLRW cosmology is based on the assumption that there is spatial homogeneity of the universe on large scales, not only for the matter and radiation densities ρ_M and ρ_R, but also for the scale factor a. All spatial gradients are prescribed to be zero; there is only evolution. The large-scale metric is assumed to be diagonal and isotropic. In the following, we will add the assumption of spatial flatness, which makes the expressions simpler and more transparent. As observations show that the spatial curvature is zero within the observational uncertainties [15], there is no need to complicate the expressions by including the curvature term.

With these assumptions $g_{tt} = -1$ and $g_{xx} = g_{yy} = g_{zz} = a^2(t)$ in terms of Cartesian coordinates that are orthogonal to the time axis. The scale factor is thus prescribed to be independent of the spatial coordinates and depends exclusively on proper time t. The same metric can also be expressed in terms of spherical coordinates r, θ, and ϕ, as is common in FLRW cosmological modeling. Then, with spatial flatness, $g_{rr} = a(t)^2$, $g_{\theta\theta} = a(t)^2 r^2$, and $g_{\phi\phi} = a(t)^2 r^2 \sin^2\theta$.

4.1. Spatial Perturbations of the FLRW Metric

In the FLRW formulation, the 4D problem is, by assumption, reduced to a 1D problem: The large-scale evolution is fully described by a scale factor $a(t)$ that only varies with the proper time t that can be recorded by local, comoving clocks. Because there is no dependence on distance in the direction orthogonal to the time axis, the $a(t)$ function can be readily obtained as a 1D solution of the Einstein equation while accounting for energy conservation for matter and radiation.

Our aim now is to determine the scale factor as a function of distance r in the observable universe (along the past light cone). The $a(r)$ function relates the scale factor between the positions of non-local objects, which are separated from the comoving observer by different distances within a bounded r domain. The FLRW prescription for a metric of 4D spacetime that only depends on local, unbounded proper time is too restrictive and limited to be of use to account for the bounded distance scale. It becomes necessary to relax the FLRW requirement of zero spatial gradients by allowing for small-amplitude vacuum perturbations of the metric in the form of a "probe" field $h(r)$.

The metric of our minimally extended, perturbed version of FLRW has the following form (in units where $c = 1$):

$$ds^2 = -dt^2 + a(t)^2\,[1 + h(r)]\,[\,dr^2 + r^2\,(\,d\theta^2 + \sin^2\theta\,d\phi^2\,)\,], \tag{6}$$

where $h(r) \ll 1$ represents the small-amplitude probe field needed to define consistent boundary conditions, which relate to the static 3D subspace that constitutes the observable universe. h is, by design, isotropic with respect to the observer. The size of the 3D subspace in which the probe field is confined determines the scale of the fluctuations. The $a(t)$

function, which governs the evolution of the CMB spectrum and is not constrained by any boundaries, is not affected.

4.2. Solution of the Einstein Equation in the Presence of Spatial Perturbations

It is convenient to express the Einstein equation in the form

$$R_{\mu\nu} - g_{\mu\nu}\Lambda = 8\pi\,G\,S_{\mu\nu} \tag{7}$$

in units where $c = 1$.

$$S_{\mu\nu} = T_{\mu\nu} - \tfrac{1}{2}\,g_{\mu\nu}T \tag{8}$$

contains the energy–momentum sources. The metric is represented by the Ricci tensor $R_{\mu\nu}$ and the cosmological Λ term. For the signature of the metric we use the convention $(-+++)$.

For our purpose of exploring the relation between Λ and the spatial boundary conditions, it is sufficient to focus our attention on the $R_{\theta\theta}$ component of the Ricci tensor, because the expression for the other components is analogous. In first order in the perturbation h one finds

$$R_{\theta\theta} - g_{\theta\theta}\,\Lambda \approx g_{\theta\theta}\left[\frac{\ddot{a}}{a} + 2\left(\frac{\dot{a}}{a}\right)^2\right] - \tfrac{1}{2}r^2\,\nabla^2 h - g_{\theta\theta}\,\Lambda = 4\pi\,G(\rho - p)\,g_{\theta\theta}\,. \tag{9}$$

The expression for $S_{\theta\theta} = \tfrac{1}{2}(\rho - p)\,g_{\theta\theta}$ has been used, where ρ is the mass density and p the pressure.

Because the spatial and temporal coordinates in this 4D spacetime representation are orthogonal to each other, the t-dependent terms assume their values independent of the spatial $\nabla^2 h$ term. There is no boundary-induced integration constant that could be the source of Λ, because the t and r coordinates in infinite 4D spacetime are unbounded. If one refrains from postulating the existence of some "dark energy" field, then Λ in Equation (9) is zero. The time-dependent terms decouple to form an equation that is independent of the spatial perturbation:

$$\frac{\ddot{a}}{a} + 2\left(\frac{\dot{a}}{a}\right)^2 = \frac{1}{a^3}\frac{\mathrm{d}^2 a}{\mathrm{d}\eta^2} + \frac{1}{a^4}\left(\frac{\mathrm{d}a}{\mathrm{d}\eta}\right)^2 = 4\pi\,G(\rho - p)\,. \tag{10}$$

It represents the evolutionary equation for the unperturbed background—a standard Friedmann universe without a cosmological constant. The time dependence of the scale factor has been expressed in two equivalent forms, one with respect to proper time t, the other with respect to conformal time η, which is defined using $\mathrm{d}\eta \equiv \mathrm{d}t/a$.

Let us now rewrite Equation (9) by rearranging the terms and expanding the expression for $g_{\theta\theta}$:

$$(1+h)\left[\frac{1}{a^3}\frac{\mathrm{d}^2 a}{\mathrm{d}\eta^2} + \frac{1}{a^4}\left(\frac{\mathrm{d}a}{\mathrm{d}\eta}\right)^2 - 4\pi\,G(\rho - p)\right] = \frac{1}{2a(t)^2}\left(\nabla^2 h + 2a(t)^2\Lambda h\right) + \Lambda\,. \tag{11}$$

The left-hand side vanishes because of Equation (10). While this implies that the right-hand side must also vanish, it does not lead to any constraint on the magnitude of Λ. Since the 4D representation is unbounded, it does not offer the possibility of useful boundary conditions.

The situation changes when we impose the restriction to the light cone and thereby reduce the domain from infinite 4D spacetime to the bounded 3D subspace that represents the observable universe.

4.3. Restriction to the Light Cone

The observer is by definition located at the spacetime tip of the past light cone. The observable universe represents the static and bounded 3D hypersurface of this null cone. It is bounded between $r = 0$ (where the redshift $z = 0$) and $r = r_c$ (where the redshift z

becomes infinite). The distance r_c to the causal horizon increases with time, which makes r_c a function of scale factor a. For any given value of a, the coordinate distance r along a hypersurface of constant a has a limiting value $r_c(a)$ that defines the causally connected domain. Any disturbance that originates in the Big Bang can spread to a maximal distance $r_c(a)$ by the time t when the scale factor has reached the value $a(t)$.

On the light cone, all spacetime intervals ds vanish (which can be seen as an expression of the property that proper time does not flow along light rays). This implies a differential relation between the r and η coordinates. According to Equation (6), the d$s^2 = 0$ restriction implies that

$$\mathrm{d}\eta^2 = (1+h)\,\mathrm{d}r^2 \tag{12}$$

in the radial direction. Implementing this in Equation (11) gives

$$\frac{1}{a^3}\frac{\mathrm{d}^2 a}{\mathrm{d}r^2} + \frac{1}{a^4}\left(\frac{\mathrm{d}a}{\mathrm{d}r}\right)^2 - 4\pi\, G(\rho - p) - \Lambda = \frac{1}{2a(t)^2}\left[\nabla^2 h(r) + 2a(t)\,\Lambda\, h(r)\right]. \tag{13}$$

We have not included the term $-4\pi\, G(\rho - p)\, h(r)$, because h is treated as a probe field that represents vacuum fluctuations of the metric and therefore does not couple to the material sources.

Note that because d$s^2 = 0$ is a differential constraint, it only transforms the differential expressions on the left-hand side, but not the proper time t in $a(t)$ on the right-hand side. Since proper time and distance are independent coordinates, the zero points of their scales are unrelated to each other. Only when we introduce an observer (by definition at $r = 0$), who exists when the age of the universe is $t = t_0$ and the scale factor is $a(t_0)$, the equation becomes physically well-defined. Distances r only have a meaning relative to an observer.

Proper time does not vary with r in contrast to the formal "look-back time," which does not represent a dynamically relevant time scale. Therefore the scale factor a is allowed to vary with distance r on the left-hand side of the equation, while $a(t)$ on the right-hand side remains constant because t is kept fixed to represent the proper time of the tip of the light cone, which defines the r scale that is used for the left-hand side.

In the limit of vanishing spatial perturbations, h, the right-hand side of Equation (13), is zero, and we are left with an equation for the unperturbed background. It shows how the scale factor a varies with distance r from the perspective of the observer, who is located at $r = 0$ at proper time t_0. From this vantage point in spacetime, the observable universe represents a time-frozen "snapshot," a bounded 3D subspace that is embedded in the unbounded 4D universe.

The right-hand side of Equation (13) serves both to define the light cone to which the left-hand side refers and to constrain the value of Λ. The proper time t there defines the temporal location for the tip of the light cone (while the spatial location of the tip and the observer is arbitrary). For time $t = t_0$ (the present age of the universe, when human observers exist), the scale factor has the value $a_0 = a(t_0)$. The standard convention in cosmological modeling is to normalize the scale factors so that $a(t_0) \equiv 1$.

For this particular value of a, which represents our time in cosmic history, the perturbation h is confined within the bounded distance interval $0 \leq r \leq r_c(a_0)$. Equation (13) is only satisfied if the right-hand side of the equation with $a = a_0$ is zero, i.e., when

$$\nabla^2 h(r) + 2\Lambda\, h(r) = 0\,. \tag{14}$$

This equation is subject to a boundary condition, which constrains the value of Λ. Equation (9) did not admit such a possibility, because it represents unbounded 4D spacetime. It is the light cone restriction that generates a bounded 3D subspace in which a non-zero Λ can get induced. This is the reason why Equation (13) for the $a(r)$ function can contain a non-zero cosmological constant, while Equation (10) for the $a(t)$ function cannot.

Equation (14) is a 3D equation for a spherical coordinate system, but since there is only dependence on the r coordinate because of isotropy, it is more convenient to express it in a 1D form that eliminates the coordinate singularity at $r = 0$ and makes the perturbation symmetric with respect to the two boundaries (0 and r_c). This is accomplished if we replace h with ψ, where

$$\psi \equiv r\,h\,. \tag{15}$$

Inserting this in Equation (14) we get

$$\frac{\mathrm{d}^2\psi}{\mathrm{d}r^2} + 2\Lambda\,\psi = 0\,. \tag{16}$$

4.4. Implementation of the Spatial Boundary Condition to Obtain the Value for Λ

Equation (16) is satisfied for Fourier modes with wave number k_Λ and wavelength r_Λ, which are related to the cosmological constant through

$$\sqrt{2\Lambda} = k_\Lambda \equiv \frac{2\pi}{r_\Lambda}\,. \tag{17}$$

The wavelength r_Λ defines a length scale, which can be expected to be related to the only available cosmological length scale, the size r_c of the bounded 3D subspace that represents the observable universe. We may therefore expect r_Λ to be proportional to r_c with a value for the proportionality constant that gets fixed by a boundary condition.

At present the choice of boundary condition needs to be based on rather general symmetry arguments, because it does not follow from any yet identified equation. Regardless of its plausibility, a given choice will only be acceptable if it can be validated using observational data. An obvious choice is that the perturbation ψ vanishes at both boundaries. It vanishes by definition at the location of the observer ($r = 0$) because of Equation (15). A symmetric condition would then require that ψ vanishes at the other boundary as well, at $r = r_c$, the beginning of time.

The modes with the longest wavelengths (and lowest energies) that satisfy this condition are the ones for which the size r_c of the bounded r domain equals either half or a single wavelength. As shown in the next subsection, only the single-wavelength mode satisfies the observational constraints, but it does so with high precision (within 2 % or 2σ). This implies that

$$r_\Lambda = r_c\,. \tag{18}$$

In contrast, the half-wavelength choice would give a value for Λ that is four times smaller than the observed value. Other half-integer multiples would similarly be rejected using the observations by a huge margin. The observational data unambiguously direct us to the single-wavelength choice of Equation (18).

When the boundary condition that is defined by Equation (18) is inserted in Equation (17), it gives a definite, predicted value for the cosmological constant:

$$\Lambda = \Lambda_0/r_c^2\,, \tag{19}$$

where

$$\Lambda_0 = 2\,\pi^2\,. \tag{20}$$

The value of Λ is constant throughout the observable universe, for the entire domain $0 < r < r_c$, and therefore corresponds to an equation of state $w = -1$ if interpreted as an energy density. Λ varies with epoch of the observer (but not with look-back time), because r_c is a function of proper time t. This variation is, however, unobservable (because human time scales are insignificant compared to the age of the universe). If we use dimensionless spatial coordinates by expressing r in units of r_c, then Λ becomes equal to Λ_0 and remains constant, not only for all of observable space but also throughout all of cosmic history.

4.5. Observational Validation

Equation (19) represents a definite theoretical prediction for the magnitude of the cosmological constant Λ, which needs to be validated through comparison with the observationally determined value. To make this comparison explicit we first need to convert Λ to the more convenient dimensionless form Ω_Λ that is used for model fits of cosmological data and provide an explicit expression for the limiting distance scale r_c (the radius of the particle horizon).

Mathematically, the solution $a(r)$ for the r dependence of the scale factor is obtained in the same way as the solution for $a(t)$, and is

$$H(r)^2 = \left(\frac{8\pi G}{3\,c^2}\right)\left[\rho_M/a(r)^3 + \rho_R/a(r)^4\right] + \frac{c^2\Lambda}{3}\,. \tag{21}$$

The difference is that the Hubble constant is not represented by $\dot{a}/a$, but by $c\,dz/dr$ as in Equation (4) or in terms of a as in Equation (5). Further, the Λ term only appears in the expression that governs $a(r)$, not in the expression for $a(t)$, because the proper time scale is not bounded, in contrast to the r scale in Equation (21).

The next step is to introduce dimensionless parameters $\Omega_{M,R,\Lambda}$ through the definitions

$$\Omega_{M,R} \equiv \frac{8\pi G}{3\,c^2 H_0^2}\,\rho_{M,R} \tag{22}$$

and

$$\Omega_\Lambda \equiv \frac{c^2\Lambda}{3\,H_0^2}\,. \tag{23}$$

This gives us

$$H = H_0\,E(z)\,, \tag{24}$$

where

$$E(z) \equiv \sqrt{\Omega_M(1+z)^3 + \Omega_R(1+z)^4 + \Omega_\Lambda}\,. \tag{25}$$

Equations (24) and (25) imply the normalization of the $\Omega_{M,R,\Lambda}$ coefficients

$$\Omega_M + \Omega_R + \Omega_\Lambda = 1\,. \tag{26}$$

The definition of the Hubble constant as $c\,dz/dr$ (cf. Equation (4)) then allows us to calculate the distance r to an object at a given redshift z:

$$r(z) = \int dr = \frac{c}{H_0}\int_0^z \frac{dz'}{E(z')}\,. \tag{27}$$

The limiting distance r_c, the causal horizon radius, is where the redshift z becomes infinite:

$$r_c = r(\infty)\,. \tag{28}$$

It is convenient to express r_c in dimensionless form x_c, in units of the Hubble radius c/H_0:

$$x_c \equiv r_c\,H_0/c\,. \tag{29}$$

From Equation (27) we see that

$$x_c = \int_0^\infty \frac{dz}{E(z)}\,. \tag{30}$$

Inserting the expression for Λ from Equation (19) into the defining Equation (23) for Ω_Λ and using the definition for x_c from Equation (29), we get

$$\Omega_\Lambda = \frac{2}{3}\left(\frac{\pi}{x_c}\right)^2\,. \tag{31}$$

The identical expression has been obtained previously via more heuristic derivations [19,20], which are nevertheless based on the same kind of boundary condition that is being used as the source of an integration constant in the form of the Λ term.

Since x_c depends on Ω_Λ according to Equation (30), the solution cannot be expressed in closed algebraic form but needs to be obtained numerically. The apparent dependence on the density parameters Ω_M and Ω_R is not a real dependence, because the value of Ω_R is fixed by the observed temperature of the CMB radiation. The value of Ω_M then follows from the spatial flatness condition of Equation (26) for a given choice of Ω_Λ:

$$\Omega_M = 1 - \Omega_R - \Omega_\Lambda. \tag{32}$$

These relations imply that the $E(z)$ function in Equation (30) only depends on Ω_Λ but not on $\Omega_{M,R}$.

Equation (31), therefore, defines a unique value for Λ, a predicted value that is obtained without any adjustable parameters. The solution that follows from our boundary condition is $x_c \approx 3.13$ and $\Omega_\Lambda \approx 0.671$. This is to be compared with the observationally based values of $\Omega_\Lambda = 0.685 \pm 0.007$ according to Planck Collaboration et al. [15] and 0.669 ± 0.038 from analysis of supernovae data by the Dark Energy Survey project [21]. We note that the theoretical prediction agrees with the Planck value within 2 % or 2σ and is in even closer agreement with the DES value.

5. Implications beyond the Standard Model

The nearly perfect agreement between the observed and predicted values for the cosmological constant can be seen as a validation of the theory and a resolution of the long-standing problem of the origin of the Λ term. At the same time the cosmological coincidence problem is made irrelevant. Our epoch in cosmic history is not special because of the particular value that the cosmological constant is observed to have. There is no violation of the Copernican principle.

The theory has further implications beyond the standard model. Here we highlight two of them: the age of the universe and a resolution of the H_0 tension.

5.1. Age of the Universe

It is commonly believed that the age of the Universe that is given by standard cosmology, 13.8 Gyr, is known with high accuracy and is not in doubt. However, this bypasses the fundamental fact that this age value is model dependent, and that the few direct measurements that are currently available actually slightly favor a larger value.

The age of the universe is readily obtained from the solution of the scale factor a as a function of proper time t that can be measured by a comoving clock. If one uses the notations and concepts of standard cosmology, then the equation that governs this $a(t)$ function looks the same as Equation (21) that governs $a(r)$. In our case, however, the cosmological constant is only relevant for the $a(r)$ function that represents our bounded and static observable 3D subspace, not for $a(t)$, which represents the evolution of a comoving region in the 4D universe. Furthermore, it is conceptually incorrect to use notation H to represent $\dot{a}/a$, because the Hubble constant depends on $a(r)$ for our observable universe but not on proper time t, as explained in Section 3.3.

With these conceptual differences the equation that governs $a(t)$ can be expressed as

$$\dot{a}(t)^2 = a(t)^2 \left(\frac{8\pi G}{3 c^2}\right)\left[\rho_M/a(t)^3 + \rho_R/a(t)^4\right]. \tag{33}$$

The density parameters $\rho_{M,R}$ have the same meanings and identical magnitudes as the corresponding parameters in Equation (21) for the $a(r)$ function. They represent the mass densities of matter and radiation at the present epoch t_0. As before, the scale factor is assumed to be normalized to the value that it has at present: $a(t_0) \equiv 1$.

The present age of the universe, t_0, is then obtained through straightforward integration of Equation (33):

$$t_0 = \int_0^{t_0} \mathrm{d}t = \left(\frac{3\,c^2}{8\pi\,G}\right)^{1/2} \int_0^1 \frac{\mathrm{d}a}{a\,(\rho_M/a^3 + \rho_R/a^4)^{1/2}}. \tag{34}$$

The solution of this integral expression can be obtained more conveniently after the ρ parameters have been converted into the dimensionless Ω forms via the defining Equation (22). The result is

$$t_0 = \frac{1}{H_0} \int_0^1 \frac{\mathrm{d}a}{a\,(\Omega_M/a^3 + \Omega_R/a^4)^{1/2}} \approx \frac{1}{H_0\sqrt{1-\Omega_\Lambda}} \int_0^1 \sqrt{a}\,\mathrm{d}a = \frac{2}{3H_0\sqrt{1-\Omega_\Lambda}}. \tag{35}$$

Note that the present Hubble constant H_0 appears in this expression, but not because the Hubble constant has any conceptual relation to t_0. It is exclusively because H_0 is used in Equation (22) as a mathematical parameter that defines the conversion relation between the $\rho_{M,R}$ and $\Omega_{M,R}$ parameters.

The second, approximate equality in Equation (35) has been obtained by disregarding the contribution Ω_R due to radiation, which is $\ll \Omega_M$. The factor $\sqrt{1-\Omega_\Lambda}$ then appears as a consequence of Equation (32). The approximation allows us to obtain a simple algebraic expression for the age t_0 in terms of the Hubble time $1/H_0$ and the cosmological constant. Without a cosmological constant, one recovers the standard result that the age is 2/3 of the Hubble time when the universe is matter dominated. The presence of the Λ term changes this result by the factor $1/\sqrt{1-\Omega_\Lambda}$.

In our actual numerical evaluation, we have not made use of this approximation but retained the Ω_R term with its observed magnitude (obtained from the measured CMB temperature). However, the effect of Ω_R on the value of t_0 turns out to be smaller than one per mille and is thus insignificant.

The numerical solution gives $t_0 = 15.52\,\mathrm{Gyr}$. The theoretical value $\Omega_\Lambda = 0.671$ that comes from the solution of Equation (31) has been used together with the supernovae value for H_0, $73.2\,\mathrm{km\,s^{-1}\,Mpc^{-1}}$ [13,14], because it has been derived in a way that is consistent with the present cosmological framework (within which Λ is an integration constant), in contrast to the value that has been obtained through CMB parameter fitting with the standard model. This is clarified in the next subsection that deals with the resolution of the H_0 tension.

The new age value is to be compared with the age $13.80\,\mathrm{Gyr}$ that has been derived using Planck Collaboration et al. [15]. It is based on using the Planck values $\Omega_\Lambda = 0.685$ and $H_0 = 67.4 \pm 0.5\,\mathrm{km\,s^{-1}\,Mpc^{-1}}$ together with the dark-energy assumption of standard cosmology that the $a(t)$ and $a(r)$ functions are governed by the same Ω_Λ term. Including the Ω_Λ term in the solution for $a(t)$ adds an exponential contribution to the cosmic expansion that takes effect when one comes close to the present epoch. This accelerated expansion shortens the time scale and leads to a (formal) age value that is $1.7\,\mathrm{Gyr}$ shorter than the proper, dynamically relevant age.

The larger age relieves some existing tension between the age of the universe and the ages of the oldest stars. For a careful age comparison, one needs to account for the time that it took to form the first stars after the Big Bang. First-generation stars, so-called pop. III stars with essentially zero metallicity, have not yet been identified. The oldest, low-metallicity stars that have been observed are second-generation pop. II stars in the halo of our galaxy. Numerical simulations indicate that the rate of GC (globular cluster) formation may have peaked about 0.4–0.6 Gyr after the Big Bang [22,23].

A reasonable estimate for the upper stellar age limit is, therefore, $t_0 - 0.4\,\mathrm{Gyr}$ instead of t_0 by itself. It is 13.4 Gyr in the case of standard cosmology. The existence of a single star with a significantly larger reliable age would prove standard cosmology to be inconsistent.

The most prominent stellar case that seems to violate the age limit of standard cosmology is represented by the so-called "Methuselah" star HD 140283. It is classified as a pop. II halo subgiant at a distance of 202 ly with a metallicity that is 250 times less than that of the Sun. Stellar evolution modeling with isochrones gives the age $14.46 \pm 0.8\,$Gyr [24]. It violates the mentioned age limit by 1.1 Gyr, but in terms of the error bars this is little more than one σ.

It has been claimed that the ages of globular clusters (GC) support the conventional age value for the universe. A GC age of $13.32 \pm 0.5\,$Gyr has been reported [25], which represents an average over 38 clusters that have metallicity [Fe/H] < -1.5, but it was obtained after applying an ad hoc prior that excludes values larger than 15 Gyr [25]. Because there is an age distribution within the used subset, which exceeds what may be expected from the quoted measurement uncertainties, the ad hoc exclusion of values above 15 Gyr artificially truncates the intrinsic distribution. The average value that is extracted from such a distribution is therefore not representative of the upper age limit of the clusters. This gives reason to expect that the upper GC age limit lies significantly higher than the quoted value. Depending on the school of thought with respect to GC formation [26,27] the GC age limit may indeed be consistent with a significantly larger age of the universe.

These examples demonstrate that the current uncertainties in stellar age values are not yet small enough to conclusively discriminate between the two cosmological frameworks. The error bars are, however, expected to come down in the future.

A new and independent avenue for the determination of the ages of individual stars is that of asteroseismology. The detections by the CoRoT and Kepler satellite missions of oscillating modes for many thousands of stars across the HRD have provided a rich database for surveys of the age structure of the Milky Way [28]. It has been used to determine the vertical age gradient for the red giant stars in the Galactic disk [29]. While the majority of stars in the sample have ages in the range 1–6 Gyr, the tail of the age distribution extends well beyond 14 Gyr, although the sample is believed to be representative of the Galactic disk and not of the halo. We do not yet know whether the extended tail is exclusively an artefact of the uncertainties.

Fortunately, there is another testing ground, the enigmatic H_0 tension, which already allows an unambiguous discrimination between the two cosmological frameworks. The H_0 tension represents an inconsistency of standard cosmology, which disappears within our alternative framework, as explained in the next subsection.

5.2. H_0 Tension

While the tension in the Hubble constant has received much attention in recent years, it has been pointed out [16] that the H_0 tension should not be considered in isolation, because there is an anti-correlated tension in the parameter r_d. This parameter represents the radius of the sound horizon at the epoch when the baryons decouple from the photon drag force and become free to cluster and form galactic structures. There is an imprint of the r_d scale on the observed galaxy distribution because of the Baryon Acoustic Oscillations (BAO). As $r_d \approx 1.02\, r_\star$ it is closely linked to the radius $r_\star$ of the sound horizon at the epoch of hydrogen recombination, which governs the angular scale of the acoustic peaks in the CMB spectrum.

The value of $r_\star$ (as well as that of r_d) depends only on the physics of the early universe, which is almost independent of the cosmological model, because all effects of Λ and spatial curvature were insignificant before the r_d and $r_\star$ epochs. The value of $r_\star$ is therefore predominantly constrained by the standard model of particle physics and is found to be $r_\star \approx 144.4 \pm 0.3\,$Mpc, while the closely related scale $r_d \approx 147.2 \pm 0.3\,$Mpc [15]. With these values the Planck CMB modeling results in a Hubble constant $H_0 = 67.4 \pm 0.5\,\mathrm{km\,s^{-1}\,Mpc^{-1}}$, significantly smaller than $H_0 = 73.2 \pm 1.3\,\mathrm{km\,s^{-1}\,Mpc^{-1}}$ that is obtained from direct distance measurements in the nearby universe [13,14]. It is this discrepancy between H_0 as determined from the near and distant universe that is referred to as the H_0 tension.

There is, however, no significant tension for the product $H_0\, r_d$ [16]. If one allows r_d to be a free parameter rather than being fixed by known physics, the product $H_0\, r_d$ is constrained by the observed angular scale in the CMB or the BAO, but the individual factors H_0 and r_d are not. This degeneracy can be broken by combining BAO with supernovae (SN) observations. The result is a low value for r_d, typically $r_d \approx 136 \pm 3\,\text{Mpc}$ (depending somewhat on the choice of joint data sets), together with the high supernovae value for H_0. The same kind of conclusion would be obtained from analysis of CMB data, if one would allow $r_\star$ to be treated as a free parameter.

Attempts to find the origin for such a low value of r_d in terms of new physics have been unsuccessful. It has been pointed out [16] that late-time modifications (like dark energy with a varying equation of state) only affects H_0 but not r_d and therefore only makes things worse. All proposed early-time modifications (before the CMB epoch) through the introduction of new kinds of fields or modifications of general relativity can be shown to be insufficient for a resolution of the tension [16,17]. The H_0 tension may be slightly reduced by raising the value of the $\Omega_M h^2$ parameter, but the potential of this change is limited, because it would exacerbate the tension in the S_8 parameter that has been revealed by the weak lensing surveys [30,31]. $S_8 \equiv \sigma_8 \, (\Omega_M/0.3)^{1/2}$, where σ_8 is the matter clustering amplitude on the scale $8h^{-1}$ Mpc (where h is the Hubble constant in units of $100\,\text{km}\,\text{s}^{-1}\,\text{Mpc}^{-1}$).

The main CMB observable that has been used to constrain the value of the Hubble constant H_0 is the angular scale of the acoustic peaks in the observed CMB spectrum. This scale can be characterized in terms of an angle parameter $\theta_\star$ that has been determined from observations with an accuracy of about 0.03 % [15]. It is one of the best known parameters in cosmology.

$\theta_\star$ is an angular measure of the anisotropy of the CMB radiation field, in which the observer is immersed. It is a local property of the radiation field in a comoving region. The angular scale is governed by basic physics that is directly related to the radius $r_\star$ of the sound horizon at the epoch $t_\star$ of hydrogen recombination.

Parameters $\theta_\star$ and $r_\star$ do not contain information on the present value H_0 of the Hubble constant. The H_0 dependence enters when $\theta_\star$ and $r_\star$ are brought in relation with the so-called angular diameter distance $D_\star$ between the observer and the surface of last scattering via the defining equation

$$\theta_\star \equiv \frac{r_\star}{D_\star}, \tag{36}$$

where

$$D_\star = c \int_{t_\star}^{t_0} \frac{\mathrm{d}t}{a(t)}. \tag{37}$$

So far there is no difference between our treatment and that of standard cosmology. The difference enters in the explicit expression for the $a(t)$ function, which determines how $D_\star$ is calculated.

The anisotropy of the ambient radiation field in a comoving region depends on the evolution and dynamics of the local region. The $a(t)$ function in Equation (37) is therefore the scale factor in its dependence on the proper time that can be measured with a comoving clock. It does not contain any contribution from a Λ term, because such a term only enters as an integration constant when we enforce the light-cone restriction to transform the unbounded 4D problem to a problem for a bounded and static 3D subspace. When the dynamics and evolution of a comoving region are calculated, the light-cone restriction is not applied or relevant. The H_0 tension arises when Λ is not treated as an integration constant that is only relevant for the static 3D subspace but is instead treated as a new field (dark energy) that also contributes to the scale factor evolution $a(t)$ of a comoving region.

The explicit evaluation of the $D_\star$ integral is done in the same way as in Section 5.1 for the calculation of age t_0. The starting point is Equation (33), which defines the solution for the $a(t)$ function that is used in Equation (37). The explicit expression for $D_\star$ then becomes

$$D_\star = \frac{c}{H_0} \int_{a_\star}^{1} \frac{\mathrm{d}a}{a^2 \, (\Omega_M/a^3 + \Omega_R/a^4)^{1/2}}, \tag{38}$$

which is similar to that of t_0 in Equation (35) except for the extra factor a in the denominator, the factor c, and the lower integration limit $a_\star$ (at the epoch of last scattering) instead of zero (Big Bang). Note also (as in the t_0 case) that the appearance of the Hubble constant in this expression is not because H_0 has any conceptual relation to $D_\star$, but because it comes from Equation (22), where it is used to define the conversion between the ρ and Ω parameters.

If we disregard Ω_R, because it is $\ll \Omega_M$, then we get a simple approximate algebraic expression for $D_\star$ (similar to Equation (35)):

$$D_\star \approx \frac{c}{H_0\sqrt{1-\Omega_\Lambda}} \int_{a_\star}^{1} \frac{da}{\sqrt{a}} \approx \frac{2c}{H_0\sqrt{1-\Omega_\Lambda}} . \tag{39}$$

Our numerical evaluation, however, does not make use of this approximation but retains the Ω_R contribution and uses the full expression of Equation (38).

It is convenient to introduce the dimensionless parameter $x_\star$, which represents $D_\star$ in units of the Hubble radius c/H_0:

$$D_\star \equiv x_\star \frac{c}{H_0} . \tag{40}$$

Converting for mathematical convenience the a scale into a z scale via Equation (1), we get

$$x_\star = \int_0^{z_\star} \frac{dz}{[\Omega_M(1+z)^3 + \Omega_R(1+z)^4]^{1/2}} . \tag{41}$$

Note that this is a purely formal conversion, because no redshift observations are involved in the CMB analysis, and $z_\star$ is not an observable. z is here a mathematical parameter that does not represent redshift.

With Equations (36), (40) and (41) it is possible to derive the value of H_0 from the observed anisotropies $\theta_\star$ in the CMB radiation field and the radius $r_\star$ of the sound horizon through

$$H_0 = c\, x_\star\, \theta_\star / r_\star . \tag{42}$$

This results in a value for H_0 from CMB data that agrees with the value that has been derived from the redshift—brightness relation of supernovae observations (as shown explicitly below). Thus, with the current formalism, there is no H_0 tension.

Equation (42) also makes it clear why there must be an anti-correlation between the tensions in H_0 and the radius of the sound horizon $r_\star$ (or r_d), as noticed in [16]. Since $\theta_\star$ is fixed by the observations, the model dependence of the $H_0\, r_\star$ product is contained in the dimensionless $x_\star$ factor, which only depends on Ω_Λ if we assume spatial flatness. For a given value of Λ, the variations in the H_0 and $r_\star$ parameters are, therefore, anti-correlated.

The reason why CMB analysis with standard cosmology gives a different answer that is at odds with the supernovae results is that a different expression $x_{\text{st. cosm.}}$ for the dimensionless parameter $x_\star$ is used. It is based on the implicit assumption of standard cosmology that the cosmological constant Λ is a physical field that governs the behavior of both the $a(t)$ and $a(r)$ functions, rather than being a boundary condition that only affects the $a(r)$ function that represents the bounded 3D subspace.

The $\theta_\star$ and $r_\star$ parameters are governed by the local evolution of the sound waves and the anisotropy of the electromagnetic radiation field, not by a distance-related function that depends on the location of an observer. The physics is therefore determined by the $a(t)$ function as in Equation (37), not by the $a(r)$ function. For this reason, $x_\star$, as given by Equation (41), does not contain any Ω_Λ contribution. In contrast standard cosmology uses

$$x_{\text{st. cosm.}} = \int_0^{z_\star} \frac{dz}{[\Omega_M(1+z)^3 + \Omega_R(1+z)^4 + \Omega_\Lambda]^{1/2}} . \tag{43}$$

When this expression is inserted in Equation (42) instead of the expression (41) for $x_\star$, one obtains a different value for the Hubble constant:

$$H_{0,\,\text{st. cosm.}} = c\, x_{\text{st. cosm.}}\, \theta_\star / r_\star \,. \tag{44}$$

The interpretation of the redshift—brightness relation for supernovae (SN) as standard candles must, on the other hand, be based on the $a(r)$ function for scale factor vs. distance along light rays rather than on the $a(t)$ function for comoving regions. It will therefore be affected by a boundary-induced integration constant that appears in the form of a Λ term. As the existence of such a non-zero term is demanded by the observed SN relation between redshift and apparent brightness and is, in fact, being used for the interpretation of the observed relation, it is the supernovae value $H_{0,\,\text{SN}}$ that represents the correctly determined present Hubble constant. This allows us to make the identification

$$H_{0,\,\text{SN}} \equiv H_0 \,, \tag{45}$$

where H_0 is the correct value that appears in expression (42).

Because Ω_R is known from the observed temperature of the CMB radiation field, and Ω_M and Ω_Λ are related via Equation (26), both $x_\star$ and $x_{\text{st. cosm.}}$ are effectively functions of only Ω_Λ (if we disregard the minor model dependence on $z_\star$, the integration limit that represents the surface of last scattering). Dividing Equations (42) and (44) with each other and using the identification of Equation (45), we obtain an expression for the H_0 tension:

$$\frac{H_{0,\,\text{SN}}}{H_{0,\,\text{st. cosm.}}} = \frac{x_\star(\Omega_\Lambda)}{x_{\text{st. cosm.}}(\Omega_\Lambda)} \,. \tag{46}$$

With the theoretical value $\Omega_\Lambda \approx 0.671$, which was obtained through numerical solution of expression (31) for Ω_Λ, the evaluation of the full integral expressions (41) and (43) for $x_\star$ and $x_{\text{st. cosm.}}$ gives

$$\left(\frac{H_{0,\,\text{SN}}}{H_{0,\,\text{st. cosm.}}} \right)_{\text{theory}} \approx 1.099 \,. \tag{47}$$

This is to be compared with the observed value for the tension. According to the supernovae observations, $H_{0,\,\text{SN}} = 73.2 \pm 1.3\,\text{km}\,\text{s}^{-1}\,\text{Mpc}^{-1}$ [13,14], while CMB analysis in the framework of standard cosmology gives $H_{0,\,\text{st. cosm.}} = 67.4 \pm 0.5\,\text{km}\,\text{s}^{-1}\,\text{Mpc}^{-1}$ [15]. The ratio between these two values represents the observed H_0 tension:

$$\left(\frac{H_{0,\,\text{SN}}}{H_{0,\,\text{st. cosm.}}} \right)_{\text{obs}} \approx 1.086 \pm 0.021 \,. \tag{48}$$

The theoretical prediction is thus well within 1σ of the observed value.

6. Conclusions

Standard cosmology assumes that the cosmological constant represents some unknown form of physical field, commonly referred to as "dark energy." To explain the redshift—brightness observations of standard candles such a hypothetical field needs to have an equation of state that is nearly the same as the vacuum energy of quantum fields and a value, which in energy density units "happens" to be of the same order as the current matter energy density. This would suggest that our epoch in cosmic history is very special, a "cosmic coincidence." Another problem is that standard cosmology is inconsistent with respect to the Hubble constant H_0. The derived value for H_0 depends on the type of data that are being used, standard candles in the form of supernovae for the nearby universe, or CMB data for the early universe.

Here we show that these stubborn enigmas and inconsistencies go away if one abandons the assumption that Λ represents a physical field, and instead treats it as a covariant integration constant. This constant comes from a boundary condition that applies to the $a(r)$ function, the dependence of the scale factor a on distance r along light rays. While the r scale is constrained to the bounded 3D hypersurface of the past light cone, which represents the observable universe, the proper time scale t that can be measured in 4D spacetime by comoving clocks is unrelated to light cones. Consequently the $a(t)$ function cannot contain any boundary-induced integration constant.

The integration constant has the dimension of the inverse square of a length scale. The only available characteristic length scale of the system is the limiting size r_c of the observable universe. It is therefore not surprising that the value of the cosmological constant is linked to the r_c scale that characterizes the universe now. This means that our epoch is not special, there is no violation of the Copernican principle.

When the distinction between the $a(r)$ and $a(t)$ functions is accounted for, the H_0 tension goes away. While the Hubble constant that has previously been determined from the redshift—brightness observations of supernovae as standard candles remains unaffected, our revised CMB analysis gives a value for H_0 that now agrees with the supernovae value, in contrast to the CMB value that is obtained within the interpretational framework of standard cosmology.

Another consequence is that the proper age of the universe is increased by 1.7 Gy, from the 13.8 Gyr of standard cosmology to 15.5 Gyr. This is consistent with other age determinations, and, in fact, relieves some tension with ages that have been determined for the oldest halo stars in the Milky Way. The error bars of stellar ages, however, need to come down before they can be used reliably for the discrimination between the cosmological frameworks.

So is the universe accelerating or not, according to the observational data? Our answer is no, it is decelerating, because the $a(t)$ function does not contain any cosmological constant. It is the Λ-free proper time scale t that is relevant for the dynamics. Standard cosmology answers yes, because it interprets the $a(r)$ function that contains a non-zero Λ term in terms of dynamics. However, while distances r can formally be expressed in time units as "look-back time," this time scale is not relevant for dynamics and, therefore, not for the question of an accelerated expansion.

Funding: This research received no external funding.

Data Availability Statement: Not applicable.

Acknowledgments: I am grateful to Åke Nordlund for many helpful suggestions.

Conflicts of Interest: The author declares no conflict of interest.

References

1. Riess, A.G.; Filippenko, A.V.; Challis, P.; Clocchiatti, A.; Diercks, A.; Garnavich, P.M.; Gilliland, R.L.; Hogan, C.J.; Jha, S.; Kirshner, R.P.; et al. Observational Evidence from Supernovae for an Accelerating Universe and a Cosmological Constant. *Astron. J.* **1998**, *116*, 1009–1038. [CrossRef]
2. Perlmutter, S.; Aldering, G.; Goldhaber, G.; Knop, R.A.; Nugent, P.; Castro, P.G.; Deustua, S.; Fabbro, S.; Goobar, A.; Groom, D.E.; et al. Measurements of Ω and Λ from 42 High-Redshift Supernovae. *Astrophys. J.* **1999**, *517*, 565–586. [CrossRef]
3. Frieman, J.A.; Turner, M.S.; Huterer, D. Dark energy and the accelerating universe. *Ann. Rev. Astron. Astrophys.* **2008**, *46*, 385–432. [CrossRef]
4. Durrer, R. What do we really know about dark energy? *Philos. Trans. R. Soc. Lond. Series A* **2011**, *369*, 5102–5114. [CrossRef]
5. Binétruy, P. Dark energy and fundamental physics. *Astron. Astrophys. Rev.* **2013**, *21*, 67. . [CrossRef]
6. Joyce, A.; Lombriser, L.; Schmidt, F. Dark Energy Versus Modified Gravity. *Annu. Rev. Nucl. Part. Sci.* **2016**, *66*, 95–122. [CrossRef]
7. Amendola, L.; Appleby, S.; Avgoustidis, A.; Bacon, D.; Baker, T.; Baldi, M.; Bartolo, N.; Blanchard, A.; Bonvin, C.; Borgani, S.; et al. Cosmology and fundamental physics with the Euclid satellite. *Living Rev. Relativ.* **2018**, *21*, 2. [CrossRef]
8. Sabulsky, D.O.; Dutta, I.; Hinds, E.A.; Elder, B.; Burrage, C.; Copeland, E.J. Experiment to Detect Dark Energy Forces Using Atom Interferometry. *Phys. Rev. Lett.* **2019**, *123*, 61102. [CrossRef]
9. Martin, J. Everything you always wanted to know about the cosmological constant problem (but were afraid to ask). *Comptes Rendus Phys.* **2012**, *13*, 566–665. [CrossRef]
10. Padmanabhan, T.; Padmanabhan, H. Cosmic information, the cosmological constant and the amplitude of primordial perturbations. *Phys. Lett. B* **2017**, *773*, 81–85. [CrossRef]
11. Lombriser, L. On the cosmological constant problem. *Phys. Lett. B* **2019**, *797*, 134804. . [CrossRef]

12. Weinberg, S. Anthropic bound on the cosmological constant. *Phys. Rev. Lett.* **1987**, *59*, 2607–2610. PhysRevLett.59.2607. [CrossRef] [PubMed]
13. Riess, A.G.; Casertano, S.; Yuan, W.; Macri, L.; Anderson, J.; MacKenty, J.W.; Bowers, J.B.; Clubb, K.I.; Filippenko, A.V.; Jones, D.O.; et al. New Parallaxes of Galactic Cepheids from Spatially Scanning the Hubble Space Telescope: Implications for the Hubble Constant. *Astrophys. J.* **2018**, *855*, 136. [CrossRef]
14. Riess, A.G.; Casertano, S.; Yuan, W.; Bowers, J.B.; Macri, L.; Zinn, J.C.; Scolnic, D. Cosmic Distances Calibrated to 1% Precision with Gaia EDR3 Parallaxes and Hubble Space Telescope Photometry of 75 Milky Way Cepheids Confirm Tension with ΛCDM. *Astrophys. J. Lett.* **2021**, *908*, L6. [CrossRef]
15. Aghanim, N. et al. [Planck Collaboration] Planck 2018 results. VI. Cosmological parameters. *Astron. Astrophys.* **2020**, *641*, A6. [CrossRef]
16. Arendse, N.; Wojtak, R.; Agnello, A.; Chen, G.-F.; Fassnacht, C.; Sluse, D.; Hilbert, S.; Millon, M.; Bonvin, V.; Wong, K.; et al. Cosmic dissonance: Are new physics or systematics behind a short sound horizon? *Astron. Astrophys.* **2020**, *639*, A57. [CrossRef]
17. Jedamzik, K.; Pogosian, L.; Zhao, G.B. Why reducing the cosmic sound horizon alone can not fully resolve the Hubble tension. *Commun. Phys.* **2021**, *4*, 123. [CrossRef]
18. Einstein, A. Kosmologische Betrachtungen zur allgemeinen Relativitätstheorie. In *Sitzungsberichte der Königlich Preußischen Akademie der Wissenschaften*; Verlag der Königlichen Akademie der Wissenschaften: Berlin, Germany, 1917; pp. 142–152.
19. Stenflo, J.O. Dark energy as an emergent phenomenon. *arXiv* **2018**, arXiv:1901.01317. Available online: http://xxx.lanl.gov/abs/1901.01317 (accessed on 24 December 2018).
20. Stenflo, J.O. Cosmological constant caused by observer-induced boundary condition. *J. Phys. Commun.* **2020**, *4*, 105001. [CrossRef]
21. Abbott, T.M.C. et al. [DES Collaboration] First Cosmology Results using Type Ia Supernovae from the Dark Energy Survey: Constraints on Cosmological Parameters. *Astrophys. J. Lett.* **2019**, *872*, L30. [CrossRef]
22. Trenti, M.; Padoan, P.; Jimenez, R. The Relative and Absolute Ages of Old Globular Clusters in the LCDM Framework. *Astrophys. J. Lett.* **2015**, *808*, L35. [CrossRef]
23. Naoz, S.; Noter, S.; Barkana, R. The first stars in the Universe. *Mon. Not. R. Astron. Soc.* **2006**, *373*, L98–L102. [CrossRef]
24. Bond, H.E.; Nelan, E.P.; VandenBerg, D.A.; Schaefer, G.H.; Harmer, D. HD 140283: A Star in the Solar Neighborhood that Formed Shortly after the Big Bang. *Astrophys. J. Lett.* **2013**, *765*, L12. [CrossRef]
25. Valcin, D.; Bernal, J.L.; Jimenez, R.; Verde, L.; Wandelt, B.D. Inferring the age of the universe with globular clusters. *J. Cosmol. Astropart. Phys.* **2020**, *2020*, 2. [CrossRef]
26. Forbes, D.A.; Bastian, N.; Gieles, M.; Crain, R.A.; Kruijssen, J.M.D.; Larsen, S.S.; Ploeckinger, S.; Agertz, O.; Trenti, M.; Ferguson, A.M.N.; et al. Globular cluster formation and evolution in the context of cosmological galaxy assembly: Open questions. *Proc. R. Soc. Lond. Ser. A* **2018**, *474*, 20170616. [CrossRef] [PubMed]
27. Krumholz, M.R.; McKee, C.F.; Bland-Hawthorn, J. Star Clusters Across Cosmic Time. *Ann. Rev. Astron. Astrophys.* **2019**, *57*, 227–303. [CrossRef]
28. Silva Aguirre, V.; Serenelli, A.M. Asteroseismic age determination for dwarfs and giants. *Astron. Nachrichten* **2016**, *337*, 823. [CrossRef]
29. Casagrande, L.; Silva Aguirre, V.; Schlesinger, K.J.; Stello, D.; Huber, D.; Serenelli, A.M.; Schönrich, R.; Cassisi, S.; Pietrinferni, A.; Hodgkin, S.; et al. Measuring the vertical age structure of the Galactic disc using asteroseismology and SAGA. *Mon. Not. R. Astron. Soc.* **2016**, *455*, 987–1007. [CrossRef]
30. Abbott, T.M.C.; Abdalla, F.B.; Alarcon, A.; Aleksic, J.; Allam, S.; Allen, S.; Amara, A.; Annis, J.; Asorey, J.; Avila, S.; et al. Dark Energy Survey year 1 results: Cosmological constraints from galaxy clustering and weak lensing. *Phys. Rev. D* **2018**, *98*, 43526. [CrossRef]
31. Asgari, M.; Lin, C.-A.; Joachimi, B.; Giblin, B.; Heymans, C.; Hildebrandt, H.; Kannawadi, A.; Stölzner, B.; Tröster, T.; van den Busch, J.L.; et al. KiDS-1000 cosmology: Cosmic shear constraints and comparison between two point statistics. *Astron. Astrophys.* **2021**, *645*, A104. [CrossRef]

Article

Quantum Fractionary Cosmology: K-Essence Theory

J. Socorro [1,*,†] and J. Juan Rosales [2,*,†]

1   Department of Physics, Division of Science and Engineering, University of Guanajuato, Campus León, León 37150, Mexico
2   Department of Electrical Engineering, Engineering Division Campus Irapuato-Salamanca, University of Guanajuato, Salamanca 36885, Mexico
*   Correspondence: socorro@fisica.ugto.mx (J.S.); rosales@ugto.mx (J.J.R.)
†   These authors contributed equally to this work.

Abstract: Using a particular form of the quantum K-essence scalar field, we show that in the quantum formalism, a fractional differential equation in the scalar field variable, for some epochs in the Friedmann–Lemaȷ̈tre–Robertson–Walker (FLRW) model (radiation and inflation-like epochs, for example), appears naturally. In the classical analysis, the kinetic energy of scalar fields can falsify the standard matter in the sense that we obtain the time behavior for the scale factor in all scenarios of our Universe by using the Hamiltonian formalism, where the results are analogous to those obtained by an algebraic procedure in the Einstein field equations with standard matter. In the case of the quantum Wheeler–DeWitt (WDW) equation for the scalar field ϕ, a fractional differential equation of order $\beta = \frac{2\alpha}{2\alpha-1}$ is obtained. This fractional equation belongs to different intervals, depending on the value of the barotropic parameter; that is to say, when $\omega_X \in [0,1]$, the order belongs to the interval $1 \leq \beta \leq 2$, and when $\omega_X \in [-1,0)$, the order belongs to the interval $0 < \beta \leq 1$. The corresponding quantum solutions are also given.

Keywords: fractional derivative; fractional quantum cosmology; K-essence formalism

Citation: Socorro, J.; Rosales, J.J.
Quantum Fractionary Cosmology:
K-Essence Theory. *Universe* **2023**, *9*,
185. https://doi.org/10.3390/
universe9040185

Academic Editor: Kazuharu Bamba

Received: 8 March 2023
Revised: 6 April 2023
Accepted: 10 April 2023
Published: 13 April 2023

1. Introduction

In general, fractional calculus (FC) is the natural generalization of ordinary calculus [1]. That is, FC considers integrals and derivatives of non-integer order. Despite the fact that there is no fully accepted physical and geometric interpretation of fractional derivatives, during the last three decades, FC has been the subject of intense theoretical and applied studies, because different types of fractional derivatives have emerged in the scientific literature, each with its advantages and disadvantages [2]. There are many works that investigate fractional calculus and its applications [3], being a powerful mathematical tool for describing complex processes, such as the tautochrone problem [4], models based on memory mechanism [5], anomalous diffusion [6], linear capacitor theory [7], non-local description of quantum mechanics [8], processing of medical images [9,10], and so on. FC has been quite successful in many areas of science and engineering [11–13].

With the exception of some fractional models that arise naturally, for example, in ref. [7], the vast majority of models starts from an ordinary differential equation corresponding to a certain physical model. Then, it is fractionated; that is, the derivatives of the system are taken as fractional by applying any of the definitions: Riemann–Liouville, Caputo, Caputo–Fabrizio, and Atangana–Baleanu.

Recently, the FC has been applied to the general theory of relativity, as in ref. [14], and in particular, to the Friedmann–Lemaȷ̈tre–Robertson–Walker model, with interesting results in cosmology [15–29], and quantum mechanics to quantum cosmology [30,31]. These have been featured recently in the chapter "Fractional Quantum Cosmology in Challenging Routes" in a Quantum Cosmology book [32], and in the publication revising fractional cosmology [33]. In a previous paper [34], we mentioned that in the quantum formalism, applied to different epochs for the K-essence theory, we would get a fractional Wheeler–DeWitt equation in the scalar field component. Now, we report our results in this direction. Additionally, one point worth mentioning is that by employing the classical information on the barotropic parameter of the scalar field, we present a relation for using a real parameter, which can reproduce the different epochs of our Universe (inflation-like phenomenon, radiation era, stiff and dust matter) for particular values. With this parameter, we found that the scale factor is analogous to those found in a previous work for the standard matter in the same scenarios [35,36]. In this sense, we can introduce the idea that the kinetic energy of the scalar field should falsify the standard matter by employing the K-essence formalism.

Usually, K-essence models are restricted to the Lagrangian density of the form [37–42]

$$S = \int d^4x \, \sqrt{-g} \, [f(\phi) \, \mathcal{G}(X)], \tag{1}$$

where the canonical kinetic energy is given by $\mathcal{G}(X) = X = -\frac{1}{2}\nabla_\mu\phi\nabla^\mu\phi$, $f(\phi)$ is an arbitrary function of the scalar field ϕ, and g is the determinant of the metric. K-essence was originally proposed as a model for inflation, and then as a model for dark energy, along with explorations of unifying dark energy and dark matter [38,43,44]. Another motivation to consider this type of Lagrangian originates from string theory [45,46]. For more details on K-essence applied to dark energy, one can see ref. [47] and references therein.

On another front, the quantized version of this theory has not been constructed, perhaps due to the difficulties in building up the ADM formalism for it. Thus, we transform this theory to a conventional one, where the dimensionless scalar field is obtained from an energy-momentum tensor as an exotic matter component; and in this sense, we can use this structure for the quantization program, where the ADM formalism is well-known for different classes of matter [48].

This work is arranged as follows: in Section 1, we give some definitions of the fractional calculus that we employ in this work and the main ideas over the K-essence formalism, applied to obtain the classical solution to the scalar field, including the fractional parameter defined in a general way. In Section 2, we construct the Lagrangian and Hamiltonian densities for the plane FLRW cosmological model, considering a barotropic perfect fluid for the scalar field in the variable X, and present the general case. Next, the radiation particular scenario, where fractional momenta appear in the scalar field, will be used in Section 3, where we present the quantum regime for several cases of interest. Finally, Section 4 is devoted to discussions.

1.1. Basic Definitions from Fractional Calculus

Fractional derivatives are defined by means of analytical continuation of the Cauchy's formula for the multiple integral of an integer order as a single integral with a power-law kernel into the field of real order $\gamma > 0$, [1,11]. The fractional integral of order γ is written as:

$$I^\gamma f(t) = \frac{1}{\Gamma(\gamma)} \int_0^t \frac{f(\tau)}{(t-\tau)^{1-\gamma}} d\tau, \tag{2}$$

recovering an ordinary integral when $\gamma \to 1$. The Caputo fractional derivative of order $\gamma \geq 0$ of a function $f(t)$ is defined as the fractional order integral (2) of the integer order derivative (in the following, in the conventional notation, the sub-index 0 corresponds to the definition domain (0,x) by example, where other notations appear as (a,b), a+ and b-. In other words, they are the derivation limits).

$$\frac{d^\gamma}{dt^\gamma} f(t) = {}^C_0 D_t^\gamma f(t) = I^{n-\gamma} {}_0 D_t^n f(t) = \frac{1}{\Gamma(n-\gamma)} \int_0^t \frac{f^{(n)}(\tau)}{(t-\tau)^{\gamma-n+1}} d\tau, \tag{3}$$

with $n - 1 < \gamma \leq n \in \mathbb{N} = 1, 2, \ldots$, and $\gamma \in \mathbb{R}$ is the order of the fractional derivative and $f^{(n)}$ are the ordinary integer derivatives, and $\Gamma(x) = \int_0^\infty e^{-t} t^{x-1} dt$ is the gamma function. The Caputo derivative satisfies the following relations:

$$
\begin{aligned}
{}^C_0 D_t^\gamma [f(t) + g(t)] &= {}^C_0 D_t^\gamma f(t) + {}^C_0 D_t^\gamma g(t), && (4) \\
{}^C_0 D_t^\gamma c &= 0, \quad \text{where } c \text{ is a constant.} && (5)
\end{aligned}
$$

The Laplace transform of the function $f(t)$ is defined as

$$
\mathbb{L}[f(t)] = \int_0^\infty f(t) e^{-st} dt = F(s). \tag{6}
$$

Then, the Laplace transform of the Caputo fractional derivative (3) has the form [1],

$$
\mathbb{L}[{}^C_0 D_t^\gamma f(t)] = s^\gamma F(s) - \sum_{k=0}^{n-1} s^{\gamma-k-1} f^{(k)}(0), \tag{7}
$$

where $f^{(k)}$ is the ordinary derivative.

Another definition which will be used is the Mittag–Leffler generalized function [49,50] (and references therein), defined as the series expansion; in a general way, under a Maclaurin series, with $z \in \mathbb{C}$

$$
E_{\chi,\sigma}(z) = \sum_{n=0}^\infty \frac{z^n}{\Gamma(\chi n + \sigma)}, \quad (\chi > 0, \ \sigma > 0), \tag{8}
$$

and for $\sigma = 1$, we have one parametric Mittag–Leffler function

$$
E_\chi(z) = E_{\chi,1}(z) = \sum_{n=0}^\infty \frac{z^n}{\Gamma(\chi n + 1)}, \quad \chi > 0. \tag{9}
$$

Some special cases are [49,50]:

$$
\begin{aligned}
E_1(\pm z) &= e^{\pm z}, \quad E_2(z) = Cosh(\sqrt{z}), \quad E_2(-z^2) = Cos(z), \\
E_{2,2}(z^2) &= \frac{Sinh(z)}{z}, \quad E_{2,2}(-z^2) = \frac{Sin(z)}{z}.
\end{aligned} \tag{10}
$$

Laplace transform of the Mittag–Leffler function is given by Equation (1)

$$
\int_0^\infty e^{-st} t^{\chi m + \sigma - 1} E_{\chi,\sigma}^{(m)}(\pm a t^\chi) dt = \frac{m! \, s^{\chi - \sigma}}{(s^\chi \mp a)^{m+1}}. \tag{11}
$$

Consequently, the inverse Laplace transform is

$$
\mathbb{L}^{-1}\left[\frac{m! \, s^{\chi - \sigma}}{(s^\chi \mp a)^{m+1}} \right] = t^{\chi m + \sigma - 1} E_{\chi,\sigma}^{(m)}(\pm a t^\chi). \tag{12}
$$

This expression is very useful for obtaining analytical solutions.

1.2. K-Essence Theory

One of the simplest K-essence Lagrangian densities is

$$
\mathcal{L}_{geo} = (R + f(\phi)\mathcal{G}(X)), \tag{13}
$$

where R is the scalar of curvature, $f(\phi)$ and $\mathcal{G}(X)$ have been defined before. Then, the field equations are given by

$$G_{\mu\nu} + f(\phi)\left[\mathcal{G}_X \phi_{,\mu} \phi_{,\nu} + \mathcal{G} g_{\mu\nu}\right] = T_{\mu\nu}, \tag{14}$$

$$f(\phi)\left[\mathcal{G}_X \phi^{\nu}_{;\nu} + \mathcal{G}_{XX} X_{;\nu} \phi^{\nu}\right] + \frac{df}{d\phi}\left[\mathcal{G} - 2X\mathcal{G}_X\right] = 0, \tag{15}$$

where we have assumed the units with $8\pi G = 1$ and, as usual, the semicolon means a covariant derivative, and the subscript X denotes differentiation with respect to X. (Equations (14) and (15) are deduced in Appendix A).

The same set of Equations (14) and (15) is obtained if we consider the scalar field $X(\phi)$ as part of the matter content, to say $\mathcal{L}_{X,\phi} = f(\phi)\mathcal{G}(X)$, with the corresponding energy-momentum tensor

$$\mathcal{T}_{\mu\nu} = f(\phi)\left[\mathcal{G}_X \phi_{,\mu} \phi_{,\nu} + \mathcal{G}(X) g_{\mu\nu}\right]. \tag{16}$$

Additionally, considering the energy-momentum tensor of a barotropic perfect fluid,

$$T_{\mu\nu} = (\rho + P)u_\mu u_\nu + P g_{\mu\nu}, \tag{17}$$

with u_μ being the four-velocity satisfying the relation $u_\mu u^\mu = -1$, ρ the energy density, and P the pressure of the fluid. For simplicity, we consider a co-moving perfect fluid. The pressure and energy density, corresponding to the energy momentum tensor of the field X, are

$$P(X) = f(\phi)\mathcal{G}, \qquad \rho(X) = f(\phi)\left[2X\mathcal{G}_X - \mathcal{G}\right]; \tag{18}$$

thus, the barotropic parameter $\omega_X = \frac{P(X)}{\rho(X)}$ for the equivalent fluid is

$$\omega_X = \frac{\mathcal{G}}{2X\mathcal{G}_X - \mathcal{G}}. \tag{19}$$

Notice that the case of a constant barotropic index ω_X, (with the exception when $\omega_X = 0$) can be obtained by the $\mathcal{G}$ function

$$\mathcal{G} = X^{\frac{1+\omega_X}{2\omega_X}}. \tag{20}$$

Choosing the barotropic parameter as

$$\omega_X = \frac{2\kappa - 1}{2\kappa + 1}, \qquad \rightarrow \qquad \mathcal{G} = X^\alpha, \tag{21}$$

where the α parameter

$$\alpha = \frac{2\kappa}{2\kappa - 1}, \tag{22}$$

is relevant in our approach. Thus, we can write the barotropic parameter in terms of $\omega_X = \frac{1}{2\alpha - 1}$, when $\kappa = \frac{\alpha}{2(\alpha - 1)}$. With this, we can write the states in the evolution of our Universe as:

$$\left\{ \begin{array}{llll}
\text{stiff matter}: & \kappa \to \infty, & \omega_X = 1, \to \mathcal{G}(X) = X. \\
\text{Radiation}: & \kappa = 1, & \omega_X = \frac{1}{3}, \to \mathcal{G}(X) = X^2. \\
\text{Radiation like}: & \kappa = \frac{5}{4}, & \omega_X = \frac{2}{3}, \to \mathcal{G}(X) = X^{5/2}. \\
\text{such as dust like}: & \kappa \to \frac{1}{2}, & \omega_X \to 0, \to \mathcal{G}(X) = X^m, \quad m \to \infty. \\
\text{inflation}: & \kappa = 0, & \omega_X = -1, \to \mathcal{G}(X) = 1, \ f(\phi) = \Lambda = constant. \\
\text{inflation such as} & \kappa = \frac{1}{4}, & \omega_X = -\frac{1}{3}, \to \mathcal{G}(X) = \frac{1}{X} \\
& \kappa = \frac{1}{10}, & \omega_X = -\frac{2}{3}, \to \mathcal{G}(X) = \frac{1}{\sqrt[4]{X}}.
\end{array} \right.$$

The classical and quantum solutions for the stiff matter case $\omega_X = 1$ with the function $\mathcal{G}(X) = X$ were treated in the refs. [34,37], considering anisotropic cosmologies, which are the standard quintessence, such as $f(\phi) = constant$. For the inflation phenomenon, we chose the particular value for the cosmological constant function $f(\phi) = \Lambda$. The original Einstein field Equations (14) were reduced to the traditional problem with the cosmological constant with exponential time behavior for the scale factor [35].

It is clear that the stiff-matter case falls into the traditional treatment of quintessence cosmology, and in the other cases, a Hamiltonian density with a fractional momentum in the scalar field; then, the quantum Wheeler–DeWitt equation appears as a fractional differential equation. In the ref. [51], the authors present the classical analysis of the radiation era by using dynamic systems and obtaining rebound solutions.

In the following, by choosing the generic formula of the barotropic parameter $\omega_X = \frac{1}{2\alpha-1}$, we obtain the classical and quantum solutions.

1.3. Classical Cosmological FLRW Model, $f(\phi) = Constant$

The space-time background to be considered is the spatially flat FLRW with element

$$ds^2 = -N(t)^2 dt^2 + A^2(t)\left[dr^2 + r^2(d\theta^2 + sin^2\theta d\phi^2)\right], \tag{23}$$

where $N(t)$ represents the lapse function, $A(t) = e^{\Omega(t)}$ is the scale factor in the Misner parametrization, and Ω is a scalar function, whose interval is $(-\infty, \infty)$. If we consider the cosmological FLRW model, then the Equation (15) is written as (we use $\prime = \frac{d}{d\tau} = \frac{d}{Ndt}$, so $g_{\tau\tau} = -1$),

$$[\mathcal{G}_X + 2X\mathcal{G}_{XX}]X' + 6\frac{A'}{A}X\mathcal{G}_X = 0, \tag{24}$$

with the exact solution

$$X\mathcal{G}_X^2 = \eta A^{-6}, \tag{25}$$

where A is the scale factor of the cosmological FLRW model, and η is an integration constant, which is linked to the parameters of matter in the Universe epoch in study. This solution has been known for some time and was found by different authors [41,52,53]. (Equations (24) and (25) have been deduced in Appendix B).

In the following, we present the generic case, $\omega_X = \frac{1}{2\alpha-1}$, given by $\mathcal{G} = X^\alpha$ and substituting into (25), we have $\alpha^2 X^{2\alpha-1} = \eta A^{-6}$ with $X = \frac{1}{2}\left(\frac{d\phi}{d\tau}\right)^2$, obtaining for the scalar field ϕ the equation

$$\frac{d\phi}{d\tau} = \sqrt{2}\left[\left(\frac{\eta}{\alpha^2}\right)^{\frac{1}{2(2\alpha-1)}} A^{-\frac{3}{2\alpha-1}}\right] = \sqrt{2}\left[\left(\frac{\eta}{\alpha^2}\right)^{\frac{1}{2(2\alpha-1)}} e^{-\frac{3}{2\alpha-1}\Omega}\right], \tag{26}$$

whose solution is

$$\Delta\phi = \sqrt{2}\left(\frac{\eta}{\alpha^2}\right)^{\frac{1}{2(2\alpha-1)}} \int A^{-\frac{3}{2\alpha-1}} d\tau = \sqrt{2}\left(\frac{\eta}{\alpha^2}\right)^{\frac{1}{2(2\alpha-1)}} \int e^{-\frac{3}{2\alpha-1}\Omega} d\tau, \tag{27}$$

which is dependent on the scale factor; it was obtained by using the relationship between the momenta and the Hamiltonian density constraint from the Lagrangian density, in the usual way.

2. Hamiltonian Cosmological Models

Introducing the line element (23) in Misner's parametrization, the Ricci scalar becomes $R = -6\frac{\ddot{\Omega}}{N^2} - 12\left(\frac{\dot{\Omega}}{N}\right)^2 + 6\frac{\dot{\Omega}\dot{N}}{N^3}$ and $\sqrt{-g} = Ne^{3\Omega}$; then, the total Lagrangian density (13) for the generic-like Universe is

$$\mathcal{L} = e^{3\Omega}\left[-6\frac{\ddot{\Omega}}{N} - 12\frac{(\dot{\Omega})^2}{N} + 6\frac{\dot{\Omega}\dot{N}}{N^2} - \left(\frac{1}{2}\right)^{\alpha}(\dot{\phi})^{2\alpha}N^{-2\alpha+1}\right]. \tag{28}$$

Thus, by using the total time derivative $\left(-6e^{3\Omega}\frac{\dot{\Omega}}{N}\right)^{\bullet} = -6e^{3\Omega}\frac{\ddot{\Omega}}{N} - 18\,e^{3\Omega}\frac{(\dot{\Omega})^2}{N} + 6e^{3\Omega}\frac{\dot{\Omega}\dot{N}}{N^2}$ in (28), we obtain

$$\mathcal{L} = e^{3\Omega}\left[6\frac{\dot{\Omega}^2}{N} - \left(\frac{1}{2}\right)^{\alpha}(\dot{\phi})^{2\alpha}N^{-2\alpha+1}\right]. \tag{29}$$

Using the standard definition of the momenta $\Pi_{q^{\mu}} = \frac{\partial \mathcal{L}}{\partial \dot{q}^{\mu}}$, where $q^{\mu} = (\Omega, \phi)$, we obtain

$$\Pi_{\Omega} = \frac{12}{N}e^{3\Omega}\dot{\Omega}, \quad \rightarrow \quad \dot{\Omega} = \frac{N}{12}e^{-3\Omega}\Pi_{\Omega}, \tag{30}$$

$$\Pi_{\phi} = -\left(\frac{1}{2}\right)^{\alpha}\frac{2\alpha}{N^{2\alpha-1}}e^{3\Omega}\dot{\phi}^{2\alpha-1}, \quad \rightarrow \quad \dot{\phi} = -N\left[\frac{2^{\alpha-1}}{\alpha}e^{-3\Omega}\Pi_{\phi}\right]^{\frac{1}{2\alpha-1}}, \tag{31}$$

and introducing them into the Lagrangian density, we obtain the canonical Lagrangian $\mathcal{L}_{canonical} = \Pi_{q^{\mu}}\dot{q}^{\mu} - N\mathcal{H}$ as

$$\mathcal{L}_{canonical} = \Pi_{q^{\mu}}\dot{q}^{\mu} - \frac{N}{24}e^{-\frac{3}{2\alpha-1}\Omega}\left\{e^{-\frac{6(\alpha-1)}{2\alpha-1}\Omega}\Pi_{\Omega}^2 - \frac{12(2\alpha-1)}{\alpha}\Pi_{\phi}^{\frac{2\alpha}{2\alpha-1}}\right\}. \tag{32}$$

Performing the variation with respect to the lapse function N, $\delta\mathcal{L}_{canonical}/\delta N = 0$, the Hamiltonian constraint $\mathcal{H} = 0$ is obtained, where the classical density is written as

$$\mathcal{H} = \frac{1}{24}e^{-\frac{3}{2\alpha-1}\Omega}\left\{e^{-\frac{6(\alpha-1)}{2\alpha-1}\Omega}\Pi_{\Omega}^2 - \frac{12(2\alpha-1)}{\alpha}\left(\frac{2^{\alpha-1}}{\alpha}\right)^{\frac{1}{2\alpha-1}}\Pi_{\phi}^{\frac{2\alpha}{2\alpha-1}}\right\}. \tag{33}$$

In this point, we return to the equation of the scalar field (27) writing $d\tau = Ndt$

$$\Delta\phi = \sqrt{2}\left(\frac{\eta}{\alpha^2}\right)^{\frac{1}{2(2\alpha-1)}}\int e^{-\frac{3}{2\alpha-1}\Omega}Ndt, \tag{34}$$

and considering the gauge $N = 24e^{\frac{3}{2\alpha-1}\Omega}$; the classical scalar field goes like

$$\phi(t) = \phi_i(t_i) + 24\sqrt{2}\left(\frac{\eta}{\alpha^2}\right)^{\frac{1}{2(2\alpha-1)}}(t - t_i), \tag{35}$$

where t_i is the initial time for generic epoch and $\phi(t_i)$ is the scalar field evaluated in this time. In this way, the scalar field is present in the following epochs in our Universe. However, when we use the equation for momentum (31) in time τ, we have $\frac{d\phi}{d\tau} = -\left(\frac{2^{\alpha-1}}{\alpha}e^{-3\Omega}\Pi_{\phi}\right)^{\frac{1}{2\alpha-1}}$, and using the first time derivative of the scalar field (26), we obtain $\Pi_\phi^{\frac{1}{2\alpha-1}} = -\sqrt{2}\left(\frac{\eta}{2^{2(\alpha-1)}}\right)^{\frac{1}{2(2\alpha-1)}}$. Plugging this back into the Hamiltonian constraint, we fiind that the momenta in the variable Ω become

$$\Pi_\Omega = 2\sqrt{\frac{6(2\alpha-1)}{\alpha}}\left(\frac{2^{\alpha-1}}{\alpha}\right)^{\frac{1}{2(2\alpha-1)}}\left(\frac{\eta}{2^{2(\alpha-1)}}\right)^{\frac{\alpha}{2(2\alpha-1)}} e^{\frac{3(\alpha-1)}{2\alpha-1}\Omega}.$$

Now, using Equation (30) at time τ, we find that the scale factor becomes

$$A(\tau) = \left[\frac{\alpha}{2(2\alpha-1)}\sqrt{\frac{6(2\alpha-1)}{\alpha}}\left(\frac{2^{\alpha-1}}{\alpha}\right)^{\frac{1}{2(2\alpha-1)}}\left(\frac{\eta}{2^{2(\alpha-1)}}\right)^{\frac{\alpha}{2(2\alpha-1)}}(\tau-\tau_i)\right]^{\frac{2\alpha-1}{3\alpha}}, \tag{36}$$

which is consistent with the result obtained in the ref. [35], Equations (6) and (34), in the time τ, for ordinary matter $p = \omega\rho$. When we substitute the barotropic parameter $\omega = \frac{1}{2\alpha-1}$ in Equation (34) of the paper [35], we obtain the power law in the time τ, resulting in the same behavior as in (36). In this sense, we mention that the kinetic energy of the scalar field in the k-essence formalism falsifies the standard matter.

In the following, we will place all our effort in solving the quantum fractionary Wheeler–DeWitt equation.

3. Quantum Regime

The WDW equation for these models is obtained by making the usual substitution $\Pi_{q^\mu} = -i\hbar\partial_{q^\mu}$ into (33) and promoting the classical Hamiltonian density in the differential operator, applied to the wave function $\Psi(\Omega,\phi)$, $\hat{\mathcal{H}}\Psi = 0$. Then, we have

$$-\hbar^2 e^{-\frac{6(\alpha-1)}{2\alpha-1}\Omega}\frac{\partial^2\Psi}{\partial\Omega^2} - \frac{12(2\alpha-1)}{\alpha}\hbar^{\frac{2\alpha}{2\alpha-1}}\left(\frac{2^{\alpha-1}}{\alpha}\right)^{\frac{1}{2\alpha-1}}\frac{\partial^{\frac{2\alpha}{2\alpha-1}}}{\partial\phi^{\frac{2\alpha}{2\alpha-1}}}\Psi = 0. \tag{37}$$

We noted that the fractional differential equation with degree $\beta = \frac{2\alpha}{2\alpha-1}$ belongs to different intervals, depending on the value of the barotropic parameter; that is, when $\omega_X \in [0,1]$, the degree belongs to the interval $[1,2]$, and when $\omega_X \in [-1,0)$, the degree belongs to the interval $[0,1)$, for the scalar field ϕ (for this calculation, we remember that $\alpha = \frac{1}{2}\left(1 + \frac{1}{\omega_X}\right)$). It is well-known in standard quantum cosmology that the best candidates for quantum solutions are those that have a damping behavior with respect to the scale factor; then, we use this conjecture in this formalism.

For simplicity, the factor $e^{-\frac{6(\alpha-1)}{2\alpha-1}\Omega}$ may be the factor ordered with $\hat{\Pi}_\Omega$ in many ways. Hartle and Hawking [54] suggested what might be called semi-general factor ordering, which, in this case, would order the terms $e^{-\frac{6(\alpha-1)}{2\alpha-1}\Omega}\hat{\Pi}_\Omega^2$ as $-e^{-(\frac{6(\alpha-1)}{2\alpha-1}-Q)\Omega}\partial_\Omega e^{-Q\Omega}\partial_\Omega = -e^{-\frac{6(\alpha-1)}{2\alpha-1}\Omega}\partial_\Omega^2 + Q\,e^{-\frac{6(\alpha-1)}{2\alpha-1}\Omega}\partial_\Omega$, where Q is any real constant that measures the ambiguity in the factor ordering in the variables Ω and its corresponding momenta. We will assume in the following that this factor ordering for the Wheeler–DeWitt equation becomes

$$-\hbar^2 e^{-\frac{6(\alpha-1)}{2\alpha-1}\Omega}\frac{\partial^2\Psi}{\partial\Omega^2} + Q\hbar^2 e^{-\frac{6(\alpha-1)}{2\alpha-1}\Omega}\frac{\partial\Psi}{\partial\Omega} - \frac{12(2\alpha-1)}{\alpha}\hbar^{\frac{2\alpha}{2\alpha-1}}\left(\frac{2^{\alpha-1}}{\alpha}\right)^{\frac{1}{2\alpha-1}}\frac{\partial^{\frac{2\alpha}{2\alpha-1}}}{\partial\phi^{\frac{2\alpha}{2\alpha-1}}}\Psi = 0, \tag{38}$$

which when written in terms of the β parameter, becomes

$$-\hbar^2 e^{-3(2-\beta)\Omega}\frac{\partial^2\Psi}{\partial\Omega^2} + Q\hbar^2 e^{-3(2-\beta)\Omega}\frac{\partial\Psi}{\partial\Omega} - \frac{24}{\beta}\left(\frac{2^{\alpha-1}}{\alpha}\right)^{\frac{1}{2\alpha-1}}\hbar^\beta\frac{\partial^\beta}{\partial\phi^\beta}\Psi = 0. \tag{39}$$

By employing the separation variables method for the wave function $\Psi = \mathcal{A}(\Omega)\,\mathcal{B}(\phi)$, we have the following two differential equations for (Ω, ϕ)

$$\frac{d^2\mathcal{A}}{d\Omega^2} - Q\frac{d\mathcal{A}}{d\Omega} \mp \frac{\mu^2}{\hbar^2}e^{3(2-\beta)\Omega}\mathcal{A} = 0, \tag{40}$$

$$\frac{d^\beta\mathcal{B}_\pm}{d\phi^\beta} \pm \left(\frac{\alpha}{2^{\alpha-1}}\right)^{\frac{1}{2\alpha-1}}\frac{\mu^2\,\beta}{24\hbar^\beta}\mathcal{B}_\pm = 0, \tag{41}$$

where $\mathcal{B}_\pm$ considers the sign in the differential equation. The fractional differential Equation (41) can be given in the fractional frameworks, following [55,56] and identifying $\gamma = \frac{\beta}{2} = \frac{\alpha}{2\alpha-1}$, where now, γ is the order of the fractional derivative taking values in $0 < \gamma \le 1$; then, we can write

$$\frac{d^{2\gamma}\mathcal{B}_\pm}{d\phi^{2\gamma}} \pm \left(\frac{\alpha}{2^{\alpha-1}}\right)^{\frac{1}{2\alpha-1}}\frac{\gamma\mu^2}{12\hbar^{2\gamma}}\mathcal{B}_\pm = 0, \qquad 0 < \gamma \le 1, \tag{42}$$

the solution of the Equation (42) with a positive sign may be obtained by applying direct and inverse Laplace transforms [56], providing

$$\mathcal{B}_+(\phi,\gamma) = \mathbb{E}_{2\gamma}\left(-z^2\right), \qquad z = \left(\frac{\alpha}{2^{\alpha-1}}\right)^{\frac{1}{2(2\alpha-1)}}\frac{\sqrt{\gamma}\mu}{2\sqrt{3}\hbar^\gamma}\phi^\gamma, \qquad 0 < \gamma \le 1. \tag{43}$$

In the ordinary case, $\gamma = 1$; then, the solution is [56],

$$\mathcal{B}_+(\phi,1) = \mathbb{E}_2\left[-\left(\frac{\mu}{2\sqrt{3}\hbar}(\phi - \phi_0)\right)^2\right] = cos\left(\frac{\mu}{2\sqrt{3}\hbar}(\phi - \phi_0)\right), \tag{44}$$

which is in agreement with the Equation (10), employing the Taylor series.

Following the book of Polyanin [57] (page 179.10), we discovered the solution for the first equation, considering different values in the factor ordering parameter (we take the corresponding sign minus in the constant μ^2)

$$\mathcal{A} = A_0\,e^{\frac{Q\Omega}{2}}\,Z_\nu\left[\frac{2\mu}{3\hbar(2-\beta)}\sqrt{-1}\,e^{\frac{3(2-\beta)}{2}\Omega}\right] = A_0\,e^{\frac{Q\Omega}{2}}\,K_\nu\left[\frac{\mu}{3\hbar(1-\gamma)}e^{3(1-\gamma)\Omega}\right], \tag{45}$$

with order $\nu = \pm\frac{Q}{6(1-\gamma)}$, where we had written the second expression in terms of the fractional order $\gamma = \frac{\beta}{2}$, and the solutions which become dependent on the sign of its argument; when $\sqrt{1}$ (for $\mathcal{B}_-$), the Bessel function Z_ν becomes the ordinary Bessel function J_ν. When $\sqrt{-1}$ (for $\mathcal{B}_+$), this becomes the modified Bessel function K_ν. For the particular values $\beta = 2$ ($\gamma = 1$), it will be necessary to solve the original differential equation for this variable.

Then, we have the probability density $|\Psi|^2$ by considering only $\mathcal{B}_+$, $\gamma \neq 1$,

$$|\Psi|^2 = \psi_0^2\,e^{Q\Omega}\,\mathbb{E}_{2\gamma}^2\left(-z^2\right)\,K_\nu\left[\frac{\mu}{3\hbar(1-\gamma)}e^{3(1-\gamma)\Omega}\right], \qquad z = \left(\frac{\alpha}{2^{\alpha-1}}\right)^{\frac{1}{2(2\alpha-1)}}\frac{\sqrt{\gamma}\mu}{2\sqrt{3}\hbar^\gamma}\phi^\gamma. \tag{46}$$

On the other hand, it is well-known that in standard quantum cosmology, the wave function is unnormalized. There is no systematic method to do this, as the Hamiltonian density is not Hermitian. In particular cases, wave packets can be constructed, and from these wave packets we can construct a normalized wave function. In this work, we could not construct these wave packets. We hope to be able to do it in future studies.

In the following, we present particular cases in the evolution of the Universe and some plots by employing the Equation (46), and for better viewing in the plots, we introduce by hand particular values to the constant ψ_0.

1. Radiation epoch, $\omega_X = \frac{1}{3}$, $\alpha = 2$, $\rightarrow \beta = \frac{4}{3} \rightarrow \gamma = \frac{2}{3}$.

When we choose the radiation case, (46) is written as

$$|\Psi|^2 = \psi_0^2\, e^{Q\Omega}\, \mathbb{E}_{\frac{4}{3}}^2\left(-z^2\right)\, K_{\frac{Q}{2}}^2\left[\frac{\mu}{\hbar}e^{\Omega}\right], \quad z = \frac{\mu}{3\sqrt{2}\hbar^{\frac{2}{3}}}\phi^{\frac{2}{3}}. \tag{47}$$

In the following Figure 1, we take the probability density (47); in the first and second Figures, and for better viewing in the plots, we take the constant $\psi_0 = \frac{1}{10}$, and in the third Figure the value becomes 1. In all Figures, the behavior of the probability density, in both variables (Ω, ϕ), has the appropriate decadent behavior. The range of the variable equals to $\phi \in [0, 3000], [0, 200]$, and $[0, 40]$, respectively.

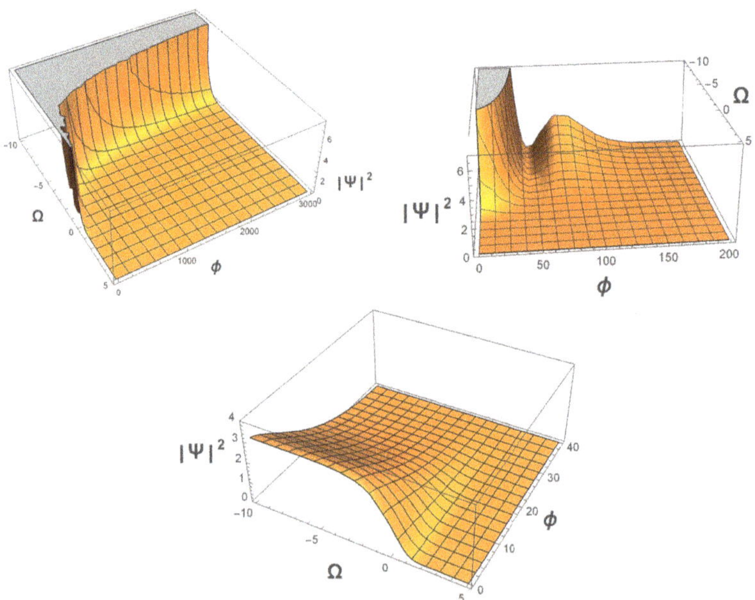

Figure 1. In the radiation era, we plot the Equation (47), considering different values in the order parameter $Q = -1, 1, 0$, from top to bottom, respectively. We consider the value for the parameter $\mu = 0.5$; we discard other values in the Q parameter.

2.   Solution to $\omega_X = \frac{2}{3}, \alpha = \frac{5}{4}, \rightarrow \beta = \frac{5}{3} \rightarrow \gamma = \frac{5}{6}$.

The probability density of the wave function becomes (here, $z = \left(\frac{5}{4\sqrt[4]{2}}\right)^{\frac{1}{3}} \frac{\sqrt{5}\mu}{6\sqrt{2}\hbar^{\frac{5}{6}}}\phi^{\frac{5}{6}}$)

$$\Psi^2 = \psi_0^2\, e^{Q\Omega}\mathbb{E}_{\frac{5}{3}}^2\left(-z^2\right)\, K_Q^2\left[\frac{\mu}{\hbar}e^{\frac{1}{2}\Omega}\right]. \tag{48}$$

In the Figure 2, we take the probability density (48); in the first and second Figures, and for better viewing in the plots, we take the constant $\psi_0 = \frac{1}{\sqrt{10}}$, and in the third Figure the value becomes 1. In all Figures, the behavior of the probability density, in both variables (Ω, ϕ), has the appropriate decadent behavior, and it presents an oscillatory behavior when $\omega_X \rightarrow 1$, since that is the behavior according to the Equation (44). Only for $Q = -1$, the probability density has a moderate increase in the direction where the scalar field evolves.

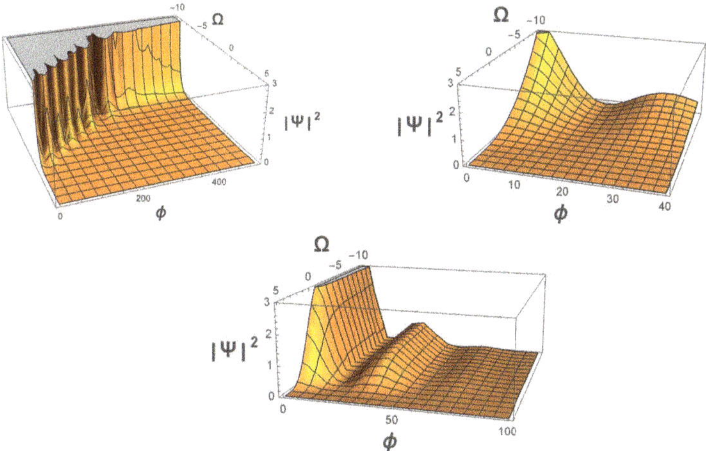

Figure 2. In the radiation-like era, we plot different combinations of the Equation (48), considering different values in the order parameter $Q = -1, 0, 1$, from top to bottom, respectively. We consider the value for the parameter $\mu = 0.5$; we discard other values in the Q parameter.

3. Dust era, $\omega_X = 0$, $\alpha \to \infty$; thus, $\beta = 1 \to \gamma = \frac{1}{2}$. In the dust case, the solution for the scale factor becomes

$$\mathcal{A} = A_0 \, e^{\frac{Q\Omega}{2}} \, Z_\nu \left[\frac{\mu}{\hbar} \sqrt{\pm 1} e^{\frac{3}{2}\Omega} \right], \qquad \nu = \pm \frac{Q}{2}. \tag{49}$$

In this case, the fractional differential Equation (41) for the scalar field is reduced to the first-order differential equation (for both signs in μ^2)

$$\frac{d\mathcal{B}_\mp}{d\phi} \mp \left(\frac{\alpha}{2^{\alpha-1}} \right)^{\frac{1}{2\alpha-1}} \frac{\mu^2}{24\hbar} \mathcal{B}_\mp = 0, \qquad \to \qquad \mathcal{B}_\mp = \beta_\mp \, e^{\pm \left(\frac{\alpha}{2^{\alpha-1}} \right)^{\frac{1}{(2\alpha-1)}} \frac{\mu^2}{24\hbar} \Delta\phi}.$$

Then, the probability density of the wave function becomes

$$\Psi^2 = \psi_0^2 \begin{cases} e^{\left(Q\Omega + \left(\frac{\alpha}{2^{\alpha-1}} \right)^{\frac{1}{(2\alpha-1)}} \frac{\mu^2}{12\hbar} \Delta\phi \right)} J_{\frac{Q}{3}}^2 \left[\frac{\mu}{\hbar} e^{\frac{3}{2}\Omega} \right] \\[2ex] e^{\left(Q\Omega - \left(\frac{\alpha}{2^{\alpha-1}} \right)^{\frac{1}{(2\alpha-1)}} \frac{\mu^2}{12\hbar} \Delta\phi \right)} K_{\frac{Q}{3}}^2 \left[\frac{\mu}{\hbar} e^{\frac{3}{2}\Omega} \right] \end{cases} \tag{50}$$

In the following Figure 3, we present the behavior of the probability density Ψ^2 by using the Equation (50) and taking the values for the order parameter $Q = -1, 0, 1$, because with these values, the probability density presents a structure well-defined for this era. In some of them, one structure did not appear; thus, we gave it a profile for the probability density for particular values in the scalar field. In these cases, the behavior of our Universe is quite selective in this formalism. Additionally, we can notice that the probability density has a moderate increase in the direction where the scalar field evolves. Similar results were reported in other formalisms [58–60].

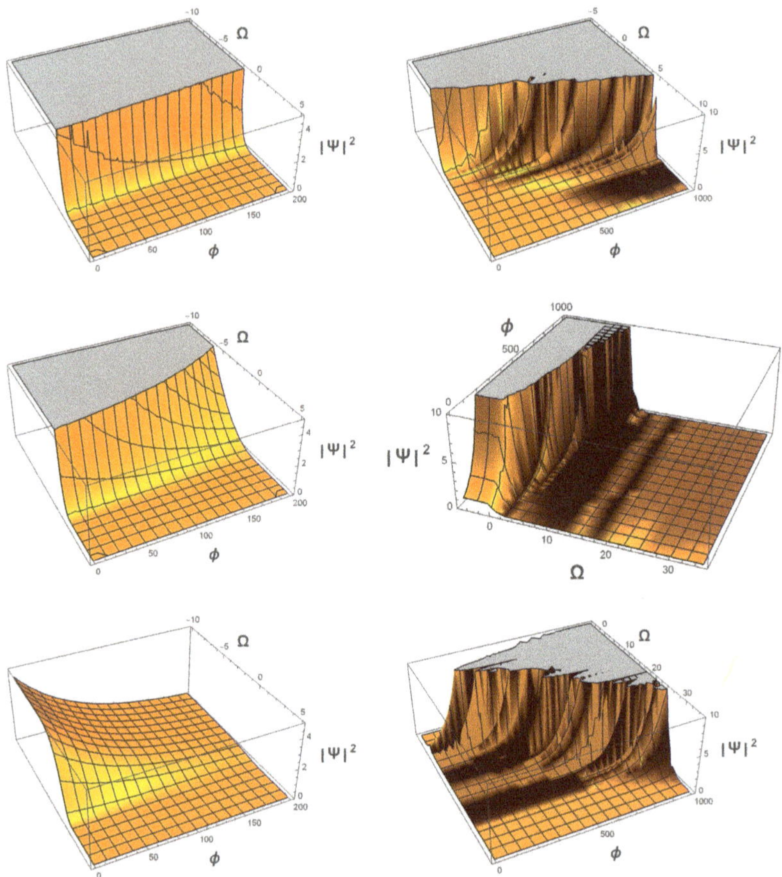

Figure 3. In the dust era, we have the corresponding solution $\mathcal{B}_+$ with the modified Bessel function K_ν and $\mathcal{B}_-$, the ordinary Bessel function of the Equation (50), considering different values in the order parameter $Q = -1, 0, 1$, from top to bottom, respectively. We consider the value for the parameter $\mu = 0.5$; we discard other values in the Q parameter.

4.   inflation such as $\omega_X = -\frac{1}{3}$, $\alpha = -1$; thus, $\beta = \frac{2}{3} \to \gamma = \frac{1}{3}$.
For this particular case, (46) is written as

$$\Psi^2 = \psi_0^2 \, e^{Q\Omega} \mathbb{E}_{\frac{2}{3}}^2(-z^2) \; K_{\frac{Q}{4}}^2\left[\frac{\mu}{\hbar}2e^\Omega\right],; \tag{51}$$

however, the argument in the Mittag–Leffer function is complex, being

$$z = (0.6873648184993014 - 0.396850262992049841)\frac{\mu}{6\hbar^{\frac{1}{3}}}\phi^{\frac{1}{3}}, \tag{52}$$

and the corresponding graph of the probability density can be made based on this function, taking the Re[z] or Im[z] parts.

5.   inflation such as $\omega_X = -\frac{2}{3}$, $\alpha = -\frac{1}{4}$; thus, $\beta = \frac{1}{3} \to \gamma = \frac{1}{6}$.
For this particular case, (46) is written as

$$\Psi^2 = \psi_0^2 \, e^{Q\Omega} \mathbb{E}_{\frac{1}{3}}^2(-z^2) \; K_{\frac{Q}{5}}^2\left[\frac{\mu}{\hbar}e^{\frac{5}{2}\Omega}\right]. \tag{53}$$

and the argument of the Mittag–Leffer function, in this case, is equal to the previous case, complex

$$z = (0.5946035575013603 - 1.0298835719535588I)\frac{\mu}{6\hbar^{\frac{1}{6}}}\phi^{\frac{1}{6}}. \tag{54}$$

In a general way, the behavior of the probability density for both inflation-like scenarios is similar, in the Re[z] or Im[z] parts, over a wide range of values in the scalar field, as it appears in Figure 4. For the behavior for both inflation-like cases in the value of Ω, the behavior is appropriate.

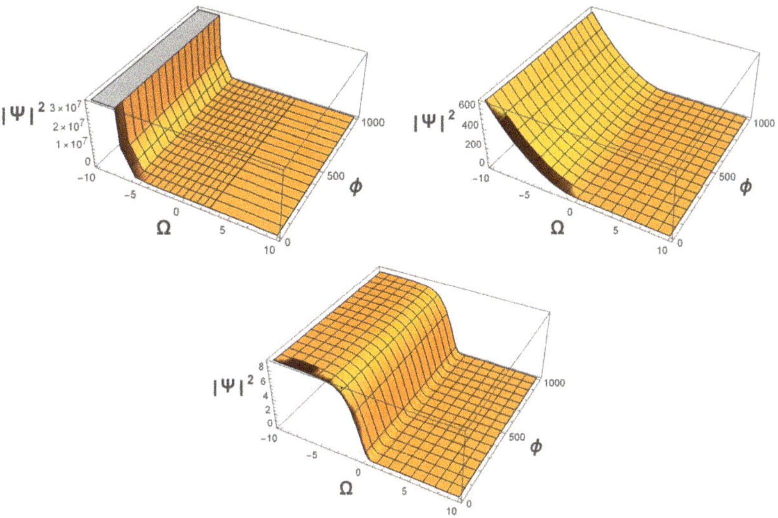

Figure 4. In the second case of the inflation-like scenario, corresponding to Equation (53), the behavior for the probability density in terms of the Mittag–Leffler function with complex values in its argument z is shown, for $Q = -1, 0, 1$. From top to bottom, respectively, we consider the value for the parameter $\mu = 0.5$. In these graphs, we use the Re[z] only; however, the plots with Im[z] are similar.

4. Final Remarks

There are different formalisms to incorporate fractional derivatives to cosmology. One of them starts from the variational principle of the action of general relativity with a fractional kernel; another way is starting from a particular configuration, for example, the FLRW model, and changing the ordinary derivatives with fractional one [14–20]. Unlike the previous formalisms, in the present work, we employed an action that contains a Lagrangian and a fractional parameter; in this way, equations of non-integer order in cosmology are obtained.

Unlike the previous formalism, in the present work, we employed a barotropic equation with perfect fluid for the energy momentum tensor in the K-essence scalar field into the Lagrangian and Hamiltonian formalism, obtaining the momentum of the scalar field with fractional numbers. However, the momentum of the scale factor appeared in the usual way. We obtained the classical solutions for different scenarios in the Universe, which are similar to those which were obtained for standard matter 16 years ago, see ref. [35] in Equations (6) and (34) in the time τ. In this sense, we can introduce the idea that the kinetic energy of the scalar field should falsify the standard matter by employing the K-essence formalism. In the quantum regime, we found a fractional differential equation for the scalar field, where the Mittag–Leffler function is the novel solution in many scenarios with real or complex values in its argument z. With this in mind, we visualized two alternatives in our analysis; the first one is within the traditional expectation over the behavior of the probability density, where the best candidates for quantum solutions are those that have a damping behavior with respect to the scale factor, which appear in all scenarios under our study, without saying anything about the scalar field. The other alternative scenario is when we keep the scale factor scenario, and we consider the values of the scalar field as significant in the quantum regime, appearing in various scenarios in the behavior of the Universe. This is mainly in those where the Universe shows huge behavior, for example, in the inflation-like scenario, see Figure 4 and the actual epoch, Figure 3 or Figure 5, where the scalar field appears as a background. In other words, the interpretation of probability density of the unnormalized wave function, is given when we demand that Ψ does not diverge when the scale factor A (or Ω) goes to infinity, and the scalar field is arbitrary. However, the evolution with the scalar field is now important in this class of theory and others, as it appears in some stages of evolution of our Universe, intended to serve as a a background for the evolution of the Universe in the classical world. The quantum regime appears with big values in the corresponding figures (see the corresponding (3), (5) and (4) plots). However, it is interesting to mention that in the radiation-like scenario, Figures 1 and 2, this behavior over the scalar field is less significant in the formation of atoms and close to the stiff matter scenarios, where an oscillatory behavior takes place in the scalar field. We briefly illustrate the main results in this work.

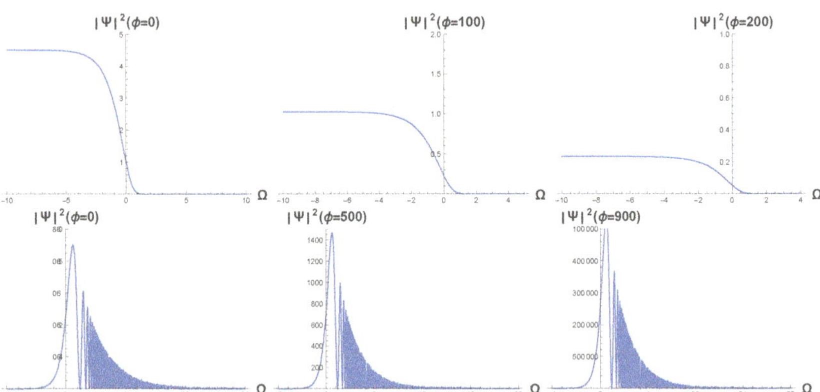

Figure 5. These graphs represent a break-off for the defined value in the scalar field ϕ, for the case in the factor ordering Q = 1, taking into account both solutions (50) given in Figure 3. The first line corresponds to the solution with the modified Bessel function, and the second line employs the ordinary Bessel function. This behavior appears in a similar way for different values in the factor ordering parameter, $Q = -1, 0, 1$. We can see that the values in the amplitude of the probability density for big values in the scalar field, have very large growth, acting as a background in the classical level.

1. Using the K-essence formalism in a general way, applied to the Friedmann–Lemaître–Robertson–Walker cosmological model, we found the Hamiltonian density in the scalar field momenta raised to a power with non-integers. This produces in the quantum scheme a fractional differential equation in a natural way, such as in this variable with order $\beta = \frac{2\alpha}{2\alpha-1}$, where $\alpha \in (-1, \infty)$, which was solved for different scenarios of our Universe.
2. We found in the classical scheme that the time evolution τ of the scale factor for ordinary matter was found 16 years ago by one of us; this time, behavior is reproduced in the K-essence formalism, see Equation (36) in this work, which is consistent with the result obtained in the ref. [35], Equations (6) and (34), with ordinary matter.
3. In the quantum regime, the novel solution at the fractional differential equation in the scalar field was found in terms of the Mittag–Leffler function, with a real or complex argument, and we can see that this function appears in several scenarios of

our Universe in this work. This function is reported in particular work dealing with different disciplines of cosmology.

4. In one of our analyses presented on the probability density, we considered the values of the scalar field as significant in the quantum regime, appearing in various scenarios in the behavior of the Universe; mainly in those where the Universe has huge behavior. For example, in the inflation-like scenario and the actual epoch, where the scalar field appears as a background, the quantum regime appears with big values, but it presents a moderate development in other scenarios with a different ordering parameter Q.

Author Contributions: Conceptualization, J.S. and J.J.R.; Methodology, J.S. and J.J.R.; Writing—Original Draft, J.S. and J.J.R.; Writing—Review and Editing, J.S. and J.J.R.; Visualization, J.S. and J.J.R. All authors have read and agreed to the published version of the manuscript.

Funding: J.S. was partially supported by PROMEP grants UGTO-CA-3. Both authors were partially supported by SNI-CONACyT. J.J. Rosales is supported by the UGTO-CA-20 nonlinear photonics and Department of Electrical Engineering.

Institutional Review Board Statement: Not applicable.

Informed Consent Statement: Not applicable.

Data Availability Statement: Not applicable.

Acknowledgments: Authors thank anonymous referees, since by answering their criticisms we understood our problem better. This work is part of the collaboration within the Instituto Avanzado de Cosmología and Red PROMEP: Gravitation and Mathematical Physics, under project Quantum aspects of gravity in cosmological models, phenomenology, and geometry of space-time. Many calculations were done by Symbolic Program REDUCE 3.8.

Conflicts of Interest: The authors declare no conflict of interest.

Appendix A. Obtaining Equations of Motion

We take the variation over the fields components in the action for K-essence theory coupled with gravity

$$S = \int \sqrt{-g}[R + f(\phi)\mathcal{G}(X)]d^4x, \tag{A1}$$

where R is the Ricci scalar, g is the determinant to the metric, $f(\phi)$ is a function of the scalar field, and $\mathcal{G}[X]$ is a functional depending of the kinetic energy $X(\phi, g^{\mu\nu}) = -\frac{1}{2}g^{\mu\nu}\nabla_\mu\phi\nabla_\nu\phi$. The variation of the fields $(g^{\mu\nu}, \phi)$ in the action (A1) becomes

$$
\begin{aligned}
\delta S &= \int \delta[\sqrt{-g}R]d^4x + \int \delta\sqrt{-g}[f(\phi)\mathcal{G}(X)]d^4x \\
&\quad + \int \sqrt{-g}[\delta f(\phi)\mathcal{G}(X) + f(\phi)\delta\mathcal{G}(X)]d^4x, \\
&= \int \sqrt{-g}\, G_{\lambda\theta}\delta g^{\lambda\theta}d^4x + \int \sqrt{-g}\frac{1}{2}[-f(\phi)\mathcal{G}(X)]g_{\lambda\theta}\delta g^{\lambda\theta}d^4x + \\
&\quad + \int \sqrt{-g}\left[\frac{\partial f(\phi)}{\partial \phi}\delta\phi\mathcal{G}(X) + f(\phi)\frac{\partial\mathcal{G}(X)}{\partial X}\delta X\right]d^4x,
\end{aligned}
\tag{A2}
$$

where the variation of the functional $\mathcal{G}(X)$ is over the kinetic energy

$$
\begin{aligned}
\delta X(\phi) &= -\frac{1}{2}\nabla_\mu\phi\nabla_\nu\phi\delta g^{\mu\nu} - \frac{1}{2}g^{\mu\nu}\nabla_\mu\delta\phi\nabla_\nu\phi - \frac{1}{2}g^{\mu\nu}\nabla_\mu\phi\nabla_\nu\delta\phi, \\
&= -\frac{1}{2}\nabla_\mu\phi\nabla_\nu\phi\delta g^{\mu\nu} - \nabla^\nu\phi\nabla_\nu\delta\phi;
\end{aligned}
$$

introducing into the last equation, we have

$$\delta S = \int \sqrt{-g}\left[G_{\lambda\theta} - \frac{1}{2}f(\phi)\left(\mathcal{G}(X)g_{\lambda\theta} + \frac{\partial \mathcal{G}(X)}{\partial X}\nabla_\mu\phi\nabla_\nu\phi\right)\right]\delta g^{\lambda\theta}d^4x$$
$$+ \int \sqrt{-g}\left[\frac{\partial f(\phi)}{\partial \phi}\delta\phi\mathcal{G}(X) + f(\phi)\frac{\partial \mathcal{G}(X)}{\partial X}\{-\nabla^\nu\phi\nabla_\nu\delta\phi\}\right]d^4x. \tag{A3}$$

However, we know that the total derivative

$$\nabla_\nu(f(\phi)G_X\nabla^\nu\phi\delta\phi) = \frac{df(\phi)}{d\phi}\nabla_\nu\phi\nabla^\nu\phi\,G_X\delta\phi + f(\phi)G_{XX}X_{,\nu}\nabla^\nu\delta\phi + f(\phi)G_X\nabla^\nu_{;\nu}\phi\delta\phi$$
$$+ f(\phi)G_X\nabla^\nu\phi\nabla_\nu\delta\phi. \tag{A4}$$

Thus,

$$-f(\phi)G_X\nabla^\nu\phi\nabla_\nu\delta\phi = \frac{df(\phi)}{d\phi}\nabla_\nu\phi\nabla^\nu\phi\,G_X\delta\phi + f(\phi)G_{XX}X_{,\nu}\nabla^\nu\delta\phi + f(\phi)G_X\nabla^\nu_{;\nu}\phi\delta\phi$$
$$- \nabla_\nu(f(\phi)G_X\nabla^\nu\phi\delta\phi), \tag{A5}$$

and reinserting into (A3), we have

$$\delta S = \int \sqrt{-g}\left[G_{\lambda\theta} - \frac{1}{2}f(\phi)\left(\mathcal{G}(X)g_{\lambda\theta} + \frac{\partial \mathcal{G}(X)}{\partial X}\nabla_\mu\phi\nabla_\nu\phi\right)\right]\delta g^{\lambda\theta}d^4x$$
$$+ \int \sqrt{-g}\left\{\frac{\partial f(\phi)}{\partial \phi}\delta\phi\mathcal{G}(X) + \frac{df(\phi)}{d\phi}\underbrace{\nabla_\nu\phi\nabla^\nu\phi}_{-2X}\,G_X\delta\phi + f(\phi)G_{XX}X_{,\nu}\nabla^\nu\phi\,\delta\phi\right.$$
$$\left. + f(\phi)G_X\nabla^\nu_{;\nu}\phi\delta\phi - \nabla_\nu(f(\phi)G_X\nabla^\nu\phi\delta\phi)\right\}d^4x,$$
$$= \int \sqrt{-g}\left[G_{\lambda\theta} - \frac{1}{2}f(\phi)\left(\mathcal{G}(X)g_{\lambda\theta} + \frac{\partial \mathcal{G}(X)}{\partial X}\nabla_\mu\phi\nabla_\nu\phi\right)\right]\delta g^{\lambda\theta}d^4x$$
$$+ \int \sqrt{-g}\left\{\frac{df(\phi)}{d\phi}[\mathcal{G}(X) - 2XG_X] + f(\phi)[G_{XX}X_{,\nu}\nabla^\nu + G_X\nabla^\nu_{;\nu}\phi]\right\}\delta\phi d^4x, \tag{A6}$$

where we have eliminated the integral over the total derivative.

The variation over the scalar field gives the equation of motion for this field, being

$$\frac{df(\phi)}{d\phi}[\mathcal{G}(X) - 2XG_X] + f(\phi)[G_{XX}X_{,\nu}\nabla^\nu + G_X\nabla^\nu_{;\nu}\phi] = 0, \tag{A7}$$

which corresponds to Equation (15). For obtaining the Einstein field-like equations, we take the variation on the metric $g^{\mu\nu}$,

$$G_{\mu\nu} = \frac{1}{2}f(\phi)\left[\nabla_\mu\phi\nabla_\nu\phi\frac{\partial \mathcal{G}(X)}{\partial X} + g_{\mu\nu}\mathcal{G}(X)\right], \tag{A8}$$

where the energy-momentum tensor becomes

$$T_{\mu\nu}(\phi) = +\frac{1}{2}f(\phi)\left[\nabla_\mu\phi\nabla_\nu\phi\frac{\partial \mathcal{G}(X)}{\partial X} + g_{\mu\nu}\mathcal{G}(X)\right], \tag{A9}$$

and considering the energy-momentum tensor of a barotropic perfect fluid for the scalar fields

$$T_{\mu\nu}(\phi) = (\rho + P)u_\mu(\phi)u_\nu(\phi) + P g_{\mu\nu}, \tag{A10}$$

we have that the pressure P and the energy density ρ of the scalar fields become

$$P(\phi) = \frac{1}{2}f(\phi)\mathcal{G}, \qquad \rho(\phi) = \frac{1}{2}f\left[2X\frac{\partial \mathcal{G}}{\partial X} - \mathcal{G}\right], \tag{A11}$$

the four-velocity becomes $u_\mu u_\nu = \frac{\nabla_\mu \phi \nabla_\nu \phi}{2X}$ and the barotropic index ω_X is

$$\omega_X = \frac{f(\phi)\mathcal{G}}{f\left[2X\frac{\partial \mathcal{G}}{\partial X} - \mathcal{G}\right]}. \tag{A12}$$

Appendix B. Obtaining the Equations of Motion with Particular Metric

We have rewritten the line element in the time $\tau = Ndt = $ /,

$$\begin{aligned}
ds^2 &= -N(t)^2 dt^2 + A^2(t)\left[dr^2 + r^2(d\theta^2 + \sin^2\theta d\phi^2)\right], \tag{A13} \\
&= -d\tau^2 + A^2(\tau)\left[dr^2 + r^2(d\theta^2 + \sin^2\theta d\phi^2)\right], \tag{A14}
\end{aligned}$$

where the metric element $g_{\tau\tau} = -1$ implies that $\Gamma^\tau_{\tau\tau} = 0$ and $\Gamma^j_{j\tau} = \Omega', j = r, \theta, \phi = 1, 2, 3$. When the function $f(\phi)$ is constant, the Equation (15) is reduced to

$$\mathcal{G}_X \phi^{;\nu}_{;\nu} + \mathcal{G}_{XX} X_{;\nu} \phi^{;\nu} = 0; \tag{A15}$$

using the metric (A14), we obtain that the different parameters into the Equation (A15) are

$$X = \frac{1}{2}(\phi')^2, \quad \rightarrow \quad (\phi')^2 = 2X, \qquad X' = \phi'\phi'', \quad \rightarrow \quad \phi'' = \frac{X'}{\phi'}$$

$$\phi^{;\nu}_{;\nu} = \phi^{;\nu}_{,\nu} + \Gamma^\nu_{\nu\rho}\phi^\rho = \phi'' + \left(\Gamma^\tau_{\tau\tau} + \Gamma^1_{1\tau} + \Gamma^2_{2\tau} + \Gamma^3_{3\tau}\right)\phi' = \phi'' + 3\Omega'\phi'. \tag{A16}$$

Thus, the Equation (A15) is rewritten as

$$\mathcal{G}_X\left(\phi'' + 3\Omega'\phi'\right) + \mathcal{G}_{XX}X'\phi' = \mathcal{G}_X\left(\frac{X'}{\phi'} + 3\Omega'\phi'\right) + \mathcal{G}_{XX}X'\phi' = 0; \tag{A17}$$

multiplying by ϕ', we have

$$[\mathcal{G}_X + 2X\mathcal{G}_{XX}]X' + 6\Omega' X\mathcal{G}_X = 0, \tag{A18}$$

where we had used the previous relations; this equation correspond to (24).

Dividing between $X\mathcal{G}_X$ the last equation, we have

$$\begin{aligned}
\frac{X'}{X} + 2\frac{\mathcal{G}_{XX}}{\mathcal{G}_X}X' + 6\frac{A'}{A} &= \frac{d}{d\tau}\left(LnX + Ln\mathcal{G}_X^2 + LnA^6\right) \\
&= \frac{d}{d\tau}Ln\left(A^6 X\mathcal{G}_X^2\right) = 0, \quad \rightarrow \quad A^6 X\mathcal{G}_X^2 = \eta = constant, \tag{A19}
\end{aligned}$$

obtaining that

$$X\mathcal{G}_X^2 = \eta A^{-6}, \tag{A20}$$

which is the Equation (25).

When $\mathcal{G} = X^\alpha$ and substituting into (A20), we have $\alpha^2 X^{2\alpha-1} = \eta A^{-6}$ with $X = \frac{1}{2}\left(\frac{d\phi}{d\tau}\right)^2$, obtaining for the scalar field ϕ the equation

$$\frac{d\phi}{d\tau} = \sqrt{2}\left[\left(\frac{\eta}{\alpha^2}\right)^{\frac{1}{2(2\alpha-1)}} A^{-\frac{3}{2\alpha-1}}\right] = \sqrt{2}\left[\left(\frac{\eta}{\alpha^2}\right)^{\frac{1}{2(2\alpha-1)}} e^{-\frac{3}{2\alpha-1}\Omega}\right], \tag{A21}$$

whose solution in the time τ is

$$\Delta\phi = \sqrt{2}\left(\frac{\eta}{\alpha^2}\right)^{\frac{1}{2(2\alpha-1)}}\int A^{-\frac{3}{2\alpha-1}}d\tau = \sqrt{2}\left(\frac{\eta}{\alpha^2}\right)^{\frac{1}{2(2\alpha-1)}}\int e^{-\frac{3}{2\alpha-1}\Omega}d\tau, \qquad (A22)$$

Appendix C. Equivalence between Lagrangian Densities

The canonical Lagrangian density $\mathcal{L}_{canonical}(q_i,\Pi_i,t)$ (32) in gravitation theories is obtained from the usual Lagrangian density $\mathcal{L}(q_i,\dot{q}_i,t)$ (29), rewritten the velocities $\dot{q}_i$ in term of the momenta $\Pi_i = \frac{\partial\mathcal{L}}{\partial\dot{q}_i}$ to the corresponding coordinate field q_i. With this procedure, the canonical Lagrangian density appear directly written as a Lagrangian density in constrained systems, where the Lagrangian multiplier is the lapse function $N(t)$, being the corresponding gauge parameter in this theory. This is equivalent to using the canonical transformation where the hamiltonian density is $\mathcal{H} = \Pi_j\dot{q}^j - \mathcal{L}$. However, from this point of view, in this expression the $\mathcal{H}$ must be interpreted as $\mathcal{H} = N\mathcal{H}_{canonical}$ where the lapse function N appears as a lagrangian multiplier. In the following we realize this calculation, employing the usual canonical transformation. We have the momenta

$$\Pi_\Omega = \frac{12}{N}e^{3\Omega}\dot{\Omega}, \quad \rightarrow \quad \dot{\Omega} = \frac{N}{12}e^{-3\Omega}\Pi_\Omega, \qquad (A23)$$

$$\Pi_\phi = -\left(\frac{1}{2}\right)^\alpha\frac{2\alpha}{N^{2\alpha-1}}e^{3\Omega}\dot{\phi}^{2\alpha-1}, \quad \rightarrow \quad \dot{\phi} = -N\left[\frac{2^\alpha}{2\alpha}e^{-3\Omega}\Pi_\phi\right]^{\frac{1}{2\alpha-1}}, \qquad (A24)$$

substituting into the canonical transformation between the Hamiltonian density and Lagrangian density

$$\mathcal{H} = \Pi_j\dot{q}^j - \mathcal{L}$$

$$\begin{aligned}
\mathcal{H} &= \Pi_\Omega\dot{\Omega} + \Pi_\phi\dot{\phi} - e^{3\Omega}\left[6\frac{\dot{\Omega}^2}{N} - \left(\frac{1}{2}\right)^\alpha(\dot{\phi})^{2\alpha}N^{-2\alpha+1}\right] \\
&= \Pi_\Omega\left(\frac{N}{12}e^{-3\Omega}\Pi_\Omega\right) + \Pi_\phi\left(-N\left[\frac{2^\alpha}{2\alpha}e^{-3\Omega}\Pi_\phi\right]^{\frac{1}{2\alpha-1}}\right) \\
&\quad - e^{3\Omega}\left[\frac{6}{N}\left(\frac{N}{12}e^{-3\Omega}\Pi_\Omega,\right)^2 - \left(\frac{1}{2}\right)^\alpha\left(-N\left[\frac{2^\alpha}{2\alpha}e^{-3\Omega}\Pi_\phi\right]^{\frac{1}{2\alpha-1}}\right)^{2\alpha}N^{-2\alpha+1}\right] \\
&= Ne^{3\Omega}\Pi_\Omega^2\left(\frac{1}{12} - \frac{1}{24}\right) - Ne^{-\frac{3\Omega}{2\alpha-1}}\left(\frac{2^{\alpha-1}}{\alpha}\right)^{\frac{1}{2\alpha-1}}\Pi_\phi^{\frac{2\alpha}{2\alpha-1}} \\
&\quad + Ne^{-\frac{3\Omega}{2\alpha-1}}\frac{1}{2\alpha}\left(\frac{2^{\alpha-1}}{\alpha}\right)^{\frac{1}{2\alpha-1}}\Pi_\phi^{\frac{2\alpha}{2\alpha-1}} \\
&= N\frac{e^{-3\Omega}}{24}\Pi_\Omega^2 - Ne^{-\frac{3\Omega}{2\alpha-1}}\left(\frac{2^{\alpha-1}}{\alpha}\right)^{\frac{1}{2\alpha-1}}\left[\frac{1}{1-2\alpha}\right]\Pi_\phi^{\frac{2\alpha}{2\alpha-1}} \\
&= N\frac{e^{-3\Omega}}{24}\Pi_\Omega^2 - Ne^{-\frac{3\Omega}{2\alpha-1}}\frac{2\alpha-1}{2\alpha}\left(\frac{2^{\alpha-1}}{\alpha}\right)^{\frac{1}{2\alpha-1}}\Pi_\phi^{\frac{2\alpha}{2\alpha-1}} \\
&= N\frac{e^{-\frac{3\Omega}{2\alpha-1}}}{24}\left(e^{-\frac{6(\alpha-1)}{2\alpha-1}\Omega}\Pi_\Omega^2 - \frac{12(2\alpha-1)}{\alpha}\left(\frac{2^{\alpha-1}}{\alpha}\right)^{\frac{1}{2\alpha-1}}\Pi_\phi^{\frac{2\alpha}{2\alpha-1}}\right), \qquad (A25)
\end{aligned}$$

corresponding to Equation (33).

References

1. Podlubny, I. *Fractional Differential Equations*; Academic Press: New York, NY, USA, 1999.
2. Ortigueira, M.D.; Tenreiro Machado, J.A. What is a fractional derivative? *J. Comput. Phys.* **2015**, *293*, 4–13. [CrossRef]

3. Rosu, H.C.; Madueño, A.L.; Socorro, J. Transform of Riccati equation of constant coefficients through fractional procedure. *J. Phys. A Math. Gen.* **2003**, *36*, 1087–1093. [CrossRef]
4. Abel, N.H. Résolution d'un probléme de mécanique. In *Oeuvres Complètes de Niels Henrik Abel: Nouvelle Édition (Cambridge Library Collection—Mathematics)*; Sylow, L., Lie, S., Eds.; Cambridge University Press: Cambridge, UK, 2012; pp. 97–101. [CrossRef]
5. Caputo, M.; Mainardi, F. A new dissipation model based on memory mechanism. *Pure Appl. Geophys.* **1971**, *91*, 134–137. [CrossRef]
6. Wyss, W. Fractional diffusion equation. *J. Math. Phys.* **1986**, *27*, 2782–2785. [CrossRef]
7. Westerlund, S. Capacitor theory. *IEEE Trans. Dielectr. Electr. Insul.* **1994**, *1*, 826–839. [CrossRef]
8. Hermann, R. *Fractional Calculus*; World Scientific: Singapore, 2011.
9. Cruz-Duarte, J.M.; Rosales-García, J.; Correa-Cely, C.R.; García-Perez, A.; Avina-Cervantes, J.G. A closed form expression for the Gaussian-based Caputo-Fabrizio fractional derivative for signal processing applications. *Commun. Nonlinear Sci. Numer. Simulat.* **2018**, *61*, 138–148. [CrossRef]
10. Martínez-Jiménez, L.; Cruz-Duarte, J.M.; Rosales-García, J.J.; Cruz-Aceves, I. Enhancement of vessels in coronary angiograms using a Hessian matrix based on Grunwald–Letnikov fractional derivative. In Proceedings of the 8th International Conference on Biomedical Engineering and Technology (ICBET '18), Bali, Indonesia, 23–25 April 2018; pp. 51–54.
11. Uchaikin, V. *Fractional Derivatives for Physicists and Engineers*; Springer: Berlin/Heidelberg, Germany, 2013.
12. Tarasov, V.E. *Fractional Dynamics: Applications of Fractional Calculus to Dynamics of Particles, Fields and Media*; Springer: Berlin/Heidelberg, Germany, 2010.
13. Magin, R.L. *Fractional Calculus in Bioengineering*; Begell House Publisher: Rodding, Denmark, 2006.
14. Roberts, M.D. Fractional Derivative Cosmology. *SOP Trans. Theor. Phys.* **2014**, *1*, 310. [CrossRef]
15. El-Nabulsi, R.A. Gravitons in fractional action cosmology. *Int. J. Theor. Phys.* **2012**, *51*, 3978–3992. [CrossRef]
16. Jamil, M.; Momeni, D.; Rashid, M.A. Fractional Action Cosmology with Power Law Weight Function. *J. Phys. Conf. Ser.* **2012**, *354*, 012008. [CrossRef]
17. Debnath, U.; Jamil, M.; Chattopadhyay, S. Fractional Action Cosmology: Emergent, Logamediate, Intermediate, Power Law Scenarios of the Universe and Generalized Second Law of Thermodynamics. *Int. J. Theor. Phys.* **2012**, *51*, 812–837. [CrossRef]
18. El-Nabulsi, R.A. Nonstandard fractional exponential Lagrangians, fractional geodesic equation, complex general relativity, and 915 discrete gravity. *Can. J. Phys.* **2013**, *91*, 618–622. [CrossRef]
19. El-Nabulsi, R.A. Non-minimal coupling in fractional action cosmology. *Indian J. Phys.* **2013**, *87*, 835–840. [CrossRef]
20. Debnath, U.; Chattopadhyay, S.; Jamil, M. Fractional action cosmology: Some dark energy models in emergent, logamediate, and intermediate scenarios of the Universe. *J. Theor. Appl. Phys.* **2013**, *7*, 25. [CrossRef]
21. Rami, E.N.A. Fractional action oscillating phantom cosmology with conformal coupling. *Eur. Phys. J. Plus* **2015**, *130*, 102. [CrossRef]
22. El-Nabulsi, R.A. A Cosmology Governed by a Fractional Differential Equation and the Generalized Kilbas-Saigo-Mittag–Leffler Function. *Int. J. Theor. Phys.* **2016**, *55*, 625–635. [CrossRef]
23. El-Nabulsi, R.A. Implications of the Ornstein-Uhlenbeck-like fractional differential equation in cosmology. *Rev. Mex. FíSica* **2016**, *62*, 240–250.
24. El-Nabulsi, R.A. Fractional Action Cosmology with Variable Order Parameter. *Int. J. Theor. Phys.* **2017**, *56*, 1159–1182. [CrossRef]
25. El-Nabulsi, R.A. Wormholes in fractional action cosmology. *Can. J. Phys.* **2017**, *95*, 605–609. [CrossRef]
26. García-Aspeitia, M.A.; Fernandez-Anaya, G.; Hernández-Almada, A.; Leon, G.; Magaña, J. Cosmology under the fractional calculus approach. *Mon. Not. R. Astron. Soc.* **2022**, *517*, 4813–4826. [CrossRef]
27. Rasouli, S.M.M.; Jalalzadeh, S.; Moniz, P.V. Broadening quantum cosmology with a fractional whirl. *Mod. Phys. Lett.* **2021**, *36*, 2140005. [CrossRef]
28. Jalalzadeh, S.; Costa, E.W.O.; Moniz, P.V. De Sitter fractional quantum cosmology. *Phys. Rev.* **2022**, *105*, L121901. [CrossRef]
29. Rasouli, S.M.M.; Costa, E.W.O.; Moniz, P.V.; Jalalzadeh, S. Inflation and fractional quantum cosmology. *Fractal Fract.* **2022**, *6*, 655. [CrossRef]
30. Moniz, P.V.; Jalalzadeh, S. From Fractional Quantum Mechanics to Quantum Cosmology: An Overture. *Mathematics* **2020**, *8*, 313. [CrossRef]
31. Jalalzadeh, S.; da Silva, F.R.; Moniz, P.V. Prospecting black hole thermodynamics with fractional quantum mechanics. *Eur. Phys. J.* **2021**, *81*, 632. [CrossRef]
32. Jalalzadeh, S.; Moniz, P.V. *Challenging Routes in Quantum Cosmology*; World Scientific: Singapore, 2023.
33. Micolta-Riascos, B.; Millano, A.D.; Leon, G.; Erices, C.; Paliathanasis, A. Revisiting Fractional Cosmology. *Fractal Fract.* **2023**, *7*, 149. [CrossRef]
34. Socorro, J.; Pimentel, L.O.; Espinoza García, A. Classical Bianchi type I cosmology in K-essence theory. *Adv. High Energy Phys.* **2014**, *2014*, 805164. [CrossRef]
35. Berbena, S.R.; Arellano, A.V.; Socorro, J.; Pimentel, L.O. The Einstein-Hamilton-Jacobi equation: Searching the classical solution for barotropic FRW. *Rev. Mex. Física* **2007**, *53*, 115–119.
36. Cota, J.C. *Konstanzer Dissertationen, Induced Gravity and Cosmology*; Hartung-Corre: Konstanz, Germany, 1996.
37. Espinoza-GarcÃa, A.; Socorro, J.; Pimentel, L.O. Quantum Bianchi type IX cosmology in K-essence theory. *Int. J. Theor. Phys.* **2014**, *53*, 3066–3077. [CrossRef]
38. De Putter, R.; Linder, E.V. Kinetic k-essence and Quintessence. *Astropart. Phys.* **2007**, *28*, 263–272. [CrossRef]
39. Chiba, T.; Dutta, S.; Scherrer, R.J. Slow-roll k-essence. *Phys. Rev. D* **2009**, *80*, 043517. [CrossRef]

40. Bose, N.; Majumdar, A.S. A k-essence model of inflation, dark matter and dark energy. *Phys. Rev. D* **2009**, *79*, 103517. [CrossRef]
41. Arroja, F.; Sasaki, M. A note on the equivalence of a barotropic perfect fluid with a k-essence scalar field. *Phys. Rev. D* **2010**, *81*, 107301. [CrossRef]
42. García, L.A.; Tejeiro J.M.; Castañeda, L. K-essence scalar field as dynamical dark energy. *arXiv* **2012**, arXiv:1210.5259.
43. Bilic, N.; Tupper, G.; Viollier, R. Unification of dark matter and dark energy: The inhomogeneous Chaplygin gas. *Phys. Lett. B* **2002**, *535*, 17–21. [CrossRef]
44. Bento, M.; Bertolami, O.; Sen, A. Dynamics of dark energy. *Phys. Rev. D* **2002**, *66*, 043507. [CrossRef]
45. Armendariz-Picon, C.; Damour, T.; Mukhanov, V. k-Inflation. *Phys. Lett. B* **1999**, *458*, 209–218. [CrossRef]
46. Garriga, J.; Mukhanov, V. Perturbations in k-inflation. *Phys. Lett. B* **1999**, *458*, 219–225. [CrossRef]
47. Copeland, E.J.; Sami, M.; Tsujikawa, S. Dynamics of dark energy. *Int. J. Mod. Phys. D* **2006**, *15*, 1753–1936. [CrossRef]
48. Ryan, M.P. *Hamiltonian Cosmology*; Springer: Berlin, Germany, 1972.
49. Erdélyi, A.; Magnus, W.; Oberhettinger, F.; Tricomi, F.G. *Higher Transcendental Functions*; McGraw-Hill: New York, NY, USA, 1955; Volume 3.
50. Haubold, H.J.; Mathai, A.M.; Saxena, R.K. Mittag–Leffler functions and their applications. *J. Appl. Math.* **2011**, *2011*, 298628. [CrossRef]
51. De-Santiago, J.; Cervantes-Cota, J.L. Generalizing a unified model of dark matter, dark energy, and inflation with a noncanonical kinetic term. *Phys. Rev. D* **2011**, *83*, 063502. [CrossRef]
52. Chimento, L.P. Extended tachyon field, Chaplygin gas, and solvable k-essence cosmologies. *Phys. Rev. D* **2004**, *69*, 123517. [CrossRef]
53. Scherrer, R.J. Purely Kinetic k Essence as Unified Dark Matter. *Phys. Rev. Lett.* **2004**, *93*, 011301. [CrossRef]
54. Hartle, J.B.; Hawking, S.W. Wave function of the Universe. *Phys. Rev. D* **1983**, *28*, 2960–2975. [CrossRef]
55. Rosales, J.J.; Gómez, J.F.; Guía, M.; Tkach, V.I. Fractional electromagnetic waves. In Proceedings of the 11th International Conference on Laser and Fiber-Optical Networks Modeling (LFNM), Kharkov, Ukraine, 5–9 September 2011. [CrossRef]
56. Gómez Aguilar, J.F.; Rosales, J.J.; Bernal Alvarado, J.J.; Cordova Fraga, T.; Guzmán Cabrera, R. Fractional mechanics oscillators. *Rev. Mex. Física* **2012**, *58*, 348–352.
57. Polyanin, A.C.; Zaitsev, V.F. *Handbook of Exact Solutions for Ordinary Differential Equations*, 2nd ed.; Chapman & Hall/CRC: Boca Raton, FL, USA, 2003.
58. Socorro, J.; Pérez-Payan, S.; Hernández-Jiménez, R.; Espinoza-García, A.; Díaz-Barrón, L.R. Classical and quantum exact solutions for a FRW in chiral like cosmology. *Class. Quantum Grav.* **2021**, *38*, 135027. [CrossRef]
59. Socorro, J.; Pérez-Payán, S.; Hernández-Jiménez, R.; Espinoza-García, A.; Díaz-Barrón, L.R. Quintom fields from chiral K-essence cosmology. *Universe* **2022**, *8*, 548. [CrossRef]
60. Socorro, J.; Pérez-Payán, S.; Hernández-Jiménez, R.; Espinoza-García, A.; Díaz-Barrón, L.R. Quintom fields from chiral anisotropic cosmology. *arXiv* **2022**, arXiv:2210.01186.

Fitting Type Ia Supernova Data to a Cosmological Model Based on Einstein–Newcomb–De Sitter Space

Vladimir N. Yershov 

Formerly Mullard Space Science Laboratory, University College London, Holmbury St.Mary, Dorking RH5 6NT, UK; vyershov@moniteye.co.uk

Abstract: Einstein–Newcomb–de Sitter (ENdS) space is de Sitter's modification of spherical space used by Einstein in his first cosmological model paper published in 1917. The modification by de Sitter incorporated the topological identification of antipodal points in space previously proposed by Newcomb in 1877. De Sitter showed that space topologically modified in this way (called elliptical or projective space) satisfies Einstein's field equations. De Sitter also found that in a space with constant positive curvature, spectral lines of remote galaxies would be red-shifted (called the de Sitter effect). However, de Sitter's formulae relating distances to red shifts do not satisfy observational data. The likely reason for this mismatch is that de Sitter mainly focused on space curvature and ignored the identification of antipodal points. Herein, we demonstrate that it is this particular feature that allows an almost perfect fit of the ENdS-based cosmological model to observational data. We use 1701 sources from the type Ia supernovae data sample called *Pantheon+*, which was previously used to fit the ΛCDM model. ΛCDM and ENdS diverge in their predictions for red shifts exceeding $z \sim 2.3$. Since there are no available type Ia supernovae (SNe) data for higher red shifts, both models can be validated by using an additional sample of 193 gamma-ray bursts (GRBs) spanning red shifts up to $z \sim 8$. This validation shows that the minimum χ^2 for the SNe+GRBs sample is about 2.7% smaller for the ENdS space model than for the ΛCDM model.

Keywords: type Ia supernovae; elliptical space; wormholes; nonlocality; Schwartzschild metric; gravitational red shift; cosmological red shift

Citation: Yershov, V.N. Fitting Type Ia Supernova Data to a Cosmological Model Based on Einstein–Newcomb–De Sitter Space. *Universe* **2023**, *9*, 204. https://doi.org/10.3390/universe9050204

Academic Editor: Kazuharu Bamba

Received: 31 March 2023
Revised: 21 April 2023
Accepted: 22 April 2023
Published: 25 April 2023

1. Introduction

The first cosmological model was introduced in 1917 by A. Einstein [1]. It was based on static space with constant positive curvature. At that time, Einstein was not concerned with the cosmological red-shift problem because there was no observational evidence for such a phenomenon at the time, and the universe was commonly believed to be static. Therefore, Einstein introduced to his model a fine-tuned cosmological constant to make space static.

Even so, based on Einstein's static universe model alone, it was already possible to foresee the existence of cosmological red shift. This was described by W. de Sitter [2], who analysed various cosmological models based on positively curved three-manifolds of spherical ($\mathbb{S}^3$) and elliptical shapes, the latter having been previously studied by S. Newcomb [3]. For brevity, we use the acronym ENdS (Einstein–Newcomb–de Sitter) when referring to this particular version of de Sitter's models.

Elliptical space (also called projective space ($\mathbb{P}^3$)) differs from $\mathbb{S}^3$ by identification of antipodal points, which is schematically shown in Figure 1 in the form of an embedding diagram, in which spherical curvature is neglected for simplicity. Two antipodal points of space are separated from each other by the maximal possible distance corresponding to the projective angle ($\chi = \pi$) (the solid and dot-dashed line along the manifold in this diagram). Topological identification is indicated by the dashed line connecting two points in the perpendicular direction. According to de Sitter, elliptical space ($\mathbb{P}^3$) is preferable for modelling the physical world because when $\mathbb{P}^3$ is projected to the Euclidean ($\mathbb{E}^3$) or hyperbolic ($\mathbb{H}^3$) spaces, it covers them once, whereas $\mathbb{S}^3$ covers them twice. The projection corresponds to the coordinate transformation

$$r = R \tan \chi, \tag{1}$$

where R^{-2} is the constant positive curvature of $\mathbb{S}^3$ or $\mathbb{P}^3$, and χ is the projective angle. Locally, $\mathbb{S}^3$ and $\mathbb{P}^3$ are identical to $\mathbb{E}^3$. However, de Sitter noted that since velocity and energy are related to different reference frames, they change when observed from one or another reference frame. The time component of the elliptical space metric is

$$g_{tt} = \cos^2 \chi. \tag{2}$$

Therefore, from the point of view of a remote observer, all physical processes slow down, including chemical and atomic reactions, which leads to the reduced frequencies of electromagnetic waves emitted in these reactions. This time-dilatation effect (called the de Sitter effect) was regarded by E. Hubble as one of the main possible physical mechanisms explaining the distance–red shift relationship [4].

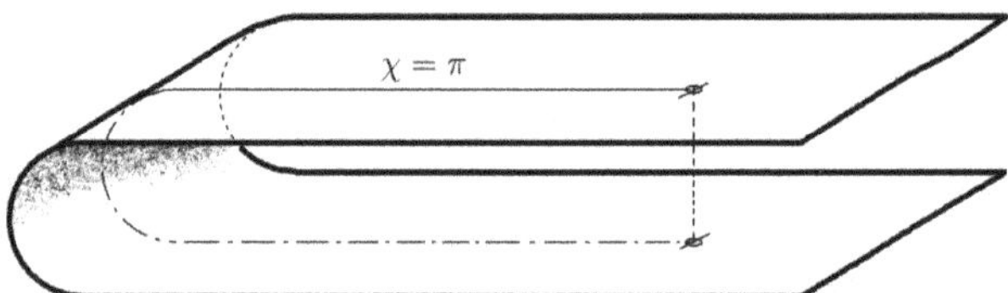

Figure 1. Embedding diagram of Einstein–Newcomb–de Sitter elliptical space with two antipodal points topologically identified (the dashed line). The radius of curvature (R) is neglected in this plot for simplicity. The distance between two antipodal points is measured along the spatial coordinate (thin solid and dot-dashed line) by the projective angle (χ), which is equal to π for the maximal possible separation in the elliptical space.

An alternative interpretation of the distance–red shift relationship due to space expansion was widely debated by many authors, including A. Friedmann [5], G. Lemaître [6], H.P. Robertson [7] and A. G. Walker [8] (FLRW), whose approach became a basis for the standard FRLW cosmological model. The 1998 discovery of type Ia SNe excessive dimming at $z \simeq 1$ [9–11] was interpreted as evidence in favour of dark energy—repulsive gravity or the Λ term in Einstein's equations, which was incorporated into the standard cosmological model, together with cold dark matter (CDM) detected via its gravitational effects. Among other expanding universe models, the standard ΛCDM cosmological model is the best to fit observational data, including the accurately measured distance moduli of type Ia supernovae and the fluctuations of the cosmic microwave background (CMB).

In recent years, the standard cosmological model has been subject to some problems. For example, extremely remote galaxies were found to host very high-mass supermassive black holes, with not enough time for their formation since the beginning of the universe [12]. Galaxies were also discovered whose sizes and surface brightness contradicted the predictions of ΛCDM [13]. More recently, observations made using the James Webb Space Telescope have demonstrated that there are many well-evolved galaxies with red shifts ($z > 15$) that must have been formed in an impossibly short time span of $\sim$230 Myr available since the beginning of the universe according to ΛCDM (see a review by N. Lovyagin [14] and references therein). Static cosmological models provide much more time for the evolution of those galaxies, which would explain the unexpected JWST results.

All this suggests that the old debate on the nature of the cosmological red shift needs to be revived. This debate has lasted since the 1930s [15–17] through the 1970s–1990s [18–21], continuing to very recent years [22,23]. Herein, we contribute to this debate but without exploring the de Sitter effect in its original form [2], or the Zwicky's photon-energy dissipation mechanism [15], as neither matches the modern standard-candle data based on the type Ia supernovae distance moduli.

Instead, we focus on a different aspect of ENdS elliptical space, namely the topological identification of antipodal points. This aspect was neglected by de Sitter when he associated distances with red shifts in his theory, likely due to the oversight that de Sitter's formulae, (1) and (2), do not match observations.

In order to relate red shifts to distances in the elliptical space, the mathematical concept of identical antipodal points needs to be interpreted in the form of a physical model. We propose such an interpretation in the next section. In Section 2.2, we derive a formalism relating the luminosity distances of remote sources with their cosmological red shifts, which is needed to obtain parameter values for our model using accurately calibrated observational data of type Ia supernovae (Section 2.3). In Section 3, we compare the theoretical predictions of the ΛCDM and ENdS models with the distance moduli based on type Ia supernova data from the *Pantheon+* sample. In Section 4, we validate our model via an external χ^2 criterion for red shifts exceeding $z \simeq 2.3$ using gamma-ray burst (GRB) distance moduli provided by L. Amati et al. [24].

2. Materials and Methods

2.1. Physical Interpretation of Identified Antipodal Points

The topological identification of antipodal points of a manifold is a mathematical abstraction. In order to interpret this in terms of physics, we would need to find a physical object, preferably one described in a generally relativistic way, which allows widely separated points of space to be connected. Such an object was found and described in 1935 within the framework of the theory of general relativity by the very author of this theory [25]. It is called an Einstein–Rosen bridge or, more frequently, a "wormhole" because it can connect two different spaces (universes).

Later on, it was found that wormholes may connect widely separated regions of the same space, which was viewed as a possibility for interstellar travel [26]. Thus, wormholes can also connect antipodal points of ENdS space, which is what we require. This is schematically shown in Figure 2 in the form of an embedding diagram, where the topological connection previously indicated as a dashed line in Figure 1 is now replaced with a wormhole's throat. In Figure 2, the observer (o) and the observed source (s) are at distances of r_o and r_s, respectively, from the far end of the wormhole's throat, where r_g is the wormhole's gravitational radius. The source is at a distance of $d = r_o - r_s$ from the observer, and we need to relate this distance to the observed red shift (z) of the source (s) in order to compare our model with observations.

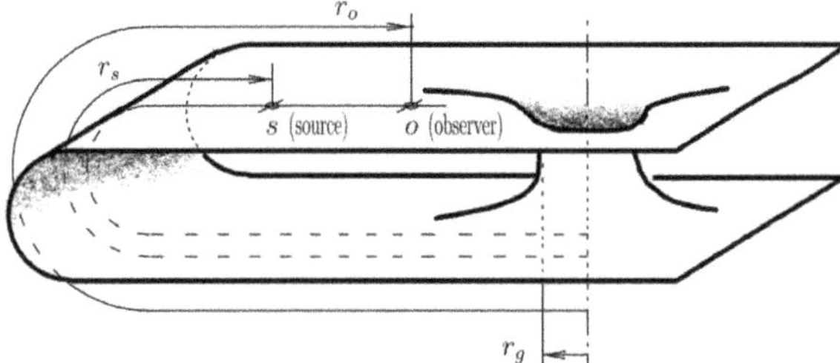

Figure 2. Embedding diagram depicting a possible physical realisation of the topological connection between two distant (antipodal) regions of space via a wormhole structure. Here, r_o and r_s are distances of the observer (o) and source (s) from the wormhole's far throat, opposite to the near throat in the vicinity of the observer; and r_g is the gravitational radius of the wormhole.

M.S. Morris, K.S. Thorne and U. Yurtsever highlighted [27] that wormhole creation is accompanied by extremely large space-time curvature, which corresponds to microscopic structures on a scale length of the order of the Planck–Wheeler length, $\sqrt{G\hbar/c^3} = 1.3 \times 10^{-33}$ cm. Therefore, in principle, wormholes can be microscopic. There might be an arbitrarily large number (n) of microscopic wormholes, and when $n \to \infty$, physical space becomes approximately equivalent to its mathematical idealisation, called elliptical space.

The fact that two sides of the wormhole's throat can be located at a very large distance from each other can be (and is) used to solve the Einstein–Podolsky–Rosen (EPR) paradox [28] and to explain the phenomenon of quantum entanglement of subatomic particles separated by large distances (see, e.g., [29]).

Einstein referred to the possibility of quantum entanglement as to a "spooky action at distance" [30]. However, surprisingly, it is his own wormhole theory [25] that provides a solution to the quantum entanglement and nonlocality puzzle [1]. This implies that quantum mechanics, in order to be consistent, require general relativity, which is not a welcome idea to most modern physicists. However, the matter stands.

Herein, we do not discuss the possibility of solving the EPR paradox using the connectivity between distant regions of space via wormholes. Instead, we focus on another aspect of this nonlocal connectivity, namely the possibility of perceiving the event horizons of local microscopic wormholes (near the observer) as horizons in the far distance from the observer, as these horizons have an effect on both the source and the observer remote from the far-throats of the same wormholes.

These remote horizons of local microscopic wormholes are spherically symmetric around the observer. This follows from the topology of the elliptical space. For example, when looking at the near throat of a microscopic wormhole, the observer localises it within a very narrow solid angle (a zero-aperture solid angle in the case of the mathematical elliptical space). However, when the observer looks at the far throat of the same microscopic wormhole, the solid angle spans 4π steradians (a sphere around the observer), even in the case of an ideal elliptical space when the far throat of the wormhole at $\chi = \pi$ is point-like.

As the remote horizons are at extremely large distances from the observer, their deviations from spherical symmetry due to different locations of neighbouring points around the observer are negligible. Furthermore, the average global collective horizon around any point of ENdS space is practically ideally spherically symmetric, making all points of this space equivalent to each other from the observer's perspective.

As with any Schwartzschild event horizon, there is an associated gravitational red shift, which is calculable when we know the distances from the horizon to the source and to the observer. Thus, in this particular setup, the observed cosmological red shift is gravitational in nature. However, in principle, this gravitational red shift can be mixed with the red shift caused by the changing scale factor of the FLRW metric, as the concept of space expansion can equally be applied to ENdS space. Therefore, the resulting cosmological red shift might be a mix of the gravitational red shift with the red shift caused by a changing scaling factor. We shall discuss this option later elsewhere.

2.2. Red Shift–Distance Relationship

Consistent with ENdS, we adopt the Schwarzschild metric related to the global remote horizon. In our setup (Figure 2), we include unknown distances (r_s, r_o), the source-to-observer distance ($d = r_o - r_s$) and the unknown gravitational radius (r_g). Both the source and observer are within the Schwartzschild metric, which corresponds to the following spacetime interval in spherical coordinates (r, θ, ϕ):

$$ds^2 = g_{tt}c^2dt^2 - g_{rr}dr^2 - r^2(d\theta^2 + \sin^2\theta d\varphi^2), \tag{3}$$

where $g_{tt} = 1 - \frac{r_g}{r}$ and $g_{rr} = g_{tt}^{-1}$. The source's red shift with respect to the observer is

$$z = \sqrt{g_{tt}^o / g_{tt}^s} - 1, \tag{4}$$

or, if we define $r_g := 1$ as the distance unit,

$$(1+z)^2 = \left(1 - r_o^{-1}\right)\left(1 - r_s^{-1}\right)^{-1}. \tag{5}$$

Then, the source-to-observer distance reads

$$d(z) = r_o - \left[1 - (1 - r_o^{-1})(1+z)^{-2}\right]^{-1} \text{ [in units of } r_g\text{]}, \tag{6}$$

which has to be multiplied by the scaling factor $((1+z)^2)$ in order to obtain the luminosity distance

$$d_L(z) = \left\{ r_o - \left[1 - (1 - r_o^{-1})(1+z)^{-2}\right]^{-1} \right\}(1+z)^2, \tag{7}$$

with one of the $(1+z)$ factors accounting for the decrease in the number of incoming photons due to time dilatation in the Schwartzschild metric (the g_{tt} metric coefficient) and another accounting for the photon path distortion (the g_{rr} metric coefficient). Equation (7) is the required red shift–luminosity relationship, permitting us to compare the ENdS-based model with observational data.

2.3. Comparison with Observations

In order to compare the theoretical luminosity distances with observational data (e.g., distance moduli of the type Ia supernovae), the distances need to be scaled and converted into magnitude values comparable with the observed source magnitudes.

For our comparison, we use the well-calibrated sample of 1701 type Ia SNe called *Pantheon+* [32,33]. The uncertainty of the parameters of the standard ΛCDM cosmological model were recently substantially reduced using this *Pantheon+* sample [34]. For example, the uncertainty in the H_0 parameter was reduced to ± 1 km/s/Mpc.

Although the ΛCDM model was previously fitted to the *Pantheon+* sample by D. Brout et al. [33], we repeat that fit here to ensure that our calculation algorithms, when applied to both ΛCDM and ENdS, remain the same in order to consistently intercompare these two models [2].

Starting with the ΛCDM model, the luminosity distance in this model is calculate as a function of red shift (z) from

$$D_L(z) = D_A(z)(1+z)^2, \tag{8}$$

where the scaling factor $((1+z)^2)$ is the same as in Equation (7), and D_A is the angular diameter distance:

$$D_A(z) = \frac{c}{H_0}\frac{1}{1+z}\int_0^z \frac{dz'}{\sqrt{1 + \Omega_{\mathrm{M}}[(1+z')^3 - 1]}}, \tag{9}$$

as calculated for a flat cosmology $(\Omega_k = 0)$. On the other hand, the luminosity distance is defined as the relationship between the bolometric flux and luminosity of a source, which is encoded in the source distance moduli provided by the *Pantheon+* sample. If D_L is expressed in Mpc, the distance modulus is

$$\mu_{\Lambda\mathrm{CDM}} = 5\log D_L + 25. \tag{10}$$

3. Results

By fitting the theoretical values (10) to the observationally determined distance moduli of type Ia supernovae, we can find the values of the ΛCDM parameters. In the flat ΛCDM, there are two free parameters–H_0 and Ω_M– and a fixed parameter, $\Omega_\Lambda = 1 - \Omega_M$. The fit can be achieved by minimising Pearson's χ^2 [33]:

$$\chi^2 = \Delta\mathbf{D}^T C^{-1}\Delta\mathbf{D}, \tag{11}$$

where C is the covariance matrix, and $\Delta \mathbf{D}$ is the vector of SN distance–modulus residuals

$$\Delta \mathbf{D}_i = \mu_{\Lambda\text{CDM}}(z_i) - \mu_i, \tag{12}$$

the length of which is $N = 1701$ for the *Pantheon+* sample.

Since here, we are only interested in comparing the goodness of fit of two different cosmological models, we do not need to reach out for the correct cosmological parameters via these fits. Thus, we can use a simplified statistic

$$\chi^2 = (\text{diag}\, C^T \Delta \mathbf{D})^2 = \sum_{i=1}^{N} \frac{\Delta \mathbf{D}_i^2}{\sigma_{\mu_i}^2}, \tag{13}$$

where $\sigma_{\mu_i}^2$ are the uncertainties of μ_i as determined from the diagonal of the covariance matrix (see, e.g., [36], § IIIc for theoretical work or [37] for practical examples of using (13) and the *Pantheon* sample [38] for comparison of various cosmological models between each other). For our comparison, we use μ_{SH0ES} from [33], which are the corrected distance moduli where fiducial type Ia SNe magnitudes (M) were determined from SH0ES 2022 Cepheid host absolute distances [34]. This minimisation of χ^2 gives

$$\begin{aligned} H_0 &= 72.429^{+0.116}_{-0.109}\,[\text{km/s/Mpc}]; \\ \Omega_M &= 0.389^{+0.010}_{-0.007}, \end{aligned} \tag{14}$$

which differ, as expected, from those based on the full covariance matrix ($H_0 = 73.6 \pm 1.1\,[\text{km/s/Mpc}]$; $\Omega_M = 0.334 \pm 0.018$ [33]). This is acceptable, as we are interested in the goodness of fit characterised by the minimal value

$$\chi^2_{\Lambda\text{CDM}} = 881.15. \tag{15}$$

Minimisation is achieved by the global descent method with consecutive iterations, and the confidence limits in (14) are estimated by using the calculated parameter values corresponding to Pearson's probability of 68.3% divided by the square root of the number of degrees of freedom ($\sqrt{N - n_{\text{p}}}$), where the number of free parameters is $n_{\text{p}} = 2$.

In the case of ENdS, besides its free parameter (r_0), the expression (10) requires an extra free parameter, such as s_g (a scaling factor), in order to match the theoretical μ_{ENdS} with the observationally determined μ from the type Ia SNe:

$$\mu_{\text{ENdS}} = 5\log(s_g d_L) + 25, \tag{16}$$

because d_L is expressed in units of r_g, and we need to scale it to Mpc. Thus, the ENdS model, like the flat ΛCDM, also has two free parameters, the χ^2-minimised values of which are

$$\begin{aligned} r_0 - 1 &= (9.91^{+0.02}_{-0.01}) \cdot 10^{-8}; \\ s_g &= (2.13^{+0.14}_{-0.13}) \cdot 10^{10}\,[\text{Mpc}], \end{aligned} \tag{17}$$

with the minimal $\chi^2_{\text{ENdS}} = 887.56$.

The results of these two χ^2 fits are graphically presented in Figure 3, where the distance moduli (μ) from the type Ia SNe *Pantheon+* sample are plotted as red points, and the minimum χ^2-fitted theoretical curves for the flat ΛCDM model and the ENdS model are plotted with solid and dashed curves, respectively.

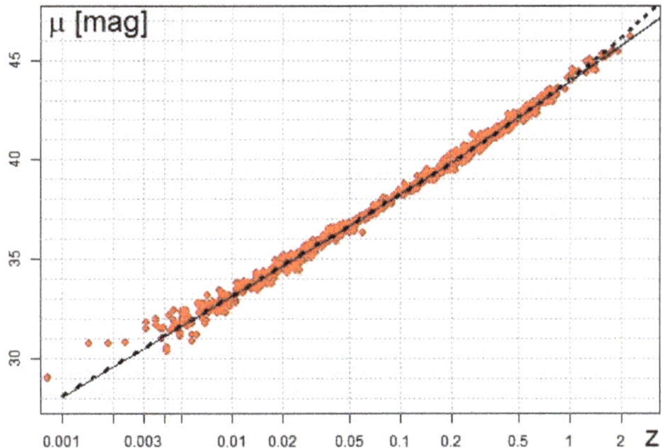

Figure 3. Distance moduli (μ) from the type Ia SNe *Pantheon+* sample (red points) as a function of red shift (z), with the minimal χ^2-fitted theoretical curves for the flat ΛCDM model (thin solid curve) and the ENdS model (dashed curve).

4. Discussion

4.1. Validation Using a Gamma-Ray Burst Sample

By comparing the minimal values ($\chi^2_{\Lambda\text{CDM}} = 881.15$ and $\chi^2_{\text{ENdS}} = 887.56$), we conclude that, according to observational evidence (the type Ia SNe distance moduli), both the ΛCDM and ENdS cosmological models compete on an equal footing with respect to the prediction of distance moduli.

We see that within the red-shift range of $0 < z < 0.7$, the two models are practically identical. However, for larger red shifts, the ENdS model predicts slightly larger distance moduli (fainter SNe) than those expected within the ΛCDM framework (see the upper-right corner of Figure 3).

The difference between the model predictions within the red-shift range of $1 < z < 2$ is not significant (a fraction of magnitude). On the other hand, the small number of available SNe within this red-shift range and the scatter of their magnitudes do not allow for confident selection one model over the other.

For a robust comparison of these two models, we would need accurately calibrated type Ia SNe with red shifts of $z > 3$. Unfortunately, they have not yet been discovered. However, they are expected to be discovered in a few years by the James Webb Space Telescope (JWST). Should the future newly discovered SNe with $z > 3$ be fainter than what is expected in ΛCDM cosmology, then our cosmological model based on Einstein–Newcomb–de Sitter space would be robustly confirmed.

The 1998 discovery of type Ia SNe excessive dimming at $z \simeq 1$ [9–11] was interpreted as evidence in favour of dark energy (repulsive gravity or the Λ term in Einstein's equations). In physics, dark energy is an unknown entity, and it can only be viably physically interpreted in terms of vacuum energy. Experimental evidence from particle physics suggests that the vacuum energy density (due to quantum fluctuations) must be large enough that it is discrepant by the order of 10^{120} from what is currently deduced from type Ia SNe observations.

In contrast, the competing model based on ENdS space discussed here is based on the experimentally observed effect of gravitational red shift. In addition, the ENdS model prediction can be appropriately validated in the near future via the expected aforementioned JWST discoveries.

As we cannot yet do so due to the lack of standard candle data for $z > 3$, we can get a hint of what these data might be by using a proxy for standard candle data in the form of gamma-ray burst distance moduli (μ_{GRB}) obtained via the Amati relation [24]. These GRB distance moduli (μ_{GRB}) are extremely noisy in comparison with the distance moduli (μ)

of the type Ia SNe, including a low-red shift systematic bias of μ_{GRB}. We calculated this systematic bias to be +0.258 (mag) by minimising χ^2 for the 27 μ_{GRB} values for $z < 0.7$ (as the ΛCDM and ENdS models are identical within this red-shift range). The distance moduli of both type Ia SNe (red points) and GRBs (blue points) are plotted in Figure 4, with the GRB red shifts projected to $z \simeq 8$. As in the previous plot, the ΛCDM-based theoretical distance moduli are indicated by a thin, solid curve, and the ENdS-based distance moduli are indicated by a thicker dashed curve.

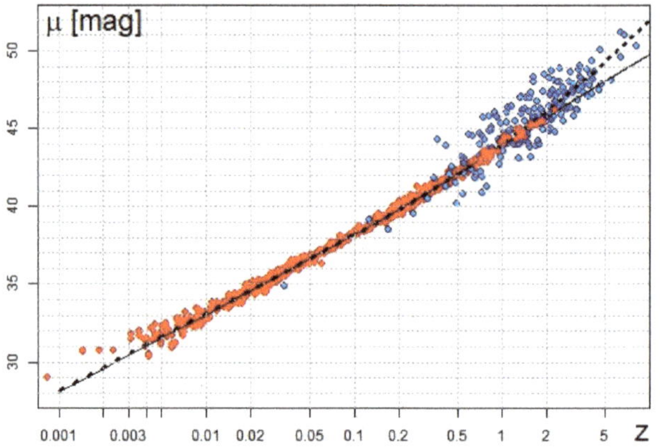

Figure 4. Distance moduli μ from the *Pantheon+* type Ia SNe sample (red points), together with the distance moduli from the GRB sample (blue points) calibrated using the Amati relation. The thin, solid and thick, dashed curves indicate the minimum χ^2-fitted theoretical curves for the flat ΛCDM and ENdS models, respectively.

A large sample of quasars spanning approximately the same red-shift range as GRBs is also available [39]. This sample is thought to be useful as yet another proxy for cosmological standard candles. However, the sample of quasars is strongly affected by the observational selection effect for $z > 3$, as found by Raikov, Lovyagin & Yershov [40]. This means that this sample cannot be used for robust validation of cosmological models. Some other authors, e.g., M. López-Corredoira [41] and N. Khadka & B. Ratra [42], arrived at the same conclusion, which is why we do not use quasars in our study.

4.2. Experimental Challenges of Static and Dynamic Cosmological Models

Our analysis favours a static rather than expanding cosmological model. Static models were largely superseded by the standard cosmological model of the expanding universe because it appeared to explain many observations in an elegant way, including the formation of light elements, the cosmic microwave background and the power spectrum of matter overdensities in the local universe. The standard ΛCDM cosmology has therefore been almost universally adopted because of its strength in explaining observational evidence.

However, the same observational evidence has, in the past, also been used to support static universe models. As we previously mentioned in Section 1, the cosmological red shift was predicted by de Sitter for Einstein's static model well before such a phenomenon was experimentally discovered [2]. A thermalised background radiation with $T = 3\,\mathrm{K}$ (CMB) for a static universe was predicted in 1926 by A.S. Eddington [43], and with $T = 2.8\,\mathrm{K}$ by W. Nernst in 1937 [44]. These predictions were also made well before G. Gamow's CMB prediction in 1953 for the expanding universe [45]. Further detailed explanations of CMB properties in the framework of a static universe model were made by several authors, including Yu. Baryshev [46], M. Cirkovic [23] and others. As for the temperature (energy) of CMB photons arriving at the observer from remote distances, it scales with red shift in ENdS space in exactly the same way as in expanding universe models,

$$T(z) = T_0(1 + z), \tag{18}$$

where $T_0 \approx 2.7\,\text{K}$. It follows from the fact that both models reproduce the Hubble law, i.e., photons arriving from further away (including remote background photons) have reduced energies compared with the energies of photons emitted nearer to the observer. The possibility that the cosmic background has a local origin in static universe models is confidently excluded by measurements of excitation lines in absorption features of quasar spectra (for example, see [47]) or by measuring the imprint of galaxy clusters on the cosmic background via the Sunyaev–Zeldovich effect, as in [48]. However, for a cosmic background of a distant origin, the temperature–red shift relation in static universe models follows Equation (18). As mentioned above in Section 2.1, ENdS space evolves over time. Therefore, in contrast with any purely static universe model, the origin of the remote cosmic background in the model based on ENdS space can include an evolutionary component similar to expanding universe models.

The abundances of light elements in a static universe were explained by G.R. Burbidge and F. Hoyle [18,19], R. Salvaterra and A. Ferrara [49] and others. However, there are some unresolved issues for static universe models. For example, according to the standard cosmological model, deuterium (^{2}H) was created exclusively during the Big Bang nucleosynthesis stage, after which it cannot be produced and can only be destroyed in stars [50]. Therefore, its observed abundance is gradually diminishing. Similarly, lithium (^{7}Li) is also regarded as having been produced during the Big Bang nucleosynthesis. However, observations suggest its continuous enrichment due to cosmic-ray spallation [51].

There are other elements (e.g., beryllium and boron) that cannot be produced in stars. Their existence can be explained by a solution in the form of cosmic-ray spallation/fusion reactions [52]. This solution has been extensively studied and discussed by many authors, e.g., [53–55]. It opens an alternative way to model light element formation, which is required by static and semistatic cosmological models [19]. This mechanism also provides the possibility of producing ^{2}H, as well as of replenishment of ^{1}H that has been burned in nuclear reactions in stars. Since the energies of cosmic rays can be as high as 10^{19} eV, they can produce spallation fragments, even from ^{4}He [56]. ^{4}He spallation processes can also involve highly energetic neutrinos [57]. Reactions of this kind are absolutely real, as they are regularly observed in laboratory experiments [58].

Other alternatives to Big Bang nucleosynthesis are the synthesis of light elements in massive objects within the nuclear regions of galaxies [18] and the creation of practically the whole periodic table of the elements in extreme processes involving relativistic compact objects, e.g., neutron stars [59,60]. As all the abovementioned alternatives are extensively discussed in literature, we do not need any deeper discussion here.

However, there is one issue that does require attention when considering a static universe model, which is the question of cosmic structure formation, which is resolved by the ΛCDM model via quantum fluctuations during the universe's inflation stage and by baryonic acoustic oscillations in the early universe. Overdensities are then modelled by performing perturbative expansions with respect to the background FLRW geometry.

A.S. Eddington [61] demonstrated that homogeneous and isotropic perturbations cause Einstein's static universe to be unstable, as a result of which Einstein abandoned his static cosmological model in 1931. Much later, N. Rosen proved the opposite—that Einstein's static model was, in fact, stable [62]. This opened up the question as to whether the structure formation issue could be resolved in static universe models.

While discussing the observational challenges for static universe models, it would be unfair to ignore the problems accumulated in the past decade for the ΛCDM model. First, it is worth mentioning that the standard Big Bang nucleosynthesis theory is not without serious problems [63–65]. Secondly, four years ago, the Hubble constant tension issue became so acute within the standard cosmology framework, which was regarded as a crisis in modern cosmology [66].

Despite the successes of ΛCDM's perturbative approach to correctly predict structure-formation parameters, this approach is still challenged by the fact that on scales below 100 Mpc, the matter distribution in the universe is extremely inhomogeneous. This problem was recently alleviated by considering quasistatic models containing black hole lattices [67]. Structure formation in such models is quite complicated, but this provides a prospect for solving the structure-formation issue in static universe models, especially in those based on black holes or wormholes, such as the model discussed here.

For simplicity, herein, we discussed a static version of ENdS space. However, mentioned above in Section 2.1, the concept of space expansion/contraction can equally be applied to ENdS space. Such a mixed model needs a third free parameter, which would result in the model's slightly better fit to the *Pantheon+* SNe data. This is a work in progress, and we shall discuss it in a separate paper. However, we can already note that mixed ENdS space allows for the exploitation of the existing structure-formation and light-element-synthesis formalisms that are in use for expanding universe models.

In 2012, we found that the CMB data obtained by the Wilkinson Microwave Anisotropy Probe were contaminated by irreducible distant (intergalactic) foreground [68]. That contamination was later confirmed to persist in the *Plank* space mission measurements [69,70]. We also found that the inhomogeneities in the *Planck* CMB maps are fractal [71], which is consistent with the well-known fractal distribution of matter in the universe [72–74].

There is statistical evidence that matter overdensities and underdensities imprint on the CMB as hot and cold spots [75,76]. For example, the CMB cold spot, which is inexplicable within the ΛCDM framework, was found to be physically related to the Eridanus supervoid [77,78]. These results support Eddington's idea that CMB radiation could be either partially or entirely explained as thermalised radiation from matter.

Since CMB data contamination is caused by a foreground whose influence is currently impossible to take into account, it is highly likely that the CMB-derived ΛCDM cosmological parameters cannot be trusted.

Recently, JWST observations of highly red-shifted galaxies [79,80] have shown them to be similar in appearance to the fully developed galaxies found in the late universe, despite their impossibly short ages as calculated using ΛCDM. This inconsistency casts doubt on one of the foundational pillars of the standard cosmological model, which is accurate prediction of geometrical and evolutionary structure.

In turn, this fact leads us to question whether the whole bulk of observational evidence supporting ΛCDM is, in fact, untrustworthy. It is this observational evidence that became the main reason for building the alternative cosmological model discussed here.

5. Conclusions

The comparison between the standard ΛCDM cosmological model and the cosmological model proposed here demonstrates that these two models are practically identical in terms of their predictions for the distance moduli of the available standard candles (type Ia SNe) within the red-shift range of $0 < z < 2.3$.

For higher red shifts, the model based on ENdS space predicts longer-distance moduli (fainter type Ia SNe) than those calculated within the framework of the standard ΛCDM cosmological model. This theoretical prediction can be experimentally verified in the future, as new discoveries of type Ia SNe with $z \simeq 3$ are expected within a few years by the JWST.

Funding: This research received no external funding.

Institutional Review Board Statement: Institutional ethical review and approval are not applicable to this study, as it does not involve animals or humans.

Informed Consent Statement: Not applicable.

Data Availability Statement: The data used to prepare this manuscript are available at https://pantheonplussh0es.github.io (accessed on 24 April 2023) and in Table 1 from [24].

Acknowledgments: The author made use of the following archives for this research: (1) The *Pantheon+* type Ia supernova distance moduli from [32–34]. (2) The Gamma-ray burst distance moduli prepared by [24]. The author acknowledges the use of the R software package developed by Andrew Harris and available at github.com/cran/cosmoFns (accessed on 24 April 2023). I would like to thank Paul Kuin, Leslie Morrison, Alice Breeveld, and Mat Page for useful discussions on the matters investigated in this paper. I am also grateful to two anonymous reviewers who drew my attention to some missing formulae and suggested highlighting various important points in the Discussion section.

Conflicts of Interest: The author declares no conflict of interest.

Abbreviations

The following abbreviations are used in this manuscript:

CMB	Cosmic microwave background (radiation)
ENdS	Enstein–Newcomb–de Sitter (space)
EPR	Einstein–Podolsky–Rosen (paradox)
FLRW	Friedmann–Lemaitre–Robertson–Walker (metric)
GRB	Gamma-ray burst
JWST	James Webb Space Telescope
ΛCDM	Lambda cold–dark matter (cosmological model)
SN	supernova.

Notes

1   In fact, this is the only viable way of understanding what matter is and what particles of matter are made of: "there is nothing in the world except empty curved space" [31]. It is also a way of answering the question as to the origin of particle species and the pattern of three generations of fundamental fermions. These are incorporated into the Standard Model of particle physics as something given to us by nature, setting aside the question, "why do we have this particular set of fundamental particles and not something else?". This is a related but different question, discussed by the author elsewhere.

2   There might be small differences in algorithms and software used by different research groups, so our results might also be slightly different from other previously published results. We conduct our calculations here using the R package *cosmoFns* [35] and the formulae presented above.

References

1. Einstein, A. Kosmologische betrachtungen zur allgemeinen Relativitätstheorie. *Sitz. Preuss. Akad. Wiss Phys.* **1917**, *VL*, 142–152.
2. de Sitter, W. On Einstein's theory of gravitation, and its astronomical consequences. Third paper. *Mon. Not. R. Astron. Soc.* **1917**, *78*, 3–28. [CrossRef]
3. Newcomb, S. Elementary theorems relating to the geometry of a space of three dimensions and of uniform positive curvature in the fourth dimension. *J. Für Die Reine Und Angew. Math.* **1877**, *LXXXIII*, 293–299.
4. Hubble, E. A relation between distance and radial velocity among extragalactic nebulae. *Proc. Natl. Acad. Sci. USA* **1929**, *15*, 168–173. [CrossRef]
5. Friedmann, A. Über die Krümmung des Raumes. *Z. Phys. A* **1922**, *10*, 377–386. [CrossRef]
6. Lemaître, G. Un univers homogène de masse constante et de rayon croissant rendant compte de la vitesse radiale des nébuleuses extra-galactiques. *Ann. Soc. Sci. Brux. A* **1927**, *47*, 49–59.
7. Robertson, H.P. Kinematics and world structure. *Astrophys. J.* **1935**, *82*, 284–301. [CrossRef]
8. Walker, A.G. On Milne's theory of world-structure. *Proc. Lond. Math. Soc. Ser. 2* **1937**, *42*, 90–127. [CrossRef]
9. Riess, A.G.; Filippenko, A.V.; Challis, P.; Clocchiatti, A.; Diercks, A.; Garnavich, P.M.; Gillil, R.L.; Hogan, C.J.; Jha, S.; Kirshner, R.P.; et al. Observational evidence from supernovae for an accelerating Universe and cosmological constant. *Astron. J.* **1998**, *116*, 1009–1038. [CrossRef]
10. Schmidt, B.P.; Suntzeff, N.B.; Phillips, M.M.; Schommer, R.A.; Clocchiatti, A.; Kirshner, R.P.; Garnavich, P.; Challis, P.; Leibundgut, B.R.; Spyromilio, J.; et al. The high-z supernovae search: Measuring cosmic deceleration and global. *Astrophys. J.* **1998**, *507*, 46–63. [CrossRef]
11. Perlmutter, S.; Aldering, G.; Goldhaber, G.; Knop, R.A.; Nugent, P.; Castro, P.G.; Deustua, S.; Fabbro, S.; Goobar, A.; Groom, D.E.; et al. Measurement of Ω and Λ from 42 high-redshift supernovae. *Astrophys. J.* **1999**, *517*, 565–586. [CrossRef]
12. Dolgov, A.D. Massive and supermassive black holes in the contemporary and early Universe and problems in cosmology and astrophysics. *Phys. Uspekhi* **2018**, *61*, 115–132. [CrossRef]
13. Lerner, E.J. Observations contradict galaxy size and surface brightness predictions that are based on the expanding universe hypothesis. *Mon. Not. R. Astron. Soc.* **2018**, *477*, 3185–3196. [CrossRef]

14. Lovyagin, N.; Raikov, A.; Yershov, V.; Lovyagin, Y. Cosmological model tests with JWST. *Galaxies* **2022**, *10*, 108. [CrossRef]
15. Zwicky, F. On the redshifts of spectral lines through interstellar space. *Proc. Natl. Acad. Sci. USA* **1929**, *15*, 773–779. [CrossRef]
16. Tolman, R.C. On the estimation of distances in a curved universe with a non-static line element. *Proc. Nat. Acad. Sci. USA* **1930**, *16*, 511–520. [CrossRef] [PubMed]
17. Hubble, E.; Tolman, R.C. Two methods of investigating the nature of the nebular redshift. *Astrophys. J.* **1935**, *82*, 302–337. [CrossRef]
18. Burbidge, G.R. Was there really a Big Bang? *Nature* **1971**, *233*, 36–40. [CrossRef] [PubMed]
19. Burbidge, G.R.; Hoyle, F. The origin of helium and the other light elements. *Astrophys. J.* **1998**, *509*, L1–L3. [CrossRef]
20. Baryshev, Y. The hierarchical structure of metagalaxy a review of problems. *Rep. Space Astrophys. Obs. Russ. Acad. Sci.* **1981**, *14*, 24–43.
21. Troitskij, V.S. A static model of the universe. *Astrophys. Space Sci.* **1995**, *229*, 89–104. [CrossRef]
22. Baryshev, Y.; Teerikorpi, P. *Fundamental Questions of Practical Cosmology*; Springer: Dordrecht, The Netherlands, 2012; p. 332.
23. Cirkovic, M.M.; Perovic, S. Alternative explanations of the Cosmic Microwave Background: A historical and an epistemological perspective. *Stud. Hist. Philos. Mod. Phys.* **2018**, *62*, 1–18. [CrossRef]
24. Amati, L.; D'Agostino, R.; Luongo, O.; Muccino, M.; Tantalo, M. Addressing the circularity problem in the $E_p - E_{ISO}$ correlation of gamma-ray bursts. *Mon. Not. R. Astron. Soc. Lett.* **2019**, *486*, L46–L51. [CrossRef]
25. Einstein, A.; Rosen, N. The Particle Problem in the General Theory of Relativity. *Phys. Rev.* **1935**, *48*, 73–77. [CrossRef]
26. Morris, M.S.; Thorne, K.S. Wormholes in spacetime and their use for interstellar travel: A tool for teaching general relativity. *Am. J. Phys.* **1988**, *56*, 395–412. [CrossRef]
27. Morris, M.S.; Thorne, K.S.; Yurtsever, U. Wormholes, time machines, and the weak energy condition. *Phys. Rev. Lett.* **1988**, *61*, 1446–1449. [CrossRef]
28. Einstein, A.; Podolsky, B.; Rosen, N. Can quantum-mechanical description of physical reality be considered complete? *Phys. Rev.* **1935**, *47*, 777–780. [CrossRef]
29. Tamburini, F.; Licata, I. General relativistic wormhole connections from Planck-scales and the ER = EPR conjecture. *Entropy* **2020**, *22*, 3. [CrossRef]
30. Einstein, A. *The Born-Einstein Letters: Correspondence between Albert Einstein and Max and Hedwig Born from 1916–1955, with Commentaries by Max Born*; Walker & Company Publ.: New York, NY, USA, 1971; p. 158.
31. Misner, C.; Wheeler, J.A. Classical physics as geometry. *Ann. Phys.* **1957**, *2*, 525–603. [CrossRef]
32. Scolnic, D.; Brout, D.; Carr, A.; Riess, A.G.; Davis, T.M.; Dwomoh, A.; Jones, D.O.; Ali, N.; Charvu, P.; Chen, R.; et al. The Pantheon+ Analysis: The Full Dataset and Light-Curve Release. *Astrophys. J.* **2022**, *938*, 113. [CrossRef]
33. Brout, D.; Scolnic, D.; Popovic, B.; Riess, A.G.; Carr, A.; Zuntz, J.; Kessler, R.; Davis, T.M.; Hinton, S.; Jones, D.; et al. The Pantheon+ Analysis: Cosmological Constraints. *Astrophys. J.* **2022**, *938*, 110. [CrossRef]
34. Riess, A.G.; Yuan, W.; Macri, L.M.; Scolnic, D.; Brout, D.; Casertano, S.; Jones, D.O.; Murakami, Y.; An, G.S.; Breuval, L.; et al. A comprehensive measurement of the local value of the Hubble constant with 1 km/s/Mpc uncertainty from the Hubble Space Telescope and the SH0ES team. *Astrophys. J.* **2022**, *934*, L7. [CrossRef]
35. Harris, A. *cosmoFns* Functions for cosmological distances, times, luminosities, etc. *R J.* **2012**, *4*, 90.
36. Cash, W. Parameter estimation in astronomy through application of likelihood ratio. *Astrophys. J.* **1979**, *228*, 939–947. [CrossRef]
37. López-Corredoira, M.; Calvo-Torel, J.L. Fitting of supernovae without dark energy. *Int. J. Mod. Phys. D* **2022**, *31*, 2250104. [CrossRef]
38. Scolnic, D.M.; Jones, D.O.; Rest, A.; Pan, Y.C.; Chornock, R.; Foley, R.J.; Huber, M.E.; Kessler, R.; Narayan, G.; Riess, A.G.; et al. The complete light-curve sample of spectroscopically confirmed SNe Ia from Pan-STARRS1 and cosmological constraints from the combined Pantheon sample. *Astrophys. J.* **2018**, *859*, 101. [CrossRef]
39. Civano, F.; Eggleston, L.; Elvis, M.; Fabbiano, G.; Gilli, R.; Marconi, A.; Paolillo, M.; Piedipalumbo, E.; Salvestrini, F.; Signorini, M.; et al. Quasars as standard candles. *A&A* **2020**, *642*, A150.
40. Raikov, A.; Lovyagin, N.; Yershov, V. Superluminous quasars and mesolensing. In *Aastronomy at the Epoch of Multimessenger Studies. VAK-2021 Proceedings*; Cherepashchuk, A.M., Emelyanov, N.V., Fedorova, A.A., Gayazov, I.S., Ipatov, A.V., Ivanchik, A.V., Malkov, O.Y., Obridko, V.N., Rastorguev, A.S., Samus, N.N., et al., Eds.; Janus-K Publ.: Moscow, Russia, 2022; pp. 392–394.
41. López-Corredoira, M. Pending Problems in QSOs. *Int. J. Astron. Astropys.* **2011**, *1*, 73–82. [CrossRef]
42. Khadka, N.; Ratra, B. Do quasar X-ray and UV flux measurements provide a useful test of cosmological models? *Mon. Not. Roy. Astron. Soc.* **2022**, *510*, 2753–2772. [CrossRef]
43. Eddington, A.S. *Internal Constitution of the Stars*; Cambridge University Press: Cambridge, UK, 1926; p. 407.
44. Nernst, W. Weitere prüfung der annahme lines stationären zustandes im weltall. *Z. Phys.* **1937**, *106*, 633–661. [CrossRef]
45. Gamow, G. The expanding universe and the origin of galaxies. *Kgl. Dan. Vidensk. Selsk. Mat. Fys. Medd.* **1953**, *27*, 3–15.
46. Baryshev, Y.V.; Raikov, A.A.; Tron, A.A. Microwave background radiation and cosmological large numbers. *Astron. Astroph. Trans.* **1996**, *10*, 135–138. [CrossRef]
47. Muller, S.; Beelen, A.; Black, J.H.; Curran, S.J.; Horellou, C.; Aalto, S.; Combes, F.; Guélin, M.; Henkel, C. A precise and accurate determination of the cosmic microwave background temperature at $z = 0.89$. *Astron. Astrophys.* **2013**, *551*, A109. [CrossRef]

48. Luzzi, G.; Shimon, M.; Lamagna, L.; Rephaeli, Y.; De Petris, M.; Conte, A.; De Gregori, S.; Battistelli, E.S. Redshift dependence of the cosmic microwave background temperature from Sunyaev-Zeldovich measurements. *Astron. J.* **2009**, *705*, 1122–1128. [CrossRef]

49. Salvaterra, R.; Ferrara, A. Is primordial ^{4}He truly from the Big Bang? *Mon. Not. R. Astron. Soc.* **2003**, *340*, L17–L20. [CrossRef]

50. Pagel, B.E.J. Abundances of elements of cosmological interest. *Philos. Trans. R. Soc. Lond. A* **1982**, *307*, 19–35.

51. Spite, F.; Spite, M. Abundances of Lithium in unevolved halo stars and old disk stars: Interpretations and consequences. *Astron. Astrophys.* **1982**, *115*, 357–366.

52. Reeves, H.; Fpwler, W.A.; Hoyle, F. Galactic cosmic ray origin of Li, Be and B in stars. *Nature* **1970**, *226*, 727–729. [CrossRef]

53. Austin, S.M. The creation of the light elements—Cosmic rays and cosmology. *Prog. Part. Nucl. Phys.* **1981**, *7*, 1–46. [CrossRef]

54. Silverberg, R.; Tsao, C.H. Spallation processes and nuclear interaction products of cosmic rays. *Phys. Rep.* **1990**, *191*, 351–408. [CrossRef]

55. Olive, K.A.; Schramm, D.N. Astrophysical ^7Li as a product of Big Bang nucleosynthesis and galactic cosmic-ray spallation. *Nature* **1992**, *360*, 439–442. [CrossRef]

56. Yamanaka, M.; Jittoh, T.; Kazunori Kohri, K.; Koike, M.; Sato, J.; Sugai, K.; Yazaki, K. Big-bang nucleosynthesis with a long-lived CHAMP including He4 spallation process. *J. Phys. Conf. Ser.* **2014**, *485*, 012020. [CrossRef]

57. Meyer, B.S.; Woosley, S.E.; Hoffman, R.D.; Mathews, G.J.; Wilson, J.R. Neutrino spallation reactions on ^{4}He and the r-process. *AIP Conf. Proc.* **1995**, *327*, 441–445.

58. Oliver, B.M.; James, M.R.; Garner, F.A.; Maloy, S.A. Helium and hydrogen generation in pure metals irradiated with high-energy protons and spallation neutrons in LANSCE. *J. Nucl. Mat.* **2002**, *307–311*, 1471–1477. [CrossRef]

59. Abbott, B.P. et al. (LIGO, Virgo and other collaborations). Multi-messenger Observations of a Binary Neutron Star Merger. *Astrophys. J.* **2017**, *848*, L12. [CrossRef]

60. Gompertz, B.P.; Ravasio, M.E.; Nicholl, M.; Levan, A.J.; Metzger, B.D.; Oates, S.R.; Lamb, G.P.; Fong, W.F.; Malesani, D.B.; Rastinejad, J.C.; et al. The case for a minute-long merger-driven gamma-ray burst from fast-cooling synchrotron emission. *Nat. Astron.* **2022**, *7*, 67–79. [CrossRef]

61. Eddington, A.S. On the instability of Einstein's spherical world. *Mon. Not. R. Astron. Soc.* **1930**, *90*, 668-678. [CrossRef]

62. Rosen, N. Static universe and cosmic field. *Ann. Math. Pure Appl.* **1970**, *14*, 305–308. [CrossRef]

63. Sargent, W.L.W.; Searle, L. The interpretation of the helium weakness in halo stars. *Astrophys. J.* **1967**, *150*, L33–L37. [CrossRef]

64. Terlevich, E.; Terlevich, R.; Skillman, E.; Stepanian, J.; Lipovetskii, V. The extremely low He abundance of SBS:0335-052. In *Elements and the Cosmos*; Edmunds, M.G., Terlevich, R., Eds.; Cambridge University Press: Cambridge, UK, 2010; pp. 21–27.

65. Izotov, Y.I.; Thuan, T.X. The primordial abundance of ^{4}He: Evidence for non-standard Big Bang nucleosynthesis. *Astrophys. J.* **2010**, *710*, L67–L71. [CrossRef]

66. Di Valentino, E.; Melchiorri, A.; Silk, J. Planck evidence for a closed Universe and a possible crisis for cosmology. *Nat. Astron.* **2020**, *4*, 196–203. [CrossRef]

67. Durk, J.; Clifton, T. A quasi-static approach to structure formation in black hole universes. *J. Cosmol. Astropart. Phys.* **2017**, *10*, 12. [CrossRef]

68. Yershov, V.N.; Orlov, V.V.; Raikov, A.A. Correlation of supernova redshifts with temperature fluctuations of the cosmic microwave background. *Mon. Not. R. Astron. Soc.* **2012**, *423*, 2147–2152. [CrossRef]

69. Yershov, V.N.; Orlov, V.V.; Raikov, A.A. Possible signature of distant foreground in the *Planck* data. *Mon. Not. R. Astron. Soc.* **2014**, *445*, 2440–2445. [CrossRef]

70. Yershov, V.N.; Raikov, A.A.; Lovyagin, N.Y.; Kuin, N.P.M.; Popova, E.A. Distant foreground and the Planck-derived Hubble constant. *Mon. Not. R. Astron. Soc.* **2020**, *492*, 5052–5056. [CrossRef]

71. Mylläri, A.A.; Raikov, A.A.; Orlov, V.V.; Tarakanov, P.A.; Yershov, V.N.; Yezhkov, M.Y. Fractality of isotherms of the Cosmic Microwave Background based on data from the *Planck* spacecraft. *Astrophysics* **2016**, *59*, 31–37. [CrossRef]

72. Coleman, P.H.; Pietronero, L. The fractal structure of the universe. *Phys. Rep.* **1992**, *213*, 311–389. [CrossRef]

73. Gaite, J. The fractal geometry of the cosmic web and its formation. *Adv. Astron.* **2019**, *2019*, 6587138. [CrossRef]

74. Teles, S.; Lopes, A.R.; Ribeiro, M.B. Galaxy distributions as fractal systems. *Eur. Phys. J. C* **2021**, *82*, 896. [CrossRef]

75. Granett, B.R.; Neyrinck, M.C.; Szapudi, I. An imprint of superstructures on the microwave background due to the integrated Sachs-Wolfe effect. *Astrophys. J.* **2008**, *683*, L99–L102. [CrossRef]

76. Cai, Y.-C.; Neyrinck, M.C.; Szapudi, I.; Cole, S.; Frenk, C.S. A possible cold imprint of voids on the microwave background radiation. *Astrophys. J.* **2014**, *786*, 110. [CrossRef]

77. Kovács, A.; García-Bellido, J. Cosmic troublemakers: The Cold Spot, the Eridanus supervoid, and the Great Walls. *Mon. Not. R. Astron. Soc.* **2016**, *462*, 1882–1893. [CrossRef]

78. Kovács, A.; Jeffrey, N.; Gatti, M.; Chang, C.; Whiteway, L.; Hamaus, N.; Lahav, O.; Pollina, G.; Bacon, D.; Kacprzak, T.; et al. The DES view of the Eridanus supervoid and the CMB cold spot. *Mon. Not. R. Astron. Soc.* **2021**, *510*, 216–229. [CrossRef]

79.   Labbé, I.; van Dokkum, P.; Nelson, E.; Bezanson, R.; Suess, K.A.; Leja, J.; Brammer, G.; Whitaker, K.; Mathews, E.; Stefanon, M.; et al. A population of red candidate massive galaxies $\sim$ 600 Myr after the Big Bang. *Nature* **2023**, *616*, 266–269. [CrossRef]
80.   Yan, H.; Ma, Z.; Ling, C.; Cheng, C.; Huang, J.-S. First batch of $z \approx$ 11–20 candidate objects revealed by the James Webb Space Telescope early release observations on SMACS 0723-73. *Astrophys. J. Leet.* **2023**, *942*, L9. [CrossRef]

Article

Revise the Phase-Space Analysis of the Dynamical Spacetime Unified Dark Energy Cosmology

Andronikos Paliathanasis [1,2]

1 Institute of Systems Science, Durban University of Technology, P.O. Box 1334, Durban 4000, South Africa; anpaliat@phys.uoa.gr
2 Departamento de Matemáticas, Universidad Católica del Norte, Avda. Angamos 0610, Casilla, Antofagasta 1280, Chile

Abstract: We analyze the phase-space of an alternate scalar field cosmology that aims to combine the concepts of dark energy and the dark sector. The investigation focuses on stationary points within this phase-space, considering different functional forms of the two potential functions. Our findings indicate that a de Sitter universe is achievable solely when at the asymptotic limit the potential function is constant. For constant potential function, the de Sitter universe is recovered in the finite regime; however, for the exponential potential, the de Sitter universe exists at the infinity regime. The cosmological viability of the present theory is discussed.

Keywords: cosmology; asymptotic dynamics; unified dark energy

Citation: Paliathanasis, A. Revise the Phase-Space Analysis of the Dynamical Spacetime Unified Dark Energy Cosmology. *Universe* **2023**, *9*, 406. https://doi.org/10.3390/universe9090406

Academic Editor: Kazuharu Bamba

Received: 22 August 2023
Revised: 1 September 2023
Accepted: 4 September 2023
Published: 5 September 2023

1. Introduction

The analysis of the observational data shows that our universe is dominated by dark energy and dark matter. Collectively, these components constitute approximately 96% of the total energy density of the universe, with dark matter making up approximately 28% and dark energy around 68%. Dark matter is a hypothetical form of matter that does not interact with radiation, and it can explain the gravitational phenomena in compact objects [1,2]. On the other hand, the impact of dark energy is observable at large scales. Dark energy has been proposed to explain the rapid acceleration of the universe [3–5], for that dark energy has negative pressure that induces repulsive gravitational forces [6].

The origin of the universe's acceleration is unknown. Over the last few decades, various models have been proposed in the literature. These models can be grouped into two broad categories, those belonging to modified theories of gravity [7–13] and those classified as "dark energy" models [14–19]. The first category offers a geometric explanation for the universe's acceleration. It introduces new geometric invariants into the Einstein–Hilbert Action, resulting in new terms in the field equations that account for the acceleration. On the contrary, "dark energy" models are defined within General Relativity, where an exotic matter source is introduced related to the cosmic acceleration. As far as dark matter is concerned in cosmological studies, it is usually introduced as a pressureless fluid source in the field equations.

There exists a distinct category of models that offer a unified approach to describing the dark sector of the universe [20]. Chaplygin gas and its modifications are a simple mechanism for the unification of dark matter and dark energy [21–25]. Another category of approaches comprises the bulk viscosity models [26,27] and the scalar field theories [28–31]. Within the unified models, both dark matter and dark energy are attributed to particular dynamical components of the effective fluid. In the unified models, there exists an interaction between dark matter and dark energy. This interaction can explain various problems related to cosmological observations [32–35].

In [36], a new mechanism was found to introduce dark matter components in cosmological dynamics. This mechanism is based on two scalar field models, a quintessence scalar field and a second scalar field coupled in the Action Integral to the energy momentum tensor of the quintessence. This approach was used in [37] to propose unified dark energy models. From the analysis of the phase space and it was found that the model can describe various eras of the cosmological history. It was found that for the exponential potential, the phase-space has asymptotics which can describe the matter epoch, the de Sitter universe and another scaling solution. Similar results were found for another functional of the potential in [38], while the theory has been used to explain the cosmological observations.

In this study, we revise the phase-space analysis presented in [37]. We demonstrate that for the exponential potential put forth in the same paper, the only asymptotic solutions correspond to the matter-dominated era. Notably, this model lacks any asymptotic solution that describes acceleration. The emergence of a de Sitter universe is solely achieved when the scalar field potential assumes the role of a cosmological constant at the asymptotic limit. For these latter two scenarios, we present analytic solutions for the field equations. Furthermore, acceleration is observed when the second potential function is not constant.

Indeed, the phase-space analysis is a powerful method for the study of proposed gravitational theories [39–41] and it can be applied for the construction of constraints [42–44]. The overview of the paper is as follows.

In Section 2 we outline the cosmological model under consideration and provide the field equations for the homogeneous and isotopic spatially flat Friedmann–Lemaître–Robertson–Walker (FLRW) geometry. Section 3 delves into the investigation of closed-form analytic solutions. We both validate previous findings and establish a novel general solution within the theory. The core analysis is presented in Section 4, where we meticulously examine the phase-space properties of the unified dark energy model. We explore three distinct cosmological scenarios, revealing that the de Sitter expansion only emerges as an attractor when all potential functions within the theory remain constant. In Section 5, we study the asymptotics when one of the variables reaches infinity, where we find that there exist two de Sitter solutions; the first is always a source and the second is always an attractor. Finally, our conclusions are presented in Section 6.

2. Field Equations

We follow the approach outlined in [37] and consider the gravitational Action Integral

$$S = \int d^4x \sqrt{-g} \left(R + \chi_{\mu;\nu} T^{\mu\nu}_{(\chi)} - \frac{1}{2} g^{\mu\nu} \phi_{;\mu} \phi_{;\nu} - V(\phi) \right). \tag{1}$$

Here, ϕ represents a scalar field that inherits the symmetries of the background space, along with the potential function $V(\phi)$, while χ_μ denotes a dynamical space-time vector; $T^{\mu\nu}_{(\chi)}$ is the energy-momentum tensor [45] defined as

$$T^{\mu\nu}_{(\chi)} = -\frac{1}{2} g^{\mu\kappa} g^{\nu\lambda} \phi_{;\kappa} \phi_{;\lambda} + U(\phi) g^{\mu\nu} \tag{2}$$

where $U(\phi)$ is now a second potential function.

Indeed, when the covariant derivative is defined by the Levi–Civita connection, then variation of (1) with respect to the vector field χ_μ provides the equation $T^{\mu\nu}_{(\chi)\ ;\nu} = 0$. The introduction of the term $\chi_{\mu;\nu}$ in the Action Integral is related to the "two measures theory" which has been proposed to solve the cosmological constant problem [46–48]. A similar vector field was introduced in [49]. For a classical analogue, we refer the reader in the discussion presented in [50]. Indeed, the term $\chi_{\mu;\nu} T^{\mu\nu}_{(\chi)}$ in the Action integral indicates the introduction of the constraint equation $T^{\mu\nu}_{(\chi)\ ;\nu} = 0$ for the theory.

For the spatially flat FLRW geometry

$$ds^2 = -dt^2 + a^2(t) \left(dx^2 + dy^2 + dz^2 \right),$$

with scale factor $a(t)$, Hubble function $H = \frac{\dot{a}}{a}$, and for a homogeneous time-like vector field $\chi^\mu = \chi_0 \delta_t^\mu$, the gravitational field equations read [37]

$$3H^2 = \rho_{eff} \tag{3}$$

$$-2\dot{H} - 3H^2 = p_{eff} \tag{4}$$

where the effective energy density ρ_{eff} and pressure components p_{eff} are defined as

$$\rho_{eff} = \dot{\phi}^2 \left(\dot{\chi}_0 \left(1 - \frac{3}{2}H \right) - \frac{1}{2} \right) + V(\phi) + \dot{\phi}\dot{\chi}_0 (U_{,\phi} + \ddot{\phi}), \tag{5}$$

$$p_{eff} = \frac{1}{2}\dot{\phi}^2 (\dot{\chi}_0 - 1) - V(\phi) - \chi_0 \dot{\phi} U_{,\phi}. \tag{6}$$

Furthermore, the equations of motion for the two scalar fields, namely ϕ and χ_0, are expressed as

$$\ddot{\phi} + \frac{3}{2}H\dot{\phi} + U_{,\phi} = 0, \tag{7}$$

$$\ddot{\phi}(\dot{\chi}_0 - 1) + \dot{\phi}(\ddot{\chi} + 3H(\dot{\chi}_0 - 1)) - U_{,\phi}(\dot{\chi}_0 + 3H\chi_0) + V_{,\phi} = 0. \tag{8}$$

This gravitational model was initially proposed as a unified dark energy model in [37]. Subsequently, in [38], the model was employed to elucidate cosmological observations. The findings indicated a slight deviation of the model from Λ-cosmology.

3. Analytic Solutions

In this Section we proceed with the derivation of analytical solutions for the field Equations (3)–(8). We explore two cosmological scenarios based on the two potential functions: (A) $U(\phi) = U_0$, $V(\phi) = V_0$ and (B) $U(\phi) = U_0$ and $V(\phi) = V_0 e^{-\lambda\phi}$.

3.1. Case A: $U(\phi) = U_0$ and $V(\phi) = V_0$

This particular case was previously examined in [37]. We now reconstruct the analytical solution. In fact, by considering (7), we deduce that

$$\phi(t) = \phi_0 + \int \phi_1 a^{-\frac{3}{2}} dt. \tag{9}$$

By replacing in (8) it follows that

$$\chi_0(t) = \chi_0^0 + \int \left(1 + \chi_0^1 a^{-\frac{3}{2}} \right) dt, \tag{10}$$

Therefore, the analytic solution for the Hubble function is given by Equation (3), that is,

$$3H^2 = V_0 + \frac{\phi_1^2}{2}a^{-3} + \phi_1^2 \chi_0^1 a^{-\frac{9}{2}}. \tag{11}$$

We note that the scalar field ϕ introduces the dark matter component a^{-3}, and furthermore, from the second scalar field χ_0, the component $a^{-\frac{9}{2}}$ emerges. This latter component corresponds to an ideal gas with an equation of state parameter of $\frac{1}{2}$.

3.2. Case B: $U(\phi) = U_0$ and $V(\phi) = V_0 e^{-\lambda\phi}$

Assume now that the scalar field potential $V(\phi) = V_0 e^{-\lambda\phi}$ and $U(\phi) = U_0$. Thus, from (8), the solution (9) follows. Moreover, for the second scalar field we derive

$$\dot{\chi}_0(t) = \frac{2}{\phi_1} \exp(-f_1(t)) \int \exp(f_1(t)) \left(\frac{\lambda \phi_1^2}{a^{\frac{3}{2}}} + 6\lambda a^{\frac{3}{2}} H^2 + 3\phi_1 H \right) dt, \tag{12}$$

where

$$f_1(t) = \ln(a)^{\frac{3}{2}} + \phi_1 \int \frac{2\lambda}{a^{\frac{3}{2}}} \, dt.$$ (13)

Therefore, the Hubble function is calculated:

$$3H^2 = -\frac{\phi_1}{2}a^{-3} + V_0 \exp\left(-\lambda\left(\phi_0 + \int \phi_1 a^{-\frac{3}{2}} dt\right)\right)$$
$$+ \left(a^{-3}\frac{2}{\phi_1}\exp(-f_1(t))\int \exp(f_1(t))\left(\frac{\lambda\phi_1^2}{a^{\frac{3}{2}}} + 6\lambda a^{\frac{3}{2}}H^2 + 3\phi_1 H\right)dt\right).$$ (14)

We remark that when $\lambda = 0$, the solution of Case A is recovered.

4. Asymptotic Solutions

We proceed with the analysis of the equilibrium points for the field Equations (3)–(8). We select three distinct cases for the analysis: (I) $U(\phi) = U_0$, $V(\phi) = 0$, (II) $U(\phi) = U_0$, $V(\phi) = V_0 e^{-\lambda\phi}$ and (III) $U(\phi) = U_0 e^{-\kappa\phi}$, $V(\phi) = V_0 e^{-\lambda\phi}$.

4.1. Case I: $U(\phi) = U_0$ and $V(\phi) = 0$

For the first case with $U(\phi) = U_0$ and $V(\phi) = 0$, we apply the H-normalization approach and we introduce the new variables

$$x = \frac{\dot{\phi}}{\sqrt{6}H} \ , \ \delta = \dot{\chi}_0 - 1 \ , \ d\tau = Hdt.$$ (15)

The field Equations (3)–(8) are expressed as the following first-order dynamical system

$$\frac{dx}{d\tau} = \frac{3}{2}\delta x^3$$ (16)

$$\frac{d\delta}{d\tau} = -\frac{3}{2}\delta$$ (17)

with the constraint equation

$$1 - x^2(1 + 2\delta) = 0.$$ (18)

In the new variables, the effective equation of the state parameter for the cosmological fluid $w_{eff} = -1 - \frac{2}{3}\frac{\dot{H}}{H^2}$ reads

$$w_{eff}(x,\delta) = x^2\delta.$$ (19)

We observe that the dynamical system is symmetric in the change of variables $x \to -x$. Hence, without loss of generality, we choose to work in the range $x \geq 0$. Therefore, there exists a unique stationary point $A = (x(A), \delta(A))$ for the latter dynamical system at the finite regime with coordinates $A_1^{\pm} = (\pm 1, 0)$. These two points describe matter-dominated eras, that is, $w_{eff}(A_1^{\pm}) = 0$.

From (18) we replace in (16) and we find that

$$\frac{dx}{d\tau} = \frac{3}{4}x\left(1 - x^2\right).$$ (20)

Thus, it is easy to observe that points A_1^2 are always attractors.

4.2. Case II: $U(\phi) = U_0$ and $V(\phi) = V_0 e^{-\lambda\phi}$

For the second case, that is, for the exponential potential $V(\phi) = V_0 e^{-\lambda\phi}$, $\lambda \neq 0$, we introduce the new variables

$$x = \frac{\dot{\phi}}{\sqrt{6}H} \ , \ \delta = \dot{\chi}_0 - 1 \ , \ y^2 = \frac{V(\phi)}{3H^2} \ , \ d\tau = Hdt.$$ (21)

Therefore, the field equations read

$$1 - x^2(1 + 2\delta) - y^2 = 0, \tag{22}$$

$$\frac{dx}{d\tau} = \frac{3}{2}\left(\delta x^3 - xy^2\right), \tag{23}$$

$$\frac{d\delta}{d\tau} = \frac{\sqrt{6}}{2}\lambda\frac{y^2}{x} - \frac{3}{2}\delta, \tag{24}$$

$$\frac{dy}{d\tau} = -\frac{y}{2}\left(\sqrt{6}\lambda x + 3\left(y^2 - 3x^2\delta - 1\right)\right). \tag{25}$$

Furthermore, the effective equation of the state parameter is derived

$$w_{eff}(x, y, \delta) = -y^2 + x^2\delta.$$

The stationary points $B = (x(B), \delta(B), y(B))$ of the dynamical system (22)–(25) are

$$B_1^{\pm} = (\pm 1, 0, 0), \; B_2 = \left(\frac{\sqrt{6}}{2}\frac{1}{\lambda}, \frac{2\lambda^2 - 3}{9}, \frac{\sqrt{2\lambda^2 - 3}}{\sqrt{6}\lambda}\right).$$

Points $B_1^{\pm}$ exhibit the same physical characteristics as points $A_1^{\pm}$, representing matter-dominated eras. Additionally, for point B_2, we calculate $w_{eff}(B_2) = 0$, implying that if B_2 exists, i.e., when $\lambda^2 > \frac{3}{2}$, it corresponds to a matter-dominated epoch.

The derivation of point B_2 was previously presented in [37], identified as point D therein. Upon substituting δ from (22) into (25), we obtain expression (33) from [37], indicating its vanishing nature at point D rather than a non-zero constant. As a result, the assertion that point B_2 signifies acceleration is incorrect.

We repeat part of the calculations outlined in [37]. From (22), it follows that

$$x^2(1 + 2\delta) = 1 - y^2, \tag{26}$$

and for $x \neq 0$,

$$\delta = \frac{1}{2}\left(-1 + \frac{1 - y^2}{x^2}\right). \tag{27}$$

It is clear that x can not take the value zero, otherwise the following system is not defined. Indeed, the dynamical system (22)–(25) can be reduced to the following two-dimensional system (with $x \neq 0$)

$$\frac{dx}{d\tau} = -\frac{3}{4}x\left(x^2 + 3y^2 - 1\right), \tag{28}$$

$$\frac{dy}{d\tau} = -\frac{1}{4}y\left(2\sqrt{6}x + 3\left(x^2 - 3y^2 - 3\right)\right), \tag{29}$$

The stationary points within the defined range of parameters x and y consist of $B_1^{\pm}$ and B_2. Consequently, we deduce that the point C discussed in [37], characterized by $x = 0$ and $y = 1$, is nonexistent at the finite regime.

The eigenvalues of the two-dimensional system (28), (29) are

$$e_1(B_1^{\pm}) = -\frac{3}{2}, \; e_2(B_1^{\pm}) = \frac{1}{2}\left(3 \mp \sqrt{6}\lambda\right),$$

and

$$e_1(B_2) = \frac{3}{4}\left(-1 + \frac{\sqrt{5\lambda^2 - 6}}{\lambda}\right), \; e_2(B_2) = \frac{3}{4}\left(-1 - \frac{\sqrt{5\lambda^2 - 6}}{\lambda}\right),$$

From the above we conclude that point B_1^+ is an attractor for $\lambda > \sqrt{\frac{3}{2}}$, while point B_1^- is an attractor for $\lambda < -\sqrt{\frac{3}{2}}$. Finally, as far as the point B_2 is concerned, we infer that when it exists, is always a saddle point.

In Figures 1–3 we present phase-portraits in the two- and three-dimensional spaces for the dynamical system (22)–(25) for different values of parameter λ. From the plots, it is clear that the trajectories go in the area where $\delta \to \infty$; this means that it seems that stationary points exist at the infinity regime. In the following analysis, specifically in Section 5, we study the existence of stationary points for very large values of δ.

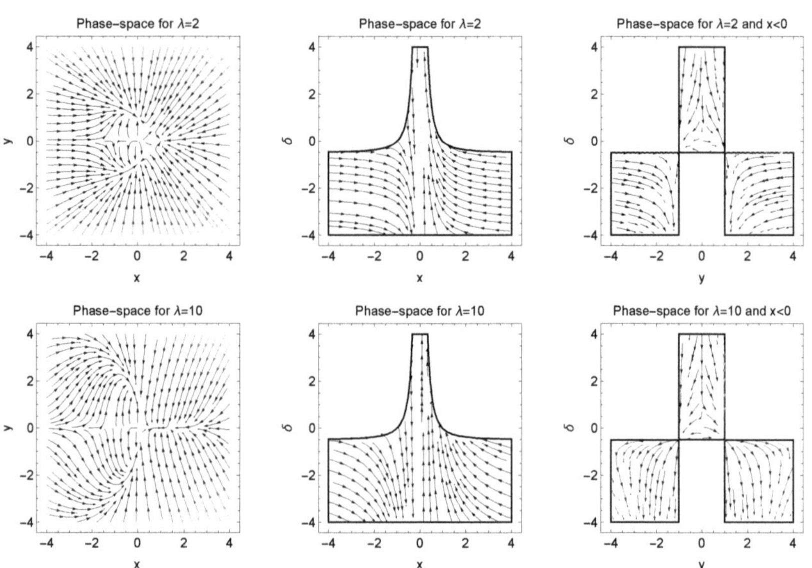

Figure 1. Phase-space portraits of the dynamical system (22), (25) on the two-dimensional planes (x, y), (x, δ) and (y, δ) for $\lambda = 2$ and $\lambda = 10$.

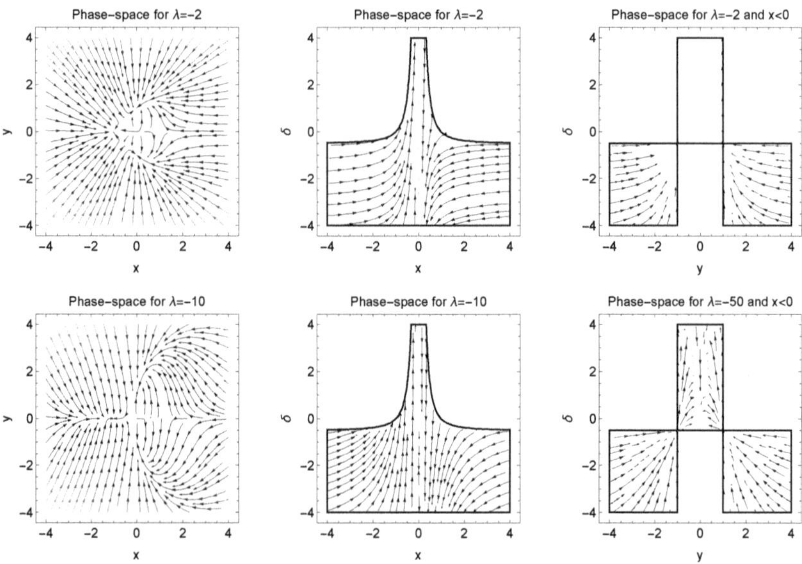

Figure 2. Phase-space portraits of the dynamical system (22), (25) on the two-dimensional planes (x, y), (x, δ) and (y, δ) for $\lambda = -2$ and $\lambda = -10$.

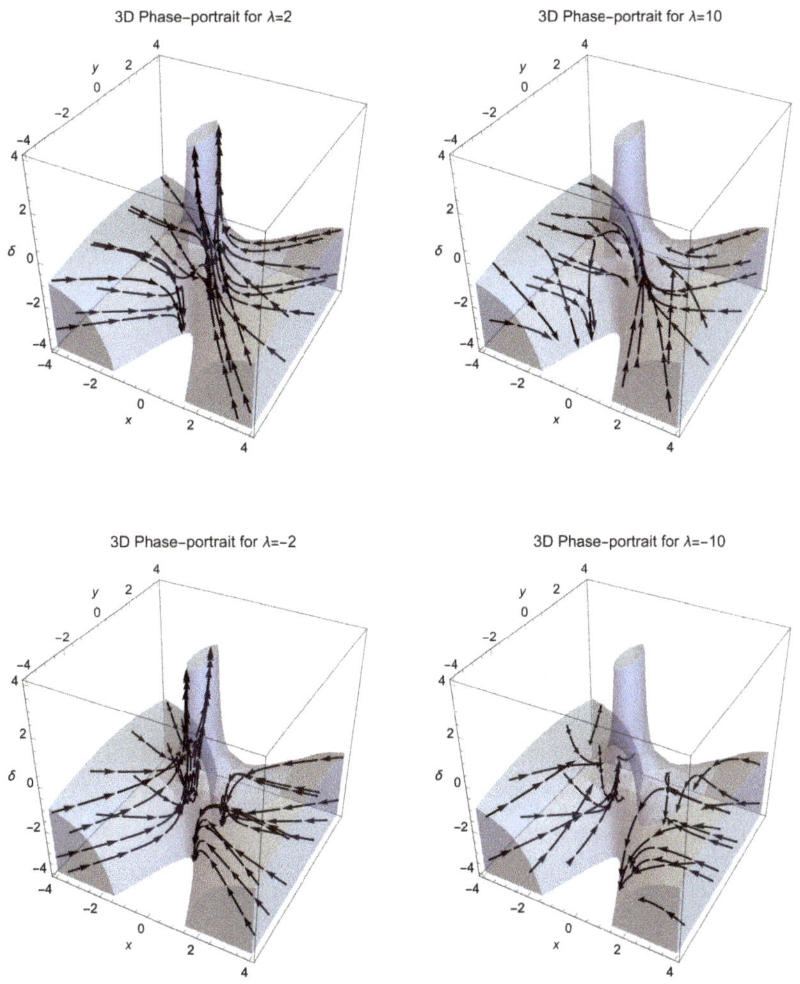

Figure 3. 3D phase-space portraits of the dynamical system (22), (25) for $\lambda = 2$, $\lambda = 10$, $\lambda = -2$ and $\lambda = -10$.

Subcase $\lambda = 0$

Consider the specific scenario where $\lambda = 0$, and the scalar field potential $V(\phi)$ assumes the role of the cosmological constant, i.e., $V(\phi) = V_0$.

Indeed, the dynamical system (23)–(25) becomes

$$\frac{dx}{d\tau} = \frac{3}{2}\left(\delta x^3 - xy^2\right),\tag{30}$$

$$\frac{d\delta}{d\tau} = -\frac{3}{2}\delta,\tag{31}$$

$$\frac{dy}{d\tau} = -\frac{y}{2}\left(3\left(y^2 - 3x^2\delta - 1\right)\right).\tag{32}$$

The stationary points are $B_1^{\pm} = (\pm 1, 0, 0)$ and $B_3 = (0, 0, 1)$. Point B_3 characterizes the de Sitter solution with $w_{eff}(B_3) = -1$, and it consistently acts as an attractor due to the negative eigenvalues of the linearized system.

In Figure 4 we present the three-dimensional phase-space for the latter dynamical system with $\lambda = 0$. We observe that B_3 is the future attractor of the dynamical system.

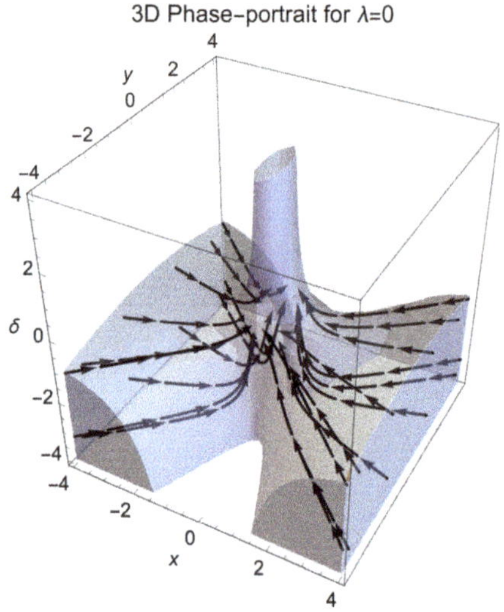

Figure 4. 3D phase-space portraits of the dynamical system (22), (25) for $\lambda = 0$, where B_3 is the unique attractor.

4.3. Case III: $U(\phi) = U_0 e^{-\kappa \phi}$ *and* $V(\phi) = V_0 e^{-\lambda \phi}$

Let us assume now the case where $U(\phi) = U_0 e^{-\kappa \phi}$ and $V(\phi) = V_0 e^{-\lambda \phi}$ and $\kappa \neq \lambda$. For this cosmological model, we introduce the variables

$$x = \frac{\dot{\phi}}{\sqrt{6}H} \ , \ \delta = \dot{\chi}_0 - 1 \ , \ y^2 = \frac{V(\phi)}{3H^2} \ , \tag{33}$$

$$z^2 = \frac{U(\phi)}{H^2} \ , \ M = \chi_0 H \ , \ d\tau = H dt. \tag{34}$$

Thus, the field equations are expressed by the following algebraic-differential system

$$1 - x^2(1 + 2\delta) - y^2 = 0, \tag{35}$$

$$\frac{dx}{d\tau} = \frac{1}{6}\left(9x\left(x^2\delta - y^2\right) + \sqrt{6}\kappa z^2\left(1 + 3x^2 M\right) \right), \tag{36}$$

$$\frac{d\delta}{d\tau} = -\frac{1}{6x}\left(\sqrt{6}\kappa z^2(1 + 3M + 2\delta) - 3\sqrt{6}\lambda y^2 + 9x\delta \right), \tag{37}$$

$$\frac{dy}{d\tau} = \frac{1}{2}y\left(3\left(1 - y^2 + x^2\delta\right) + \sqrt{6}x(\kappa z M - \lambda) \right), \tag{38}$$

$$\frac{dM}{d\tau} = 1 + \delta - \frac{1}{2}M\left(3\left(1 - y^2 + x^2\delta\right) + \sqrt{6}\kappa z^2 x M \right), \tag{39}$$

$$\frac{dz}{d\tau} = \frac{1}{2}z\left(3\left(1 - y^2 + x^2\delta\right) + \sqrt{6}\kappa x\left(M z^2 - 1\right) \right), \tag{40}$$

and

$$w_{eff} = -y^2 + x^2\delta + \frac{\sqrt{6}}{3}\kappa M x z^2. \tag{41}$$

The stationary points $C = (x(C), \delta(C), y(C), M(C), z(C))$ of the dynamical system (36)–(40) which satisfy the constraint Equation (35) are

$$C_1^{\pm} = \left(\pm 1, 0, 0, \frac{2}{3}, 0\right), \quad C_2 = \left(\frac{\sqrt{6}}{2}\frac{1}{\lambda}, \frac{2\lambda^2 - 3}{9}, \frac{\sqrt{2\lambda^2 - 3}}{\sqrt{6}\lambda}, \frac{4}{27}(\lambda^2 + 3), 0\right)$$

$$C_3 = \left(x_3, \frac{1}{2}\left(\frac{1}{x_3} - 1\right), 0, \left(\frac{1}{2} - \frac{1}{6x_3^2} - \frac{\sqrt{6}}{x_3}\kappa\right), \sqrt{\frac{x^2\left(\sqrt{17 + 10x_3^2 + 9x_3^4} - 5 - x_3^2\right)}{4(1 + x^2)}}\right)$$

where x_3 is a real solution of the algebraic equation

$$4x_3\kappa = \sqrt{\frac{3}{2}}\left(1 - 3x_3^2 + \sqrt{17 + 10x_3^2 + 9x_3^4}\right). \tag{42}$$

Points $C_1^{\pm}$ and C_2 are the extensions of points $B_1^{\pm}$ in B_2 for the nonconstant potential $U(\phi)$ in the five-dimensional space. Point C_3 is new and describes a universe where potential $U(\phi)$ contributes in the cosmological fluid. C_3 is real and physical, accepted for $\sqrt{17 + 10x_3^2 + 9x_3^4} - 5 - x_3^2 \geq 0$, that is, $x^2 \geq 1$. The effective equation of state parameter $w_{eff}(C_3) = \frac{1}{4}\left(\sqrt{17 + 10x_3^2 + 9x_3^4} - 3(1 + x_3^2)\right)$. We observe that $w_{eff}(C_3) \geq -\frac{1}{3}$, which means that point C_3 can not describe acceleration. In Figure 5, the $\kappa = \kappa(x_3)$ is given as it is presented in Equation (42).

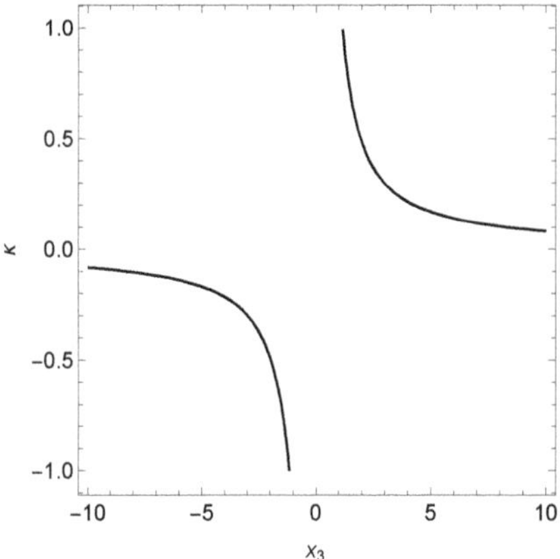

Figure 5. We plot the $\kappa = \kappa(x_3)$ as it is given by the algebraic equation $4x_3\kappa = \sqrt{\frac{3}{2}}\left(1 - 3x_3^2 + \sqrt{17 + 10x_3^2 + 9x_3^4}\right)$.

We observe that in this case where $U(\phi)$ contributes in the cosmological fluid there is not any stationary point which can describes inflation. We omit the presentation of the stability properties and we focus in the special case where $\kappa = \lambda$.

The results of this Section are summarized in Table 1.

Table 1. Stationary points and physical properties at the finite regime.

Model	$U(\phi)$	$V(\phi)$	Point	w_{eff}	Acceleration?
I	U_0	0			
			$A_1^{\pm}$	0	No
II	U_0	$V_0 e^{-\lambda\phi}$			
			$B_1^{\pm}$	0	No
			B_2	0	No
$II_{(\lambda=0)}$	U_0	V_0			
			$B_1^{\pm}$	0	No
			B_3	-1	Yes
III	$U_0 e^{-\kappa\phi}$	$V_0 e^{-\lambda\phi}$			
			$C_1^{\pm}$	0	No
			C_2	0	No
			C_3	$w_{eff}(C_3) \geq -\frac{1}{3}$	No

5. Analysis at Infinity

Until now we have seen that the de Sitter universe exists only for the asymptotic solutions with $V(\phi)$ a constant, that is, $\lambda = 0$. We have investigated the existence of asymptotic solutions at the finite regime; however, the dynamical variables are constrained by the algebraic equation $1 - x^2(1 + 2\delta) - y^2 = 0$, which means that they can take values at infinity. In order to study the existence of asymptotic solutions at infinity we should introduce Poincare variables.

In case II, we focus on $U(\phi) = U_0$ and $V(\phi) = V_0 e^{-\lambda\phi}$. We introduce the new variable $z = x^2\delta$, and the two-dimensional system in the plane $\{x, z\}$ reads

$$\frac{dz}{d\tau} = -\frac{3}{2}\left(1 - x^2 - 3z\right), \tag{43}$$

$$\frac{dz}{d\tau} = \frac{1}{2}\left(\sqrt{6}\lambda x\left(1 - x^2 - 2z\right) - 3z\left(3 - 2x^2 - 6z\right)\right). \tag{44}$$

Indeed, there exist the stationary points $B^* = (x(B^*), z(B^*))$ with coordinates $B_1^* = \left(\frac{\sqrt{6}}{2}\frac{1}{\lambda}, \frac{2\lambda^2-3}{9}\right)$, $B_2^* = \left(0, \frac{1}{2}\right)$ and $B_3^* = (0,0)$. Point B_1^* is B_1, while points B_2^* and B_3^* describe de Sitter solutions with $w_{eff}(B_2^*) = -1$ and $w_{eff}(B_3^*) = -1$.

As far as the stability properties are concerned, the resulting eigenvalues for point B_2^* are $\left\{\frac{9}{2}, \frac{3}{4}\right\}$ and for point B_3^* we derive $\left\{-\frac{9}{2}, -\frac{3}{2}\right\}$. We infer that B_2^* is always a source and B_3^* is always an attractor.

Because of the constraint equation $y^2 = 1 - x^2 - 2z$, at the new stationary points we calculate $y(B_2^*) = 0$ and $y(B_3^*) = 1$. Thus, point B_3^* is the de Sitter solution described by point C in [37].

In Figure 6, we present the two-dimensional phase-space portrait for the dynamical system (43), (44), from where it is clear that the de Sitter solution described by B_3^* is a future attractor for the dynamical system.

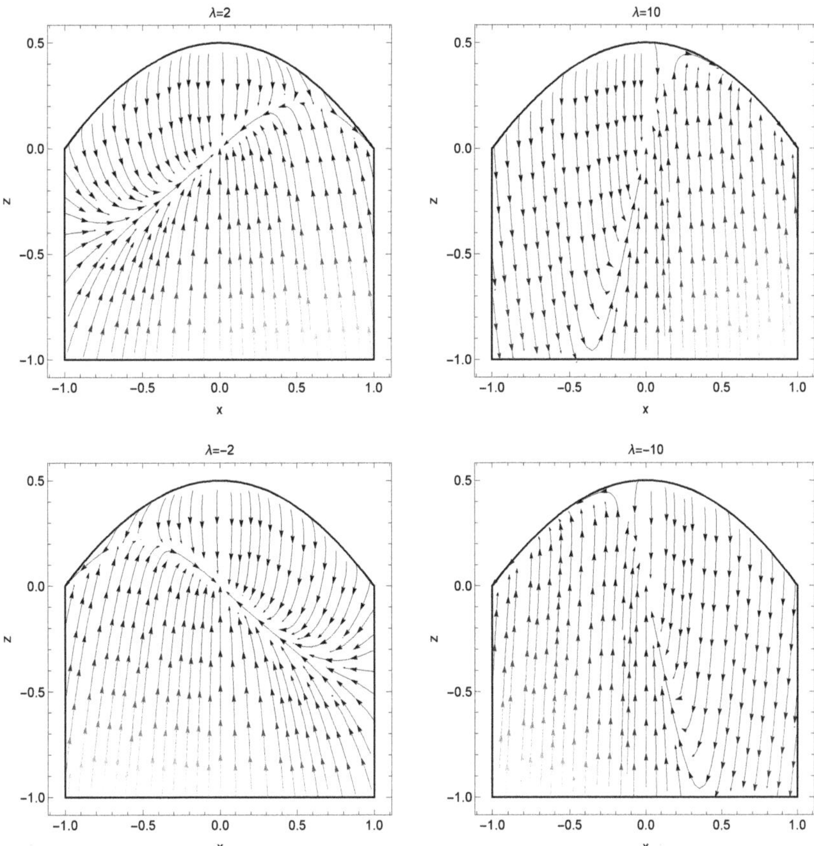

Figure 6. Phase-space portraits of the dynamical system (43), (44) on the two-dimensional planes (x, z) for $\lambda = 2$, $\lambda = 10$, $\lambda = -2$ and $\lambda = -10$.

6. Conclusions

In this study, we conducted a thorough phase-space analysis of the cosmological model, considering different forms of the potential functions. Our findings indicate that the cosmological model does not exhibit the de Sitter universe as an attractor, except in the case where the potential function $V(\phi)$ represents the cosmological constant. Specifically, we investigated the following cosmological scenarios for the two potential functions: (I) $U(\phi) = U_0$, $V(\phi) = 0$, (II) $U(\phi) = U_0$, $V(\phi) = V_0 e^{-\lambda \phi}$ and (III) $U(\phi) = U_0 e^{-\kappa \phi}$, $V(\phi) = V_0 e^{-\lambda \phi}$.

Referring to Table 1, for the analysis at the finite regime, we observe that for Model I, two stationary points emerge, characterizing the era of matter dominance. In contrast, Model II, which involves an exponential potential, reveals three stationary points that pertain to matter-dominated epochs. This finding stands in contrast to the outcomes previously reported in [37,38]. However, in the scenario where the exponent λ becomes zero, leading to the cosmological constant limit, a de Sitter solution emerges. A natural question which arises is if there are other forms for the scalar field potential $V(\phi)$ and $U(\phi) = U_0$ in which the de Sitter universe exists as an asymptotic solution. For a general potential function $V(\phi)$ parameter $\lambda = -\frac{V_{,\phi}}{V}$ is not a constant but a dynamical parameter which satisfies the equation

$$\frac{d\lambda}{d\tau} = -\sqrt{6}x\lambda^2(\Gamma(\lambda) - 1), \; \Gamma(\lambda) = \frac{V(\phi(\lambda))V_{,\phi\phi}(\phi(\lambda))}{\left(V_{,\phi}(\phi(\lambda))\right)^2}. \tag{45}$$

Thus, the stationary points depend on the value of λ which solves the algebraic equation $\lambda^2(\Gamma(\lambda) - 1) = 0$. Consequently, a de Sitter point exist when $\lambda = 0$ is a root of the algebraic equation $\lambda^2(\Gamma(\lambda) - 1) = 0$. For the potential $V(\phi) = V_0 e^{-\sigma_0\phi} + \Lambda$, it follows that $\Gamma(\lambda) = \frac{\sigma_0}{\lambda}$, which means that the algebraic equation reads $\lambda(\sigma_0 - \lambda) = 0$, with solutions $\lambda = \sigma_0$ and $\lambda = 0$. As a result, on the surface $\lambda = 0$, the de Sitter point exist. On the other hand, for the power-law potential $V(\phi) = V_0\phi^{-\alpha}$, it follows $\Gamma(\lambda) = \frac{\alpha+1}{\alpha}$, and the algebraic equation reads $\lambda^2 = 0$, that is, the de Sitter solution exist. Of course the stability properties of the de Sitter solution depend on the function form of $V(\phi)$. It is important to note that the asymptotic solution with $\lambda = 0$ leads to a scalar field potential $V(\phi)$ which remains constant.

For the third cosmological scenario, Model III, the absence of solutions capable of describing acceleration is notable. An exception is found only in the special case where $\kappa = \lambda$. In this circumstance, a collection of stationary points corresponding to scaling solutions manifests. Although accelerated solutions are present, they do not provide a de Sitter universe at the finite regime.

However, because the dynamical variables of the field equations are not constrained, they can reach infinity. Specifically for Model II, we investigate the existence of stationary points when $\delta \to \infty$. We found that, at infinity, there exist two stationary points which correspond to two de Sitter solutions. The one solution is always a source, while the other is always an attractor.

The current investigation underscores the potency of phase-space analysis as a robust tool for assessing the viability of gravitational models and constraining proposed theories. Importantly, this analysis can also serve as a classification and selection rule for the initial value problem. For instance, Model II, for $|\lambda| > \sqrt{\frac{3}{2}}$, has two attractors, one of the points $B_1^{\pm}$ and the stationary point B_3^*. On the other hand, for $|\lambda| < \sqrt{\frac{3}{2}}$, B_3^* is the unique attractor, and the de Sitter universe is the unique future solution.

Funding: This work was partially financially supported by the National Research Foundation of South Africa (Grant Numbers 131604). The author thanks the support of Vicerrectoría de Investigación y Desarrollo Tecnológico (Vridt) at Universidad Católica del Norte through Núcleo de Investigación Geometría Diferencial y Aplicaciones, Resolución Vridt No—098/2022.

Data Availability Statement: No data available.

Conflicts of Interest: The author declares no conflict of interest.

References

1. Persic, M.; Salucci, P.; Stel, F. The universal rotation curve of spiral galaxies—I. The dark matter connection. *Mon. Not. R. Astron. Soc.* **1996**, *281*, 27–47. [CrossRef]
2. Weinberg, D.H.; Colombi, S.; Davé, R.; Katz, N. Baryon Dynamics, Dark Matter Substructure, and Galaxies. *Astrophys. J.* **2008**, *678*, 6.
3. Riess, A.G.; Filippenko, A.V.; Challis, P.; Clocchiatti, A.; Diercks, A.; Garnavich, P.M.; Gilliland, R.L.; Hogan, C.J.; Jha, S.; Kirshner, R.P.; et al. Observational evidence from Supernovae for an accelerating universe and cosmological constant. *Astron. J.* **1998**, *116*, 1009.
4. Tegmark, M. et al. [SDSS Collaboration]. The 3D power spectrum of galaxies from the SDSS. *Astrophys. J.* **2004**, *606*, 702–740.
5. Kowalski, M.; Rubin, D.; Aldering, G.; Agostinho, R.J.; Amadon, A.; Amanullah, R.; Balland, C.; Barbary, K.; Blanc, G.; Challis, P.J.; et al. Improved Cosmological Constraints from New, Old, and Combined Supernova Data Sets. *Astrophys. J.* **2008**, *686*, 749.
6. Yoo, J.; Watanabe, Y. Theoretical models of dark energy. *Int. J. Mod. Phys. D* **2012**, *21*, 1230002.
7. Clifton, T.; Ferreira, P.G.; Padilla, A.; Skordis, C. Modified Gravity and Cosmology. *Phys. Rep.* **2012**, *513*, 1–189.
8. Nojiri, S.; Odintsov, S.D.; Oikonomou, V. Modified gravity theories on a nutshell: Inflation, bounce and late-time evolution. *Phys. Rep.* **2017**, *692*, 1–104.
9. Ferraro, R.; Fiorini, F. Modified teleparallel gravity. *Phys. Rev. D* **2007**, *75*, 084031. [CrossRef]
10. Paliathanasis, A. Dynamical analysis of f(Q) -cosmology. *Phys. Dark Univ.* **2023**, *41*, 101255.

11. Krssak, M.; van den Hoogen, R.J.; Pereira, J.G.; Boehmer, C.G.; Coley, A.A. Teleparallel Theories of Gravity. *Class. Quantum Grav.* **2019**, *36*, 183001.
12. Rani, S.; Jawad, A.; Bamba, K.; Malik, I.-U. Cosmological Consequences of New Dark Energy Models in Einstein-Aether Gravity. *Symmetry* **2019**, *11*, 509.
13. Li, B.; Barrow, J.D.; Mota, D.F. Cosmology of modified Gauss-Bonnet gravity. *Phys. Rev. D* **2007**, *76*, 044027.
14. Ratra, B.; Peebles, P.J.E. Cosmological consequences of a rolling homogeneous scalar field. *Phys. Rev. D* **1988**, *37*, 3406.
15. Hordenski, G.W. Second-order scalar-tensor field equations in a four-dimensional space. *Int. J. Theor. Phys.* **1975**, *10*, 363–384.
16. Socorro, J.; Pérez-Payán, S.; Hernández-Jiménez, R.; Espinoza-García, A.; Díaz-Barrón, L.R. Classical and quantum exact solutions for a FRW in chiral like cosmology. *Class. Quantum Grav.* **2021**, *38*, 135027.
17. von Marttens, R.; Barbosa, D.; Alcaniz, J. One-parameter dynamical dark-energy from the generalized Chaplygin gas. *J. Cosmol. Astropart. Phys.* **2023**, *2023*, 052.
18. Akrami, Y.; Sasaki, M.; Solomon, A.R.; Vardanyan, V. Multi-field dark energy: Cosmic acceleration on a steep potential. *Phys. Lett. B* **2021**, *819*, 136427.
19. Mamon, A.A.; Paliathanasis, A.; Saha, S. An extended analysis for a generalized Chaplygin gas model. *Eur. Phys. J. C* **2022**, *82*, 232.
20. Cardone, V.F.; Troisi, A.; Capozziello, S. Unified dark energy models: A phenomenological approach. *Phys. Rev. D* **2004**, *69*, 083517.
21. Bento, M.C.; Bertolami, O.; Sen, A.A. Generalized Chaplygin gas, accelerated expansion, and dark-energy-matter unification. *Phys. Rev. D* **2009**, *70*, 083519. [CrossRef]
22. Perrotta, F.; Matarrese, S.; Torki, M. Instability of Chaplygin gas trajectories in unified dark matter models. *Phys. Rev. D* **2004**, *70*, 121304. [CrossRef]
23. Wu, Y.-B.; Li, S.; Lu, J.-B.; Yang, X.-Y. The modified Chaplygin gas as a unified dark sector model. *Mod. Phys. Lett. A* **2007**, *22*, 783–790. [CrossRef]
24. Gorini, V.; Kamenshchik, A.Y.; Moschella, U.; Piattella, O.F.; Starobinsky, A.A. More about the Tolman-Oppenheimer-Volkoff equations for the generalized Chaplygin gas. *Phys. Rev. D* **2009**, *80*, 104038. [CrossRef]
25. Zhu, Z.-H. Generalized Chaplygin gas as a unified scenario of dark matter/energy: Observational constraints. *Astron. Astrophys.* **2004**, *423*, 421–426.
26. Li, B.; Barrow, J.D. Does bulk viscosity create a viable unified dark matter model? *Phys. Rev. D* **2009**, *79*, 103521.
27. Atreya, A.; Bhatt, J.R.; Mishra, A.K. Viscous self interacting dark matter cosmology for small redshift. *J. Cosmol. Astropart. Phys.* **2019**, *2019*, 045.
28. Bertacca, D.; Matarrese, S.; Pietroni, M. Unified Dark Matter in Scalar Field Cosmologies. *Mod. Phys. Lett. A* **2007**, *22*, 2893–2907.
29. Bertacca, D.; Bartolo, N.; Matarrese, S. Unified Dark Matter Scalar Field Models. *Adv. Astron.* **2010**, *2010*, 904379.
30. Paliathanasis, A. Dynamics of Chiral Cosmology. *Class. Quantum Grav.* **2020**, *37*, 195014.
31. Leon, G.; Paliathanasis, A.; Saridakis, E.N.; Basilakos, S. Unified dark sectors in scalar-torsion theories of gravity. *Phys. Rev. D* **2022**, *106*, 024055. [CrossRef]
32. Amendola, L. Coupled quintessence. *Phys. Rev. D* **2000**, *62*, 043511. [CrossRef]
33. Wetterich, C. An asymptotically vanishing time-dependent cosmological "constant". *Astron. Astrophys.* **1995**, *301*, 321.
34. Di Valentino, E.; Melchiorri, A.; Mena, O.; Pan, S.; Yang, W. Interacting Dark Energy in a closed universe. *Mon. Not. R. Astron. Soc. Lett.* **2021**, *502*, L23–L28. [CrossRef]
35. Bonilla, A.; Kumar, S.; Nunes, R.C.; Pan, S. Reconstruction of the dark sectors' interaction. *Mon. Not. R. Astron. Soc.* **2022**, *512*, 4231–4238. [CrossRef]
36. Benisty, D.; Guendelman, E.I. Interacting diffusive unified dark energy and dark matter from scalar fields. *Eur. Phys. J. C* **2017**, *77*, 396. [CrossRef]
37. Benisty, D.; Guendelman, E.I. Unified dark energy and dark matter from dynamical spacetime. *Phys. Rev. D* **2018**, *98*, 023506. [CrossRef]
38. Anagnostopulos, F.K.; Benisty, D.; Basilakos, S.; Guendelman, E.I. Dark energy and dark matter unification from dynamical space time: Observational constraints and cosmological implications. *J. Cosmol. Astropart. Phys.* **2019**, *2019*, 003. [CrossRef]
39. Gonzales, T.; Leon, G.; Quiros, I. Dynamics of quintessence models of dark energy with exponential coupling to dark matter. *Class. Quantum Grav.* **2006**, *23*, 3165. [CrossRef]
40. Tot, J.; Yildirim, B.; Coley, A.; Leon, G. The dynamics of scalar-field Quintom cosmological models. *Phys. Dark Univ.* **2023**, *39*, 101155. [CrossRef]
41. Millano, A.D.; Leon, G.; Paliathanasis, A. Global dynamics in Einstein-Gauss-Bonnet scalar field cosmology with matter. *Mathematics* **2023**, *11*, 1408. [CrossRef]
42. Coley, A.A.; van den Hoogen, R.J. Dynamics of multi-scalar-field cosmological models and assisted inflation. *Phys. Rev. D* **2000**, *62*, 023517. [CrossRef]
43. Amendola, L.; Gannouji, R.; Polarski, D.; Tsujikawa, S. Conditions for the cosmological viability of f(R) dark energy models. *Phys. Rev. D* **2007**, *75*, 083504. [CrossRef]
44. Khyllep, W.; Dutta, J.; Saridakis, E.N.; Yesmakhanova, K. Cosmology in f(Q) gravity. *Phys. Rev. D* **2023**, *107*, 044022. [CrossRef]

45. Gao, C.; Kunz, M.; Liddle, A.R.; Parkinson, D. Unified dark energy and dark matter from a scalar field different from quintessence. *Phys. Rev. D* **2010**, *81*, 043520. [CrossRef]
46. Guendelman, E.I.; Kaganovich, A.B. Dark energy, dark matter and fermion families in the two measures theory. *Int. J. Mod. Phys A* **2004**, *19*, 5325–5332. [CrossRef]
47. Guendelman, E.I.; Kaganovich, A.B. Exotic Low Density Fermion States in the Two Measures Field Theory: Neutrino Dark Energy. *Int. J. Mod. Phys. A* **2006**, *21*, 4373–4406. [CrossRef]
48. Gronwald, F.; Muench, U.; Macias, A.; Hehl, F.W. Volume elements of spacetime and a quartet of scalar fields. *Phys. Rev. D* **1998**, *58*, 084021. [CrossRef]
49. Comelli, D. A Way to Dynamically Overcome the Cosmological Constant Problem. *Int. J. Mod. Phys. A* **2008**, *23*, 4133–4143. [CrossRef]
50. Guendelman, E.I. Gravitational Theory with a Dynamical Time. *Int. J. Mod. Phys. A* **2010**, *15*, 4081–4099. [CrossRef]

Article

An Effective Sign Switching Dark Energy: Lotka–Volterra Model of Two Interacting Fluids

Yen Chin Ong [1,2]

1 Center for Gravitation and Cosmology, College of Physical Science and Technology, Yangzhou University, 180 Siwangting Road, Yangzhou 225002, China; ycong@yzu.edu.cn
2 Shanghai Frontier Science Center for Gravitational Wave Detection, School of Aeronautics and Astronautics, Shanghai Jiao Tong University, No. 800, Dongchuan Rd, Minhang, Shanghai 200240, China

Abstract: One of the recent attempts to address the Hubble and S_8 tensions is to consider that the Universe started out not as a de Sitter-like spacetime, but rather anti-de Sitter-like. That is, the Universe underwent an "AdS-to-dS" transition at some point. We study the possibility that there are two dark energy fluids, one of which gave rise to the anti-de Sitter-like early Universe. The interaction is modeled by the Lotka–Volterra equations commonly used in population biology. We consider "competition" models that are further classified as "unfair competition" and "fair competition". The former involves a quintessence in competition with a phantom, and the second involves two phantom fluids. Surprisingly, even in the latter scenario it is possible for the overall dark energy to cross the phantom divide. The latter model also allows a constant w "AdS-to-dS" transition, thus evading the theorem that such a dark energy must possess a singular equation of state. We also consider a "conversion" model in which a phantom fluid still manages to achieve "AdS-to-dS" transition even if it is being converted into a negative energy density quintessence. In these models, the energy density of the late time effective dark energy is related to the coefficient of the quadratic self-interaction term of the fluids, which is analogous to the resource capacity in population biology.

Keywords: Hubble tension; interacting dark energy; sign-switching dark energy; Lotka–Volterra equations

Citation: Ong, Y.C. An Effective Sign Switching Dark Energy: Lotka–Volterra Model of Two Interacting Fluids. *Universe* **2023**, *9*, 437. https://doi.org/10.3390/universe9100437

Academic Editors: Joan Solà Peracaula and Aharon Davidson

Received: 20 August 2023
Revised: 15 September 2023
Accepted: 26 September 2023
Published: 30 September 2023

1. Introduction: Cosmology with Sign Switching Dark Energy

The Hubble tension [1–4] and the S_8 tension [5–7] in cosmology continue to be highly debated [8–17]. The former is the mismatch between the locally measured expansion rate and the inferred rate via the cosmic microwave background (CMB), while the latter concerns the measurement of galaxy clusters on a scale of $8\,h^{-1}$Mpc, which revealed that matter has not clumped as much as expected assuming the concordance ΛCDM cosmology and its parameters constraints given by the CMB data (for a review on these issues, as well as other challenges facing ΛCDM cosmology, see [18]; see [19] for an introduction to various dark energy scenarios). If these effects are real, they could be due to modified gravity or other new physics [20,21].

Note that in the ΛCDM model, the Hubble parameter as a function of the redshift $H(z)$ is specified by two constant fitting parameters [22] (H_0, Ω_m), or equivalently (A, B), as follows:

$$H(z)^2 = H_0^2\Big[1 - \Omega_m + \Omega_m(1+z)^3\Big] \tag{1}$$
$$= A + B(1+z)^3,$$

where A is the term associated with dark energy. The aforementioned tensions could mean that ΛCDM is not correct and thus the "constant" fitting parameters could evolve with redshift (or equivalently, with time). See also [23,24].

Indeed, one of the possible ways to ameliorate these tensions is to consider the possibility that the Universe was originally more anti-de Sitter-(AdS)-like than de Sitter-(dS)-like [25–41]. That is, the physics of dark energy (DE) might be more complicated than we initially expected[1]. The main idea is to reduce the tension between the higher value of H_0 obtained from CMB with the lower value obtained by local measurements, by changing cosmological models either at the recombination epoch or at late time [11]. For example, a phantom energy at late time would accelerate cosmic expansion faster than a cosmological constant would. In addition, Baryon Oscillation Spectroscopic Survey (BOSS) of SDSS-III probing the Lyα forest of quasars also indicates preference for a positive dark energy density at late time but a negative one at early time [48]. Future observations such as SKA [49], BINGO [50,51], and Euclid [52] could potentially further constrain this possibility. Another observational support for negative energy density comes from Pantheon+ data of high redshift supernovae [53].

Such a possibility can be realized by simply promoting the cosmological constant Λ to[2] $\Lambda_s := \Lambda_{s0}\mathrm{sgn}[z_\dagger - z]$, where Λ_{s0} is the present value of the cosmological constant, and $z_\dagger$ is the value of redshift at which the sign switching abruptly happened [34]. See also [54]. Another model considers a "graduated dark energy" with energy density and pressure satisfying $\rho + p \propto \rho^\lambda$, which provides a continuous transition (controlled by the parameter λ) from AdS-like to dS-like Universe [28]. Other options include the possibility that the dark energy sector could consist of a negative cosmological constant and a phantom dark energy [27] or a quintessence [35] (see, however, [55]). A different approach based on fractal modification to the entropy via a running Barrow index [56] could also give rise to an effective sign-changing dark energy [38].

In this work, we shall consider what happens if instead of a scalar field on top of a negative cosmological constant, we have either (1) a quintessence with negative energy density, which competes with a phantom dark energy with a positive energy density[3], or (2) a phantom with a negative energy density that competes with another phantom with a positive energy density, or (3) a phantom with positive energy density being converted into a negative energy density quintessence. We shall refer to these scenarios, respectively, as "unfair competition", "fair competition", and "conversion" models. This is inspired by the interacting models between dark matter (DM) and dark energy [59–64], as well as from biological species interactions. In fact, the connection between these two subjects has been noticed in the literature. For example, [65] Perez et al., as well as Aydiner [66], pointed out that the DM–DE interaction can be re-written as the Lotka–Volterra equation, which is commonly used in population biology to model the interactions between various species. In cosmological contexts, the Lotka–Volterra equation was also studied in [67,68]. In the population model, it is of course required that the numbers of the species involved are non-negative, whereas in our model, the corresponding quantities are the energy densities of the DE fluids, which can be negative by assumption. Indeed, one might expect that if DM is converted into DE at late time, H_0 can be increased, while the matter density Ω_m can be decreased, so that $S_8 \propto \sqrt{\Omega_m}$ also decreases. This is indeed true for some models, but in others, the S_8 tension can in fact worsen [69]. In [70], it is shown that generically, allowing interactions between DE and DM does not allow both the Hubble and S_8 tensions to be ameliorated. Somewhat surprisingly, in the same work, it is argued that phantom dark energy with energy flow from the opposite direction, i.e., from DE to DM, is slightly preferred. In this work, we consider the alternative of non-minimal interaction between two DE fields to obtain one effective DE that exhibits sign-switching energy density, which might have a better chance at addressing both tensions simultaneously (as illustrated in [71] with the DE sector consisting of multiple interacting axion-like particle species). Our work is only meant to be an illustration of concept with the simplest models. Some assumptions would need to be relaxed or improved before a more realistic model can be used for data fitting the actual universe.

We will work out the conditions on the interactions between two dark energy fluids ("DE-DE interaction") in order to obtain a late time accelerated expansion with a very small but positive energy density. In our models, unlike the single fluid models in the literature, neither fluid exhibits any singular behavior in their equation of state, although if the phantom divide is crossed, the combined effective dark energy does exhibit such a singular behavior during the AdS-to-dS transition. Remarkably, we found that in the fair competition model, it is possible for the effective dark energy to cross the phantom divide despite both component fluids satisfying $w < -1$. In addition, in this scenario, there are evolutions that allow AdS-to-dS transition without crossing the phantom divide, which therefore is free of singular behavior in its equation of state. Finally, it is often said that the fact that the dark energy density is extremely small is "unnatural". We shall see that in the two-fluid model, this value is related to the coefficient of the quadratic self-interaction term of the fluids, which mathematically plays the same role as the resource capacity of a biological population.

2. The Unfair Competition Model

In [65,66], the authors considered a model in which dark energy is being converted into dark matter via

$$\begin{cases} \dot{\rho}_{\text{DE}} + 3H(\rho_{\text{DE}} + p_{\text{DE}}) = -\gamma\rho_{\text{DE}}\rho_{\text{DM}}, \\ \dot{\rho}_{\text{DM}} + 3H(\rho_{\text{DM}} + p_{\text{DM}}) = \gamma\rho_{\text{DE}}\rho_{\text{DM}}, \end{cases} \tag{2}$$

where $\gamma > 0$. Likewise, dark matter can be converted into dark energy with $\gamma < 0$. See [72] for generalizations.

The equations of state considered in [66] are $w_{\text{DM}} \geqslant 0$ and $w_{\text{DE}} < -1$, i.e., dark matter is a "normal matter" while dark energy is a phantom fluid. Thus, one can define two positive quantities:

$$\begin{cases} R_1 := -3H(1 + w_{\text{DE}}) > 0, \\ R_2 := 3H(1 + w_{\text{DM}}) > 0. \end{cases} \tag{3}$$

Furthermore, assuming that the Hubble parameter is slowly varying (so that R_1 and R_2 are both approximately constant) and upon introducing[4] $x_1 := \gamma\rho_{\text{DE}}/r_2$ and $x_2 := \gamma\rho_{\text{DM}}/r_1$, Equation (2) can be re-written as

$$\begin{cases} \dfrac{\mathrm{d}x_1}{\mathrm{d}t} = R_1 x_1 - R_1 x_1 x_2, \\ \dfrac{\mathrm{d}x_2}{\mathrm{d}t} = R_2 x_1 x_2 - R_2 x_2, \end{cases} \tag{4}$$

which is explicitly a Lotka–Volterra equation that describes a predator–prey dynamic with x_1 being the "prey" and x_2 the "predator". The system thus displays an oscillatory behavior. Such an interacting model could potentially help to resolve the coincidence problem.

In our case, we have two interacting dark energy fluids, which we will denote by $\Lambda_1 < 0$ and $\Lambda_2 > 0$ (despite the notation, they are not constant; the notation is meant to remind us that they are mimicking cosmological constants). Their energy densities would be denoted by ρ_{Λ_1} and ρ_{Λ_2}, respectively. The transition from an early time AdS-like Universe to a late time dS-like spacetime thus amounts to Λ_1 becoming subdominant to Λ_2. Let us first consider an unrealistic model (to be improved upon later) that is analagous to Equation (2), so we can point out the differences:

$$\begin{cases} \dot{\rho}_{\Lambda_1} + 3H(\rho_{\Lambda_1} + p_{\Lambda_1}) = -\gamma\rho_{\Lambda_1}\rho_{\Lambda_2}, \\ \dot{\rho}_{\Lambda_2} + 3H(\rho_{\Lambda_2} + p_{\Lambda_2}) = \gamma\rho_{\Lambda_1}\rho_{\Lambda_2}, \end{cases} \tag{5}$$

where $\gamma > 0$. As before, we assume that the late time dark energy is a phantom; thus, $w_{\Lambda_2} < -1$. On the other hand[5], $w_{\Lambda_1} > -1$ but with $\rho_{\Lambda_1} < 0$. This is justified by [74,75], in which it was argued that solving both the Hubble and S_8 tensions require the overall effective dark energy to cross the phantom divide (if we assume that Newton's gravitational constant is not varying). See also [76]. This is also similar to the model in [28].

Thus, we define, analogous to Equation (3), two parameters

$$\begin{cases} r_1 := 3H(1 + w_{\Lambda_1}) > 0, \\ r_2 := -3H(1 + w_{\Lambda_2}) > 0. \end{cases} \tag{6}$$

We assume that both w_{Λ_1} and w_{Λ_2} are constant. Upon introducing the dimensionless quantities $x := \gamma \rho_{\Lambda_1}/r_2$ and $y := \gamma \rho_{\Lambda_2}/r_1$, we obtain the Lotka–Volterra equation of the form

$$\begin{cases} \dfrac{\mathrm{d}x}{\mathrm{d}t} = -r_1 x - r_1 xy, \\ \dfrac{\mathrm{d}y}{\mathrm{d}t} = r_2 xy + r_2 y, \end{cases} \tag{7}$$

where some of the signs differ from Equation (4) that describes the interacting DM–DE model. In addition, here, we have $x < 0, y > 0$. This system does not oscillate, but rather there is an attractor $x \to 0^-$ and $y \to \infty$. Note that in contrast to the DM–DE interaction model, it is not quite right to say that Λ_1 is being "converted" into Λ_2 here (a true conversion model will be studied in Section 4), since $x < 0$ implies that the interaction term is negative for both x and y. In other words, the two fluids are competing, but unlike two competing species whose birth rates are both positive, x is itself diminishing exponentially due to the "death rate" term $-r_1 x$, hence the name "unfair competition". It is clear that the phantom fluid thus dominates over the quintessence. That is, ρ_{Λ_1} is asymptotically vanishing while ρ_{Λ_2} becomes large (and eventually diverges) at late time. This is not a desired property since we know from observation that dark energy density is only of the order of 10^{-30} g/cm^3.

We can improve upon this model by modifying the $\mathrm{d}y/\mathrm{d}t$ term so that

$$\begin{cases} \dfrac{\mathrm{d}x}{\mathrm{d}t} = -r_1 x - r_1 xy, \\ \dfrac{\mathrm{d}y}{\mathrm{d}t} = r_2 xy + r_2 y\left(1 - \dfrac{y}{K}\right), \end{cases} \tag{8}$$

where $K > 0$ is a constant. The attractor is then $x \to 0^-$ and $y \to K$, so we can prescribe to K the observed value. At this point, this seems rather ad hoc, but later on we will give it a physical interpretation.

The analogy to population biology can also be made here: with x small at late time, in the absence of y/K term, what we have is analogous to an exponential growth population $\mathrm{d}y/\mathrm{d}t = r_2 y$, whereas Equation (8) corresponds to the logistic model with a resource capacity (or "carrying capacity") K, so that the actual population size cannot diverge but rather asymptotes to a constant value. Note that since $x \to 0^-$ anyway, there is no good reason to add the resource term to $\mathrm{d}x/\mathrm{d}t$ in this model. A typical phase diagram is given in Figure 1.

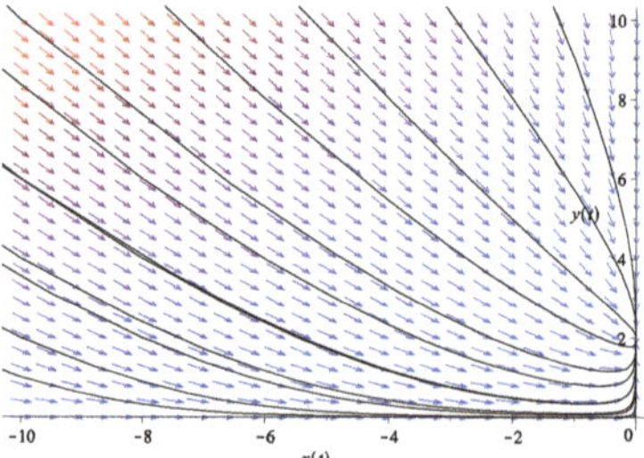

Figure 1. The phase diagram of the two dark energy fluid model in Equation (8). Here, we use $r_1 = r_2 = 1$ and $K = 2$. The system exhibits a fixed point at $(x, y) = (0, K)$.

Note that the effective dark energy density is the sum $\rho_{\Lambda_1} + \rho_{\Lambda_2}$. Thus, in order that this quantity starts out negative, we need the initial condition to satisfy $y(0) < |x(0)|$. Then, since the attractor is $(x, y) = (0, K)$ with $K > 0$, it follows that the continuity $x(t) + y(t)$ must cross over to $y > 0$ at some point. The exact profile of $\rho_{\Lambda_1}(t) + \rho_{\Lambda_2}(t)$ or the re-scaled equation $x(t) + y(t)$ would depend on the initial conditions, but the transition from an overall negative energy density to a positive one can be smoother than the model in [28]. One example is given in Figure 2.

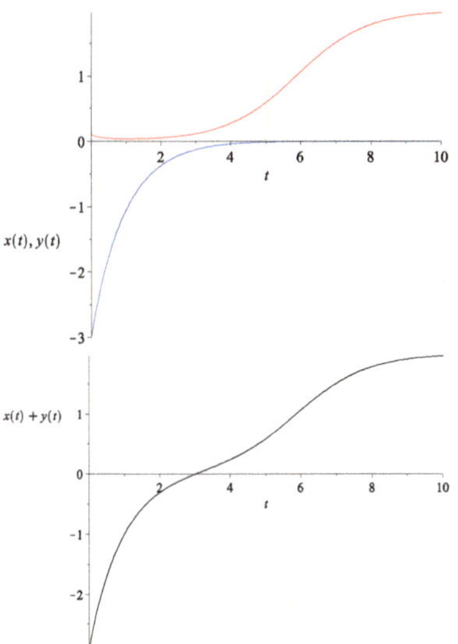

Figure 2. Top: The evolution of $x(t)$ (bottom curve in blue) and $y(t)$ (top curve in red) with the initial condition set to be $x(0) = -3$, $y(0) = 0.1$, and with $r_1 = 1 = r_2$, $K = 2$. **Bottom**: The evolution of $x(t) + y(t)$, which is essentially the re-scaled version of $\rho_{\Lambda_1}(t) + \rho_{\Lambda_2}(t)$ (in this example, they are in fact equal).

If we know what kind of profile is desired from observational constraints, this would in turn provide us with a mean to choose the coefficients r_1 and r_2, as well as the initial conditions of the Lotka–Volterra equation. We also note that the equation of states of both dark energy components are never singular, though the combined effective dark energy density has to pass through $\rho_{\Lambda_1}(t) + \rho_{\Lambda_2}(t) = 0$, and the effective equation of the state is singular at that point [77]. This can be seen as follows: if we consider the combined fluid to still satisfy the equation of state of the form $p = w\rho$, where $\rho := \rho_{\Lambda_1} + \rho_{\Lambda_2}$ and $p := p_{\Lambda_1} + p_{\Lambda_2}$, then

$$\rho = \rho_{\Lambda_1} + \rho_{\Lambda_2} = w^{-1}(p_{\Lambda_1} + p_{\Lambda_2}),\tag{9}$$

which implies that the effective varying w is the weighted average:

$$w = \frac{w_{\Lambda_1}\rho_{\Lambda_1} + w_{\Lambda_2}\rho_{\Lambda_2}}{\rho_{\Lambda_1} + \rho_{\Lambda_2}} = \frac{w_1 r_2 x_1 + w_2 r_1 x_2}{r_2 x_1 + r_1 x_2}.\tag{10}$$

Therefore, evidently $w \to \pm\infty$ when the denominator is zero. Again, the situation is similar to the "graduated dark energy" model of [28], though in that case there is only one dark energy fluid, whose equation of state becomes singular. An example of the evolution of w is shown in Figure 3. Note that the values of w_{Λ_1} and w_{Λ_1} are not freely prescribed since they are constrained by Equation (6), in the sense that once we fix r_1 and r_2, the relations between w_{Λ_1} and w_{Λ_2} are also determined. In our example, choosing $r_1 = r_2 = 1$ implies $w_{\Lambda_1} + w_{\Lambda_2} = -2$.

We remark that the sign of $x(t) + y(t)$ is not necessarily the same as the sign of $\rho_{\Lambda_1}(t) + \rho_{\Lambda_2}(t)$. In fact,

$$x(t) + y(t) = \frac{\gamma}{r_1 r_2}\left(r_1 \rho_{\Lambda_1}(t) + r_2 \rho_{\Lambda_2}(t)\right). \tag{11}$$

Thus, the sign of $x(t) + y(t)$ is the same as the sign of $r_1 \rho_{\Lambda_1}(t) + r_2 \rho_{\Lambda_2}(t)$. For simplicity, our example deals with $r_1 = r_2$, and the two expressions do have the same sign; furthermore, w is singular when $x + y = 0$ or equivalently at $r_1 \rho_{\Lambda_1} + r_2 \rho_{\Lambda_2} = 0$.

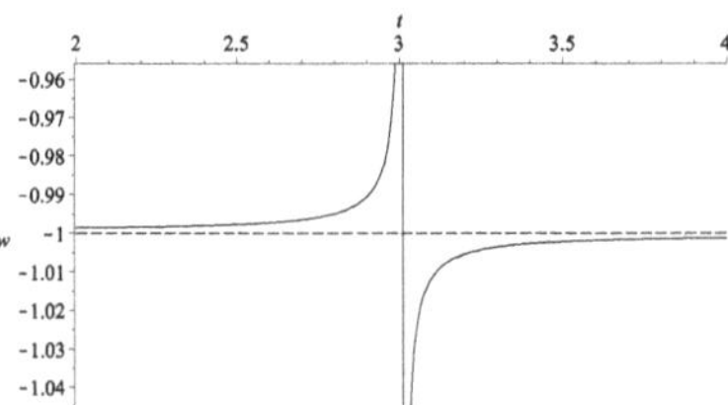

Figure 3. The evolution of the overall effective w of the example in Figure 2 with $w_{\Lambda_2} = -0.999$ and $w_{\Lambda_1} = -1.001$.

3. The Fair Competition Model

In population biology, two species that are very similar (i.e., fulfilling the same ecological niche) will compete for the same resources. To model such a situation, we consider both species to have a positive "birth rate", so that in this sense, the competition is fair. In place of Equation (8), we have:

$$\begin{cases} \dfrac{\mathrm{d}x}{\mathrm{d}t} = r_1 x\left(1 + \dfrac{x}{K_1}\right) - r_1 xy, \\ \dfrac{\mathrm{d}y}{\mathrm{d}t} = r_2 xy + r_2 y\left(1 - \dfrac{y}{K_2}\right), \end{cases} \tag{12}$$

where now $r_1 := -3H(1 + w_{\Lambda_2}) > 0$. Note the coefficient pre-multiplying r_1 is now 1 instead of -1. Since $x < 0$, we consider the resource term to be $1 + x/K_1$ instead of $1 - x/K_1$, keeping $K_1 > 0$. Note again that "competition" means that the interaction is harmful for both species, so the interaction term is negative for both fluids ($r_2 xy < 0$ because $x < 0$). For a fair competition, we also include the carrying capacities K_1 and K_2 for both fluids, with $\mathcal{O}(K_1) = \mathcal{O}(K_2)$, both being positive. A typical phase diagram is shown in Figure 4. Two families of flows are observed: those that flow towards K_1 and those that flow towards K_2. We are concerned with the latter.

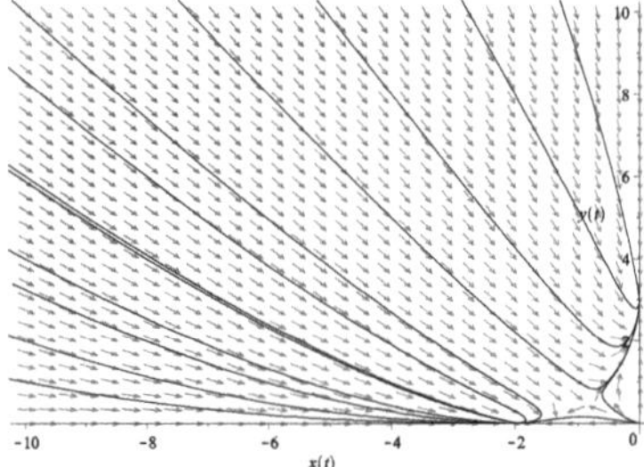

Figure 4. The phase diagram of the two dark energy fluid model in Equation (12). Here, we use $r_1 = 1$, $K_1 = 2$, and $r_2 = 1.5$, $K_2 = 3$. The system exhibits trivial fixed points at $(x, y) = (0, K_2)$ and $(x, y) = (K_1, 0)$, but also notice the existence of a saddle point that "separates" the two families of flow.

An explicit example of the rescaled energy density x and y as well as their sum are provided in Figure 5. Here, we let $r_1 = 1$ and $r_2 = 1.5$. We can see that $x(t) + y(t)$ changes sign in Figure 6. Now, in this example, since $r_1 \neq r_2$, we cannot directly compare the sign change of $x(t) + y(t)$ to that of the overall dark energy density. However, we note that initially, $\mathrm{sgn}(x + y) = \mathrm{sgn}(\rho_{\Lambda_1} + 1.5\rho_{\Lambda_2}) = -1$. Since $\rho_{\Lambda_2} > 0$, it follows that $\rho_{\Lambda_1} + \rho_{\Lambda_2} < \rho_{\Lambda_1} + 1.5\rho_{\Lambda_2} < 0$. Thus, it follows that $\mathrm{sgn}(\rho_{\Lambda_1} + \rho_{\Lambda_2}) = -1$ initially. On the other hand, asymptotically, we have $\mathrm{sgn}(x + y) = \mathrm{sgn}(y) = \mathrm{sgn}(K_2) = +1$ as $x \to 0^-$. Equivalently, at late time $\mathrm{sgn}(\rho_{\Lambda_1} + \rho_{\Lambda_2}) = \mathrm{sgn}(\rho_{\Lambda_2}) = +1$. Thus, we see that $\rho_{\Lambda_1} + \rho_{\Lambda_2}$ does indeed change the sign., i.e., the Universe transits from AdS-like to dS-like.

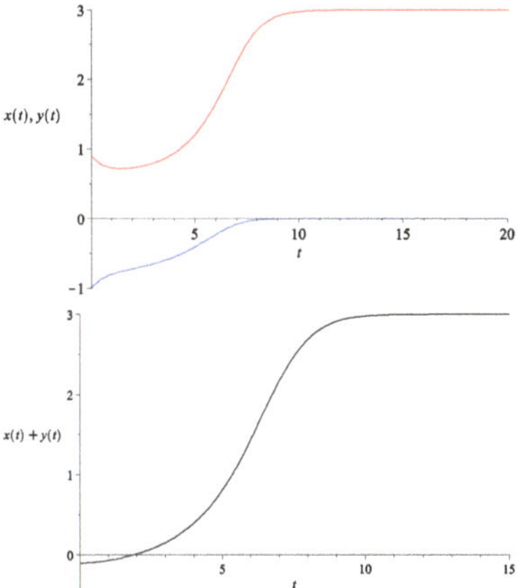

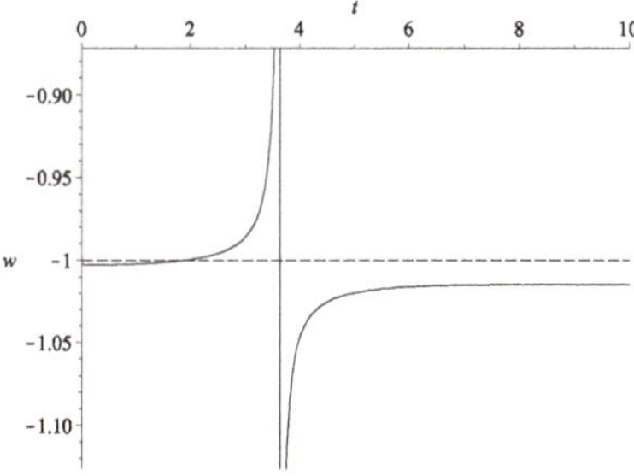

Figure 5. Top: The evolution of $x(t)$ (bottom curve in blue) and $y(t)$ (top curve in red) with the initial condition set to be $x(0) = -1, y(0) = 0.9$, and with $r_1 = 1$, $r_2 = 1.5$; $K_1 = 2$, $K_2 = 3$. **Bottom**: The evolution of $x(t) + y(t)$, which is essentially the re-scaled version of $\rho_{\Lambda_1}(t) + \rho_{\Lambda_2}(t)$.

Figure 6. The evolution of the overall effective w of the example in Figure 5 with $w_{\Lambda_1} = -1.01$ and $w_{\Lambda_1} = -2.01$. There are two phantom crossings. The first crossing is smooth and corresponds to the time when $x + y = 0$. The second crossing is singular and corresponds to $\rho_{\Lambda_1} + \rho_{\Lambda_2} = 0$. These two conditions are not the same since $r_1 \neq r_2$.

Incidentally, we also note that if $x(t) + y(t)$ is monotonically increasing, then we can give a bound on $\rho_{\Lambda_1}(t) + \rho_{\Lambda_2}(t)$. To see this, simply observe that

$$\frac{\mathrm{d}(x(t) + y(t))}{\mathrm{d}t} > 0 \tag{13}$$

is equivalent to

$$\frac{\gamma}{r_1 r_2}(r_1\dot{\rho}_{\Lambda_1} + r_2\dot{\rho}_{\Lambda_2}) > 0. \tag{14}$$

Given that $\gamma, r_1, r_2 > 0$, this means $r_1 \dot{\rho}_{\Lambda_1} + r_2 \dot{\rho}_{\Lambda_2} > 0$. Inserting a few terms that cancel each other yields:

$$r_1 \dot{\rho}_{\Lambda_1} + (r_1 \dot{\rho}_{\Lambda_2} - r_1 \dot{\rho}_{\Lambda_2}) + (r_2 \dot{\rho}_{\Lambda_1} - r_2 \dot{\rho}_{\Lambda_1}) + r_2 \dot{\rho}_{\Lambda_2} > 0. \tag{15}$$

Therefore,

$$(r_1 + r_2)(\dot{\rho}_{\Lambda_1} + \dot{\rho}_{\Lambda_2}) > r_2 \dot{\rho}_{\Lambda_1} + r_1 \dot{\rho}_{\Lambda_2}$$
$$= r_2^2 \frac{\dot{x}}{\gamma} + r_1^2 \frac{\dot{y}}{\gamma}. \tag{16}$$

If $r_2 > r_1$, we can write the last equation as

$$\gamma^{-1}[r_1^2(\dot{x} + \dot{y}) + (r_2^2 - r_1^2)\dot{x}]. \tag{17}$$

Likewise, if $r_1 > r_2$, we can write

$$\gamma^{-1}[r_2^2(\dot{x} + \dot{y}) + (r_1^2 - r_2^2)\dot{y}]. \tag{18}$$

Thus, for example, if $r_2 > r_1$ (and hence $r_2^2 > r_1^2$ – recall that r_1, r_2 are positive), and we observe that $x + y$ and x are both monotonicaly increasing, then so must $\rho_{\Lambda_1} + \rho_{\Lambda_2}$:

$$(r_1 + r_2)(\dot{\rho}_{\Lambda_1} + \dot{\rho}_{\Lambda_2}) > \gamma^{-1}[r_1^2(\dot{x} + \dot{y}) + (r_2^2 - r_1^2)\dot{x}] > 0 \tag{19}$$

The phantom divide can be crossed as shown in Figure 6, where w is given by Equation (10). The values of w_{Λ_1} and w_{Λ_2} are constrained by the defining equations $r_1 = -3H(1 + w_{\Lambda_1})$ and $r_2 = -3H(1 + w_{\Lambda_2})$. With $r_1 = 1$ and $r_2 = 1.5$, if we take $w_{\Lambda_1} = -1.01$, say, then $w_{\Lambda_1} = -1.015$. Note that there are two phantom crossings[6] here: the first one occurs without any singularity. From Equation (10), it can be seen that if the denominator is not zero, then such a smooth crossing occurs precisely when $x + y = 0$. This cannot happen for the unfair competition model as the condition would be $x - y = 0$ instead (which cannot occur since $x < 0$ and $y > 0$). Note that despite the fact that there are two phantom crossings, there is only one AdS-to-dS transition in this example.

Even more surprising is the fact that the overall $w < -1$ can stay constant, yet there is still a AdS-to-dS transition. To achieve this we simply need to choose $r_1 = r_2$. The plots of $x(t), y(t)$ and their sum is qualitatively the same as Figure 5 and are thus not shown. However, from the defining relations of r_1 and r_2 we would have $w_1 = w_2$, and so in Equation (10), we obtain

$$w = \frac{w_{\Lambda_1}\rho_{\Lambda_1} + w_{\Lambda_2}\rho_{\Lambda_2}}{\rho_{\Lambda_1} + \rho_{\Lambda_2}} = \frac{w_1(\rho_{\Lambda_1} + \rho_{\Lambda_2})}{\rho_{\Lambda_1} + \rho_{\Lambda_2}} = w_1. \tag{20}$$

Strictly speaking, during the transition point $\rho_{\Lambda_1} + \rho_{\Lambda_2} = 0$, which otherwise would give rise to a singular behavior, we obtain an indeterminate form $0/0$, but both the left and right limit are well defined and equal to w_1, so physically it makes sense to say that $w \equiv w_1 = w_2$ for all t. As such, this evades the recent theorem that a sign-changing dark energy must have a singular equation of state [77]. The reason this does not really violate the theorem therein is because the proof in [77] is strictly for DE fluids that obey the usual continuity equation $\dot{\rho} + 3H(\rho + p) = 0$, whereas in our model it can be checked that the combined DE does not satisfy the continuity equation; the "carrying capacity" term K breaks the continuity equation. This is equivalent to saying that $\nabla_\mu T_{\mathrm{DE}}^{\mu\nu} \neq 0$. This is not surprising—as we will see in the Section 5, our models have a nontrivial nonlinear self-interaction term that acts as a source term for the continuity equation.

4. The Conversion Model

Given the results above, one might wonder whether the AdS-to-dS transition can still happen if we restrict the growth of Λ_2 by converting it into Λ_1, or equivalently, by giving Λ_1 an advantage. This is achieved by the following model involving a quintessence Λ_1 and a phantom Λ_2:

$$\begin{cases} \dfrac{\mathrm{d}x}{\mathrm{d}t} = -r_1 x\left(1 - \dfrac{x}{K_1}\right) + r_1 xy, \\ \dfrac{\mathrm{d}y}{\mathrm{d}t} = r_2 xy + r_2 y\left(1 - \dfrac{y}{K_2}\right), \end{cases} \tag{21}$$

in which we note that the second term of $\mathrm{d}x/\mathrm{d}t$ is now $+r_1 xy$ instead of $-r_1 xy$. The interaction is therefore beneficial to x but harmful to y. This sounds like the complete opposite of what we wish to achieve (to have Λ_2 being the dominant term at late time). Surprisingly, even in such a scenario, it is possible to have a phantom crossing. The only difference being the attractor is now a stable spiral centered at

$$(x, y) = \left(-\frac{K_1(K_2 - 1)}{K_1 K_2 + 1},\, \frac{K_2(K_1 + 1)}{K_1 K_2 + 1}\right), \tag{22}$$

as can be seen in the example depicted in Figure 7.

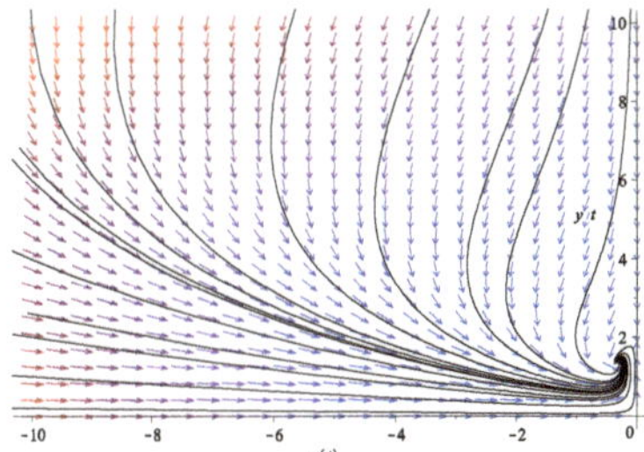

Figure 7. The phase diagram of the two dark energy fluid model in Equation (21). Here, we use $r_1 = 1 = r_2$, $K_1 = 1$, and $K_2 = 2$. The system exhibits an attractor fixed point at $(x, y) = (-1/3, 3/2)$.

What happens is that, despite the conversion term, Λ_2 can still dominate at late time. After all, we do not need Λ_2 to grow too big. The evolution of $x(t)$ and $y(t)$, as well as their sum (which in this example is the same as $\rho_{\Lambda_1} + \rho_{\Lambda_2}$), is shown in Figure 8. The phantom crossing is shown in Figure 9.

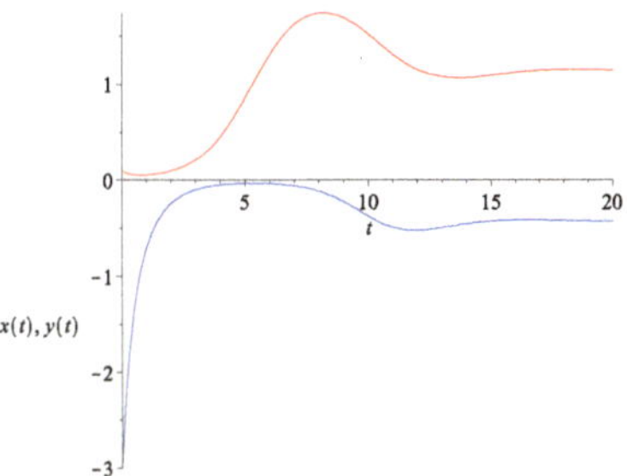

Figure 8. *Cont.*

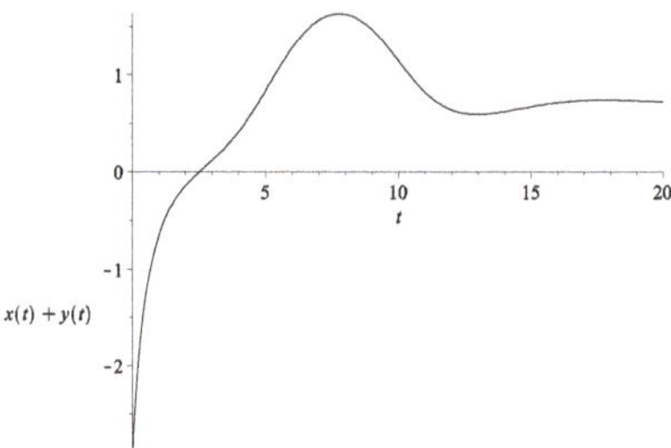

Figure 8. Top: The evolution of $x(t)$ (bottom curve in blue) and $y(t)$ (top curve in red) with the initial condition set to be $x(0) = -3$, $y(0) = 0.1$, and with $r_1 = 1 = r_2$; $K_1 = 1, K_2 = 2$. **Bottom**: The evolution of $x(t) + y(t)$, which in this case is the same as the overall ρ.

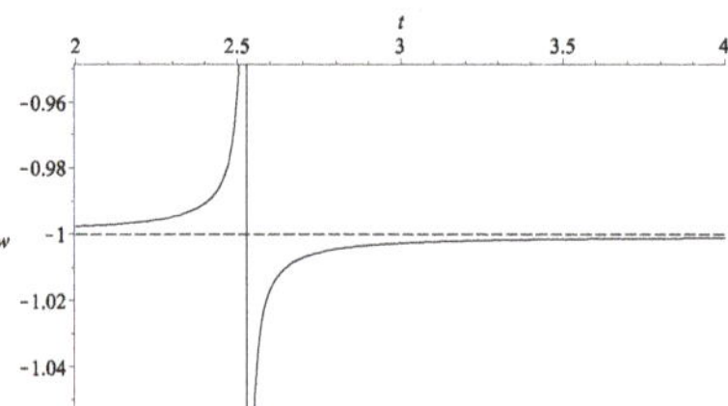

Figure 9. The evolution of the overall effective w of the example in Figure 8 with $w_{\Lambda_1} = -0.999$ and $w_{\Lambda_1} = -1.001$.

The reason we consider a resource term $1 - x/K_1$ instead of $1 + x/K_1$ as in the previous section (incidentally, this puts the fixed point $x = K_1 > 0$ outside the physical phase space) is that otherwise, with $1 + x/K_1$ and the coefficient pre-multiplying r_1 being negative (quintessence) instead of positive (phantom), we observe that for "most" initial conditions,

$$\frac{\mathrm{d}|x|}{\mathrm{d}t} \sim \frac{r_1}{K_1} \frac{|x|^2}{2}, \tag{23}$$

and thus the magnitude of x is increasing and $x \to -\infty$ instead of 0, which is not the behavior that we want. This can be seen in Figure 10.

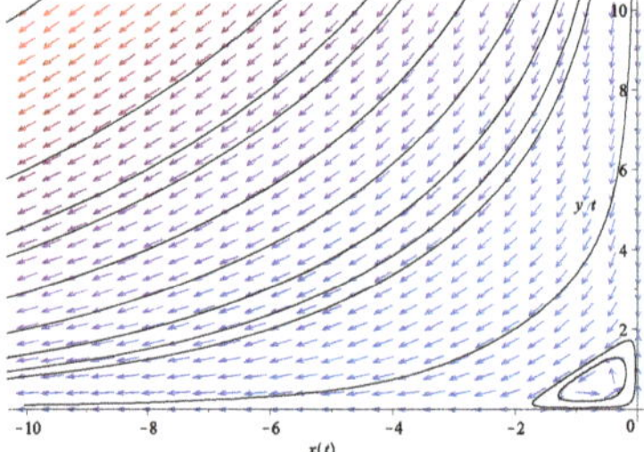

Figure 10. The phase diagram of the two dark energy fluid model in Equation (21) but with the resource term changed to $1 - x/K_1$ instead of $1 + x/K_1$. Here, we use $r_1 = 1 = r_2$, $K_1 = 2 = K_2$. Most trajectories flow towards $x \to -\infty$ and $y \to 0$. Note, however, the presence of a center surrounded by cyclic flows.

However, even in this scenario, there is one interesting feature worth mentioning. In the neighborhood of the origin, there exists a center around which the flows are cyclic. This implies both x and y, as well as their sum, are oscillatory. As a result, there are multiple (infinitely many) phantom crossings, and infinitely many transitions between AdS-like and dS-like cosmology. These are shown in Figures 11 and 12. Indeed, multiple transition scenarios have been considered in the literature [39].

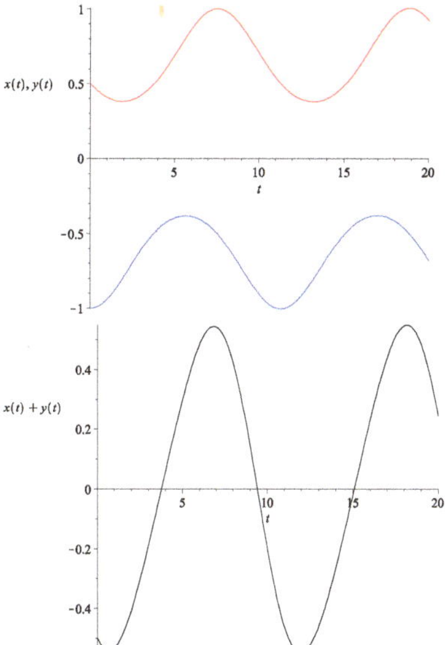

Figure 11. Top: The evolution of $x(t)$ (bottom curve in blue) and $y(t)$ (top curve in red) with the initial condition set to be $x(0) = -1, y(0) = 0.5$, for the model that corresponds to Figure 10. **Bottom**: The evolution of $x(t) + y(t)$, which in this case is the same as the overall ρ.

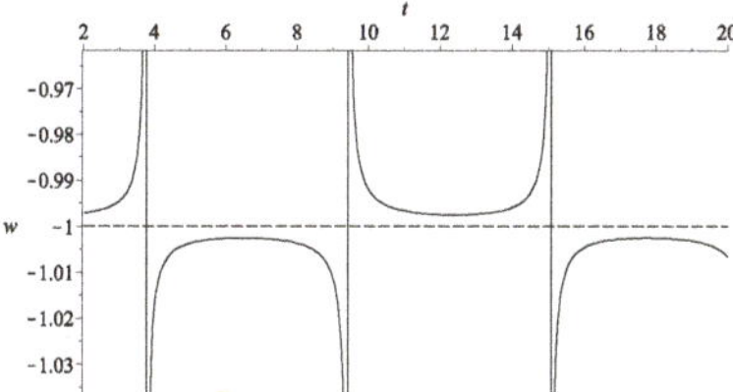

Figure 12. The evolution of the overall effective w of the example in Figure 11 with $w_{\Lambda_1} = -0.999$ and $w_{\Lambda_1} = -1.001$.

5. Discussion: Sign Switching Dark Energy and Naturalness

One of the longstanding questions about dark energy density is why its value is so small, which is some 10^{-120} times smaller than the natural scale for a quantum vacuum energy if it is indeed a cosmological constant (for a dynamical field, the problem translates into an extremely light mass of the field). Of course, it is debatable whether this is indeed a problem [78]. In any case, it would be interesting to see what this value corresponds to in the Lotka–Volterra equations in these models.

Take for example, the unfair competition model. We note that the evolution equation for $y(t)$, namely

$$\frac{\mathrm{d}y}{\mathrm{d}t} = r_2 xy + r_2 y\left(1 - \frac{y}{K}\right), \tag{24}$$

is equivalent to the following fluid equation:

$$\frac{\mathrm{d}\rho_{\Lambda_2}}{\mathrm{d}t} + 3H(\rho_{\Lambda_2} + p_{\Lambda_2}) = \gamma \rho_{\Lambda_1}\rho_{\Lambda_2} + \frac{\gamma}{K}\left(\frac{1 + w_{\Lambda_2}}{1 + w_{\Lambda_1}}\right)\rho_{\Lambda_2}^2. \tag{25}$$

In other words, the "resource term" in the Lotka–Volterra equation corresponds to a quadratic self-interaction term. How might one interpret this term?

Such a term was also considered in [66,79]. As commented therein, pressure and density may not be linearly related in more complicated and more realistic systems. If we assume $p = p(\rho)$ for any barotropic fluid to be an analytic function, we can consider equation of state of the form $p = p_0 + A_1\rho + A_2\rho^2 + \mathcal{O}(\rho^3)$. This is a Taylor expansion of $p = p(\rho)$ about $\rho = 0$, or upon re-grouping of terms, the expansion about the present energy density [79,80]. If this is the correct interpretation, then the self-interaction term in Equation (25) can be interpreted as the result of the first order non-linear term in the expansion. However, in a series expansion, typically, the coefficients of the subsequent terms are roughly of the same order of magnitude, so the "natural" expectation is that $\frac{\gamma}{K}\left(\frac{1+w_{\Lambda_2}}{1+w_{\Lambda_1}}\right) \sim \mathcal{O}(\gamma)$. Even if w_{Λ_1} and w_{Λ_2} can be very close to -1, generically we would have the ratio $(1 + w_{\Lambda_2})/(1 + w_{\Lambda_1})$ to be of order 1. This means that K being small $\mathcal{O}(\varepsilon) \ll 1$ would typically result in the quadratic coefficient being unnaturally large and dominate over the linear term, which in turn suggests that we should not, in fact, interpret this term as a term in a Taylor series expansion of $p_{\Lambda_2} = p_{\Lambda_2}(\rho_{\Lambda_2})$. Note that if we do not interpret ρ_{Λ_2} and $\rho_{\Lambda_2}^2$ term as part of a Taylor series, we can still absorb the $\rho_{\Lambda_2}^2$ term as part of the pressure so that $p_{\Lambda_2} = w_2\rho_{\Lambda_2} + \mathrm{const.}\rho_{\Lambda_2}^2$. In which case, K is related to the mass scale M of Λ_2 via $K \sim M^4$, see [81,82]; so this is just the aforementioned fact that in the case of dark energy being dynamical, the naturalness problem is its small mass scale. In the conversion model, the situation is similar. The attractor of the spiral is given in Equation (22), in which we see that y and x are both small if K_1 and K_2 are small. Obviously, our models do not solve the naturalness problem, unless one could explain dynamically why the attractor has such a small value. Perhaps a fundamental understanding of the nature of the phantom fluid or an entropic argument could provide such a mechanism (in the context of cosmological constant, it was argued in [83] that gravitational entropy is maximized by $\Lambda \to 0^+$.)

The DM–DE interacting model also has naturalness problem. In this scenario, in addition to the aforementioned mass scale issue, the coincidence problem also enters the discussion. The coincidence problem is the following: why has the DE energy density only recently become comparable to DM energy density? If the two dark sectors interact dynamically and one can convert into another, then it is conceivable that this problem can be resolved. In practice, however, this is not easily achieved and some amount of fine-tuning may be required [84] (though more complicated models like DE coupling to DM inhomogeneities can better accommodate the coincidence problem, as well as other issues like fine-tuning of initial conditions [85]). Interestingly, even in the DM–DE interacting scenario, it was shown in the same work [84] that at least in a general class of models they discussed, DE with a negative energy density (hence AdS-like in our language) in the past is better at resolving the coincidence problem. Indeed, we note that the coincidence problem may be resolved by considering interactions between two DE components. An example is provided by the so-called "cosmon" model [86,87], in which a running cosmological "constant" interacts with another effective DE fluid (the cosmon). Here, it is worth mentioning another interesting proposal, the "running vacuum model" [88], that could potentially solve the Hubble tension, the $S8$ tension, along with the naturalness problem and the coincidence problem.

To conclude, in this work, motivated by the idea that a sign switching dark energy from an early time AdS-like Universe to a late time dS-like Universe can help to ameliorate the Hubble tension and the S_8 tension, we consider a scenario in which the dark energy sector consists of two interacting fluids. We found that AdS-to-dS transition can happen under various models, even if both fluids are phantom, and even if the combined effective dark energy has a constant w. Of course, these are only toy models serve to illustrate the qualitative features. The profiles of these fluids need to be constrained by observations. Still, the possibility that the dark energy sector may contain various interacting components deserves a closer look (see [82,89] for other aspects of self-interacting dark energy) as it can realize many different interesting features.

For generalizations, one could also consider interactions between the two dark energy components with dark matter and/or dark radiation in a more complicated model (a quintom model was considered in [90], with the phantom component interacting with dark matter, but not with the quintessence sector; see also [91]). Another possibility is to consider other forms of interaction terms in place of $\pm \gamma \rho_{\Lambda_1} \rho_{\Lambda_2}$; see, for example, [92] and the references therein for the case of DM–DE interaction. Most importantly, more realistic models need to go beyond the assumption that the Hubble parameter is slowly varying when setting up the Lotka–Volterra equations.

Funding: This research was funded by the National Natural Science Foundation of China grant number 11922508.

Data Availability Statement: This research is a theoretical study and has no associated data.

Acknowledgments: Y.C.O. thanks Brett McInnes for useful discussions.

Conflicts of Interest: The author declares no conflict of interest.

Notes

1. Indeed, such a possibility was already considered from other perspectives before the Hubble tension became a serious issue [42–46]. See also the recent work [47].

2. Here sgn is the sign function (i.e., it is $+1$ for positive argument and -1 for negative argument).

3. The idea that different dark energy components might interact with each other is not new. For example, models in which a quintessence interacts with a Chaplygin gas was considered in [57], in an attempt to explain the coincidence problem [58].

4. One can check that in the units in which the speed of light $c = 1$, R_1 and R_2 have physical dimension $[\text{time}]^{-1}$, while γ has dimension $[\text{density} \cdot \text{time}]^{-1}$; while x_1 and x_2 are dimensionless.

5. In the cosmological case, AdS spacetime has $w = -1$, which is the same as dS spactime. Unlike dS spacetime, however, in AdS spacetime the negative cosmological constant corresponds to a negative energy density but with a positive pressure. Cosmological evolution with negative energy densities was previously studied in details in [73].

6. This phantom crossing is achieved by exhibiting a pole/singularity in their equation of state parameter, which is quite different from the more well-known quintom models.

References

1. Bernal, J.L.; Verde, L.; Riess, A.G. The Trouble With H_0. *arXiv* **2016**, arXiv:1607.05617.
2. Verde, L.; Treu, T.; Riess, A.G. Tensions Between the Early and Late Universe. *Nat. Astron.* **2019**, *3*, 891–895. [CrossRef]
3. Valentino, E.D.; Anchordoqui, L.A.; Akarsu, O.; Ali-Haimoud, Y.; Amendola, L.; Arendse, N.; Asgari, M.; Ballardini, M.; Basilakos, S.; Battistelli, E.; et al. Snowmass2021—Letter of Interest Cosmology Intertwined II: The Hubble Constant Tension. *Astropart. Phys.* **2021**, *131*, 102605.
4. Di Valentino, E.; Mena, O.; Pan, S.; Visinelli, L.; Yang, W.; Melchiorri, A.; Mota, D.F.; Riess, A.G.; Silk, J. In the Realm of the Hubble Tension—A Review of Solutions. *Class. Quantum Grav.* **2021**, *38*, 153001.
5. Battye, R.A.; Charnock, T.; Moss, A. Tension Between the Power Spectrum of Density Perturbations Measured on Large and Small Scales. *Phys. Rev. D* **2015**, *91*, 103508. [CrossRef]
6. Benisty, D. Quantifying the S_8 Tension With the Redshift Space Distortion Data Set. *Phys. Dark Univ.* **2021**, *1*, 100766.
7. Di Valentino, E.; Anchordoqui, L.A.; Akarsu, Ö.; Ali-Haimoud, Y.; Amendola, L.; Arendse, N.; Asgari, M.; Ballardini, M.; Basilakos, S.; Battistelli, E.; et al. Cosmology Intertwined III: $f\sigma8$ and S8. *Astropart. Phys.* **2021**, *131*, 102604.
8. Knox, L.; Millea, M. The Hubble Hunter's Guide. *Phys. Rev. D* **2020**, *101*, 043533. [CrossRef]
9. Rameez, M.; Sarkar, S. Is There Really a Hubble Tension? *Class. Quantum Grav.* **2021**, *38*, 154005.
10. Vagnozzi, S. New Physics in Light of the H_0 Tension: An Alternative View. *Phys. Rev. D* **2020**, *102*, 023518. [CrossRef]
11. Benevento, G.; Hu, W.; Raver, M. Can Late Dark Energy Transitions Raise the Hubble Constant? *Phys. Rev. D* **2020**, *101*, 103517. [CrossRef]
12. Colgáin, E.Ó.; Sheikh-Jabbari, M.M.; Solomon, R.; Bargiacchi, G.; Capozziello, S.; Dainotti, M.G.; Stojkovic, D. Revealing Intrinsic Flat ΛCDM Biases with Standardizable Candles. *Phys. Rev. D* **2022**, *106*, L041301. [CrossRef]
13. Abdalla, E.; Abellán, G.F.; Aboubrahim, A.; Agnello, A.; Akarsu, Ö.; Akrami, Y.; Alestas, G.; Aloni, D.; Amendola, L.; Anchordoqui, L.A.; et al. Cosmology Intertwined: A Review of the Particle Physics, Astrophysics, and Cosmology Associated with the Cosmological Tensions and Anomalies. *J. High Energy Astrophys.* **2022**, *34*, 49–211.
14. Mazo, B.Y.D.V.; Romano, A.E.; Quintero, M.A.C. H_0 Tension or M Overestimation? *Eur. Phys. J. C* **2022**, *82*, 610. [CrossRef]

15. Escudero, H.G.; Kuo, J.L.; Keeley, R.E.; Abazajian, K.N. Early or Phantom Dark Energy, Self-Interacting, Extra, or Massive Neutrinos, Primordial Magnetic Fields, or a Curved Universe: An Exploration of Possible Solutions to the H_0 and σ_8 Problems. *Phys. Rev. D* **2022**, *106*, 103517. [CrossRef]

16. de Sá, R.; Benetti, M.; Graef, L.L. An Empirical Investigation Into Cosmological Tensions. *Eur. Phys. J. Plus* **2022**, *137*, 1129. [CrossRef]

17. Schöneberg, N.; Verde, L.; Gil-Marín, H.; Brieden, S. BAO+BBN Revisited—Growing the Hubble Tension With a 0.7km/s/Mpc Constraint. *J. Cosmol. Astropart. Phys.* **2022**, *11*, 039. [CrossRef]

18. Perivolaropoulos, L.; Skara, F. Challenges for ΛCDM: An Update. *New Astron. Rev.* **2022**, *95*, 101659.

19. Bamba, K.; Capozziello, S.; Nojiri, S.; Odintsov, S.D. Dark Energy Cosmology: The Equivalent Description via Different Theoretical Models and Cosmography Tests. *Astrophys. Space Sci.* **2012**, *342*, 155. [CrossRef]

20. Dainotti, M.G.; Simone, B.D.; Schiavone, T.; Montani, G.; Rinaldi, E.; Lambiase, G. On the Hubble Constant Tension in the SNe IA Pantheon Sample. *Astrophys. J.* **2021**, *912*, 150. [CrossRef]

21. Dainotti, M.G.; Simone, B.D.; Schiavone, T.; Montani, G.; Rinaldi, E.; Lambiase, G.; Bogdan, M.; Ugale, S. On the Evolution of the Hubble Constant With the SNe IA Pantheon Sample and Baryon Acoustic Oscillations: A Feasibility Study for GRB-Cosmology in 2030. *Galaxies* **2022**, *10*, 24. [CrossRef]

22. Colgáin, E.Ó.; Sheikh-Jabbari, M.M.; Solomon, R.; Dainotti, M.G.; Stojkovic, D. Putting Flat ΛCDM In The (Redshift) Bin. *arXiv* **2022**, arXiv:2206.11447.

23. Colgáin, E.Ó.; Sheikh-Jabbari, M.M.; Solomon, R. High Redshift ΛCDM Cosmology: To Bin or not to Bin? *Phys. Dark Univ.* **2023**, *40*, 101216. [CrossRef]

24. Jia, X.D.; Hu, J.P.; Wang, F.Y. The Evidence for a Decreasing Trend of Hubble Constant. *Astron. Astrophys.* **2023**, *674*, A45. [CrossRef]

25. Dutta, K.; Ruchika; Roy, A.; Sen, A.A.; Sheikh-Jabbari, M.M. Beyond ΛCDM with Low and High Redshift Data: Implications for Dark Energy. *Gen. Rel. Grav.* **2020**, *52*, 15. [CrossRef]

26. Peracaula, J.S.; Gomez-Valent, A.; Perez, J.D. Signs of Dynamical Dark Energy in Current Observations. *Phys. Dark Univ.* **2019**, *25*, 100311. [CrossRef]

27. Visinelli, L.; Vagnozzi, S.; Danielsson, U. Revisiting a Negative Cosmological Constant From Low-Redshift Data. *Symmetry* **2019**, *11*, 1035. [CrossRef]

28. Akarsu, Ö.; Barrow, J.D.; Escamilla, L.A.; Vazquez, J.A. Graduated Dark Energy: Observational Hints of a Spontaneous Sign Switch in the Cosmological Constant. *Phys. Rev. D* **2020**, *101*, 063528. [CrossRef]

29. Ye, G.; Piao, Y.-S. Is the Hubble Tension a Hint of AdS Phase Around Recombination? *Phys. Rev. D* **2020**, *101*, 083507. [CrossRef]

30. Valentino, E.D.; Linder, E.V.; Melchiorri, A. H_0 Ex Machina: Vacuum Metamorphosis and Beyond H_0. *Phys. Dark Univ.* **2020**, *30*, 100733. [CrossRef]

31. Calderón, R.; Gannouji, R.; L'Huillier, B.; Polarski, D. Negative Cosmological Constant in the Dark Sector? *Phys. Rev. D* **2021**, *103*, 023526. [CrossRef]

32. Lin, W.; Chen, X.; Mack, K.J. Early Universe-Physics Insensitive and Uncalibrated Cosmic Standards: Constraints on Ω_M and Implications for the Hubble Tension. *Astrophys. J.* **2021**, *920*, 159. [CrossRef]

33. Cai, R.-G.; Guo, Z.-K.; Wang, S.-J.; Yu, W.-W.; Zhou, Y. No-Go Guide for the Hubble Tension: Late-Time Solutions. *Phys. Rev. D* **2022**, *105*, L021301. [CrossRef]

34. Akarsu, Ö.; Kumar, S.; Ozulker, E.; Vazquez, J.A. Relaxing Cosmological Tensions With a Sign Switching Cosmological Constant. *Phys. Rev. D* **2021**, *104*, 123512. [CrossRef]

35. Sen, A.A.; Adil, S.A.; Sen, S. Do Cosmological Observations Allow a Negative Λ? *Mon. Not. R. Astron. Soc.* **2022**, *518*, 1098. [CrossRef]

36. Cai, R.-G.; Guo, Z.-K.; Wang, S.-J.; Yu, W.-W.; Zhou, Y. No-Go Guide for Late-Time Solutions to the Hubble Tension: Matter Perturbations. *Phys. Rev. D* **2022**, *106*, 063519. [CrossRef]

37. Hu, J.-P.; Wang, F. Revealing the Late-Time Transition of H_0: Relieve the Hubble Crisis. *Mon. Not. R. Astron. Soc.* **2022**, *517*, 576–581. [CrossRef]

38. Gennaro, S.D.; Ong, Y.C. Sign Switching Dark Energy from a Running Barrow Entropy. *Universe* **2022**, *8*, 541. [CrossRef]

39. Moshafi, H.; Firouzjahi, H.; Talebian, A. Multiple Transitions in Vacuum Dark Energy and H_0 Tension. *Astrophys. J.* **2022**, *940*, 2–121. [CrossRef]

40. Akarsu, Ö.; Kumar, S.; Ozulker, E.; Vazquez, J.A.; Yadav, A. Relaxing Cosmological Tensions With a Sign Switching Cosmological Constant: Improved Results With Planck, BAO and Pantheon Data. *Phys. Rev. D* **2023**, *108*, 023513. [CrossRef]

41. Antonini, S.; Simidzija, P.; Swingle, B.; Raamsdonk, M.V.; Waddell, C. Accelerating Cosmology From $\Lambda < 0$ Gravitational Effective Field Theory. *J. High Energy Phys.* **2023**, *2023*, 203.

42. Kallosh, R.; Kratochvil, J.; Linde, A.; Linder, E.V.; Shmakova, M. Observational Bounds on Cosmic Doomsday. *J. Cosmol. Astropart. Phys.* **2003**, *2003*, 015. [CrossRef]

43. McInnes, B. Quintessential Maldacena-Maoz Cosmologies. *J. High Energy Phys.* **2004**, *2004*, 036. [CrossRef]

44. Prokopec, T. Negative Energy Cosmology and the Cosmological Constant. *arXiv* **2011**, arXiv:1105.0078.

45. Biswas, T.; Koivisto, T.; Mazumdar, A. Could Our Universe Have Begun With Negative Lambda? *arXiv* **2011**, arXiv:1105.2636.

46. Banerjee, S.; Danielsson, U.; Dibitetto, G.; Giri, S.; Schillo, M. Emergent de Sitter Cosmology from Decaying AdS. *Phys. Rev. Lett.* **2018**, *121*, 261301. [CrossRef] [PubMed]

47. Raamsdonk, M.V. Cosmology Without Time-Dependent Scalars Is Like Quantum Field Theory Without RG Flow. *arXiv* **2022**, arXiv:2211.12611.

48. Delubac, T. et al. [BOSS Collaboration]. Baryon Acoustic Oscillations in the Lyα Forest of Boss DR11 Quasars. *Astron. Astrophys.* **2015**, *574*, A59. [CrossRef]

49. Bacon, D.J. et al. [SKA Collaboration]. Cosmology With Phase 1 of the Square Kilometre Array Red Book 2018: Technical Specifications and Performance Forecasts. *Publ. Astron. Soc. Austral.* **2020**, *37*, e007.

50. Abdalla, E. et al. [BINGO Collaboration]. The BINGO Project I: Baryon Acoustic Oscillations from Integrated Neutral Gas Observations. *Astron. Astrophys.* **2022**, *664*, A14. [CrossRef]

51. Costa, A.A. et al. [BINGO Collaboration]. The BINGO Project VII: Cosmological Forecasts from 21cm Intensity Mapping. *Astron. Astrophys.* **2022**, *664*, A20. [CrossRef]

52. Amendola, L.; Appleby, S.; Avgoustidis, A.; Bacon, D.; Baker, T.; Baldi, M.; Bartolo, N.; Blanchard, A.; Bonvin, C.; Borgani, S.; et al. Cosmology and Fundamental Physics with the Euclid Satellite. *Living Rev. Rel.* **2018**, *21*, 2. [CrossRef] [PubMed]

53. Malekjani, M.; Conville, R.M.; Colgáin, E.Ó.; Pourojaghi, S.; Sheikh-Jabbari, M.M. Negative Dark Energy Density from High Redshift Pantheon+ Supernovae. *arXiv* **2023**, arXiv:2301.12725.

54. Acquaviva, G.; Akarsu, O.; Katirci, N.; Vazquez, J.A. Simple-Graduated Dark Energy and Spatial Curvature. *Phys. Rev. D* **2021**, *104*, 023505. [CrossRef]

55. Lee, B.-H.; Lee, W.; Colgáin, E.Ó.; Sheikh-Jabbari, M.M.; Thakur, S. Is Local H_0 At Odds With Dark Energy EFT? *J. Cosmol. Astropart. Phys.* **2022**, *2022*, 004. [CrossRef]

56. Barrow, J.D. The Area of a Rough Black Hole. *Phys. Letts. B* **2020**, *808*, 135643. [CrossRef]

57. Farooq, M.U.; Jamil, M.; Debnath, U. Dynamics of Interacting Phantom and Quintessence Dark Energies. *Astrophys. Space Sci.* **2011**, *334*, 243. [CrossRef]

58. Velten, H.E.S.; von Marttens, R.; Zimdahl, W. Aspects of the Cosmological 'Coincidence Problem'. *Eur. Phys. J. C* **2014**, *74*, 3160. [CrossRef]

59. Gonzalez, T.; Quiros, I. Exact Models With Non-minimal Interaction Between Dark Matter and (Either Phantom or Quintessence) Dark Energy. *Class. Quant. Grav.* **2008**, *25*, 175019. [CrossRef]

60. Pu, B.-Y.; Xu, X.-D.; Wang, B.; Abdalla, E. Early Dark Energy and Its Interaction With Dark Matter. *Phys. Rev. D* **2015**, *92*, 123537. [CrossRef]

61. Wang, B.; Abdalla, E.; Atrio-Barandela, F.; Pavon, D. Dark Matter and Dark Energy Interactions: Theoretical Challenges, Cosmological Implications and Observational Signatures. *Rept. Prog. Phys.* **2016**, *79*, 096901. [CrossRef]

62. Wang, H.; Piao, Y.-S. A Fraction of Dark Matter Faded With Early Dark Energy? *arXiv* **2022**, arXiv:2209.09685.

63. Yang, W.; Pan, S.; Mena, O.; Valentino, E.D. On the Dynamics of a Dark Sector Coupling. *arXiv* **2022**, arXiv:2209.14816.

64. Bernui, A.; Valentino, E.D.; Giarè, W.; Kumar, S.; Nunes, R.C. Exploring the H0 Tension and the Evidence for Dark Sector Interactions From 2D BAO Measurements. *Phys. Rev. D* **2023**, *107*, 103531. [CrossRef]

65. Perez, J.; Füzfa, A.; Carletti, T.; Mélot, L.; Guedezounme, L. The Jungle Universe. *Gen. Rel. Grav.* **2014**, *46*, 1753. [CrossRef]

66. Aydiner, E. Chaotic Universe Model. *Sci. Rep.* **2018**, *8*, 721. [CrossRef]

67. Simon-Petit, A.; Yap, H.-H.; Perez, J. Refinements in the Jungle Universes. *arXiv* **2016**, arXiv:1603.02267.

68. Haba, Z.; Stachowski, A.; Szydlowski, M. Dynamics of the Diffusive DM–DE Interaction–Dynamical System Approach. *J. Cosmol. Astropart. Phys.* **2016**, *2016*, 024. [CrossRef]

69. Gariazzo, S.; Valentino, E.D.; Mena, O.; Nunes, R.C. Late Time Interacting Cosmologies and the Hubble Constant Tension. *Phys. Rev. D* **2022**, *106*, 023530. [CrossRef]

70. Bhattacharyya, A.; Alam, U.; Pandey, K.L.; Das, S.; Pal, S. Are H_0 and σ_8 Tensions Generic to Present Cosmological Data? *Astrophys. J.* **2019**, *876*, 143. [CrossRef]

71. Mawas, E.; Street, L.; Gass, R.; Wijewardhana, L.C.R. Interacting Dark Energy Axions in Light of the Hubble Tension. *arXiv* **2021**, arXiv:2108.13317.

72. Cruz, M.; Lepe, S.; Morales-Navarrete, G. Qualitative Description of the Universe in the Interacting Fluids Scheme. *Nucl. Phys. B* **2019**, *943*, 114623. [CrossRef]

73. Saharian, A.A.; Avagyan, R.M.; de Mello, E.R.B.; Kotanjyan, V.K.; Petrosyan, T.A.; Babujyan, H.G. Cosmological Evolution With Negative Energy Densities. *Astrophysics* **2022**, *65*, 427. [CrossRef]

74. Heisenberg, L.; Villarrubia-Rojo, H.; Zosso, J. Simultaneously Solving the H_0 and σ_8 Tensions With Late Dark Energy. *Phys. Dark Univ.* **2023**, *39*, 101163. [CrossRef]

75. Heisenberg, L.; Villarrubia-Rojo, H.; Zosso, J. Can Late-Time Extensions Solve the H_0 and σ_8 Tensions? *Phys. Rev. D* **2022**, *106*, 043503. [CrossRef]

76. Valentino, E.D.; Mukherjee, A.; Sen, A.A. Dark Energy With Phantom Crossing and the H_0 Tension. *Entropy* **2021**, *23*, 404. [CrossRef] [PubMed]

77. Ozulker, E. Is the Dark Energy Equation of State Parameter Singular? *Phys. Rev. D* **2022**, *106*, 063509. [CrossRef]

78. Bianchi, E.; Rovelli, C. Why All These Prejudices Against a Constant? *arXiv* **2010**, arXiv:1002.3966.

79. Ananda, K.N.; Bruni, M. Cosmo-Dynamics and Dark Energy With Non-linear Equation of State: A Quadratic Model. *Phys. Rev. D* **2006**, *74*, 023523. [CrossRef]
80. Visser, M. Jerk, Snap, and the Cosmological Equation of State. *Class. Quant. Grav.* **2004**, *21*, 2603. [CrossRef]
81. Arkani-Hamed, N.; Cheng, Hs.; Luty, M.A.; Mukohyama, S.; Wiseman, T. Dynamics of Gravity in a Higgs Phase. *J. High Energy Phys.* **2007**, *2007*, 036. [CrossRef]
82. Bhattacharya, G.; Mukherjee, P.; Saha, A. On the Self-Interaction of Dark Energy in a Ghost-Condensate Model. *arXiv* **2013**, arXiv:1301.4746.
83. Boyle, L.; Turok, N. Thermodynamic Solution of the Homogeneity, Isotropy and Flatness Puzzles (And a Clue to the Cosmological Constant). *arXiv* **2022**, arXiv:2210.01142.
84. Quartin, M.; Calvao, M.O.; Joras, S.E.; Reis, R.R.R.; Waga, I. Dark Interactions and Cosmological Fine-Tuning. *J. Cosmol. Astropart. Phys.* **2008**, *2008*, 007. [CrossRef]
85. Marra, V. Coupling Dark Energy to Dark Matter Inhomogeneities. *Phys. Dark Univ.* **2016**, *13*, 25. [CrossRef]
86. Grande, J.; Sola, J.; Stefancic, H. LXCDM: A Cosmon Model Solution to the Cosmological Coincidence Problem? *J. Cosmol. Astropart. Phys.* **2006**, *2006*, 011. [CrossRef]
87. Grande, J.; Pelinson, A.; Sola, J. Dark Energy Perturbations and Cosmic Coincidence. *Phys. Rev. D* **2009**, *79*, 043006. [CrossRef]
88. Peracaula, J.S. The Cosmological Constant Problem and Running Vacuum in the Expanding Universe. *Philos. Trans. R. Soc. Lond. A* **2022**, *380*, 20210182. [CrossRef]
89. Gonzalez, M.C.; Trodden, M. Field Theories and Fluids for an Interacting Dark Sector. *Phys. Rev. D* **2018**, *97*, 043508; Erratum in *Phys. Rev. D* **2020**, *101*, 089901. [CrossRef]
90. Panpanich, S.; Burikham, P.; Ponglertsakul, S.; Tannukij, L. Resolving Hubble Tension with Quintom Dark Energy Model. *Chin. Phys. C* **2021**, *45*, 015108. [CrossRef]
91. Wu, P.; Zhang, S.N. Cosmological Evolution of Interacting Phantom (Quintessence) Model in Loop Quantum Gravity. *J. Cosmol. Astropart. Phys.* **2008**, *2008*, 007. [CrossRef]
92. Bouali, A.; Albarran, I.; Bouhmadi-Lopez, M.; Errahmani, A.; Ouali, T. Cosmological Constraints of Interacting Phantom Dark Energy Models. *Phys. Dark Univ.* **2021**, *34*, 100907. [CrossRef]

Article

Geodesic Structure of Generalized Vaidya Spacetime through the K-Essence

Bivash Majumder [1], **Maxim Khlopov** [2,3,4], **Saibal Ray** [5] **and Goutam Manna** [6,7,*]

1   Department of Mathematics, Prabhat Kumar College, Contai 721404, India; bivashm@pkcollegecontai.ac.in
2   Institute of Physics, Southern Federal University, 194 Stachki, Rostov-on-Don 344090, Russia; khlopov@apc.in2p3.fr
3   Department of Physics, National Research Nuclear University, MEPHI, Moscow 115409, Russia
4   Virtual Institute of Astroparticle Physics, Rue Garreau, 75018 Paris, France
5   Centre for Cosmology, Astrophysics and Space Science (CCASS), GLA University, Mathura 281406, India; saibal.ray@gla.ac.in
6   Department of Physics, Prabhat Kumar College, Contai 721404, India
7   Institute of Astronomy, Space and Earth Science (IASES), Kolkata 700054, India
*   Correspondence: goutamphs@pkcollegecontai.ac.in

Abstract: This article investigates the radial and non-radial geodesic structures of the generalized K-essence Vaidya spacetime. Within the framework of K-essence geometry, it is important to note that the metric does not possess conformal equivalence to the conventional gravitational metric. This study employs a non-canonical action of the Dirac–Born–Infeld kind. In this work, we categorize the generalized K-essence Vaidya mass function into two distinct forms. Both the forms of the mass functions have been extensively utilized to analyze the radial and non-radial time-like or null geodesics in great detail inside the comoving plane. Indications of the existence of wormholes can be noted during the extreme phases of spacetime, particularly in relation to black holes and white holes, which resemble the Einstein–Rosen bridge. In addition, we have also detected a distinctive indication of the quantum tunneling phenomenon around the singularity ($r \to 0$). Furthermore, we have found that for certain types of solutions, there exist circular orbits through the event horizon as well as quasicircular orbits. Also, we have noted that there is no central singularity in our spacetime where both r and t tend towards zero. The existence of a central singularity is essential for any generalized Vaidya spacetime. This indicates that spacetime can be geodesically complete, which correlates with the findings of Kerr's recent work (2023).

Keywords: geodesic structure; Euler–Lagrange equation; K-essence geometry; Vaidya spacetime

Citation: Majumder, B.; Khlopov, M.; Ray, S.; Manna, G. Geodesic Structure of Generalized Vaidya Spacetime through the K-Essence. *Universe* **2023**, *9*, 510. https://doi.org/10.3390/universe9120510

Academic Editor: Kazuharu Bamba

Received: 6 November 2023
Revised: 5 December 2023
Accepted: 6 December 2023
Published: 8 December 2023

1. Introduction

Chandrasekhar extensively analyzed the time-like and null geodesic features of the Schwarzschild spacetime in his book [1]. In addition, he examined the orbital configurations of both the confined and unconfined trajectories using graphical representations. In addition, the authors of [2] examined the geodesic structures of the Schwarzschild anti-de Sitter spacetime. The researchers assessed both radial and non-radial paths for time-like and null geodesics. Additionally, they demonstrated that the geodesic structures of this black hole exhibit distinct forms of motion that are not permitted by the Schwarzschild spacetime. The geometric framework of the Schwarzschild spacetime is also examined in [3]. The Jacobi metric for time-like geodesics in static spacetimes was examined in Ref. [4]. They demonstrated that the unrestricted movement of large particles in stationary spacetimes is determined by the geodesics of a Riemannian metric that depends on the particle's energy. This metric is similar to Jacobi's metric in classical dynamics. When the mass of an object approaches zero, Jacobi's metric becomes identical to the Fermat or optical metric, which does not depend on energy. In addition, they provided a detailed account of the characteristics of the Jacobi metric pertaining to the motion of heavy particles beyond the event horizon of a Schwarzschild black hole. The authors of [5] derived the Jacobi metric for different stationary metrics and developed the Jacobi–Maupertuis metric for time-dependent metrics by using the Eisenhart–Duval lift [6,7]. The authors of [8] documented the remarkable characteristics of the time-like geodesic structure when dark energy is present in an emergent gravity framework, specifically for the Barriola–Vilenkin metric [9]. The K-essence emergent gravity metric is precisely correlated with the Barriola–Vilenkin (BV) metric for the Schwarzschild background, specifically for a certain form of K-essence scalar field [10]. The researchers analyzed the various paths that time-like geodesics can take in the presence of dark energy in the Barriola–Vilenkin spacetime [8], which is equivalent to the Schwarzschild spacetime in terms of its fundamental structure. However, the permissible ranges for the maximum and minimum distances from the central object are significantly distinct. For a constant dark energy density, the orbits, both bound and unbound, were graphed.

In 1951, Vaidya proposed the first relativistic line element that properly represented the spacetime of a conceivable star [11]. It extended the specific solution of Schwarzschild by depicting the emission of radiation for a mass that is not in a static state. The Schwarzschild solution describes the geometry of spacetime around a spherically symmetric, non-rotating, black object with a constant mass. Therefore, it is clear that the model is incapable of accurately depicting spacetime outside the confines of a star. The solution proposed by Vaidya [11], known as the Vaidya spacetime or the radiating Schwarzschild metric, was introduced as a possible explanatory framework. The main distinction between the two metrics is that the Vaidya metric adds a time-dependent mass parameter, whereas the Schwarzschild metric uses a constant mass value. As a result, the spacetime in the Vaidya metric evolves with time. The Vaidya metric is primarily used to investigate gravitational collapse. The occurrence of gravitational collapse is widely acknowledged in the disciplines of general relativity and astrophysics, as demonstrated by the research conducted by Joshi et al. [12–19]. It plays a vital role in understanding several astrophysical aspects of our cosmos. The phenomenon of gravitational collapse provides useful insights into several elements of astronomy, including the evolution of structures, the features of stars, the genesis of black holes, and the construction of white dwarfs or neutron stars, among other events. Gravitational collapse refers to the phenomenon in which a star collapses as a result of its mass. The outcome of this collapse might vary depending on the exact beginning mass conditions, leading to distinct stages of collapse. Papapetrou [20] was the first to demonstrate that the solution of a null dust fluid with spherical symmetry in gravitational collapse can lead to the creation of naked singularities. This statement presents a counterexample of the cosmic censorship hypothesis (CCH) as proposed by Penrose [21,22]. The authors of [14,23] provided a detailed account of the causal paths that connect the singularities in the continuing Vaidya scenario. Furthermore, a comprehensive classification of the non-space-like geodesics that link the naked singularity in the past is presented, offering a rather thorough discussion of the restrictions involved. It is subsequently demonstrated to be a robust curvature singularity in a more significant manner.

The Vaidya solution, as a generalization, encompasses all the established solutions of Einstein's field equations that include a mix of type-I and type-II matter fields [24–29]. The composition of this work is attributed to Husain [30] and Wang and Wu [31]. The extension of the Vaidya solution is sometimes referred to as the generalized Vaidya spacetime. The work performed in [32] examines the gravitational collapse of the generalized Vaidya spacetime within the framework of the cosmic censorship theory. They demonstrated that the categories of generalized Vaidya mass functions emerged in the situation, suggesting the end of collapse with a locally visible central singularity. The authors computed the magnitude of these singularities. A comprehensive mathematical framework was created to examine the requirements for the mass function for non-space-like geodesics going towards the future to end at the singularity in the past. Furthermore, they demonstrated that, when considering a certain generalized Vaidya mass function, the ultimate outcome of the collapse can be precisely defined as either a black hole or a naked singularity. The work by Patil [33] examined the phenomenon of gravitational collapse in higher dimensions within the context of the charged Vaidya spacetime. It was demonstrated that singularities occur in a charged null fluid in a higher dimension. These singularities consistently lacked any form of covering, hence contradicting the strong CCH. This idea does not specifically pertain to weak cosmic censorship. The Vaidya metric has received significant attention in scholarly research, with several major contributions to our comprehension of this subject. The authors of the study [34] examined the geometric properties of Vaidya's spacetime while considering a white hole that undergoes a decrease in mass. They found that the white hole can either stabilize and transform into a black hole within a limited or indefinite amount of time, or entirely evaporate. The researchers focused specifically on the scenario of total evaporation over an indefinite period of time. They successfully demonstrated the presence of an asymptotic light-like singularity in the conformal curvature, which connects both the past space-like singularity and the future time-like infinity. Vertogradov [35] conducted a study on the structure of the generalized Vaidya spacetime, specifically focusing on the case when the matter field of type-II follows the equation of state $P = \rho$. The findings of the study revealed the presence of an eternal naked singularity in this spacetime, which meets all energy conditions. Once formed, the singularity will remain perpetually uncovered by the apparent horizon. Nevertheless, the formation of the apparent horizon leads to the emergence of a white hole. Solanki et al. [36] derived precise mathematical equations that describe the changes in the photon sphere and the angular radius of the shadow in a certain Vaidya spacetime. The mass function $m(v)$ was seen as a function of time that either increases or decreases linearly. The initial scenario can function as a basic representation of a black hole that is accumulating matter, whereas the subsequent scenario can be seen as an illustration of a black hole that is emitting radiation, as theorized by Hawking.

In the realm of K-essence geometry, Manna et al. [37] were the first to establish a link between K-essence geometry and Vaidya spacetime. They achieved this by introducing a new definition of the generalized Vaidya mass function, which directly depends on the kinetic energy of the K-essence scalar field. Subsequently, Manna et al. [38] demonstrated that the K-essence emergent gravity metric bears a strong resemblance to the recently found generalized Vaidya metrics for the collapse of a null fluid. This similarity arises from the presence of a K-essence emergent mass function. Notably, Manna's analysis exclusively considers the K-essence scalar field as a function of either the advanced or the retarded time. The recently developed K-essence model, known as the K-essence emergent Vaidya spacetime, has successfully met all the necessary energy conditions. The presence of the centrally exposed singularity and the intensity and stability of the singularities in the K-essence emergent Vaidya metric yield intriguing results in their research. The evaporation of the dynamical horizon with the Hawking temperature in the K-essence Vaidya Schwarzschild spacetime was investigated by Manna et al. in [39]. This study uses the dynamical horizon equation to quantify the reduction in mass caused by Hawking radiation. Additionally, the tunneling formalism, namely, the Hamilton–Jacobi technique, is utilized to compute the Hawking temperature. In addition, Sawayama's revised explanation of the dynamical horizon [40] is utilized to demonstrate that the results obtained differ from the conventional Vaidya spacetime geometry. The authors establish, using analytical measures, that the mass of the black hole, denoted as $m(v,r)$, in the K-essence emergent Schwarzschild–Vaidya spacetime, consistently decreases over time but does not fully evaporate.

The K-essence theory is a scalar field model that deviates from the canonical form. In this theory, the dominant energy component of the field is its kinetic energy, rather than its potential energy. This concept and related others have been extensively studied by several researchers [41–52]. The distinctions between the K-essence theory employing a non-canonical Lagrangian and the relativistic field theories utilizing a canonical Lagrangian are found in the sophisticated dynamical solutions of the K-essence equation of motion. These solutions not only spontaneously violate Lorentz invariance but also alter the metric for the perturbations around them. The disturbances propagate in the emergent or analogous curved spacetime, characterized by a metric distinct from the gravitational metric. The non-canonical Lagrangian may be expressed as $\mathcal{L}(X) = -V(\phi)F(X)$, where $X = \frac{1}{2}g_{\mu\nu}\nabla^\mu\phi\nabla^\nu\phi$, ϕ is the K-essence scalar field, and $V(\phi)$ is the potential term. An alternate form of the Lagrangian, as described by Tian [53], may be represented as $\mathcal{L} = [1 + f(y)]X + [1 + g(y)]V_{exp}$, where $V_{exp} = V_0\, exp(-\lambda\phi)$, V_0 and λ are constants, $y = X/V_{exp}$, and $f(y)$ and $g(y)$ are arbitrary functions. The functions $f(y)$ and $g(y)$ are unrestricted and can have any form. Furthermore, it is important to mention that there exist examples of K-essence theories that are not minimally linked, as mentioned in Refs. [54–56]. Nevertheless, this article only addresses the minimally coupled K-essence theory, as investigated in Refs. [41–50]. In a general sense, the Lagrangian has the capacity to depend on any functions of ϕ and X. The K-essence theory offers the benefit of circumventing both the fine-tuning problem and the coincidence problem [57] of the current universe. Additionally, it generates the negative pressure required for the universe's acceleration only through the kinetic energy of the field. The kinetic term of the field dominates over the potential term. The article [41] presents attractor solutions where the dynamics of the cosmos are governed by the scalar field of the models. During the radiation-dominated phase, the K-essence field mimics the equation of state of radiation and has a constant ratio to the radiation density. The K-essence field is unable to replicate the dust-like equation of state (EOS) due to dynamical limits during the time dominated by dust. However, it rapidly reduces its energy value by many orders of magnitude and eventually reaches a constant value. Subsequently, over a period approximately equivalent to the current age of the universe, the density of matter is diminished by the K-essence field, leading to the commencement of cosmic acceleration. The equation of state (EOS) of the K-essence theory ultimately converges to a value within the range of 0 to -1. Although, in theory, it has the potential to extend beyond -1. Another intriguing aspect of the K-essence idea is its potential to generate a type of dark energy where the speed of sound is consistently slower than that of light. This feature may mitigate the cosmic microwave background (CMB) disruptions on large angular scales [58–60]. In this specific situation, Manna and coworkers [8,10,37–39,61–66] have developed a fascinating emergent gravity metric, referred to as $\bar{G}_{\mu\nu}$. This metric possesses distinct attributes in contrast to the standard gravitational metric $g_{\mu\nu}$ and is derived from the notions of the Dirac–Born–Infeld (DBI)-type action, as outlined in Refs. [67–70]. Dirac et al. [70] proposed a non-canonical Lagrangian in order to eliminate the infinite self-energy of the electron, as described in their work. The specific reasons and objectives for selecting the non-canonical theory, such as the K-essence theory, may be found in Refs. [71,72]. The Planck collaborations' findings, as shown in Refs. [73–75], have examined the empirical evidence supporting the concept of K-essence with a DBI-type non-canonical Lagrangian, along with other modified theories. Furthermore, it has been noted that the K-essence theory may be applied in a model that combines dark energy and dark matter [8,10,45,61–64], as well as from a purely gravitational perspective [37–39,65,66].

This article is organized as follows: in Section 2, we provided a concise explanation of the K-essence geometry and its connection to the conventional generalized Vaidya spacetime, which leads to the construction of a new generalized K-essence Vaidya spacetime. Section 3 offers a comprehensive analysis of the geodesic structures observed in the generalized K-essence Vaidya spacetime. This analysis considers two forms of mass function while ensuring that the condition on the kinetic energy of the K-essence scalar field is maintained. This section also provides a detailed analysis of the radial and non-radial geodesics used to examine the structure of time-like and null geodesics in the given spacetime. This is achieved by solving the Euler–Lagrange equations. A graphical and numerical analysis is also performed in this section. In Section 4, we wrap up both the discussion and conclusions.

2. Summary of the Relation between K-Essence and Generalized Vaidya Spacetime

This section offers a short introduction to the geometry of K-essence and the generalized Vaidya spacetime. Initially, we present a brief summary of the geometric aspects related to the K-essence, as extensively explored in many scholarly references [41–50]. The action performed by the minimally couple K-essence geometry is

$$S_k[\phi, g_{\mu\nu}] = \int d^4x \sqrt{-g} \mathcal{L}(X, \phi), \tag{1}$$

where the expression $X = \frac{1}{2} g^{\mu\nu} \nabla_\mu \phi \nabla_\nu \phi$ represents the canonical kinetic term, whereas $\mathcal{L}(X, \phi)$ denotes the non-canonical Lagrangian. In this scenario, the conventional gravitational metric $g_{\mu\nu}$ has formed a minimum coupling with the K-essence scalar field (ϕ).

The energy–momentum tensor that corresponds solely to the K-essence scalar field is

$$T_{\mu\nu} \equiv \frac{-2}{\sqrt{-g}} \frac{\delta S_k}{\delta g^{\mu\nu}} = -2 \frac{\partial \mathcal{L}}{\partial g^{\mu\nu}} + g_{\mu\nu} \mathcal{L}$$
$$= -\mathcal{L}_X \nabla_\mu \phi \nabla_\nu \phi + g_{\mu\nu} \mathcal{L}, \tag{2}$$

where $\mathcal{L}_X = \frac{d\mathcal{L}}{dX}$, $\mathcal{L}_{XX} = \frac{d^2\mathcal{L}}{dX^2}$, $\mathcal{L}_\phi = \frac{d\mathcal{L}}{d\phi}$ and ∇_μ is the covariant derivative defined with respect to the gravitational metric $g_{\mu\nu}$.

The equation of motion (EOM) for the K-essence scalar field is

$$-\frac{1}{\sqrt{-g}} \frac{\delta S_k}{\delta \phi} = \tilde{G}^{\mu\nu} \nabla_\mu \nabla_\nu \phi + 2X \mathcal{L}_{X\phi} - \mathcal{L}_\phi = 0, \tag{3}$$

where

$$\tilde{G}^{\mu\nu} \equiv \frac{c_s}{\mathcal{L}_X^2} [\mathcal{L}_X g^{\mu\nu} + \mathcal{L}_{XX} \nabla^\mu \phi \nabla^\nu \phi], \tag{4}$$

with $1 + \frac{2X\mathcal{L}_{XX}}{\mathcal{L}_X} > 0$ and $c_s^2(X, \phi) \equiv (1 + 2X \frac{\mathcal{L}_{XX}}{\mathcal{L}_X})^{-1}$.

Following [8,10,37,38,61], the inverse metric can be written as

$$\bar{G}_{\mu\nu} = g_{\mu\nu} - \frac{\mathcal{L}_{XX}}{\mathcal{L}_X + 2X\mathcal{L}_{XX}} \nabla_\mu \phi \nabla_\nu \phi. \tag{5}$$

Equations (4) and (5) have physical relevance when $\mathcal{L}_X$ is non-zero, assuming a positive definite c_s^2. Equation (5) states that the emergent metric, represented as $\bar{G}_{\mu\nu}$, differs in its conformal properties from the metric $g_{\mu\nu}$ when considering non-trivial configurations of the scalar field ϕ. Like canonical scalar fields, the variable ϕ exhibits diverse local causal structural properties. It also differs from those that are defined using $g_{\mu\nu}$. The EOM, as stated in Equation (3), is valid even when taking into account the implicit relationship between L and ϕ. Then, the EOM Equation (3) is

$$\frac{1}{\sqrt{-g}} \frac{\delta S_k}{\delta \phi} = \bar{G}^{\mu\nu} \nabla_\mu \nabla_\nu \phi = 0. \tag{6}$$

This study addresses the Dirac–Born–Infeld (DBI)-type non-canonical Lagrangian, which is represented as $\mathcal{L}(X, \phi) \equiv \mathcal{L}(X)$ [8,10,37,67–72]

$$\mathcal{L}(X) = 1 - \sqrt{1 - 2X}. \tag{7}$$

The K-essence paradigm posits that the prevalence of kinetic energy over potential energy results in the exclusion of the potential term in the Lagrangian Equation (7) [67,71,72]. The squared speed of sound, represented as c_s^2, is determined by the expression $(1 - 2X)$. Therefore, Equation (5) for the effective emergent metric is expressed as

$$\bar{G}_{\mu\nu} = g_{\mu\nu} - \nabla_\mu\phi\nabla_\nu\phi = g_{\mu\nu} - \partial_\mu\phi\partial_\nu\phi, \tag{8}$$

since ϕ is a scalar.

The Christoffel symbol, corresponding to the emergent gravity metric given by Equation (8), can be written as [8,10,71,72]

$$\bar{\Gamma}^\alpha_{\mu\nu} = \Gamma^\alpha_{\mu\nu} - \frac{1}{2(1-2X)}\left[\delta^\alpha_\mu\partial_\nu + \delta^\alpha_\nu\partial_\mu\right]X, \tag{9}$$

where $\Gamma^\alpha_{\mu\nu}$ is the usual Christoffel symbol associated with the gravitational metric $g_{\mu\nu}$.

Hence, the geodesic equation governing the K-essence geometry may be expressed as

$$\frac{d^2x^\alpha}{d\lambda^2} + \bar{\Gamma}^\alpha_{\mu\nu}\frac{dx^\mu}{d\lambda}\frac{dx^\nu}{d\lambda} = 0, \tag{10}$$

where λ is an affine parameter.

The covariant derivative D_μ [48,71,72] linked with the emergent metric $\bar{G}_{\mu\nu}$ $(D_\alpha\bar{G}^{\alpha\beta}=0)$ gives

$$D_\mu A_\nu = \partial_\mu A_\nu - \bar{\Gamma}^\lambda_{\mu\nu}A_\lambda, \tag{11}$$

and the inverse emergent metric is $\bar{G}^{\mu\nu}$, such as $\bar{G}_{\mu\lambda}\bar{G}^{\lambda\nu} = \delta^\nu_\mu$.

Therefore, considering the extensive behavior that defines the dynamics of K-essence and general relativity [47,71,72], the emergent Einstein equation (EEE) may be formulated as

$$\bar{\mathcal{G}}_{\mu\nu} = \bar{R}_{\mu\nu} - \frac{1}{2}\bar{G}_{\mu\nu}\bar{R} = \kappa\bar{T}_{\mu\nu}, \tag{12}$$

where $\kappa = 8\pi G$ is constant, $\bar{R}_{\mu\nu}$ is the Ricci tensor, and $\bar{R}$ $(= \bar{R}_{\mu\nu}\bar{G}^{\mu\nu})$ is the Ricci scalar. Moreover, the energy–momentum tensor $\bar{T}_{\mu\nu}$ is linked to this emergent spacetime.

Now, we would like to provide a concise overview of the K-essence emergent generalized Vaidya spacetime. In the cited work [38], the author introduced the concept of K-essence emergent generalized Vaidya spacetime. This framework considers the background gravitational metric to be the typical generalized Vaidya metric [30,31], while also satisfying the necessary energy requirements. The line element for the emergent generalized Vaidya metric in K-essence theory is as follows:

$$\begin{aligned}
dS^2 &= -\left[1 - \frac{2m(t,r)}{r} - \phi_t^2\right]dt^2 + 2dtdr + r^2d\Omega^2 \\
&= -\left[1 - \frac{2\mathcal{M}(t,r)}{r}\right]dt^2 + 2dtdr + r^2d\Omega^2,
\end{aligned} \tag{13}$$

with $d\Omega^2 = d\theta^2 + \sin^2\theta d\Phi^2$.

They defined the K-essence emergent Vaidya mass function:

$$\mathcal{M}(t,r) = m(t,r) + \frac{r}{2}\phi_t^2, \tag{14}$$

where $m(t,r)$ is the usual generalized Vaidya mass function and ϕ_t^2 $(\phi_t = \frac{\partial\phi}{\partial t})$ is the non-zero kinetic energy of the K-essence scalar field.

The above-mentioned mass function pertains to the gravitational energy associated with the K-essence emergent gravity within a specified radius r. Here, we substitute the Eddington advanced time coordinate with the conventional time coordinate, without any loss of generality, denoted as $v \to t$. In this study [38], the author examined the effective K-essence emergent metric, as denoted by Equation (8). Additionally, the author calculated all the components of the EEE (Equation (12)) and the necessary energy conditions. It is important to mention that the assumption about ϕ contradicts local Lorentz invariance since, in general, spherical symmetry only requires $\phi(x) = \phi(t,r)$. The inclusion of the assumption of the independence of ϕ, denoted as $\phi(t,r) = \phi(t)$, suggests that beyond this specific frame selection, a spherically symmetric ϕ is indeed a function of both t and r. The K-essence theory permits the occurrence of Lorentz violation due to the fact that the dynamic solutions of the K-essence equation of motion spontaneously break Lorentz invariance and alter the metric for the perturbations around these solutions.

Furthermore, the authors of [37] successfully established a connection between the geometry of K-essence and the Vaidya spacetime. The researchers developed a model of the Vaidya spacetime with generalized K-essence, which takes into account any spherically symmetric static black hole as the underlying spacetime. The line element of the new geometry is (using Equation (8))

$$dS^2 = -\left[f(r) - \phi_t^2\right]dt^2 + 2dtdr + r^2d\Omega^2 = -\left(1 - \frac{2\mathcal{M}(t,r)}{r}\right)dt^2 + 2dtdr + r^2d\Omega^2, \quad (15)$$

gives the mass function

$$\mathcal{M}(t,r) = \frac{1}{2}r\left[1 + \phi_t^2 - f(r)\right]. \quad (16)$$

In this article [37], the authors also calculated all the components of the EEE and the required energy conditions. If we consider $f(r) = (1 - 2M/r)$, i.e., the background physical spacetime is Schwarzschild spacetime, the mass function may be expressed as

$$\mathcal{M}(t,r) = M + \frac{r}{2}\phi_t^2. \quad (17)$$

Again, if we select the function $f(r) = 1 - \frac{2M}{r} + \frac{Q^2}{r^2}$, Q represents the charge of the Reissner–Nordström (RN) black hole in the physical spacetime. In this case, the related mass function [37] is modified as

$$\mathcal{M}(t,r) = M - \frac{Q^2}{2r} + \frac{r}{2}\phi_t^2. \quad (18)$$

It is important to mention that the values of ϕ_t^2 must be between 0 and 1. Otherwise, the metric (13) and (15) cannot be specified properly, and the presence of a dynamical horizon is also questionable [37,38]. In order to maintain the energy conditions, it is evident that the ϕ_t^2 must be a monotonically increasing function of t, with the condition $\phi_t^2 < 1$. The admissible configurations for the K-essence scalar field in the generalized Vaidya solution, in order to have a dynamical horizon, are subject to a highly restrictive constraint. It is important to note that the metrics mentioned above represent dynamical horizons rather than isolated or event horizons, as explained extensively in Refs. [37,38].

It is also noted that the time dependence in the given mass functions (Equations (17) and (18)) arises from the kinetic energy of the K-essence scalar field. However, in the mass function Equation (14), the time dependence comes from both the usual generalized Vaidya mass and the K-essence scalar fields. Thus, considering the above situations of the K-essence generalized Vaidya spacetime, we may conclude that the K-essence Vaidya mass function adheres to the general form specified in Equation (14). Therefore, we may conclude that the background metric can be chosen from any standard gravitational metric, with the only alteration being the replacement of their masses with a background mass, which likewise satisfies the EEE equation.

3. Geodesics for the Generalized K-Essence Vaidya Spacetime

This section focuses on analyzing the geodesic structure of the generalized K-essence Vaidya spacetime. In this context, we define our investigative metric as Equation (13), where the K-essence emergent Vaidya mass function is represented by Equation (14). For metric (13), we can write the Lagrangian as [1,8,35,36]

$$2\mathcal{L} = -\left(1 - \frac{2\mathcal{M}(t,r)}{r}\right)\dot{t}^2 + 2\dot{t}\dot{r} + r^2\dot{\theta}^2 + r^2\sin^2\theta\,\dot{\Phi}^2. \tag{19}$$

where $\dot{t} = \frac{dt}{d\tau}$, $\dot{r} = \frac{dr}{d\tau}$, $\dot{\theta} = \frac{d\theta}{d\tau}$, $\dot{\Phi} = \frac{d\Phi}{d\tau}$, τ is to be identified with the proper time.

Now, using the Euler–Lagrange equation we have

$$\frac{d}{d\tau}\left(\frac{\partial\mathcal{L}}{\partial\dot{t}}\right) = \frac{\mathcal{M}_t}{r}\dot{t}^2, \tag{20}$$

$$2r\dot{\theta} + r^2\ddot{\theta} = r^2\sin\theta\cos\theta\,\dot{\Phi}^2, \tag{21}$$

$$\text{and}\qquad r^2\sin^2\theta\,\dot{\Phi} = \text{Constant}, \tag{22}$$

$$\text{with}\qquad \frac{\partial\mathcal{L}}{\partial\dot{t}} = -\left(1 - \frac{2\mathcal{M}}{r}\right)\dot{t} + \dot{r}, \tag{23}$$

where $\mathcal{M}_t = \frac{\partial\mathcal{M}(t,r)}{\partial t}$ and we write $\mathcal{M}(t,r)$ as $\mathcal{M}$.

Because our object and metric are spherically symmetric, we can simplify everything by examining just motion on the equatorial plane $\theta = \frac{\pi}{2}$ and, therefore, $\dot{\theta} = 0$. For the above choice of equatorial plane, Equation (22) becomes

$$r^2\dot{\Phi} = \text{Constant} = L,\,(\text{say}). \tag{24}$$

Thus, by employing Equation (24) on the equatorial plane, we may write from Equation (19) that the Lagrangian is

$$2\mathcal{L} = -\left(1 - \frac{2\mathcal{M}(t,r)}{r}\right)\dot{t}^2 + 2\dot{t}\dot{r} + \frac{L^2}{r^2}. \tag{25}$$

Due to the inclusion of the generalized K-essence Vaidya mass function ($\mathcal{M}(t,r)$) in the Lagrangian formulation provided above, further analysis is not possible as it can have varying values based on the gravitational mass. Within this particular situation, we have the option to select the mass function. For our subsequent analysis, we have selected two distinct mass functions. Moreover, it is mentioned that the K-essence Vaidya mass function (14) depends on ϕ_t^2, which has values between 0 and 1. Therefore, we can select ϕ_t^2 as an explicit function of the time in order to keep the values of ϕ_t^2 throughout the article as [37]

$$\phi_t^2 = e^{-t/t_0}, \tag{26}$$

where t_0 is a positive constant.

3.1. Case-I: $\mathcal{M}(t,r) = M + \frac{r}{2}\phi_t^2$

In this subsection, we will look at the K-essence Vaidya mass function as

$$\mathcal{M}(t,r) = M + \frac{r}{2}e^{-t/t_0}, \tag{27}$$

where M is the mass of a Schwarzschild black hole.

In this scenario, the time dependence of the mass parameter is derived from the K-essence scalar field via Equation (26), which was previously discussed in the preceding section. For this mass function (27), the radii of the dynamical horizon can be written as $r_D^+ = \frac{2M}{1-e^{-t/t_0}}$ and $r_D^{++} = \frac{2M}{1-e^{-t/t_0}}\left[1 + W_0\left(\frac{1-e^{-t/t_0}}{2Me}e^{-\frac{t}{2M}}\right)\right]$ (see Appendix A). Given that the mass parameter in Equation (25) is directly influenced by the time through Equations (14) and (26), we may infer from Refs. [35,36] that the energy E can be expressed as a function of the time t:

$$\frac{\partial\mathcal{L}}{\partial\dot{t}} = E(t). \tag{28}$$

Now, using Equations (27) and (28) in Equation (20), we have

$$\dot{E}(t) = -\frac{e^{-\frac{t}{t_0}}}{2t_0}\dot{t}^2.$$

(29)

The solution of the aforementioned equation, as denoted by Equation (29), is highly intricate and cannot be solved directly. To determine the expression for $E(t)$ in the above equation, we converted our measurement to a comoving plane, where $\frac{dt}{d\tau} = 1$. This conversion is consistently maintained throughout the article. Thus, on the comoving plane ($d\tau \equiv dt$), the expression for $E(t)$ is

$$E(t) = \frac{1}{2}e^{-\frac{t}{t_0}} \equiv \frac{1}{2}\phi_t^2.$$

(30)

Thus, we can say that in our model, specifically when we select a specific form (26) for the kinetic energy of the K-essence scalar field while satisfying the imposed conditions, we have found a direct relationship between the energy of the system we have chosen and the kinetic energy of the K-essence scalar field in the comoving plane. So, the K-essence Vaidya mass function (27) can be written as

$$\mathcal{M}(t.r) = M + E(t)r.$$

(31)

It is important to point out that the Vaidya metric defines the gravitational field surrounding a massive object, often a dying star, that emits radiation in the form of null dust. The notion of the Vaidya spacetime is expanded in the generalized version to encompass a wide range of scenarios, accommodating different forms of matter and radiation. The spacetime is dynamic and undergoes evolution as matter compresses, with the metric describing the changing curvature [30,31]. Within the framework of the generalized Vaidya spacetime, employing comoving observers that satisfy the condition "$dt = d\tau$" simplifies the mathematical representation of the spacetime. This enables us to utilize a temporal reference that tracks the movement of matter as it undergoes gravitational collapse to become a black hole or emits radiation as a star. Using a time parameter that evolves with the behavior of matter is a practical approach for investigating gravitational collapse or radiating stars. It improves the intuitiveness and physical significance of describing the collapse process. While examining geodesic structures in the generalized Vaidya spacetime from the perspective of a comoving observer, our focus lies on the trajectories that objects or particles take when they deal with the changing spacetime caused by the reducing matter. These geodesics illustrate the paths that things follow as they move through spacetime, which is influenced by changes in curvature caused by the dynamics of matter. Comprehending these geodesic structures is essential for analyzing the dynamics of particles, photons, and observers in spacetime. It facilitates forecasting the movement and interaction of objects inside the gravitational field generated by collapsing matter and is a crucial component in the analysis of the physics and astrophysical phenomena occurring in these spacetimes.

Using Equations (28) and (31) in Equation (23), we have

$$\frac{dr}{dt} + E(t) = 1 - \frac{2M}{r}.$$

(32)

On the other hand, using Equation (31) in Equation (25), we obtain

$$\frac{dr}{dt} + E(t) = 2\mathcal{L} - \frac{L^2}{r^2}.$$

(33)

By substituting Equations (24) and (30) into Equation (33), we obtain the angular relation as follows:

$$r + L\Phi = 2\mathcal{L}t + t_0 E(t),\tag{34}$$

taking the integration constant to be zero.

For the non-radial geodesic, using Equations (32) and (33) we have

$$\left(\frac{dr}{dt} + E(t)\right)^2 + \left(D^2 - 2\right)\left(\frac{dr}{dt} + E(t)\right) + \left(1 - 2D^2\mathcal{L}\right) = 0,\tag{35}$$

where $D = \frac{2M}{L}$ and $L \neq 0$.

Solving the above Equation (35), we obtain

$$\frac{dr}{dt} + E(t) = \frac{1}{2}\left[(2 - D^2) \pm D\sqrt{D^2 - 4(1 - 2\mathcal{L})}\right],\tag{36}$$

where we take only positive solutions for our study.

By employing Equation (30) and performing integration on the aforementioned Equation (36), we obtain

$$r = t_0 E(t) + \frac{t}{2}\left[(2 - D^2) \pm D\sqrt{D^2 - 4(1 - 2\mathcal{L})}\right] + c,\tag{37}$$

where c is an integration constant.

Alternatively, equating the LHS of Equations (32) and (33), we also can have a solution such as

$$r_\pm = \frac{M \pm \sqrt{M^2 - L^2(1 - 2\mathcal{L})}}{1 - 2\mathcal{L}}\tag{38}$$

provided $M^2 \geq L^2(1 - 2\mathcal{L})$ and from Equation (24), we obtain

$$\Phi = \frac{Lt}{r_\pm^2}.\tag{39}$$

From Equation (38), it is evident that $r_\pm$ remains constant throughout the time, which is not a usual characteristic of any generalized Vaidya geometry. According to the observations in the generalized Vaidya spacetime, the value of r is expected to vary with the time [37]. This distinctive attribute can be explained in the following manner: According to Equation (30), it can be observed that the value of $E(t)$ is directly proportional to the kinetic energy of the K-essence scalar field. This relationship is expressed by Equation (31) through the K-essence generalized mass function $(\mathcal{M}(t,r))$. In the given mass function (31), the variation with time is determined by the function $E(t)$ through the term ϕ_t^2. This is because the background gravitational metric is Schwarzschild, which has a constant mass (M). In this solution (38), the values of $r_\pm$ may be determined for constant values of M, L, and $\mathcal{L}$, resulting in a circular orbit instead of the dynamic behavior of the orbit. So we have an event horizon instead of a dynamical horizon, which can be supported by Ishihara et al. [76] in a different context.

3.1.1. Time-like Geodesics for Case-I

In order to analyze the structure of the time-like geodesics in the specified spacetime (13) with the mass function (27), we impose the condition $2\mathcal{L} \equiv \bar{G}_{\mu\nu}\dot{x}^\mu \dot{x}^\nu = -1$. In this particular circumstance, Equation (33) is transformed as

$$\frac{dr}{dt} + E(t) = -1 - \frac{L^2}{r^2}.\tag{40}$$

First, we will analyze the radial geodesics with $L = 0$, and then we will go on to the non-radial geodesics with $L \neq 0$. These geodesics are studied from the perspective of time-like geodesics of the generalized K-essence Vaidya metric (13), considering the mass function (27), inside a comoving system.

In order to track the radialgeodesics ($\dot{\Phi} = 0$), we consider the motion of a particle with no angular momentum ($L = 0$) that starts its journey from a state of rest at a distance of $r = r_a$ and time $t = t_a$, so that the rate of change in its radial position with respect to time, $\frac{dr}{dt}$, is zero. Thus, by referring to Equation (32), we obtain

$$r_a = \frac{2M}{1 - E(t_a)}. \tag{41}$$

At $t = t_a$, we have $r_a < r_D^+ < r_D^{++}$. Using Equation (30) in Equation (40), we obtain

$$r = -t + t_0 E(t) + c_1, \tag{42}$$

where c_1 is an integration constant. Using the aforementioned two Equations (41) and (42) in conjunction with the previously mentioned radial geodesics criteria for a particle, we obtain the expression for constant c_1 as

$$c_1 = \frac{2M}{1 - E(t_a)} + t_a - t_0 E(t_a). \tag{43}$$

Hence, from Equation (42), we obtain

$$r = -t + t_0 E(t) + \frac{2M}{1 - E(t_a)} + t_a - t_0 E(t_a). \tag{44}$$

Now, we have the capability to compute the duration it takes for a particle to reach the singularity ($r = 0$) at a specific time $t = t_s$ along the radial geodesic in our given spacetime, which is

$$-t_s + t_0 E(t_s) + \frac{2M}{1 - E(t_a)} + t_a - t_0 E(t_a) = 0,$$
$$\implies t_s = t_0 \ln\left[\frac{1}{2W_0\left(\frac{1}{2}e^{\frac{p}{t_0}}\right)}\right], \tag{45}$$

where $p = \left(-\frac{2M}{1-E(t_a)} - t_a + t_0 E(t_a)\right)$ and $W_0(e^{\frac{p}{t_0}})$ is the Lambert W function [77] provided that $e^{\frac{p}{t_0}} \geq 0$. It is important to note that Equation (35) can be used to investigate radial geodesics. Because $D \to \infty$ when $L \to 0$, i.e., $\frac{1}{D} \to 0$ when $L \to 0$. So that Equation (35) is transformed to

$$\frac{dr}{dt} + E(t) = 2\mathcal{L}, \tag{46}$$

so that for radial time-like geodesics, if we substitute $2\mathcal{L} = -1$ then it the exactly same with Equation (40) for $L = 0$.

At the moment, we are tracking the non-radial scenario using the time-like geodesic framework. In this study, we use Equations (35) and (36).

Case–A: First, we look at $D = 2\sqrt{2}$ for the real root of Equation (36), and then we use Equation (30) to find

$$r - t_0 E(t) + 3t - c_2 = 0, \tag{47}$$

where c_2 is an integration constant, and from Equation (34) we obtain

$$\Phi = \frac{1}{L}\left(2t - c_2\right).\tag{48}$$

Let a particle start its journey from $r = r_a$ at time $t = t_a$, then from Equation (47), we obtain

$$c_2 = r_a - t_0 E(t_a) + 3t_a.\tag{49}$$

By observing that $\frac{t_a}{t_0} \geq 0$ and $0 \leq E(t_a) < \frac{1}{2}$, as long as all the values are finite we obtain the condition

$$c_2 \leq r_a + 3t_a < \frac{t_0}{2} + c_2.\tag{50}$$

So that Equations (47) and (48) are transformed to

$$r - t_0 E(t) + 3t - r_a + t_0 E(t_a) - 3t_a = 0,\tag{51}$$

$$\Phi = \frac{1}{L}\left(2t - r_a + t_0 E(t_a) - 3t_a\right).\tag{52}$$

If we assume that the particle is moving towards the singularity $(r \to 0)$ when time $t \to t_s$, then according to Equation (51), we obtain

$$t_s = t_0 \ln\left[\frac{1}{6W_0\left(\frac{1}{6}e^{\frac{p_1}{3t_0}}\right)}\right],\tag{53}$$

where $p_1 = -r_a + t_0 E(t_a) - 3t_a$ and $e^{\frac{p_1}{3t_0}} \geq 0$.

Hence, Equations (47) and (48) demonstrate the possibility of tracing several non-radial time-like geodesics for varying values of c_2 under the condition $D = 2\sqrt{2}$. As a result, a particle starts its journey from $r = r_a$ when $\Phi = \Phi_a = \frac{1}{L}\left(-r_a + t_0 E(t_a) - t_a\right)$ and $t = t_a$ and will approach to singularity $r \to 0$ when $\Phi = \Phi_s = \frac{1}{L}\left(2t_s - r_a + t_0 E(t_a) - 3t_a\right)$ and $t = t_s = t_s = t_0 \ln\left[\frac{1}{6W_0\left(\frac{1}{6}e^{\frac{p_1}{3t_0}}\right)}\right]$, and it will leave the singularity when $t > t_s = t_0 \ln\left[\frac{1}{6W_0\left(\frac{1}{6}e^{\frac{p_1}{3t_0}}\right)}\right]$. Given that $c_2 \leq r_a + 3t_a < \frac{t_0}{2} + c_2$, it follows that the values of r_a and t_a are finite for all finite values of c_2 and t_0. Also, since $t_s = t_0 \ln\left[\frac{1}{6W_0\left(\frac{1}{6}e^{\frac{p_1}{3t_0}}\right)}\right]$, the value of t_s is finite. By using Equations (47) and (48), we have plotted two non-radial time-like geodesics for $L = 1$ and $t_0 = 0.1$, each corresponding to distinct values of $c_2 = 0.1$ and $c_2 = 15$. These geodesics are depicted in Figures 1 and 2, respectively. To plot the geodesics in these diagrams, we have converted the coordinate system from polar coordinates (r, Φ) to Cartesian coordinates (x, y). It should be noted that there exist central singularities for both the conventional generalized Vaidya spacetime [30,31] and the generalized K-essence Vaidya spacetime [37,38], meaning that both $r \to 0$ and $t \to 0$. However, a singularity is typically defined as only $r \to 0$. In this case, we may see that the particle's track will allow a future observer to watch the particle reach $r \to 0$ at a specific time and escape the singularity. These phenomena are discussed in the Conclusions.

For circular orbits, Equations (38) and (39) can be employed. For a given value of $D = 2\sqrt{2}$, the corresponding values are $r_{\pm} = L\sqrt{2}$ and $\Phi_{\pm} = \frac{t}{L}$. Thus, there is only a single non-radial time-like circular orbit with a radius of $L\sqrt{2}$ for $D = 2\sqrt{2}$, as seen in Figure 3, when $L = 1$ and $M = \sqrt{2}$. In this regard, it is worth noting that Ishihara et al. [76] have demonstrated that a spacetime that varies with time can possess both an event horizon and a possibility for quasicircular orbits in a different context.

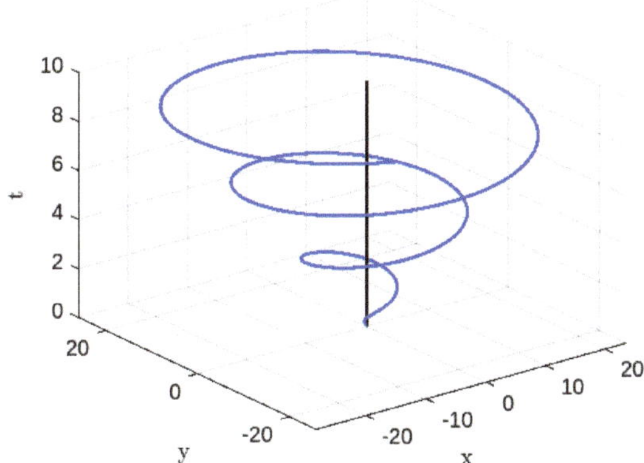

Figure 1. Time-like geodesics for $D = 2\sqrt{2}$, $L = 1$, $t_0 = 0.1$, and $c_2 = 0.1$.

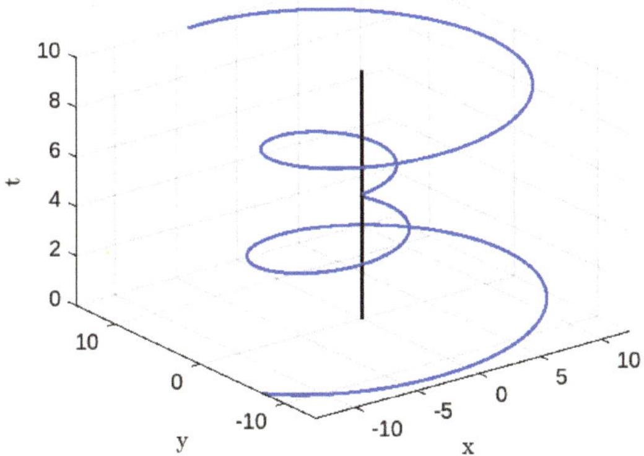

Figure 2. Time-like geodesics for $D = 2\sqrt{2}$, $L = 1$, $t_0 = 0.1$, and $c_2 = 15$.

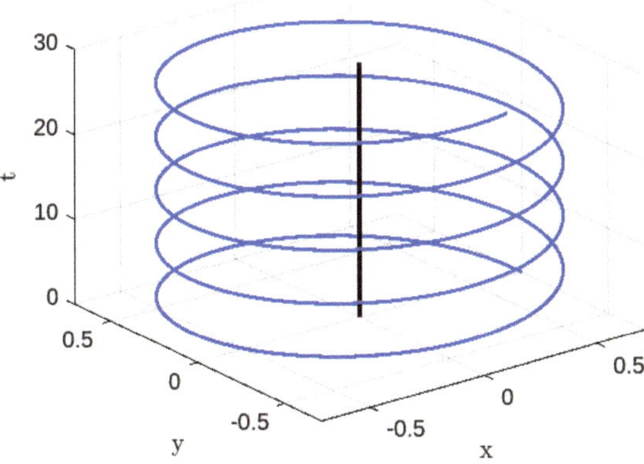

Figure 3. Circular orbit of time-like geodesics for $D = 2\sqrt{2}$, $L = 1$, and $M = \sqrt{2}$.

Case–B: To calculate the real solutions of Equation (35) for the purpose of analyzing non-radial time-like geodesics, we assume that $D > 2\sqrt{2}$. So that Equation (35) can be written as

$$\left(\tfrac{dr}{dt} + E(t) - A\right)\left(\tfrac{dr}{dt} + E(t) - B\right) = 0,$$
$$\implies \left(r - t_0 E(t) - At - c_3\right)\left(r - t_0 E(t) - Bt - c_3\right) = 0, \tag{54}$$

where

$$A = 1 - \frac{D^2}{2} + \frac{D}{2}\sqrt{D^2 - 8}, \text{ and } B = 1 - \frac{D^2}{2} - \frac{D}{2}\sqrt{D^2 - 8}, \tag{55}$$

and c_3 is an integration constant.

Equation (54) demonstrates that there are two non-radial time-like geodesics that may be followed for any finite value of c_3.

For

$$r - t_0 E(t) - At - c_3 = 0, \tag{56}$$

then from Equation (34), we obtain

$$\Phi = \frac{1}{L}\left[-t - At - c_3\right], \tag{57}$$

and for

$$r - t_0 E(t) - Bt - c_3 = 0, \tag{58}$$

we have from (34)

$$\Phi = \frac{1}{L}\left[-t - Bt - c_3\right]. \tag{59}$$

Let a particle start its journey from $r = r_a$ in the path at time $t = t_a$, then Equation (56) yields

$$c_3 = r_a - t_0 E(t_a) - At_a. \tag{60}$$

Clearly, since $\frac{t_a}{t_0} \geq 0$ and $0 \leq E(t_a) < \frac{1}{2}$ so that

$$c_3 \leq r_a - At_a < \frac{t_0}{2} + c_3, \tag{61}$$

then, Equations (56) and (57) are transformed to

$$r - t_0 E(t) - At - r_a + t_0 E(t_a) + At_a = 0, \tag{62}$$
$$\Phi = \tfrac{1}{L}\left[-t - At - r_a + t_0 E(t_a) + At_a\right]. \tag{63}$$

If we consider that the particle approaches $r \to 0$ when $t \to t_s$ on path (56), then from (62) we have

$$t_s = t_0 \ln\left[\frac{1}{-2AW_0\left(-\frac{1}{2A}e^{\frac{p_2}{At_0}}\right)}\right], \tag{64}$$

where $p_2 = r_a - t_0 E(t_a) - At_a$ and $\frac{1}{A}e^{\frac{p_2}{At_0}} < 0$.

Similarly, as before, let us consider a particle that begins its trajectory at $r = r_a$ along the path (58) at time $t = t_a$ and eventually approaches singularity at $t = t_s$. By analyzing Equations (58) and (59), we may derive the following results:

$$r - t_0 E(t) - Bt - r_a + t_0 E(t_a) + Bt_a = 0, \tag{65}$$

$$\Phi = \tfrac{1}{L}\Big[-t - Bt - r_a + t_0 E(t_a) + Bt_a \Big], \tag{66}$$

$$c_3 = r_a - t_0 E(t_a) - Bt_a, \tag{67}$$

$$c_3 \le r_a - Bt_a < \tfrac{t_0}{2} + c_3, \tag{68}$$

$$t_s = t_0 \ln \left[\frac{1}{-2BW_0\left(-\frac{1}{2B}e^{\frac{p_3}{Bt_0}}\right)} \right], \tag{69}$$

where $p_3 = r_a - t_0 E(t_a) - Bt_a$ and $\frac{1}{B}e^{\frac{p_3}{Bt_0}} \le 0$.

Therefore, Equations (56)–(59) demonstrate the possibility of tracking several non-radial time-like geodesics for distinct values of c given that $D > 2\sqrt{2}$. Hence, for path (56), a particle starts its journey from $r = r_a$ when $\Phi = \Phi_a = \tfrac{1}{L}\big[-t_a - r_a + t_0 E(t_a) \big]$ and $t = t_a$, and approach $r \to 0$ when $\Phi = \Phi_s = \tfrac{1}{L}\big[-t_s - At_s - r_a + t_0 E(t_a) + At_a \big]$ and $t = t_s = t_0 \ln \left[\frac{1}{-2AW_0\left(-\frac{1}{2A}e^{\frac{p_2}{At_0}}\right)} \right]$, and it will leave the singularity when $t > t_s = t_0 \ln \left[\frac{1}{-2AW_0\left(-\frac{1}{2A}e^{\frac{p_2}{At_0}}\right)} \right]$. Also, for path (58), a particle starts its journey from $r = r_a$ when $\Phi = \Phi_a = \tfrac{1}{L}\big[-t_a - r_a + t_0 E(t_a) \big]$ and $t = t_a$, and will arrive at the singularity $r = 0$ when $\Phi = \Phi_s = \tfrac{1}{L}\big[-t_s - Bt_s - r_a + t_0 E(t_a) + Bt_a \big]$ and $t = t_s = t_0 \ln \left[\frac{1}{-2BW_0\left(-\frac{1}{2B}e^{\frac{p_3}{Bt_0}}\right)} \right]$, and it will leave the singularity when $t > t_s = t_0 \ln \left[\frac{1}{-2BW_0\left(-\frac{1}{2B}e^{\frac{p_3}{Bt_0}}\right)} \right]$. Since $c_3 \le r_a - At_a < \frac{t_0}{2} + c_3$ and $c_3 \le r_a - Bt_a < \frac{t_0}{2} + c_3$, then for any finite values of c_3, t_0, A, and B, the values of r_a and t_a are also finite. Using Equations (56)–(59), for $D = 2.9$, $L = 1$, and $t_0 = 0.1$, we have traced two non-radial time-like for $c_3 = 0.1$ in Figure 4 and also for $c_3 = 10$ in Figure 5.

Again, for the circular orbits, we recall Equations (38) and (39), so that for $D > 2\sqrt{2}$, we have $r_\pm = \frac{M \pm \sqrt{M^2 - 2L^2}}{2}$ and $\Phi_\pm = \frac{Lt}{r_\pm^2}$. Therefore, two non-radial time-like circular orbits can be traced for $D > 2\sqrt{2}$ as shown in Figure 6 for $D = 2.9$, $L = 1$, and $M = 1.45$.

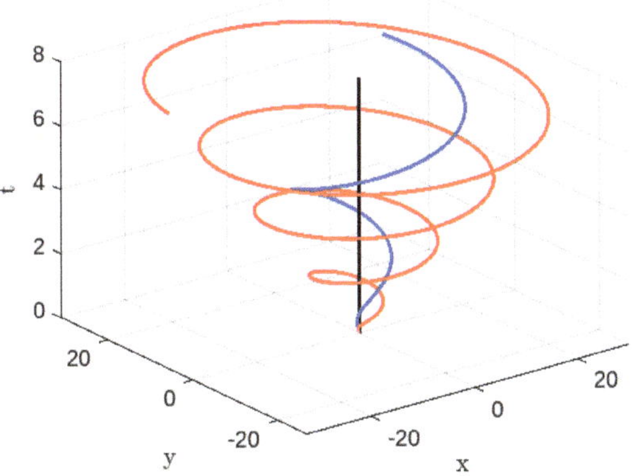

Figure 4. Time-like geodesics for $D = 2.9$, $L = 1$, $t_0 = 0.1$, and $c_3 = 0.1$. Blue-line for $r - t_0 E(t) - At - c_3 = 0$ and red-line for $r - t_0 E(t) - Bt - c_3 = 0$.

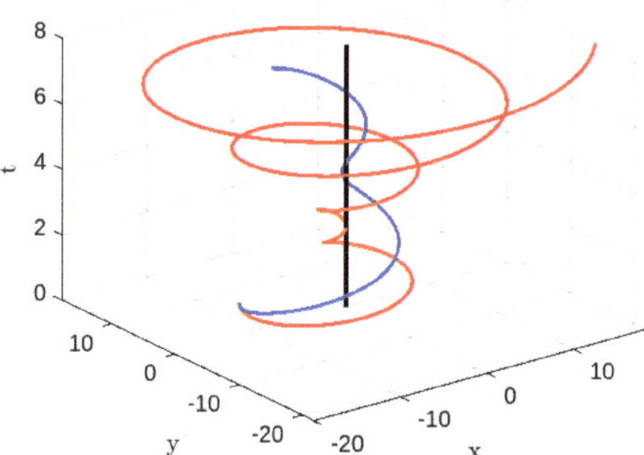

Figure 5. Time-like geodesics for $D = 2.9$, $L = 1$, $t_0 = 0.1$, and $c_3 = 10$. Blue-line for $r - t_0 E(t) - At - c_3 = 0$ and red-line for $r - t_0 E(t) - Bt - c_3 = 0$.

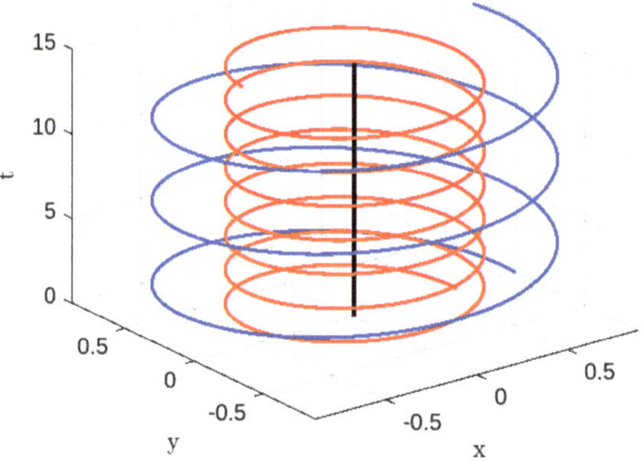

Figure 6. Circular time-like geodesics for $D = 2.9$, $L = 1$, and $M = 1.45$. Blue-line for $r_+ = \frac{M+\sqrt{M^2-2L^2}}{2}$ and red-line for $r_- = \frac{M-\sqrt{M^2-2L^2}}{2}$.

3.1.2. Null Geodesics for Case-I

In this subsection, we want to elucidate the null geodesic behavior of the generalized K-essence Vaidya spacetime, as described by Equation (13) and the mass function given by Equation (27). We assume that $2\mathcal{L} = 0$. For this study, Equations (33) and (34) reduce to

$$\frac{dr}{dt} + E(t) = -\frac{L^2}{r^2}. \tag{70}$$

$$r + L\Phi = t_0 E(t). \tag{71}$$

In order to track the radial geodesics, we assume that a particle with zero angular momentum ($L = 0$) begins its motion from a state of rest at a distance of $r = r_a$ and time $t = t_a$, such that the rate of change in r with respect to t is zero ($\frac{dr}{dt} = 0$). Thus, based on Equation (32), the value of r_a may be determined as $\frac{2M}{1-E(t_a)}$. Clearly, at $t = t_a$, we have $r_a < r_D^+ < r_D^{++}$. Furthermore, according to Equation (70),

$$r = t_0 E(t) + c_4, \tag{72}$$

where c_4 is an integration constant.

At $t = t_a$, we obtain

$$c_4 = \frac{2M}{1 - E(t_a)} - t_0 E(t_a),\tag{73}$$

so that

$$r = t_0 E(t) + \frac{2M}{1 - E(t_a)} - t_0 E(t_a).\tag{74}$$

Assuming that the particle reaches the singularity ($r = 0$) at a specific time $t = t_s$, we determine

$$t_s = t_0 \ln\left[\frac{1 - E(t_a)}{2E(t_a)\left(1 - E(t_a)\right) - (4M/t_0)}\right].\tag{75}$$

Now, we will examine the characteristics of non-radial ($L \neq 0$) geodesics inside the framework of null geodesics in the given spacetime (13), which is governed by the particular mass function (27). The requirement for the existence of real roots of Equation (35) under the assumption that $2\mathcal{L} = 0$ is that $D \geq 2$. Now, we discuss the following two cases:

Case–A: When $D = 2$, we have from Equation (36)

$$r - t_0 E(t) + t - c_5 = 0,\tag{76}$$

and from Equation (34) we obtain

$$\Phi = \frac{1}{L}\left(t - c_5\right).\tag{77}$$

Let a particle start its journey from $r = r_a$ at time $t = t_a$; then, from (76)

$$c_5 = r_a - t_0 E(t_a) + t_a.\tag{78}$$

It is evident that the inequality $\frac{t_a}{t_0} \geq 0$ is true, and we also have the condition $0 \leq E(t_a) < \frac{1}{2}$ so that

$$c_5 \leq r_a + t_a < \frac{t_0}{2} + c_5,\tag{79}$$

then, Equations (76) and (77) transform to

$$r - t_0 E(t) + t - r_a + t_0 E(t_a) - t_a = 0,\tag{80}$$
$$\Phi = \frac{1}{L}\left(t - r_a + t_0 E(t_a) - t_a\right).\tag{81}$$

If we consider that the particle reaches the singularity ($r = 0$) at time $t = t_s$, then from (80) we have

$$t_s = t_0 \ln\left[\frac{1}{2W_0\left(\frac{1}{2}e^{\frac{p_4}{t_0}}\right)}\right],\tag{82}$$

where $p_4 = -r_a + t_0 E(t_a) - t_a$ and $e^{\frac{p_4}{t_0}} \geq 0$.

Therefore, Equations (76) and (77) demonstrate that we may track several non-radial null geodesics for varying values of c_5 in the case of $D = 2$. Hence, a particle starts its trajectory at a distance of $r = r_a$ when $\Phi = \Phi_a = \frac{1}{L}\left[-r_a + t_0 E(t_a)\right]$ and $t = t_a$, and will arrive at the singularity $r = 0$ when $\Phi = \Phi_s = \frac{1}{L}\left(t_s - r_a + t_0 E(t_a) - t_a\right)$ and

$t = t_s = t_0 \ln \left[\dfrac{1}{2W_0\left(\frac{1}{2}e^{\frac{p_4}{t_0}}\right)} \right]$, and it will leave the singularity when $t > t_s = t_0 \ln \left[\dfrac{1}{2W_0\left(\frac{1}{2}e^{\frac{p_4}{t_0}}\right)} \right]$.

Given that $c_5 \leq r_a + t_a < \frac{t_0}{2} + c_5$, it follows that both r_a and t_a are finite for any finite values of c_5 and t_0. Also, since $t_s = t_0 \ln \left[\dfrac{1}{2W_0\left(\frac{1}{2}e^{\frac{p_4}{t_0}}\right)} \right]$, the value of t_s is finite. By using Equations (76) and (77), we have plotted two non-radial time-like geodesics for $L = 1$ and $t_0 = 0.1$, with two distinct values of c_5, namely, $c_5 = 1$ and $c_5 = 10$. These geodesics are depicted in Figures 7 and 8, respectively.

Now, for the null circular orbits, using Equations (38) and (39) for $D = 2$, we have $r_\pm = L$ and $\Phi_\pm = \frac{t}{L}$. Therefore, there is only one non-radial null circular orbit of the radius L which can be traced for $D = 2$, as shown in Figure 9 for $L = 1$ and $M = 1$.

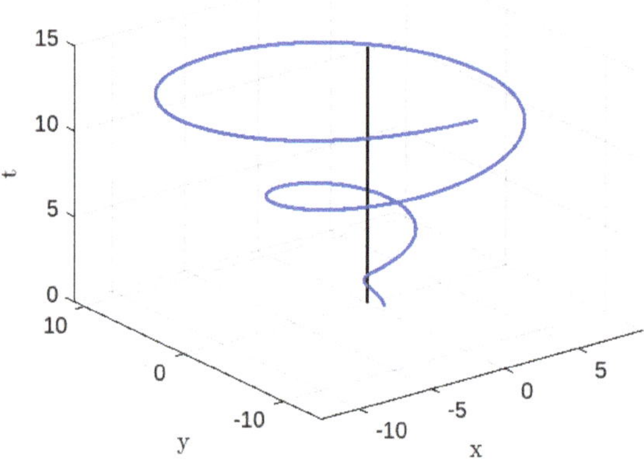

Figure 7. Null geodesics for $D = 2$, $L = 1$, $t_0 = 0.1$, and $c_5 = 1$.

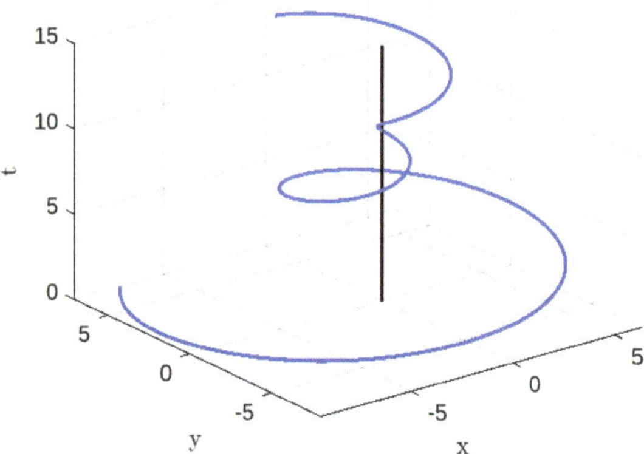

Figure 8. Null geodesics for $D = 2$, $L = 1$, $t_0 = 0.1$, and $c_5 = 10$.

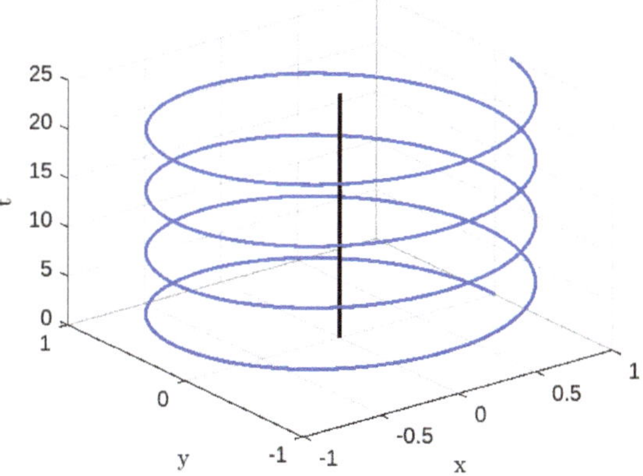

Figure 9. Circular null geodesics for $D = 2$, $L = 1$, and $M = 1$.

Case–B: When $D > 2$, we have from Equation (35)

$$\left(r - t_0 E(t) - At - c_6\right)\left(r - t_0 E(t) - Bt - c_6\right) = 0, \tag{83}$$

where

$$A = 1 - \frac{D^2}{2} + \frac{D}{2}\sqrt{D^2 - 4} \ \text{ and } \ B = 1 - \frac{D^2}{2} - \frac{D}{2}\sqrt{D^2 - 4}, \tag{84}$$

and c_6 is an integration constant.

Equation (83) demonstrates the existence of two non-radial null geodesics that may be followed for any finite value of c_6. Equation (83) is of the same form as (54) in the context of time-like geodesics, but with different constants. Therefore, based on this closeness, we may infer Equations (77) and (83) as

$$r - t_0 E(t) - At - c_6 = 0, \tag{85}$$

$$\Phi = \tfrac{1}{L}\left(- At - c_6\right). \tag{86}$$

Also,

$$r - t_0 E(t) - Bt - c_6 = 0, \tag{87}$$

$$\Phi = \tfrac{1}{L}\left(- Bt - c_6\right). \tag{88}$$

Let a particle start its journey from $r = r_a$ in path (85) or (87) at time $t = t_a$; then, from Equation (85), we have

$$c_6 = r_a - t_0 E(t_a) - At_a, \tag{89}$$

$$\text{and} \quad c_6 \leq r_a - At_a < \tfrac{t_0}{2} + c_6. \tag{90}$$

Also, from Equation (87) we may obtain

$$c_6 = r_a - t_0 E(t_a) - Bt_a, \tag{91}$$

$$\text{and} \quad c_6 \leq r_a - Bt_a < \tfrac{t_0}{2} + c_6. \tag{92}$$

Therefore, we have

$$r - t_0 E(t) - At - r_a + t_0 E(t_a) + At_a = 0, \tag{93}$$

$$\Phi = \tfrac{1}{L}\Big[-At - r_a + t_0 E(t_a) + At_a \Big], \tag{94}$$

and also

$$r - t_0 E(t) - Bt - r_a + t_0 E(t_a) + Bt_a = 0, \tag{95}$$

$$\Phi = \tfrac{1}{L}\Big[-Bt - r_a + t_0 E(t_a) + Bt_a \Big]. \tag{96}$$

If we consider that the particle approaches the singularity ($r \to 0$) at time $t = t_s$ along path (85) or (87), then we may deduce from Equation (93)

$$t_s = t_0 \ln \left[\frac{1}{-2AW_0\left(-\frac{1}{2A} e^{\frac{p_5}{At_0}} \right)} \right], \tag{97}$$

and from (95)

$$t_s = t_0 \ln \left[\frac{1}{-2BW_0\left(-\frac{1}{2B} e^{\frac{p_6}{Bt_0}} \right)} \right], \tag{98}$$

where $p_5 = r_a - t_0 E(t_a) - At_a$, $p_6 = r_a - t_0 E(t_a) - Bt_a$, $\frac{1}{A} e^{\frac{p_5}{At_0}} \leq 0$, and $\frac{1}{B} e^{\frac{p_6}{Bt_0}} \leq 0$.

As the outcome of Equations (85)–(88), we may trace distinct non-radial null geodesics for different values of c_6 when $D > 2$. As a result, given the path (85), a particle begins its journey from $r = r_a$ when $\Phi = \Phi_a = \tfrac{1}{L}\big[-r_a + t_0 E(t_a) \big]$ and $t = t_a$ and approaches singularity $r \to 0$ when $\Phi = \Phi_s = \tfrac{1}{L}\big[-At_s - r_a + t_0 E(t_a) + At_a \big]$ and $t = t_s = t_0 \ln \left[\frac{1}{-2AW_0\left(-\frac{1}{2A} e^{\frac{p_5}{At_0}} \right)} \right]$, and it will leave from the singularity when $t > t_s = t_0 \ln \left[\frac{1}{-2AW_0\left(-\frac{1}{2A} e^{\frac{p_5}{At_0}} \right)} \right]$.

Also, for path (87), a particle starts its journey from $r = r_a$ when $\Phi = \Phi_a = \tfrac{1}{L}\big[-r_a + t_0 E(t_a) \big]$ and $t = t_a$ and will approaches singularity $r \to 0$ when $\Phi = \Phi_s = \tfrac{1}{L}\big[-Bt_s - r_a + t_0 E(t_a) + Bt_a \big]$ and $t = t_s = t_0 \ln \left[\frac{1}{-2BW_0\left(-\frac{1}{2B} e^{\frac{p_6}{Bt_0}} \right)} \right]$, and it will leave the singularity when $t > t_s = t_0 \ln \left[\frac{1}{-2BW_0\left(-\frac{1}{2B} e^{\frac{p_6}{Bt_0}} \right)} \right]$.

Since $c_6 \leq r_a - At_a < \frac{t_0}{2} + c_6$ and $c_6 \leq r_a - Bt_a < \frac{t_0}{2} + c_6$, then for any finite values of c_6, t_0, A and B, the values of r_a and t_a are also finite.

By substituting the values $D = 2.1$, $L = 1$, and $t_0 = 0.1$ into Equations (85)–(88), we have obtained two non-radial nulls for $c_6 = 0.1$ in Figure 10 and also for $c_6 = 10$ in Figure 11.

Again, for the circular orbits, we recall Equations (38) and (39), so that for $D > 2$, we have $r_{\pm} = M \pm \sqrt{M^2 - L^2}$ and the corresponding $\Phi_{\pm} = \frac{Lt}{r_{\pm}^2}$. Therefore, two non-radial null circular orbits can be traced for $D > 2$, as shown in Figure 12 for $D = 2.1$, $L = 1$, and $M = 1.05$.

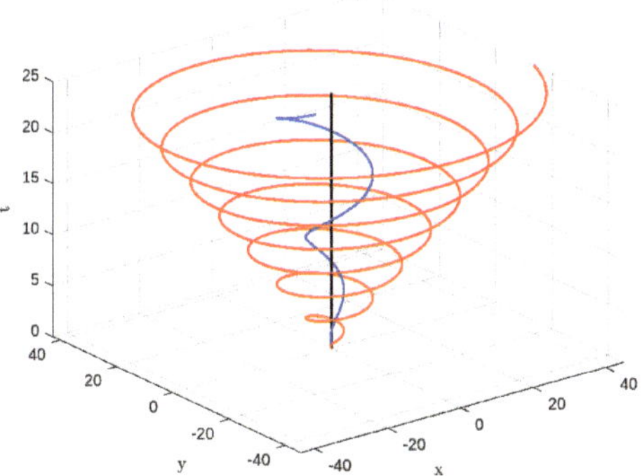

Figure 10. Null geodesics for $D = 2.1$, $L = 1$, $t_0 = 0.1$, and $c_6 = 0.1$. Blue-line for $r - t_0 E(t) - At - c_6 = 0$ and red-line for $r - t_0 E(t) - Bt - c_6 = 0$.

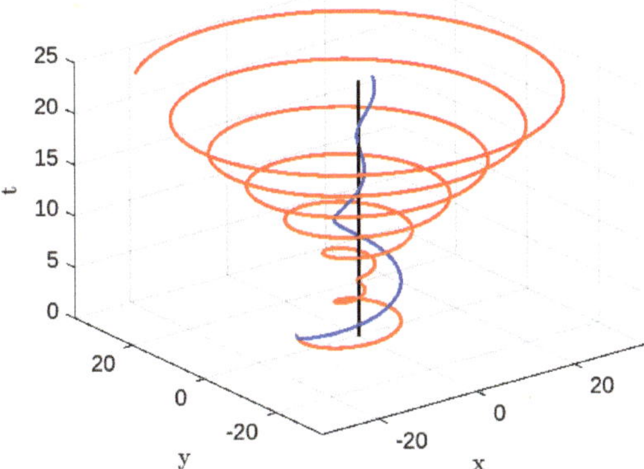

Figure 11. Null geodesics for $D = 2.1$, $L = 1$, $t_0 = 0.1$, and $c_6 = 10$. Blue-line for $r - t_0 E(t) - At - c_6 = 0$ and red-line for $r - t_0 E(t) - Bt - c_6 = 0$.

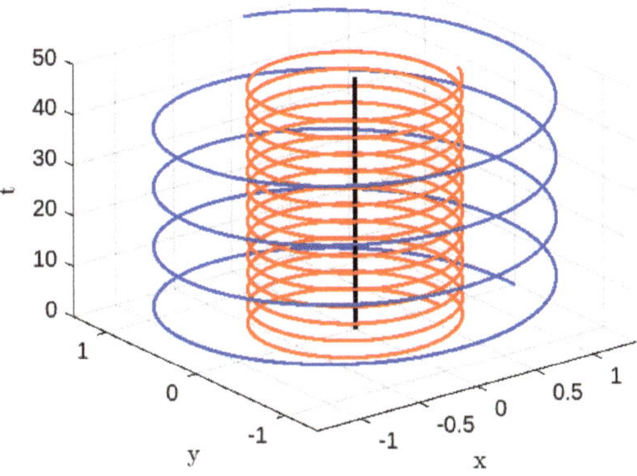

Figure 12. Circular null geodesics for $D = 2.1$, $L = 1$, and $M = 1.05$. Blue-line for $r_+ = M + \sqrt{M^2 - L^2}$ and red-line for $r_- = M - \sqrt{M^2 - L^2}$.

3.2. Case-II: $\mathcal{M}(t,r) = \mu t + \frac{r}{2}\phi_t^2$

In this part, we consider the generalized K-essence Vaidya mass function (14) with the assumption (26) as

$$\mathcal{M}(t,r) = \mu t + \frac{r}{2}e^{-\frac{t}{t_0}}, \tag{99}$$

where we take the usual generalized Vaidya mass function $m(t,r) \equiv m(t) = \mu t$, i.e., is a linear function of t [35,36,78] and μ is a positive constant. In this case, the radii of the dynamical horizons are $r_D^+ = \frac{2\mu t}{1 - e^{-t/t_0}}$ and $r_D^{++} = \frac{2\mu t}{1 - e^{-t/t_0}}\left[1 + W_0\left(\frac{1 - e^{-t/t_0}}{2\mu te}e^{-\frac{1}{2\mu}}\right)\right]$ (see Appendix A).

From the above mass function (99), we have

$$\frac{\bar{\mathcal{M}}_t}{r} = -\frac{e^{-\frac{t}{t_0}}}{2t_0}, \tag{100}$$

where we redefine the first derivative of the given mass function with respect to time as $\bar{\mathcal{M}}_t \equiv \mathcal{M}_t - \mu$.

Following the same procedure as for Case-I, and using Equations (20)–(25) and (28), and again assuming the comoving plane $(d\tau \equiv dt)$, we obtain

$$E(t) = \tfrac{1}{2}e^{-\frac{t}{t_0}}, \tag{101}$$

$$\text{and} \quad \mathcal{M}(t.r) = \mu t + E(t)r. \tag{102}$$

Again, continuing the similar procedure as for Case-I, we found the following relations:

$$\frac{dr}{dt} + E(t) = 2\mathcal{L} - \frac{L^2}{r^2}, \tag{103}$$

$$\frac{dr}{dt} + E(t) = 1 - \frac{2\mu t}{r}, \tag{104}$$

$$r + L\Phi = 2\mathcal{L}t + t_0 E(t). \tag{105}$$

Again, proceeding on as before, we have another solution from Equations (103) and (104):

$$r_{\pm} = \frac{\mu t \pm \sqrt{\mu^2 t^2 - L^2(1 - 2\mathcal{L})}}{1 - 2\mathcal{L}}, \tag{106}$$

provided $\mu^2 t^2 \geq L^2(1 - 2\mathcal{L})$ and from Equation (24), we obtain

$$\dot{\Phi}_{\pm} = \frac{L}{r_{\pm}^2}. \tag{107}$$

Under this situation, it is evident that the radial part, as indicated by (106), exhibits time dependence. This characteristic is expected in any generalized Vaidya spacetime [30,31,37], given that the background generalized Vaidya mass function is defined as $m(t,r) \equiv \mu t$.

3.2.1. Time-like Geodesics for Case-II

In this subsection, we analyze the behavior of time-like geodesics in the generalized K-essence Vaidya spacetime, as described by Equation (13) with the mass function given by Equation (99). For the sake of this investigation, we assume $2\mathcal{L} = -1$.

Following from the previous analysis for Case-I, we will now focus on tracking the radial geodesics. Specifically, we examine a particle with zero angular momentum $(L = 0)$ that begins its motion from a state of rest at a distance $r = r_a$ and time $t = t_a$, so that

$$r_a = \frac{2\mu t_a}{1 - E(t_a)}. \tag{108}$$

In this case, at $t = t_a$, we also have $r_a < r_D^+ < r_D^{++}$, which is the same type of condition as in the first Case-I, and using Equation (40), we obtain

$$r = -t + t_0 E(t) + c_7, \tag{109}$$

$$c_7 = \frac{2\mu t_a}{1 - E(t_a)} + t_a - t_0 E(t_a), \tag{110}$$

$$r = -t + t_0 E(t) + \frac{2\mu t_a}{1 - E(t_a)} + t_a - t_0 E(t_a). \tag{111}$$

Similarly, if the particle approaches the singularity ($r \to 0$) at time $t = t_s$, the same type of consideration applies; we obtain

$$t_s = t_0 \ln \left[\frac{1}{2 W_0 \left(\frac{1}{2} e^{\frac{p_7}{t_0}} \right)} \right], \tag{112}$$

provided $e^{\frac{p_7}{t_0}} \geq 0$, where $p_7 = -\frac{2\mu t_a}{1 - E(t_a)} - t_a + t_0 E(t_a)$.

When studying non-radial time-like geodesics, we obtain the same expression as Equation (35), but with a different value for the parameter D ($D = \frac{2\mu t}{L}$). In order to trace the geodesics, it is necessary to satisfy the given requirement

$$D \geq 2\sqrt{2} \implies t \geq \frac{\sqrt{2} L}{\mu}. \tag{113}$$

In this non-radial study, the scenario where $D = 2\sqrt{2}$ cannot be taken into consideration. This is because if $D = 2\sqrt{2}$, then $t = \frac{\sqrt{2}L}{\mu}$, resulting in a constant value for time. As a result, the spacetime is not like the generalized K-essence Vaidya type. So, in this non-radial time-like geodesic investigation, we only look into $D > 2\sqrt{2}$, ensuring that the spacetime is time-dependent, which is an essential characteristic of both the normal Vaidya spacetime and the generalized K-essence Vaidya spacetime. As before, we solve Equation (35) for $D > 2\sqrt{2}$, where $D = \frac{2\mu t}{L}$, and we obtain

$$\left(r - t_0 E(t) - \bar{A}(t) - c_8 \right) \left(r - t_0 E(t) - \bar{B}(t) - c_8 \right) = 0, \tag{114}$$

where

$$\bar{A}(t) = t - \frac{2\mu^2 t^3}{3L^2} + \frac{2\mu^2}{3L^2} \left(t^2 - \frac{2L^2}{\mu^2} \right)^{\frac{3}{2}}, \tag{115}$$

$$\bar{B}(t) = t - \frac{2\mu^2 t^3}{3L^2} - \frac{2\mu^2}{3L^2} \left(t^2 - \frac{2L^2}{\mu^2} \right)^{\frac{3}{2}}. \tag{116}$$

Clearly, Equation (114) shows that there are two non-radial time-like geodesics that can be traced for any finite value of c_8. Therefore, we have two non-radial time-like geodesics, either

$$r - t_0 E(t) - \bar{A}(t) - c_8 = 0, \tag{117}$$

$$\Phi = \frac{1}{L} \left[-t - \bar{A}(t) - c_8 \right], \tag{118}$$

or

$$r - t_0 E(t) - \bar{B}(t) - c_8 = 0, \tag{119}$$

$$\Phi = \frac{1}{L} \left[-t - \bar{B}(t) - c_8 \right]. \tag{120}$$

Let a particle start its journey from $r = r_a$ in path (117) at time $t = t_a > \frac{\sqrt{2}L}{\mu}$; we have

$$c_8 = r_a - t_0 E(t_a) - \bar{A}(t_a). \tag{121}$$

Since $\frac{t_a}{t_0} \geq 0$ and $0 \leq E(t_a) < \frac{1}{2}$

$$c_8 \leq r_a - \bar{A}(t_a) < \frac{t_0}{2} + c_8, \tag{122}$$

then, Equations (117) and (118) are transformed to

$$r - t_0 E(t) - \bar{A}(t) - r_a + t_0 E(t_a) + \bar{A}(t_a) = 0, \tag{123}$$

$$\Phi = \frac{1}{L}\left[-t - \bar{A}(t) - r_a + t_0 E(t_a) + \bar{A}(t_a)\right], \tag{124}$$

then, at $t = t_a$

$$\Phi = \Phi_a = \frac{1}{L}\left[-t_a - r_a + t_0 E(t_a)\right]. \tag{125}$$

If we assume that the particle reaches the singularity $(r \to 0)$ at time $t = t_s$ along path (117), then we can deduce from Equation (123) the following:

$$t_0 E(t_s) + \bar{A}(t_s) + c_8 = 0,$$

$$E(t_s) - \frac{2\mu^2 t_s^3}{3 t_0 L^2} + \frac{c_8}{t_0} = -\frac{t_s}{t_0} - \frac{2\mu^2}{3 t_0 L^2}\left(t_s^2 - \frac{2L^2}{\mu^2}\right)^{\frac{3}{2}}. \tag{126}$$

In this scenario, the precise expression for t_s cannot be determined due to the presence of a distinct mass function (99). However, an expression for the transcendental Equation (116) is available, allowing for numerical and graphical analysis to obtain the finite time for any given values of μ and L. Now, let $f(t_s) = E(t_s) - \frac{2\mu^2 t_s^3}{3 t_0 L^2} + \frac{c}{t_0}$ and $g(t_s) = -\frac{t_s}{t_0} - \frac{2\mu^2}{3 t_0 L^2}\left(t_s^2 - \frac{2L^2}{\mu^2}\right)^{\frac{3}{2}}$. Then, the only possibility for a finite value of t_s is if and only if $f(t_s) = g(t_s)$.

To start with let us consider $\mu = 0.05$ and $L = 1$, then $t_a > 28.2843$. Let $t_a = 28.3$, and then from (115), $\bar{A}(t_a) \approx -9.4739$. Now, if we consider $t_0 = 0.1$, then for $c_8 = 0.1$, $r_a \approx -9.3739$ and $\Phi_a \approx -18.9261$, and also for $c_8 = 30$, $r_a \approx 20.5261$ and $\Phi_a \approx -48.8261$ by using (121) and (122). Figure 13 shows that for $c_8 = 0.1$ Equation (126) has no solution for t_s since the curves $y = f(t_s)$ and $y = g(t_s)$ do not intersect each other. So, for $c_8 = 0.1$, the particle which starts from $r = r_a$ at time $t = t_a > 28.2843$ will never reach the singularity, as shown in Figure 14 (blue-line) using Equations (117) and (118). However, for $c_8 = 30$, Figure 15 shows that it has a solution for t_s since the curves $y = f(t_s)$ and $y = g(t_s)$ intersect each other. For $c_8 = 30$, the particle initially located at $r = r_a$ at time $t = t_a > 28.2843$ will eventually reach the singularity $(r = 0)$ at time $t = t_s$, as depicted by the red-line in Figure 14. The particle will then depart from the singularity $(r = 0)$, which is shown graphically using Equations (117) and (118).

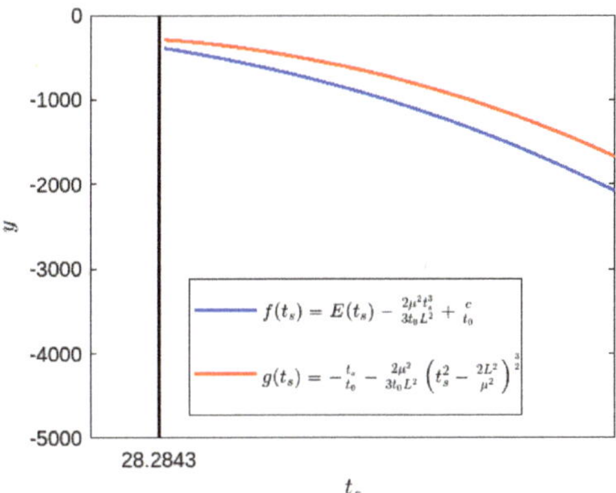

Figure 13. $\mu = 0.05$, $L = 1$, $t_0 = 0.1$, and $c_8 = 0.1$. Blue-line for $f(t_s)$ and red-line for $g(t_s)$.

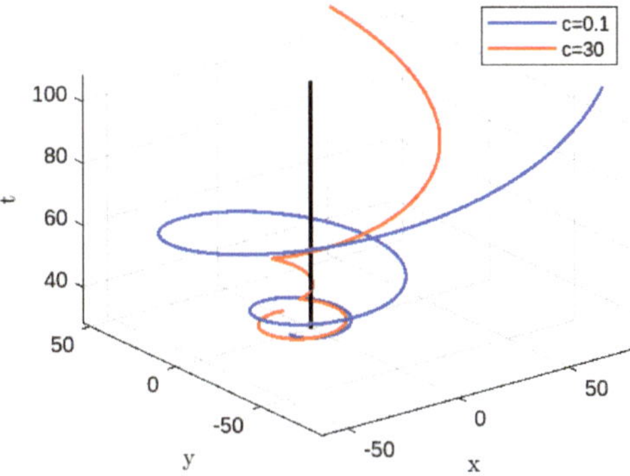

Figure 14. Time-like geodesics for $r - t_0 E(t) - \bar{A} - c_8 = 0$ when $\mu = 0.05$, $L = 1$, and $t_0 = 0.1$. Blue-line for $c_8 = 0.1$ and red-line for $c_8 = 30$.

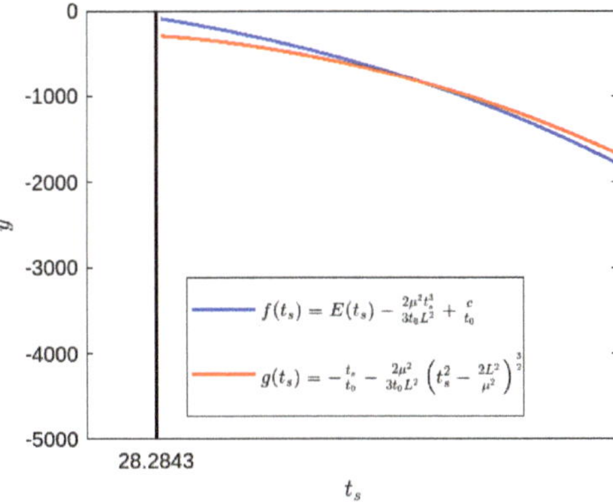

Figure 15. $\mu = 0.05$, $L = 1$, $t_0 = 0.1$, and $c_8 = 30$. Blue-line for $f(t_s)$ and red-line for $g(t_s)$.

Again, if we consider a particle starts its journey from $r = r_b$ in path (119) at time $t = t_b > \frac{\sqrt{2}L}{\mu}$, we have

$$c_8 = r_b - t_0 E(t_b) - \bar{B}(t_b). \tag{127}$$

In this path, the condition is $\frac{t_b}{t_0} \geq 0$ and $0 \leq E(t_b) < \frac{1}{2}$

$$c_8 \leq r_b - \bar{B}(t_b) < \frac{t_0}{2} + c_8, \tag{128}$$

then, Equations (119) and (120) are transformed to

$$r - t_0 E(t) - \bar{B}(t) - r_b + t_0 E(t_b) + \bar{B}(t_b) = 0, \tag{129}$$

$$\Phi = \frac{1}{L}\left[-t - \bar{B}(t) - r_b + t_0 E(t_b) + \bar{B}(t_b)\right], \tag{130}$$

then, at $t = t_b$

$$\Phi = \Phi_b = \frac{1}{L}\left[-t_b - r_b + t_0 E(t_b)\right]. \tag{131}$$

Once more, assuming it is possible, let us contemplate the scenario in which the particle moves towards the singularity $(r \to 0)$ at time $t = t_s$ along trajectory (119). Consequently, based on Equation (129), we may establish the following relation:

$$E(t_s) - \frac{2\mu^2 t_s^3}{3t_0 L^2} + \frac{c_8}{t_0} = -\frac{t_s}{t_0} + \frac{2\mu^2}{3t_0 L^2}\left(t_s^2 - \frac{2L^2}{\mu^2}\right)^{\frac{3}{2}}. \tag{132}$$

Once again, this equation is transcendental, meaning it can be analyzed numerically and in pictures for any finite values of μ and L. Now, let $f(t_s) = E(t_s) - \frac{2\mu^2 t_s^3}{3t_0 L^2} + \frac{c_8}{t_0}$ and $g(t_s) = -\frac{t_s}{t_0} + \frac{2\mu^2}{3t_0 L^2}\left(t_s^2 - \frac{2L^2}{\mu^2}\right)^{\frac{3}{2}}$. Then, the only possibility for a finite value of t_s is if and only if $f(t_s) = g(t_s)$.

To start with, let us consider $\mu = 0.05$ and $L = 1$, then $t_b > 28.2843$. Again, let $t_b = 28.3$ and then, from (116), $\bar{B}(t_b) \approx -9.4767$. Now, if we consider $t_0 = 0.1$, then for $c_8 = 0.1$, $r_b \approx -9.3767$ and $\Phi_b \approx -18.9233$, and also for $c_8 = 30$, $r_b \approx 20.5233$ and $\Phi_b \approx -48.8233$ by using (127) and (131). Now, Figure 16 demonstrates that when $c_8 = 0.1$, Equation (132) does not have a solution for t_s because the curves $y = f(t_s)$ and $y = g(t_s)$ do not cross. So, for $c_8 = 0.1$, the particle which starts from $r = r_b$ at time $t = t_b > 28.2843$ will never reach the singularity as shown in Figure 17 (blue-line) using Equations (119) and (120). However, Figure 18 shows that when $c_8 = 30$, Equation (132) has a solution for t_s since the curves $y = f(t_s)$ and $y = g(t_s)$ intersect. For a value of c_8 equal to 30, the particle begins at $r = r_b$ at a time $t = t_b$ greater than 28.2843. It will eventually reach the singularity at $r = 0$ at time $t = t_s$, as depicted by the red-line in Figure 17. The particle will then depart from the location $r = 0$, which can be determined from Figure 17 using Equations (119) and (120).

Now, we recall Equations (106) and (107), and for $D > 2\sqrt{2}$ we have $r_{\pm} = \frac{\mu t \pm \sqrt{\mu^2 t^2 - 2L^2}}{2}$ and the corresponding $\Phi_{\pm} = \frac{4L}{\mu\sqrt{1+2L^2}} \tanh^{-1}\left[\frac{\pm\sqrt{\mu t - L\sqrt{2}} - L\sqrt{2}\sqrt{\mu t + L\sqrt{2}}}{\sqrt{2\mu t}\sqrt{1+2L^2}}\right]$. Therefore, two non-radial time-like geodesics can be traced for $D > 2\sqrt{2}$, as shown in Figures 19 and 20 for $L = 1$ and $\mu = 0.05$. Figure 19 shows that the orbit $r_{+} = \frac{\mu t + \sqrt{\mu^2 t^2 - 2L^2}}{2}$ starts from a finite distance $\frac{\mu t_a + \sqrt{\mu^2 t_a^2 - 2L^2}}{2}$ from the singularity $(r \to 0)$ at time $t = t_a > \frac{L\sqrt{2}}{\mu}$ and then departs to infinity since $r \to \infty$ for a large t, and Figure 20 shows that the orbit $r_{-} = \frac{\mu t - \sqrt{\mu^2 t^2 - 2L^2}}{2}$ starts from a finite distance $\frac{\mu t_a - \sqrt{\mu^2 t_a^2 - 2L^2}}{2}$ from the singularity at time $t = t_a > \frac{L\sqrt{2}}{\mu}$ and then gradually plunges to singularity. Assuming $t = t_a$ as our reference point, we observe from Figure 19 that the particle trajectory moves away from this reference point. Similarly, from Figure 20 we see that the particle trajectory moves towards $r \to 0$ from the reference point. This phenomenon again indicates the potential existence of a membrane-like wormhole or a quantum tunneling effect. It is also mentioned that the path of the particle trajectories of the above two scenarios (Figures 19 and 20) are quasicircular [76].

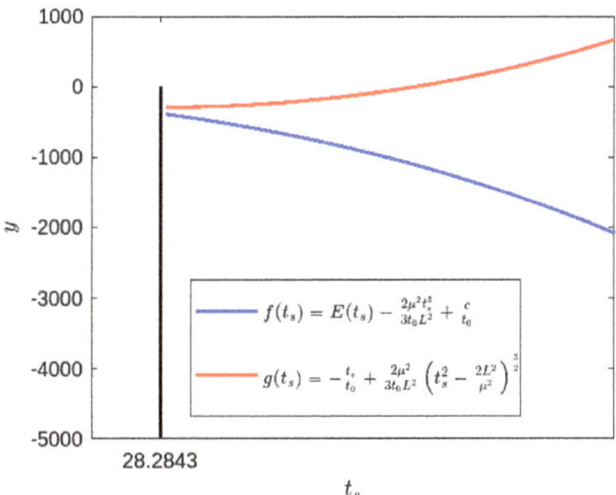

Figure 16. $\mu = 0.05$, $L = 1$, $t_0 = 0.1$, and $c_8 = 0.1$. Blue-line for $f(t_s)$ and red-line for $g(t_s)$.

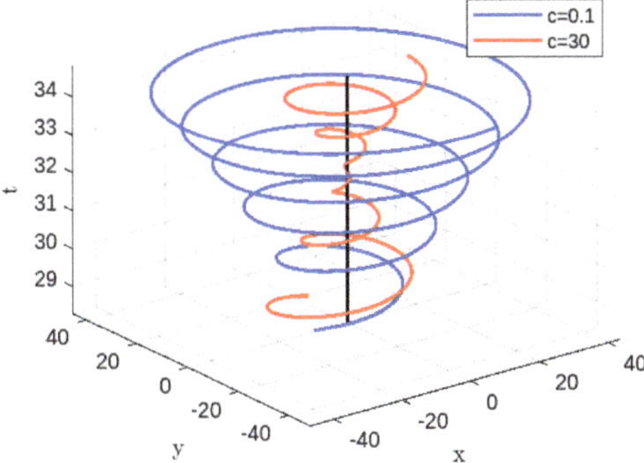

Figure 17. Time-like geodesics for $r - t_0 E(t) - \bar{B} - c_8 = 0$ when $\mu = 0.05$, $L = 1$, and $t_0 = 0.1$. Blue-line for $c_8 = 0.1$ and red-line for $c_8 = 30$.

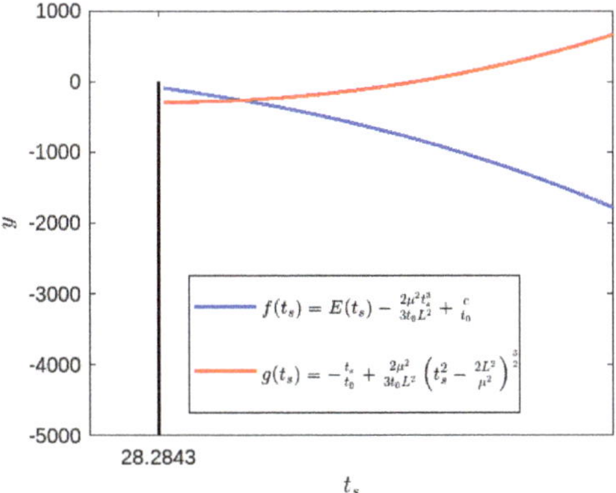

Figure 18. $\mu = 0.05$, $L = 1$, $t_0 = 0.1$, and $c_8 = 30$. Blue-line for $f(t_s)$ and red-line for $g(t_s)$.

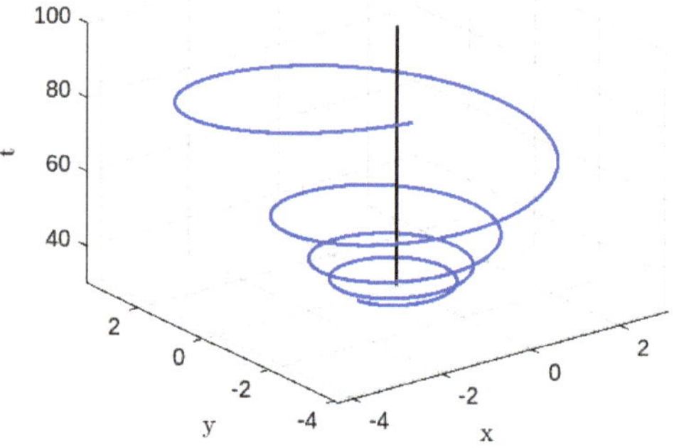

Figure 19. Time-like geodesics for $r_+ = \frac{\mu t + \sqrt{\mu^2 t^2 - 2L^2}}{2}$ for $\mu = 0.05$ and $L = 1$.

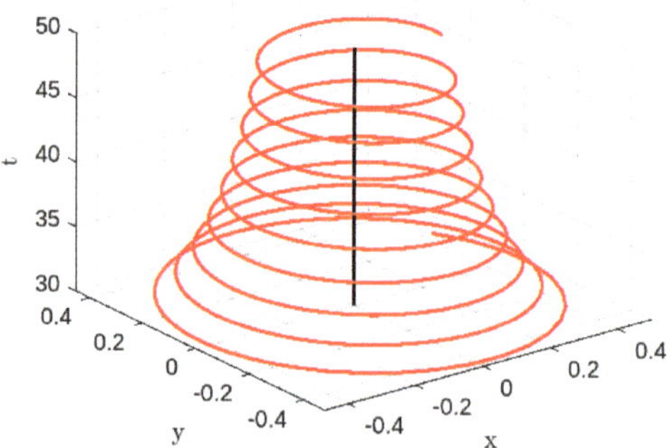

Figure 20. Time-like geodesics for $r_- = \frac{\mu t - \sqrt{\mu^2 t^2 - 2L^2}}{2}$ for $\mu = 0.05$ and $L = 1$.

3.2.2. Null Geodesics for Case-II

Within this part, we study the characteristics of both radial ($L = 0$) and non-radial ($L \neq 0$) null geodesics in the generalized K-essence Vaidya spacetime (13) with the specific mass function (99).

By tracing the radial null geodesic using the same reasoning as previously, we obtain the identical value of r_a as stated in Equation (108). In this case, at $t = t_a$, we also have $r_a < r_D^+ < r_D^{++}$. Continuing similarly as before, we have

$$r = t_0 E(t) + \frac{2\mu t_a}{1 - E(t_a)} - t_0 E(t_a).$$
(133)

Again, we are able to calculate the time $t = t_s$ for a particle approaching to singularity $r \to 0$; we obtain

$$t_s = t_0 \ln \left[\frac{1 - E(t_a)}{2 E(t_a)[1 - E(t_a)] - \frac{4\mu t_a}{t_0}} \right].$$
(134)

To trace the characteristics of non-radial ($L \neq 0$) null geodesics for the given spacetime (13) through Equation (35) we consider $D > 2 \implies t \geq \frac{L}{\mu}$ ($D = \frac{2\mu t}{L}$). Similarly, we exclude the situation when $D = 2$ due to the aforementioned rationale for non-radial time-like geodesics. Therefore, when $D > 2$, Equation (35) can be written as

$$\left(r - t_0 E(t) - \bar{A}(t) - c_9\right)\left(r - t_0 E(t) - \bar{B}(t) - c_9\right) = 0. \tag{135}$$

where

$$\bar{A}(t) = t - \frac{2\mu^2 t^3}{3L^2} + \frac{2\mu^2}{3L^2}\left(t^2 - \frac{L^2}{\mu^2}\right)^{\frac{3}{2}}$$

$$\implies \bar{A}(t) = \frac{2\mu^2}{3L^2}\mathcal{O}\left(\frac{1}{t}\right), \tag{136}$$

$$\text{and } \bar{B}(t) = t - \frac{2\mu^2 t^3}{3L^2} - \frac{2\mu^2}{3L^2}\left(t^2 - \frac{L^2}{\mu^2}\right)^{\frac{3}{2}}$$

$$\implies \bar{B}(t) = 2t - \frac{4\mu^2 t^3}{3L^2} - \frac{2\mu^2}{3L^2}\mathcal{O}\left(\frac{1}{t}\right). \tag{137}$$

Now, Equation (135) indicates the existence of two non-radial null geodesics that may be followed for any finite value of c_9. Proceeding as before, we have

$$r = t_0 E(t) + \frac{2\mu^2}{3L^2}\mathcal{O}\left(\frac{1}{t}\right) + c_9, \tag{138}$$

$$\text{with } \Phi = \frac{1}{L}\left[-\frac{2\mu^2}{3L^2}\mathcal{O}\left(\frac{1}{t}\right) - c_9\right], \tag{139}$$

$$\text{and } r = t_0 E(t) + 2t - \frac{4\mu^2 t^3}{3L^2} - \frac{2\mu^2}{3L^2}\mathcal{O}\left(\frac{1}{t}\right) + c_9, \tag{140}$$

$$\text{with } \Phi = \frac{1}{L}\left[-2t + \frac{4\mu^2 t^3}{3L^2} + \frac{2\mu^2}{3L^2}\mathcal{O}\left(\frac{1}{t}\right) - c\right]. \tag{141}$$

Again, we consider a particle that starts its journey from $r = r_a$ in path (138) at time $t = t_a > \frac{L}{\mu}$; we obtain

$$c_9 = r_a - t_0 E(t_a) - \bar{A}(t_a). \tag{142}$$

For this scenario, the condition imposed on the particle trajectory is as follows:

$$c_9 \leq r_a - \bar{A}(t_a) < \frac{t_0}{2} + c_9, \tag{143}$$

since $\frac{t_a}{t_0} \geq 0$ and $0 \leq E(t_a) < \frac{1}{2}$.

Hence, Equations (138) and (139) become

$$r - t_0 E(t) - \bar{A}(t) - r_a + t_0 E(t_a) + \bar{A}(t_a) = 0, \tag{144}$$

$$\Phi = \frac{1}{L}\left[-\bar{A}(t) - r_a + t_0 E(t_a) + \bar{A}(t_a)\right], \tag{145}$$

then, at $t = t_a$

$$\Phi = \Phi_a = \frac{1}{L}\left[-r_a + t_0 E(t_a)\right]. \tag{146}$$

If it is achievable, let us assume that the particle travels along path (138) and reaches the singularity $(r = 0)$ at time $t = t_s$. Then, from Equation (144) we obtain

$$E(t_s) - \frac{2\mu^2 t_s^3}{3t_0 L^2} + \frac{c}{t_0} = -\frac{t_s}{t_0} - \frac{2\mu^2}{3t_0 L^2}\left(t_s^2 - \frac{L^2}{\mu^2}\right)^{\frac{3}{2}}. \tag{147}$$

Again, this is also a transcendental equation and it can be analyzed numerically and graphically for any finite value of μ and L. As before, let $f(t_s) = E(t_s) - \frac{2\mu^2 t_s^3}{3t_0 L^2} + \frac{c_9}{t_0}$ and $g(t_s) = -\frac{t_s}{t_0} - \frac{2\mu^2}{3t_0 L^2}\left(t_s^2 - \frac{L^2}{\mu^2}\right)^{\frac{3}{2}}$. Then, the only possibility for a finite value of t_s is if and only if $f(t_s) = g(t_s)$. To start with let us consider $\mu = 0.01$ and $L = 3$, then $t_a > 300$. Let $t_a = 301$, then from (136), $\bar{A}(t_a) \approx 99.1025$. Now, if we consider $t_0 = 0.1$, then for $c_9 = 0.1$, $r_a \approx 99.2025$ and $\Phi_a \approx -33.0675$, and also for $c_9 = 20$, $r_a \approx 119.1025$ and $\Phi_a \approx -39.7008$ by using (142) and (144). Again, Figures 21 and 22 show that for $c_9 = 0.1$ and $c_9 = 20$, Equation (147) does not have a solution for finite t_s since the curves $y = f(t_s)$ and $y = g(t_s)$ do not intersect each other. From Equations (138) and (139), we obtain $r = t_0 E(t) + \frac{2\mu^2}{3L^2}\mathcal{O}\left(\frac{1}{t}\right) + c_9$ and $\Phi = \frac{1}{L}\left[-\frac{2\mu^2}{3L^2}\mathcal{O}\left(\frac{1}{t}\right) - c_9\right]$, so that for large t, we have $r \to c_9$ and $\Phi \to -\frac{c_9}{L}$. So there is no way to trace any null geodesics for non-zero values of c_9 like $c = 0.1$ and $c = 20$, as shown in Figure 23 using Equations (144) and (145). But if we take $c_9 = 0$, a particle starts its journey from $r_a \approx 99.1025$ and $\Phi_a \approx -33.0342$, it finally plunges to the singularity, as shown in Figure 24.

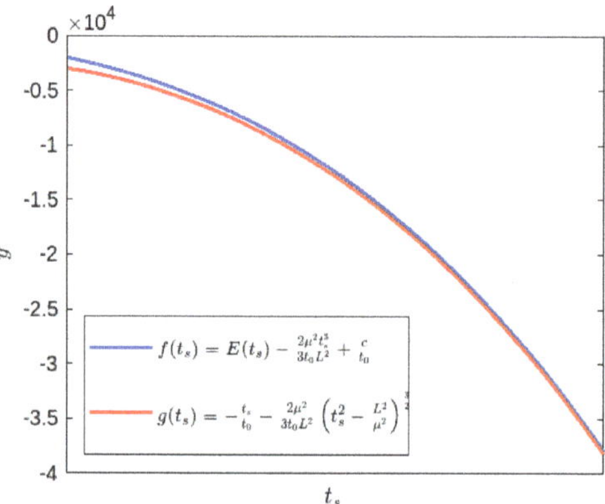

Figure 21. $\mu = 0.01$, $L = 3$, $t_0 = 0.1$, and $c_9 = 0.1$. Blue-line for $f(t_s)$ and red-line for $g(t_s)$.

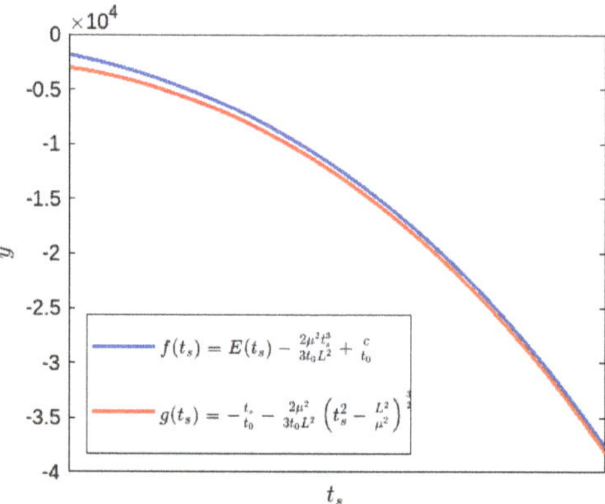

Figure 22. $\mu = 0.01$, $L = 3$, $t_0 = 0.1$, and $c_9 = 20$. Blue-line for $f(t_s)$ and red-line for $g(t_s)$.

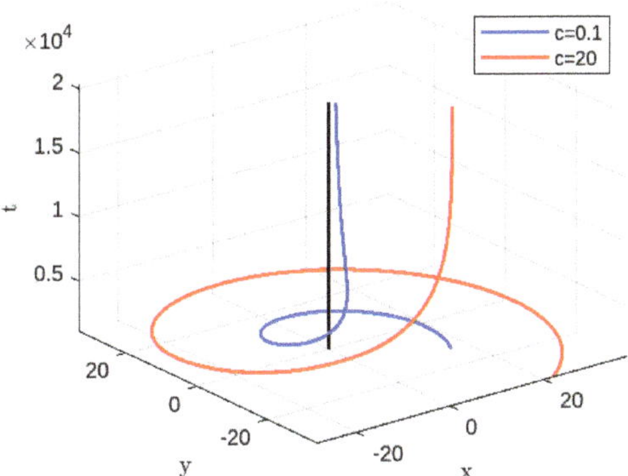

Figure 23. Null geodesics for $r - t_0 E(t) - \bar{A} - c_9 = 0$ when $\mu = 0.01$, $L = 3$, and $t_0 = 0.1$. Blue-line for $c_9 = 0.1$ and red-line for $c_9 = 20$.

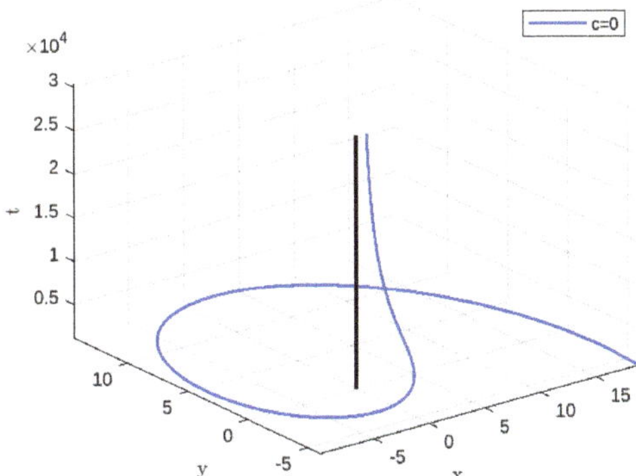

Figure 24. Null geodesics for $r - t_0 E(t) - \bar{A} - c_9 = 0$ when $\mu = 0.01$, $L = 3$, and $t_0 = 0.1$. Blue-line for $c_9 = 0$.

Similar to the previous example, if we look at a particle that begins its journey from $r = r_b$ in the path described in Equation (140) at time $t = t_b > \frac{L}{\mu}$, we obtain the following relations with the same type of meaning from Equations (140), (141) and (137):

$$c_9 = r_b - t_0 E(t_b) - \bar{B}(t_b), \tag{148}$$

$$c_9 \leq r_b - \bar{B}(t_b) < \frac{t_0}{2} + c_9, \tag{149}$$

$$r - t_0 E(t) - \bar{B}(t) - r_b + t_0 E(t_b) + \bar{B}(t_b) = 0, \tag{150}$$

$$\Phi = \frac{1}{L}\left[-\bar{B}(t) - r_b + t_0 E(t_b) + \bar{B}(t_b) \right], \tag{151}$$

$$\Phi = \Phi_b = \frac{1}{L}\left[-r_b + t_0 E(t_b) \right], \tag{152}$$

$$E(t_s) - \frac{2\mu^2 t_s^3}{3 t_0 L^2} + \frac{c_9}{t_0} = -\frac{t_s}{t_0} + \frac{2\mu^2}{3 t_0 L^2}\left(t_s^2 - \frac{L^2}{\mu^2} \right)^{\frac{3}{2}}. \tag{153}$$

Again, as before, Equation (153) is a transcendental equation, so it can be analyzed numerically and graphically for any finite value of μ and L. Now, let $f(t_s) = E(t_s) - \frac{2\mu^2 t_s^3}{3 t_0 L^2} + \frac{c_9}{t_0}$ and $g(t_s) = -\frac{t_s}{t_0} + \frac{2\mu^2}{3 t_0 L^2}\left(t_s^2 - \frac{L^2}{\mu^2} \right)^{\frac{3}{2}}$. Then, the only possibility for a finite value of t_s is if and only if $f(t_s) = g(t_s)$.

To start with let us consider $\mu = 0.05$ and $L = 1$, then $t_b > 20$. Let $t_b = 21$, then from (137), $\bar{B}(t_a) \approx 5.1275$. Now, if we consider $t_0 = 0.1$, then for $c_9 = 0.1$, $r_b \approx 5.2275$ and $\Phi_b \approx -5.2275$, and also for $c_9 = 20$, $r_b \approx 25.1275$ and $\Phi_b \approx -25.1275$ by using (140) and (141). Figure 25 shows that for $c_9 = 0.1$, Equation (153) has a solution for t_s since the curves $y = f(t_s)$ and $y = g(t_s)$ intersect each other. Also, Figure 26 shows that for $c_9 = 20$, Equation (153) has a solution for t_s since the curves $y = f(t_s)$ and $y = g(t_s)$ intersect each other. So, for $c_9 = 0.1$ and $c_9 = 20$, the particle which starts from $r = r_b$ at time $t = t_b > 20$ will reach the singularity ($r = 0$) at $t = ts$, as shown in Figure 27 using Equations (140) and (141).

Lastly, from Equations (106) and (107), for $D > 2$, we have $r_\pm = \mu t \pm \sqrt{\mu^2 t^2 - L^2}$ and the corresponding $\Phi_\pm = \frac{L\sqrt{2}}{\mu\sqrt{1+L^2}} \tanh^{-1}\left[\frac{\pm\sqrt{\mu t - L} - L\sqrt{\mu t + L}}{\sqrt{2\mu t}\sqrt{1+L^2}}\right]$. Therefore, two non-radial null geodesics can be traced for $D > 2$, as shown in Figures 28 and 29 for $L = 1$ and $\mu = 0.05$. Figure 28 shows that the orbit $r_+ = \mu t + \sqrt{\mu^2 t^2 - L^2}$ starts from a finite distance $\mu t_a + \sqrt{\mu^2 t_a^2 - L^2}$ from the singularity at time $t = t_a > \frac{L}{\mu}$ and then departs to infinity, since $r \to \infty$ for large t, and Figure 29 shows that the orbit $r_- = \mu t - \sqrt{\mu^2 t^2 - L^2}$ starts from a finite distance $\mu t_a - \sqrt{\mu^2 t_a^2 - L^2}$ from the singularity at time $t = t_a > \frac{L}{\mu}$ and then gradually plunges to singularity. By analyzing Figures 28 and 29, we can identify a recurrence of the previously observed phenomena, namely, the indication of a probable possibility of a membrane-like wormhole or a quantum tunneling effect.

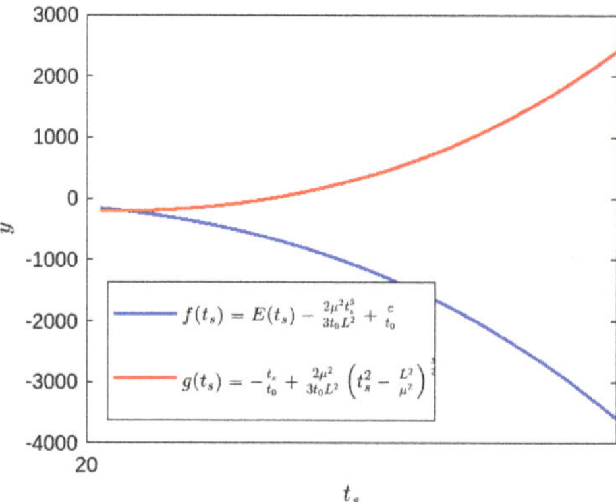

Figure 25. $\mu = 0.05$, $L = 1$, $t_0 = 0.1$, and $c_9 = 0.1$. Blue-line for $f(t_s)$ and red-line for $g(t_s)$.

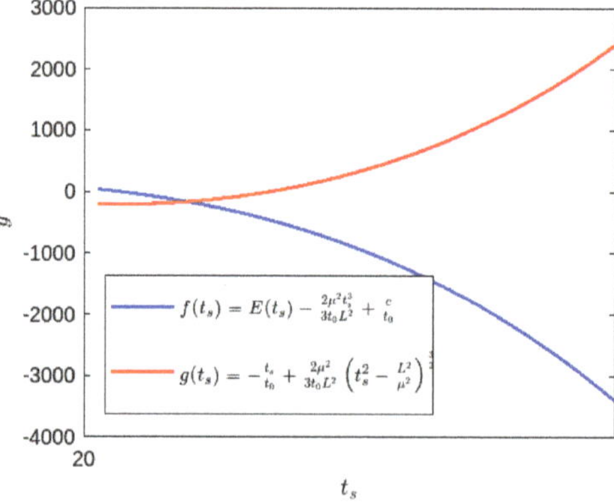

Figure 26. $\mu = 0.05$, $L = 1$, $t_0 = 0.1$, and $c_9 = 20$. Blue-line for $f(t_s)$ and red-line for $g(t_s)$.

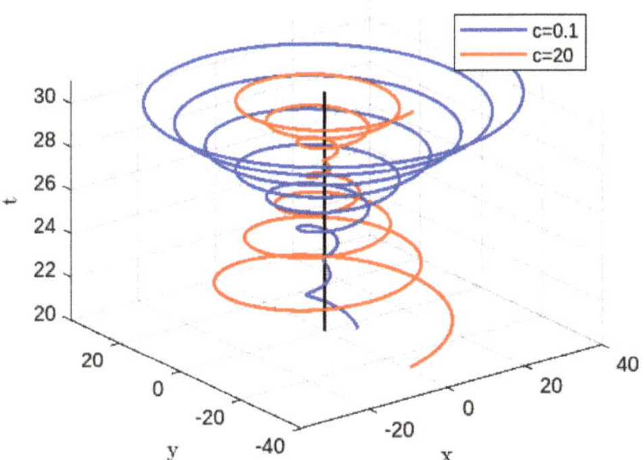

Figure 27. Null geodesics for $r - t_0 E(t) - \bar{B} - c_9 = 0$ when $\mu = 0.05$, $L = 1$, $t_0 = 0.1$. Blue-line for $c_9 = 0.1$ and red-line for $c_9 = 20$.

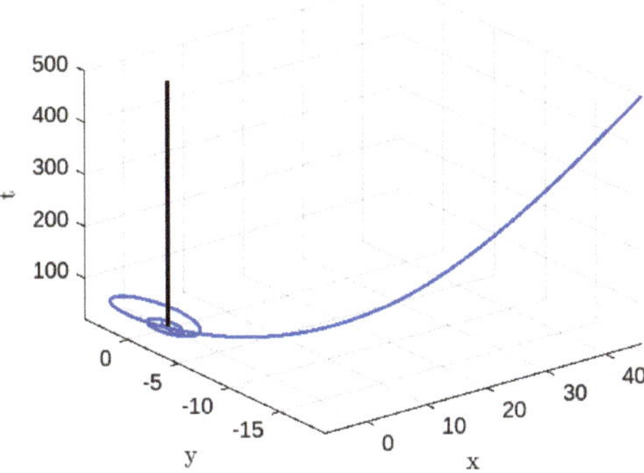

Figure 28. $\mu = 0.05$ and $L = 1$ for $r_+ = \mu t + \sqrt{\mu^2 t^2 - L^2}$.

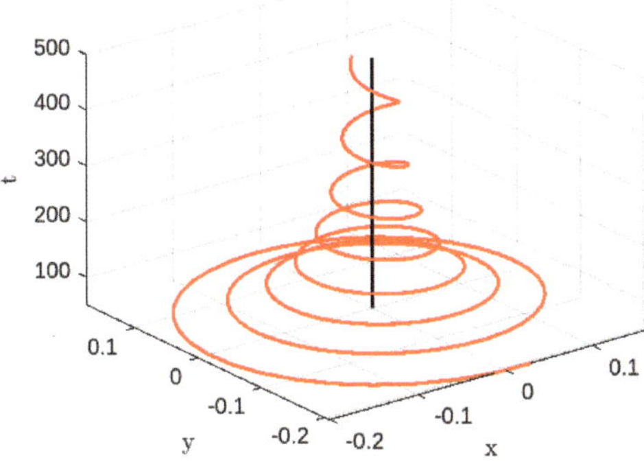

Figure 29. $\mu = 0.05$ and $L = 1$ for $r_- = \mu t - \sqrt{\mu^2 t^2 - L^2}$.

4. Discussion and Conclusions

In the present investigation, we have traced time-like geodesics and null geodesics for two different K-essence Vaidya mass functions $\mathcal{M}(t,r) = M + \frac{r}{2}e^{-\frac{t}{t_0}}$ and $\mathcal{M}(t,r) = \mu t + \frac{r}{2}e^{-\frac{t}{t_0}}$ on equatorial plane $\theta = \frac{\pi}{2}$ by using Euler–Lagrange equations. For both the cases, we observe that $\mathcal{M}(t,r)$ can be written in terms of the energy $E(t)$. We have seen that for the radial time-like and radial null geodesics for both the K-essence Vaidya masses, a particle having zero angular momentum (L) starting from a finite distance takes a finite time to reach the singularity $(r \to 0)$ but not in the central singularity, where both the time and space approach to zero, i.e., $r \to 0$ and $t \to 0$. For the first K-essence Vaidya mass function, there are two types of families for the time-like case that can be traced, one for $M = \sqrt{2}L$ and the other for $M > \sqrt{2}L$, and similarly for the null case one for $M = L$ and the other for $M > L$. But a bit of difference was observed for the second K-essence Vaidya mass function. There exists only one type of family for the time-like case when $\mu t > \sqrt{2}L$ and for the null case $\mu t > L$. For all the above-mentioned cases, for different finite value of arbitrary constants c, we have traced different kinds of orbits for both the generalized K-essence Vaidya mass function, and the conclusions are as follows.

Based on Figure 1, it may be inferred that a future observer will observe the particle approaching to the singularity within a finite amount of time, as previously stated. According to Figures 2–5 it is evident that a future observer can watch the particle reaching the singularity within a finite period and thereafter departing from it. Based on the information provided in Figures 1–5, for non-radial time-like geodesics of the generalized K-essence Vaidya spacetime (13) with the first mass function (17) or (27), it can be inferred that there is a worry with the presence of a central singularity (where both r and t approach to 0). It is evident that the central singularity in the generalized Vaidya spacetime is defined by the limits $r \to 0$ and $t \to 0$, as shown in [19,30,32,33,38]. However, based on our analysis of the given data presented in Figures 1–5, we may conclude that when r approaches zero, t does not tend to zero. This phenomenon is also observed in the non-radial null geodesics with the same mass function but with varying finite time for different situations, as depicted in Figures 7–11. These phenomena can be explained as follows: a future observer is expected to see the particle approaching the singularity and then moving away from it for a limited period of time, (53), (64), or (69) (non-radial time-like geodesic) and (82), (97), or (98) (non-radial null geodesic) in different situations. This observation could potentially indicate the presence of a wormhole during the extreme stages of spacetime, specifically black holes and white holes, similar to the concept of the Einstein–Rosen bridge [79–89].

Regarding particles, it is clear that their signature is changed by the singularity. In this interesting scenario, it seems that the presence of a wormhole allows for the possibility of gravitational collapse resulting in either a naked singularity or a black hole and a white hole in the generalized *K*-essence Vaidya spacetime with the given mass function (27). Based on the analysis of collapsing scenarios in the generalized Vaidya spacetime, it has been established that there can be either a naked singularity or a black hole [30,38]. This is supported by the potential presence of a dynamical horizon, as discussed in [90–93]. On the other hand, Vertogradov [35] has also addressed the issue and showed the presence of naked singularities as well as white holes within the framework of the usual generalized Vaidya spacetime in a different setting. However, in our case, the results are a bit different as we have already mentioned that there is the possibility of either a naked singularity or a black hole and a white hole. If we look at all the aforementioned figures with a closer view, then it actually reveals the presence of a wormhole situation also in between the black hole and white hole scenarios. But here, the conditions being extremal ($r \to 0$ and $t \neq 0$), we have a membrane-type wormhole instead of a tunnel as is usually suggested. Alternatively, the previous explanation can also be explained as follows: Given that the particle reaches the point at $r = 0$ and then escapes from there within a certain period, it may be possible that a quantum tunneling event occurs at the vicinity of the central singularity. Since there is a finite time at the typical singular region at $r = 0$, it follows that there is likewise a finite probability of escaping the region at $r = 0$. We also have finite times (45) and (75) to approach the singularity for a particle in radial time-like and null geodesic situations. In these instances, it can be possible to compute the probability amplitude for a time-dependent spacetime, such as the generalized K-essence Vaidya geometry (13). It is important to note that Ellis's recent study [94] has demonstrated that quantum behavior may occur in time-dependent spacetimes, such as the Vaidya geometry, which is based on Hawking's recent work [95]. However, it is to be noted that the quantum behavior of generalized K-essence Vaidya geometry cannot be discussed under the present purview.

On the other hand, we have the characteristics of the radial and non-radial time-like or null geodesics structures for the case of the second mass function (99). Through an analysis of the radial properties of the time-like and null geodesics, we have determined that the particle ultimately approaches the singularity at $r = 0$ within a limited amount of time, as shown by Equations (112) and (134), respectively. To investigate the non-radial geodesics of the specified spacetime (13) with the mass function (99), we have not found a finite expression for time for a particle to reach the singularity at $r = 0$. However, we have obtained a transcendental equation for different possibilities, namely, Equation (126) or (131) for time-like geodesics and Equation (147) or (153) for null geodesics. But from these transcendental equations, we have analyzed the non-radial time-like or null geodesics through numerical and graphical analyses. Based on Figures 13–17, our investigation of non-radial Time-like geodesics reveals that when $c_8 = 0.1$ the particle is unable to reach the singularity at $r = 0$ (blue-line). This phenomenon may be explained as follows: The presence of a black hole singularity with an event horizon might result in the spectator being unable to perceive the particle's journey towards the singularity. Based on the aforementioned figures (red-line), it has been observed that when the constant c_8 is set to 30, the particle approaches to the singularity at $r = 0$ after a finite amount of time and subsequently moves away from it. Here, we also see the occurrence of a wormhole-like physical phenomenon or quantum tunneling effect near the singularity at $r = 0$.

Regarding the non-radial null geodesic characteristics of the aforementioned spacetime and mass function (99), we note that the transcendental Equations (147) and (153) exhibit the same kind of solutions as Equation (135). In this situation also, we are as yet unable to determine the expression for the time it takes to reach the singularity at $r = 0$. After solving the transcendental equations using numerical analysis, we find the graphical solutions through Figures 21–27. From Figure 23, we have not found any tracing of the non-radial null geodesic structure for two choices of the constants, viz., $c_9 = 0.1$ and $c_9 = 20$, though it is available for the choice of $c_9 = 0$; see Figures 24 and 25. No traces of the non-radial null geodesic structure were noticed in Figure 23 for two different values of the constants, namely, $c_9 = 0.1$ and $c_9 = 20$. However, by setting $c_9 = 0$, as seen in Figure 24, it is evident that the particle originates from a specific location and ultimately drops into the singularity. Alternatively, we have explored an additional solution to Equation (135) and its associated Equation (153). We have analyzed the non-radial null geodesic structure and depicted it in Figure 27. In this scenario, we note that for two specific values of the constants, namely, $c_9 = 0.1$ and $c_9 = 20$, the particle begins its trip from a fixed position and eventually reaches the singularity within a limited amount of time. After reaching the singularity, the particle then escapes from it. This once again indicates the presence of a wormhole or the quantum tunneling effect.

We have found that instead of the dynamical horizon, the circular orbits via the event horizons (38) provide alternative solutions for Equations (32) and (33) in the case of the first mass function (27). The corresponding graphical representations of the circular orbits can be observed in Figures 3, 6, 9 and 12. These phenomena can be supported by the study conducted by Ishihara et al. [76] on a time-dependent spacetime in a different setting. On the other hand, for the second mass function (99), Equations (103) and (104) have alternative solutions, which are (106) and (107). By observing solution (106), it is evident that the values of $r_\pm$ vary with time, a characteristic that is inherent in all forms of generalized Vaidya spacetime. This suggests the possible presence of a dynamical horizon rather than an event horizon, which is seen in the case of the generalized K-essence Vaidya spacetime. Based on solutions (106) and (107), we obtain four plots: Figures 19, 20, 28 and 29. Based on these plots, we observe evidence of the possible presence of a membrane-like wormhole or a quantum tunneling phenomenon, as well as hints of a quasicircular orbit along the particle trajectory.

Therefore, under the aforementioned discussion, we can conclude that in our specific model, we have investigated the existence of a wormhole during the extreme stages of spacetime, specifically black holes and white holes, similar to the concept of the Einstein–Rosen bridge or quantum tunneling effect near the central singularity. This investigation was carried out in the comoving frame for the generalized K-essence Vaidya spacetime using two types of K-essence Vaidya mass functions. It is possible to argue that extra interactions between the K-essence scalar field and the usual gravity are the cause of this occurrence. Furthermore, it should be noted that in our specific scenario, there is no presence of a central singularity, i.e., both r and t approach to zero, which is a need for any form of generalized Vaidya geometry. Usually, the singularity means $r \to 0$, at which point geodesics are incomplete. It might be argued that the presence of singularity is uncertain, as has been shown very recently by Kerr [96], where he demonstrates the lack of evidence supporting the existence of singularities within black holes formed by real physical spacetime. In a nutshell, we may state that a physical spacetime does not exhibit geodesic incompleteness. Geodesic completeness relates to the concept that these trajectories should continue endlessly in time inside a well-behaved and physically plausible spacetime. Penrose [97] and Hawking [98] demonstrated that geodesics can be incomplete under specific circumstances, indicating that they may only persist for a finite time before meeting a singularity, a location where the curvature of spacetime becomes indefinitely large. In this context, we may state that in the present investigation our spacetime (13) is geodesically complete under certain perspectives.

At this juncture, we would like to point out that the wormhole geometries given by the present solutions may or may not be traversable. This is because, though K-essence inherently admits dark energy, we have not directly included any exotic component that is essential for forming a wormhole traversable via its throat [99–105]. But here, we have not used the K-essence model as a dark energy model, rather we have adapted it to a purely gravitational perspective [37,38]. However, in the case of possible quantum tunneling, there is no need of such a type of traversable structure where the quantum field theoretic effect becomes predominant.

Moreover, based on the investigation of the present work, especially the observations in Figures 2–5, one may raise the issue of black holes and baby universes [106], and whether there is any possibility of such situations under the generalized K-essence Vaidya spacetime. This obviously draws special attention and can be considered as a potential topic in the near future.

Author Contributions: Conceptualization and writing—original draft preparation, G.M.; writing—original draft preparation, B.M.; writing—review and editing, M.K. and S.R. All authors have read and agreed to the published version of the manuscript.

Funding: G.M. acknowledges the DSTB, Government of West Bengal, India for financial support through Grant Nos. 322(Sanc.)/ST/P/S&T/16G-3/2018 dated 6 March 2019. The research by M.K. was carried out at Southern Federal University with financial support from the Ministry of Science and Higher Education of the Russian Federation (State Contract GZ0110/23-10-IF).

Data Availability Statement: There are no assoiated data with this artile as such no new data were generated or analyzed in support of this research.

Acknowledgments: S.R. is thankful to the Inter-University Centre for Astronomy and Astrophysics (IUCAA), Pune, India, for providing the Visiting Associateship under which a part of this work was carried out. S.R. also acknowledges the facilities under ICARD at CCASS, GLA University, Mathura, India. Additionally, the authors express their gratitude to the referees for providing insightful comments to enhance the quality of the work. Also, all the authors are grateful to P. Panchadhyayee of P.K. College, Contai, West Bengal, India, for fruitful discussions.

Conflicts of Interest: The authors declare no conflict of interest.

Appendix A. Radius of the Dynamical Horizon

To start with, let us consider

$$1 - \frac{2\mathcal{M}(t,r)}{r} = F(t,r) \tag{A1}$$

and then generalized K-essence Vaidya spacetime (13) can be expressed as

$$dS^2 = -F(t,r)dt^2 + 2dtdr + r^2 d\Omega^2. \tag{A2}$$

Setting $t = t' + r^*$, where r^* is a tortoise coordinate with $dr^* = \frac{dr}{F(t,r)}$, i.e.,

$$r^* = K(t,r) \tag{A3}$$

where $K_r(t,r) = \frac{1}{F(t,r)}$, which implies $K(t,r) = \frac{1}{N(t)}\left[r + \frac{2m(t)}{N(t)} \ln\left(r - \frac{2m(t)}{N(t)}\right)\right]$. Here we have used $\mathcal{M}(t,r) = m(t) + \frac{r}{2}\phi_t^2$, where $m(t,r) = m(t)$ and $N(t) = 1 - \phi_t^2$. Following [39,40], the null vectors can be written as $l^\alpha = \begin{bmatrix} -a^{-1} \\ a^{-1} \\ 0 \\ 0 \end{bmatrix}$ and $n^\alpha = \begin{bmatrix} -a^{-1} \\ -\frac{F}{F-2a}a^{-1} \\ 0 \\ 0 \end{bmatrix}$ and the corresponding expansions of the two null vectors l^a and n^a, $\Theta_{(l)}$ and $\Theta_{(n)}$ [39,40], are

$$\Theta_{(l)} = \frac{1}{r}(2F - a) \tag{A4}$$

and

$$\Theta_{(n)} = \frac{1}{ar}\left(\frac{-2F^2 + aF - 2a^2}{-F + 2a}\right) \tag{A5}$$

where $a = \frac{dr}{dr^*}$.

One can note that

$$\Theta_l = 0 \implies 2F(t,r) - a = 0, \tag{A6}$$

and the other null expansion $\Theta_{(n)}$ is strictly negative, which are the required conditions for the dynamical horizon [40,90–92].

So, we can proceed by using the dynamical horizon equation.

Now, from Equation (A3), we have

$$a = \frac{1 - K_t(t,r)}{K_r(t,r)}. \tag{A7}$$

Therefore, the dynamical horizon radii are $r_D^+ = \frac{2m(t)}{N(t)}$ and $r_D^{++} = \frac{2m(t)}{N(t)}\left[1 + W_0\left(\frac{N(t)}{2em(t)}e^{-\frac{t}{2m(t)}}\right)\right]$, where $W_0(q)$ is the solution of the equation $xe^x = q$ when q is a non-negative real number. It is also noted that the Lambert W_0 function [77] is a single-valued function.

Detail Derivation

$$F(t,r) = 1 - \frac{2\mathcal{M}(t,r)}{r} \implies F(t,r) = 1 - \frac{2m(t)}{r} - \phi_t^2,$$

$$\text{since } \mathcal{M}(t,r) = m(t) + \frac{r}{2}\phi_t^2, \text{ where } m(t,r) = m(t)$$

$$\implies F(t,r) = N(t)\left[1 - \frac{B(t)}{r}\right] \text{ where we consider } N(t) = 1 - \phi_t^2 \text{ and } B(t) = \frac{2m(t)}{1-\phi_t^2}, \tag{A8}$$

$$K_r(t,r) = \frac{1}{F(t,r)} = \frac{1}{N(t)}\left[\frac{r}{r - B(t)}\right] = \frac{1}{N(t)}\left[1 + \frac{B(t)}{r - B(t)}\right]$$

$$\implies K(t,r) = \frac{1}{N(t)}\left[r + B(t)\ln\left(r - B(t)\right)\right], \tag{A9}$$

$$r^* = K(t,r) \implies 1 = \frac{dK(t,r)}{dr^*}$$

$$\implies 1 = K_t(t,r)\frac{dt}{dr^*} + K_r(t,r)\frac{dr}{dr^*} = K_t(t,r) + aK_r(t,r)$$

$$\implies a = \frac{1 - K_t(t,r)}{K_r(t,r)} = F(t,r)\left[1 - K_t(t,r)\right]. \tag{A10}$$

On the other hand, at the dynamical horizon ($r = r_D$), we can write

$$2F(t,r_D) - a = 0 \implies 2F(t,r_D) = F(t,r_D)\left[1 - K_t(t,r_D)\right]$$

$$\implies F(t,r_D)\left[1 + K_t(t,r_D)\right] = 0. \tag{A11}$$

So that the above Equation (A11) has two solutions; one is

$$F(t,r_D) = 0 = N(t)\left[1 - \frac{B(t)}{r_D}\right] = 0$$

$$\implies r_D = \frac{2m(t)}{N(t)} \equiv r_D^+ \ (Say), \quad \text{since } N(t) \neq 0 \tag{A12}$$

whereas the other one is

$$1 + K_t(t,r_D) = 0 \implies K(t,r_D) = -t$$

$$\implies \frac{1}{N(t)}\left[r_D + B(t)\ln\left(r_D - B(t)\right)\right] = -t$$

$$\implies r_D + B(t)\ln\left(r_D - B(t)\right) = -tN(t)$$

$$\implies e^{r_D}\left(r_D - B(t)\right)^{B(t)} = e^{-tN(t)} \implies e^{\frac{r_D}{B(t)}}\left(r_D - B(t)\right) = e^{-\frac{tN(t)}{B(t)}}$$

$$\implies B(t)e^{\frac{r_D}{B(t)}}\left(\frac{r_D}{B(t)} - 1\right) = e^{-\frac{tN(t)}{B(t)}} \implies e^{\left(\frac{r_D}{B(t)}-1\right)}\left(\frac{r_D}{B(t)} - 1\right) = \frac{1}{eB(t)}e^{-\frac{tN(t)}{B(t)}}$$

$$\implies \frac{r_D}{B(t)} - 1 = W_0\left[\frac{1}{eB(t)}e^{-\frac{tN(t)}{B(t)}}\right] \implies r_D = B(t) + B(t)W_0\left[\frac{1}{eB(t)}e^{-\frac{tN(t)}{B(t)}}\right]$$

$$\implies r_D = \frac{2m(t)}{N(t)}\left[1 + W_0\left(\frac{N(t)}{2em(t)}e^{-\frac{t}{2m(t)}}\right)\right] \equiv r_D^{++} \ (Say). \tag{A13}$$

It is important to mention that the procedure for deriving the dynamical horizon differs from our previous work [39], while the outcome is identical.

References

1. Chandrasekhar, S. *The Mathematical Theory of Black Holes*; Indian Edition 2010; Oxford University Press: Oxford, UK, 1992.
2. Cruz, N.; Olivares, M.; Saavedra, J.; Villanueva, J.R. The geodesic structure of the Schwarzschild anti-de Sitter black hole. *Class. Quantum Grav.* **2005**, *22*, 1167–1190. [CrossRef]
3. Berti, E. A Black-Hole Primer: Particles, Waves, Critical Phenomena and Superradiant Instabilities. *arXiv* **2014**, arXiv:1410.4481.
4. Gibbons, G.W. The Jacobi metric for timelike geodesics in static spacetimes. *Class. Quantum Grav.* **2016**, *33*, 025004. [CrossRef]
5. Chanda, S.; Gibbons, G.W.; Guha, P. Jacobi-Maupertuis-Eisenhart metric and geodesic flows. *J. Math. Phys.* **2017**, *58*, 032503. [CrossRef]
6. Eisenhart, L.P. Dynamical Trajectories and Geodesics. *Ann. Math. Second Ser.* **1928**, *30*, 591–606. [CrossRef]
7. Duval, C.; Burdet, G.; Künzle, H.P.; Perrin, M. Bargmann structures and Newton-Cartan theory. *Phy. Rev. D* **1985**, *31*, 1841. [CrossRef]
8. Majumder, B.; Manna, G.; Das, A. Time-like geodesic structure for the emergent Barriola–Vilenkin type spacetime. *Class. Quantum Grav.* **2020**, *37*, 115002. [CrossRef]
9. Barriola, M.; Vilenkin, A. Gravitational field of a global monopole. *Phys. Rev. Lett.* **1989**, *63*, 341. [CrossRef]
10. Gangopadhyay, D.; Manna, G. The Hawking temperature in the context of dark energy. *Europhys. Lett.* **2012**, *100*, 49001. [CrossRef]
11. Vaidya, P.C. The gravitational field of a radiating star. *Proc. Indian Acad. Sci. Sect. A* **1951**, *33*, 264. [CrossRef]
12. Joshi, P.S. Gravitational collapse: The story so far. *Pramana J. Phys.* **2000**, *55*, 529. [CrossRef]
13. Joshi, P.S. *Global Aspects in Gravitation and Cosmology*; Clarendon: Oxford, UK, 1993.
14. Joshi, P.S.; Dwivedi, I.H. Naked singularities in spherically symmetric inhomogeneous Tolman-Bondi dust cloud collapse. *Phys. Rev. D* **1993**, *47*, 5357. [CrossRef] [PubMed]
15. Joshi, P.S.; Dwivedi, I.H. The structure of naked singularity in self-similar gravitational collapse: II. *Commun. Math. Phys.* **1992**, *146*, 333. [CrossRef]
16. Joshi, P.S.; Singh, T.P. Role of initial data in the gravitational collapse of inhomogeneous dust. *Phys. Rev. D* **1995**, *51*, 6778. [CrossRef] [PubMed]
17. Oppenheimer, J.R.; Snyder, H. On Continued Gravitational Contraction. *Phys. Rev.* **1939**, *56*, 455. [CrossRef]
18. Malafarina, D. Classical Collapse to Black Holes and Quantum Bounces: A Review. *Universe* **2017**, *3*, 48. [CrossRef]
19. Dwivedi, I.H.; Joshi, P.S. On the nature of naked singularities in Vaidya spacetimes: II. *Class. Quantum Gravit.* **1989**, *6*, 1599. [CrossRef]
20. Papapetrou, A. *A Random Walk in Relativity and Cosmology*; Dadhich, N., Rao, J.K., Narlikar, J.V., Vishveshwara, C.V., Eds.; John Wiley and Sons: New York, NY, USA, 1985; p. 184.
21. Penrose, R. Gravitational Collapse: The Role of General Relativity. *Riv. Nuovo Cim.* **1969**, *1*, 252. Available online: https://ui.adsabs.harvard.edu/abs/1969NCimR...1..252P/abstract (accessed on 3 April 2023).
22. Penrose, R. "Golden Oldie": Gravitational Collapse: The Role of General Relativity. *Gen. Rel. Grav.* **2002**, *34*, 1141. [CrossRef]
23. Vertogradov, V. Gravitational collapse of Vaidya spacetime. *Int. J. Mod. Phys. Conf. Ser.* **2016**, *41*, 1660124. [CrossRef]
24. Vaidya, P.C. Nonstatic Solutions of Einstein's Field Equations for Spheres of Fluids Radiating Energy. *Phys. Rev.* **1951**, *83*, 1. [CrossRef]
25. Vaidya, P.C.; Pandya, I.M. Gravitational Field of a Radiating Spheroid. *Prog. Theor. Phys.* **1966**, *35*, 1. [CrossRef]
26. Vaidya, P.C. Nonstatic Analogs of Schwarzschild's Interior Solution in General Relativity. *Phys. Rev.* **1968**, *174*, 5. [CrossRef]
27. Vaidya, P.C. An Analytical Solution for Gravitational Collapse with Radiation. *APJ* **1966**, *144*, 943. [CrossRef]
28. Vaidya, P.C. The External Field of a Radiating Star in General Relativity. *Gen. Rel. Grav.* **1999**, *31*, 119. [CrossRef]
29. Griffiths, J.B.; Podolsky, J. *Exact Space-Times in Einstein's General Relativity*; Cambridge Monographs on Mathematical Physics; Cambridge University Press: Cambridge, UK, 2009.
30. Husain, V. Exact solutions for null fluid collapse. *Phys. Rev. D* **1996**, *53*, 4. [CrossRef] [PubMed]
31. Wang, A.; Wu, Y. LETTER: Generalized Vaidya Solutions. *Gen. Rel. Grav.* **1999**, *31*, 107. Available online: https://www.springer.com/journal/10714 (accessed on 3 April 2023). [CrossRef]
32. Mkenyeleye, M.D.; Goswami, R.; Maharaj, S.D. Collapsing spherical stars in f(R) gravity. *Phys. Rev. D* **2014**, *90*, 064034. [CrossRef]
33. Patil, K.D. Gravitational collapse in higher-dimensional charged-Vaidya space-time. *Pramana—J. Phys.* **2003**, *60*, 423. [CrossRef]
34. Coudray, A.; Nicolas, J.P. Geometry of Vaidya spacetimes. *Gen. Rel. Gravit.* **2021**, *53*, 73. [CrossRef]
35. Vertogradov, V. The structure of the generalized Vaidya space–time containing the eternal naked singularity. *Int. J. Mod. Phys. A* **2022**, *37*, 2250185. [CrossRef]
36. Solanki, J.; Perlick, V. Photon sphere and shadow of a time-dependent black hole described by a Vaidya metric. *Phys. Rev. D* **2022**, *105*, 064056. [CrossRef]
37. Manna, G.; Majumdar, P.; Majumder, B. k-essence emergent spacetime as a generalized Vaidya geometry. *Phys. Rev. D* **2020**, *101*, 124034. [CrossRef]
38. Manna, G. Gravitational collapse for the K-essence emergent Vaidya spacetime. *Eur. Phys. J. C* **2020**, *80*, 813. [CrossRef]
39. Majumder, B.; Ray, S.; Manna, G. Evaporation of Dynamical Horizon with the Hawking Temperature in the K-essence Emergent Vaidya Spacetime. *Fortschr. Phys.* **2023**, *71*, 2300133. [CrossRef]
40. Sawayama, S. Evaporating dynamical horizon with the Hawking effect in Vaidya spacetime. *Phys. Rev. D* **2006**, *73*, 064024. [CrossRef]

41. Armendariz-Picon, C.; Damour, T.; Mukhanov, V. k-Inflation. *Phys. Lett. B* **1999**, *458*, 209–218. [CrossRef]
42. Armendariz-Picon, C.; Mukhanov, V.; Steinhardt, P.J. Dynamical Solution to the Problem of a Small Cosmological Constant and Late-Time Cosmic Acceleration. *Phys. Rev. Lett.* **2000**, *85*, 4438. [CrossRef]
43. Armendariz-Picon, C.; Mukhanov, V.; Steinhardt, P.J. Essentials of k-essence. *Phys. Rev. D* **2001**, *63*, 103510. [CrossRef]
44. Visser, M.; Barcelo, C.; Liberati, S. Analogue Models of and for Gravity. *Gen. Rel. Grav.* **2002**, *34*, 1719. [CrossRef]
45. Scherrer, R.J. Purely Kinetic k Essence as Unified Dark Matter. *Phys. Rev. Lett.* **2004**, *93*, 011301. [CrossRef]
46. Chimento, L.P. Extended tachyon field, Chaplygin gas, and solvable k-essence cosmologies. *Phys. Rev. D* **2004**, *69*, 123517. [CrossRef]
47. Vikman, A. K-Essence: Cosmology, Causality and Emergent Geometry. Doctoral Dissertation, der Ludwig-Maximilians-Universitat Munchen, Munchen, Germany, 2007.
48. Babichev, E.; Mukhanov, V.; Vikman, A. k-Essence, superluminal propagation, causality and emergent geometry. *J. High Energy Phys.* **2008**, *2*, 101. [CrossRef]
49. Babichev, E.; Mukhanov, V.; Vikman, A. Looking beyond the horizon. Looking beyond the Horizon. *arXiv* **2007**, arXiv:0704.3301.
50. Chimento, L.P. Internal space structure generalization of the quintom cosmological scenario. *Phys. Rev. D* **2009**, *79*, 043502. [CrossRef]
51. Singh, T.; Chaubey, R.; Singh, A. k-essence cosmologies in Kantowski–Sachs and Bianchi space–times. *Can. J. Phys.* **2015**, *93*, 1319. [CrossRef]
52. Singh, A.; Raushan, R.; Chaubey, R.; Mandal, S.; Mishra, K.C. Lagrangian formulation and implications of barotropic fluid cosmologies. *Int. J. Geom. Meth. Mod. Phys.* **2022**, *19*, 2250107. [CrossRef]
53. Tian, S.X.; Zhu, Z.H. Early dark energy in k-essence. *Phys. Rev. D* **2021**, *103*, 043518. [CrossRef]
54. Myrzakulov, R.; Sebastiani, L. k-Essence Non-Minimally Coupled with Gauss–Bonnet Invariant for Inflation. *Symmetry* **2016**, *8*, 57. [CrossRef]
55. Sen, A.A.; Devi, N.C. Cosmology with non-minimally coupled k-field. *Gen. Relativ. Gravit.* **2010**, *42*, 821. [CrossRef]
56. Chatterjee, A.; Hussain, S.; Bhattacharya, K. Dynamical stability of the k-essence field interacting nonminimally with a perfect fluid. *Phys. Rev. D* **2021**, *104*, 103505. [CrossRef]
57. Velten, H.E.S.; Vom Marttens, R.F.; Zimdahl, W. Aspects of the cosmological "coincidence problem". *Eur. Phys. J. C* **2014**, *74*, 3160. [CrossRef]
58. Erickson, J.K.; Caldwell, R.R.; Steinhardt, P.J.; Armendariz-Picon, C.; Mukhanov, V. Measuring the Speed of Sound of Quintessence. *Phys. Rev. Lett.* **2002**, *88*, 121301. [CrossRef] [PubMed]
59. DeDeo, S.; Caldwell, R.R.; Steinhardt, P.J. Effects of the Sound Speed of Quintessence on the Microwave Background and Large Scale Structure. *Phys. Rev. D* **2003**, *67*, 103509. [CrossRef]
60. Bean, R.; Dore, O. Probing dark energy perturbations: The dark energy equation of state and speed of sound as measured by WMAP. *Phys. Rev. D* **2004**, *69*, 083503. [CrossRef]
61. Manna, G.; Gangopadhyay, D. The Hawking temperature in the context of dark energy for Reissner–Nordstrom and Kerr background. *Eur. Phys. J. C* **2014**, *74*, 2811. [CrossRef]
62. Manna, G.; Majumder, B. The Hawking temperature in the context of dark energy for Kerr–Newman and Kerr–Newman–AdS backgrounds. *Eur. Phys. J. C* **2019**, *79*, 553. [CrossRef]
63. Manna, G.; Majumder, B.; Das, A. Quantum corrections to the accretion onto a Schwarzschild black hole in the background of quintessence. *Eur. Phys. J. Plus* **2020**, *135*, 107. [CrossRef]
64. Manna, G.; Panda, A.; Karmakar, A.; Ray, S. $f(\bar{R}, L(X))$-gravity in the context of dark energy with power law expansion and energy conditions*. *Chin. Phys. C* **2023**, *47*, 025101. [CrossRef]
65. Ray, S.; Panda, A.; Majumder, B.; Islam, M.R.; Manna, G. Collapsing scenario for the k-essence emergent generalized Vaidya spacetime in the context of massive gravity's rainbow. *Chin. Phys. C* **2022**, *46*, 125103. [CrossRef]
66. Das, S.; Panda, A.; Manna, G.; Ray, S. Raychaudhuri Equation in K-essence Geometry: Conditional Singular and Non-Singular Cosmological Models. *Fortschr. Phys.* **2023**, *2023*, 2200193. [CrossRef]
67. Mukohyama, S.; Namba, R.; Watanabe, Y. Is the DBI scalar field as fragile as other k-essence fields? *Phys. Rev. D* **2016**, *94*, 023514. [CrossRef]
68. Born, M.; Infeld, L. Foundations of the new field theory. *Proc. R. Soc. Lond. A* **1934**, *144*, 425. [CrossRef]
69. Heisenberg, W. Production of Meson Showers. *Zeit. Phys.* **1939**, *113*, 61. [CrossRef]
70. Dirac, P.A.M. An extensible model of the electron. *Proc. R. Soc. Lond. A* **1962**, *268*, 57. [CrossRef]
71. Panda, A.; Manna, G.; Ray, S.; Khlopov, M.; Islam, M.R. Gravitational collapse in generalized K-essence emergent Vaidya spacetime via $f(\bar{R}, \bar{T})$ gravity. *arXiv* **2023**, arXiv:2308.13574.
72. Panda, A.; Das, S.; Manna, G.; Ray, S.; Islam, M.R.; Ranjit, C. $f(\bar{R}, \bar{T})$ gravity in a Non-canonical theory. *arXiv* **2022**, arXiv:2206.14808.
73. Ade, P.A.R.; Aghanim, N.; Arnaud, M.; Ashdown, M.; Aumont, J.; Baccigalupi, C.; Banday, A.J.; Barreiro, R.B.; Bartolo, N.; Battaner, E.; et al. [Planck Collaboration]. XIV. Dark energy and modified gravity. *Astron. Astrophys.* **2016**, *594*, A14. [CrossRef]
74. Aghanim, N.; Akrami, Y.; Ashdow, M.; Aumont, J.; Baccigalupi, C.; Ballardini, M.; Banday, A.J.; Barreiro, R.B.; Bartolo, N.; Basak, S.; et al. [Planck Collaboration]. VI. Cosmological parameters. *Astron. Astrophys.* **2020**, *641*, A6. [CrossRef]

75. Aghanim, N.; Akrami, Y.; Arroja, F.; Ashdow, M.; Aumon, J.; Baccigalupi, C.; Ballardini, M.; Banday, A.J.; Barreir, R.B.; Bartolo, N.; et al. [Planck Collaboration]. I. Overview and the cosmological legacy of Planck. *Astron. Astrophys.* **2020**, *641*, A1. [CrossRef]

76. Ishihara, H.; Kimura, M.; Matsuno, K. Charged black strings in a five-dimensional Kasner universe. *Phys. Rev. D* **2016**, *93*, 024037. [CrossRef]

77. Corless, R.M.; Gonnet, G.H.; Hare, D.E.G.; Jeffrey, D.J.; Knuth, D.E. On the LambertW function. *Adv. Comp. Math.* **1996**, *5*, 329. [CrossRef]

78. Blau, M. Lecture Notes on General Relativity. 8 October 2022. Available online: http://www.blau.itp.unibe.ch/GRLecturenotes.html (accessed on 3 April 2023).

79. Einstein, A.; Rosen, N. The Particle Problem in the General Theory of Relativity. *Phys. Rev.* **1935**, *48*, 73. [CrossRef]

80. Fuller, R.W.; Wheeler, J.A. Causality and Multiply Connected Space-Time. *Phys. Rev.* **1962**, *128*, 919. [CrossRef]

81. Ellis, H.G. Ether flow through a drainhole: A particle model in general relativity. *J. Math. Phys.* **1973**, *14*, 104. [CrossRef]

82. Morris, M.S.; Thorne, K.S. Wormholes in spacetime and their use for interstellar travel: A tool for teaching general relativity. *Am. J. Phys.* **1988**, *56*, 395. [CrossRef]

83. Visser, M. Traversable wormholes: Some simple examples. *Phys. Rev. D* **1989**, *39*, 3182. [CrossRef]

84. Hayward, S.A. Wormhole dynamics in spherical symmetry. *Phys. Rev. D* **2009**, *79*, 124001. [CrossRef]

85. Bronnikov, K.A.; Lipatova, L.N.; Novikov, I.D.; Shatskiy, A.A. Example of a stable wormhole in general relativity. *Gravit. Cosmol.* **2013**, *19*, 269. [CrossRef]

86. Garattini, R. Casimir wormholes. *Eur. Phys. J. C* **2019**, *79*, 951. [CrossRef]

87. Mishra, B.; Agrawal, A.S.; Tripathy, S.K.; Ray, S. Wormhole solutions in f(R) gravity. *Int. J. Mod. Phys. D* **2021**, *30*, 2150061. [CrossRef]

88. Mishra, B.; Agrawal, A.S.; Tripathy, S.K.; Ray, S. Traversable wormhole models in f(R) gravity. *Int. J. Mod. Phys. A* **2022**, *37*, 2250010. [CrossRef]

89. Mustafa, G.; Maurya, S.K.; Ray, S. On the Possibility of Generalized Wormhole Formation in the Galactic Halo Due to Dark Matter Using the Observational Data within the Matter Coupling Gravity Formalism. *Astrophys. J.* **2022**, *941*, 170. [CrossRef]

90. Ashtekar, A.; Krishnan, B. Dynamical horizons and their properties. *Phys. Rev. D* **2003**, *68*, 104030. [CrossRef]

91. Ashtekar, A.; Krishnan, B. Dynamical Horizons: Energy, Angular Momentum, Fluxes, and Balance Laws. *Phys. Rev. Lett.* **2002**, *89*, 261101. [CrossRef] [PubMed]

92. Ashtekar, A.; Krishnan, B. Isolated and Dynamical Horizons and Their Applications. *Living Rev. Relativ.* **2004**, *7*, 10. [CrossRef] [PubMed]

93. Hayward, S.A. General laws of black-hole dynamics. *Phys. Rev. D* **1994**, *49*, 6467. [CrossRef]

94. Firouzjaee, J.T.; Ellis, G.F.R. Particle creation rate for dynamical black holes. *Eur. Phys. J. C* **2016**, *76*, 620. [CrossRef]

95. Hawking, S.W.; Perry, M.J.; Strominger, A. Soft Hair on Black Holes. *Phys. Rev. Lett.* **2016**, *116*, 231301. [CrossRef] [PubMed]

96. Kerr, R. Do Black Holes have Singularities? *arXiv* **2023**, arXiv:2312.00841. [CrossRef].

97. Penrose, R. Gravitational Collapse and Space-Time Singularities. *Phys. Rev. Lett.* **1965**, *14*, 57. [CrossRef]

98. Hawking, S.W. Black holes in general relativity. *Commun. Math. Phys.* **1972**, *25*, 152–166. [CrossRef]

99. Hochberg, D.; Visser, M. Geometric structure of the generic static traversable wormhole throat. *Phys. Rev. D* **1997**, *56*, 4745. [CrossRef]

100. Krasnikov, S. Traversable wormhole. *Phys. Rev. D* **2000**, *62*, 084028. [CrossRef]

101. Matos, T.; Nunez, D. Rotating scalar field wormhole. *Class. Quantum Gravit.* **2006**, *23*, 4485. [CrossRef]

102. Rahaman, F.; Islam, S.; Kuhfittig, P.K.F.; Ray, S. Searching for higher-dimensional wormholes with noncommutative geometry. *Phys. Rev. D* **2012**, *86*, 106010. [CrossRef]

103. Rahaman, F.; Karmakar, S.; Karar, I.; Ray, S. Wormhole inspired by non-commutative geometry. *Phys. Lett. B* **2015**, *746*, 73. [CrossRef]

104. Chew, X.Y.; Kleihaus, B.; Kunz, J. Geometry of spinning Ellis wormholes. *Phys. Rev. D* **2016**, *94*, 104031. [CrossRef]

105. Mustafa, G.; Maurya, S.K.; Ray, S.; Javed, F. Construction of thin-shell around new wormhole solutions via solitonic quantum wave dark matter. *Ann. Phys.* **2014**, *460*, 169551. [CrossRef]

106. Hawking, S. *Black Holes and Baby Universes and Other Essays*; Bantam Dell Publishing Group: New York, NY, USA, 1993.

Article

MeV Dark Energy Emission from a De Sitter Universe

Yasmín B. Alcántara-Pérez [1], Miguel. A. García-Aspeitia [2,*], Humberto Martínez-Huerta [3] and Alberto Hernández-Almada [4]

1 Departamento de Física, DCI, Campus León, Universidad de Guanajuato, León 37150, Guanajuato, Mexico; yasbe21@yahoo.com.mx
2 Departamento de Física y Matemáticas, Universidad Iberoamericana, Prolongación Paseo de la Reforma 880, Mexico 01219, Mexico
3 Departamento de Física y Matemáticas, Universidad de Monterrey, Av. Morones Prieto 4500, San Pedro Garza García 66238, Nuevo León, Mexico; humberto.martinezhuerta@udem.edu
4 Facultad de Ingeniería, Universidad Autónoma de Querétaro, Centro Universitario Cerro de las Campanas, Santiago de Querétaro 76010, Mexico; ahalmada@uaq.mx
* Correspondence: angel.garcia@ibero.mx

Abstract: The evolution of a de Sitter Universe is the basis for both the accelerated Universe and the late-stationary Universe. So, how do we differentiate between both universes? In this paper, we state that it is not possible to design an experiment using luminous or angular distances to distinguish between the two cases because they are the same during the de Sitter phase. However, this equivalence allows us prediction of the signal of *a constant dark energy emission* with a signal peak around 29.5 MeV, in which, according to our astrophysical test of survival probability, the radiation must be *non-standard photons*. Remarkably, experiments by EGRET and COMPTEL have observed an excess of gamma photons in this predicted region, coming from a possible decay process of dark energy emission, which may constitute the smoking gun of a late-stationary Universe with the continuous creation of non-standard radiation, an alternative approach to understanding the current stages of the Universe's evolution.

Keywords: cosmological constant; cosmology

Citation: Alcántara-Pérez, Y.B.; García-Aspeitia, M.A.; Martínez-Huerta, H.; Hernández-Almada, A. MeV Dark Energy Emission from a De Sitter Universe. *Universe* **2023**, *9*, 513. https://doi.org/10.3390/universe 9120513

Academic Editors: Kazuharu Bamba and Jean-Pierre Gazeau

Received: 11 October 2023
Revised: 13 November 2023
Accepted: 28 November 2023
Published: 13 December 2023

1. Introduction

The late acceleration of the Universe—which was first observed by the teams dedicated to study the supernovae Type Ia (SNIa) [1,2] and was later confirmed by the WMAP and Planck satellites through cosmic microwave background radiation (CMB) [3]—is today an undisputed fact. According to the general theory of relativity (GR), and assuming a homogeneous and isotropic line element and a perfect fluid energy-momentum tensor, the scale factor for an accelerated Universe ($z \lesssim 0.6$) is expressed as $a(t) = a_0 \exp[\Lambda(t - t_0)]$, where Λ is the cosmological constant at $z = 0$, representing a de Sitter evolution. Additionally, the equation of state (EoS) of the fluid responsible for the acceleration must fulfill inequality $\omega < -1/3$. According to these demands, the cosmological constant (CC) is one of the explanations, and through this approach, it is possible to construct the standard paradigm for cosmology, also known as the Λ cold dark matter (ΛCDM) model, which is in agreement with modern observations. However, we should not lose sight of the profound unsolved problems afflicting the CC [4,5], which can be attributed to possible modifications to the GR or to a misinterpretation of the CC from the quantum field theory point of view, in which the ultraviolet divergence of the energy density arises.

From the experimental approach, we know that we are in an accelerated phase because the observed luminous distances from SNIa (and confirmed by CMB [3]) coincide with the theoretical expression expected for an accelerated Universe according to the background model. The luminous distance expression is $d_L(z) \approx H_0^{-1}\Omega_\Lambda^{-1/2}(1 + z)z$, where Ω_Λ is the density parameter of the CC, and it is valid only when Ω_Λ dominates over the other components (matter and radiation). This assumption is valid because $\Omega_\Lambda = 0.68$ according to [3], and eventually $\Omega_\Lambda = 1$, while the other components tend to dilute. A similar situation happens, for example, with other observations, like Baryon Acoustic Oscillations (BAO), when we explore the final stages of the Universe's evolution, where Ω_Λ dominates. In this case, the angular distance can be reduced to $d_A(z) \approx z H_0^{-1} \Omega_\Lambda^{-1/2} (z+1)^{-1}$. This confirms that the observations and theory fit only if the Universe is considered to be in an accelerated stage nowadays. Both these measurements (and others) are important evidence that the Universe is transiting to a de Sitter phase in its last stages, and two of its main physical observables are the luminous and angular distances.

On the other hand, the steady-state model (SSM) demands a *perfect cosmological principle* (PCP) [6], generalizing the isotropy not only in all directions at cosmological scales but also at all times. Indeed, the SSM demands $\dot{a}/a = H_0$, where H_0 is the Hubble constant, implying $a = a_0 \exp[H_0(t - t_0)]$ with a deceleration parameter $q \simeq -1$ and a luminous and angular distance that coincides precisely with the one obtained from the de Sitter evolution in the ΛCDM model in its last stages. We note that the SSM was initially discarded because of its inability to predict CMB radiation and its black-body spectrum [7]. Additionally, when the SSM was proposed, there was no observational evidence regarding the transition of our Universe into an accelerated stage; thus, the best at that epoch was an universe dominated by matter at $z \sim 0$, which is no longer the case today. Nowadays, SSM has evolved into the so-called *matter/radiation creation model* [8–12], in which the de Sitter evolution is not caused by a cosmological constant; instead, a continuous creation of matter/radiation is produced through a diffusion term in the continuity equations driving the de Sitter evolution observed.

Evidence shows that our Universe is an evolving system rising from a Big Bang 13.7 billion years ago, with different transitions, but in particular with a transition at $z \lesssim 0.6$ that coincides with the mathematical description of both an accelerated and a *late steady-state universe* (LSS). Additionally, based on the ΛCDM model, the Universe eventually tend to $q \to -1$ when $z \to -1$, i.e., in the far future. The word *late* is used because the steady state is not valid for all epochs in the Universe evolution; prior to this epoch (accelerated/stationary), the ΛCDM model is still the cornerstone. This affirmation arises from the mathematical equivalence between models. Thus, how can we be certain of which model for the Universe is the ideal interpretation for $z \leqslant 0.6$? Are both conditions equivalent, and is there no way to differentiate between an accelerated and a steady-state Universe for $z \leqslant 0.6$? If there is an equivalence, maybe the interpretation under one condition is easier than under the other (accelerated $\leftrightarrow$ LSS), as it happens with the equivalence principle or the equivalence between anti-de Sitter and the conformal field theory (AdS/CFT).

Thus, this paper is dedicated to tackling these questions and following up on all of their consequences. The outline of the paper is as follows: Section 2 is dedicated to discussing the LSS model; it presents the predictions related to the continuous emission of radiation at MeV energy. Section 3 tackles the consequences of MeV emission at astrophysical scales and concludes that this radiation must belong to the dark sector. In Section 4, the underlying classical field theory and the consequences in cosmology are revisited. Finally, in Section 5, we present our discussions and conclusions.

2. The LSS Model

LSS requires the continuous creation of matter and radiation in order to have a steady-state condition, implying the violation of the conservation of the energy-momentum tensor $\nabla^\mu T_{\mu\nu} \neq 0$, which is incompatible with GR and, thus, we need to have matter creation at a rate of $\sim 3H$ per the existing accumulation of matter in the Universe, i.e., we need to maintain $\dot{\rho} = \dot{p} = 0$ for the dominant components. On the other hand, we note that the CC shows similar complications: in this case, we also need to fulfill condition $\dot{\rho}_\Lambda = \dot{p}_\Lambda = 0$, pushing GR to its limits due to the continuous creation of energy/matter to keep CC constant. This is a consequence of introducing quantum–vacuum fluctuations to account for this continuous creation of energy to maintain $\rho_\Lambda = cte$ [13]. In the SSM scenario, Hoyle [14,15] introduced the C-field in the Einstein equations as $G_{\mu\nu} + C_{\mu\nu} = 8\pi G T_{\mu\nu}$ in order to have a solution for the conundrum of the continuous creation. From a more recent perspective, this C-field can be expressed as $G_{\mu\nu} + \frac{1}{4}(R + 8\pi G T)g_{\mu\nu} = 8\pi G T_{\mu\nu}$ using unimodular gravity (UG) [16–18], which contains an extra term equivalent to the C-field proposed by Hoyle but in a natural deduction. We keep in mind that UG is a model that naturally emerges from a Lagrangian, only demanding the invariance in its volume $\sqrt{-g} = \zeta$, where $\zeta = 1$ is a constant. In this context, the vacuum energy has no direct gravitational effects, and Λ is only an integration constant, implying that it is possible to choose a small value (or even $\Lambda = 0$) as demanded in studies like [19].

Connecting all these arguments, we raise the following proposition: *we are not in a position to know whether the universe is already transiting to an accelerated or a steady-state phase using the redshift drift. (The detection of the radiation exposed hereafter could be evidence of LSS instead of an accelerated Universe. However, the radiation acts as an equivalent to CC in the standard way.) Indeed, even if we use the luminous distance, $d_L(z) = H^{-1}z(z+1)$, or the angular distance, $d_A(z) = H^{-1}z(z+1)^{-1}$, for $z \lesssim 0.6$, there is no experiment/observation that would allow us to conclude if the Universe is transiting to an accelerated or a late steady-state Universe.*

Thus, under the previous state of accelerated and LSS conditions, it is possible to follow Universe consequences.

The traditional view for CC needs to deal with quantum vacuum fluctuations afflicted by the ultraviolet divergences that grow at k^4 when the energy density is calculated (see [13] and references therein). The reinterpretation consists in the use of a continuous creation of radiation in concordance with previous calculations for SSM, which can be deduced through expression

$$\dot{\delta}(t) = -\sigma\left(\nu_1 \frac{a(t_1)}{a(t)}, t\right)\delta(t), \tag{1}$$

where $\delta(t)$ is the number of particles whose solution takes the form $\delta(t_0) = e^{-\tau}\delta(t_1)$ in terms of optical depth τ, dot represents a time derivative, $\sigma(\nu)$ is the absorption rate of a radiation of frequency ν, $a(t)$ is the scale factor related with the line element $ds^2 = -dt^2 + a(t)^2(dr^2 + r^2 d\Omega^2)$, $d\Omega^2$ is the solid angle and it is considered null curvature $k = 0$ according to recent observations [3]. Due to that, we assume that our particles follow Bose statistic; with relation $8\pi\nu^2\zeta(t,\nu) = \Gamma(\nu,t)$, we can generalize $\delta(t)$ as the following differential equation:

$$\frac{\dot{\delta}(t)}{\delta(t)} = -\sigma\left(\nu_1 \frac{a(t_1)}{a(t)}, t\right) + \zeta\left(\nu_1 \frac{a(t_1)}{a(t)}, t\right), \tag{2}$$

where $\zeta(\nu)$ is the emission rate. After some straightforward calculations, we arrive to (see details in [7])

$$n(\nu) = 8\pi\nu^2 \int_{t_1}^{t_0} \exp\left\{ -\int_{t}^{t_0} \left[\sigma\left(\nu\frac{a(t_0)}{a(t')}, t'\right) - \zeta\left(\nu\frac{a(t_0)}{a(t')}, t'\right)\right]dt'\right\} \times$$
$$\zeta\left(\nu\frac{a(t_0)}{a(t)}, t\right)dt, \tag{3}$$

where $n(\nu)$ is the number function and $8\pi\nu^2\zeta(\nu)d\nu$ is the emission rate per unit volume of radiation between frequency ν and $\nu + d\nu$, with t_0 and t_1 arbitrary chosen. We notice that Equation (3) is restricted to the scale factor, which in turn is coupled with a field theory of gravitation which in principle is GR but not restricted to possible extensions in which the scale factor is involved.

Thus, our starting point is Equation (3), and we also consider that the radiation could be standard photons but not restricted to them (i.e., other particles with similar behavior such as axions and dark photons, among others, are allowed).

In the de Sitter phase, Expression (3) can be written as [7]

$$n(v) = 8\pi v^2 \int_{-\infty}^{t_0} \exp\left\{ -\int_t^{t_0} \left[\sigma(v\exp(H_0[t_0 - t'])) - \zeta(v\exp(H_0[t_0 - t']))\right] dt' \right\} \times$$
$$\zeta(v\exp(H_0[t_0 - t]))dt, \tag{4}$$

where it is assumed that, during this de Sitter phase, $\dot{a}/a \approx H_0$ (this assumption is only valid for $z \leq 0.6$). With an appropriate change in variables to avoid dependence on t_0, we have

$$n(v) = 8\pi v^2 \int_v^\infty \frac{\zeta(v')}{H_0 v'} \exp\left(-\int_v^{v'} \frac{dv^*}{H_0 v^*}\left[\sigma(v^*) - \zeta(v^*)\right]\right) dv', \tag{5}$$

and differentiating with respect to v, we arrive to the following expression:

$$\zeta(v) = \frac{n(v)\sigma(v)}{8\pi v^2 + n(v)} + \left[2n(v) - v\frac{dn(v)}{dv}\right]\frac{H_0}{8\pi v^2 + n(v)}, \tag{6}$$

where we separate terms that depend on Hubble constant H_0. Thus, if we stand by our hypothesis of assuming that the radiation (standard or dark emission) follows a Planckian number distribution, we have homogeneity and isotropy

$$n(v) = 8\pi v^2[\exp(v/\mathcal{T}) - 1]^{-1}, \tag{7}$$

where $\mathcal{T}$ is a fixed temperature. Therefore, we obtain, from Equation (6),

$$\zeta(v) = \exp(-v/\mathcal{T})\sigma(v) + \zeta(v)_{\text{CMV}}, \tag{8}$$

where

$$\zeta(v)_{\text{CMV}} = H_0\left(\frac{v}{\mathcal{T}}\right)[\exp(v/\mathcal{T}) - 1]^{-1}. \tag{9}$$

We notice that $\zeta(v)_{\text{CMV}}$ is a constant emission of radiation that is independent of absorption $\sigma(v)$ and can be expressed as a function of the Hubble parameter, where subscript CMV indicates the Cosmic MeV emission. The first term in (8) (r.h.s) is the classical behavior emission–absorption term for radiation implying interactions with baryonic matter and astrophysical background lights, such as the extragalactic background light (EBL) and the CMB. (This interaction depends on the particle we are dealing with; for example, in the dark sector, interactions are weaker than those associated with standard photons.)

We also notice that even in the case $v \to 0$, we always have a non-negligible emission in the form $\zeta(0)_{\text{CMV}} = H_0$. After an integration in the form of $\rho = 8\pi \int_0^\infty v^2 \zeta(v)_{\text{CMV}} dv$ (which is convergent) for energy density, we conclude

$$\rho_{\text{CMV}} = \frac{\pi^2}{15} H_0 \mathcal{T}^3. \tag{10}$$

Due to the equivalence between an accelerated and a stationary Universe, it is possible to propose identification $\rho_{\text{CMV}} = \rho_\Lambda \simeq 2.46 \times 10^{-11}\text{eV}^4$, where the last number is the value expected for the cause of Universe acceleration. Consequently, we arrive to a fluid with energy

$$E_{\text{CMV}} \simeq 29.5\,\text{MeV}. \tag{11}$$

The energy is obtained by using the current value for the Hubble constant reported in [20] ($E_{\text{CMV}} \simeq 28.8\,\text{MeV}$ using a local measurement of H_0 [21]). This energy region for standard photons is a minimum in the photon interaction cross-section, and the transition from Compton scattering to pair production as the dominant process makes a piece of new physics evidence in this region particularly challenging. This would explain why it has not been detected yet. Consequently, in the following sections, we investigate the possibility of detection through astrophysical tests.

3. Astrophysical Tests

In a CMB-like scenario, Universe acceleration forecasts an isotropic and homogeneous cosmic background of standard light radiation, such as the known CMB [3,22]. Thus, if E_{CMV} is a CMB-like radiation, it follows the energy density distribution given by Equation (7) as black-body radiation with a mean energy density given by Equation (11). As a background light, it is likely to interact with gamma rays and annihilate by the photon pair-production process ($\gamma \gamma \rightarrow e^+ e^-$). The astrophysical expected outcome of this process is attenuating the expected astrophysical photon flux; for a CMB energy range, such attenuation is expected on the ultra-high-energy gamma rays. Lighter background photons as the EBL attenuates the TeV gamma-ray flux of extragalactic point sources, such as blazars [23–26]. In this line of thought, we find the survivable probability ($P_{surv} = e^{-\tau}$) of standard photons, such as gamma rays, by finding the optical depth (τ) but considering the presence of the E_{CMV} radiation as a background light. The survivable probability is given by

$$P_{surv} = \exp\left[-\int_0^z \frac{c\,dz}{H_0(1+z)h(z)} \int_{-1}^1 d\cos\theta \frac{1-\cos\theta}{2}\right.$$
$$\left. \times \int_{\epsilon_{th}}^{\infty} d\epsilon\, n(\epsilon, z)\sigma(E_\gamma, \epsilon, z)\right], \tag{12}$$

where H_0 stands for the Hubble constant in the present time, $h(z) = \sqrt{\Omega_m(1+z)^3 + \Omega_{\text{CMV}}}$ is the distance element in an expanding universe, $\sigma(E_\gamma, \epsilon, z)$ is the Breit–Wheeler cross-section for pair production process $\gamma \gamma \rightarrow e^+ e^-$ [27], and $n(\epsilon, z)$ is the background density given by Equation (7), with $\epsilon = \nu$.

The outcome of these is that if E_{CMV} radiation is CMB-like, any astrophysical photon above the keV energy range and from a distance beyond 10^{-26} Mpc will be attenuated by the presence of E_{CMV} emission. In Figure 1, we show the survival probability of gamma rays from different z's to us on Earth. For comparison, we include the gamma-ray survival probability due to CMB; the shadow area represents the attenuated region for gamma rays from $z = 1$, from which CMB allows ultra-high-energy photon propagation from $z = 1$ to Earth. Once again, we conclude that non-standard photons of the keV order would survive E_{CMV} radiation, which contradicts the observations. Therefore, if CMV emission exists, it cannot be made by standard photons.

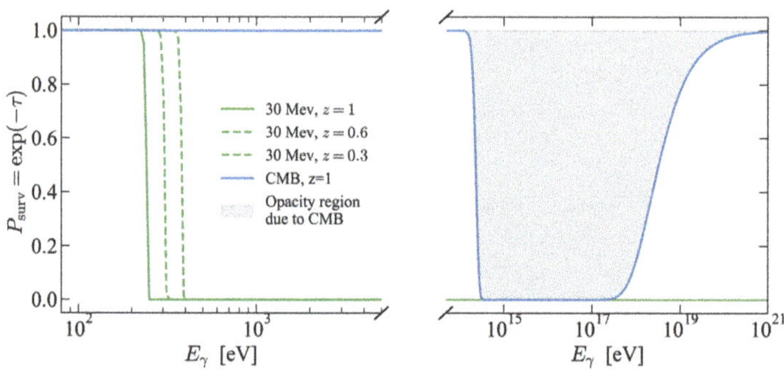

Figure 1. Photon survival probability due to interaction with the 30 MeV radiation from a given z as standard photons. For comparison, survival probability due to CMB is included. Everything above such lines is attenuated; we illustrate this as the shadow area in the CMB case.

Thus, considering the unknown nature of the CMV and assuming that it can decay to some standard photons, it may be proven through astrophysical tests. Experiments like the Compton Telescope (COMPTEL) [28] and the Energetic Gamma-ray Experiment Telescope (EGRET) [29] have reported measurements of photon flux at the energy region of our interest [30].

Therefore, in order to test the predicted signal, we model EGRET and COMPTEL data extracted from [28,29] of the energy flux in the energy range $1.12 < E < 15.28 \times 10^3$ MeV. Based on [31], we consider the background modeled by the sum of two components, $\mathrm{Bkg}_1(E) + \mathrm{Bkg}_2(E)$, where

$$\mathrm{Bkg}_{1,2}(E) = \frac{C_{1,2}}{(E/E_b)^{\Gamma_{1,3}} + (E/E_b)^{\Gamma_{2,4}}}, \tag{13}$$

where $C_{1,2}$ are normalization constants, E_b is the energy peak, and Γ_1 and Γ_2 are constants. For the Bkg_1 component, we fix $\Gamma_1 = 1.32$, $\Gamma_2 = 2.88$, $E_b = 25\,\mathrm{keV}$, and for the Bkg_2 component, $\Gamma_3 = 1.0$, $\Gamma_4 = 2.41$ and $E_b = 20\,\mathrm{MeV}$ [31], and we allow variation of both C_1 and C_2. The full model is obtained by adding the signal presented in Equation (9), which is named Bkg+Signal. Figure 2 shows the fit obtained, the solid blue line corresponds to Signal+Bkg and the solid red line is the total background.

We compare the two models, Bkg and Bkg+Signal, statistically through the Akaike information criterion corrected (AICc) [32,33] for small samples defined as $\mathrm{AICc} = \chi^2_{min} + 2k + (2k^2 + 2k)/(N - k - 1)$ where χ^2_{min} is the minimum of the χ^2-function, k is the number of degrees of freedom and N is the size of the sample. In this criterion, the model with the lower value of AICc is the one preferred by data. When difference ΔAICc between a given model and the best one is ΔAICc < 4, both models are equally supported by the data. For the range $4 < \Delta$AICc < 10, the data still support the given model but less than the preferred one, and if ΔAICc > 10, the given model is not supported. Our result gives ΔAICc $= \mathrm{AICc}(\mathrm{Bkg} + \mathrm{Signal}) - \mathrm{AICc}(\mathrm{Bkg}) = -118.8$, which indicates that the model Bkg+Signal is preferred by EGRET and COMPTEL datasets. Additionally, we find signal amplitude $A = 0.151 \pm 0.014\,\mathrm{MeVcm}^{-2}\mathrm{s}^{-1}\mathrm{sr}^{-1}$.

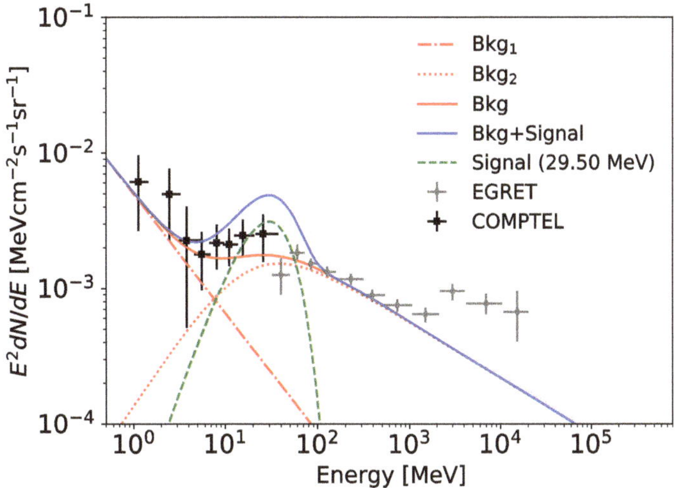

Figure 2. Energy flux vs energy. Black and grey markers are COMPTEL and EGRET data, respectively. The solid blue line corresponds to Bkg+Signal fit, and the solid red line to Bkg is composed of Bkg_1 (dot-dashed red line) and Bkg_2 (dotted red line).

As can be seen, the expected energy is in the extreme of both data sets, in the maximum of COMPTEL and in the minimum of EGRET, causing difficulties in its detection through the recent data compilations (see Figure 2). Future experiments could improve the signal-to-noise detection in the MeV region, such as the GRAMS Project [34] and GammaTPC [35]. They might be able to confirm whether CMV emission exists, and if it does, the cause of the current de Sitter transition might be explained.

4. The Underneath Field Theory, a Revision

The continuous creation of energy, in principle, is incompatible with GR because $\nabla^\mu T_{\mu\nu} \neq 0$ unless we accept fluids with the $w < -1/3$ equation of state. One of the best approaches under a field equation to have a non-conservation of the energy–momentum tensor and compatible with the fluid discussed in Section 2 is through the equations of unimodular gravity (UG) given by

$$R_{\mu\nu} - \frac{1}{4}g_{\mu\nu}R = 8\pi G\left(T_{\mu\nu} - \frac{1}{4}g_{\mu\nu}T\right),\tag{14}$$

which clearly is traceless [16], $R_{\mu\nu}$, R are the Ricci tensor and scalar, respectively, $T_{\mu\nu}$ is the energy–momentum tensor, $T = g^{\mu\nu}T_{\mu\nu}$ is the energy–momentum scalar and G is the Newton gravitational constant. In this context, a CC is unnecessary, and traditional fluids can produce evolution $a(t) = \exp \Lambda(t - t_0)$ only under the restriction that $\dot{\rho}^{fluid} = 0$, caused naturally by

$$32\pi G\nabla^\mu T_{\mu\nu} = \nabla^\mu(R + 8\pi GT)g_{\mu\nu},\tag{15}$$

where the general covariance is not demanded. We notice that this approach is similar to the stationary model [6,14,15] in which a continuous creation of matter/energy produces a de Sitter behavior, indistinguishable from the scale factor for an accelerated Universe. Some studies [36,37] suggest that a continuous creation of radiation (relativistic particles, axions, dark photons) under the UG approach could resolve the problem of the observed de Sitter phase, without the necessity of $w < -1/3$, demanded in standard GR.

Assuming a Friedmann–Lemaitre–Robertson–Walker (FLRW) metric and a perfect fluid energy–momentum tensor, we arrive at the following equation:

$$\dot{H} = -4\pi G\sum_i(\rho_i + p_i),\tag{16}$$

where ρ_i, p_i are the density and pressure of the fluids, respectively, $H \equiv \dot{a}/a$ is the Hubble parameter, a is the scale factor, and dot stands for time derivative. On the other hand, resolving for Equation (15), we have

$$\sum_i[(\dot{\rho}_i + \dot{p}_i) + 3H(\rho_i + p_i)] = \frac{H^3}{4\pi G}(1 - j),\tag{17}$$

where $j \equiv \dddot{a}/aH^3$ is known as the jerk parameter [38] where the dots stand for third-order derivatives. The third-order derivatives that this theory contains are tamed through the cosmographic jerk parameter, which also acts as a source term. Thus, after some manipulation using Equations (16) and (17), the dynamical equations for the cosmology in this context can be presented as follows:

$$H^2 = \frac{8\pi G}{3}\sum_i\rho_i + H^2_{UG},\tag{18}$$

$$\frac{\ddot{a}}{a} = -\frac{4\pi G}{3}\sum_i(\rho_i + 3p_i) + H^2_{UG},\tag{19}$$

$$H^2_{UG} = \frac{8\pi G}{3}\sum_i p_i + \frac{2}{3}\int_{a_{ini}}^a H(a')^2[j(a') - 1]\frac{da'}{a'},\tag{20}$$

where subscript UG refers to Unimodular Gravity. According to [36,37], the best selection of j, to mimic the ΛCDM model but taking into account the clues about the causative of the Universe's acceleration, is $j = \frac{9}{2}(1 + w)wE(z)^{-2}\Omega_{0i}(z + 1)^{3(w+1)} + 1$, where w is the fluid equation of state and Ω_{0i} is the density parameter, $z = a^{-1} - 1$ is the redshift and $E(z) \equiv H(z)/H_0$. The election, naturally, allows us to decouple matter with the standard equation of state while radiation depends on the j parameter. Therefore, radiation plays a central role in the Universe's acceleration due to the coupling with j. Thus, demanding a continuity equation for new dark energy term gives us the following Friedmann equation:

$$E(z)^2 = \Omega_{0m}(z + 1)^3 + \Omega_{0r}(z + 1)^4 + \omega_{\mathrm{CMV}}\Omega_{0\mathrm{CMV}}(z_{ini} + 1)^4, \tag{21}$$

where the last term acts like a cosmological constant but with an origin based in the term $\frac{1}{4}(R + 8\pi GT)g_{\mu\nu}$. This constant acts like a diffusion parameter and can be interpreted as a CMV component plus the effects of the UG dictated by the z_{ini} free parameter (see [36,37] for details), where Ω_{0m} and $\Omega_{0\mathrm{CMV}}$ are the matter and CMV parameters, respectively, $\omega_{\mathrm{CMV}} = 1/3$ and $z_{ini} = 11.473^{+0.074}_{-0.073}$ constrained through recent observations (see [36,37]) and interpreted as the region where the term that generates the de Sitter expansion emerges (but not dominates).

5. Discussion and Conclusions

As we discussed previously, at the moment, it is impossible to differentiate between an accelerated or late stationary Universe because the observables (d_L and d_A) in terms of the redshift drift of all the experiments that measure the Universe evolution are equally compatible with both approaches. We emphasize that this argument is only valid for $z \lesssim 0.6$, where the data confirms the transition.

On the other hand, it is well known that both models (accelerated and stationary) need a continuous creation of energy/matter to maintain the energy density constant of some species in the Universe, which, in turn, pushes the GR's limits in its current form and suggests that changes are required to obtain, for instance, a non-conservation to the energy–momentum tensor. This equivalence suggests a new strategy to simultaneously tackle the energy density problem of the CC and the interpretation of the emission in the regime of $\sim$29.5 MeV, which, in principle, could be detected through decaying to standard photons. *Confirming this signal is crucial, so we encourage further astrophysical analysis in this scope.* Suggestively, the MeV region is also connected to several new physics proposals, such as Primordial Black Holes (PBH) [39–41] emitting radiation through Hawking effect, self-annihilation of DM particles candidates with MeV masses that can produce gamma radiation [42,43], and some quantum gravity effects (like Continuous Spontaneous Localization, compatible with UG [18] or the Diosi–Penrose model, DP [44–46]) could also produce radiation emission at the same energy scale of tens of MeV. Identifying the event that causes the excess radiation is vital because the theory presented here cannot identify the event that causes the previously mentioned excess. For example, PBH as DM implies decaying radiation at MeV generating the observed accelerated/stationary Universe, sustaining a unified framework that relates several events. In this case, the excess of radiation is a transitory effect; therefore, the accelerated/stationary Universe is also a transitory process. For example, in the case where the cause of the excess of radiation is the DP model under charged particles, the collapse of the wave function by gravitational effects has significant consequences for the evolution of the Universe and its current stationary/accelerated stage. The radiation energy for this case is in the range of $\Delta E = (10 - 10^5)$ keV, with CMV inside the expected energy region. Other particles like axions [47] or dark photons [48] could be possible candidates. Therefore, models could prove that we are dealing with an LSS instead of an accelerated Universe.

We need to remark that a stationary Universe is not a natural state (at least under the Friedmann–Lemaitre–Robertson–Walker line element) because maintaining constant energy density requires the violation of the conservation of the energy–momentum tensor. As a consequence, we expect that this state will eventually stop, finishing this particular condition. However, to respond to the question of how to differentiate between the two universes, we need to know what the source of the CMV radiation is, if it exists. In this work, we showed that the survival probability of standard photons due to the CMV predicts an impossible opacity region, not allowing the detection of any >keV astrophysical photons on Earth. This strongly constrains the hypothesis that this radiation exists as standard photons. Hence, we also studied the possibility of an indirect signal of standard photons derived from a potential non-standard component of the CMV emission at 29.5 MeV. Our findings suggest that the model Bkg+Signal is preferred over the only Bkg scenario, using EGRET and COMPTEL datasets.

Finally, we must remember that a change in variables to the de Sitter line element produces a stationary metric as $ds^2 = -(1 - r^2/\alpha^2)dt^2 + (1 - r^2/\alpha^2)^{-1}dr^2 + r^2 d\Omega^2$, where α is a nonzero constant of the hyperboloid of one sheet. Remarkably, a de Sitter evolution is analogous to a stationary Universe, as discussed throughout this paper. Additionally, we emphasize that when using standard GR, a fluid with $w < -1/3$ EoS is necessary in order to have a de Sitter Universe. We also do not have the physics of DE, and thus it is impossible to know whether the particles decay in an energy region in order to be detected by some experiment. Under these conditions, it is not possible to have the same predictions presented in this paper.

We encourage further and novel astrophysical experiments and studies to unravel the 29.5 MeV emission in order to fully comprehend the current de Sitter stage of our Universe.

Author Contributions: Conceptualization, Y.B.A.-P.; Methodology, M.A.G.-A., H.M.-H. and A.H.-A. All authors have read and agreed to the published version of the manuscript.

Funding: YBAP acknowledges the Ph.D. grant provided by CONACYT. MAGA acknowledges support from cátedra Marcos Moshinsky and Universidad Iberoamericana for the support with the National Research System (SNI) grant.

Data Availability Statement: Data are contained within the article.

Acknowledgments: We thank the anonymous referees for their thoughtful remarks and suggestions. The numerical analysis was carried out by *Numerical Integration for Cosmological Theory and Experiments in High-energy Astrophysics* (Nicte Ha) cluster at IBERO University, acquired through cátedra MM support. A.H.-A. thank for the support from Luis Aguilar, Alejandro de León, Carlos Flores, and Jair García of the Laboratorio Nacional de Visualización Científica Avanzada. A.H.-A. and M.A.G.-A. acknowledge partial support from project ANID Vinculación Internacional FOVI220144. We are also grateful to Verónica Motta for revising this paper, suggesting changes, and correcting misinterpretations in the text. This paper is in memory of my father Miguel Ángel.

Conflicts of Interest: The authors declare no conflict of interest.

References

1. Riess, A.G.; Filippenko, A.V.; Challis, P.; Clocchiatti, A.; Diercks, A.; Garnavich, P.M.; Gilliland, R.L.; Hogan, C.J.; Jha, S.; Kirshner, R.P.; et al. Observational Evidence from Supernovae for an Accelerating Universe and a Cosmological Constant. *Astron. J.* **1998**, *116*, 1009. [CrossRef]
2. Perlmutter, S.; Aldering, G.; Goldhaber, G.; Knop, R.A.; Nugent, P.; Castro, P.G.; Deustua, S.; Fabbro, S.; Goobar, A.; Groom, D.E.; et al. Measurements of Ω and Λ from 42 High-Redshift Supernovae. *Astrophys. J.* **1999**, *517*, 565. [CrossRef]
3. Aghanim, N.; Akrami, Y.; Ashdown, M.; Aumont, J.; Baccigalupi, C.; Ballardini, M.; Banday, A.J.; Barreiro, R.B.; Bartolo, N.; Basak, S.; et al. *Planck*/2018 results. VI. Cosmological parameters. *Astron. Astrophys.* **2020**, *641*, A6. [CrossRef]
4. Zeldovich, Y.B. The cosmological constant and the theory of elementary particles. *Sov. Phys. Uspekhi* **1968**, *11*, 381. [CrossRef]
5. Weinberg, S. The cosmological constant problem. *Rev. Mod. Phys.* **1989**, *61*, 1. [CrossRef]
6. Bondi, H.; Gold, T. The Steady-State Theory of the Expanding Universe. *Mon. Not. R. Astron. Soc.* **1948**, *108*, 252–270. [CrossRef]
7. Weinberg, S. *Gravitation and Cosmology*; Wiley: Hoboken, NJ, USA, 1972.
8. Trevisani, S.R.G.; Lima, J.A.S. Gravitational matter creation, multi-fluid cosmology and kinetic theory. *Eur. Phys. J. C* **2023**, *83*, 244. [CrossRef]
9. Cárdenas, V.H.; Cruz, M.; Lepe, S. Cosmic expansion with matter creation and bulk viscosity. *Phys. Rev. D* **2020**, *102*, 123543. [CrossRef]
10. Lima, J.A.S.; Santos, R.C.; Cunha, J.V. Is ΛCDM an effective CCDM cosmology? *J. Cosmol. Astropart. Phys.* **2016**, *3*, 027. [CrossRef]

11. Vargas dos Santos, M.; Waga, I.; Ramos, R.O. Degeneracy between CCDM and ΛCDM cosmologies. *Phys. Rev. D* **2014**, *90*, 127301. [CrossRef]
12. Ramos, R.O.; Vargas dos Santos, M.; Waga, I. Matter creation and cosmic acceleration. *Phys. Rev. D* **2014**, *89*, 083524. [CrossRef]
13. Carroll, S.M. The Cosmological constant. *Living Rev. Rel.* **2001**, *4*, 1. [CrossRef] [PubMed]
14. Hoyle, F. A New Model for the Expanding Universe. *Mon. Not. R. Astron. Soc.* **1948**, *108*, 372–382. [CrossRef]
15. Hoyle, F. On the Cosmological Problem. *Mon. Not. R. Astron. Soc.* **1949**, *109*, 365–371. [CrossRef]
16. Einstein, A. Do gravitational fields play an essential role in the structure of elementary particles of matter. *Siz. Preuss. Acad. Scis.* **1919**, *1919*, 349.
17. Ellis, G.F.R.; van Elst, H.; Murugan, J.; Uzan, J.P. On the trace-free Einstein equations as a viable alternative to general relativity. *Class. Quantum Gravity* **2011**, *28*, 225007. [CrossRef]
18. Josset, T.; Perez, A.; Sudarsky, D. Dark Energy from Violation of Energy Conservation. *Phys. Rev. Lett.* **2017**, *118*, 021102. [CrossRef] [PubMed]
19. Mitras, A. Interpretational conflicts between the static and non-static forms of the de Sitter metric. *Sci. Rep.* **2012**, *2*, 923. [CrossRef]
20. Aghanim, N.; Akrami, Y.; Ashdown, M.; Aumont, J.; Baccigalupi, C.; Ballardini, M.; Banday, A.J.; Barreiro, R.B.; Bartolo, N.; Basak, S.; et al. Planck 2018 results. V. CMB power spectra and likelihoods. *Astron. Astrophys.* **2020**, *641*, A5. [CrossRef]
21. Riess, A.G.; Casertano, S.; Yuan, W.; Bowers, J.B.; Macri, L.; Zinn, J.C.; Scolnic, D. Cosmic Distances Calibrated to 1% Precision with Gaia EDR3 Parallaxes and Hubble Space Telescope Photometry of 75 Milky Way Cepheids Confirm Tension with ΛCDM. *Astrophys. J. Lett.* **2021**, *908*, L6. [CrossRef]
22. Penzias, A.A.; Wilson, R.W. A Measurement of Excess Antenna Temperature at 4080 Mc/s. *Astrophys. J.* **1965**, *142*, 419–421. [CrossRef]
23. De Angelis, A.; Galanti, G.; Roncadelli, M. Transparency of the Universe to gamma rays. *Mon. Not. R. Astron. Soc.* **2013**, *432*, 3245–3249. [CrossRef]
24. Gilmore, R.; Somerville, R.; Primack, J.; Dominguez, A. Semi-analytic modeling of the EBL and consequences for extragalactic gamma-ray spectra. *Mon. Not. R. Astron. Soc.* **2012**, *422*, 3189. [CrossRef]
25. Franceschini, A.; Rodighiero, G.; Vaccari, M. The extragalactic optical-infrared background radiations, their time evolution and the cosmic photon-photon opacity. *Astron. Astrophys.* **2008**, *487*, 837. [CrossRef]
26. Domínguez, A.; Primack, J.R.; Rosario, D.J.; Prada, F.; Gilmore, R.C.; Faber, S.M.; Koo, D.C.; Somerville, R.S.; Pérez-Torres, M.A.; Pérez-González, P.; et al. Extragalactic Background Light Inferred from AEGIS Galaxy SED-type Fractions. *Mon. Not. R. Astron. Soc.* **2011**, *410*, 2556. [CrossRef]
27. Breit, G.; Wheeler, J.A. Collision of Two Light Quanta. *Phys. Rev.* **1934**, *46*, 1087–1091. [CrossRef]
28. Weidenspointner, G.; Varendorff, M.; Kappadath, S.C.; Bennett, K.; Bloemen, H.; Diehl, R.; Hermsen, W.; Lichti, G.G.; Ryan, J.; Schönfelder, V. The cosmic diffuse gamma-ray background measured with COMPTEL. *AIP Conf. Proc.* **2000**, *510*, 467–470. [CrossRef]
29. Strong, A.W.; Moskalenko, I.V.; Reimer, O. A new determination of the extragalactic diffuse gamma-ray background from egret data. *Astrophys. J.* **2004**, *613*, 956–961. [CrossRef]
30. Ajello, M.; Greiner, J.; Sato, G.; Willis, D.R.; Kanbach, G.; Strong, A.W.; Diehl, R.; Hasinger, G.; Gehrels, N.; Markwardt, C.B.; et al. Cosmic X-ray Background and Earth Albedo Spectra with Swift BAT. *Astrophys. J.* **2008**, *689*, 666. [CrossRef]
31. Oberlack, U. Extragalactic diffuse gamma-ray emission at high energies. *Physics* **2010**, *3*, 21. [CrossRef]
32. Akaike, H. A New Look at the Statistical Model Identification. *IEEE Trans. Autom. Control* **1974**, *19*, 716–723. [CrossRef]
33. Sugiura, N. Further analysts of the data by akaike's information criterion and the finite corrections. *Commun. Stat.-Theory Methods* **1978**, *7*, 13–26. [CrossRef]
34. Aramaki, T.; Adrian, P.O.H.; Karagiorgi, G.; Odaka, H. Dual MeV gamma-ray and dark matter observatory—GRAMS Project. *Astropart. Phys.* **2020**, *114*, 107–114. [CrossRef]
35. Shutt, T.; Akerib, D.; Breur, S.; Buuck, M.; Dragone, A.; Digel, S.; Haller, G.; Hitchcock, O.; Linehan, R.; Luitz, S.; et al. A Next-Generation LAr TPC-Based MeV Gamma Ray Instrument. Available online: https://www.snowmass21.org/docs/files/summaries/CF/SNOWMASS21-CF7CF1-NF7NF10-IF8IF0Shutt-224.pdf (accessed on 13 November 2023).
36. García-Aspeitia, M.A.; Martínez-Robles, C.; Hernández-Almada, A.; Magaña, J.; Motta, V. Cosmic acceleration in unimodular gravity. *Phys. Rev.* **2019**, *D99*, 123525. [CrossRef]
37. García-Aspeitia, M.A.; Hernández-Almada, A.; Magaña, J.; Motta, V. The Universe acceleration from the Unimodular gravity view point: Background and linear perturbations. *Phys. Dark Univ.* **2021**, *32*, 100840. [CrossRef]
38. Zhang, M.J.; Li, H.; Xia, J.Q. What do we know about cosmography. *Eur. Phys. J.* **2017**, *C77*, 434. [CrossRef]
39. Zel'dovich, Y.B.; Novikov, I.D. The Hypothesis of Cores Retarded during Expansion and the Hot Cosmological Model. *Astron. Zhurnal* **1966**, *43*, 758.
40. Hawking, S. Gravitationally collapsed objects of very low mass. *Mon. Not. R. Astron. Soc.* **1971**, *152*, 75. [CrossRef]
41. Carr, B.J.; Hawking, S.W. Black Holes in the Early Universe. *Mon. Not. R. Astron. Soc.* **1974**, *168*, 399–415 [CrossRef]
42. Gonzalez-Morales, A.X.; Profumo, S.; Reynoso-Córdova, J. Prospects for indirect MeV Dark Matter detection with Gamma Rays in light of Cosmic Microwave Background Constraints. *Phys. Rev. D* **2017**, *96*, 063520. [CrossRef]

43. Boudaud, M.; Lacroix, T.; Stref, M.; Lavalle, J. Robust cosmic-ray constraints on *p*-wave annihilating MeV dark matter. *Phys. Rev. D* **2019**, *99*, 061302. [CrossRef]
44. Diósi, L. A universal master equation for the gravitational violation of quantum mechanics. *Phys. Lett. A* **1987**, *120*, 377–381. [CrossRef]
45. Diósi, L. Models for universal reduction of macroscopic quantum fluctuations. *Phys. Rev. A* **1989**, *40*, 1165–1174. [CrossRef] [PubMed]
46. Penrose, R. On Gravity's role in Quantum State Reduction. *Gen. Relativ. Gravit.* **1996**, *28*, 581–600. [CrossRef]
47. Peccei, R.D. The Strong CP problem and axions. *Lect. Notes Phys.* **2008**, *741*, 3–17. [CrossRef]
48. Essig, R. Working Group Report: New Light Weakly Coupled Particles. In Proceedings of the Snowmass 2013: Snowmass on the Mississippi, Minneapolis, MN, USA, 29 July–6 August 2013; p. 10.

Article

Effective $f(R)$ Actions for Modified Loop Quantum Cosmologies via Order Reduction

Ana Rita Ribeiro [1], **Daniele Vernieri [2,3,4]** and **Francisco S. N. Lobo [1,5,]**

[1] Instituto de Astrofísica e Ciências do Espaço, Faculdade de Ciências da Universidade de Lisboa, Edifício C8, Campo Grande, 1749-016 Lisbon, Portugal; anaritaribeiro95@hotmail.com

[2] Dipartimento di Fisica "E. Pancini", Università di Napoli "Federico II", Complesso Universitario di Monte Sant'Angelo, Edificio G, Via Cinthia, 80126 Napoli, Italy; daniele.vernieri@unina.it

[3] Istituto Nazionale di Fisica Nucleare (INFN) Sezione di Napoli, Complesso Universitario di Monte Sant'Angelo, Edificio G, Via Cinthia, 80126 Napoli, Italy

[4] Scuola Superiore Meridionale, Largo San Marcellino 10, 80138 Napoli, Italy

[5] Departamento de Física, Faculdade de Ciências, Universidade de Lisboa, Edifício C8, Campo Grande, 1749-016 Lisbon, Portugal

[] Correspondence: fslobo@fc.ul.pt

Abstract: General Relativity is an extremely successful theory, at least for weak gravitational fields; however, it breaks down at very high energies, such as in correspondence to the initial singularity. Quantum Gravity is expected to provide more physical insights in relation to this open question. Indeed, one alternative scenario to the Big Bang, that manages to completely avoid the singularity, is offered by Loop Quantum Cosmology (LQC), which predicts that the Universe undergoes a collapse to an expansion through a bounce. In this work, we use metric $f(R)$ gravity to reproduce the modified Friedmann equations which have been obtained in the context of modified loop quantum cosmologies. To achieve this, we apply an order reduction method to the $f(R)$ field equations, and obtain covariant effective actions that lead to a bounce, for specific models of modified LQC, considering a massless scalar field.

Keywords: modified gravity; loop quantum cosmology; bouncing cosmologies

Citation: Ribeiro, A.R.; Vernieri, D.; Lobo, F.S.N. Effective $f(R)$ Actions for Modified Loop Quantum Cosmologies via Order Reduction. *Universe* **2023**, *9*, 181. https://doi.org/10.3390/universe9040181

Academic Editor: Kazuharu Bamba

Received: 9 March 2023
Revised: 3 April 2023
Accepted: 4 April 2023
Published: 11 April 2023

1. Introduction

1.1. Motivations: Loop Quantum Cosmology and Its Modifications

At the order of the Planck length l_p, where General Relativity (GR) breaks down, a quantum theory of gravity is expected to provide insight on the behaviour of gravity at quantum scales [1], and, consequently, in addressing fundamental open questions [2], such as the initial singularity problem. Loop Quantum Gravity (LQG) [3–5] is one of the main candidates, which predicts that classical differential geometry, at small space-time curvatures, is replaced by a discrete quantum geometry at the Planck scale. Indeed, Loop Quantum Cosmology (LQC), which is described as a symmetry-reduced model of LQG [6], provides a more complete understanding of the initial singularity, and through an effective Hamiltonian description yields the following modified Friedmann equation [7–13]:

$$H^2 = \frac{1}{3}\kappa\rho\left(1 - \frac{\rho}{\rho_c}\right), \tag{1}$$

where $H \equiv \dot{a}/a$ is the Hubble function, given in terms of the scale factor $a(t)$, and the critical energy density ρ_c is defined as $\rho_c = \sqrt{3}/(32\pi^2\gamma^3)$, where $\gamma \approx 0.2375$ is the Barbero–Immirzi parameter [14]. Throughout this work, we use the geometrized system of natural units in which $c = \hbar = G = 1$ and, accordingly, $\kappa \equiv 8\pi$. The resulting equation of motion is just the usual Friedmann equation, for a spatially flat universe with a cosmological constant equal to zero, with a modified source, which has no extra degrees of freedom compared to

GR. The modified Friedmann Equation (1) dictates that the Big Bang singularity is replaced by a quantum bounce that occurs at the critical density ρ_c, determined by the underlying quantum geometry, which is independent of the matter content of the universe, when $H = 0$ and $\ddot{a} > 0$.

Formally, in LQG, the elementary classical phase space variables for the gravitational sector are the Ashtekar–Barbero connection A_a^i and the conjugate triad E_i^a. When considering a spatially flat space-time, which is the case in standard LQC, the only relevant constraint is the Hamiltonian constraint, given by [15,16]

$$\mathcal{H} = (\mathcal{C}_M + \mathcal{C}_{grav})/8\pi G,\tag{2}$$

whose vanishing yields the equations of motion. $\mathcal{C}_M$ is the matter sector constraint and $\mathcal{C}_{grav}$ is the gravitational constraint, given by [15,16]

$$\mathcal{C}_{grav} = \mathcal{C}_{grav}^{(E)} - (1 + \gamma^2)\mathcal{C}_{grav}^{(L)},\tag{3}$$

where $\mathcal{C}_{grav}^{(E)}$ is the Euclidean term and $\mathcal{C}_{grav}^{(L)}$ is the Lorentzian term. The Euclidean part is given by [15,16]

$$\mathcal{C}_{grav}^{(E)} = \frac{1}{2}\int d^3x \epsilon_{ijk} F_{ab}^i \frac{E^{aj}E^{bk}}{\sqrt{\det(q)}},\tag{4}$$

where F_{ab}^i is the field strength of the connection A_a^i and $\det(q)$ is the determinant of the spatial metric compatible with the triads. The Lorentzian part, on the other hand, is given by [15,16]

$$\mathcal{C}_{grav}^{(L)} = \int d^3x K_{[a}^j K_{b]}^k \frac{E^{aj}E^{bk}}{\sqrt{\det(q)}},\tag{5}$$

where K_a^i is the extrinsic curvature. Unlike in the framework of full LQG, in the standard LQC scenario, the Euclidean and Lorentzian terms are treated in the same way, due to a classical symmetry reduction, which makes the Lorentzian term a multiple of the Euclidean one. A quantization is then performed on the combination of these two terms, leading to the Hamiltonian constraint in LQC [15]. In this sense, this procedure does not grasp the totality of the full theory of LQG, since the Lorentzian and the Euclidean terms need to be quantized in different ways, leading to a different Hamiltonian constraint. Thus, due to the quantization ambiguities in constructing the Hamiltonian constraint operator in isotropic loop quantum cosmology, this has the important implication that different treatments may arrive at different Planck scale physics. One of the first attempts to understand this problem and to construct a Hamiltonian constraint that is actually similar to the construction in LQG, and inherits more features from the full theory, was undertaken in Ref. [17], in which Thiemann's regularization was used. Indeed, the Lorentz term of the gravitational Hamiltonian constraint in the spatially flat Friedmann–Lemaître–Robertson–Walker (FLRW) model was quantized by two approaches different from that of the Euclidean term. One of the approaches is very similar to the treatment of the Lorentz part of the Hamiltonian in loop quantum gravity and, therefore, inherits more features from the full theory. Two symmetric Hamiltonian constraint operators were constructed, respectively, in the improved scheme, and it was shown that both have the correct classical limit through the semi-classical analysis. Nevertheless, there is still no systematic derivation of LQC as a cosmological sector of LQG and, as a consequence, different paths to obtain LQG-like Hamiltonians from LQC followed [18–21]. Due to these quantization ambiguities in constructing the effective Hamiltonian from LQG, it is pertinent to understand how different quantization methods affect the physical predictions. More specifically, there are still great possibilities for the expanding universe to recollapse due to quantum gravity effects.

Throughout this work, we will focus on two specific modified loop quantum cosmology (mLQC) models [15,17], which differ from standard LQC in the way the Lorentzian term in the Hamiltonian constraint is treated [22]. One of the main goals of this study was to verify whether the Big Bang was still replaced by a quantum bounce. Moreover, numerical solutions were shown to be reliable approximations in order to extract phenomenological implications, particularly in the bounce regime.

1.2. Modified Loop Quantum Cosmology—I: mLQC-I

As emphasized in Ref. [15], in LQC it is not necessary to eliminate the Lorentzian term explicitly in order to express the Hamiltonian constraint. In fact, if the Lorentzian term is maintained, different effective Hamiltonians are obtained. In this formal sense, the model derived in Ref. [15], which here we denote by mLQC-I, by treating the Lorentzian term in the Hamiltonian constraint separately and using Thiemann's regularization [17] yielded a more fundamental difference between LQC and mLQC-I, in that in the latter the evolution of the universe is described by two different branches, b_+ and b_-, and, therefore, it is asymmetric with respect to the bounce.

Considering the b_- branch, the modified Friedmann equation takes the form [15,22]

$$H_-^2 = \frac{\kappa \rho}{3}\left(1 - \frac{\rho}{\rho_c^I}\right)\left[1 + \frac{\gamma^2 \rho/\rho_c^I}{(\gamma^2 + 1)\left(1 + \sqrt{1 - \rho/\rho_c^I}\right)^2}\right],\tag{6}$$

where $\rho_c^I \equiv \rho_c/[4(1 + \gamma^2)]$ is the critical density at which the quantum bounce occurs, that is suppressed by a factor of $4(1 + \gamma^2)$ with respect to the standard model of LQC. For $\rho/\rho_c^I \ll 1$, the standard Friedmann equation $H^2 \approx \kappa \rho/3$ is recovered.

Concerning the b_+ branch, the modified Friedmann equation takes the form [15,22]

$$H_+^2 = \frac{\kappa \alpha \rho_\Lambda}{3}\left(1 - \frac{\rho}{\rho_c^I}\right)\left[1 + \frac{\rho\left(1 - 2\gamma^2 + \sqrt{1 - \rho/\rho_c^I}\right)}{4\gamma^2 \rho_c^I\left(1 + \sqrt{1 - \rho/\rho_c^I}\right)}\right],\tag{7}$$

where $\alpha \equiv (1 - 5\gamma^2)/(\gamma^2 + 1)$, $\rho_\Lambda \equiv 3/[\kappa \lambda^2(1 + \gamma^2)^2]$, and $\lambda^2 \equiv 4\sqrt{3}\pi\gamma$. For this branch, the limit of Equation (7) when $\rho/\rho_c \ll 1$, is given by $H^2 \approx 8\pi G_\alpha(\rho + \rho_\Lambda)/3$, where $G_\alpha := \alpha G$. It can be shown that the classical energy conservation law still holds in this modified model of LQC, concerning both branches.

Taking the current cosmological constraints into account, the evolution of the universe must start from the b_+ branch, in the pre-bounce phase, and afterwards the evolution will be switched to the one described by the b_- branch. In LQC, we only have one branch due to the fact that the Lorentzian term is not explicit in the Hamiltonian. The two branches coincide at the bounce, $\rho = \rho_c^I$ [15]. Contrary to the previous branch, in the b_+ branch, the quantum geometric effects are still present in the above limit. Comparing with the standard Friedmann equation, we see that these effects are present in the form of a modified Newtonian constant, G_α, and an emergent positive cosmological constant, ρ_Λ. These pose phenomenological problems, if this branch lies in the post-bounce universe in which we live [15].

1.3. Modified Loop Quantum Cosmology—Ii: mLQC-II

The second modification of LQC that we consider in this work, outlined in Ref. [17], was derived by using Thiemann's regularization, similar to the previous model, while using the proportionality of Ashtekar's connection and extrinsic curvature [16]. The corresponding modified Friedmann equation takes the form [17]

$$H^2 = \frac{2\kappa \rho}{3}\left(1 - \frac{\rho}{\rho_c^{II}}\right)\left[\frac{1 + 4\gamma^2(\gamma^2 + 1)\rho/\rho_c^{II}}{1 + 2\gamma^2 \rho/\rho_c^{II} + \sqrt{1 + 4\gamma^2(1 + \gamma^2)\rho/\rho_c^{II}}}\right],\tag{8}$$

where we find that the quantum bounce occurs at the critical density $\rho_c^{II} \equiv 4(\gamma^2 + 1)\rho_c$. Unlike the mLQC-I model, the bounce in this model is perfectly symmetric, as in LQC.

However, the modified dynamics in the Planck regime are less trivial, similarly to the mLQC-I model. Another similarity between LQC and mLQC-II is their classical limit. The limit of Equation (8), for $\rho/\rho_c^{II} \ll 1$, is given by $H^2 \approx \kappa\rho/3$ [16]. Once again, using the modified equations that this model provides, the energy–momentum conservation law still holds without any change in the properties of the equation of state (EoS).

1.4. Order Reduction

Now, the only covariant action that leads to second-order equations, under metric variation, is the Einstein–Hilbert (EH) action, and any change in this action leads to higher-order equations. Considering an action other than the one from GR appears to be unjustified given that it implies a departure of a framework that has proven to be successful in describing several physical phenomena. However, considering such theories might be useful, especially if they are close to the theory which we believe to be presently more correct. In fact, given that the modified Friedmann Equations (6)–(8) are obtained from modified sources, a reasonable question to ask is whether it is possible to derive them from an effective action, other than the EH action. An interesting analysis was considered in the Palatini approach of $f(R)$ gravity in Ref. [23]. In this context, the functional Palatini $f(R)$ is a function of the trace of the energy–momentum tensor T, so that the modified Friedmann equation in Palatini $f(R)$ gravity does not involve any higher derivatives of the geometrical quantities and is just a function of the matter sources [24]. Indeed, this result provided new insights on the continuum properties of the discrete structure of quantum geometry. In this work, in order to deduce effective actions, we use metric $f(R)$ gravity [25–31], where the full field equations in this framework are of fourth-order. One advantage of using $f(R)$ gravity is that this theory seems to be the only one which can avoid the Ostrogradski instability [29,32]. We can either consider this theory as an exact theory, meaning that their field equations are considered as genuinely higher-order, or we can treat it as an effective field theory by only considering the solutions which are perturbatively close to GR as the physical ones and the rest as spurious [33–35]. A method that provides the latter approach is that of covariant order reduction, which allows one to find solutions that are perturbatively close to GR. A further advantage of this procedure consists in the fact that it may also serve as a model selection approach. The idea of considering only the solutions which are perturbatively close to GR is not new in cosmology. For example, an inflation model with a polynomial expression for $f(R)$ was phenomenologically investigated in [36]. Another example is presented in [37], where the $R + R^2 + R\Box R$ inflationary model was considered by treating the $R\Box R$ term as a small perturbation.

Indeed, it was shown that, at least for isotropic models in LQC, using a covariant order reduction then a covariant effective action can be found if one considers higher-order theories of gravity but faithfully follows effective field theory techniques [32]. In this context, these techniques were applied within bouncing cosmologies, which can be envisaged as candidates for solving the Big Bang initial singularity problem. In Ref. [38], bouncing solutions in a modified Gauss–Bonnet gravity theory were explored, of the type $R + f(G)$, where R is the Ricci scalar, G is the Gauss–Bonnet term, and f is some function of it [38]. In finding such a bouncing solution, the order reduction technique was used to reduce the order of the differential equations of the $R + f(G)$ theory to second order equations. As GR is a theory whose equations are of second order, this order reduction technique enables one to find solutions which are perturbatively close to GR. Furthermore, the covariant action of the order reduced theory was also obtained. The analysis explored in Ref. [38] was extended in the context of $f(R, G)$ gravity by using the same order reduction technique [39]. Indeed, several covariant gravitational actions leading to a bounce are directly selected by demanding that the Friedmann equation derived within such gravity theories coincides with the one emerging from LQC. The same approach has also been implemented in the context of $f(Q)$ symmetric teleparallel gravity theories, where Q is the non-metricity scalar [40]. Further cases were explored in Refs. [41,42].

1.5. Outline of the Paper

In this work, we are interested in extending the analysis outlined in Ref. [32], by applying the method of covariant order reduction to the modified Friedmann Equations (6)–(8), within the modified loop quantum cosmologies considered above. To this effect, we write out the Lagrangian density of $f(R)$ gravity as the sum of the gravitational Lagrangian of GR and a deviation term, and present the final form of the reduced modified Friedmann equations. In Section 2, we present the covariant order reduction technique in some detail and outline a strategy to deduce the effective actions. In Section 3, we obtain the effective actions for the modified loop quantum cosmological models considered above. Finally, in Section 4, we conclude and discuss our results.

2. Covariant Order Reduction Technique

2.1. Reduced Equations in $f(R)$ Gravity

In this work, we consider metric $f(R)$ gravity, whose corresponding action is given by

$$S = \frac{1}{2\kappa} \int_V d^4 x \sqrt{-g} f(R) + S_M(g_{\mu\nu}, \Psi),\tag{9}$$

where g is the determinant of the metric tensor $g_{\mu\nu}$, S_M is the matter action defined as $S_M = \int_V d^4 x \sqrt{-g}\, \mathcal{L}_M(g_{\mu\nu}, \Psi)$, being $\mathcal{L}_M$ the matter Lagrangian density, in which matter is minimally coupled to the metric $g_{\mu\nu}$, and Ψ collectively denotes the matter fields. Without loss of generality, we parametrize the Lagrangian density $f(R)$ as the sum of the gravitational Lagrangian of GR and a deviation term, as follows

$$f(R) = R + 2\Lambda + \epsilon \varphi(R),\tag{10}$$

where Λ is the cosmological constant, ϵ is a dimensionless parameter, and $\epsilon \varphi(R)$ represents the deviation from GR. Substituting Equation (10) in Equation (9), and varying the action with respect to the metric yields the following gravitational field equations

$$G_{\mu\nu} - g_{\mu\nu}\Lambda + \epsilon\left[-\frac{1}{2}g_{\mu\nu}\varphi(R) + \varphi'(R)R_{\mu\nu} - \left(\nabla_\mu \nabla_\nu - g_{\mu\nu}\Box\right)\varphi'(R)\right] = \kappa T_{\mu\nu}.\tag{11}$$

The energy–momentum tensor $T_{\mu\nu}$ is defined as

$$T_{\mu\nu} = -\frac{2}{\sqrt{-g}}\frac{\delta(\sqrt{-g}\mathcal{L}_M)}{\delta g^{\mu\nu}}.\tag{12}$$

We now apply the order reduction technique to Equation (11). In the present case, this amounts to replacing the Ricci scalar and the Ricci tensor, in the terms of order ϵ, by the expression we obtain from them, from the $\epsilon = 0$ version of the same equations. Thus, setting $\epsilon = 0$ in Equation (11), yields the following

$$R^T_{\mu\nu} = \frac{1}{2}g_{\mu\nu}(R_T + 2\Lambda) + \kappa T_{\mu\nu},\tag{13}$$

as the reduced expression of the Ricci tensor, where R_T is the order reduced Ricci scalar. Taking into account the trace of Equation (13), the latter provides $R_T = -\kappa T - 4\Lambda$, and after substituting it into Equation (13), we arrive at

$$R^T_{\mu\nu} = \kappa T_{\mu\nu} - \frac{1}{2}g_{\mu\nu}\kappa T - \Lambda g_{\mu\nu}.\tag{14}$$

Replacing R and $R_{\mu\nu}$, in the ϵ-order terms in Equation (11), with the above expressions, we finally obtain

$$G_{\mu\nu} - g_{\mu\nu}\Lambda + \epsilon\left[-\frac{1}{2}g_{\mu\nu}\varphi(R_T) + \varphi'(R_T)\left(\kappa T_{\mu\nu} - \frac{1}{2}g_{\mu\nu}\kappa T - \Lambda g_{\mu\nu}\right) - \left(\nabla_\mu \nabla_\nu - g_{\mu\nu}\Box\right)\varphi'(R_T)\right] = \kappa T_{\mu\nu}.\tag{15}$$

These are reduced field equations in the sense that we are not using the complete field equations to determine expressions for R and $R_{\mu\nu}$. We should notice that this approximation is only valid when

$$|\epsilon\varphi(R)| \ll |R|. \tag{16}$$

After a solution $\varphi(R)$ is determined, it has to be verified a posteriori for which interval of values of R this is valid.

Throughout this work, we consider an isotropic and homogeneous universe described by the FLRW metric with a $(-,+,+,+)$ signature, which, in spherical coordinates, is given by

$$ds^2 = -dt^2 + a(t)^2\left[\frac{dr^2}{1-kr^2} + r^2d\theta^2 + r^2\sin^2\theta d\phi^2\right], \tag{17}$$

where $a(t)$ is the scale factor and k is the curvature of the universe, which can be set equal to $-1, 0$, or $+1$, depending if one is considering a hyperspherical, spatially flat, or hyperbolic universe, respectively. We assume a perfect fluid description for the content of the universe, given by the energy–momentum tensor, $T_{\mu\nu} = (\rho + p)U_\mu U_\nu + pg_{\mu\nu}$, where U^μ is the four-velocity field of an observer comoving with the fluid, defined in such a way that $U_\mu U^\mu = -1$, and $\rho = \rho(t)$ and $p = p(t)$ are the fluid's energy density and isotropic pressure, respectively.

Now, taking into account the FLRW metric (17), then Equation (15) provides the following modified Friedmann equation

$$H^2 = \frac{1}{3}\kappa\rho - \frac{k}{a^2} - \frac{\Lambda}{3} - \frac{\epsilon}{3}\left[\frac{1}{2}(3w+1)\varphi_T'\kappa\rho + \varphi_T'\Lambda + \frac{1}{2}\varphi_T + 9H^2\varphi_T''(1+w)(3w-1)\kappa\rho\right], \tag{18}$$

by taking into account a barotropic EoS $p = w\rho$, with $-1 \leq w \leq 1$. Note that we recover the standard Friedmann equation for $\epsilon = 0$.

In the context of the order reduction method, we substitute the H^2 term within the ϵ term on the right hand side of Equation (18) by the standard Friedmann equation, which yields [32]

$$
\begin{aligned}
H^2 = {} & \frac{1}{3}\kappa\rho - \frac{k}{a^2} - \frac{\Lambda}{3} - \frac{\epsilon}{3}\left[\frac{1}{2}(3w+1)\varphi'(R_T)\kappa\rho + \varphi'(R_T)\Lambda + \frac{1}{2}\varphi(R_T)\right. \\
& \left. - \left(3\kappa\rho - \frac{9k}{a^2} - 3\Lambda\right)\varphi''(R_T)(1+w)(1-3w)\kappa\rho\right],
\end{aligned}
\tag{19}
$$

as our final form of the reduced modified Friedmann equation in $f(R)$ gravity. It is clear now, that the reduced differential equation is effectively a second-order one. With this result, we aim to find a function $\varphi(R_T)$, such that Equation (19) is the same as some Friedmann equation with a modified source.

2.2. Strategy to Obtain Effective Actions

Here, we specify the conditions in which we will be using Equation (19). In this context, we consider a spatially flat universe ($k = 0$) with a zero cosmological constant ($\Lambda = 0$), so that $R_T = -\kappa\rho(3w - 1)$. Taking these aspects into account, then Equation (19) becomes

$$H^2 = \frac{1}{3}\kappa\rho - \frac{\epsilon}{3}\left[\frac{1}{2}(3w+1)\varphi'(R_T)\kappa\rho + \frac{1}{2}\varphi(R_T) - 3\kappa^2\rho^2\varphi''(R_T)(1+w)(1-3w)\right]. \tag{20}$$

In the following sections, we will compare Equation (20) to the modified Friedmann Equations (6)–(8) considered in the modified loop quantum cosmologies, outlined in the Introduction. However, we reinforce the idea that there are two kinds of modified Friedmann equations at play. One of them is Equation (20), which was derived in the context of metric $f(R)$ gravity, and the other one was derived in the context of LQC. Thus, our main goal is to determine $\varphi(R_T)$, such that Equation (20) is the same as Equations (6)–(8). In doing so, we will determine an effective action that leads to the quantum bounce which characterizes those models. Thus, the modified Friedmann equation, in the context of LQC and its modifications, can be written, in a general way, as

$$H^2 = \frac{1}{3}\kappa\rho + \Psi(\rho), \tag{21}$$

where $\Psi(\rho)$ is some algebraic function, which will depend on the model. Comparing with Equation (20), the requirement is that $\varphi(R_T)$ satisfies

$$-\frac{\epsilon}{3}\left[\frac{1}{2}(3w+1)\varphi'(R_T)\kappa\rho + \frac{1}{2}\varphi(R_T) - 3\kappa^2\rho^2\varphi''(R_T)(1+w)(1-3w)\right] = \Psi(\rho), \tag{22}$$

for a given $\Psi(\rho)$. The same equation, written in terms of R_T, using $R_T = -\kappa\rho(3w-1)$, is given by

$$-\frac{\epsilon}{3}\left[\frac{1}{2}\varphi(R_T) - \frac{(3w+1)}{2(3w-1)}R_T\,\varphi'(R_T) + \frac{3(1+w)}{(3w-1)}R_T^2\,\varphi''(R_T)\right] = \Psi(R_T). \tag{23}$$

For each model, mLQC-I and mLQC-II, we will be considering the scenario of a massless scalar field ($w = 1$). In this context, an effective action for LQC was already determined in Ref. [32].

3. Effective Actions for Modified Loop Quantum Cosmology Models

In order to be in agreement with the approach leading to Equation (1), we consider a massless scalar field, i.e., we set the EoS parameter to be $w = 1$. Therefore, Equation (20) simplifies to

$$H^2 = \frac{1}{3}\kappa\rho - \frac{\epsilon}{3}\left[2\varphi'(R_T)\kappa\rho + \frac{1}{2}\varphi(R_T) + 12\kappa^2\rho^2\varphi''(R_T)\right], \tag{24}$$

and, as a consequence, Equation (22) reduces to

$$-\frac{\epsilon}{3}\left[2\varphi'(R_T)\kappa\rho + \frac{1}{2}\varphi(R_T) + 12\kappa^2\rho^2\varphi''(R_T)\right] = \Psi(\rho), \tag{25}$$

with

$$R_T = -2\kappa\rho. \tag{26}$$

Note that R_T is negative, as the energy density ρ is positive. Finally, we write out Equation (25) in terms of R_T as

$$-\frac{\epsilon}{3}\left[\frac{1}{2}\varphi(R_T) - R_T\varphi'(R_T) + 3R_T^2\varphi''(R_T)\right] = \Psi(R_T). \tag{27}$$

Independently of the model of LQC that we are considering, at a given moment, the solution to the non-homogeneous Equation (27) consists of the sum of the solution to the corresponding homogeneous equation and its particular solution,

$$\varphi(R_T) = \varphi_h(R_T) + \varphi_p(R_T). \tag{28}$$

As such, before specifying any of the models, by substituting $\Psi(R_T)$ in Equation (27) for each case, let us consider the homogeneous equation first, which is common to all cases and is given by

$$\frac{1}{2}\varphi(R_T) - R_T\varphi'(R_T) + 3R_T^2\varphi''(R_T) = 0. \tag{29}$$

This equation is a homogeneous second order Cauchy–Euler equation, having the form

$$x^2 y''(x) + a x y'(x) + b y(x) = 0. \tag{30}$$

Assuming a trial solution, $y(x) = x^m$ gives

$$x^2 m(m-1)x^{m-2} + a x m x^{m-1} + b x^m = 0. \tag{31}$$

We now see that $y = x^m$ was a rather natural choice since we obtained a common factor x^m that we can drop obtaining the characteristic equation for m, given by $m^2 + (a-1)m + b = 0$. Through the roots of this equation, we obtain a solution for Equation (30) whose form depends on the number of roots, and if they are real or complex. In the present case, the characteristic equation for the homogeneous Equation (29) is $m^2 - 4m/3 + 1/6 = 0$ and, solving for m, one obtains two real roots, $m = \frac{1}{6}(4 \pm \sqrt{10})$, meaning that the solution to Equation (29) is given by

$$\varphi_h(R) = c_1 R^{\frac{1}{6}(4-\sqrt{10})} + c_2 R^{\frac{1}{6}(4+\sqrt{10})}. \tag{32}$$

This solution does not contain analytic functions of R, meaning it is not locally given by a convergent power series. For this reason, and also for the fact that $\varphi_h(R)$ does not contribute to the full Equation (27), we can set $c_1 = c_2 = 0$, without loss of generality [32]. As such, the solutions to Equation (27) are simply given by the particular solution in question

$$\varphi(R_T) = \varphi_p(R_T), \tag{33}$$

depending on $\Psi(R_T)$.

3.1. Effective Action for LQC

Here, for self-completeness and self-consistency, we briefly present the particular scenario studied in Ref. [32], where an effective action for LQC was found, requiring matter to be a scalar field, in order to be in agreement with the approach leading to Equation (1) [13]. From Equation (21), for LQC we have that $\Psi(\rho) = -\kappa \rho^2/(3\rho_c)$ or, as a function of R_T, using $R_T = -2\kappa\rho$, we have $\Psi(R_T) = -R_T^2/(12\kappa\rho_c)$. Therefore, in this case, Equation (27) is given by

$$\frac{1}{2}\varphi(R_T) - R_T\varphi'(R_T) + 3R_T^2\varphi''(R_T) = \frac{R_T^2}{4\kappa\rho_c\epsilon}. \tag{34}$$

Since we already covered the solution to Equation (29), we now focus on its particular solution. Looking at Equation (34), the particular solution should be of the form $\varphi_p(R_T) = A R_T^2$, where A is a constant. Plugging this in Equation (34) and solving for A, one obtains $A = 1/(18\epsilon\kappa\rho_c)$. We therefore have [32]:

$$\epsilon\varphi_p(R) = \frac{R^2}{18\kappa\rho_c}, \tag{35}$$

where we have dropped the subscript T since, to ϵ order, $f(R)$ and $f(R_T)$ are the same. Thus, the $f(R)$ function for the effective action, that leads to Equation (1), is [32]

$$f(R) = R + \epsilon\varphi_p(R) = R + \frac{R^2}{18\kappa\rho_c} + ..., \tag{36}$$

where the dots are an indication of an infinite series. Notice that the case of a quadratic model $f(R) \propto R^2$, in the context of LQC, was also considered in Ref. [43]. An effective brane-world and the LQC background expansion histories were also reproduced from a modified gravity perspective in terms of a quadratic term in the Ricci scalar, in a theory non-minimally coupled with the matter Lagrangian [44]. We can neglect the rest of the terms, as they are subdominant with respect to the quadratic term. According to the order reduction method, this solution is only valid for a certain range of curvatures, given by Equation (16), or, equivalently, for a certain range of energy density values. In this case, the solution is valid for

$$R \gg -18\kappa\rho_c \quad \Rightarrow \quad \rho \ll 9\rho_c . \tag{37}$$

Thus, a metric $f(R)$ action was found in Ref. [32] which, when treated as an effective action, leads to the modified Friedmann equation, provided by LQC. Here, we follow an analogous procedure for the modified models of LQC.

3.2. Effective Actions for mLQC-I

As described in the Introduction, mLQC-I is divided in two branches. As such, we consider each of them separately and find an effective action for both.

3.2.1. Effective Action for the b_- Branch

Considering Equation (6) for the b_- branch, we find that Equation (21), in this case, is given by

$$\Psi(\rho) = -\frac{\kappa\rho^2}{3\rho_c^I} + \frac{\kappa\rho}{3}\left(1 - \frac{\rho}{\rho_c^I}\right)\frac{\gamma^2\rho/\rho_c^I}{(\gamma^2+1)(1+\sqrt{1-\rho/\rho_c^I})^2} \tag{38}$$

or, using Equation (26),

$$\Psi(R_T) = -\frac{R_T^2}{12\rho_c^I\kappa} + \frac{R_T}{6}\left(1 + \frac{R_T}{2\kappa\rho_c^I}\right)\frac{\gamma^2 R_T/(2\kappa\rho_c^I)}{(\gamma^2+1)(1+\sqrt{1+R_T/(2\kappa\rho_c^I)})^2} . \tag{39}$$

Therefore, in this case, Equation (27) provides

$$-\frac{\epsilon}{3}\left[\frac{1}{2}\varphi(R_T) - R_T\varphi'(R_T) + 3R_T^2\varphi''(R_T)\right] = -\frac{R_T^2}{12\rho_c^I\kappa} + \frac{R_T}{6}\frac{\left(1 + \frac{R_T}{2\kappa\rho_c^I}\right)\gamma^2 R_T/(2\kappa\rho_c^I)}{(\gamma^2+1)\left(1 + \sqrt{1+R_T/(2\kappa\rho_c^I)}\right)^2} . \tag{40}$$

Using *Wolfram Mathematica*, we obtain the following solution

$$\begin{aligned}
\epsilon\varphi(R) = {} & c_1 R^{\frac{1}{6}(4-\sqrt{10})}\epsilon + c_2 R^{\frac{1}{6}(\sqrt{10}+4)}\epsilon + \frac{1}{90(\gamma^2+1)}\left\{5\left(-72\gamma^2\kappa\rho_c^I + \frac{R^2}{\kappa\rho_c^I} + 54\gamma^2 R\right)\right. \\
& + \frac{18\gamma^2}{\sqrt{10}+10}\left[3\left(4\left(5-4\sqrt{10}\right)\kappa\rho_c^I + \left(\sqrt{10}+10\right)R\right){}_2F_1\left[-\frac{1}{2},\frac{1}{6}\left(-\sqrt{10}-4\right);\frac{1}{6}\left(2-\sqrt{10}\right);-\frac{R}{2\kappa\rho_c^I}\right]\right. \\
& + 3\left(12\left(2\sqrt{10}+5\right)\kappa\rho_c^I + \left(\sqrt{10}+10\right)R\right){}_2F_1\left[-\frac{1}{2},\frac{1}{6}\left(\sqrt{10}-4\right);\frac{1}{6}\left(\sqrt{10}+2\right);-\frac{R}{2\kappa\rho_c^I}\right] \\
& + (2\kappa\rho_c^I + R)\left(9\sqrt{10}\,{}_2F_1\left[\frac{1}{2},\frac{1}{6}\left(-\sqrt{10}-4\right);\frac{1}{6}\left(2-\sqrt{10}\right);-\frac{R}{2\kappa\rho_c^I}\right]\right. \\
& \left.\left.\left. - \left(11\sqrt{10}+20\right){}_2F_1\left[\frac{1}{2},\frac{1}{6}\left(\sqrt{10}-4\right);\frac{1}{6}\left(\sqrt{10}+2\right);-\frac{R}{2\kappa\rho_c^I}\right]\right)\right]\right\},
\end{aligned} \tag{41}$$

where, again, we have dropped the subscript T since, to ϵ order, $f(R)$ and $f(R_T)$ are the same.

As expected, the first two terms in Equation (41) correspond to Equation (32), and the remaining terms correspond to the particular solution to Equation (40). This solution contains hypergeometric functions[1], denoted by ${}_2F_1$, and its domain is $-2\kappa\rho_c^I \leq R$. Along with the fact that it only makes physical sense that $R \leq 0$, we have that our physical solution is defined in the range

$$-2\kappa\rho_c^I \leq R \leq 0 . \tag{44}$$

Thus, a metric $f(R)$ action was found which, when treated as an effective action, leads to the modified Friedmann Equation (7), provided by the b_- branch of mLQC-I. Setting $c_1 = c_2 = 0$, without loss of generality, the effective action is given by

$$
\begin{aligned}
f(R) = {} & R + \frac{1}{90(\gamma^2+1)}\left\{5\left(-72\gamma^2\kappa\rho_c^I + \frac{R^2}{\kappa\rho_c^I} + 54\gamma^2 R\right) + \frac{18\gamma^2}{\sqrt{10}+10}\left[3\left(4\left(5-4\sqrt{10}\right)\kappa\rho_c^I\right.\right.\right. \\
& + \left(\sqrt{10}+10\right)R\right)\,_2F_1\left[-\frac{1}{2},\frac{1}{6}\left(-\sqrt{10}-4\right);\frac{1}{6}\left(2-\sqrt{10}\right);-\frac{R}{2\kappa\rho_c^I}\right] \\
& + 3\left(12\left(2\sqrt{10}+5\right)\kappa\rho_c^I + \left(\sqrt{10}+10\right)R\right)\,_2F_1\left[-\frac{1}{2},\frac{1}{6}\left(\sqrt{10}-4\right);\frac{1}{6}\left(\sqrt{10}+2\right);-\frac{R}{2\kappa\rho_c^I}\right] \\
& + \left(2\kappa\rho_c^I + R\right)\left(9\sqrt{10}\,_2F_1\left[\frac{1}{2},\frac{1}{6}\left(-\sqrt{10}-4\right);\frac{1}{6}\left(2-\sqrt{10}\right);-\frac{R}{2\kappa\rho_c^I}\right]\right. \\
& \left.\left.\left.- \left(11\sqrt{10}+20\right)\,_2F_1\left[\frac{1}{2},\frac{1}{6}\left(\sqrt{10}-4\right);\frac{1}{6}\left(\sqrt{10}+2\right);-\frac{R}{2\kappa\rho_c^I}\right]\right)\right]\right\}.
\end{aligned}
\tag{45}
$$

Using *Wolfram Mathematica*, a Taylor expansion of Equation (45), about $R = 0$ and up to third order, was computed and is given by

$$
f(R) = R + \epsilon\varphi(R) = R + \frac{(4+3\gamma^2)R^2}{72(1+\gamma^2)\kappa\rho_c^I} + \frac{\gamma^2 R^3}{992(1+\gamma^2)(\kappa\rho_c^I)^2} + \dots ,
\tag{46}
$$

where the critical density for this model is given by

$$
\rho_c^I \equiv \frac{\sqrt{3}}{128\pi^2\gamma^3(1+\gamma^2)}.
\tag{47}
$$

In order to better convey the magnitude of the deviation from GR in this case, we provide two different plots in Figure 1. In Figure 1a, we present the deviation of the effective Lagrangian density from the one we find in GR, as a function of R. In this plot, we are able to compare the deviation in the case of LQC, given by Equation (35), with that of mLQC-I, in the case of b_- branch. Notice that we are plotting Equation (41), with $c_1 = c_2 = 0$, not the corresponding series expansion. In Figure 1b, we compare the effective $f(R)$ functions of LQC and the b_- branch of mLQC-I, given by Equations (36) and (45), respectively, with that of GR, with $\Lambda = 0$.

3.2.2. Effective Action for the b_+ Branch

Considering the b_+ branch, with the modified Friedmann Equation (6), we find that Equation (21) is given by

$$
\Psi(\rho) = -\frac{\kappa\rho}{3} + \frac{1}{3}\alpha\kappa\rho_\Lambda\left(1-\frac{\rho}{\rho_c^I}\right)\left[1+\frac{\rho\left(1-2\gamma^2+\sqrt{1-\rho/\rho_c^I}\right)}{4\gamma^2\rho_c^I\left(1+\sqrt{1-\rho/\rho_c^I}\right)}\right],
\tag{48}
$$

or, using Equation (26),

$$
\Psi(R_T) = \frac{R_T}{6} + \frac{1}{3}\alpha\kappa\rho_\Lambda\left(1+\frac{R_T}{2\kappa\rho_c^I}\right)\left[1-\frac{R_T\left(1-2\gamma^2+\sqrt{1+R_T/(2\kappa\rho_c^I)}\right)}{8\gamma^2\kappa\rho_c^I\left(1+\sqrt{1+R_T/(2\kappa\rho_c^I)}\right)}\right].
\tag{49}
$$

Therefore, in this case Equation (27) becomes

$$
-\frac{\epsilon}{3}\left[\frac{1}{2}\varphi_T - R_T\varphi_T' + 3R_T^2\varphi_T''\right] = \frac{R_T}{6} + \frac{1}{3}\alpha\kappa\rho_\Lambda\left(1+\frac{R_T}{2\kappa\rho_c^I}\right)\left[1-\frac{R_T\left(1-2\gamma^2+\sqrt{1+R_T/(2\kappa\rho_c^I)}\right)}{8\gamma^2\kappa\rho_c^I\left(1+\sqrt{1+R_T/(2\kappa\rho_c^I)}\right)}\right],
\tag{50}
$$

where $\varphi_T = \varphi(R_T)$. As before, the solution to Equation (50) was obtained using *Wolfram Mathematica*, and is given by

$$
\begin{aligned}
\epsilon\varphi(R) = {}& c_1 R^{\frac{1}{6}(4-\sqrt{10})}\epsilon + c_2 R^{\frac{1}{6}(\sqrt{10}+4)}\epsilon - \alpha\kappa\rho_\Lambda + R + \frac{\alpha\rho_\Lambda R\left(18(2\gamma^2-1)\kappa\rho_c^I + R\right)}{72\gamma^2\kappa(\rho_c^I)^2} \\
&+\frac{\alpha\rho_\Lambda}{20\rho_c^I}\left\{ -3\left(R - 2\left(\sqrt{10}-2\right)\kappa\rho_c^I\right){}_2F_1\left[-\frac{1}{2},\frac{1}{6}\left(-\sqrt{10}-4\right);\frac{1}{6}\left(2-\sqrt{10}\right);-\frac{R}{2\kappa\rho_c^I}\right] \right. \\
&-3\left(2\left(\sqrt{10}+2\right)\kappa\rho_c^I + R\right){}_2F_1\left[-\frac{1}{2},\frac{1}{6}\left(\sqrt{10}-4\right);\frac{1}{6}\left(\sqrt{10}+2\right);-\frac{R}{2\kappa\rho_c^I}\right] \\
&+(2\kappa\rho_c^I + R)\left(-\left(\sqrt{10}-1\right){}_2F_1\left[\frac{1}{2},\frac{1}{6}\left(-\sqrt{10}-4\right);\frac{1}{6}\left(2-\sqrt{10}\right);-\frac{R}{2\kappa\rho_c^I}\right]\right. \\
&\left.\left.+\left(\sqrt{10}+1\right){}_2F_1\left[\frac{1}{2},\frac{1}{6}\left(\sqrt{10}-4\right);\frac{1}{6}\left(\sqrt{10}+2\right);-\frac{R}{2\kappa\rho_c^I}\right]\right)\right\},
\end{aligned}
\tag{51}
$$

where we have dropped the subscript T since, to ϵ order, $f(R)$ and $f(R_T)$ are the same. Once again, the first two terms correspond to Equation (32), and the remaining ones correspond to the particular solution of Equation (50). As in the previous branch, the solution contains hypergeometric functions and its domain is given by $-2\kappa\rho_c^I \leq R$. As such, since it only makes physical sense that $R \leq 0$, Equation (51) is defined, as a physical solution, for the range

$$
-2\kappa\rho_c^I \leq R \leq 0.
\tag{52}
$$

Taking the same argument as in the previous calculations, we can set $c_1 = c_2 = 0$ and, as a consequence, the effective action is given by

$$
\begin{aligned}
f(R) = {}& -\alpha\kappa\rho_\Lambda + 2R + \frac{\alpha\rho_\Lambda R\left(18(2\gamma^2-1)\kappa\rho_c^I + R\right)}{72\gamma^2\kappa(\rho_c^I)^2} \\
&+\frac{\alpha\rho_\Lambda}{20\rho_c^I}\left\{ -3\left[R - 2\left(\sqrt{10}-2\right)\kappa\rho_c^I\right]{}_2F_1\left[-\frac{1}{2},\frac{1}{6}\left(-\sqrt{10}-4\right);\frac{1}{6}\left(2-\sqrt{10}\right);-\frac{R}{2\kappa\rho_c^I}\right] \right. \\
&-3\left(2\left(\sqrt{10}+2\right)\kappa\rho_c^I + R\right){}_2F_1\left[-\frac{1}{2},\frac{1}{6}\left(\sqrt{10}-4\right);\frac{1}{6}\left(\sqrt{10}+2\right);-\frac{R}{2\kappa\rho_c^I}\right] \\
&+(2\kappa\rho_c^I + R)\left(-\left(\sqrt{10}-1\right){}_2F_1\left[\frac{1}{2},\frac{1}{6}\left(-\sqrt{10}-4\right);\frac{1}{6}\left(2-\sqrt{10}\right);-\frac{R}{2\kappa\rho_c^I}\right]\right. \\
&\left.\left.+\left(\sqrt{10}+1\right){}_2F_1\left[\frac{1}{2},\frac{1}{6}\left(\sqrt{10}-4\right);\frac{1}{6}\left(\sqrt{10}+2\right);-\frac{R}{2\kappa\rho_c^I}\right]\right)\right\}.
\end{aligned}
\tag{53}
$$

A Taylor expansion of Equation (53) up to third order, about $R = 0$, was computed using *Wolfram Mathematica*, resulting in

$$
f(R) = -2\alpha\kappa\rho_\Lambda + \left(\frac{\alpha(5\gamma^2-1)\rho_\Lambda}{4\gamma^2\rho_c^I}+2\right)R + \frac{\alpha(4-3\gamma^2)\rho_\Lambda R^2}{288\gamma^2\kappa(\rho_c^I)^2} + \frac{\alpha\rho_\Lambda R^3}{3968\kappa^2(\rho_c^I)^3} + \dots .
\tag{54}
$$

We observe that, for this particular branch, $f(0) \neq 0$. We recall that ρ_Λ was interpreted as the energy density of an emergent positive cosmological constant in Ref. [15].

Thus, a metric $f(R)$ action was also found for the b_+ branch of mLQC-I which, when treated as an effective action, leads to the modified Friedmann Equation (7). Finally, using the same definitions we used before, to provide plots for the b_- branch, we also provide the same kind of plots for this branch, in Figure 2. In Figure 2a, we can observe that, for mLQC-II, $\epsilon\varphi(R)$ is an increasing function, contrary to all the other models we are considering.

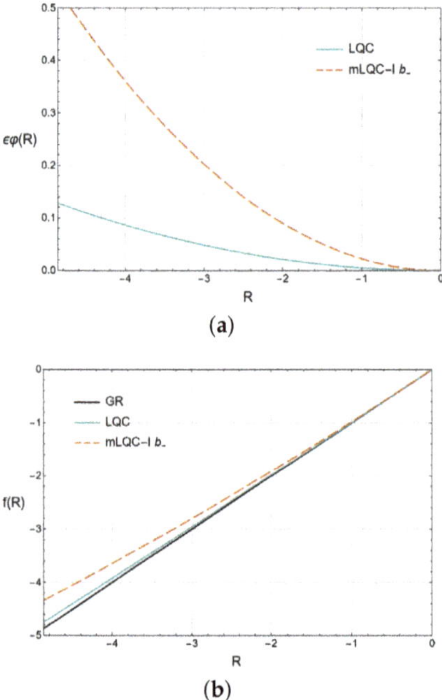

Figure 1. The plot (**a**) (Deviation, from GR, of the effective Lagrangian: a comparison between the b_- branch of mLQC-I and LQC) presents the comparison between Equation (41), for the b_- branch of mLQC-I (dashed dark orange line), and Equation (35), for LQC (solid blue line), with $c_1 = c_2 = 0$, for $-2\kappa\rho_c^I \leq R \leq 0$. Although the deviation is higher for the case of the b_- branch of this modified model, both of them are dominated by the quadratic term. The plot (**b**) (Effective $f(R)$ function: a comparison between the b_- branch of mLQC-I, LQC, and GR) illustrates the differences between the Lagrangian density of GR (solid black line), with $\Lambda = 0$, and the effective Lagrangian densities given by Equation (36), in the case of LQC (solid blue line) and Equation (45), for the b_- branch of mLQC-I (dashed dark orange line), for the interval $-2\kappa\rho_c^I \leq R \leq 0$.

3.3. Effective Action for mLQC-II

We now consider the mLQC-II model, in which the modified Friedmann equation is given by Equation (8). In this case, Equation (21) is given by

$$\Psi(\rho) = \frac{2\kappa\rho\left(1 - \rho/\rho_c^{II}\right)\left(1 + 4(\gamma^2 + 1)\gamma^2\rho/\rho_c^{II}\right)}{3\left(1 + 2\gamma^2\rho/\rho_c^{II} + \sqrt{1 + 4(\gamma^2 + 1)\gamma^2\rho/\rho_c^{II}}\right)} - \frac{\kappa\rho}{3}, \tag{55}$$

or, using Equation (26),

$$\Psi(R_T) = \frac{R_T}{6} - \frac{R_T\left(1 + R_T/(2\kappa\rho_c^{II})\right)\left(1 - 2\gamma^2(\gamma^2 + 1)R_T/(\kappa\rho_c^{II})\right)}{3\left(1 - \gamma^2 R_T/(\kappa\rho_c^{II}) + \sqrt{1 - 2\gamma^2(\gamma^2 + 1)R_T/(\kappa\rho_c^{I})}\right)}. \tag{56}$$

Therefore, in this case Equation (27) is given by

$$-\frac{\epsilon}{3}\left[\frac{1}{2}\varphi(R_T) - R_T\varphi'(R_T) + 3R_T^2\varphi''(R_T)\right] = \frac{R_T}{6} - \frac{R_T\left(1 + \frac{R_T}{2\kappa\rho_c^{II}}\right)\left(1 - \frac{2\gamma^2(\gamma^2 + 1)R_T}{\kappa\rho_c^{II}}\right)}{3\left(1 - \frac{\gamma^2 R_T}{\kappa\rho_c^{II}} + \sqrt{1 - \frac{2\gamma^2(\gamma^2 + 1)R_T}{\kappa\rho_c^{II}}}\right)}. \tag{57}$$

The solution to Equation (57) was obtained using *Wolfram Mathematica*, and leads to

$$\epsilon\varphi(R) = c_1 R^{\frac{1}{6}(4-\sqrt{10})}\epsilon + c_2 R^{\frac{1}{6}(\sqrt{10}+4)}\epsilon + \frac{1}{90\gamma^4}\left\{270\left(\gamma^4+\gamma^2\right)R + \frac{20\left(\gamma^6+\gamma^4\right)R^2}{\kappa\rho_c^{II}} + 90\kappa\rho_c^{II}\right.$$

$$+9\left[3\left(\left(\sqrt{10}-2\right)\kappa\rho_c^{II} + 2\left(\gamma^4+\gamma^2\right)R\right)\,_2F_1\left[-\frac{1}{2},\frac{1}{6}\left(-\sqrt{10}-4\right);\frac{1}{6}\left(2-\sqrt{10}\right);\frac{2R\gamma^2\left(\gamma^2+1\right)}{\kappa\rho_c^{II}}\right]\right.$$

$$+\left(6\left(\gamma^4+\gamma^2\right)R - 3\left(\sqrt{10}+2\right)\kappa\rho_c^{II}\right)\,_2F_1\left[-\frac{1}{2},\frac{1}{6}\left(\sqrt{10}-4\right);\frac{1}{6}\left(\sqrt{10}+2\right);\frac{2R\gamma^2\left(\gamma^2+1\right)}{\kappa\rho_c^{II}}\right]$$

$$+\left(2\left(\gamma^4+\gamma^2\right)R - \kappa\rho_c^{II}\right)\left(\left(\sqrt{10}-1\right)\,_2F_1\left[\frac{1}{2},\frac{1}{6}\left(-\sqrt{10}-4\right);\frac{1}{6}\left(2-\sqrt{10}\right);\frac{2R\gamma^2\left(\gamma^2+1\right)}{\kappa\rho_c^{II}}\right]\right.$$

$$\left.\left.\left.-\left(\sqrt{10}+1\right)\,_2F_1\left[\frac{1}{2},\frac{1}{6}\left(\sqrt{10}-4\right);\frac{1}{6}\left(\sqrt{10}+2\right);\frac{2R\gamma^2\left(\gamma^2+1\right)}{\kappa\rho_c^{II}}\right]\right)\right]\right\},\tag{58}$$

where we have dropped the subscript T since, to order ϵ, $f(R)$, and $f(R_T)$ are the same. As usual, the first two terms correspond to Equation (32) and the remaining terms correspond to the particular solution of Equation (57). For this model, the solution also contains hypergeometric functions but with a different argument and its domain is given by $R \leq \kappa\rho_c^{II}/(2\gamma^2(\gamma^2+1))$. As such, in this case the solution only has physical meaning for the interval

$$R \leq 0.\tag{59}$$

Thus, a metric $f(R)$ action was found which, when treated as an effective action, leads to the modified Friedmann Equation (8), provided by mLQC-II. Setting $c_1 = c_2 = 0$ this effective action is given by

$$f(R) = R + \frac{1}{90\gamma^4}\left\{270\left(\gamma^4+\gamma^2\right)R + \frac{20\left(\gamma^6+\gamma^4\right)R^2}{\kappa\rho_c^{II}} + 90\kappa\rho_c^{II} + 9\left[3\left(\left(\sqrt{10}-2\right)\kappa\rho_c^{II}\right.\right.\right.$$

$$+2\left(\gamma^4+\gamma^2\right)R\right)\,_2F_1\left[-\frac{1}{2},\frac{1}{6}\left(-\sqrt{10}-4\right);\frac{1}{6}\left(2-\sqrt{10}\right);\frac{2R\gamma^2\left(\gamma^2+1\right)}{\kappa\rho_c^{II}}\right]$$

$$+\left(6\left(\gamma^4+\gamma^2\right)R - 3\left(\sqrt{10}+2\right)\kappa\rho_c^{II}\right)\,_2F_1\left[-\frac{1}{2},\frac{1}{6}\left(\sqrt{10}-4\right);\frac{1}{6}\left(\sqrt{10}+2\right);\frac{2R\gamma^2\left(\gamma^2+1\right)}{\kappa\rho_c^{II}}\right]$$

$$+\left(2\left(\gamma^4+\gamma^2\right)R - \kappa\rho_c^{II}\right)\left(\left(\sqrt{10}-1\right)\,_2F_1\left[\frac{1}{2},\frac{1}{6}\left(-\sqrt{10}-4\right);\frac{1}{6}\left(2-\sqrt{10}\right);\frac{2R\gamma^2\left(\gamma^2+1\right)}{\kappa\rho_c^{II}}\right]\right.$$

$$\left.\left.\left.-\left(\sqrt{10}+1\right)\,_2F_1\left[\frac{1}{2},\frac{1}{6}\left(\sqrt{10}-4\right);\frac{1}{6}\left(\sqrt{10}+2\right);\frac{2R\gamma^2\left(\gamma^2+1\right)}{\kappa\rho_c^{II}}\right]\right)\right]\right\}.\tag{60}$$

Using *Wolfram Mathematica*, a Taylor expansion of Equation (60), about $R = 0$ and up to third order, was computed and is given by

$$f(R) = R + \frac{\left(-3\gamma^4 - 2\gamma^2 + 1\right)R^2}{18\kappa\rho_c^{II}} - \frac{\gamma^2\left(\gamma^2+1\right)^3 R^3}{62\left(\kappa\rho_c^{II}\right)^2} + \dots,\tag{61}$$

where the critical density for this model is

$$\rho_c^{II} \equiv \frac{\sqrt{3}\left(\gamma^2+1\right)}{8\pi^2\gamma^3}.\tag{62}$$

This solution for the mLQC-II model is depicted in Figure 3. In Figure 3a, we present the deviation of the effective Lagrangian density from the general relativistic counterpart, and compare them to the the deviation in the case of LQC, given by Equation (35), with that of mLQC-II, given by Equation (41) with $c_1 = c_2 = 0$. In Figure 3b, we compare the effective $f(R)$ functions of LQC and mLQC-II, given by Equation (36) and Equation (60), respectively, with that of GR, with $\Lambda = 0$.

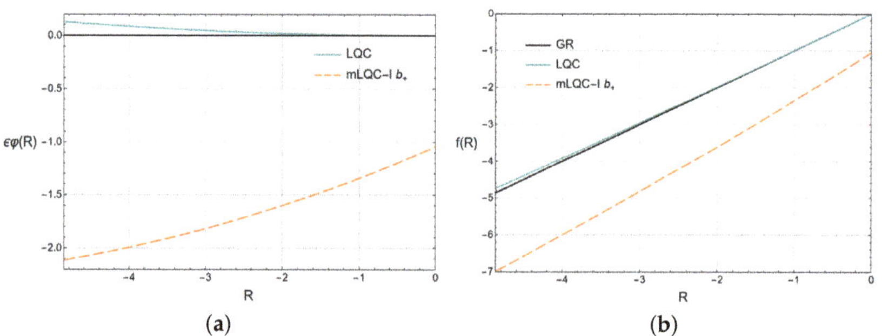

Figure 2. The plot (**a**) (Deviation, from GR, of the effective Lagrangian: a comparison between the b_+ branch of mLQC-I and LQC) presents the comparison between Equation (51), for the b_+ branch of mLQC-I (dashed light orange line), and Equation (35), for LQC (solid blue line), with $c_1 = c_2 = 0$, for $-2\kappa\rho_c^I \leq R \leq 0$. Contrary to the previous case, for this branch the correction is dominated by the linear term. As a result, since $R \leq 0$, the deviation is predominantly negative. The plot (**b**) (Effective $f(R)$ function: a comparison between the b_+ branch of mLQC-I, LQC, and GR) illustrates the differences between the Lagrangian density of GR (solid black line), with $\Lambda = 0$, and the effective Lagrangian densities given by Equation (36), in the case of LQC (solid blue line), and Equation (53), for the b_+ branch of mLQC-I (dashed light orange line), for the interval $-2\kappa\rho_c^I \leq R \leq 0$.

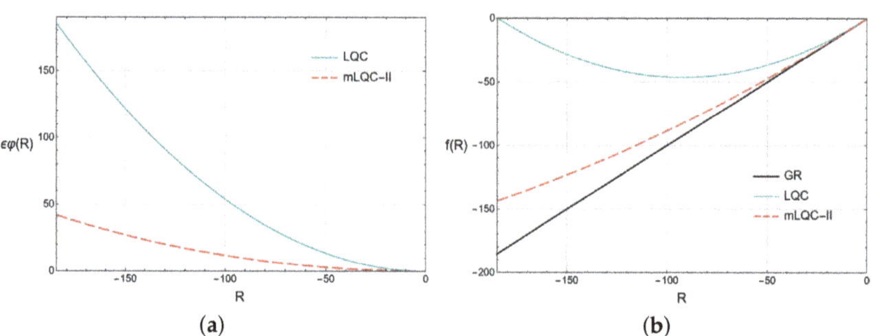

Figure 3. The plot (**a**) (Deviation, from GR, of the effective Lagrangian: a comparison between mLQC-II and LQC) presents the comparison between Equation (58), for the case of mLQC-II (dashed pink line), and Equation (35), for LQC (solid blue line), with $c_1 = c_2 = 0$, for $-18\kappa\rho_c \leq R \leq 0$. Both of the corrections are dominated by the quadratic term. The plot (**b**) (Effective $f(R)$ function: a comparison between mLQC-II, LQC, and GR) illustrates the differences between the Lagrangian density of GR (solid black line), with $\Lambda = 0$, and the effective Lagrangian densities, in the case of LQC (solid blue line), and Equation (60), for mLQC-II (dashed pink line), for the interval $-18\kappa\rho_c \leq R \leq 0$.

Finally, we also present in Figure 4 two plots with all the solutions. In Figure 4a, we compare all the deviations with respect to the Lagrangian density of GR. In Figure 4b, all the effective metric $f(R)$ functions are displayed. With the exception of the b_+ branch of mLQC-I, all deviations from GR are dominated by the quadratic term and the remaining terms become more subdominant, with increasing order, as is the case in LQC. In the case of the b_+ branch, the dominant term is linear, due to the presence of the constant on the right-hand side of Equation (50), which is not the case in any of the other equations we have solved.

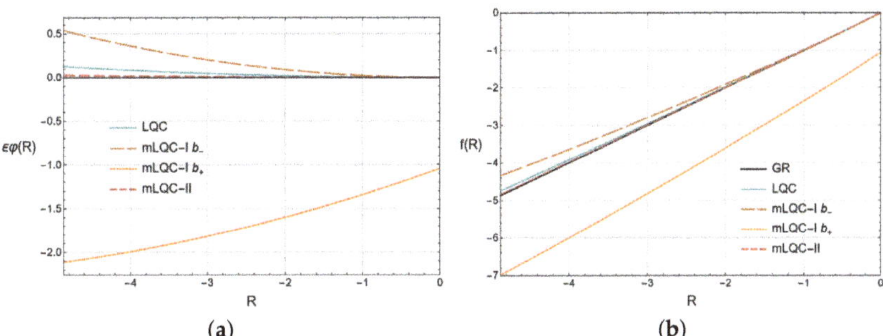

Figure 4. The plot (**a**) (Deviation, from GR, of the effective Lagrangian: a comparison between all models) contains Equation (35), for LQC (solid blue line), Equation (41), for the b_- branch of mLQC-I (dashed dark orange line), Equation (51), for the b_+ branch of mLQC-I (dashed light orange line) and Equation (58), for mLQC-II (dashed pink line), with $c_1 = c_2 = 0$, for $-2\kappa\rho_c^I \leq R \leq 0$. The plot (**b**) (Effective $f(R)$ function: a comparison between all models) presents the effective $f(R)$ functions for all models, including GR (solid black line), with $\Lambda = 0$. These are Equation (36), for LQC (solid blue line), Equation (45), for the b_- branch of mLQC-I (dashed dark orange line), Equation (53), for b_+ branch of mLQC-I (dashed light orange line) and Equation (60), for mLQC-II (dashed pink line), for $-2\kappa\rho_c^I \leq R \leq 0$.

4. Conclusions

In this work, we have addressed the initial singularity problem that is present in the ΛCDM model, in which the Universe emerges from a single point with infinite density, and considered a possible resolution to this question, as proposed by LQC, in which the Big Bang is replaced by a quantum bounce. Through an effective Hamiltonian description, this bounce is manifested by an effective Friedmann equation that is similar to the one found in GR, but with a modified source, given by Equation (1). It is clear from this equation that the energy density reaches a finite maximum value, a critical density, contrary to the Big Bang scenario. In this regard, LQC proposes a scenario in which the Universe undergoes a collapse to an expansion through a bounce.

Since this result comes from a different field in physics, it is relevant to ask if it is possible to replicate it in the framework of GR and its modifications. In this context, in order to obtain the modified Friedmann Equation (1), we considered the class of metric $f(R)$ gravity, where the Ricci scalar in the EH action, is substituted by a general function of R. Using this particular modification of GR, an effective action that leads to Equation (1) was determined in Ref. [32], assuming a massless scalar field. Motivated by this approach, we considered two modifications of standard LQC, which also yield a quantum bounce that occurs for a critical density, specific to each model, which we denoted by mLQC-I and mLQC-II, as outlined in the Introduction. These modifications came to be formulated as a result of a departure of LQC from LQG. More specifically, in LQC, which is a symmetry-reduced model of LQG, the different components of the Hamiltonian are treated as multiples of each other. This is not the case in the full theory of LQG. As such, mLQC-I and mLQC-II are two attempts to incorporate more aspects of LQG in LQC, by means of similar treatments of the Hamiltonian, with respect to the treatment given in LQG.

Furthermore, we applied the covariant order reduction method by obtaining a reduced version of the full field equations of $f(R)$ gravity, in the sense that they are second-order equations and give solutions perturbatively close to GR, and deduced the modified Friedmann equations. This equation depends on the $f(R)$ function, which we chose to parametrize as $f(R) = R + \epsilon\varphi(R)$. Motivated by the example given in Ref. [32], in which a function $\varphi(R)$ was found, we applied this procedure to the mLQC-I and mLQC-II models. The first model, mLQC-I, is divided in two branches, denoted by b_- and b_+, providing the modified Friedmann equations given by Equations (6) and (7), respectively. The second model, mLQC-II, gives the modified Friedmann Equation (8). Using the software *Wolfram Mathematica*, we then found a function $\varphi(R)$, such that Equations (6)–(8) are the same as Equation (20), for $w = 1$. As such, specific effective covariant actions were found, in the context of metric $f(R)$ gravity, which provide Equations (6)–(8), for $|R| \ll \rho_c$, according to Equation (16). These effective actions satisfy $f'(R) > 0$, allowing for a positive effective gravitational coupling, and $f''(R) > 0$, avoiding the Dolgov–Kawasaki instability [25].

The approach followed in this article breaks down in regions of very high energy densities, which are the most interesting regions for physics and phenomenology. Indeed, the approximation that we use is valid for $|R| \ll \rho_c \sim l_p^{-2}$, breaking down at ρ_c, since $|R| \sim \rho_c \sim l_p^{-2}$. Thus, the argument that the effective Lagrangians deduced in this work yield Equations (6)–(8) is not quite a rigorous statement. However, our claim is that at stages close enough to the bounce, such as just before and after it, the Lagrangians found in this work may effectively and correctly describe the collapsing universe. Moreover, it should be noted that the modified Friedmann equations derived in this paper are not guaranteed to respect general covariance, a problem that remains open in LQC models that involve quantum bounces [45].

We would like to point out that there is a simpler approach compared to the one presented in this paper, namely $f(\chi)$ gravity [46], where χ is an invariant, such as the torsion, the intrinsic curvature, a combination of the Carminati–McLenaghan, or the invariant coming from mimetic gravity which, at the background level, mimics standard LQC [47]. This approach could, in principle, be applied to the modifications of LQC, providing simpler formulations than those obtained in the present paper. Moreover, in some works dealing with $f(\chi)$ gravity, it has been reported that the corresponding equations do not coincide with the ones of LQC and, in fact, they can lead to the Ostrogradski instability, as opposed to $f(R)$ gravity theories [29,32], or gradient instability, as well as to the appearance of ghost fields. A similar conclusion could be obtained for modified LQC at the perturbations level.

A successful theory that combines GR and quantum mechanics is yet to be found. In order to predict and describe the beginning of the Universe, we need physical laws that are valid in that regime. If GR is the correct theory in describing the Universe in that period, then the singularity theorem would show that, in the beginning, the Universe was contained in a single point, with infinite density and infinite curvature. However, what the theorem really shows is that, in the beginning of time, the magnitude of the gravitational interaction was so strong, that quantum gravitational effects were, most likely, relevant. As such, it is expected that a quantum theory of gravity will allow for a proper description of the beginning of the Universe. In this regard, the ΛCDM model is incomplete and, for this reason, it is pertinent to attempt modifications of GR, in order to accommodate scenarios in which the Big Bang singularity is non-existent, such as the covariant effective actions determined in this work. In a forthcoming paper we will extend our analysis to arbitrary values of w for mLQC–II.

Author Contributions: Formal analysis, A.R.R., D.V. and F.S.N.L.; Investigation, A.R.R., D.V. and F.S.N.L.; Writing original draft, A.R.R., D.V. and F.S.N.L. All the authors have substantially contributed to the present work. All authors have read and agreed to the published version of the manuscript.

Funding: This research was funded by the Fundação para a Ciência e a Tecnologia (FCT) from the research grants UIDB/04434/2020, UIDP/04434/2020 and CERN/FIS-PAR/0037/2019 and PTDC/FIS-AST/0054/2021.

Institutional Review Board Statement: Not applicable.

Informed Consent Statement: Not applicable.

Data Availability Statement: Not applicable.

Acknowledgments: D.V. acknowledges support from the *Istituto Nazionale di Fisica Nucleare* (INFN) (*iniziativa specifica* TEONGRAV). F.S.N.L. acknowledges support from the Fundação para a Ciência e a Tecnologia (FCT) Scientific Employment Stimulus contract with reference CEECINST/00032/2018, and funding from the research grants UIDB/04434/2020, UIDP/04434/2020 and CERN/FIS-PAR/ 0037/2019. D.V. and F.S.N.L. also acknowledge support from the grant PTDC/FIS-AST/0054/2021.

Conflicts of Interest: The authors declare no conflict of interest.

Note

¹ The Gaussian, or ordinary hypergeometric function, is a special function defined by the hypergeometric series

$$
{}_2F_1[a,b;c;z] = \sum_{n=0}^{\infty} \frac{(a)_n (b)_n}{(c)_n} \frac{z^n}{n!} \tag{42}
$$

on the disk $|z| < 1$, where $(a)_n$, $(b)_n$, and $(c)_n$ are the Pochhammer symbols defined as

$$
(q)_n = \begin{cases} 1, & n = 0, \\ q(q+1)...(q+n-1), & n > 0, \end{cases} \tag{43}
$$

and by analytic continuation elsewhere. ${}_2F_1[a,b;c;z]$ has a branch cut discontinuity in the complex plane running from 1 to ∞. The hypergeometric series is convergent for arbitrary a, b, and c for real $-1 < z < 1$ and for $z \pm 1$ if $c > a + b$.

References

1. Amelino-Camelia, G. Quantum-Spacetime Phenomenology. *Living Rev. Relativ.* **2013**, *16*, 1–137. [CrossRef]
2. Carlip, S. Quantum gravity: A Progress report. *Rep. Prog. Phys.* **2001**, *64*, 885. [CrossRef]
3. Rovelli, C. Loop quantum gravity. *Living Rev. Relativ.* **2008**, *11*, 1–69. [CrossRef]
4. Thiemann, T. *Modern Canonical Quantum General Relativity*; Cambridge University Press: Cambridge, UK, 2008.
5. Ashtekar, A.; Lewandowski, J. Background independent quantum gravity: A Status report. *Class. Quantum Gravity* **2004**, *21*, R53. [CrossRef]
6. Bojowald, M. Loop quantum cosmology. *Living Rev. Relativ.* **2008**, *11*, 1–131. [CrossRef] [PubMed]
7. Bojowald, M. Absence of singularity in loop quantum cosmology. *Phys. Rev. Lett.* **2001**, *86*, 5227. [CrossRef] [PubMed]
8. Ashtekar, A.; Pawlowski, T.; Singh, P. Quantum Nature of the Big Bang: An Analytical and Numerical Investigation. *Phys. Rev. D* **2006**, *73*, 124038. [CrossRef]
9. Ashtekar, A.; Pawlowski, T.; Singh, P. Quantum Nature of the Big Bang: Improved dynamics. *Phys. Rev. D* **2006**, *74*, 84003. [CrossRef]
10. Taveras, V. Corrections to the Friedmann Equations from LQG for a Universe with a Free Scalar Field. *Phys. Rev. D* **2008**, *78*, 64072. [CrossRef]
11. Banerjee, K.; Calcagni, G.; Martin-Benito, M. Introduction to loop quantum cosmology. *SIGMA* **2012**, *8*, 16. [CrossRef]
12. De Haro, J.; Amorós, J. Bouncing cosmologies via modified gravity in the ADM formalism: Application to Loop Quantum Cosmology. *Phys. Rev. D* **2018**, *97*, 64014. [CrossRef]
13. Singh, P. Loop cosmological dynamics and dualities with Randall-Sundrum braneworlds. *Phys. Rev. D* **2006**, *73*, 63508. [CrossRef]
14. Meissner, K.A. Black hole entropy in loop quantum gravity. *Class. Quantum Gravity* **2004**, *21*, 5245. [CrossRef]
15. Li, B.F.; Singh, P.; Wang, A. Towards Cosmological Dynamics from Loop Quantum Gravity. *Phys. Rev. D* **2018**, *97*, 84029. [CrossRef]
16. Li, B.F.; Singh, P.; Wang, A. Qualitative dynamics and inflationary attractors in loop cosmology. *Phys. Rev. D* **2018**, *98*, 66016. [CrossRef]
17. Yang, J.; Ding, Y.; Ma, Y. Alternative quantization of the Hamiltonian in loop quantum cosmology II: Including the Lorentz term. *Phys. Lett. B* **2009**, *682*, 1–7. [CrossRef]
18. Dapor, A.; Liegener, K. Cosmological Effective Hamiltonian from full Loop Quantum Gravity Dynamics. *Phys. Lett. B* **2018**, *785*, 506–510. [CrossRef]
19. Dapor, A.; Liegener, K. Cosmological coherent state expectation values in loop quantum gravity I. Isotropic kinematics. *Class. Quantum Gravity* **2018**, *35*, 135011. [CrossRef]
20. Alesci, E.; Barrau, A.; Botta, G.; Martineau, K.; Stagno, G. Phenomenology of Quantum Reduced Loop Gravity in the isotropic cosmological sector. *Phys. Rev. D* **2018**, *98*, 106022. [CrossRef]
21. Bilski, J.; Marcianò, A. Critical Insight into the Cosmological Sector of Loop Quantum Gravity. *Phys. Rev. D* **2020**, *101*, 66026. [CrossRef]
22. Li, B.F.; Singh, P.; Wang, A. Genericness of pre-inflationary dynamics and probability of the desired slow-roll inflation in modified loop quantum cosmologies. *Phys. Rev. D* **2019**, *100*, 63513. [CrossRef]

23. Olmo, G.J.; Singh, P. Effective Action for Loop Quantum Cosmology a la Palatini. *J. Cosmol. Astropart. Phys.* **2009**, *2009*, 30. [CrossRef]
24. Olmo, G.J. Palatini Approach to Modified Gravity: $f(R)$ Theories and Beyond. *Int. J. Mod. Phys. D* **2011**, *20*, 413–462. [CrossRef]
25. Sotiriou, T.P.; Faraoni, V. $f(R)$ Theories Of Gravity. *Rev. Mod. Phys.* **2010**, *82*, 451. [CrossRef]
26. Lobo, F.S.N. The Dark side of gravity: Modified theories of gravity. *arXiv* **2008**, arXiv:0807.1640.
27. De Felice, A.; Tsujikawa, S. $f(R)$ theories. *Living Rev. Relativ.* **2010**, *13*, 1–161. [CrossRef] [PubMed]
28. Nojiri, S.; Odintsov, S.D. Unified cosmic history in modified gravity: From $f(R)$ theory to Lorentz non-invariant models. *Phys. Rep.* **2011**, *505*, 59–144. [CrossRef]
29. Clifton, T.; Ferreira, P.G.; Padilla, A.; Skordis, C. Modified Gravity and Cosmology. *Phys. Rep.* **2012**, *513*, 1–189. [CrossRef]
30. Capozziello, S.; De Laurentis, M. Extended Theories of Gravity. *Phys. Rep.* **2011**, *509*, 167–321. [CrossRef]
31. Harko, T.; Lobo, F.S.N. Extensions of $f(R)$ Gravity: Curvature-Matter Couplings and Hybrid Metric-Palatini Theory. In *Cambridge Monographs on Mathematical Physics*; Cambridge University Press: Cambridge, UK, 2018.
32. Sotiriou, T.P. Covariant Effective Action for Loop Quantum Cosmology from Order Reduction. *Phys. Rev. D* **2009**, *79*, 44035. [CrossRef]
33. Bel, L.; Zia, H.S. Regular reduction of relativistic theories of gravitation with a quadratic Lagrangian. *Phys. Rev. D* **1985**, *32*, 3128–3135. [CrossRef] [PubMed]
34. Simon, J.Z. Higher Derivative Lagrangians, Nonlocality, Problems and Solutions. *Phys. Rev. D* **1990**, *41*, 3720. [CrossRef]
35. Simon, J.Z. No Starobinsky inflation from selfconsistent semiclassical gravity. *Phys. Rev. D* **1992**, *45*, 1953–1960. [CrossRef]
36. Huang, Q.G. A polynomial $f(R)$ inflation model. *J. Cosmol. Astropart. Phys.* **2014**, *2*, 35. [CrossRef]
37. Castellanos, A.R.R.; Sobreira, F.; Shapiro, I.L.; Starobinsky, A.A. On higher derivative corrections to the $R + R^2$ inflationary model. *J. Cosmol. Astropart. Phys.* **2018**, *12*, 7. [CrossRef]
38. Terrucha, I.; Vernieri, D.; Lemos, J.P.S. Covariant action for bouncing cosmologies in modified Gauss–Bonnet gravity. *Ann. Phys.* **2019**, *404*, 39–46. [CrossRef]
39. Barros, B.J.; Teixeira, E.M.; Vernieri, D. Bouncing cosmology in $f(R, \mathcal{G})$ gravity by order reduction. *Ann. Phys.* **2020**, *419*, 168231. [CrossRef]
40. Bajardi, F.; Vernieri, D.; Capozziello, S. Bouncing Cosmology in f(Q) Symmetric Teleparallel Gravity. *Eur. Phys. J. Plus* **2020**, *135*, 912. [CrossRef]
41. Miranda, M.; Vernieri, D.; Capozziello, S.; Lobo, F.S.N. Effective actions for loop quantum cosmology in fourth-order gravity. *Eur. Phys. J. C* **2021**, *81*, 975. [CrossRef]
42. Miranda, M.; Vernieri, D.; Capozziello, S.; Lobo, F.S.N. Bouncing Cosmology in Fourth-Order Gravity. *Universe* **2022**, *8*, 161. [CrossRef]
43. Amorós, J.; de Haro, J.; Odintsov, S.D. $R + \alpha R^2$ Loop Quantum Cosmology. *Phys. Rev. D* **2014**, *89*, 104010. [CrossRef]
44. Olmo, G.J.; Rubiera-Garcia, D. "Brane-world and loop cosmology from a gravity–matter coupling perspective. *Phys. Lett. B* **2015**, *740*, 73–79. [CrossRef]
45. Bojowald, M. Noncovariance of the dressed-metric approach in loop quantum cosmology. *Phys. Rev. D* **2020**, *102*, 23532. [CrossRef]
46. Bernal, T.; Capozziello, S.; Hidalgo, J.C.; Mendoza, S. Recovering MOND from extended metric theories of gravity. *Eur. Phys. J. C* **2011**, *71*, 1794. [CrossRef]
47. de Haro, J.; Aresté Saló, L.; Pan, S. Limiting curvature mimetic gravity and its relation to Loop Quantum Cosmology. *Gen. Relativ. Gravit.* **2019**, *51*, 49. [CrossRef]

 *universe*

MDPI

Article

Cosmological Fluctuations in Delta Gravity

Jorge Alfaro [1,*,†], Carlos Rubio [2,†] and Marco San Martín [3,†]

1   Instituto de Física, Pontificia Universidad Católica de Chile, Av. Vicuña Mackenna 4860, Santiago 7820436, Chile
2   Facultad de Ingeniería y Ciencias, Universidad Adolfo Ibañez, Av. Diagonal Las Torres 2640, Peñalolén, Santiago 7941169, Chile; carlos.rubio@edu.uai.cl
3   Escuela de Ingeniería Mecánica, Pontificia Universidad Católica de Valparaíso, Av. Los Carrera 1567, Quilpué 2430120, Chile; mlsanmartin@uc.cl
*   Correspondence: jalfaro@uc.cl; Tel.: +569-6843-1908
†   These authors contributed equally to this work.

Abstract: About 70% of the Universe is Dark Energy, but the physics community still does not know what it is. Delta gravity (DG) is an alternative theory of gravitation that could solve this cosmological problem. Previously, we studied the Universe's accelerated expansion, where DG was able to explain the SNe-Ia data successfully. In this work, we computed the cosmological fluctuations in DG that give rise to the CMB through a hydrodynamic approximation. We calculated the gauge transformations for the metric and the perfect fluid to present the equations of the evolution of cosmological fluctuations. This provided the necessary equations to solve the scalar TT power spectrum in a semi-analytical way. These equations are useful for comparing the DG theory with astronomical observations and thus being able to constrain the DG cosmology.

Keywords: cosmology; modified gravity; cosmic microwave background; cosmological perturbations; dark energy

Citation: Alfaro, J.; Rubio, C.; San Martín, M. Cosmological Fluctuations in Delta Gravity. *Universe* **2023**, *9*, 315. https://doi.org/10.3390/universe9070315

Academic Editor: Kazuharu Bamba

Received: 6 April 2023
Revised: 27 June 2023
Accepted: 29 June 2023
Published: 30 June 2023

1. Introduction

In the past decade, there has been a surge of interest in cosmology due to the increasingly precise observational constraints that can help elucidate the physics underlying the Universe. Despite the mounting evidence for cosmological phenomena such as the acceleration of the Universe attributed to Dark Energy (DE) and the presence of non-visible matter known as Dark Matter (DM) [1–3], the physics community has yet to provide a comprehensive explanation for their nature.

The standard cosmological model, known as ΛCDM, describes the composition of the Universe, where 69% of the energy density corresponds to DE, 26% corresponds to DM, and the remaining 5% is composed of ordinary matter and light [1]. This model has successfully accounted for various observations, including those of Type Ia Supernovae (SNe-Ia) and the cosmic microwave background (CMB), and it has been validated through cosmological simulations that depict the formation of large-scale structures [4,5].

However, the ΛCDM model exhibits inconsistencies between its description of the early and late Universe [6]. These inconsistencies manifest in different cosmological parameters, such as the Hubble constant [7,8], the curvature [9,10], and the S_8 tension [11].

The Planck team measured the local expansion rate through the cosmic microwave background (CMB) radiation and obtained a value of $H_0 = 67.37 \pm 0.54$ km/s/Mpc, which aligns with a flat ΛCDM model [1]. However, the SH0ES collaboration independently measured a higher value of $H_0 = 73.52 \pm 1.62$ km/s/Mpc for the local Universe [7], creating a discrepancy with the Planck value exceeding 3.5σ. Importantly, this tension between the early and late Universe persists even without considering the Planck CMB data or the SH0ES distance ladder [6].

Furthermore, the H0LiCOW collaboration derived a direct measurement of $H_0 = 72.5^{+2.1}_{-2.3}$ km/s/Mpc based on lensing time delays, which exhibits a moderate tension with the Planck value [12]. Additionally, a constraint obtained from the Big Bang nucleosynthesis (BBN) combined with baryon acoustic oscillation (BAO) data yielded $H_0 = 66.98 \pm 1.18$ km/s/Mpc, which is inconsistent with the SH0ES measurement [6].

Other studies have attempted to explain the discrepancy by suggesting that the Hubble constant determined from nearby SNe-Ia may differ from that measured from the CMB due to cosmic variance, with a potential difference of ± 0.8 percent at 1σ statistical significance. However, this variation does not account for the observed discrepancy between SNe-Ia and CMB measurements [13]. In an extreme case, observers situated in the centers of vast cosmic voids might measure a Hubble constant biased high by 5 percent from SNe-Ia.

Since the initial publication highlighting the H_0 tension [2], numerous inquiries have emerged regarding the source of this discrepancy. One suggestion is that errors in the calibration of Cepheids, which contribute to systematic errors, could be responsible. However, this potential error has been thoroughly discussed and dismissed by Riess et al. [14].

Other publications have explored possible solutions to the observed acceleration of the Universe, including anisotropies at local scales. Wang et al. [15], using SNe-Ia data, found evidence of anisotropies associated with the direction and amplitude of the bulk flow. Nonetheless, the impact of dipolar distribution of dark energy cannot be ruled out at high redshifts. Similarly, another publication [16] suggests that the anisotropies in cosmic acceleration may be linked to the nature of Dark Energy, implying that the perceived cosmic acceleration deduced from supernovae could be an artifact of our non-Copernican perspective rather than evidence of a dominant "dark energy" component in the Universe. Sun et al. [17] conclude that even in the presence of anisotropy, Dark Energy cannot be entirely ruled out. While such proposals could potentially explain variations in local measurements, including different values for the local Hubble constant, they could contradict the analyses conducted by Planck using the ΛCDM model. The Dark Energy component is crucial for the evolution of CMB photons from the last scattering surface until the present, and altering the sum over Ω for each component in the Universe would lead to significant changes. Various other suggestions concerning discrepancies have emerged not only related to SNe-Ia measurements but also within the Planck data itself. The presence of anisotropies in these measurements has been a subject of debate due to high uncertainties and inconsistent results. Hypotheses proposing the possibility of a Universe with less Dark Energy [18] have also been put forward.

Another potential source of error in local measurements could be the inhomogeneity in local density [19,20]. However, in this scenario, the presence of local structures does not seem to impede the possibility of measuring the Hubble constant with a precision of 1%, and there is no evidence of a Hubble constant change corresponding to an inhomogeneity.

Today, there are different methods to obtain the Hubble constant, including the use of SNe-II, ref. [21]. In this research, SNe-II were employed as standard candles to obtain an independent measurement of the Hubble constant. The resulting value was $H_0 = 75.8^{+5.2}_{-4.9}$ km /s/Mpc. The local H_0 value is higher than the value derived from the early Universe, with a confidence level of 95%. The researchers concluded that *there is no evidence that SNe-Ia are the source of the H_0 tension*. In another publication analyzing SNe-Ia as standard candles in the near-infrared, it was concluded that $H_0 = 72.8 \pm 1.6$ (statistical) ± 2.7 (systematic) km/s/Mpc. This study also suggested that the tension in the competing H_0 distance ladders is likely not a result of supernova systematics.

Other proposals have tried to reconcile Planck and SNe-Ia data, including modifications to the physics of the DE. In other words, introducing an equation of state of the interacting dark energy component, where w is allowed to vary freely, could solve the H_0 tension [22]. Additionally, a decaying dark matter model has been proposed to alleviate the H_0 and σ_8 anomalies [23]; in their work, they reduce the tension for both measurements when only consider Planck CMB data and the local SH0ES prior on H_0. However, when BAOs and the JLA supernova dataset are included, their model is weakened.

Other disagreements are related to inconsistencies with curvature (and other parameters needed to describe the CMB) [10], or they are related to the tension between measurements of the amplitude of the power spectrum of density perturbations (inferred using CMB) and directly measured by large-scale structure (LSS) on smaller scales [11]. Extensions of ΛCDM models have been considered [24] in an attempt to solve the tension of H_0. However, they concluded that none of these extended models can convincingly resolve the H_0 tension. For a full scope of the Hubble tension, please see [25]. Through the time, the tension between Planck and SNe-Ia persists [1,14], where the H_0 is the most significant tension. Furthermore, the Universe is composed principally by DE, but we still do not know what it is.

Over the past decades, various proposals have been made to explain the observed acceleration of the Universe. These proposals involve the inclusion of additional fields in approaches such as Quintessence, Chameleon, Vector Dark Energy or Massive Gravity, the addition of higher-order terms in the Einstein–Hilbert action, such as $f(R)$ theories and Gauss–Bonnet terms, and the introduction of extra dimensions for a modification of gravity on large scales ([26]). Other interesting possibilities include the search for non-trivial ultraviolet fixed points in gravity (asymptotic safety, [27]) and the notion of induced gravity ([28–31]). The first possibility uses exact renormalization-group techniques ([32,33]) together with a lattice and numerical techniques, such as Lorentzian triangulation analysis ([34]). Induced gravity proposes that gravitation is a residual force produced by other interactions.

Delta gravity (DG) is an extension of General Relativity (GR) where new fields are added to the Lagrangian through a new symmetry [35–38]. The main properties of this model at the classical level follow: (a) It agrees with GR outside the sources and with adequate boundary conditions. In particular, the causal structure of delta gravity in a vacuum is the same as in General Relativity, satisfying all standard tests automatically. (b) When studying the evolution of the Universe, it predicts acceleration without a cosmological constant or additional scalar fields. The Universe ends in a BigRip, which is similar to the scenario considered in [39]. (c) The scale factor agrees with the standard cosmology at early times and show acceleration only at late times. Therefore, we expect that density perturbations should not have large corrections at the moment of last scattering (denoted by t_{ls}).

It was noticed in [40] that the Hamiltonian of delta models is not bounded from below. Phantom cosmological models [39,41] also have this property. The present model could provide an arena to study the quantum properties of a phantom field, since the model has a finite quantum effective action. In this respect, the advantage of the present model is that being a gauge model, it could give us the possibility of solving the problem of lack of unitarity using standard techniques of gauge theories such as the BRST method ([36]). However, we are not concerned about this feature in this work, because we are considering DG as a phenomenological model that interpolates the observations of the early with the late Universe.

This theory predicts an accelerating Universe without a cosmological constant Λ and a Hubble parameter $H_0 = 74.47 \pm 1.63$ Km/s/Mpc [42] when fitting SN-Ia Data, which is in agreement with SH0ES.

On the other hand, temperature correlations provide us with information about the constituents of the Universe, including baryonic and dark matter. Typically, these calculations are performed using software such as CMBFast [43,44] or CAMB[1] [45]. These codes employ Boltzmann equations for the fluids and their interactions, yielding well-established results that are consistent with Planck measurements [1].

Nevertheless, one can obtain a good approximation of this complex problem [46,47]. In this work, we use an analytical method that consists of two steps instead of studying the evolution of the scalar perturbations using Boltzmann equations. First, we use a hydrodynamic approximation, which assumes photons and baryonic plasma as a fluid in thermal equilibrium at recombination time when there is a high rate of collisions between free electrons and photons. Second, we study the propagation of photons [35] by radial geodesics from the moment when the Universe switches from opaque to transparent at time t_{ls} until now.

In this research, we develop the theory of scalar perturbations at first order. We discuss the gauge transformations in an extended Friedmann–Lemaître–Robertson–Walker (FRLW) Universe. Then, we show how to obtain an expression for temperature fluctuations, and we demonstrate that they are gauge invariant, which is crucial from a theoretical point of view. With this result, we derive a formula for the scalar contribution to temperature multipole coefficients. This formula is useful to test the theory, and it could indicate the physical consequence of the "delta matter" introduced in this theory. This work has been incorporated as a part of the Ph.D. thesis [48], where more details can be found.

The CMB provides cosmological constraints crucial for testing a model. Many cosmological parameters can be obtained directly from the CMB power spectrum, such as $h^2\Omega_b, h^2\Omega_c, 100\theta, \tau, A_s$ and n_s [1], while others can be derived from constraining CMB observation with SNe-Ia or BAOs. By studying the CMB anisotropies, we can address two aspects: the compatibility between the CMB power spectrum and DG fluctuations and the compatibility between CMB and SNe-Ia in the DG theory. In [49], we already fitted Planck satellite's data with a DG model using Markov Chain Monte Carlo analysis. We also studied the compatibility between SNe Ia and CMB observation in this framework. We obtained the scalar CMB TT power spectrum and the fitted parameters needed to explain both SNe-Ia data and CMB measurements. The results are in reasonable agreement with both observations considering the analytical approximation. We also discussed whether the Hubble constant and the accelerating Universe are in concordance with the observational evidence in the DG context. With this in mind, the aim of this work is to present the full theoretical scheme for the scalar perturbation theory.

The paper is organized as follows: In Section 2, we introduce the definition of DG and its equations of motion. We then review some implications of the first law of thermodynamics, which will allow us to interpret the physical quantities of DG. Before finishing this section, we state the ansatz that the moment of equality between matter and radiation was equal in both DG and GR, and we discuss its implications. In Section 3, we study the gauge transformation for small perturbations of both geometrical and matter fields. We choose a gauge and present the gauge-invariant equations of motion for small perturbations. In Section 4, we study the evolution of cosmological perturbations, solving the equations partially when the Universe is dominated by radiation and when it is dominated by matter. In Section 5, we derive the formula for temperature fluctuation. Here, we find that this fluctuation can be expressed in three independent and gauge-invariant terms. In Section 6, we obtain a formula for temperature multipole coefficients for scalar modes. We also present preliminary numerical results for the power spectrum of the CMB. Finally, we provide conclusions and remarks.

For notation, we will use the Riemann tensor:

$$R^{\alpha}{}_{\beta\mu\nu} = \partial_\mu \Gamma^{\alpha}_{\nu\beta} - \partial_\nu \Gamma^{\alpha}_{\mu\beta} + \Gamma^{\alpha}_{\mu\gamma}\Gamma^{\gamma}_{\nu\beta} - \Gamma^{\alpha}_{\nu\gamma}\Gamma^{\gamma}_{\mu\beta},\tag{1}$$

where the Ricci Tensor is given by $R_{\mu\nu} = R^{\alpha}{}_{\mu\alpha\nu}$, the Ricci scalar $R = g^{\mu\nu}R_{\mu\nu}$ and:

$$\Gamma^{\alpha}_{\mu\nu} = \frac{1}{2}g^{\alpha\beta}(\partial_\nu g_{\beta\mu} + \partial_\mu g_{\nu\beta} - \partial_\beta g_{\mu\nu})\tag{2}$$

is the usual Christoffel symbol. Finally, the covariant derivative is given by:

$$D_\nu A_\mu \equiv A_{\mu;\nu} = A_{\mu,\nu} - \Gamma^{\alpha}_{\mu\nu}A_\alpha.\tag{3}$$

So, it is defined with the usual metric $g_{\mu\nu}$.

2. Definition of Delta Gravity

In this section, we will present the action as well all the symmetries of the model and derive the equations of motion. These approaches are based on the application of a variation called $\tilde{\delta}$, and it has the usual properties of a variation such as:

$$
\begin{aligned}
\tilde{\delta}(AB) &= (\tilde{\delta}A)B + A(\tilde{\delta}B), \\
\tilde{\delta}\delta A &= \delta\tilde{\delta}A, \\
\tilde{\delta}(\Phi_\mu) &= (\tilde{\delta}\Phi)_\mu,
\end{aligned}
\tag{4}
$$

where δ is another variation. The main point of this variation is that when it is applied on a field (function, tensor, etc), it produces new elements that we define as $\tilde{\delta}$ fields, which we treat as an entirely new independent object from the original, $\tilde{\Phi} = \tilde{\delta}(\Phi)$. We use the convention that a tilde tensor is equal to the $\tilde{\delta}$ transformation of the original tensor when all its indexes are covariant. (For more detail about $\tilde{\delta}$, please see Appendix A.1.)

Now, we will present the $\tilde{\delta}$ prescription for a general action. The extension of the new symmetry is given by:

$$
S_0 = \int d^n x \mathcal{L}_0(\phi, \partial_i \phi) \to S = \int d^n x (\mathcal{L}_0(\phi, \partial_i \phi) + \tilde{\delta}\mathcal{L}_0(\phi, \partial_i \phi)),
\tag{5}
$$

where S_0 is the original action and S is the extended action in Delta Gauge Theories. When we apply this formalism to the Einstein–Hilbert action of GR, we obtain [35]

$$
S = \int d^4 x \sqrt{-g}\left(\frac{R}{2\varkappa} + L_M - \frac{1}{2\varkappa}\left(G^{\alpha\beta} - \varkappa T^{\alpha\beta}\right)\tilde{g}_{\alpha\beta} + \tilde{L}_M\right),
\tag{6}
$$

where $\varkappa = \frac{8\pi G}{c^4}$ (hereafter, we set $c = 1$), $\tilde{g}_{\mu\nu} = \tilde{\delta}g_{\mu\nu}$, L_M is the matter Lagrangian and:

$$
T^{\mu\nu} = \frac{2}{\sqrt{-g}}\frac{\delta}{\delta g_{\mu\nu}}\left[\sqrt{-g}L_M\right],
\tag{7}
$$

$$
\tilde{L}_M = \tilde{\phi}_I \frac{\delta L_M}{\delta \phi_I} + (\partial_\mu \tilde{\phi}_I)\frac{\delta L_M}{\delta(\partial_\mu \phi_I)},
\tag{8}
$$

where $\tilde{\phi} = \tilde{\delta}\phi$ are the $\tilde{\delta}$ matter fields or "delta matter" fields. The equations of motion are given by the variation of $g_{\mu\nu}$ and $\tilde{g}_{\mu\nu}$. It is easy to see that we obtain the usual Einstein's equations varying the action (6) with respect to $\tilde{g}_{\mu\nu}$. On the other hand, variations with respect to $g_{\mu\nu}$ give the equations for $\tilde{g}_{\mu\nu}$:

$$
\begin{aligned}
F^{(\mu\nu)(\alpha\beta)\rho\lambda} D_\rho D_\lambda \tilde{g}_{\alpha\beta} \;+\;& \frac{1}{2}R^{\alpha\beta}\tilde{g}_{\alpha\beta}g^{\mu\nu} + \frac{1}{2}R\tilde{g}^{\mu\nu} - R^{\mu\alpha}\tilde{g}_\alpha^\nu - R^{\nu\alpha}\tilde{g}_\alpha^\mu + \frac{1}{2}\tilde{g}_\alpha^\alpha G^{\mu\nu} \\
=\;& \frac{\varkappa}{\sqrt{-g}}\frac{\delta}{\delta g_{\mu\nu}}\left[\sqrt{-g}\left(T^{\alpha\beta}\tilde{g}_{\alpha\beta} + 2\tilde{L}_M\right)\right],
\end{aligned}
\tag{9}
$$

with:

$$
\begin{aligned}
F^{(\mu\nu)(\alpha\beta)\rho\lambda} &= P^{((\rho\mu)(\alpha\beta))}g^{\nu\lambda} + P^{((\rho\nu)(\alpha\beta))}g^{\mu\lambda} - P^{((\mu\nu)(\alpha\beta))}g^{\rho\lambda} - P^{((\rho\lambda)(\alpha\beta))}g^{\mu\nu}, \\
P^{((\alpha\beta)(\mu\nu))} &= \frac{1}{4}\left(g^{\alpha\mu}g^{\beta\nu} + g^{\alpha\nu}g^{\beta\mu} - g^{\alpha\beta}g^{\mu\nu}\right),
\end{aligned}
\tag{10}
$$

where $(\mu\nu)$ denotes the totally symmetric combination of μ and ν. It is possible to simplify (9) (see [35]) to obtain the following system of equations:

$$
G^{\mu\nu} = \varkappa T^{\mu\nu},
\tag{11}
$$

$$
F^{(\mu\nu)(\alpha\beta)\rho\lambda} D_\rho D_\lambda \tilde{g}_{\alpha\beta} + \frac{1}{2}g^{\mu\nu}R^{\alpha\beta}\tilde{g}_{\alpha\beta} - \frac{1}{2}\tilde{g}^{\mu\nu}R = \varkappa \tilde{T}^{\mu\nu},
\tag{12}
$$

where $\tilde{T}^{\mu\nu} = \tilde{\delta}T^{\mu\nu}$. The energy momentum conservation now is given by

$$D_\nu T^{\mu\nu} = 0,$$
(13)

$$D_\nu \tilde{T}^{\mu\nu} = \frac{1}{2}T^{\alpha\beta}D^\mu \tilde{g}_{\alpha\beta} - \frac{1}{2}T^{\mu\beta}D_\beta \tilde{g}^\alpha_\alpha + D_\beta(\tilde{g}^\beta_\alpha T^{\alpha\mu}).$$
(14)

Then, we are going to work with Equations (11)–(14). However, as the perturbation theory in the standard sector is well known (see [47]), we will focus on the DG sector.

One important result of DG is that photons follow geodesic trajectories given by the effective metric $\mathbf{g}_{\mu\nu} = g_{\mu\nu} + \tilde{g}_{\mu\nu}$ [35], and for an FRLW Universe, these metrics take the form (with constant curvature parameter $k = 0$)

$$\bar{g}_{\mu\nu}dx^\mu dx^\nu = -dt^2 + a^2(t)(dx^2 + dy^2 + dz^2),$$
(15)

and

$$\tilde{\bar{g}}_{\mu\nu}dx^\mu dx^\nu = -3F(t)dt^2 + F(t)a^2(t)(dx^2 + dy^2 + dz^2),$$
(16)

where $F(t)$ is a time-dependent function which is determined by the solution of the unperturbed equations system, and $a(t)$ is the standard scale factor, which in Section 4 we will show is no longer the physical scale factor of the Universe. To obtain the form of $\bar{g}_{\mu\nu}$ and $\tilde{\bar{g}}_{\mu\nu}$, first we impose isotropy and homogeneity, and then, we apply the harmonic gauge $g^{\mu\nu}\Gamma^\alpha_{\mu\nu} = 0$ and its tilde version (for details, see [37]). One of the implications of this effective metric is that geometry is now described by a new tridimensional metric given by [35][2] (latin indexes run from 1 to 3)

$$dl^2 = \gamma_{ij}dx^i dx^j,$$
(17)

$$\gamma_{ij} = \frac{g_{00}}{g_{00}}\left(g_{ij} - \frac{g_{i0}g_{j0}}{g_{00}}\right),$$

while the proper time is defined by $g_{\mu\nu}$. In this case, t is the cosmic time.

2.1. DG and Thermodynamics

Now, we will study some implications of thermodynamics in cosmology for DG. Equation (17) defines the modified scale factor of this theory:

$$a_{DG}(t) = a(t)\sqrt{\frac{1+F(t)}{1+3F(t)}}.$$
(18)

Then, the volume of a cosmological sphere is now

$$V = \frac{4}{3}\pi r^3 a_{DG}^3.$$

Any physical fluid has a density given by

$$\rho_{DG} = \frac{U}{V},$$
(19)

where U is the internal energy and V is the volume. From the first law of thermodynamics, we have

$$\frac{dU}{dt} = T\frac{dS}{dt} - P_{DG}\frac{dV}{dt}.$$
(20)

We will assume that the Universe evolved adiabatically; this means $\dot{S} = 0$. Then, we obtain the well-known relation for the energy conservation

$$\dot{\rho}_{DG} = -3H_{DG}(\rho_{DG} + P_{DG}),$$
(21)

with $H_{DG} = \dot{a}_{DG}/a_{DG}$. In order to know the evolution of ρ, we need an equation of state $P(\rho)$. In [38], they showed that $H_{DG}(t)$ replaces the first Friedmann equation. Now, we know that the second Friedmann equation is the thermodynamics statement that the Universe evolves adiabatically, so the physical densities must satisfy Equation (21). If we assume $P = \omega\rho$, we found

$$\rho_{DG} a_{DG}^{3(1+\omega)} = \rho_{DG\,0} a_{DG\,0}^{3(1+\omega)}, \tag{22}$$

where ρ_0 is the density at present. A crucial point in this theory is that the GR field Equations (11) and (13) are valid; then, we also have a similar relation for the densities of GR but with the standard scale factor $a(t)$, explicitly

$$\rho_{GR} a^{3(1+\omega)} = \rho_{GR\,0} a_0^{3(1+\omega)}. \tag{23}$$

Then, we can relate both densities by the ratio between them

$$\frac{\rho_{DG}}{\rho_{GR}} \left(\sqrt{\frac{1+F(t)}{1+3F(t)}} \right)^{3(1+\omega)} = constant(\omega). \tag{24}$$

This ratio will be vitally important when we study the perturbations of the system. Because we will study the evolution of fractional perturbations at the last-scattering time defined as

$$\delta_{GR\,\alpha} = \frac{\delta\rho_{GR\,\alpha}}{\bar{\rho}_{GR\,\alpha} + \bar{p}_{GR\,\alpha}}, \tag{25}$$

where α runs between γ, ν, B and D (photons, neutrinos, baryons and dark matter, respectively). If we consider the results from [38], at the moment of last-scattering ($T \sim 3000$ K), we obtain

$$\sqrt{\frac{1+F(t_{ls})}{1+3F(t_{ls})}} \sim 1. \tag{26}$$

This mean that at that moment, the physical density was proportional to the densities of GR, and without a loss of generality, we can take

$$\delta_{DG\,\alpha}(t_{ls}) = \delta_{GR\,\alpha}(t_{ls}) \equiv \delta_\alpha(t_{ls}), \tag{27}$$

as it will be introduced in Section 4. In fact, Equation (26) is valid for a wide range of times, from the beginning of the Universe ($z \to \infty$) until $z \sim 10$, so this approximation is valid in the study of primordial perturbations in DG when using the equations of GR. On the other hand, the number density (number of photons over the volume) at equilibrium with matter at temperature T is

$$n_T(\nu)d\nu = \frac{8\pi\nu^2 d\nu}{e^{\frac{h\nu}{k_B T}} - 1}; \tag{28}$$

After decoupling, photons travel freely from the surface of last scattering to us. So, the number of photons is conserved

$$dN = n_{T_{ls}}(\nu_{ls})d\nu_{ls}dV_{ls} = n_T(\nu)d\nu dV, \tag{29}$$

as frequencies are redshifted by $\nu = \nu_{ls} a_{DG}(t_{ls})/a_{DG}$, and the volume $V = V_{ls} a_{DG}^3/a_{DG}^3(t_{ls})$. We find that in order to keep the form of a black body distribution, temperature in the number density should evolve as $T = T_{ls} a_{DG}(t_{ls})/a_{DG}$.

2.2. Equality Time t_{EQ}

After concluding this section, there is an ansatz that we need to propose in order to be completely consistent when solving the cosmological perturbation theory in the next section. This is about when the radiation was equal to the non-relativistic matter. We state that the moment when radiation and matter were equal at some t_{EQ} is the same in GR as in DG. The implication of this statement is the following: let us consider the ratio of the matter and radiation densities of GR (23)

$$\frac{\rho_{GR\,M}}{\rho_{GR\,R}} = \frac{Y}{C},$$ (30)

We remind that $C = \Omega_R/\Omega_M$. Then, the moment of equality in GR corresponds to $Y_{EQ} = C$. On the other hand, if we consider the same ratio but now between the physical densities using (22), we obtain

$$\frac{\rho_{DG;M}}{\rho_{DG\,R}} = \frac{Y_{DG}}{C_{DG}},$$ (31)

where $C_{DG} = \Omega_{DG\,R}/\Omega_{DG\,M}$. Then, in the equality, we need to impose $Y_{DG}(Y_{EQ}) = C_{DG}$, explicitly

$$C_{DG} = C\frac{\sqrt{\frac{1+F(C)}{1+3F(C)}}}{\sqrt{\frac{1+F(1)}{1+3F(1)}}},$$ (32)

if we take the value from [42] (they used L_2 instead of L, but these are the same quantity also), $C \sim 10^{-4}$ and $L \sim 0.45$ implies $F(C) \sim 10^{-3} << 1$ and $F(1) \sim -L/3$, then

$$C_{DG} = C\sqrt{\frac{1-L}{1-L/3}}.$$ (33)

This means that the total density of matter and radiation today depend explicitly on the geometry measured with L [42].

3. Perturbation Theory

Now, we perturbed the metric as the following

$$g_{\mu\nu} = \bar{g}_{\mu\nu} + h_{\mu\nu},$$ (34)
$$\tilde{g}_{\mu\nu} = \bar{\tilde{g}}_{\mu\nu} + \tilde{h}_{\mu\nu}.$$ (35)

Then, we follow the standard method, known as Scalar–Vector–Tensor decomposition [50]. This decomposition depends on four scalar functions (A, B, E and H), two vector functions (C_i and G_i), and one tensor function (D_{ij}) (with their respective delta part). This process allows us to study those sectors independently. Therefore, the perturbations are

$$h_{00} = -E \quad h_{i0} = a\left[\frac{\partial H}{\partial x^i} + G_i\right] \quad h_{ij} = a^2\left[A\delta_{ij} + \frac{\partial^2 B}{\partial x^i \partial x^j} + \frac{\partial C_i}{\partial x^j} + \frac{\partial C_j}{\partial x^i} + D_{ij}\right],$$ (36)

where

$$\frac{\partial C_i}{\partial x^i} = \frac{\partial G_i}{\partial x^i} = 0 \quad \frac{\partial D_{ij}}{\partial x^j} = 0 \quad D_{ii} = 0.$$ (37)

This decomposition must be equivalent for $\tilde{h}_{\mu\nu}$ (by group theory):

$$\tilde{h}_{00} = -\tilde{E} \quad \tilde{h}_{i0} = a\left[\frac{\partial \tilde{H}}{\partial x^i} + \tilde{G}_i\right] \quad \tilde{h}_{ij} = a^2\left[\tilde{A}\delta_{ij} + \frac{\partial^2 \tilde{B}}{\partial x^i \partial x^j} + \frac{\partial \tilde{C}_i}{\partial x^j} + \frac{\partial \tilde{C}_j}{\partial x^i} + \tilde{D}_{ij}\right],$$ (38)

with

$$\frac{\partial \tilde{C}_i}{\partial x^i} = \frac{\partial \tilde{G}_i}{\partial x^i} = 0 \quad \frac{\partial \tilde{D}_{ij}}{\partial x^j} = 0 \quad \tilde{D}_{ii} = 0 . \tag{39}$$

If we replace perturbations in (11)–(14), we obtain the equations for the perturbations. However, there are degrees of freedom that we have to take into account to have physical solutions. In the next subsection, we show how to choose a gauge to delete the nonphysical solutions.

3.1. Choosing a Gauge

Under a space–time coordinate transformation, the metric perturbations transform as [3]

$$\Delta h_{\mu\nu}(x) = -\bar{g}_{\lambda\nu}(x)\frac{\partial \epsilon^\lambda}{\partial x^\mu} - \bar{g}_{\mu\lambda}(x)\frac{\partial \epsilon^\lambda}{\partial x^\nu} - \frac{\partial \bar{g}_{\mu\nu}}{\partial x^\lambda}\epsilon^\lambda . \tag{40}$$

In more detail,

$$\Delta h_{ij} = -\frac{\partial \epsilon_i}{\partial x^j} - \frac{\partial \epsilon_j}{\partial x^i} + 2a\dot{a}\delta_{ij}\epsilon_0, \tag{41}$$

$$\Delta h_{i0} = -\frac{\partial \epsilon_i}{\partial t} - \frac{\partial \epsilon_0}{\partial x^i} + 2\frac{\dot{a}}{a}\epsilon_i, \tag{42}$$

$$\Delta h_{00} = -2\frac{\partial \epsilon_0}{\partial t} . \tag{43}$$

For delta perturbations, we obtain

$$\Delta \tilde{h}_{\mu\nu} = -\tilde{\bar{g}}_{\mu\lambda}\frac{\partial \epsilon^\lambda}{\partial x^\nu} - \tilde{\bar{g}}_{\lambda\nu}\frac{\partial \epsilon^\lambda}{\partial x^\mu} - \frac{\partial \tilde{\bar{g}}_{\mu\nu}}{\partial x^\lambda}\epsilon^\lambda - \bar{g}_{\mu\lambda}\frac{\partial \tilde{\epsilon}^\lambda}{\partial x^\nu} - \bar{g}_{\lambda\nu}\frac{\partial \tilde{\epsilon}^\lambda}{\partial x^\mu} - \frac{\partial \bar{g}_{\mu\nu}}{\partial x^\lambda}\tilde{\epsilon}^\lambda . \tag{44}$$

In more detail,

$$\Delta \tilde{h}_{ij} = -F\frac{\partial \epsilon_i}{\partial x^j} - F\frac{\partial \epsilon_j}{\partial x^i} - \frac{\partial \tilde{\epsilon}_j}{\partial x^i} - \frac{\partial \tilde{\epsilon}_i}{\partial x^j} + \left[\epsilon_0\left(2F a\dot{a} + \dot{F}a^2\right) + 2\tilde{\epsilon}_0 a\dot{a}\right]\delta_{ij} , \tag{45}$$

$$\Delta \tilde{h}_{i0} = -F\frac{\partial \epsilon_i}{\partial t} - 3F\frac{\partial \epsilon_0}{\partial x^i} - \frac{\partial \tilde{\epsilon}_i}{\partial t} - \frac{\partial \tilde{\epsilon}_0}{\partial x^i} + 2F\frac{\dot{a}}{a}\epsilon_i + 2\frac{\dot{a}}{a}\tilde{\epsilon}_i , \tag{46}$$

$$\Delta \tilde{h}_{00} = -3\epsilon_0\dot{F} - 6F\frac{\partial \epsilon_0}{\partial t} - 2\frac{\partial \tilde{\epsilon}_0}{dt} , \tag{47}$$

where ϵ and $\tilde{\epsilon} = \delta\epsilon$ define the coordinates transformation. In addition, we raised and lowered the index using $\bar{g}_{\mu\nu}$, so $\epsilon^0 = -\epsilon_0$, $\tilde{\epsilon}^0 = -\tilde{\epsilon}_0$, $\epsilon^i = a^{-2}\epsilon_i$ and $\tilde{\epsilon}^j = a^{-2}\tilde{\epsilon}_j$. Following the standard procedure, we decompose the spatial part of ϵ^μ and $\tilde{\epsilon}^\mu$ into the gradient of a spatial scalar plus a divergenceless vector:

$$\epsilon_i = \partial_i \epsilon^S + \epsilon_i^V , \quad \partial_i \epsilon^V = 0 , \tag{48}$$

$$\tilde{\epsilon}_i = \partial_i \tilde{\epsilon}^S + \tilde{\epsilon}_i^V , \quad \partial_i \tilde{\epsilon}^V = 0 . \tag{49}$$

Thus, we can compare equations (36) and (38) with (41)–(43) and (45)–(47) to obtain the gauge transformations of the metric components:

$$\Delta A = \frac{2\dot{a}}{a}\epsilon_0 , \quad \Delta B = -\frac{2}{a^2}\epsilon^S ,$$

$$\Delta C_i = -\frac{1}{a^2}\epsilon_i^V , \quad \Delta D_{ij} = 0 , \quad \Delta E = 2\dot{\epsilon}_0 , \tag{50}$$

$$\Delta H = \frac{1}{a}\left(-\epsilon_0 - \dot{\epsilon}^S + \frac{2\dot{a}}{a}\epsilon^S\right) , \quad \Delta G_i = \frac{1}{a}\left(-\dot{\epsilon}_i^V + \frac{2\dot{a}}{a}\epsilon_i^V\right) ,$$

and

$$\Delta\tilde{A} = \left(\frac{2\dot{a}F}{a} + \dot{F}\right)\epsilon_0 + 2\frac{\dot{a}}{a}\tilde{\epsilon}_0, \quad \Delta\tilde{B} = -\frac{2}{a^2}\left(F\epsilon^S + \tilde{\epsilon}^S\right),$$

$$\Delta\tilde{C}_i = -\frac{1}{a^2}\left(F\epsilon_i^V + \tilde{\epsilon}_i^V\right), \quad \Delta\tilde{D}_{ij} = 0, \quad \Delta\tilde{E} = 6F\dot{\epsilon}_0 + 3\dot{F}\epsilon_0 + 2\dot{\tilde{\epsilon}}_0,$$

$$\Delta\tilde{H} = \frac{1}{a}\left(-3F\epsilon_0 - \tilde{\epsilon}_0 - F\dot{\epsilon}^S - \dot{\tilde{\epsilon}}^S + \frac{2F\dot{a}}{a}\epsilon^S + \frac{2\dot{a}}{a}\tilde{\epsilon}^S\right),$$

$$\Delta\tilde{G}_i = \frac{1}{a}\left(-F\dot{\epsilon}_i^V - \dot{\tilde{\epsilon}}_i^V + \frac{2F\dot{a}}{a}\epsilon_i^V + \frac{2\dot{a}}{a}\tilde{\epsilon}_i^V\right). \tag{51}$$

There are different scenarios in which we can continue with the calculations when we impose conditions on the parameters ϵ_μ and $\tilde{\epsilon}_\mu$. However, before discussing this, we will study the gauge transformation of energy-momentum tensors $T_{\mu\nu}$ and $\tilde{T}_{\mu\nu}$.

3.2. $T_{\mu\nu}$ and $\tilde{T}_{\mu\nu}$

Now, we will decompose the energy-momentum tensors $T_{\mu\nu}$ and $\tilde{T}_{\mu\nu}$ in the same way. For a perfect fluid, we would have (for more details, see [37])

$$T_{\mu\nu} = pg_{\mu\nu} + (\rho + p)u_\mu u_\nu, \tag{52}$$

while for $\tilde{T}_{\mu\nu}$ (here $u_\mu^T = e_{\mu\alpha}\tilde{u}^\alpha$, where $e_{\mu\alpha}$ is the Vierbein $g_{\mu\nu} = e_{\mu\alpha}e_{\nu\beta}\eta^{\alpha\beta}$, with η being the Minkowski metric); for a full derivation, please see [35,37]):

$$\tilde{T}_{\mu\nu} = \tilde{p}g_{\mu\nu} + p\tilde{g}_{\mu\nu} + (\tilde{\rho} + \tilde{p})u_\mu u_\nu + (\rho + p)\left(\frac{1}{2}(\tilde{g}_{\mu\alpha}u_\nu u^\alpha + \tilde{g}_{\nu\alpha}u_\mu u^\alpha) + u_\mu^T u_\nu + u_\mu u_\nu^T\right), \tag{53}$$

where

$$g^{\mu\nu}u_\mu u_\nu = -1, \tag{54}$$

$$g^{\mu\nu}u_\mu u_\nu^T = 0. \tag{55}$$

The tensors $g_{\mu\nu}$ and $\tilde{g}_{\mu\nu}$ are defined in (15) and (16), respectively; besides, we consider

$$\begin{aligned}
p &= \bar{p} + \delta p, \\
\rho &= \bar{\rho} + \delta\rho, \\
u_\mu &= \bar{u}_\mu + \delta u_\mu, \\
\tilde{p} &= \bar{\tilde{p}} + \delta\tilde{p}, \\
\tilde{\rho} &= \bar{\tilde{\rho}} + \delta\tilde{\rho}, \\
u_\mu^T &= \bar{u}_\mu^T + \delta u_\mu^T.
\end{aligned} \tag{56}$$

Usually, the equation of state is given by $p(\rho)$, so we could reduce this system. For now, we will work in the generic case. When we work in the frame $\bar{u}_\mu = (-1, 0, 0, 0)$, we have $\bar{u}_\mu^T = 0$, and the normalization conditions (54) and (55) give

$$\begin{aligned}
\delta u^0 &= \delta u_0 = \frac{h_{00}}{2} \\
\delta u_0^T &= \delta u_T^0 = 0
\end{aligned} \tag{57}$$

while δu_i and δu_i^T are independent dynamical variables (note that $\delta u^\mu \equiv \delta(g^{\mu\nu}u_\nu)$ is not given by $\bar{g}^{\mu\nu}\delta u_\nu$. The same is true for δu_T^μ). Then, the first-order perturbations for both energy-momentum tensors (a perfect fluid) are

$$\delta T_{\mu\nu} = \bar{p}h_{\mu\nu} + \delta p\bar{g}_{\mu\nu} + (\bar{p} + \bar{\rho})(\bar{u}_\mu\delta u_\nu + \delta u_\mu u_\nu) + (\delta p + \delta\rho)\bar{u}_\mu\bar{u}_\nu, \tag{58}$$

Therefore,

$$\delta T_{ij} = \bar{p}h_{ij} + a^2\delta_{ij}\delta p\,, \quad \delta T_{i0} = \bar{p}h_{i0} - (\bar{p}+\bar{\rho})\delta u_i\,, \quad \delta T_{00} = -\bar{\rho}h_{00} + \delta\rho\,. \tag{59}$$

While

$$
\begin{aligned}
\delta\tilde{T}_{\mu\nu} &= \bar{\tilde{p}}h_{\mu\nu} + \delta\bar{\tilde{p}}\bar{g}_{\mu\nu} + \bar{p}\tilde{h}_{\mu\nu} + \delta p\tilde{\bar{g}}_{\mu\nu} + (\bar{\tilde{\rho}}+\bar{\tilde{p}})(\bar{u}_\mu\delta u_\nu + \delta u_\mu\bar{u}_\nu) \\
&\quad + (\delta\bar{\tilde{\rho}} + \delta\bar{\tilde{p}})\bar{u}_\mu\bar{u}_\nu + (\bar{\rho}+\bar{p})\left\{\frac{1}{2}\left[\tilde{\bar{g}}_{\mu\alpha}(\bar{u}_\nu\delta u^\alpha + \delta u_\nu\bar{u}^\alpha) + \tilde{h}_{\mu\alpha}\bar{u}_\nu\bar{u}^\alpha\right.\right. \\
&\quad + \left. \tilde{\bar{g}}_{\nu\alpha}(\bar{u}_\mu\delta u^\alpha + \delta u_\mu\bar{u}^\alpha) + \tilde{h}_{\nu\alpha}\bar{u}_\mu\bar{u}^\alpha\right] + \bar{u}_\mu^T\delta u_\nu + \delta u_\mu^T\bar{u}_\nu + \bar{u}_\mu\delta u_\nu^T + \delta u_\mu\bar{u}_\nu^T\Big\} \\
&\quad + (\delta\rho + \delta p)\left\{\frac{1}{2}\left[\tilde{\bar{g}}_{\mu\alpha}\bar{u}_\nu\bar{u}^\alpha + \tilde{\bar{g}}_{\nu\alpha}\bar{u}_\mu\bar{u}^\alpha\right] + \bar{u}_\mu^T\bar{u}_\nu + \bar{u}_\mu\bar{u}_\nu^T\right\},
\end{aligned}
\tag{60}
$$

and

$$
\begin{aligned}
\delta\tilde{T}_{00} &= -\bar{\tilde{\rho}}h_{00} - \bar{\rho}\tilde{h}_{00} + 3F\delta\rho + \delta\tilde{\rho}, \\
\delta\tilde{T}_{i0} &= \bar{\tilde{p}}h_{i0} + \bar{p}\tilde{h}_{i0} - (\bar{\tilde{\rho}}+\bar{\tilde{p}})\delta u_i + (\bar{\rho}+\bar{p})\left\{\frac{1}{2}[Fh_{i0} - \tilde{h}_{i0} - 4F\delta u_i] - \delta u_i^T\right\}, \\
\delta\tilde{T}_{ij} &= \bar{\tilde{p}}h_{ij} + \delta\bar{\tilde{p}}a^2\delta_{ij} + \bar{p}\tilde{h}_{ij} + \delta p F a^2\delta_{ij},
\end{aligned}
\tag{61}
$$

where we used $\delta u^\alpha \equiv \delta(g^{\alpha\beta}u_\beta) = \bar{g}^{\alpha\beta}\delta u_\beta - h^{\alpha\beta}\bar{u}_\beta$.

Generally, we decompose $\delta u_i\,(\delta u_i^T)$ into the gradient of a scalar velocity potential $\delta u\,(\delta\tilde{u})$ and a divergenceless vector $\delta u_i^V\,(\delta\tilde{u}_i^V)$, and the dissipative corrections to the inertia tensor are added as follows

$$\delta T_{ij} = \bar{p}h_{ij} + a^2\left[\delta_{ij}\delta p + \partial_i\partial_j\pi^S + \partial_i\pi_j^V + \partial_j\pi_i^V + \pi_{ij}^T\right], \tag{62}$$

$$\delta T_{i0} = \bar{p}h_{i0} - (\bar{p}+\bar{\rho})\left(\partial_i\delta u + \delta u_i^V\right), \tag{63}$$

$$\delta T_{00} = -\bar{\rho}h_{00} + \delta\rho, \tag{64}$$

and

$$
\begin{aligned}
\delta\tilde{T}_{ij} &= \bar{\tilde{p}}h_{ij} + a^2\left[\delta_{ij}\delta\tilde{p} + \partial_i\partial_j\tilde{\pi}^S + \partial_i\tilde{\pi}_j^V + \partial_j\tilde{\pi}_i^V + \tilde{\pi}_{ij}^T\right] + \bar{p}\tilde{h}_{ij} \\
&\quad + Fa^2\left[\delta_{ij}\delta p + \partial_i\partial_j\pi^S + \partial_i\pi_j^V + \partial_j\pi_i^V + \pi_{ij}^T\right],
\end{aligned}
\tag{65}
$$

$$
\begin{aligned}
\delta\tilde{T}_{i0} &= \bar{\tilde{p}}h_{i0} + \bar{p}\tilde{h}_{i0} - (\bar{\tilde{\rho}}+\bar{\tilde{p}})(\partial_i\partial u + \partial u_i^V) \\
&\quad + (\bar{\rho}+\bar{p})\left\{\frac{1}{2}[Fh_{i0} - \tilde{h}_{i0} - 4F(\partial_i\delta u + \delta u_i^V)] - \partial_i\delta\tilde{u} + \delta\tilde{u}_i^V\right\},
\end{aligned}
\tag{66}
$$

$$\delta\tilde{T}_{00} = -\bar{\tilde{\rho}}h_{00} - \bar{\rho}\tilde{h}_{00} + 3F\delta\rho + \delta\tilde{\rho}, \tag{67}$$

where $\pi_i^V\,(\tilde{\pi}_i^V)$, $\pi_{ij}^T\,(\tilde{\pi}_{ij}^T)$ and $\delta u_i^V\,(\delta\tilde{u}_i^V)$ satisfy similar conditions to (37) and (39). These conditions are (expressed before as $C_i\,(\tilde{C}_i)$, $D_{ij}\,(\tilde{D}_{ij})$ $G_i\,(\tilde{G}_i)$):

$$\partial_i\pi_i^V = \partial_i\tilde{\pi}_i^V = \partial_i\delta u_i^V = \partial_i\delta\tilde{u}_i^V = 0 \quad \partial_i\pi_{ij}^T = \partial_i\tilde{\pi}_{ij}^T = 0, \quad \pi_{ii}^T = \tilde{\pi}_{ii}^T = 0. \tag{68}$$

3.3. Gauge Transformations for the Energy-Momentum Tensors

The gauge transformation for $T_{\mu\nu}$ is given by

$$\Delta\delta T_{\mu\nu}(x) = -\bar{T}_{\lambda\nu}(x)\frac{\partial\epsilon^\lambda}{\partial x^\mu} - \bar{T}_{\mu\lambda}(x)\frac{\partial\epsilon^\lambda}{\partial x^\nu} - \frac{\partial\bar{T}_{\mu\nu}}{\partial x^\lambda}\epsilon^\lambda, \tag{69}$$

where the components are

$$\Delta\delta T_{ij} = -\bar{p}\left(\frac{\partial\epsilon_i}{\partial x^j} + \frac{\partial\epsilon_j}{\partial x^i}\right) + \frac{\partial}{\partial t}(a^2\bar{p})\delta_{ij}\epsilon_0, \tag{70}$$

$$\Delta\delta T_{i0} = -\bar{p}\frac{\partial\epsilon_i}{\partial t} + \bar{\rho}\frac{\partial\epsilon_0}{\partial x^i} + 2\bar{p}\frac{\dot{a}}{a}\epsilon_i, \tag{71}$$

$$\Delta\delta T_{00} = 2\bar{\rho}\frac{\partial\epsilon_0}{\partial t} + \dot{\bar{\rho}}\epsilon_0. \tag{72}$$

While the gauge transformation of $\delta\tilde{T}_{\mu\nu}$ is given by

$$\Delta\delta\tilde{T}_{\mu\nu} = -\tilde{\bar{T}}_{\mu\lambda}\frac{\partial\epsilon^\lambda}{\partial x^\nu} - \tilde{\bar{T}}_{\lambda\nu}\frac{\partial\epsilon^\lambda}{\partial x^\mu} - \frac{\partial\tilde{\bar{T}}_{\mu\nu}}{\partial x^\lambda}\epsilon^\lambda - \bar{T}_{\mu\lambda}\frac{\partial\tilde{\epsilon}^\lambda}{\partial x^\nu} - \bar{T}_{\lambda\nu}\frac{\partial\tilde{\epsilon}^\lambda}{\partial x^\mu} - \frac{\partial\bar{T}_{\mu\nu}}{\partial x^\lambda}\tilde{\epsilon}^\lambda, \tag{73}$$

where the components are

$$\Delta\delta\tilde{T}_{ij} = -(\tilde{\bar{p}} + \bar{p}F)\frac{\partial\epsilon_i}{\partial x^j} - (\tilde{\bar{p}} + \bar{p}F)\frac{\partial\epsilon_j}{\partial x^i} - \bar{p}\frac{\partial\tilde{\epsilon}_j}{\partial x^i} - \bar{p}\frac{\partial\tilde{\epsilon}_i}{\partial x^j} + \left[\epsilon_0\frac{\partial}{\partial t}[a^2(\tilde{\bar{p}} + \bar{p}F)] + \frac{\partial}{\partial t}(a^2\bar{p})\tilde{\epsilon}_0\right]\delta_{ij} \tag{74}$$

$$\Delta\delta\tilde{T}_{i0} = -(\tilde{\bar{p}} + \bar{p}F)\frac{\partial\epsilon_i}{\partial t} + (\tilde{\bar{\rho}} + 3F\bar{\rho})\frac{\partial\epsilon_0}{\partial x^i} - \bar{p}\frac{\partial\tilde{\epsilon}_i}{\partial t} + \bar{\rho}\frac{\partial\tilde{\epsilon}_0}{\partial x^i} + 2(\tilde{\bar{p}} + \bar{p}F)\frac{\dot{a}}{a}\epsilon_i + 2\bar{p}\frac{\dot{a}}{a}\tilde{\epsilon}_i \tag{75}$$

$$\Delta\delta\tilde{T}_{00} = \epsilon_0\frac{\partial}{\partial t}(\tilde{\bar{\rho}} + 3F\bar{\rho}) + 2(\tilde{\bar{\rho}} + 3F\bar{\rho})\frac{\partial\epsilon_0}{\partial t} + \dot{\bar{\rho}}\tilde{\epsilon}_0 + 2\bar{\rho}\frac{\partial\tilde{\epsilon}_0}{\partial t}. \tag{76}$$

ϵ_i and $\tilde{\epsilon}_i$ were decomposed in (48) to write these gauge transformations in terms of the scalar, vector and tensor components. The transformations (41)–(43) and (45)–(47) with (70)–(71) and (74)–(76) give the gauge transformation for the pressure, energy density and velocity potential:

$$\Delta\delta p = \dot{\bar{p}}\epsilon_0, \quad \Delta\delta\rho = \dot{\bar{\rho}}\epsilon_0, \quad \Delta\delta u = -\epsilon_0. \tag{77}$$

The other ingredients of the energy-momentum tensor are gauge invariants:

$$\Delta\pi^S = \Delta\pi_i^V = \Delta\pi_{ij}^T = \Delta\delta u_i^V = 0. \tag{78}$$

Nevertheless, the other transformations are

$$\Delta\delta\tilde{\rho} = \frac{\partial}{\partial t}(\tilde{\bar{\rho}} + 3F\bar{\rho})\epsilon_0 + 2(\tilde{\bar{\rho}} + 3F\bar{\rho})\dot{\epsilon}_0 + \dot{\bar{\rho}}\tilde{\epsilon}_0 + 2\bar{\rho}\dot{\tilde{\epsilon}}_0 - \tilde{\bar{\rho}}\Delta E$$
$$- \quad 3F\bar{\rho}\Delta\tilde{E} - 3F\Delta\delta\rho, \tag{79a}$$

$$\Delta\delta\tilde{p} = \frac{1}{a^2}\frac{\partial}{\partial t}[a^2(\tilde{\bar{p}} + \bar{p}F)]\epsilon_0 + \frac{1}{a^2}\frac{\partial}{\partial t}(a^2\bar{p})\tilde{\epsilon}_0 - \tilde{\bar{p}}\Delta A - \bar{p}F\Delta\tilde{A} - F\Delta\delta p, \tag{79b}$$

$$\Delta\delta\tilde{u} = \frac{1}{(\bar{\rho} + \bar{p})}\Bigg\{(\tilde{\bar{p}} + \bar{p}F)\dot{\epsilon}^S - (\tilde{\bar{\rho}} + 3F\bar{\rho})\epsilon_0 + \bar{p}\dot{\tilde{\epsilon}}^S - \bar{\rho}\tilde{\epsilon}_0 - 2(\tilde{\bar{p}} + \bar{p}F)\frac{\dot{a}}{a}\epsilon^S$$
$$- \quad 2\bar{p}\frac{\dot{a}}{a}\tilde{\epsilon}^S + \tilde{\bar{p}}a\Delta H + \bar{p}a\Delta\tilde{H} - (\bar{\rho} + \bar{p})\left[\frac{1}{2}(1 - F)a\Delta\tilde{H} + 2F\Delta\delta u\right]\Bigg\}, \tag{79c}$$

$$\Delta\delta\tilde{u}_i^V = \frac{1}{(\bar{\rho} + \bar{p})}\Bigg\{(\tilde{\bar{p}} + \bar{p}F)\dot{\epsilon}_i^V + \bar{p}\dot{\tilde{\epsilon}}_i^V - 2(\tilde{\bar{p}} + \bar{p}F)\frac{\dot{a}}{a}\epsilon_i^V - 2\bar{p}\frac{\dot{a}}{a}\tilde{\epsilon}_i^V + \tilde{\bar{p}}a\Delta G_i$$
$$+ \quad \bar{p}a\Delta\tilde{G}_i - \frac{1}{2}(\bar{\rho} + \bar{p})(1 - F)a\Delta\tilde{G}_i\Bigg\}, \tag{79d}$$

$$\Delta\delta\tilde{\pi}^S = -\frac{2}{a^2}(\tilde{\bar{p}} + \bar{p}F)\epsilon^S - 2\frac{\bar{p}}{a^2}\tilde{\epsilon}^S - \tilde{\bar{p}}\Delta B - \bar{p}F\Delta\tilde{B}, \tag{79e}$$

$$\Delta\delta\tilde{\pi}_i^V = -\frac{1}{a^2}(\tilde{\bar{p}} + \bar{p}F)\epsilon_i^V - \frac{\bar{p}}{a^2}\tilde{\epsilon}_i^V - \tilde{\bar{p}}\Delta C_i - \bar{p}F\Delta\tilde{C}_i, \tag{79f}$$

$$\Delta\delta\tilde{\pi}_{ij} = 0. \tag{79g}$$

The results given in (50), (51) and (77) are used to obtain

$$\Delta\delta\tilde{\rho} \;=\; \dot{\tilde{\rho}}\epsilon_0 + (\dot{\rho} - 3F\bar{\rho})\tilde{\epsilon}_0, \tag{80a}$$

$$\Delta\delta\tilde{p} \;=\; \dot{\tilde{p}}\epsilon_0 + \dot{\bar{p}}\tilde{\epsilon}_0, \tag{80b}$$

$$\Delta\delta\tilde{u} \;=\; \left[(1-3F)\frac{F}{2} - \frac{(\tilde{p}+\tilde{\rho})}{(\bar{p}+\bar{\rho})}\right]\epsilon_0 - \frac{1}{2}(1+F)\tilde{\epsilon}_0 - (1-F)\frac{\dot{a}}{a}\left(F\epsilon^S + \tilde{\epsilon}^S\right)$$
$$\qquad +\; \frac{1}{2}(1-F)\left(F\dot{\epsilon}^S + \dot{\tilde{\epsilon}}^S\right), \tag{80c}$$

$$\Delta\delta\tilde{u}_i^V \;=\; \frac{1}{2}(1-F)[F\dot{\epsilon}_i^V + \dot{\tilde{\epsilon}}_i^V - 2\frac{\dot{a}}{a}F\epsilon_i^V - 2\frac{\dot{a}}{a}\tilde{\epsilon}_i^V], \tag{80d}$$

$$\Delta\delta\tilde{\pi}^S \;=\; 0, \tag{80e}$$

$$\Delta\delta\tilde{\pi}_i^V \;=\; 0, \tag{80f}$$

$$\Delta\delta\tilde{\pi}_{ij} \;=\; 0. \tag{80g}$$

As we said before, there are different choices for the ϵ and $\tilde{\epsilon}$ parameters to fix all the gauge freedoms. The most common and well-known gauges are the Newtonian gauge and synchronous gauge. The former fix ϵ^S such that $B = 0$, and we choose ϵ_0 such that $H = 0$ (in Equation (50)). In DG, this choice is extended, imposing similar conditions in (51) for $\tilde{\epsilon}^S$ and $\tilde{\epsilon}_0$, such that $\tilde{B} = \tilde{H} = 0$. There is no remaining freedom to make a gauge transformation in this scenario. Nevertheless, in this work, we will use the synchronous gauge, where we will choose ϵ_0 such that $E = 0$ and ϵ^S such that $H = 0$ (similar conditions for $\tilde{\epsilon}_0$ and $\tilde{\epsilon}^S$). In the next section, we present the perturbed equations of motion in this frame, and we discuss the suitability of this choice for our purposes.

3.4. Fields Equations and Energy Momentum Conservations in Synchronous Gauge

Under this gauge fixing, the perturbed Einstein Equation (11) reads (at first order)[4]:

$$-4\pi G(\delta\rho + 3\delta p + \nabla^2\pi^S) = \frac{1}{2}\left(3\ddot{A} + \nabla^2\ddot{B}\right) + \frac{\dot{a}}{2a}\left(3\dot{A} + \nabla^2\dot{B}\right). \tag{81}$$

where $\nabla^2 \equiv \partial_x^2 + \partial_y^2 + \partial_z^2$. The energy-momentum conservation gives

$$\delta p + \nabla^2\pi^S + \partial_0[(\bar{\rho}+\bar{p})\delta u] + \frac{3\dot{a}}{a}(\bar{\rho}+\bar{p})\delta u \;=\; 0, \tag{82}$$

$$\delta\dot{\rho} + \frac{3\dot{a}}{a}(\delta\rho + \delta p) + \nabla^2\left[a^{-2}(\bar{\rho}+\bar{p})\delta u + \frac{\dot{a}}{a}\pi^S\right] + \frac{1}{2}(\bar{\rho}+\bar{p})\partial_0\left[3A + \nabla^2 B\right] \;=\; 0. \tag{83}$$

We define

$$\Psi \equiv \frac{1}{2}\left[3A + \nabla^2 B\right], \tag{84}$$

then,

$$-4\pi Ga^2(\delta\rho + 3\delta p + \nabla^2\pi^S) = \frac{\partial}{\partial t}\left(a^2\dot{\Psi}\right), \tag{85}$$

$$\delta\dot{\rho} + \frac{3\dot{a}}{a}(\delta\rho + \delta p) + \nabla^2\left[a^{-2}(\bar{\rho}+\bar{p})\delta u + \frac{\dot{a}}{a}\pi^S\right] + \frac{1}{2}(\bar{\rho}+\bar{p})\dot{\Psi} = 0. \tag{86}$$

The unperturbed Einstein equations correspond to the Friedmann equations. In the delta sector, computations give the non-perturbed equations:

$$3\dot{F}\frac{\dot{a}}{a} = \varkappa(3F\bar{\rho} + \tilde{\rho}) \tag{87}$$

and

$$12F\frac{\ddot{a}}{a} + 6F\left(\frac{\dot{a}}{a}\right)^2 + 3\dot{F}\frac{\dot{a}}{a} - 3\ddot{F} = \varkappa(\tilde{\rho} + 3\tilde{p} + 3F\bar{\rho} + 3F\bar{p}). \tag{88}$$

The perturbed contribution (at first order) is

$$\left[2\dot{F}\frac{\dot{a}}{a} + \ddot{F}\right]\left[3A + \nabla^2 B\right] + \left[6F\frac{\dot{a}}{a} + \frac{5}{2}\dot{F}\right]\left[3\dot{A} + \nabla^2\dot{B}\right] - \left[2\frac{\dot{a}}{a}\right]\left[3\dot{A} + \nabla^2\dot{B}\right]$$

$$+3F\left[3\ddot{A} + \nabla^2\ddot{B}\right] - \left[3\dot{A} + \nabla^2\dot{B}\right] = \varkappa\left(3\delta\tilde{p} + \delta\tilde{\rho} + F\delta\rho + 3F\delta p + \nabla^2\tilde{\pi} + F\nabla^2\pi\right) \tag{89}$$

In addition, the 00 component of delta energy-momentum conservation in (14) gives

$$\delta\dot{\tilde{\rho}} + \frac{3\dot{a}}{a}(\delta\tilde{\rho} + \delta\tilde{p}) + \frac{3\dot{F}}{2}(\delta\rho + \delta p) + \nabla^2\left[\frac{(\tilde{\rho} + \tilde{p})}{a^2}\delta u + \frac{(\bar{\rho} + \bar{p})F}{a^2}\delta u + \frac{(\bar{\rho} + \bar{p})}{a^2}\delta\tilde{u}\right]$$

$$+\frac{(\tilde{\rho} + \tilde{p})}{2}\partial_0[3A + \nabla^2 B] + \frac{(\bar{\rho} + \bar{p})}{2}\partial_0[3\tilde{A} + \nabla^2\tilde{B}] - \frac{(\bar{\rho} + \bar{p})}{2}\partial_0(F[3A + \nabla^2 B]) = 0, \tag{90}$$

while the $i0$ component gives

$$\delta\tilde{p} + \partial_0[(\tilde{\rho} + \tilde{p})\delta u] + \partial_0[(\bar{\rho} + \bar{p})\delta\tilde{u}] - \partial_0[(\bar{\rho} + \bar{p})F\delta u] + 3(\bar{\rho} + \bar{p})\dot{F}\delta u$$

$$+\frac{3\dot{a}}{a}(\bar{\rho} + \bar{p})\delta\tilde{u} + \frac{3\dot{a}}{a}(\tilde{\rho} + \tilde{p})\delta u - \frac{3\dot{a}}{a}F(\bar{\rho} + \bar{p})\delta u = 0. \tag{91}$$

Analogous to the standard sector, we define

$$\tilde{\Psi} \equiv \frac{1}{2}\left[3\tilde{A} + \nabla^2\tilde{B}\right], \tag{92}$$

then the gravitational equation becomes

$$\left[2\dot{F}\frac{\dot{a}}{a} + \ddot{F}\right]a^2\Psi + \left[6F\frac{\dot{a}}{a} + \frac{5}{2}\dot{F}\right]a^2\Psi + 3Fa^2\ddot{\Psi} - \frac{d}{dt}\left(a^2\dot{\tilde{\Psi}}\right) = \frac{\varkappa}{2}(3\delta\tilde{p} + \delta\tilde{\rho} + F\delta\rho$$

$$+3F\delta p + \nabla^2\tilde{\pi} + F\nabla^2\pi). \tag{93}$$

Now, the delta energy conservation is given by

$$\delta\dot{\tilde{\rho}} + \frac{3\dot{a}}{a}(\delta\tilde{\rho} + \delta\tilde{p}) + \frac{3\dot{F}}{2}(\delta\rho + \delta p) + \nabla^2\left[\frac{(\tilde{\rho} + \tilde{p})}{a^2}\delta u + \frac{(\bar{\rho} + \bar{p})F}{a^2}\delta u + \frac{(\bar{\rho} + \bar{p})}{a^2}\delta\tilde{u}\right]$$

$$+(\tilde{\rho} + \tilde{p})\Psi + (\bar{\rho} + \bar{p})\dot{\tilde{\Psi}} - (\bar{\rho} + \bar{p})\partial_0(F\Psi) = 0. \tag{94}$$

The study of the non-perturbed sector was already treated in Alfaro et al. and applied to the supernovae observations [37,42]. We will consider these results when necessary. For now, we only need the expression for the time-dependent function $F(t)$, which is

$$F(Y) = -\frac{LY}{3}\sqrt{Y + C}, \tag{95}$$

where $Y \equiv Y(t) = a(t)/a_0$ is the quotient between the scale factor at a time t over the scale factor in the actuality (which for our purposes, we will consider equal to one). L (~ 0.45) and C ($\sim 10^{-4}$) are the new parameters of DG that are already determined by supernova data [37,42]. We have to remark that our definition of $\tilde{\Psi}$ is not the usual, since the standard definition is with the time derivative of fields A and B, respectively. In the delta sector, the combinations of these fields appear explicitly without a time derivative, so if the reader wants to compare results with other works, he or she should take into consideration this definition to analyze the gauge. In the next section, we will discuss the evolution of the cosmological fluctuations, which will help us compute the scalar contribution to the CMB.

4. Evolution of Cosmological Fluctuations

Until now, we have developed the perturbation theory in DG; now, we are interested in studying the evolution of the cosmological fluctuations to have a physical interpretation of the delta matter fields, which this theory naturally introduces. Even in the standard cosmology, the system of equations that describes these perturbations are complicated to allow analytic solutions, and there are comprehensive computer programs for this task, such as CMBfast [43,44] and CAMB [45]. However, such computer programs cannot give a clear understanding of the physical phenomena involved. Nevertheless, some good approximations allow computing the spectrum of the CMB fluctuations with a rather good agreement with these computer programs [46,47]. In particular, we are going to extend the Weinberg approach for this task. This method consists of two main aspects: first, the hydrodynamic limit, which assumes that near recombination time photons were in local thermal equilibrium with the baryonic plasma; then, photons could be treated hydro-dynamically, such as plasma and cold dark matter. Second, a sharp transition from thermal equilibrium to complete transparency at the moment t_{ls} of the last scattering.

Since we will reproduce this approach, we consider the Universe's standard components, which means photons, neutrinos, baryons, and cold dark matter. Then, the task is to understand the role of their own delta counterpart. We will also neglect both anisotropic inertia tensors and took the usual state equation for pressures and energy densities and perturbations. As we will treat photons and delta photons hydrodynamically, we will use $\delta u_\gamma = \delta u_B$ and $\delta \tilde{u}_\gamma = \delta \tilde{u}_B$. Finally, as the synchronous scheme does not completely fix the gauge freedom, one can use the remaining freedom to put $\delta u_D = 0$, which means that cold dark matter evolves at rest with respect to the Universe expansion. In our theory, the extended synchronous scheme also has extra freedom, which we will use to choose $\delta \tilde{u}_D = 0$ as its standard part. Now, we will present the equations for both sectors. However, we will provide more detail in the delta sector because Weinberg [47] already calculates the solution of Einstein's equations. Einstein's equations and its energy-momentum conservation in Fourier space are[5]

$$\frac{d}{dt}\left(a^2 \dot{\Psi}_q\right) = -4\pi G a^2 \left(\delta\rho_{Dq} + \delta\rho_{Bq} + 2\delta\rho_{\gamma q} + 2\delta\rho_{vq}\right), \quad (96)$$

$$\delta\dot{\rho}_{\gamma q} + 4H\delta\rho_{\gamma q} - (4q/3a)\bar{\rho}_\gamma \delta u_{\gamma q} = -(4/3)\bar{\rho}_\gamma \dot{\Psi}_q, \quad (97)$$

$$\delta\dot{\rho}_{Dq} + 3H\delta\rho_{Dq} = -\bar{\rho}_D \dot{\Psi}_q, \quad (98)$$

$$\delta\dot{\rho}_{Bq} + 3H\delta\rho_{Bq} - (q/a)\bar{\rho}_B \delta u_{\gamma q} = -\bar{\rho}_B \dot{\Psi}_q, \quad (99)$$

$$\delta\dot{\rho}_{vq} + 4H\delta\rho_{vq} - (4q/3a)\bar{\rho}_v \delta u_{vq} = -(4/3)\bar{\rho}_v \dot{\Psi}_q, \quad (100)$$

where $H \equiv \dot{a}/a$. It is useful to rewrite these equations in term of the dimensionless fractional perturbation

$$\delta_{\alpha q} = \frac{\delta\rho_{\alpha q}}{\bar{\rho}_\alpha + \bar{p}_\alpha}, \quad (101)$$

where α can be γ, v, B and D (photons, neutrinos, baryons and dark matter, respectively). $a^4\bar{\rho}_\gamma$, $a^4\bar{\rho}_v$, $a^3\bar{\rho}_D$, $a^3\bar{\rho}_B$ are time-independent quantities; then, (96)–(100) are

$$\frac{d}{dt}\left(a^2\dot{\Psi}_q\right) = -4\pi G a^2\left(\bar{\rho}_D\delta_{Dq} + \bar{\rho}_B\delta_{Bq} + \frac{8}{3}\bar{\rho}_\gamma\delta_{\gamma q} + \frac{8}{3}\bar{\rho}_v\delta_{vq}\right), \quad (102\text{a})$$

$$\dot{\delta}_{\gamma q} - (q^2/a^2)\delta u_{\gamma q} = -\dot{\Psi}_q, \quad (102\text{b})$$

$$\dot{\delta}_{Dq} = -\dot{\Psi}_q, \quad (102\text{c})$$

$$\dot{\delta}_{Bq} - (q^2/a^2)\delta u_{\gamma q} = -\dot{\Psi}_q, \quad (102\text{d})$$

$$\dot{\delta}_{vq} - (q^2/a^2)\delta u_{vq} = -\dot{\Psi}_q, \quad (102\text{e})$$

$$\frac{d}{dt}\left(\frac{(1+R)\delta u_{\gamma q}}{a}\right) = -\frac{1}{3a}\delta_{\gamma q}, \quad (102\text{f})$$

$$\frac{d}{dt}\left(\frac{\delta u_{vq}}{a}\right) = -\frac{1}{3a}\delta_{vq}, \quad (102\text{g})$$

where $R = 3\bar{\rho}_B/4\bar{\rho}_\gamma$. By the other side, in the delta sector, we will use a dimensionless fractional perturbation. However, this perturbation is defined as the delta transformation of (101)[6],

$$\tilde{\delta}_{\alpha q} \equiv \tilde{\delta}\delta_{\alpha q} = \frac{\delta\tilde{\rho}_{\alpha q}}{\bar{\rho}_\alpha + \bar{p}_\alpha} - \frac{\tilde{\bar{\rho}}_\alpha + \tilde{\bar{p}}_\alpha}{\bar{\rho}_\alpha + \bar{p}_\alpha}\delta_{\alpha q}\,. \tag{103}$$

In [37], they found

$$\frac{\tilde{\bar{\rho}}_R}{\bar{\rho}_R} = -2F(a) \quad and \quad \frac{\tilde{\bar{\rho}}_M}{\bar{\rho}_M} = -\frac{3}{2}F(a)\,. \tag{104}$$

We will assume that this quotient holds for every component. In addition, using the result that $a^4\tilde{\bar{\rho}}_\gamma/F$, $a^4\tilde{\bar{\rho}}_v/F$, $a^3\tilde{\bar{\rho}}_D/F$, $a^3\tilde{\bar{\rho}}_B/F$ are time independent, the equations for the delta sector are

$$\left[2\dot{F}\frac{\dot{a}}{a} + \ddot{F}\right]a^2\Psi_q + \left[6F\frac{\dot{a}}{a} + \frac{5}{2}\dot{F}\right]a^2\dot{\Psi}_q + 3Fa^2\ddot{\Psi}_q - \frac{d}{dt}\left(a^2\dot{\tilde{\Psi}}_q\right) \tag{105a}$$
$$= \frac{\varkappa}{2}a^2\left[\bar{\rho}_D\tilde{\delta}_{Dq} + \bar{\rho}_B\tilde{\delta}_{Bq} + \frac{8}{3}\bar{\rho}_\gamma\tilde{\delta}_{\gamma q} + \frac{8}{3}\bar{\rho}_v\tilde{\delta}_{vq} - \frac{F}{2}\left(\bar{\rho}_D\delta_{Dq} + \bar{\rho}_B\delta_{Bq}\right) \quad - \quad \frac{8}{3}F\left(\bar{\rho}_\gamma\delta_{\gamma q} + \bar{\rho}_v\delta_{vq}\right)\right],$$

$$\dot{\tilde{\delta}}_{\gamma q} - \frac{q^2}{a^2}\left(\delta\tilde{u}_{\gamma q} + F\delta u_{\gamma q}\right) + \dot{\tilde{\Psi}}_q - \partial_0(F\Psi_q) = 0\,, \tag{105b}$$

$$\dot{\tilde{\delta}}_{Dq} + \dot{\tilde{\Psi}}_q - \partial_0(F\Psi_q) = 0\,, \tag{105c}$$

$$\dot{\tilde{\delta}}_{Bq} - \frac{q^2}{a^2}\left(\delta\tilde{u}_{\gamma q} + F\delta u_{\gamma q}\right) + \dot{\tilde{\Psi}}_q - \partial_0(F\Psi_q) = 0\,, \tag{105d}$$

$$\dot{\tilde{\delta}}_{vq} - \frac{q^2}{a^2}\left(\delta\tilde{u}_{vq} + F\delta u_{vq}\right) + \dot{\tilde{\Psi}}_q - \partial_0(F\Psi_q) = 0\,, \tag{105e}$$

$$\frac{\tilde{\delta}_{\gamma q}}{3a} + \frac{d}{dt}\left(\frac{(1+R)\delta\tilde{u}_{\gamma q}}{a}\right) + 2F\frac{d}{dt}\left(\frac{(R-\tilde{R})\delta u_{\gamma q}}{a}\right)$$
$$-F\frac{d}{dt}\left(\frac{(1+R)\delta u_{\gamma q}}{a}\right) - 2\dot{F}(\tilde{R}-R)\frac{\delta u_{\gamma q}}{a} = 0\,, \tag{105f}$$

$$\frac{\tilde{\delta}_{vq}}{3a} + \frac{d}{dt}\left(\frac{\delta\tilde{u}_{vq}}{a}\right) - F\frac{d}{dt}\left(\frac{\delta u_{vq}}{a}\right) = 0\,, \tag{105g}$$

with $\tilde{R} = 3\tilde{\bar{\rho}}_B/4\tilde{\bar{\rho}}_\gamma$. Due to the definition of tilde fractional perturbation (103), solutions for (105a)–(105g) can be obtained easily, putting all solutions of GR equal to zero; then, the system is exactly equal to the system of Equations (102a)–(102g) and the solutions of tilde perturbations in the homogeneous system are exactly equal to the GR solutions. Then, we only need to "turn on" the GR source and find the complete solutions just like a forced-system. We will impose initial conditions to find solutions valid up to recombination time. At sufficiently early times, the Universe was dominated by radiation, and as Friedmann equations are valid in our theory (in particular the first equation), we can use a good approximation given by $a \propto \sqrt{t}$ and $8\pi G\bar{\rho}_R/3 = 1/4t^2$, while R and $\tilde{R} \ll 1$. Here

$$\bar{\rho}_M \equiv \bar{\rho}_D + \bar{\rho}_B\,, \quad \bar{\rho}_R \equiv \bar{\rho}_\gamma + \bar{\rho}_v\,. \tag{106}$$

We are interested in adiabatic solutions in the sense that all the $\delta_{\alpha q}$ and $\tilde{\delta}_{\alpha q}$ become equal at very early times. So, we make the ansatz:

$$\delta_{\gamma q} = \delta_{vq} = \delta_{Bq} = \delta_{Dq} = \delta_q\,, \quad \delta u_{\gamma q} = \delta u_{vq} = \delta u_q\,, \tag{107}$$
$$\tilde{\delta}_{\gamma q} = \tilde{\delta}_{vq} = \tilde{\delta}_{Bq} = \tilde{\delta}_{Dq} = \tilde{\delta}_q\,, \quad \delta\tilde{u}_{\gamma q} = \delta\tilde{u}_{vq} = \delta\tilde{u}_q\,. \tag{108}$$

Finally, we drop the term q^2/a^2 because we are considering very early times. Then, Equations (102a)–(102g) become

$$\frac{d}{dt}\left(t\Psi_q\right) = -\frac{1}{t}\delta_q\,, \tag{109}$$

$$\dot{\delta}_q = -\Psi_q\,, \tag{110}$$

and

$$\frac{d}{dt}\left(\frac{\delta u_q}{\sqrt{t}}\right) = -\frac{1}{t}\delta_q \, . \tag{111}$$

While Equations (105a)–(105g) become

$$\left[2\dot{F}\frac{\dot{a}}{a} + \ddot{F}\right]a^2\Psi_q + \left[6F\frac{\dot{a}}{a} + \frac{5}{2}\dot{F}\right]a^2\dot{\Psi}_q + 3Fa^2\ddot{\Psi}_q$$

$$-\frac{d}{dt}\left(a^2\dot{\tilde{\Psi}}_q\right) = \frac{a^2}{t^2}(\tilde{\delta}_q - F\delta_q) \, , \tag{112}$$

$$\dot{\tilde{\delta}}_q + \dot{\tilde{\Psi}}_q - \partial_0(F\Psi_q) = 0 \, , \tag{113}$$

$$\frac{\tilde{\delta}_q}{3a} + \frac{d}{dt}\left(\frac{\delta\tilde{u}_q}{a}\right) - F\frac{d}{dt}\left(\frac{\delta u_q}{a}\right) = 0 \, . \tag{114}$$

An inspection of Equation (95) shows that at this era, for $a \ll C$, we have $F \propto -La\sqrt{C}/3$. In addition, in DG, time can be integrated from the first Friedmann equation with only radiation and matter, and one obtains:

$$t(Y) = \frac{2\sqrt{1+C}}{3H_0}\left(\sqrt{Y+C}(Y-2C) + 2C^{\frac{3}{2}}\right) \, , \tag{115}$$

We recall that $Y = a/a_0 = a$ assuming $a_0 = 1$, $H_0 = \dot{a}_0/a_0$ is the usual Hubble parameter which we recall is no longer the physical Hubble parameter. Thus, radiation era time and $a(t)$ were related by $a(t) = (3H_0\sqrt{C}/\sqrt{1+C})^{1/2}t^{1/2}$. This complete system consisting of Equations (109)–(111) and Equations (112)–(114) has an analytical solution:

$$\delta_{\gamma q} = \delta_{Bq} = \delta_{Dq} = \delta_{vq} = \frac{q^2 t^2 \mathcal{R}_q}{a^2} \, , \tag{116}$$

$$\dot{\Psi}_q = -\frac{t q^2 \mathcal{R}_q}{a^2} \, , \tag{117}$$

$$\delta u_{\gamma q} = \delta u_{vq} = -\frac{2t^3 q^2 \mathcal{R}_q}{9a^2} \, , \tag{118}$$

where[7]

$$q^2\mathcal{R}_q \equiv -a^2 H\Psi_q + 4\pi G a^2 \delta\rho_q + q^2 H\delta u_q \, , \tag{119}$$

is a gauge-invariant quantity, which take a time-independent value for $q/a \ll H$. Here, $H = \dot{a}/a$ is the GR definition of the Hubble parameter, which we recall is no longer the physical one. On the other hand, we obtain

$$\tilde{\delta}_q = -\frac{L\sqrt{C}q^2\mathcal{R}_q t^2}{3a} \, , \tag{120}$$

$$\dot{\tilde{\Psi}}_q = \frac{L\sqrt{C}q^2\mathcal{R}_q t}{a} \, , \tag{121}$$

$$\delta\tilde{u}_q = \frac{L\sqrt{C}q^2\mathcal{R}_q t^3}{a} \, . \tag{122}$$

We will talk about these initial conditions later. Note that Equations (102b)–(102d) give

$$\frac{d}{dt}(\delta_\gamma - \delta_B) = 0 \, . \tag{123}$$

This implies that if we start from adiabatic solutions, $\delta_\gamma = \delta_B$ is true for all the Universe evolution (the same happens for its delta version from Equations (105b)–(105d)).

Matter Era

In this era, we use $a \propto t^{2/3}$, then (still using $R = \tilde{R} = 0$), we have

$$\frac{d}{dt}\left(a^2 \Psi_q\right) = -4\pi G \bar{\rho}_D a^2 \delta_{Dq} ,$$
(124a)

$$\dot{\delta}_{Dq} = -\Psi_q ,$$
(124b)

$$\frac{d}{dt}\left(\frac{(1+R)\delta u_{\gamma q}}{a}\right) = -\frac{1}{3a}\delta_{\gamma q} ,$$
(124c)

$$\frac{d}{dt}\left(\frac{\delta u_{vq}}{a}\right) = -\frac{1}{3a}\delta_{vq} .$$
(124d)

For the delta sector,

$$\left[2\dot{F}\frac{\dot{a}}{a} + \ddot{F}\right]a^2\Psi_q + \left[6F\frac{\dot{a}}{a} + \frac{5}{2}\dot{F}\right]a^2\dot{\Psi}_q + 3Fa^2\ddot{\Psi}_q,$$

$$-\frac{d}{dt}\left(a^2\dot{\tilde{\Psi}}_q\right) = \frac{2a^2}{3t^2}\left(\tilde{\delta}_{Dq} - F\frac{\delta_{Dq}}{2}\right),$$
(124e)

$$\dot{\tilde{\delta}}_{\gamma q} - \frac{q^2}{a^2}\left(\delta\tilde{u}_{\gamma q} + F\delta u_{\gamma q}\right) + \dot{\tilde{\Psi}}_q - \partial_0(F\Psi_q) = 0,$$
(124f)

$$\dot{\tilde{\delta}}_{Dq} + \dot{\tilde{\Psi}}_q - \partial_0(F\Psi_q) = 0,$$
(124g)

$$\frac{\tilde{\delta}_{\gamma q}}{3a} + \frac{d}{dt}\left(\frac{\delta\tilde{u}_{\gamma q}}{a}\right) - F\frac{d}{dt}\left(\frac{\delta u_{\gamma q}}{a}\right) = 0.$$
(124h)

where (in this era),

$$a(t) = \left(\frac{3H_0}{2\sqrt{1+C}}\right)^{2/3} t^{2/3},$$
(125)

$$F(t) \propto -\frac{L}{3}a(t)^{3/2}.$$
(126)

It is remarkable that in the GR sector, there are exact solutions given by

$$\delta_{Dq} = \frac{9q^2t^2 \mathcal{R}_q \mathcal{T}(\kappa)}{10a^2},$$
(127)

$$\dot{\Psi}_q = -\frac{3q^2 t \mathcal{R}_q \mathcal{T}(\kappa)}{5a^2},$$
(128)

$$\delta_{\gamma q} = \delta_{vq} = \frac{3\mathcal{R}_q}{5}\left[\mathcal{T}(\kappa) - \mathcal{S}(\kappa)\cos\left(q\int_0^t \frac{dt}{\sqrt{3}a} + \Delta(\kappa)\right)\right],$$
(129)

$$\delta u_{\gamma q} = \delta u_{vq} = \frac{3t\mathcal{R}_q}{5}\left[-\mathcal{T}(\kappa) + \mathcal{S}(\kappa)\frac{a}{\sqrt{3}qt}\sin\left(q\int_0^t \frac{dt}{\sqrt{3}a} + \Delta(\kappa)\right)\right].$$
(130)

where $\mathcal{T}(\kappa)$, $\mathcal{S}(\kappa)$ and $\Delta(\kappa)$ are time-independent dimensionless functions of the dimensionless re-scaled wave number

$$\kappa \equiv \frac{q\sqrt{2}}{a_{EQ}H_{EQ}}$$
(131)

.

a_{EQ} and H_{EQ} are, respectively, the Robertson–Walker scale factor and the expansion rate at matter radiation equally. These are referred to as transfer functions. (These functions can only depend on κ because they need to be independent of spatial coordinates' normalization and are dimensionless. A comprehensive analysis of the behavior of these functions can be found in [47].) Conversely, delta perturbations do not have an exact solution, and numerical calculations are necessary to determine them. However, in this work, we will not present numerical solutions and instead focus on estimating the initial conditions of the perturbations at the end of this section.

In order to obtain all transfer functions, we have to compare solutions with the full equation system (with $\rho_B = \tilde{\rho}_B = 0$). To do this task, let us make the change of variable $y \equiv a/a_{EQ} = a/C$; this means

$$\frac{d}{dt} = \frac{H_{EQ}}{\sqrt{2}} \frac{\sqrt{1+y}}{y} \frac{d}{dy}.$$ (132)

In addition, we will use the following parametrization for all perturbations

$$\delta_{Dq} = \kappa^2 \mathcal{R}_q^0 d(y)/4\,, \quad \delta_{\gamma q} = \delta_{vq} = \kappa^2 \mathcal{R}_q^0 r(y)/4\,,$$
$$\Psi_q = (\kappa^2 H_{EQ}/4\sqrt{2})\mathcal{R}_q^0 f(y)\,, \quad \delta u_{\gamma q} = \delta u_{vq} = (\kappa^2 \sqrt{2}/4H_{EQ})\mathcal{R}_q^0 g(y)\,,$$

and

$$\tilde{\delta}_{Dq} = \kappa^2 \mathcal{R}_q^0 \tilde{d}(y)/4\,, \quad \tilde{\delta}_{\gamma q} = \tilde{\delta}_{vq} = \kappa^2 \mathcal{R}_q^0 \tilde{r}(y)/4\,,$$
$$\dot{\tilde{\Psi}}_q = (\kappa^2 H_{EQ}/4\sqrt{2})\mathcal{R}_q^0 \tilde{f}(y)\,, \quad \delta\tilde{u}_{\gamma q} = \delta\tilde{u}_{vq} = (\kappa^2 \sqrt{2}/4H_{EQ})\mathcal{R}_q^0 \tilde{g}(y)\,.$$

Then, Equations (124a)–(124d) and Equations (124e)–(124h) become

$$\sqrt{1+y}\frac{d}{dy}\left(y^2 f(y)\right) = -\frac{3}{2}d(y) - \frac{4r(y)}{y}\,,$$ (133a)

$$\sqrt{1+y}\frac{d}{dy}r(y) - \frac{\kappa^2 g(y)}{y} = -yf(y)\,,$$ (133b)

$$\sqrt{1+y}\frac{d}{dy}d(y) = -yf(y)\,,$$ (133c)

$$\sqrt{1+y}\frac{d}{dy}\left(\frac{g(y)}{y}\right) = -\frac{r(y)}{3}\,,$$ (133d)

and

$$-[(1+2y)yF'(y) + y(1+y)F''(y)]d(y) + \left[6F(y) + \frac{5}{2}yF'(y)\right]y\sqrt{1+y}f(y)$$

$$+3F(y)y^2\sqrt{1+y}f'(y) - \sqrt{1+y}\frac{d}{dy}\left(y^2\tilde{f}(y)\right) = \frac{3\tilde{d}(y)}{2} + \frac{4\tilde{r}(y)}{y}\,,$$

$$-\frac{3F(y)d(y)}{4} - \frac{4F(y)r(y)}{y}$$ (133e)

$$\sqrt{1+y}\frac{d}{dy}\tilde{d}(y) = -y\tilde{f}(y) - \sqrt{1+y}\frac{d}{dy}d(y)\,,$$ (133f)

$$\sqrt{1+y}\frac{d}{dy}\tilde{r}(y) = \frac{\kappa^2}{y}[\tilde{g}(y) + F(y)g(y)] - y\tilde{f}(y) - \sqrt{1+y}\frac{d}{dy}d(y)\,,$$ (133g)

$$\sqrt{1+y}\frac{d}{dy}\left(\frac{\tilde{g}(y)}{y}\right) = -\frac{\tilde{r}(y)}{3} + \sqrt{1+y}F(y)\frac{d}{dy}\left(\frac{g(y)}{y}\right)\,.$$ (133h)

In this notation, the initial conditions are

$$
\begin{aligned}
d(y) &= r(y) \rightarrow y^2\,, \\
f(y) &\rightarrow -2\,, \\
g(y) &\rightarrow -\frac{y^4}{9}\,.
\end{aligned}
$$

For the delta sector,

$$
\begin{aligned}
\tilde{d}(y) &= \tilde{r}(y) \rightarrow -\frac{LC^{3/2}}{3}y^3\,, \\
\tilde{f}(y) &\rightarrow \sqrt{2}LC^{3/2}y\,, \\
\tilde{g}(y) &\rightarrow \frac{LC^{3/2}}{2}y^5\,.
\end{aligned}
$$

From supernovae fit, we know that $C \sim 10^{-4}$ and $L \sim 0.45$ [37,42]; thus, we can estimate that fluctuations of "delta matter" at the beginning of the Universe were much smaller than fluctuations of standard matter. For example, at $y \sim 10^{-3}$, the ratio between components of the Universe is $|\tilde{\delta}_\alpha / \delta_\alpha| \sim 10^{-10}$.

We do not show numerical solutions here because the aim of this work is to trace a guide for future work, in particular, in the numeric computation of multipole coefficients for temperature fluctuations in the CMB. However, we will derive the equations to calculate that computation.

5. Derivation of Temperature Fluctuations

It is possible to find expressions analogous to temperature fluctuations usually obtained by Boltzmann equations by studying photons propagation in FRLW-perturbed coordinates, with the condition $\tilde{g}_{i0} = 0^8$. For DG, the metric which photons follow is given by

$$
\begin{aligned}
\mathbf{g}_{00} &= -((1+3F(t)) + E(\mathbf{x},t) + \tilde{E}(\mathbf{x},t))\,, \quad \mathbf{g}_{i0} = 0\,, \\
\mathbf{g}_{ij} &= a^2(t)(1+F(t))\delta_{ij} + h_{ij}(\mathbf{x},t) + \tilde{h}_{ij}(\mathbf{x},t)\,,
\end{aligned} \tag{134}
$$

A ray of light propagating to the origin of the FRLW coordinate system, from a direction $\hat{n}$, will have a comoving radial coordinate r related with t by

$$
\begin{aligned}
0 = \tilde{\mathbf{g}}_{\mu\nu}dx^\mu dx^\nu = &-(1+3F(t) + E(r\hat{n},t) + \tilde{E}(r\hat{n},t))dt^2 \\
&+(a^2(t)(1+F(t)) + h_{rr}(r\hat{n},t) + \tilde{h}_{rr}(r\hat{n},t))dr^2\,,
\end{aligned} \tag{135}
$$

in other words,

$$
\begin{aligned}
\frac{dr}{dt} &= -\left(\frac{(1+3F(t)) + E + \tilde{E}}{a^2(t)(1+F(t)) + h_{rr} + \tilde{h}_{rr}}\right)^{1/2} \\
&\simeq -\frac{1}{a_{DG}(t)} + \frac{(h_{rr} + \tilde{h}_{rr})}{2(1+3F(t))a_{DG}^3(t)} - \frac{E + \tilde{E}}{2(1+3F(t))a_{DG}(t)}\,,
\end{aligned} \tag{136}
$$

where $a_{DG}(t)$ is the modified scale factor given by

$$
a_{DG}(t) = a(t)\sqrt{\frac{1+F(t)}{1+3F(t)}}\,. \tag{137}
$$

Now, we will use the approximation of a sharp transition between the opaque and transparent Universe at a moment t_{ls} of last scattering at red shift $z \simeq 1090$. With this approximation, the relevant term at the first order in Equation (136) is

$$r(t) = \left[s(t) + \int_{t_{ls}}^{t} \frac{dt'}{a_{DG}(t')} N\big(s(t')\hat{n}, t'\big) \right],$$ (138)

where

$$N(\mathbf{x}, t) \equiv \frac{1}{2(1+3F)} \left[\frac{h_{rr}(\mathbf{x}, t) + \tilde{h}_{rr}(\mathbf{x}, t)}{a_{DG}^2} - E(\mathbf{x}, t) - \tilde{E}(\mathbf{x}, t) \right],$$ (139)

and $s(t)$ is the zero-order solution for the radial coordinate. $s(t) = r_{ls}$ when $t = t_{ls}$:

$$s(t) = r_{ls} - \int_{t_{ls}}^{t} \frac{dt'}{a_{DG}(t')} = \int_{t}^{t_0} \frac{dt'}{a_{DG}(t')}.$$ (140)

If a ray of light arrives to $r = 0$ at a time t_0, then Equation (138) gives

$$0 = s(t_0) + \int_{t_{ls}}^{t} \frac{dt'}{a_{DG}(t')} N\big(s(t')\hat{n}, t'\big) = r_{ls} + \int_{t_{ls}}^{t_0} \frac{dt}{a_{DG}(t)} (N(s(t)\hat{n}, t) - 1).$$ (141)

A time interval δt_{ls}, between the departure of successive rays of light at time t_{ls} of last scattering, produces an interval of time δt_0 between the arrival of the rays of light at t_0, which are given by the variation of Equation (141):

$$0 = \frac{\delta t_{ls}}{a_{DG}(t_{ls})} \left[1 - N(r_{ls}\hat{n}, t_{ls}) + \int_{t_{ls}}^{t_0} \frac{dt}{a_{DG}(t)} \left(\frac{\partial N(r(t)\hat{n}, t)}{\partial r} \right)_{r=s(t)} \right]$$
$$+ \delta t_{ls}(\partial u_{\gamma}^{r}(r_{ls}\hat{n}, t_{ls}) + \partial \tilde{u}_{\gamma}^{r}(r_{ls}\hat{n}, t_{ls})) + \frac{\delta t_0}{a_{DG}(t_0)} [-1 + N(0, t_0)].$$ (142)

The velocity terms of the photon–gas or photon–electron–nucleon arise because of the variation with respect to the time of the radial coordinate r_{ls} described by Equation (141). The exchange rate of $N(s(t)\hat{n}, t)$ is

$$\frac{d}{dt} N(s(t)\hat{n}, t) = \left(\frac{\partial}{\partial t} N(r\hat{n}, t) \right)_{r=s(t)} - \frac{1}{a_{DG}(t)} \left(\frac{\partial N(r\hat{n}, t)}{\partial r} \right)_{r=s(t)},$$

then,

$$0 = \frac{\delta t_{ls}}{a_{DG}(t_{ls})} \left[1 - N(0, t_{ls}) + \int_{t_{ls}}^{t_0} dt \left(\frac{\partial N(r\hat{n}, t)}{\partial t} \right)_{r=s(t)} \right]$$
$$+ \delta t_{ls}(\partial u_{\gamma}^{r}(r_{ls}\hat{n}, t_{ls}) + \partial \tilde{u}_{\gamma}^{r}(r_{ls}\hat{n}, t_{ls})) + \frac{\delta t_0}{a_{DG}(t_0)} [-1 + N(0, t_0)].$$ (143)

This result gives the ratio between the time intervals between ray of lights that are emitted and received. However, we are interested in this ratio but for the proper time, which in DG is defined with the original metric $g_{\mu\nu}$:

$$\delta \tau_L = \sqrt{1 + E(r_{ls}, t_{ls})}\,\delta t_{ls}, \quad \delta \tau_0 = \sqrt{1 + E(0, t_0)}\,\delta t_0,$$ (144)

At first order, it gives the ratio between a received frequency and an emitted one:

$$\frac{\nu_0}{\nu_L} = \frac{\delta \tau_L}{\delta \tau_0} = \frac{a_{DG}(t_{ls})}{a_{DG}(t_0)} \left[1 + \frac{1}{2}(E(r_{ls}\hat{n}, t) - E(0, t_0)) \right.$$
$$\left. - \int_{t_{ls}}^{t_0} \left(\frac{\partial}{\partial t} N(r\hat{n}, t) \right)_{r=s(t)} dt - a_{DG}(t)(\delta u_{\gamma}^{r}(r_{ls}\hat{n}, t) + \delta \tilde{u}_{\gamma}^{r}(r_{ls}\hat{n}, t)) \right].$$ (145)

In [42], we defined the physical scale factor as $Y_{DG}(t) \equiv a_{DG}(t)/a_{DG}(t_0)$. Thus, we recover the standard expression for the redshift. The observed temperature at the present time t_0 from direction $\hat{n}$ is

$$T(\hat{n}) = \left(\frac{\nu_0}{\nu_L}\right)(\bar{T}(t_{ls}) + \delta T(r_{ls}\hat{n}, t_{ls})) \,, \tag{146}$$

In the absence of perturbations, the observed temperature in all directions should be

$$T_0 = \left(\frac{a_{DG}(t_{ls})}{a_{DG}(t_0)}\right)\bar{T}(t_{ls}) \,, \tag{147}$$

Therefore, the ratio between the observed temperature shift that comes from direction $\hat{n}$ and the unperturbed value is

$$
\begin{aligned}
\frac{\Delta T(\hat{n})}{T_0} &\equiv \frac{T(\hat{n}) - T_0}{T_0} = \frac{\nu_0 a_{DG}(t_0)}{\nu_L a_{DG}(t_{ls})} - 1 + \frac{\delta T(r_{ls}\hat{n}, t_{ls})}{\bar{T}(t_{ls})} \\
&= \frac{1}{2}(E(r_{ls}\hat{n}, t) - E(0, t_0)) - \int_{t_{ls}}^{t_0} dt \left(\frac{\partial}{\partial t}N(r\hat{n}, t)\right)_{r=s(t)} \\
&\quad - a_{DG}(t)(\delta u_\gamma^r(r_{ls}\hat{n}, t) + \delta\tilde{u}_\gamma^r(r_{ls}\hat{n}, t)) + \frac{\delta T(r_{ls}\hat{n}, t_{ls})}{\bar{T}(t_{ls})} \,.
\end{aligned}
\tag{148}
$$

For scalar perturbations in any gauge with $\mathbf{h}_{i0} = 0$, the metric perturbations are

$$
\begin{aligned}
h_{00} &= -E \,, \quad h_{ij} = (1+F)a^2\left[A\delta_{ij} + \frac{\partial^2 B}{\partial x^i \partial x^j}\right], \\
\tilde{h}_{00} &= -\tilde{E} \,, \quad \tilde{h}_{ij} = (1+F)a^2\left[\tilde{A}\delta_{ij} + \frac{\partial^2 \tilde{B}}{\partial x^i \partial x^j}\right].
\end{aligned}
\tag{149}
$$

In addition, for scalar perturbations, the radial velocity of the photon fluid and the delta versions are given in terms of the velocity potentials δu_γ and $\delta\tilde{u}_\gamma$, respectively,

$$
\begin{aligned}
\delta u_\gamma^r &= (\bar{g} + \bar{\tilde{g}})^{r\mu}\frac{\partial \delta u_\gamma}{\partial x^\mu} = \frac{1}{(1+F(t))a^2}\frac{\partial \delta u_\gamma}{\partial r} \,, \\
\delta\tilde{u}_\gamma^r &= (\bar{g} + \bar{\tilde{g}})^{r\mu}\frac{\partial \delta\tilde{u}_\gamma}{\partial x^\mu} = \frac{1}{(1+F(t))a^2}\frac{\partial \delta\tilde{u}_\gamma}{\partial r} \,.
\end{aligned}
\tag{150}
$$

Then, Equation (148) gives the scalar contribution to temperature fluctuations

$$
\begin{aligned}
\left(\frac{\Delta T(\hat{n})}{T_0}\right)^S &= \frac{1}{2}(E(r_{ls}\hat{n}, t) - E(0, t_0)) - \int_{t_{ls}}^{t_0} dt\left(\frac{\partial}{\partial t}N(r\hat{n}, t)\right)_{r=s(t)} \\
&\quad - \frac{1}{(1+3F(t))a_{DG}}\left(\frac{\partial \delta u_\gamma(r_{ls}\hat{n}, t)}{\partial r} + \frac{\partial \delta\tilde{u}_\gamma(r_{ls}\hat{n}, t)}{\partial t}\right) \\
&\quad + \frac{\delta T(r_{ls}\hat{n}, t_{ls})}{\bar{T}(t_{ls})} \,,
\end{aligned}
\tag{151}
$$

where

$$N = \frac{1}{2}\left[A + \frac{\partial^2 B}{\partial r^2} + \left(\tilde{A} + \frac{\partial^2 \tilde{B}}{\partial r^2}\right) - \frac{E}{1+3F} - \frac{\tilde{E}}{1+3F}\right]. \tag{152}$$

In the next step, we will study the gauge transformations of these fluctuations. The following identity for the fields B and $\tilde{B}$ will be useful:

$$\left(\frac{\partial^2 \dot{B}}{\partial r^2}\right)_{r=s(t)} = -\left(\frac{d}{dt}\left[a_{DG}\frac{\partial \dot{B}}{\partial r} + a_{DG}\dot{a}_{DG}\dot{B} + a_{DG}^2\ddot{B}\right] + \frac{\partial}{\partial t}\left[a_{DG}\dot{a}_{DG}\dot{B} + a_{DG}^2\ddot{B}\right]\right)_{r=s(t)}. \tag{153}$$

Then, the temperature fluctuations are described by

$$\left(\frac{\Delta T(\hat{n})}{T_0}\right)^S = \left(\frac{\Delta T(\hat{n})}{T_0}\right)^S_{early} + \left(\frac{\Delta T(\hat{n})}{T_0}\right)^S_{late} + \left(\frac{\Delta T(\hat{n})}{T_0}\right)^S_{ISW} \tag{154}$$

where

$$
\left(\frac{\Delta T(\hat{n})}{T_0}\right)^S_{early} = -\frac{1}{2}a_{DG}(t_{ls})\dot{a}_{DG}(t_{ls})\dot{B}(r_{ls}\hat{n}, t_{ls}) - \frac{1}{2}a^2_{DG}(t_{ls})\ddot{B}(r_{ls}\hat{n}, t_{ls}) + \frac{1}{2}E(r_{ls}\hat{n}, t_{ls})
$$

$$
+\frac{\delta T(r_{ls}\hat{n})}{\bar{T}(t_{ls})} - a_{DG}(t_{ls})\left[\frac{\partial}{\partial r}\left(\frac{1}{2}\dot{B}(r\hat{n}, t_{ls}) + \frac{1}{(1+3F(t_{ls}))a^2_{DG}(t_{ls})}\delta u_\gamma(r\hat{n}, t_{ls})\right)_{r=r_{ls}}\right]
$$

$$
- \left\{\left(\frac{1}{2}a_{DG}(t_{ls})\dot{a}_{DG}(t_{ls})\dot{\tilde{B}}(r_{ls}\hat{n}, t_{ls}) + \frac{1}{2}a^2_{DG}(t_{ls})\ddot{\tilde{B}}(r_{ls}\hat{n}, t_{ls})\right)\right.
$$

$$
+a_{DG}(t_{ls}) \times \left.\left[\frac{\partial}{\partial r}\left(\frac{1}{2}\dot{\tilde{B}}(r\hat{n}, t_{ls}) + \frac{1}{(1+3F(t_{ls}))a^2_{DG}(t_{ls})}\delta\tilde{u}_\gamma(r\hat{n}, t_{ls})\right)_{r=r_{ls}}\right]\right\}, \tag{155}
$$

$$
\left(\frac{\Delta T(\hat{n})}{T_0}\right)^S_{late} = \frac{1}{2}a_{DG}(t_0)\dot{a}_{DG}(t_0)\dot{B}(0, t_0) + \frac{1}{2}a^2_{DG}(t_0)\ddot{B}(0, t_0) - \frac{1}{2}E(0, t_0)
$$

$$
+ a_{DG}(t_0)\left[\frac{\partial}{\partial r}\left(\frac{1}{2}\dot{B}(r\hat{n}, t_0) + \frac{1}{(1+3F(t_0))a^2_{DG}(t_0)}\delta u_\gamma(r\hat{n}, t_0)\right)_{r=0}\right]
$$

$$
+ \left\{\left(\frac{1}{2}a_{DG}(t_0)\dot{a}_{DG}(t_0)\dot{\tilde{B}}(0, t_0) + \frac{1}{2}a^2_{DG}(t_0)\ddot{\tilde{B}}(0, t_0)\right)\right.
$$

$$
+a_{DG}(t_0) \times \left.\left[\frac{\partial}{\partial r}\left(\frac{1}{2}\dot{\tilde{B}}(r\hat{n}, t_0) + \frac{1}{(1+3F(t_0))a^2_{DG}(t_0)}\delta\tilde{u}_\gamma(r\hat{n}, t_0)\right)_{r=r_{ls}}\right]\right\}, \tag{156}
$$

$$
\left(\frac{\Delta T(\hat{n})}{T_0}\right)^S_{ISW} = -\frac{1}{2}\int_{t_{ls}}^{t_0} dt\left\{\frac{\partial}{\partial t}\left[a^2_{DG}(t)\ddot{B}(r\hat{n}, t) + a_{DG}(t)\dot{a}_{DG}(t)\dot{B}(r\hat{n}, t) + A(r\hat{n}, t)\right.\right.
$$

$$
- \frac{E(r\hat{n}, t)}{1+3F(t)} + a^2_{DG}(t)\ddot{\tilde{B}}(r\hat{n}, t) + a_{DG}(t)\dot{a}_{DG}(t)\dot{\tilde{B}}(r\hat{n}, t)
$$

$$
+ \left.\left.\tilde{A}(r\hat{n}, t) - \frac{\tilde{E}(r\hat{n}, t)}{1+3F(t)}\right]\right\}, \tag{157}
$$

The "late" term is the sum of independent direction terms and a term proportional to $\hat{n}$, which was added to represent the local anisotropies of the gravitational field and the local fluid. In GR, these terms only contribute to the multipole expansion for $l = 0$ and $l = 1$. Thus, we will ignore their contribution to our derivation of the temperature fluctuations multipoles coefficients.

5.1. Gauge Transformations

We are going to study the gauge transformations for photons propagating in the metric $\mathbf{g}_{\mu\nu}$ for a parameter ϵ_μ. Then, the transformations are

$$
\Delta A = \frac{2\dot{a}}{(1+F)a}\frac{\epsilon_0}{1+3F}, \quad \Delta B = -\frac{2}{1+F}\frac{\epsilon^S}{(1+F)a^2},
$$

$$
\Delta C_i = -\frac{1}{1+F}\frac{\epsilon^V_i}{(1+F)a^2}, \quad \Delta D_{ij} = 0, \quad \Delta E = 2\frac{\partial}{\partial t}\left(\frac{\epsilon_0}{1+3F}\right), \tag{158}
$$

$$
\Delta H = -\frac{1}{\sqrt{1+F}a}\left[a^2\frac{\partial}{\partial t}\left(\frac{\epsilon^S}{(1+F)a^2}\right) + \frac{\epsilon_0}{(1+3F)}\right], \quad \Delta G_i = -\frac{a}{\sqrt{1+F}}\frac{\partial}{\partial t}\left(\frac{\epsilon^V_i}{(1+F)a^2}\right).
$$

and

$$
\begin{aligned}
\Delta\tilde{A} &= \frac{1}{(1+F)a^2}\left[\frac{\partial}{\partial t}(Fa^2)\frac{\epsilon_0}{1+3F}\right], \quad \Delta\tilde{B} = -\frac{1}{(1+F)a^2}\left[\frac{2F}{1+F}\epsilon^S\right], \\
\Delta\tilde{C}_i &= -\frac{F}{1+F}\frac{\epsilon_i^V}{(1+F)a^2}, \quad \Delta\tilde{D}_{ij} = 0, \quad \Delta\tilde{E} = 6F\frac{\partial}{\partial t}\left(\frac{\epsilon_0}{1+3F}\right) + \frac{3\dot{F}}{1+3F}\epsilon_0 \\
\Delta\tilde{H} &= -\frac{1}{\sqrt{1+Fa}}\left[Fa^2\frac{\partial}{\partial t}\left(\frac{\epsilon^S}{(1+F)a^2}\right) + \frac{3F\epsilon_0}{(1+3F)}\right], \\
\Delta\tilde{G}_i &= -\frac{1}{\sqrt{1+Fa}}\left[Fa^2\frac{\partial}{\partial t}\left(\frac{\epsilon_i^V}{(1+F)a^2}\right)\right].
\end{aligned}
\tag{159}
$$

Now, considering the sum of the perturbations, we obtain

$$
\Delta A + \Delta\tilde{A} = \frac{1}{(1+F)a^2}\frac{\partial}{\partial t}[(1+F)a^2]\frac{\epsilon_0}{1+3F},
\tag{160a}
$$

$$
\Delta B + \Delta\tilde{B} = -\frac{2\epsilon^S}{(1+F)a^2},
\tag{160b}
$$

$$
\Delta E + \Delta\tilde{E} = 2(1+3F)\frac{\partial}{\partial t}\left(\frac{\epsilon_0}{1+3F}\right) + \frac{3\dot{F}}{1+3F}\epsilon_0,
\tag{160c}
$$

$$
\Delta H + \Delta\tilde{H} = -\frac{1}{\sqrt{1+Fa}}\left[(1+F)a^2\frac{\partial}{\partial t}\left(\frac{\epsilon^S}{(1+F)a^2}\right) + \epsilon_0\right],
\tag{160d}
$$

$$
\Delta C_i + \Delta\tilde{C}_i = -\frac{\epsilon_i^V}{(1+F)a^2},
\tag{160e}
$$

$$
\Delta G_i + \Delta\tilde{G}_i = -\frac{1}{\sqrt{1+Fa}}\left[(1+F)a^2\frac{\partial}{\partial t}\left(\frac{\epsilon_i^V}{(1+F)a^2}\right)\right].
\tag{160f}
$$

Now, we will study the gauge transformations that preserve the condition $\mathbf{g}_{i0} = g_{i0} + \tilde{g}_{i0} = 0$. This means that $\Delta H + \Delta\tilde{H} = 0$. This gives a solution for ϵ_0 given by

$$
\epsilon_0 = -(1+F)a^2\frac{\partial}{\partial t}\left(\frac{\epsilon^S}{(1+F)a^2}\right).
\tag{161}
$$

When we study how the "ISW" term transforms under this type of transformation, we found that $\Delta ISW = 0$. While for the "early" term, we should note that temperature perturbations transform as

$$
\Delta\delta T(r_{ls}\hat{n}, t) = \dot{T}(t)\frac{\epsilon_0}{1+3F},
\tag{162}
$$

With this expression and $\bar{T}a_{DG} = cte$, we finally obtain

$$
\frac{\Delta\delta T(r_{ls}\hat{n}, t)}{\bar{T}(t_{ls})} = -\frac{\dot{a}_{DG}}{a_{DG}}\frac{\epsilon_0}{1+3F}.
\tag{163}
$$

This results implies that the "early" term is invariant under this gauge transformation. Note that this gauge transformation is equivalent to the previously discussed in Section 1, because we can always take ϵ as a combination of ϵ and $\tilde{\epsilon}$. Then, we remark that temperature fluctuations are gauge invariant under scalar transformations that leave $\mathbf{g}_{i0} = 0$.

5.2. Single Modes

We will assume that since the last scattering until now, all the scalar contributions are dominated by a unique mode such that any perturbation $X(\mathbf{x}, t)$ could be written as

$$X(\mathbf{x}, t) = \int d^3 q \alpha(\mathbf{q}) e^{i\mathbf{q}\cdot\mathbf{x}} X_q(t), \tag{164}$$

where $\alpha(\mathbf{q})$ is a stochastic variable, which is normalized such that

$$\langle \alpha(\mathbf{q})\alpha^*(\mathbf{q}') \rangle = \delta^3(\mathbf{q} - \mathbf{q}'). \tag{165}$$

Then, Equations (155) and (157) become

$$\left(\frac{\Delta T(\hat{n})}{T_0}\right)^S_{early} = \int d^3 q \alpha(\mathbf{q}) e^{i\mathbf{q}\cdot\hat{n}r(t_{ls})} \left(\mathcal{F}(q) + \tilde{\mathcal{F}}(q) + i\hat{q}\cdot\hat{n}(\mathcal{G}(q) + \tilde{\mathcal{G}}(q))\right), \tag{166}$$

$$\left(\frac{\Delta T(\hat{n})}{T_0}\right)^S_{ISW} = -\frac{1}{2}\int_{t_0}^{t_1} dt \int d^3 q \alpha(\mathbf{q}) e^{i\mathbf{q}\cdot\hat{n}s(t)} \frac{d}{dt}\left[a_{DG}^2(t)\ddot{B}_q(t) + a_{DG}(t)\dot{a}_{DG}(t)\dot{B}_q(t)\right.$$

$$+ \quad A_q(t) - \frac{E_q(t)}{1 + 3F(t)} + \left(a_{DG}^2(t)\ddot{B}_q(t) + a_{DG}(t)\dot{a}_{DG}(t)\dot{B}_q(t) + \tilde{A}_q(t)\right)$$

$$- \quad \left.\frac{\tilde{E}_q(t)}{1 + 3F(t)}\right)\Bigg], \tag{167}$$

where

$$\mathcal{F}(q) = -\frac{1}{2}a_{DG}^2(t)\ddot{B}_q(t_{ls}) - \frac{1}{2}a_{DG}(t)\dot{a}_{DG}(t_{ls})\dot{B}_q(t_{ls}) + \frac{1}{2}E_q(t_{ls}) + \frac{\delta T_q(t_{ls})}{\bar{T}(t_{ls})}, \tag{168}$$

$$\tilde{\mathcal{F}}(q) = -\frac{1}{2}a_{DG}^2(t)\ddot{\tilde{B}}_q(t_{ls}) - \frac{1}{2}a_{DG}(t_{ls})\dot{a}_{DG}(t_{ls})\dot{\tilde{B}}_q(t_{ls}), \tag{169}$$

$$\mathcal{G}(q) = -q\left(\frac{1}{2}a_{DG}(t_{ls})\dot{B}_q(t_{ls}) + \frac{1}{(1 + 3F(t_{ls}))a_{DG}(t_{ls})}\delta u_\gamma(t_{ls})\right), \tag{170}$$

$$\tilde{\mathcal{G}}(q) = -q\left(\frac{1}{2}a_{DG}(t_{ls})\dot{\tilde{B}}_q(t_{ls}) + \frac{1}{(1 + 3F(t_{ls}))a_{DG}(t_{ls})}\delta\tilde{u}_\gamma(t_{ls})\right). \tag{171}$$

These functions are called form factors. We emphasize that combination given by $\mathcal{F}(q) + \tilde{\mathcal{F}}(q)$ and $\mathcal{G}(q) + \tilde{\mathcal{G}}(q)$, and the expressions inside the integral are gauge invariants under gauge transformations that preserve $\mathbf{g}_{i0}$ equal to zero.

6. Coefficients of Multipole Temperature Expansion: Scalar Modes

As an application of the previous results, we will study the contribution of the scalar modes for temperature–temperature correlation, which is given by:

$$C_{TT,l} = \frac{1}{4\pi}\int d^2\hat{n} \int d^2\hat{n}' P_l(\hat{n}\cdot\hat{n}')\langle\Delta T(\hat{n})\Delta T(\hat{n}')\rangle, \tag{172}$$

where $\Delta T(\hat{n})$ is the stochastic variable which gives the deviation of the average of observed temperature in direction $\hat{n}$, and $\langle\ldots\rangle$ denotes the average over the position of the observer. However, the observed quantity is

$$C_{TT,l}^{obs} = \frac{1}{4\pi}\int d^2\hat{n} \int d^2\hat{n}' P_l(\hat{n}\cdot\hat{n}')\Delta T(\hat{n})\Delta T(\hat{n}'), \tag{173}$$

Nevertheless, the mean square fractional difference between this equation and Equation (172) is $2/(2l + 1)$, and therefore, it may be neglected for $l \gg 1$. In order to calculate these coefficients, we use the following expansion in spherical harmonics

$$e^{i\hat{q}\cdot\hat{n}\rho} = 4\pi \sum_{l=0}^{\infty} \sum_{m=-l}^{m=l} i^l j_l(\rho) Y_l^m(\hat{n}) Y_l^{m*}(\hat{q}), \tag{174}$$

where $j_l(\rho)$ represents the spherical Bessel's functions. Using this expression in Equation (166), and replacing the factor $i\hat{q}\cdot\hat{n}$ for time derivatives of Bessel's functions, the scalar contribution of the observed T–T fluctuations in direction $\hat{n}$ is

$$(\Delta T(\hat{n}))^S = \sum_{lm} a^S_{T,lm} Y^m_l(\hat{n}) \,, \tag{175}$$

where

$$a^S_{T,lm} = 4\pi i^l T_0 \int d^3 q \, \alpha(\mathbf{q}) Y^{l*}_l(\hat{q}) \left[j_l(qr_{ls})(\mathcal{F}(q) + \tilde{\mathcal{F}}(q)) + j'_l(qr_{ls})(\mathcal{G}(q) + \tilde{\mathcal{G}}(q)) \right] \,, \tag{176}$$

and $\alpha(\mathbf{q})$ is a stochastic parameter for the dominant scalar mode. It is normalized such that

$$\langle \alpha(\mathbf{q})\alpha^*(\mathbf{q}') \rangle = \delta^3(\mathbf{q} - \mathbf{q}') \,. \tag{177}$$

Inserting this expression in Equation (172), we obtain

$$C^S_{TT,l} = 16\pi^2 T_0^2 \int_0^\infty q^2 dq \left[j_l(qr_{ls})(\mathcal{F}(q) + \tilde{\mathcal{F}}(q)) + j'_l(qr_{ls})(\mathcal{G}(q) + \tilde{\mathcal{G}}(q)) \right]^2 \,. \tag{178}$$

Now, we will consider the case $l \gg 1$. In this limit, we can use the following approximation for Bessel's functions[9]:

$$j_l(\rho) \to \begin{cases} \cos(b) \cos[\nu(\tan b - b) - \pi/4]/(\nu\sqrt{\sin b}) & \rho > \nu, \\ 0 & \rho < \nu, \end{cases} \tag{179}$$

where $\nu \equiv l + 1/2$, and $\cos b \equiv \nu/\rho$, with $0 \le b \le \pi/2$. In addition, for $\rho > \nu \gg 1$, the phase $\nu(\tan b - b)$ is a function of ρ that grows very fast; then, the derivatives of Bessel's functions only act in its phase:

$$j'_l(\rho) \to \begin{cases} -\cos(b)\sqrt{\sin b} \sin[\nu(\tan b - b) - \pi/4]/\nu & \rho > \nu, \\ 0 & \rho < \nu. \end{cases} \tag{180}$$

Using these approximations in Equation (178) and changing the variable from q to $b = \cos^{-1}(\nu/qr_{ls})$, we obtain

$$\begin{aligned}
C^S_{TT,l} &= \frac{16\pi^2 T_0^2 \nu}{r_{ls}^3} \int_0^{\pi/2} \frac{db}{\cos^2 b} \\
&\times \left[\left(\mathcal{F}\left(\frac{\nu}{r_{ls}\cos b}\right) + \tilde{\mathcal{F}}\left(\frac{\nu}{r_{ls}\cos b}\right) \right) \cos[\nu(\tan b - b) - \pi/4] \right. \\
&\left. - \sin b \left(\mathcal{G}\left(\frac{\nu}{r_{ls}\cos b}\right) + \tilde{\mathcal{G}}\left(\frac{\nu}{r_{ls}\cos b}\right) \right) \sin[\nu(\tan b - b) - \pi/4] \right]^2 \,. \end{aligned} \tag{181}$$

When $\nu \gg 1$, the functions $\cos[\nu(\tan b - b) - \pi/4]$ and $\sin[\nu(\tan b - b) - \pi/4]$ oscillate very rapidly; then, the squared average of its values are $1/2$, while the averaged cross-terms are zero. Using $l \approx \nu$, and changing the integration variable from b to $\beta = 1/\cos b$, Equation (181) becomes

$$\begin{aligned}
l(l+1)C^S_{TT,l} &= \frac{8\pi^2 T_0^2 l^3}{r_{ls}^3} \int_1^\infty \frac{\beta d\beta}{\sqrt{\beta^2 - 1}} \\
&\times \left[\left(\mathcal{F}\left(\frac{l\beta}{r_{ls}}\right) + \tilde{\mathcal{F}}\left(\frac{l\beta}{r_{ls}}\right) \right)^2 + \frac{\beta^2 - 1}{\beta^2} \left(\mathcal{G}\left(\frac{l\beta}{r_{ls}}\right) + \tilde{\mathcal{G}}\left(\frac{l\beta}{r_{ls}}\right) \right)^2 \right] \end{aligned} \tag{182}$$

Note that $d_A = r_{ls} a_{DG}(t_{ls})$ is the angular diameter distance of the last scattering surface. To calculate the CMB power spectrum, we need to know the value of $\mathring{B}_q$. We use the off-diagonal equation from the delta sector to obtain it. This gives:

$$\mathring{A}_q = \dot{A}_q F + A_q \dot{F} - 2a^2(\rho + p)\delta u_q - a^2(\tilde{\rho} + \tilde{p})\delta u_q - (\rho + p)\delta \tilde{u}_q \,, \tag{183}$$

so if we use this equation with the definition of $\mathring{\Psi}$

$$\mathring{\dot{\Psi}}_q = \frac{1}{2}(3\mathring{\dot{A}}_q - q^2 \mathring{\dot{B}}_q),\tag{184}$$

it allows us to find $\mathring{B}$. Now, we will use the approximation that perturbations of a gravitation field are dominated by perturbations of dark matter density. In this regime $\dot{A}_q(t_{ls}) = 0$ and in the synchronous gauge, the velocity perturbations for dark matter are zero; then,

$$\mathring{\dot{A}}_q(t_{ls}) = A_q(t_{ls})\dot{F}(t_{ls})\,,\tag{185}$$

and

$$\mathring{\dot{B}}_q(t_{ls}) = \frac{3}{q^2}A_q(t_{ls})\dot{F}(t_{ls}) - \frac{2\mathring{\dot{\Psi}}_q(t_{ls})}{q^2} \Rightarrow \mathring{\ddot{B}}_q(t_{ls}) = \frac{3}{q^2}A_q(t_{ls})\ddot{F}(t_{ls}) - \frac{2\mathring{\ddot{\Psi}}_q(t_{ls})}{q^2}\,,\tag{186}$$

where

$$\begin{aligned}
q^2 A_q &= 8\pi G a^2 \delta\rho_{Dq} - 2Ha^2\mathring{\Psi}_q \\
&= 3H^2 a^2 \delta_{Dq} - 2Ha^2\mathring{\Psi}_q\,.
\end{aligned}\tag{187}$$

In GR, $\mathring{\dot{B}}_q = -2\mathring{\Psi}_q/q^2$, and $\mathring{\Psi}_q \propto t^{-1/3}$ implies $\mathring{\ddot{B}}_q = 2\mathring{\Psi}_q/3tq^2$. Therefore, the usual form factors are:

$$\mathcal{F}(q) = \frac{1}{3}\delta_{\gamma q}(t_{ls}) + \frac{\mathring{\Psi}_q(t_{ls})}{q^2}\left(a_{DG}(t_{ls})\dot{a}_{DG}(t_{ls}) - \frac{2}{3}\frac{a_{DG}^2(t_{ls})}{t_{ls}}\right),\tag{188}$$

$$\mathcal{G}(q) = -q\frac{\delta u_{\gamma q}(t_{ls})}{(1+3F(t_{ls}))a_{DG}(t_{ls})} + \frac{a_{DG}(t_{ls})\mathring{\Psi}_q(t_{ls})}{q}\,.\tag{189}$$

where we have used $\delta T_q/\bar{T} = \delta\rho_{\gamma q}/4\bar{\rho}_\gamma = \delta_{\gamma q}/3$. Nevertheless, for the "delta" contribution, $\mathring{\dot{\Psi}}_q$ and $\mathring{\ddot{\Psi}}_q$ satisfy the same relation as the standard case. Due to our decomposition, the tilde expresions are

$$\begin{aligned}
\tilde{\mathcal{F}}(q) &= -\frac{3}{2}\frac{A_q(t_{ls})}{q^2}(a_{DG}^2(t_{ls})\ddot{F}(t_{ls}) + a_{DG}(t_{ls})\dot{a}_{DG}(t_{ls})\dot{F}(t_{ls})) \\
&\quad + \frac{\mathring{\dot{\Psi}}_q(t_{ls})}{q^2}\left(a_{DG}(t_{ls})\dot{a}_{DG}(t_{ls}) - \frac{2}{3}\frac{a_{DG}^2(t_{ls})}{t_{ls}}\right),
\end{aligned}\tag{190}$$

$$\tilde{\mathcal{G}}(q) = -q\frac{\delta\tilde{u}_{\gamma q}(t_{ls})}{(1+3F(t_{ls}))a_{DG}(t_{ls})} + \frac{a_{DG}(t_{ls})\mathring{\dot{\Psi}}_q(t_{ls})}{q}\,.\tag{191}$$

Unfortunately, due to all the approximations we have used, we need to add some corrections to the solutions of the GR sector. After that, we will be able to find the numerical solutions for DG perturbations. The first consideration is that in the set of equations presented in the matter era, we have used $R = 3\bar{\rho}_B/4\rho_\gamma = 0$, which is not valid in this era. Corrections to the solutions can be calculated using a WKB approximation for perturbations[10] [47]. The second consideration that we must include in the solution of photons perturbations is the so-called Silk damping[11] [52,53], which takes into account the viscosity and heat conduction of the relativistic medium. Moreover, the transition from opaque to a transparent Universe at the last scattering moment was not instantaneous, but it could be considered a Gaussian. This effect is known as Landau damping[12]. We must recall that the physical geometry now is described by $Y_{DG}(t) = a_{DG}(t)/a_{DG}(t=0)$, so the expression for both Silk and Landau effects has to be expressed in this geometry. With these considerations, the solutions of perturbations are given by:

$$\dot{\Psi}_q(t_{ls}) \;=\; -\frac{3q^2 t_{ls}\,\mathcal{R}_q^o\,\mathcal{T}(\kappa)}{5a^2(t_{ls})}\,, \tag{192}$$

$$\delta_{\gamma q}(t_{ls}) \;=\; \frac{3\mathcal{R}_q^o}{5}\left[\mathcal{T}(\kappa)(1+3R_{ls}) - (1+R_{ls})^{-1/4}e^{-q^2 d_D^2/a_{ls}^2}\right.$$

$$\left. \times\ \mathcal{S}(\kappa)\cos\left(\int_0^{t_{ls}}\frac{qdt}{\sqrt{3(1+R(t))}a(t)} + \Delta(\kappa)\right)\right], \tag{193}$$

$$\delta u_{\gamma q}(t_{ls}) \;=\; \frac{3\mathcal{R}_q^o}{5}\left[-t_{ls}\mathcal{T}(\kappa) + \frac{a(t_{ls})}{\sqrt{3}q(1+R_{ls})^{3/4}}e^{-q^2 d_D^2/a_{ls}^2}\right.$$

$$\left. \times\ \mathcal{S}(\kappa)\sin\left(\int_0^{t_{ls}}\frac{qdt}{\sqrt{3(1+R(t))}a(t)} + \Delta(\kappa)\right)\right], \tag{194}$$

Here, we used an approximation given by $a_{DG}(t_{ls}) \approx a(t_{ls}) \propto t^{2/3}$, and the error of this approximation is of the order $10^{-4}\%$.

$$\dot{\Psi}_q(t_{ls}) \;=\; -\frac{3q^2 t_{ls}\,\mathcal{R}_q^o\,\mathcal{T}(\kappa)}{5a_{DG}^2(t_{ls})}\,, \tag{195}$$

$$\delta_{\gamma q}(t_{ls}) \;=\; \frac{3\mathcal{R}_q^o}{5}\left[\mathcal{T}(\kappa)(1+3R_{ls}) - (1+R_{ls})^{-1/4}e^{-q^2 d_D^2/a_{DG}^2(t_{ls})}\right.$$

$$\left. \times\ \mathcal{S}(\kappa)\cos\left(q\int_0^{t_{ls}}\frac{dt}{\sqrt{3(1+R(t))}a_{DG}(t)} + \Delta(\kappa)\right)\right], \tag{196}$$

$$\delta u_{\gamma q}(t_{ls}) \;=\; \frac{3\mathcal{R}_q^o}{5}\left[-t_{ls}\mathcal{T}(\kappa) + \frac{a_{DG}(t_{ls})}{\sqrt{3}q(1+R_{ls})^{3/4}}e^{-q^2 d_D^2/a_{DG}^2(t_{ls})}\right.$$

$$\left. \times\ \mathcal{S}(\kappa)\sin\left(q\int_0^{t_{ls}}\frac{dt}{\sqrt{3(1+R(t))}a_{DG}(t)} + \Delta(\kappa)\right)\right], \tag{197}$$

where

$$d_D^2 = d_{Silk}^2 + d_{Landau}^2\,, \tag{198}$$

$$d_{Silk}^2 = Y_{DG}^2(t_{ls})\int_0^{t_{ls}}\frac{t_\gamma}{6Y_{DG}^2(1+R)}\left\{\frac{16}{15} + \frac{R^2}{(1+R)}\right\}dt\,, \tag{199}$$

$$d_{Landau}^2 = \frac{\sigma_t^2}{6(1+R_{ls})}\,, \tag{200}$$

where t_γ is the mean free time for photons and $R = 3\bar{\rho}_B/4\bar{\rho}_\gamma = 3h^2\Omega_B Y_{DG}/4h^2\Omega_\gamma$. In order to evaluate the Silk damping, we have

$$t_\gamma = \frac{1}{n_e \sigma_T}\,, \tag{201}$$

where n_e is the number density of electrons and σ_T is the Thomson cross-section.

On the other hand

$$q\int_0^{r_{ls}} c_s dr \;=\; q\int_0^{t_{ls}}\frac{dt}{\sqrt{3(1+R(t))}a_{DG}(t)} \equiv qr_{ls}^{SH}$$

$$=\; \frac{q}{a_{DG}(t_{ls})}\cdot(a_{DG}(t_{ls})r_{ls}^{SH}) = \frac{q}{a_{DG}(t_{ls})}\cdot d_H(t_{ls}) \tag{202}$$

where c_s is the speed of sound, r_{ls}^{SH} is the sound horizon radial coordinate and d_H is the horizon distance.

With all this approximation, the transfers functions were simplified to the following expressions:

$$\mathcal{F}(q) = \frac{1}{3}\delta_{\gamma q}(t_{ls}) + \frac{a_{DG}^2(t_{ls})\dot{\Psi}_q(t_{ls})}{3q^2 t_{ls}}, \tag{203}$$

$$\mathcal{G}(q) = -q\frac{\delta u_{\gamma q}(t_{ls})}{(1+3F(t_{ls}))a_{DG}(t_{ls})} + \frac{a_{DG}(t_{ls})\dot{\Psi}_q(t_{ls})}{q}, \tag{204}$$

where $A_q(t_{ls}) = \mathcal{R}_q^o \mathcal{T}(\kappa)$. Then, we replaced the GR solutions, and we obtain

$$\begin{aligned}
\mathcal{F}(q) &= \frac{\mathcal{R}_q^o}{5}\Big[3\mathcal{T}(qd_T/a_{DG}(t_{ls}))R_{ls} - (1+R_{ls})^{-1/4}e^{-q^2 d_D^2/a_{DG}^2(t_{ls})} \\
&\quad \times\ \mathcal{S}(qd_T/a_{DG}(t_{ls}))\cos(qd_H/a_{DG}(t_{ls}) + \Delta(qd_T/a_{DG}(t_{ls})))\Big],
\end{aligned} \tag{205}$$

$$\begin{aligned}
\mathcal{G}(q) &= \frac{\sqrt{3}\mathcal{R}_q^o}{5(1+R_{ls})^{3/4}}e^{-q^2 d_D^2/a_{DG}^2(t_{ls})} \\
&\quad \times\ \mathcal{S}(qd_T/a_{DG}(t_{ls}))\sin(qd_H/a_{DG}(t_{ls}) + \Delta(qd_T/a_{DG}(t_{ls}))),
\end{aligned} \tag{206}$$

where $\kappa = qd_T/a_{ls}$ (defined in Equation (131)) and

$$d_T(t_{ls}) \equiv \frac{\sqrt{2}a_{DG}(t_{ls})}{a_{EQ}H_{EQ}} = \frac{a_{DG}(t_{ls})\sqrt{\Omega_R}}{H_0\Omega_M} = \frac{a_{DG}(t_{ls})}{100h}\sqrt{C(C+1)}. \tag{207}$$

The final consideration that we must include is that due to the reionization of hydrogen at $z_{reion} = 10$ by ultraviolet light coming from the first generation of massive stars, photons of the CMB have a probability of being scattered $1 - \exp(-\tau_{reion})$. CMB has two contributions. The non-scattered photons provide the first contribution, where we have to correct by a factor given by $\exp(-\tau_{reion})$. The scattered photons provide the second contribution, but the reionization occurs at $z \ll z_L$ affecting only low ls. We are not interested in this effect, and therefore, we will not include it. Measurements show that in GR, $\exp(-2\tau_{reion}) \approx 0.8$.

On the other hand, we will use a standard parametrization of $\mathcal{R}_q^0$ given by

$$|\mathcal{R}_q^0|^2 = N^2 q^{-3}\left(\frac{q/R_0}{\kappa_{\mathcal{R}}}\right)^{n_s-1}, \tag{208}$$

where n_s could vary with the wave number. It is usual to take $\kappa_{\mathcal{R}} = 0.05\ \text{Mpc}^{-1}$.

Note that $d_A(t_{ls}) = r_{ls}a_{DG}(t_{ls})$ is the angular diameter distance of the last scattering surface.

$$\begin{aligned}
d_A(t_{ls}) &= a_{DG}(t_{ls})\int_{t_{ls}}^{t_0}\frac{dt'}{a_{DG}(t')} = \frac{a_{DG}(t_0)}{1+z_{ls}}\int_{t_{ls}}^{t_0}\frac{dt'}{a_{DG}(t')} = \frac{1}{1+z_{ls}}\int_{t_{ls}}^{t_0}\frac{dt'}{Y_{DG}(t')} \\
&= \frac{1}{1+z_{ls}}\int_{Y_{ls}}^{1}\frac{dY'}{Y_{DG}(Y')}\frac{dt}{dY'} = \frac{d_L(t_{ls})}{(1+z_{ls})^2}.
\end{aligned} \tag{209}$$

This is consistent with the luminosity distance definition [38]. Then, when we set $q = \beta l/r_{ls}$, we obtain

$$\begin{aligned}
|\mathcal{R}_{\beta l/r_{ls}}^0|^2 &= N^2\left(\frac{\beta l}{r_{ls}}\right)^{-3}\left(\frac{\beta l}{\kappa_{\mathcal{R}}r_{ls}}\right)^{n_s-1} = N^2\left(\frac{\beta l}{r_{ls}}\right)^{-3}\left(\frac{\beta l a_{DG}(t_{ls})}{\kappa_{\mathcal{R}}r_{ls}a_{DG}(t_{ls})}\right)^{n_s-1} \\
&= N^2\left(\frac{\beta l}{r_{ls}}\right)^{-3}\left(\frac{\beta l a_{DG}(t_{ls})}{\kappa_{\mathcal{R}}d_A(t_{ls})}\right)^{n_s-1} \equiv N^2\left(\frac{\beta l}{r_{ls}}\right)^{-3}\left(\frac{\beta l}{l_R}\right)^{n_s-1}.
\end{aligned}$$

Using similar computations for the other distances, the final form of the form factors is given by

$$\mathcal{F}(q) = \frac{\mathcal{R}_q^o}{5}\Big[3\mathcal{T}(\beta l/l_T)R_{ls} - (1+R_{ls})^{-1/4}e^{-\beta^2 l^2/l_D^2}$$
$$\times \quad \mathcal{S}(\beta l/l_T)\cos(\beta l/l_H + \Delta(\beta l/l_T))\Big], \tag{210}$$

$$\mathcal{G}(q) = \frac{\sqrt{3}\mathcal{R}_q^o}{5(1+R_{ls})^{3/4}}e^{-\beta^2 l^2/l_D^2}\mathcal{S}(\beta l/l_T)\sin(\beta l/l_H + \Delta(\beta l/l_T)), \tag{211}$$

where

$$l_R = \frac{\kappa_{\mathcal{R}}d_A(t_{ls})}{a_{DG}(t_{ls})}, \quad l_H = \frac{d_A(t_{ls})}{d_H(t_{ls})}, \quad l_T = \frac{d_A(t_{ls})}{d_T(t_{ls})}, \quad l_D = \frac{d_A(t_{ls})}{d_D(t_{ls})}. \tag{212}$$

To summarize, for reasonably large values of l (say $l > 20$), CMB multipoles are given by

$$\frac{l(l+1)C_{TT,l}^{S}}{2\pi} = \frac{4\pi T_0^2 l^3 \exp(-2\tau_{reion})}{r_{ls}^3}\int_1^\infty \frac{\beta d\beta}{\sqrt{\beta^2-1}}$$
$$\times \quad \left[\left(F\left(\frac{l\beta}{r_{ls}}\right)+\tilde{F}\left(\frac{l\beta}{r_{ls}}\right)\right)^2 + \frac{\beta^2-1}{\beta^2}\left(G\left(\frac{l\beta}{r_{ls}}\right)+\tilde{G}\left(\frac{l\beta}{r_{ls}}\right)\right)^2\right]. \tag{213}$$

The structure of Equation (213) is remarkable, where the delta sector contributes additively inside the integral. If we set all the delta sector equal to zero, we recover the result directly for scalar temperature–temperature multipole coefficients in GR given by Weinberg. Numerical solutions are needed to compute the solution for the perturbations. In a preliminary numerical solution, we obtained the temperature power spectrum of the CMB following Weinberg's approach [47]. We obtain:

$$C = 5.25 \times 10^{-4}, \quad \Omega_M = 0.13038, \quad \Omega_B = 0.02228$$

where Ω_M is the density of non-relativistic matter, and Ω_B is the baryon density. Figure 1 displays the temperature power spectrum for a specific set of cosmological parameters. These cosmological parameters were obtained through exploratory analysis without any statistical examination. In other words, preliminary values were tested to determine if there was any possibility of obtaining reasonable results for the temperature power spectrum using the delta gravity equations. Consequently, the plot does not incorporate error bars.

In the work conducted in ApJ [49], the shape of the temperature power spectrum is determined by five free parameters. These parameters were explored using a modified adaptive Metropolis MCMC algorithm. The statistical study, which aimed at determining the optimal parameter space to match the observed data of the temperature power spectrum, is complicated. The complexity arises from the involved equations of delta gravity, which encompass integrals that are computationally intensive, particularly when the MCMC algorithm performs numerous calculation cycles. Moreover, additional equations accounting for the Landau damping effect and other physical considerations during the last scattering epoch, specific to the delta gravity model, need to be included.

To overcome the computational challenges, an adaptive step method was employed for each parameter independently. Additionally, pre-generated interpolation tables for each integral involved in the calculation were utilized to reduce the computational time during each execution of the MCMC cycle. The endeavor of obtaining optimal parameters for the power spectrum involves an extensive study that significantly differs from this work in terms of physical considerations, statistical approach and numerical methods required to achieve that goal.

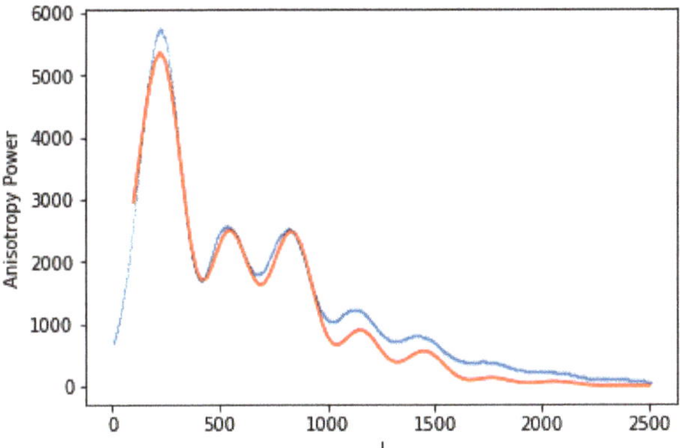

Figure 1. TT CMB spectrum was predicted by DG vs. the observed TT CMB spectrum. The blue line corresponds to the Planck observations, and the red line is the DG prediction.

7. Conclusions

We discussed the implications of the first law of thermodynamics using the modified geometry of this model. We distinguished the physical densities from the GR densities in terms of which scale factor they dilute. However, knowing the solutions of the GR sector is enough for us to know about the behavior of the physical densities. In addition, if we consider that the number of photons is conserved after the moment of decoupling, the black body distribution should keep the form, and that means that temperature is redshifted with the modified scale factor Y_{DG}. Finally, we stated the ansatz that the moment of equality between radiation and matter was the same in GR and in DG, and we showed its implications in some parameters of the theory. We had developed the theory of perturbations for delta gravity and its gauge transformations. Following Weinberg [47], we used the synchronous gauge which leaves a residual gauge transformation, which can be used to set $\delta u_D = 0$ (and also $\delta \tilde{u}_D = 0$). Then, we computed the equations for cosmological perturbations using the hydrodynamic approximation, which we solved for the radiation era, while for a matter-dominated Universe, we presented the equations with the respective initial conditions. As in GR, we found an expression for temperature fluctuations in DG, studying the photon propagation in an effective metric from the moment of the last scattering until now. We found that those temperature fluctuations can be split into three independent terms: an early term which only depends on the moment of the last scattering t_{ls}, an ISW term that includes the evolution of gravitational fields from the last scattering to the present, and a late term which depends on the actual value for those fields. We compute the gauge transformations which leaves $\mathbf{g}_{i0} = 0$, and we found that those three terms are separately gauge invariants. Then, we derived the TT multipole coefficients for scalar modes for large l ($l > 200$), where we found that DG affects additively, which could have an observational effect that could be compared with Plank results and give a physical meaning for the so-called "delta matter". As we mentioned in Section 2.1, differences between physicals and GR fluctuations will occur near the present (low z). Therefore, we should study the full scalar contribution of the multipole coefficients for low ls, where we expect relevant differences between both models. This task requires a full numerical derivation of perturbations, which is beyond the aim of this work. With the full scalar expression for the CMB power spectrum coefficients, we can find the shape of the spectrum. The full analysis and results can be found in [49]. The principal purpose of this work was to present the derivation of this expression and obtain physical insight into the cosmological fluctuations in delta gravity.

Author Contributions: The theory has been proposed by J.A. The theory of perturbation was developed by C.R. and M.S.M. The analysis and writing was completed by all the authors. All authors have read and agreed to the published version of the manuscript.

Funding: J.A. acknowledges the partial support of the Institute of Physics PUC and Fondo Gemini Astro20-0038. C.R. was supported by FONDECYT grant 3220876. M.S.M. was supported by Conicyt PhD Fellowship No. 21170604, Fondecyt 1150390 and CONICYT-PIA-ACT14177.

Data Availability Statement: Not applicable.

Conflicts of Interest: The authors declare no conflict of interest.

Appendix A. Delta Gravity

The following appendixes present details of delta gravity theory and its consequences when studying test particles in this new model. It is derived in more detail in [38].

Appendix A.1. Definition of Delta Gravity

Let us review how $\tilde{\delta}$ variation works. In Section 2, we said that we use the convention that a tilde tensor is equal to the $\tilde{\delta}$ transformation of the original tensor when all its indexes are covariant. So:

$$\tilde{S}_{\mu\nu\alpha\dots} \equiv \tilde{\delta}\left(S_{\mu\nu\alpha\dots}\right) \tag{A1}$$

and we raise and lower indexes using the metric $g_{\mu\nu}$. Therefore:

$$\begin{aligned}
\tilde{\delta}\left(S^{\mu}{}_{\nu\alpha\dots}\right) &= \tilde{\delta}(g^{\mu\rho}S_{\rho\nu\alpha\dots}) \\
&= \tilde{\delta}(g^{\mu\rho})S_{\rho\nu\alpha\dots} + g^{\mu\rho}\tilde{\delta}\left(S_{\rho\nu\alpha\dots}\right) \\
&= -\tilde{g}^{\mu\rho}S_{\rho\nu\alpha\dots} + \tilde{S}^{\mu}{}_{\nu\alpha\dots},
\end{aligned} \tag{A2}$$

where we used that $\tilde{\delta}(g^{\mu\nu}) = -\tilde{\delta}(g_{\alpha\beta})g^{\mu\alpha}g^{\nu\beta}$.

Appendix A.2. $\tilde{\delta}$ Transformation

With the previous notation in mind, we can define how the tilde elements, given by (A1), transform. In general, if we have a field Φ_i that transforms:

$$\delta\Phi_i = \Lambda_i^j(\Phi)\epsilon_j. \tag{A3}$$

Then, $\tilde{\Phi}_i = \tilde{\delta}\Phi_i$ transforms as:

$$\delta\tilde{\Phi}_i = \tilde{\Lambda}_i^j(\Phi)\epsilon_j + \Lambda_i^j(\Phi)\tilde{\epsilon}_j, \tag{A4}$$

where we used that $\tilde{\delta}\delta\Phi_i = \delta\tilde{\delta}\Phi_i = \delta\tilde{\Phi}_i$ and $\tilde{\epsilon}_j = \tilde{\delta}\epsilon_j$. Now, we consider general coordinate transformations or diffeomorphism in its infinitesimal form:

$$\begin{aligned}
x'^{\mu} &= x^{\mu} - \xi_0^{\mu}(x) \\
\delta x^{\mu} &= -\xi_0^{\mu}(x),
\end{aligned} \tag{A5}$$

where $\tilde{\delta}$ will be the general coordinate transformation from now on. Defining:

$$\xi_1^{\mu}(x) \equiv \tilde{\delta}\xi_0^{\mu}(x) \tag{A6}$$

and using (A4), we can see a few examples of how some elements transform:

(I) A scalar ϕ:

$$\begin{aligned}
\delta\phi &= \xi_0^{\mu}\phi_{,\mu} \tag{A7} \\
\delta\tilde{\phi} &= \xi_1^{\mu}\phi_{,\mu} + \xi_0^{\mu}\tilde{\phi}_{,\mu}. \tag{A8}
\end{aligned}$$

(II) A vector V_μ:

$$\bar{\delta} V_\mu = \zeta_0^\beta V_{\mu,\beta} + \zeta_{0,\mu}^\alpha V_\alpha \tag{A9}$$

$$\bar{\delta} \tilde{V}_\mu = \zeta_1^\beta V_{\mu,\beta} + \zeta_{1,\mu}^\alpha V_\alpha + \zeta_0^\beta \tilde{V}_{\mu,\beta} + \zeta_{0,\mu}^\alpha \tilde{V}_\alpha. \tag{A10}$$

(III) Rank two covariant tensor $M_{\mu\nu}$:

$$\bar{\delta} M_{\mu\nu} = \zeta_0^\rho M_{\mu\nu,\rho} + \zeta_{0,\nu}^\beta M_{\mu\beta} + \zeta_{0,\mu}^\beta M_{\nu\beta} \tag{A11}$$

$$\bar{\delta} \tilde{M}_{\mu\nu} = \zeta_1^\rho M_{\mu\nu,\rho} + \zeta_{1,\nu}^\beta M_{\mu\beta} + \zeta_{1,\mu}^\beta M_{\nu\beta} + \zeta_0^\rho \tilde{M}_{\mu\nu,\rho} + \zeta_{0,\nu}^\beta \tilde{M}_{\mu\beta} + \zeta_{0,\mu}^\beta \tilde{M}_{\nu\beta}. \tag{A12}$$

This new transformation is the basis of $\bar{\delta}$ theories. Particulary, in gravitation, we have a model with two fields. The first one is just the usual gravitational field $g_{\mu\nu}$, and the second one is $\tilde{g}_{\mu\nu}$. Then, we will have two gauge transformations associated to general coordinate transformation. We will call it extended general coordinate transformation, which is given by:

$$\bar{\delta} g_{\mu\nu} = \zeta_{0\mu;\nu} + \zeta_{0\nu;\mu} \tag{A13}$$

$$\bar{\delta} \tilde{g}_{\mu\nu}(x) = \zeta_{1\mu;\nu} + \zeta_{1\nu;\mu} + \tilde{g}_{\mu\rho}\zeta_{0,\nu}^\rho + \tilde{g}_{\nu\rho}\zeta_{0,\mu}^\rho + \tilde{g}_{\mu\nu,\rho}\zeta_0^\rho, \tag{A14}$$

where we used (A11) and (A12). With these tools, we can introduce the $\bar{\delta}$ theories, as in Section 2.

Appendix B. Test Particle

In Section 2, we present the equations of motion for $\bar{\delta}$ gravity. However, to describe some phenomenology, we need to analyze the trajectory of a particle. For this, we need to find the coupling of a test particle with the gravitational field. In this Appendix, we will separate the massive and massless particles cases.

Appendix B.1. Massive Particles

We know that in the standard case, the action for a test particle is given by:

$$S_0[\dot{x}, g] = -m \int dt \sqrt{-g_{\mu\nu} \dot{x}^\mu \dot{x}^\nu}, \tag{A15}$$

with $\dot{x}^\mu = \frac{dx^\mu}{dt}$. This action is invariant under reparametrizations, $t' = t - \epsilon(t)$. This means, in the infinitesimal form, that:

$$\delta_R x^\mu = \dot{x}^\mu \epsilon. \tag{A16}$$

In $\bar{\delta}$ gravity, we will have a new test particle action. To obtain this action, we need to evaluate (A15) in (5):

$$S[\dot{x}, y, g, \tilde{g}] = m \int dt \left(\frac{\bar{g}_{\mu\nu}\dot{x}^\mu\dot{x}^\nu + \frac{1}{2}(2g_{\mu\nu}\dot{y}^\mu\dot{x}^\nu + g_{\mu\nu,\rho}y^\rho\dot{x}^\mu\dot{x}^\nu)}{\sqrt{-g_{\alpha\beta}\dot{x}^\alpha\dot{x}^\beta}} \right), \tag{A17}$$

where $\bar{g}_{\mu\nu} = g_{\mu\nu} + \frac{1}{2}\tilde{g}_{\mu\nu}$ and $y^\mu = \bar{\delta} x^\mu$. This action is invariant under reparametrization transformations, given by (A16), plus $\bar{\delta}$ reparametrization transformations:

$$\delta_R y^\mu = \dot{y}^\mu \epsilon + \dot{x}^\mu \tilde{\epsilon}. \tag{A18}$$

The presence of y^μ suggests that we have other coordinates. Because we do not want new coordinates, we impose that $2g_{\mu\nu}\dot{y}^\mu\dot{x}^\nu + g_{\mu\nu,\rho}y^\rho\dot{x}^\mu\dot{x}^\nu = 0$ such as a gauge condition on $\bar{\delta}$ reparametrization, fixing $\tilde{\epsilon}$. With this, we eliminate this new symmetry; however, the extended general coordinate transformations as well as time reparametrizations continue to be preserved. Therefore, we can fix the gauge for $g_{\mu\nu}$ and $\tilde{g}_{\mu\nu}$ separately. Finally, (A17) is reduced to:

$$S[\dot{x}, g, \tilde{g}] = m \int dt \left(\frac{\tilde{g}_{\mu\nu}\dot{x}^{\mu}\dot{x}^{\nu}}{\sqrt{-g_{\alpha\beta}\dot{x}^{\alpha}\dot{x}^{\beta}}} \right). \tag{A19}$$

Now, if we vary (A19) with respect to x^{μ}, we obtain the equation of motion for a massive test particle. That is:

$$\hat{g}_{\mu\nu}\ddot{x}^{\nu} + \hat{\Gamma}_{\mu\alpha\beta}\dot{x}^{\alpha}\dot{x}^{\beta} = \frac{1}{4}\tilde{K}_{,\mu}, \tag{A20}$$

with:

$$\begin{aligned}
\hat{\Gamma}_{\mu\alpha\beta} &= \frac{1}{2}(\hat{g}_{\mu\alpha,\beta} + \hat{g}_{\beta\mu,\alpha} - \hat{g}_{\alpha\beta,\mu}) \\
\hat{g}_{\alpha\beta} &= \left(1 + \frac{1}{2}\tilde{K}\right)g_{\alpha\beta} + \tilde{g}_{\alpha\beta} \\
\tilde{K} &= \tilde{g}_{\alpha\beta}\dot{x}^{\alpha}\dot{x}^{\beta}
\end{aligned}$$

and, if we choose t equal to the proper time, then $g_{\mu\nu}\dot{x}^{\mu}\dot{x}^{\nu} = -1$. We can see that the equation of motion of a free massive particle is a second-order equation, but we emphasize that it is not a geodesic with an effective metric.

Appendix B.2. Massless Particles

Unfortunately, (A15) is useless for massless particles, because it is null when $m = 0$. To solve this problem, it is common practice to start from the action:

$$S_0[\dot{x}, g, v] = \frac{1}{2} \int dt \left(vm^2 - v^{-1}g_{\mu\nu}\dot{x}^{\mu}\dot{x}^{\nu} \right), \tag{A21}$$

where v is a Lagrange multiplier. From (A21), we can obtain the equation of motion for v:

$$v = -\frac{\sqrt{-g_{\mu\nu}\dot{x}^{\mu}\dot{x}^{\nu}}}{m}. \tag{A22}$$

If we substitute (A22) in (A21), we recover (A15). In other words, (A21) is a good action that includes the massless case. So, we must substitute (A21) in (5) to obtain the modified test particle action. That is:

$$S[\dot{x}, g, \tilde{g}, v, \tilde{v}] = \frac{1}{2} \int dt \left[vm^2 - v^{-1}\left(g_{\mu\nu} + \kappa_2\tilde{g}_{\mu\nu}\right)\dot{x}^{\mu}\dot{x}^{\nu} + \tilde{v}\left(m^2 + v^{-2}g_{\mu\nu}\dot{x}^{\mu}\dot{x}^{\nu}\right) \right], \tag{A23}$$

where we must discard y^{μ} for the same reason used in Appendix B.1. In addition, two Lagrange multipliers are unnecessary, so we will eliminate one of them. The equation of motion for $\tilde{v}$ is:

$$\tilde{v} = \frac{m^2 + v^{-2}\left(g_{\mu\nu} + \tilde{g}_{\mu\nu}\right)\dot{x}^{\mu}\dot{x}^{\nu}}{2v^{-3}g_{\alpha\beta}\dot{x}^{\alpha}\dot{x}^{\beta}}. \tag{A24}$$

If we now replace (A24) in (A23), we obtain our $\tilde{\delta}$ Test Particle Action:

$$S[\dot{x}, g, \tilde{g}, v] = \int dt \left(m^2 v - \frac{\left(g_{\mu\nu} + \tilde{g}_{\mu\nu}\right)\dot{x}^{\mu}\dot{x}^{\nu}}{4v} + \frac{m^2 v^3}{4g_{\alpha\beta}\dot{x}^{\alpha}\dot{x}^{\beta}}\left(m^2 + v^{-2}\tilde{g}_{\mu\nu}\dot{x}^{\mu}\dot{x}^{\nu}\right) \right). \tag{A25}$$

Therefore, we can use (A25) to represent the trajectory of a particle in the presence of a gravitational field, given by g and $\tilde{g}$, for the massless and massive case. In the previous section, we have developed the massive case, so we need to study the massless case now. Evaluating $m = 0$ in (A21) and (A25), they are, respectively:

$$S_0^{(m=0)}[\dot{x}, g, v] \;=\; -\frac{1}{2}\int dt\, v^{-1} g_{\mu\nu}\dot{x}^\mu \dot{x}^\nu \tag{A26}$$

$$S^{(m=0)}[\dot{x}, g, \tilde{g}, v] \;=\; -\frac{1}{4}\int dt\, v^{-1} \mathbf{g}_{\mu\nu}\dot{x}^\mu \dot{x}^\nu, \tag{A27}$$

with $\mathbf{g}_{\mu\nu} = g_{\mu\nu} + \tilde{g}_{\mu\nu}$. In the usual and modified case, the equation of motion for v implies that a massless particle will move in a null-geodesic. In the usual case, we have $g_{\mu\nu}\dot{x}^\mu \dot{x}^\nu = 0$, but in our model, the null-geodesic is $\mathbf{g}_{\mu\nu}\dot{x}^\mu \dot{x}^\nu = 0$.

All this means that in our theory, the equation of motion of a free massless particle is given by:

$$\mathbf{g}_{\mu\nu}\ddot{x}^\nu + \mathbf{\Gamma}_{\mu\alpha\beta}\dot{x}^\alpha \dot{x}^\beta = 0 \tag{A28}$$
$$\mathbf{g}_{\mu\nu}\dot{x}^\mu \dot{x}^\nu = 0,$$

with:

$$\mathbf{\Gamma}_{\mu\alpha\beta} = \frac{1}{2}\left(\mathbf{g}_{\mu\alpha,\beta} + \mathbf{g}_{\beta\mu,\alpha} - \mathbf{g}_{\alpha\beta,\mu}\right).$$

Appendix B.3. Proper Time

It is important to observe that the proper time is defined in terms of massive particles. So, we must define the measurement of time and distances in the model.

The equation (A20) preserves the proper time of the particle along the trajectory and along the trajectory $g_{\mu\nu}\dot{x}^\mu \dot{x}^\nu = -1$. So, we must define the proper time using the original metric $g_{\mu\nu}$. That is:

$$d\tau = \sqrt{-g_{\mu\nu}dx^\mu dx^\nu} = \sqrt{-g_{00}}\,dt. \tag{A29}$$

From here, we can observe that $g_{00} < 0$. On the other hand, in order to define dl as the interval between two infinitesimally separated events at the same time, we follow the approach outlined in [54]. They consider a scenario where a light signal is directed from point B in space to a point A located infinitely close to it and then back along the same path. By solving the geodesic equation $ds^2 = 0$ (for the metric $\mathbf{g}_{\mu\nu}$ as obtained in (A28)) with respect to dx^0, two solutions are obtained. The total time interval between the departure of the signal and its return to the original point is given by the difference between these two solutions. Consequently, the proper time interval can be obtained, as described in (A29), by multiplying by $\sqrt{-g_{00}}$, while the distance dl between the two points is obtained by multiplying by $1/2$ (keeping in mind that $c = 1$). Thus, we obtain:

$$dl^2 \;=\; \gamma_{ij}dx^i dx^j \tag{A30}$$
$$\gamma_{ij} \;=\; \frac{g_{00}}{\mathbf{g}_{00}}\left(\mathbf{g}_{ij} - \frac{\mathbf{g}_{i0}\mathbf{g}_{j0}}{\mathbf{g}_{00}}\right),$$

where $\mathbf{g}_{\mu\nu} = g_{\mu\nu} + \tilde{g}_{\mu\nu}$.

Therefore, we measure the proper time using the metric $g_{\mu\nu}$, but the space geometry is determined by both tensor fields, $g_{\mu\nu}$ and $\tilde{g}_{\mu\nu}$.

For example, in cosmology, we have (see Appendix C and Appendix D):

$$\mathbf{g}_{\mu\nu}dx^\mu dx^\nu \;=\; -(1 + 3F(t))dt^2 + a^2(t)(1 + F(t))\left(dx^2 + dy^2 + dz^2\right)$$
$$\rightarrow dl^2 \;=\; a^2(t)\frac{(1 + F(t))}{(1 + 3F(t))}\delta_{ij}dx^i dx^j$$
$$=\; a_{DG}^2(t)\delta_{ij}dx^i dx^j. \tag{A31}$$

This means that we have the same 3-geometry as in Einstein but replacing $a(t)$ by $a_{DG}(t)$. Therefore, in $\tilde{\delta}$ gravity, $a_{DG}(t)$ is the effective scale factor (it determines distances in the 3d geometry), and volume is given by $V \propto a_{DG}^3(t)$. Finally, using (A29) and (A30), we can find the relation between $a_{DG}(t)$ and redshift, z, given by (A39).

To summarize, we have analyzed the $\tilde{\delta}$ gravity. We obtained the equations of motion of $g_{\mu\nu}$ and $\tilde{g}_{\mu\nu}$, given by (11)–(14), and we know how to solve them for a perfect fluid using (52) and (53). Then, we obtained how a test particle moves when it is coupled to $g_{\mu\nu}$ and $\tilde{g}_{\mu\nu}$, which is given by (A20) or (A28) if we have a massive or massless particle, respectively. The last tool that we will need is how to fix the gauge. In the next Appendix, we will develop the harmonic gauge for the FLRW case that we want to study.

Appendix C. Harmonic Gauge

We know that Einstein's equations do not fix all degrees of freedom of $g_{\mu\nu}$. This means that if $g_{\mu\nu}$ is the solution, then other solutions $g'_{\mu\nu}$ exist given by a general coordinate transformation $x \to x'$. We can eliminate these degrees of freedom by adopting some particular coordinate system, fixing the gauge.

One particularly convenient gauge is given by the harmonic coordinate conditions. That is:

$$\Gamma^\mu \equiv g^{\alpha\beta}\Gamma^\mu_{\alpha\beta} = 0. \tag{A32}$$

Under general coordinate transformation, Γ^μ transforms:

$$\Gamma'^\mu = \frac{\partial x'^\mu}{\partial x^\alpha}\Gamma^\alpha - g^{\alpha\beta}\frac{\partial^2 x'^\mu}{\partial x^\alpha \partial x^\beta}.$$

Therefore, if Γ^α does not vanish, we can define a new coordinate system x'^μ where $\Gamma'^\mu = 0$. So, it is always possible to choose a harmonic coordinate system. For more detail about harmonic gauge see, for example, [55].

In the same form, we need to fix the gauge for $\tilde{g}_{\mu\nu}$. It is natural to choose a gauge given by:

$$\tilde{\delta}(\Gamma^\mu) \equiv g^{\alpha\beta}\tilde{\delta}\left(\Gamma^\mu_{\alpha\beta}\right) - \tilde{g}^{\alpha\beta}\Gamma^\mu_{\alpha\beta} = 0, \tag{A33}$$

where $\tilde{\delta}\left(\Gamma^\mu_{\alpha\beta}\right) = \frac{1}{2}g^{\mu\lambda}\left(D_\beta\tilde{g}_{\lambda\alpha} + D_\alpha\tilde{g}_{\beta\lambda} - D_\lambda\tilde{g}_{\alpha\beta}\right)$. So, when we refer to harmonic gauge, we will use (A32) and (A33).

Appendix C.1. FLRW

In this case, to find the harmonic coordinate system, we will change the t variable of (15) with u. So, the metric is now:

$$g_{\mu\nu}dx^\mu dx^\nu = -T^2(u)du^2 + a^2(u)\left(dx^2 + dy^2 + dz^2\right),$$

such that $T(u) = \frac{dt}{du}(u)$. In the same form, (16) is changed to:

$$\tilde{g}_{\mu\nu}dx^\mu dx^\nu = -F_b(u)T^2(u)du^2 + F_a(u)a^2(u)\left(dx^2 + dy^2 + dz^2\right).$$

Now, if we fix the harmonic gauge, we obtain that $T(u) = T_0 a^3(u)$ from (A32) and $F_b(u) = 3(F_a(u) + T_1) \equiv F(u)$ from (A33), where T_0 and T_1 are gauge constants. We use $T_0 = 1$ and $T_1 = 0$ to fix the gauge completely. So, with these conditions, the system (u,x,y,z) corresponds to harmonic coordinates. Now, we can return to the usual system where $g_{\mu\nu}$ and $\tilde{g}_{\mu\nu}$ are given by (15) and (16), where the gauge is fixed.

It is important to note that the $\tilde{\delta}$ variation defines a new independent field. In particular, $\tilde{\delta}$ can be as large as the theory allows them. This mean that they can not be considered as perturbations of the original fields. Here, as an example of this feature, after this gauge is fixed, both $g_{\mu\nu}$ and $\tilde{g}_{\mu\nu}$ evolve independently, so $\tilde{g}_{\mu\nu}$ cannot be considered as a perturbation of the $g_{\mu\nu}$ metric.

Appendix D. Photon Trajectory in a Non-Perturbed Background

When a photon emitted from a source travels to the Earth, the Universe is expanding. This means that the photon is affected by the cosmological Doppler effect. For this, we must use a null geodesic, given by (A28), in a radial trajectory from r_1 to $r = 0$. So, with (15) and (16), we have:

$$-(1 + 3F(t))dt^2 + a^2(t)(1 + F(t))dr^2 = 0.$$

In GR, we have that $dt = -a(t)dr$. So, in the $\tilde{\delta}$ gravity case, we can define the effective scale factor (Equation (18)):

$$a_{DG}(t) = a(t)\sqrt{\frac{1 + F(t)}{1 + 3F(t)}} \tag{A34}$$

such that $cdt = -a_{DG}(t)dr$ now. If we integrate this expression from r_1 to 0, we obtain:

$$r_1 = \int_{t_1}^{t_0} \frac{dt}{a_{DG}(t)}, \tag{A35}$$

where t_1 and t_0 are the emission and reception times. If a second wave crest is emitted at $t = t_1 + \Delta t_1$ from $r = r_1$, it will reach $r = 0$ at $t = t_0 + \Delta t_0$, so:

$$r_1 = \int_{t_1 + \Delta t_1}^{t_0 + \Delta t_0} \frac{dt}{a_{DG}(t)}. \tag{A36}$$

Therefore, when Δt_1, Δt_0 is small, which is appropriate for light waves, we obtain:

$$\frac{\Delta t_0}{\Delta t_1} = \frac{a_{DG}(t_0)}{a_{DG}(t_1)} \tag{A37}$$

or:

$$\frac{\Delta \nu_1}{\Delta \nu_0} = \frac{a_{DG}(t_0)}{a_{DG}(t_1)}, \tag{A38}$$

where ν_0 is the light frequency detected at $r = 0$, corresponding to a source emission at frequency ν_1. So, the redshift is now:

$$1 + z(t_1) = \frac{a_{DG}(t_0)}{a_{DG}(t_1)}. \tag{A39}$$

We see that $a_{DG}(t)$ replaces the usual scale factor $a(t)$ to compute z. This means that we need to redefine the luminosity distance, too. For this, let us consider a mirror of radius b that is receiving light from our distant source at r_1. The photons that reach the mirror are within a cone of half-angle ϵ with origin at the source.

Let us compute ϵ. The path of the light rays is given by $\vec{r}(\rho) = \rho\hat{n} + \vec{r}_1$, where $\rho > 0$ is a parameter and $\hat{n}$ is the direction of the light ray. Since the mirror is in $\vec{r} = 0$, then $\rho = r_1$ and $\hat{n} = -\hat{r}_1 + \vec{\epsilon}$, where ϵ is the angle between $-\vec{r}_1$ and $\hat{n}$ at the source, forming a cone. The proper distance is determined by the tri-dimensional metric, which is given by (see Appendix B.3):

$$\begin{aligned} dl^2 &= \gamma_{ij}dx^i dx^j \\ &= a_{DG}^2(t)\delta_{ij}dx^i dx^j \end{aligned}$$

in the cosmological case. Then, $b = a_{DG}(t_0)r_1\epsilon$ and the solid angle of the cone is:

$$\begin{aligned}
\Delta\Omega &= \int_0^{2\pi} d\phi \int_0^{\epsilon} \sin(\theta)d\theta = 2\pi(1 - \cos(\epsilon)) \\
&= \pi\epsilon^2 = \frac{A}{r_1^2 a_{DG}^2(t_0)},
\end{aligned}$$

where $A = \pi b^2$ is the proper area of the mirror. This means that $\epsilon = \frac{b}{r_1 a_{DG}(t_0)}$. So, the fraction of all isotropically emitted photons that reach the mirror is:

$$\begin{aligned}
f &= \frac{\Delta\Omega}{4\pi} \\
&= \frac{A}{4\pi r_1^2 a_{DG}^2(t_0)}.
\end{aligned}$$

We know that the apparent luminosity, denoted as l, represents the power received per unit area of the mirror. Power is defined as energy per unit time, so the received power can be expressed as $P = \frac{h\nu_0}{\Delta t_0} f$, where $h\nu_0$ represents the energy associated to the received photon. On the other hand, the total power emitted by the source is given by $L = \frac{h\nu_1}{\Delta t_1}$, where $h\nu_1$ corresponds to the energy of the emitted photon. Therefore, we can conclude that:

$$\begin{aligned}
P &= \frac{a_{DG}^2(t_1)}{a_{DG}^2(t_0)} L f \\
l &= \frac{P}{A} \\
&= \frac{a_{DG}^2(t_1)}{a_{DG}^2(t_0)} \frac{L}{4\pi r_1^2 a_{DG}^2(t_0)},
\end{aligned}$$

where we have used that $\frac{\Delta t_0}{\Delta t_1} = \frac{\nu_1}{\nu_0} = \frac{a_{DG}(t_0)}{a_{DG}(t_1)}$. In addition, we know that in an Euclidean space, the luminosity decreases with distance d_L according to $l = \frac{L}{4\pi d_L^2}$. Therefore, using (A35), the luminosity distance is:

$$\begin{aligned}
d_L &= \frac{a_{DG}^2(t_0)}{a_{DG}(t_1)} r_1 \\
&= \frac{a_{DG}^2(t_0)}{a_{DG}(t_1)} \int_{t_1}^{t_0} \frac{dt}{a_{DG}(t)}.
\end{aligned} \tag{A40}$$

On the other hand, we can define the angular diameter distance, which is denoted as d_A. Considering a light ray emitted at time t_1 and moving in the θ coordinate, our null geodesic, as given by Equation (A28), indicates that the proper distance is $s = t_1 = a_{DG}(t_1) r_1 \theta$. The angular diameter distance is defined as $\theta = \frac{s}{d_A}$, thus yielding $d_A = a_{DG}(t_1) r_1$. If we compare this expression with Equation (A40), we find that:

$$\begin{aligned}
d_A &= \frac{a_{DG}^2(t_1)}{a_{DG}^2(t_0)} d_L \\
&= \frac{d_L}{(1 + z_1)^2}.
\end{aligned} \tag{A41}$$

Therefore, the relation between d_A and d_L is the same to GR [47]. This result is important, because in other modified gravity theories, this relation is not satisfied [56].

Notes

1.   http://camb.info/ (accessed on 5 April 2023)
2.   In Appendix B.3, we discussed this derivation.

3   Here, Δ represents the gauge transformation, which affects only the field perturbations. It is defined as

$$\Delta h_{\mu\nu} \equiv g'_{\mu\nu}(x) - g_{\mu\nu}(x)$$

When applying to $\bar{h}_{\mu\nu} = \bar{\delta} h_{\mu\nu}$, we can commute the variations and apply $\bar{\delta}$ to Equation (40), obtaining Equation (44) (a full derivation can be found in Chapter 5 of [47]).

4   The calculations were made using Pytearcat [51]

5   The Fourier transform is defined as:

$$X(\mathbf{x}, t) = \int d^3 q X_q(t) e^{i\mathbf{q}\cdot\mathbf{x}}$$

6   We choose this definition because the system of equations now seems as a homogeneous system exactly equal to the GR sector (where now the variables are the tilde fields) with external forces mediated by the GR solutions. Maybe the most intuitive solution should be

$$\tilde{\delta}_{\alpha q}^{int} = \frac{\delta \tilde{\rho}_{\alpha q}}{\tilde{\rho}_\alpha + \tilde{p}_\alpha} \, ,$$

however, these definitions are related by

$$\tilde{\delta}_{\alpha q} = \frac{\tilde{\rho}_\alpha + \tilde{p}_\alpha}{\bar{\rho}_\alpha + \bar{p}_\alpha} \left(\tilde{\delta}_{\alpha q}^{int} - \delta_{\alpha q} \right) .$$

7   The definition of $\mathcal{R}_q$ is given in Section 5.4: Conservation outside the horizon, Cosmology, Weinberg.

8   See Section 7.1 : General formulas for the temperature fluctuation, Cosmology, Weinberg.

9   See, e.g., I. S. Gradsteyn & I. M. Ryzhik, *Table of Integral, Series, and Products*, translated, corrected and enlarged by A. Jeffrey (Academic Press, New York, 1980): formula 8.453.1.

10   See Section 6.3: Scalar perturbations-long wavelengths, Cosmology, Weinberg.

11   See Section 6.4: Scalar perturbations-short wavelengths, Cosmology, Weinberg.

12   See Section 7.2: Temperature multipole coefficients: Scalar modes, Cosmology, Weinberg.

References

1. Aghanim N. et al. [Planck Collaboration]. Planck 2018 results. VI. Cosmological parameters. *A&A* **2020**, *641*, A6.
2. Riess, A.G.; Macri, L.M.; Hoffmann, S.L.; Scolnic, D.; Casertano, S.; Filippenko, A.V.; Tucker, B.E.; Reid, M.J.; Jones, D.O.; Silverman, J.M.; et al. A 2.4% determination of the local value of the hubble constant. *Astrophys. J.* **2016**, *826*, 56. [CrossRef]
3. Ata, M.; Baumgarten, F.; Bautista, J.; Beutler, F.; Bizyaev, D.; Blanton, M.R.; Blazek, J.A.; Bolton, A.S.; Brinkmann, J.; Brownstein, J.R.; et al. The clustering of the SDSS-IV extended Baryon Oscillation Spectroscopic Survey DR14 quasar sample: First measurement of baryon acoustic oscillations between redshift 0.8 and 2.2. *Mon. Not. R. Astron. Soc.* **2017**, *473*, 4773–4794. [CrossRef]
4. Nelson, D.; Pillepich, A.; Genel, S.; Vogelsberger, M.; Springel, V.; Torrey, P.; Rodriguez-Gomez, V.; Sijacki, D.; Snyder, G.; Griffen, B.; et al. The illustris simulation: Public data release. *Astron. Comput.* **2015**, *13*, 12–37. [CrossRef]
5. Boylan-Kolchin, M.; Springel, V.; White, S.D.M.; Jenkins, A.; Lemson, G. Resolving cosmic structure formation with the Millennium-II Simulation. *Mnras* **2009**, *398*, 1150–1164. [CrossRef]
6. Addison, G.E.; Watts, D.J.; Bennett, C.L.; Halpern, M.; Hinshaw, G.; Weiland, J.L. Elucidating λcdm: Impact of baryon acoustic oscillation measurements on the hubble constant discrepancy. *Astrophys. J.* **2018**, *853*, 119. [CrossRef]
7. Riess, A.G.; Casertano, S.; Yuan, W.; Macri, L.; Anderson, J.; MacKenty, J.W.; Bowers, J.B.; Clubb, K.I.; Filippenko, A.V.; Jones, D.O.; et al. New parallaxes of galactic cepheids from spatially scanning the hubble space telescope: Implications for the hubble constant. *Astrophys. J.* **2018**, *855*, 136. [CrossRef]
8. Riess, A.G.; Casertano, S.; Yuan, W.; Macri, L.M.; Scolnic, D. Large magellanic cloud cepheid standards provide a 1% foundation for the determination of the hubble constant and stronger evidence for physics beyond λCDM. *Astrophys. J.* **2019**, *876*, 85. [CrossRef]
9. Handley, W. Curvature tension: Evidence for a closed universe. *arXiv* **2019**, arXiv:1908.09139.
10. Valentino, E.D.; Melchiorri, A.; Silk, J. Planck evidence for a closed Universe and a possible crisis for cosmology. *Nat. Astron.* **2020**, *4*, 196–203. [CrossRef]
11. Battye, R.A.; Charnock, T.; Moss, A. Tension between the power spectrum of density perturbations measured on large and small scales. *Phys. Rev. D* **2015**, *91*, 103508. [CrossRef]
12. Birrer, S.; Treu, T.; Rusu, C.E.; Bonvin, V.; Fassnacht, C.D.; Chan, J.H.H.; Agnello, A.; Shajib, A.J.; Chen, G.C.-F.; Auger, M.; et al. H0LiCOW – IX. Cosmographic analysis of the doubly imaged quasar SDSS 1206+4332 and a new measurement of the Hubble constant. *Mon. Not. R. Astron. Soc.* **2019**, *484*, 4726–4753. [CrossRef]
13. Wojtak, R.; Knebe, A.; Watson, W.A.; Iliev, I.T.; Heß, S.; Rapetti, D.; Yepes, G.; Gottlöber, S. Cosmic variance of the local Hubble flow in large-scale cosmological simulations. *Mon. Not. R. Astron. Soc.* **2013**, *438*, 1805–1812. [CrossRef]
14. Riess, A.G.; Yuan, W.; Casertano, S.; Macri, L.M.; Scolnic, D. The accuracy of the hubble constant measurement verified through cepheid amplitudes. *Astrophys. J.* **2020**, *896*, L43. [CrossRef]

15. Wang, J.S.; Wang, F.Y. Probing the anisotropic expansion from supernovae and grbs in a model-independent way. *Mon. Not. R. Astron. Soc.* **2014**, *443*, 1680–1687. [CrossRef]
16. Jacques, C.; Roya, M.; Mohamed, R.; Subir, S. Evidence for anisotropy of cosmic acceleration. *A&A* **2019**, *631*, L13.
17. Sun, Z.Q.; Wang, F.Y. Probing the isotropy of cosmic acceleration using different supernova samples. *Eur. Phys. J. C* **2019**, *79*, 783. [CrossRef]
18. Kang, Y.; Lee, Y.-W.; Kim, Y.-L.; Chung, C.; Ree, C.H. Early-type host galaxies of type ia supernovae. II. evidence for luminosity evolution in supernova cosmology. *Astrophys. J.* **2020**, *889*, 8. [CrossRef]
19. Kenworthy, W.D.; Scolnic, D.; Riess, A. The local perspective on the hubble tension: Local structure does not impact measurement of the hubble constant. *Astrophys. J.* **2019**, *875*, 145. [CrossRef]
20. Martín, M.S.; Rubio, C. Hubble tension and matter inhomogeneities: A theoretical perspective. *arXiv* **2021**, arXiv:2107.14377.
21. de Jaeger, T.; Stahl, B.E.; Zheng, W.; Filippenko, A.V.; Riess, A.G.; Galbany, L. A measurement of the Hubble constant from Type II supernovae. *Mon. Not. R. Astron. Soc.* **2020**, *6*, staa1801. [CrossRef]
22. Valentino, E.D.; Melchiorri, A.; Mena, O. Can interacting dark energy solve the H_0 tension? *Phys. Rev. D* **2017**, *96*, 043503. [CrossRef]
23. Pandey, K.L.; Karwal, T.; Das, S. Alleviating the H_0 and σ_8 anomalies with a decaying dark matter model. *J. Cosmol. Astropart. Phys.* **2019**, *2*, 026. [CrossRef]
24. Guo, R.-Y.; Zhang, J.-F.; Zhang, X. Can the h0 tension be resolved in extensions to ΛCDM cosmology? *J. Cosmol. Astropart. Phys.* **2019**, *2019*, 054. [CrossRef]
25. Valentino, E.D.; Mena, O.; Pan, S.; Visinelli, L.; Yang, W.; Melchiorri, A.; Mota, D.; Riess, A.G.; Silk, J. In the realm of the Hubble tension review of solutions. *Class. Quantum Gravity* **2021**, *38*, 153001. [CrossRef]
26. Tsujikawa, S. *Lectures on Cosmology*; Springer: Berlin/Heidelberg, Germany, 2010; pp. 99–145.
27. Hawking, S.; Israel, W. *General Relativity: An Einstein Centenary Survey*; Cambridge University Press: Cambridge, UK, 1979; p. 790
28. Sahni, V.; Krasiński, A.A. Republication of: The cosmological constant and the theory of elementary particles (By Ya. B. Zeldovich). *Gen Relativ Gravit* **2008**, *40*, 1557–1591. [CrossRef]
29. Sakharov, A.D. Vacuum Quantum Fluctuations in Curved Space and the Theory of Gravitation. *Sov. Phys. Dokl.* **1968**, *12*, 1040.
30. Klein, O. Generalization of Einstein's Principle of Equivalence so as to Embrace the Field Equations of Gravitation. *Phys. Scr.* **1974**, *9*, 69–72. [CrossRef]
31. Adler, S.L. Einstein gravity as a symmetry-breaking effect in quantum field theory. *Rev. Mod. Phys.* **1982**, *54*, 729. [CrossRef]
32. Litim, D.F. Fixed Points of Quantum Gravity. *Phys. Rev. Lett.* **2004**, *92*, 201301. [CrossRef]
33. Reuter, M.; Saueressig, F. Functional Renormalization Group Equations, Asymptotic Safety, and Quantum Einstein Gravity. *arXiv* **2010**, arXiv:0708.1317.
34. Ambjørn, J.; Jurkiewicz, J.; Loll, R. Nonperturbative Lorentzian Path Integral for Gravity. *Phys. Rev. Lett.* **2000**, *85*, 924. [CrossRef] [PubMed]
35. Alfaro, J. Delta-gravity and dark energy. *Phys. Lett. B* **2012**, *709*, 101–105. [CrossRef]
36. Alfaro, J.; González, P.; Avila, R. A finite quantum gravity field theory model. *Class. Quant. Grav.* **2011**, *28*, 215020. [CrossRef]
37. Alfaro, J.; Gonzalez, P. Cosmology in Delta-Gravity. *Class. Quant. Grav.* **2013**, *30*, 085002. [CrossRef]
38. Alfaro, J.; González, P. δ Gravity: Dark Sector, Post-Newtonian Limit and Schwarzschild Solution. *Universe* **2019**, *5*, 96. [CrossRef]
39. Caldwell, R.R.; Kamionkowski, M.; Weinberg, N.N. Phantom Energy and Cosmic Doomsday. *Phys. Rev. Lett.* **2003**, *91*, 071301. [CrossRef]
40. Alfaro, J.; Labraña, P. Semiclassical gauge theories. *Phys. Rev. D* **2002**, *65*, 045002. [CrossRef]
41. Caldwell, R.R. A phantom menace? Cosmological consequences of a dark energy component with super-negative equation of state. *Phys. Lett. B* **2002**, *545*, 2329. [CrossRef]
42. Alfaro, J.; Martín, M.S.; Sureda, J. An accelerating universe without lambda: Delta gravity using monte carlo. *Universe* **2019**, *5*, 51. [CrossRef]
43. Seljak, U.; Zaldarriaga, M. A line-of-sight integration approach to cosmic microwave background anisotropies. *ApJ* **1996**, *469*, 437. [CrossRef]
44. Zaldarriaga, M.; Seljak, U.; Bertschinger, E. Integral solution for the microwave background anisotropies in nonflat universes. *ApJ* **1998**, *494*, 491–502. [CrossRef]
45. Lewis, A.; Challinor, A.; Lasenby, A. Efficient computation of cosmic microwave background anisotropies in closed friedmann-robertson-walker models. *Astrophys. J.* **2000**, *538*, 473–476. [CrossRef]
46. Mukhanov, V. "CMB-Slow" or How to Determine Cosmological Parameters by Hand? *Int. J. Theor. Phys.* **2004**, *43*, 623–668. [CrossRef]
47. Weinberg, S. *Cosmology*; Cosmology OUP: Oxford, UK, 2008.
48. Rubio, C. On the Effects of the Modification of the Metric in the Gravitational Context. Ph.D Thesis, Pontificia Universidad Católica de Chile, Santiago, Chile , 2020. Available online: https://repositorio.uc.cl/xmlui/handle/11534/46088 (accessed on 26 September 2020).
49. Martín, M.S.; Alfaro, J.; Rubio, C. Observational Constraints in Delta-gravity: CMB and Supernovae. *Astrophys. J.* **2021**, *910*, 43. [CrossRef]
50. Lifshitz, E.M. On the gravitational stability of the expanding universe. *Zhurnal Eksperimentalnoi Teor. Fiz.* **1946**, *16*, 587–602.

51. Martín, M.S.; Sureda, J. Pytearcat: PYthon TEnsor AlgebRa calCulATor A python package for general relativity and tensor calculus. *Astron. Comput.* **2022**, *39*, 100572. [CrossRef]
52. Silk, J. When were Galaxies and Galaxy Clusters formed? *Nature* **1968**, *218*, 453–454. [CrossRef]
53. Kaiser, N. Small-angle anisotropy of the microwave background radiation in the adiabatic theory. *Mon. Not. R. Astron. Soc.* **1983**, *202*, 1169–1180. [CrossRef]
54. Landau, L.D.; Lifshitz, E.M. *The Classical Theory of Fields*, 4th ed.; Hamermesh, M., Translator; Butterworth-Heinemann: Oxford, UK, 2000; Volume 2.
55. Weinberg, S. *Gravitation and Cosmology: Principles and Applications of the General Theory of Relativity*; Massachusetts Institute of Technology: Cambridge, MA, USA, 1972; See Section 7.4, Chapter 8 and Chapter 9.
56. Holanda, R.F.L.; Goncalves, R.S.; Alcaniz, J.S. A test for cosmic distance duality. *JCAP* **2012**, *6*, 022. [CrossRef]

Article

Anisotropic Cosmology in the Local Limit of Nonlocal Gravity

Javad Tabatabaei [1], Abdolali Banihashemi [1], Shant Baghram [1] and Bahram Mashhoon [2,3,*]

1 Department of Physics, Sharif University of Technology, Tehran 11155-9161, Iran; smj_tabatabaei@physics.sharif.edu (J.T.); abdolali.banihashemi@sharif.edu (A.B.); baghram@sharif.edu (S.B.)
2 School of Astronomy, Institute for Research in Fundamental Sciences (IPM), Tehran 19395-5531, Iran
3 Department of Physics and Astronomy, University of Missouri, Columbia, MO 65211, USA
* Correspondence: mashhoonb@missouri.edu

Abstract: Within the framework of the local limit of nonlocal gravity (NLG), we investigate a class of Bianchi type I spatially homogeneous but anisotropic cosmological models. The modified field equations are presented in this case, and some special solutions are discussed in detail. This modified gravity theory contains a susceptibility function $S(x)$ such that general relativity (GR) is recovered for $S = 0$. In the modified anisotropic cosmological models, we explore the contribution of $S(t)$ and its temporal derivative to the local anisotropic cosmic acceleration. The implications of our results for observational cosmology are briefly discussed.

Keywords: nonlocal gravity; cosmology; cosmic acceleration

Citation: Tabatabaei, J.; Banihashemi, A.; Baghram, S.; Mashhoon, B. Anisotropic Cosmology in the Local Limit of Nonlocal Gravity. *Universe* 2023, *9*, 377. https://doi.org/10.3390/universe9090377

Academic Editor: Kazuharu Bamba

Received: 11 July 2023
Revised: 17 August 2023
Accepted: 21 August 2023
Published: 23 August 2023

1. Introduction

In the current ΛCDM model of cosmology, the energy content of the universe consists of about 70% dark energy, about 25% dark matter, and about 5% visible matter. The model is based on the standard spatially homogeneous and isotropic Friedmann–Lemaître–Robertson–Walker (FLRW) cosmological solutions of Einstein's general relativity theory [1]. The *dark* features, whose nature and origin are unknown, provide the motivation to modify and extend general relativity (GR) on galactic scales and beyond in order to account for observational data purely on the basis of the gravitational physics of the extended GR without recourse to dark ingredients.

To modify the current benchmark model of cosmology, we consider nonlocal gravity theory [2–4], a classical history-dependent generalization of GR that bears a formal resemblance to the nonlocal electrodynamics of media [5–10]. It is important to digress here and mention that there are indeed various other approaches to nonlocal gravitation. For the sake of brevity, we only refer to some examples and their cosmological implications. Nonlocally modified extensions of GR can be generated by the addition of functions of $\Box$, as in infinite derivative theories, or functions of $\Box^{-1}$ to the Einstein–Hilbert action. Here, $\Box$ is the d'Alembert-Beltrami operator. Cosmological implications of such theories have been investigated by a number of authors in connection with dynamic dark energy and accelerated expansion of the universe, for instance, see [11,12] and the references cited therein. Moreover, cosmological solutions of nonlocal infinite derivative theories have been studied that involve anisotropic bouncing models [13] or an interplay between dark matter and dark energy [14]. Quantum field theory provides the motivation for a different class of nonlocal theories of gravitation. Higher curvature nonlocal gravity theories have recently been reviewed in [15], and a generalized nonlocal quantum gravity theory has been formulated within the framework of inflationary cosmology. On the other hand, the phenomenological approach of Deser and Woodard has been based on an effective quantum gravitational action and has been designed to explain cosmic acceleration without dark energy; see [16] and the references cited therein. Furthermore, primordial bouncing cosmology and anisotropy have been investigated within the framework of the Deser–Woodard nonlocal gravity model in [17].

We now return to our classical model of nonlocal gravity that is patterned after the nonlocal electrodynamics of media. Nonlocal gravity (NLG) is a tetrad theory, where the gravitational potentials are given by the 16 components of a preferred orthonormal tetrad frame field. The extended geometric framework of NLG is based on the Weitzenböck connection [18], which renders the spacetime a parallelizable manifold. Within the framework of teleparallelism [19–22], it is possible to express GR using the Weitzenböck torsion tensor. This teleparallel equivalent of general relativity (TEGR) is a gauge theory of the Abelian group of spacetime translations [23]. The formal similarity between TEGR and electrodynamics can be employed to introduce nonlocality into GR via constitutive kernels [2,3]. In NLG, the gravitational field is local, but the theory involves an average of the field over past events resulting in 16 partial integro-differential field equations [24,25]. No exact nontrivial solution of NLG is known at present [26]; however, the linear regime of the theory has been extensively studied. Nonlocal gravity, in its Newtonian regime, simulates dark matter. It is therefore possible to account for the gravitational effects in the solar system as well as in nearby galaxies and clusters of galaxies [27–31]. A comprehensive account of these studies is contained in [4].

NLG is rather intricate, and to study its cosmological implications, we resort to its local limit, which is easier to analyze. In Section 2, we present a brief account of the modified GR field equations in the local limit of NLG. For a more detailed treatment of this limiting situation, see [32,33], where spatially homogeneous and isotropic (FLRW) cosmological models were investigated in this modified TEGR scheme. To explore anisotropy in the Hubble flow, we present, in Section 3, modified gravitational field equations for a Bianchi type I class of time-dependent spatially homogeneous but anisotropic spacetimes within the framework of the local limit of NLG. The field equations contain a susceptibility function $S(t)$ with $1 + S > 0$ and $dS/dt \neq 0$ that is characteristic of the dynamic spacetime background. For $S = 0$, we recover the GR field equations. We show that de Sitter and Kasner spacetimes are not solutions of the modified field equations unless $S(t)$ is independent of time, which is not physically reasonable. Explicit solutions of the modified field equations are studied in the next two sections. A well-known class of dynamic solutions of GR for dust with vanishing cosmological constant is extended to the local limit of NLG in Section 4. The new solutions contain the time-dependent function $S(t)$ and allow for the possibility of exploring the dependence of anisotropic acceleration on $S(t)$. These modified cosmological models are locally anisotropic but tend to the isotropic modified Einstein–de Sitter model at late times ($t \to \infty$). Similarly, we study the solution of the modified field equations for a spacetime dominated by dark energy in Section 5 and explore anisotropic cosmic acceleration in this cosmological model, which eventually becomes isotropic as well. The presence of $S(t)$ could be responsible for certain new "dark" features of accelerating bulk flows in the local universe.

Anisotropy of the Hubble flow would indicate a significant departure from the presumed large-scale spatial homogeneity and isotropy of the standard FLRW cosmology. On the other hand, there is recent observational evidence in support of *local* anisotropic cosmic acceleration [34–38]. The purpose of the present paper is to study the possible contribution of the susceptibility function $S(t)$ to anisotropic features of the Hubble flow.

2. Local Limit of NLG

We consider a spacetime manifold as in general relativity (GR). In an admissible system of coordinates x^μ, the spacetime metric can be written as

$$ds^2 = g_{\mu\nu}(x)\, dx^\mu\, dx^\nu\,. \tag{1}$$

Here, Greek indices run from 0 to 3, Latin indices run from 1 to 3, and the signature of the metric is +2; moreover, we employ units such that $c = 1$. As in GR, the world lines of free test particles and null rays are geodesics of the spacetime manifold. We assume the existence of a preferred set of observers in this gravitational field. The observers have adapted orthonormal tetrads $e^\mu{}_{\hat\alpha}(x)$,

$$g_{\mu\nu}(x)\, e^\mu{}_{\hat\alpha}(x)\, e^\nu{}_{\hat\beta}(x) = \eta_{\hat\alpha\hat\beta}\,, \tag{2}$$

where $\eta_{\alpha\beta} = \mathrm{diag}(-1,1,1,1)$ is the Minkowski metric tensor. In our convention, indices without hats are normal spacetime indices, while hatted indices indicate the tetrad axes in the local tangent space.

We employ the tetrad frame field to define the curvature-free Weitzenböck connection,

$$\Gamma^\mu{}_{\alpha\beta} = e^\mu{}_{\hat\rho}\, \partial_\alpha e_\beta{}^{\hat\rho}\,. \tag{3}$$

Let ∇ denote covariant differentiation with respect to the Weitzenböck connection; then, $\nabla_\nu e_\mu{}^{\hat\alpha} = 0$, so the preferred tetrad frames are parallel throughout the gravitational field and provide a natural scaffolding for the spacetime manifold. The spacetime is thus a parallelizable manifold by the Weitzenböck connection. In this framework of teleparallelism, two distant vectors are considered parallel if they have the same local components relative to their preferred tetrad frames. Moreover, it follows from the tetrad orthonormality relation that the Weitzenböck connection is metric compatible, namely, $\nabla_\mu g_{\alpha\beta} = 0$.

The difference between two connections on the same manifold is a tensor. We define the *torsion* tensor that corresponds to the Weitzenböck connection by

$$C_{\mu\nu}{}^\alpha = \Gamma^\alpha{}_{\mu\nu} - \Gamma^\alpha{}_{\nu\mu} = e^\alpha{}_{\hat\beta}\left(\partial_\mu e_\nu{}^{\hat\beta} - \partial_\nu e_\mu{}^{\hat\beta}\right). \tag{4}$$

In the extended GR framework, we have the Weitzenböck connection as well as the symmetric Levi–Civita connection,

$$^0\Gamma^\mu{}_{\alpha\beta} = \frac{1}{2}g^{\mu\nu}\left(g_{\nu\alpha,\beta} + g_{\nu\beta,\alpha} - g_{\alpha\beta,\nu}\right). \tag{5}$$

We use a left superscript "0" to refer to geometric quantities directly derived from the Levi–Civita connection. The *contorsion* tensor is then defined by

$$K_{\mu\nu}{}^\alpha = {}^0\Gamma^\alpha{}_{\mu\nu} - \Gamma^\alpha{}_{\mu\nu}\,, \tag{6}$$

which is related to the torsion tensor through the metric compatibility of the Weitzenböck connection. In fact,

$$K_{\mu\nu\rho} = \frac{1}{2}\left(C_{\mu\rho\nu} + C_{\nu\rho\mu} - C_{\mu\nu\rho}\right). \tag{7}$$

The Levi–Civita connection given by the Christoffel symbol is the sum of the Weitzenböck connection and the contorsion tensor. One can therefore express the Einstein tensor $^0G_{\mu\nu}$ and the gravitational field equations of GR in terms of the teleparallelism framework, resulting in the teleparallel equivalent of GR, namely, TEGR [4]. Indeed, we find

$$^0G_{\mu\nu} = \frac{\kappa}{\sqrt{-g}}\left[e_\mu{}^{\hat\gamma}\, g_{\nu\alpha}\frac{\partial}{\partial x^\beta}\,\mathfrak{H}^{\alpha\beta}{}_{\hat\gamma} - \left(C_\mu{}^{\rho\sigma}\,\mathfrak{H}_{\nu\rho\sigma} - \frac{1}{4}g_{\mu\nu}\,C^{\alpha\beta\gamma}\,\mathfrak{H}_{\alpha\beta\gamma}\right)\right] \tag{8}$$

and Einstein's field equations expressed in terms of torsion thus become the TEGR field equations

$$\frac{\partial}{\partial x^\nu}\,\mathfrak{H}^{\mu\nu}{}_{\hat\alpha} + \frac{\sqrt{-g}}{\kappa}\,\Lambda e^\mu{}_{\hat\alpha} = \sqrt{-g}\left(T_{\hat\alpha}{}^\mu + \mathbb{T}_{\hat\alpha}{}^\mu\right), \tag{9}$$

where Λ is the cosmological constant and $\kappa := 8\pi G$. Here, we define the auxiliary torsion field $\mathfrak{H}_{\mu\nu\rho}$ by means of the auxiliary torsion tensor $\mathfrak{C}_{\alpha\beta\gamma}$, namely,

$$\mathfrak{H}_{\mu\nu\rho} := \frac{\sqrt{-g}}{\kappa}\,\mathfrak{C}_{\mu\nu\rho}\,, \qquad \mathfrak{C}_{\alpha\beta\gamma} := C_\alpha\, g_{\beta\gamma} - C_\beta\, g_{\alpha\gamma} + K_{\gamma\alpha\beta}\,. \tag{10}$$

Moreover, $C_\mu := C^\alpha{}_{\mu\alpha} = -C_\mu{}^\alpha{}_\alpha$ is the torsion vector. As in GR, $T_{\mu\nu}$ is the symmetric energy-momentum tensor of matter. In Equation (9), we interpret $\mathbb{T}_{\mu\nu}$ to be the traceless energy-momentum tensor of the gravitational field

$$\mathbb{T}_{\mu\nu} := (\sqrt{-g})^{-1} \left(C_{\mu\rho\sigma}\,\mathfrak{H}_\nu{}^{\rho\sigma} - \tfrac{1}{4} g_{\mu\nu}\, C_{\rho\sigma\delta}\,\mathfrak{H}^{\rho\sigma\delta} \right). \tag{11}$$

This version of GR, namely, TEGR, is the gauge theory of the 4-parameter Abelian group of spacetime translations [23]; therefore, though nonlinear, it bears a certain resemblance to Maxwell's electrodynamics.

In analogy with the electrodynamics of media, we can consider the torsion tensor in the form $C_{\mu\nu}{}^{\hat{\alpha}} = \partial_\mu e_\nu{}^{\hat{\alpha}} - \partial_\nu e_\mu{}^{\hat{\alpha}}$ to be similar to the Faraday tensor, while the relationship between $\mathfrak{H}_{\mu\nu\rho}$ and the torsion tensor in Equation (10) can be viewed as the local constitutive relation of TEGR. Let us recall that in Maxwell's electrodynamics, the constitutive relation may change, but the field equations remain the same. We adopt the same approach for the purpose of modifying Einstein's theory. That is, we modify TEGR by introducing a tensor $N_{\mu\nu\rho} = -N_{\nu\mu\rho}$ that changes the constitutive relation of TEGR as follows:

$$\mathcal{H}_{\mu\nu\rho} = \frac{\sqrt{-g}}{\kappa} \left(\mathfrak{C}_{\mu\nu\rho} + N_{\mu\nu\rho} \right). \tag{12}$$

To obtain the field equations of modified TEGR, we simply replace $\mathfrak{H}$ in Equations (9) and (11) by $\mathcal{H}$. The gravitational field equations of extended GR based on the new tensor field $N_{\mu\nu\rho}$ now take the form

$$\frac{\partial}{\partial x^\nu} \mathcal{H}^{\mu\nu}{}_{\hat{\alpha}} + \frac{\sqrt{-g}}{\kappa} \Lambda\, e^\mu{}_{\hat{\alpha}} = \sqrt{-g}\, \left(T_{\hat{\alpha}}{}^\mu + \mathcal{T}_{\hat{\alpha}}{}^\mu \right), \tag{13}$$

where $\mathcal{T}_{\mu\nu}$ is the traceless energy-momentum tensor of the gravitational field. The gravitational energy-momentum tensor is modified by the presence of $N_{\mu\nu\rho}$; hence, we introduce a traceless tensor $Q_{\mu\nu}$ that indicates this difference, namely,

$$\kappa\,\mathcal{T}_{\mu\nu} = \kappa\,\mathbb{T}_{\mu\nu} + Q_{\mu\nu}, \tag{14}$$

where

$$Q_{\mu\nu} := C_{\mu\rho\sigma} N_\nu{}^{\rho\sigma} - \frac{1}{4} g_{\mu\nu}\, C_{\delta\rho\sigma} N^{\delta\rho\sigma}. \tag{15}$$

The total energy-momentum conservation law takes the form

$$\frac{\partial}{\partial x^\mu} \left[\sqrt{-g}\, \left(T_{\hat{\alpha}}{}^\mu + \mathcal{T}_{\hat{\alpha}}{}^\mu - \frac{\Lambda}{\kappa}\, e^\mu{}_{\hat{\alpha}} \right) \right] = 0. \tag{16}$$

It is interesting to see how $N_{\mu\nu\rho}$ modifies GR field equations; to this end, we substitute

$$\mathfrak{H}_{\mu\nu\rho} = \mathcal{H}_{\mu\nu\rho} - \frac{\sqrt{-g}}{\kappa} N_{\mu\nu\rho} \tag{17}$$

in the Einstein tensor (8) and employ modified TEGR field Equation (13) to obtain

$${}^0 G_{\mu\nu} + \Lambda g_{\mu\nu} = \kappa T_{\mu\nu} + Q_{\mu\nu} - \mathcal{N}_{\mu\nu}, \tag{18}$$

where $\mathcal{N}_{\mu\nu}$ is a tensor defined by

$$\mathcal{N}_{\mu\nu} := g_{\nu\alpha} e_\mu{}^{\hat{\gamma}} \frac{1}{\sqrt{-g}} \frac{\partial}{\partial x^\beta} \left(\sqrt{-g} N^{\alpha\beta}{}_{\hat{\gamma}} \right). \tag{19}$$

Therefore, to find the field equations of modified GR, we must add $Q_{\mu\nu} - \mathcal{N}_{\mu\nu}$ to the right-hand side of Einstein's field equations of GR.

Finally, we have to relate $N_{\mu\nu\rho} = -N_{\nu\mu\rho}$ to the torsion tensor. In NLG, the components of $N_{\mu\nu\rho}$ measured by the preferred observers of the theory with adapted tetrads $e^\mu{}_{\hat{\alpha}}$ are associated with the corresponding measured components of $X_{\mu\nu\rho}$ that are directly connected to the torsion tensor, and its expression has been discussed in detail in [4]. That is [24,25],

$$N_{\hat{\mu}\hat{\nu}\hat{\rho}}(x) = \int \mathcal{K}(x, x')\, X_{\hat{\mu}\hat{\nu}\hat{\rho}}(x')\sqrt{-g(x')}\, d^4x', \tag{20}$$

where

$$X_{\hat{\mu}\hat{\nu}\hat{\rho}} = \mathfrak{C}_{\hat{\mu}\hat{\nu}\hat{\rho}} + \check{p}\left(\check{C}_{\hat{\mu}}\, \eta_{\hat{\nu}\hat{\rho}} - \check{C}_{\hat{\nu}}\, \eta_{\hat{\mu}\hat{\rho}}\right). \tag{21}$$

Here, $\mathcal{K}(x, x')$ is the basic causal kernel of NLG that in essence must be determined via observation [4], $\check{p} \neq 0$ is a constant dimensionless parameter, and $\check{C}^\mu$ is the torsion pseudovector,

$$\check{C}_\mu = \frac{1}{3!} C^{\alpha\beta\gamma}\, E_{\alpha\beta\gamma\mu}, \tag{22}$$

where $E_{\alpha\beta\gamma\delta}$ is the Levi–Civita tensor.

Nonlocal gravity (NLG) is thus a classical extension of GR that is highly nonlinear as well. Linearized NLG has been investigated in detail [4]. Within the Newtonian regime of NLG, it appears possible to account for the rotation curves of nearby spiral galaxies as well as for the solar system data [27–31]. Beyond the linear domain, no exact solution is known except for the trivial result that in the absence of gravity we have Minkowski spacetime [26]. On the other hand, it is possible that certain nonlinear features of NLG that belong to the strong-field regimes such as those involving black holes or cosmological models may indeed survive in the local limit of the theory. It is therefore interesting to explore this limiting case of NLG.

To come up with the local limit of NLG, let us assume that the kernel in Equation (20) is proportional to the 4D Dirac delta function, namely,

$$\mathcal{K}(x, x') := \frac{S(x)}{\sqrt{-g(x)}}\, \delta(x - x'); \tag{23}$$

then, $N_{\mu\nu\rho}(x) = S(x)X_{\mu\nu\rho}$, where $S(x)$ is a dimensionless scalar function. Therefore,

$$N_{\mu\nu\rho}(x) = S(x)\left[\mathfrak{C}_{\mu\nu\rho}(x) + \check{p}\left(\check{C}_\mu\, g_{\nu\rho} - \check{C}_\nu\, g_{\mu\rho}\right)\right] \tag{24}$$

and the constitutive relation takes the form

$$\mathcal{H}_{\mu\nu\rho} = \frac{\sqrt{-g}}{\kappa}\left[(1 + S)\, \mathfrak{C}_{\mu\nu\rho} + S\,\check{p}\left(\check{C}_\mu\, g_{\nu\rho} - \check{C}_\nu\, g_{\mu\rho}\right)\right]. \tag{25}$$

Here, the susceptibility function $S(x)$ is a characteristic of the background spacetime just as $\epsilon(x)$ and $\mu(x)$ are features of the medium in electrodynamics. In general, the local electric permittivity $\epsilon(x)$ and magnetic permeability $\mu(x)$ functions are expected to preserve significant features of the electrodynamics of media such as spatial symmetries and temporal dependence. Similarly, $S(x)$ is expected to preserve the characteristics of the background spacetime. Ultimately, $S(x)$ must be determined based on observational data [32,33].

For $S(x) = 0$, we recover TEGR; otherwise, we have a natural generalization of GR that contains a new function $S(x)$. Indeed, Equation (25) implies that to have GR as a limit, we must impose the requirement that $1 + S > 0$. In this local limit of nonlocal gravity, explicit deviations from locality have vanished; however, nontrivial aspects of NLG may have survived through $S(x)$, which would be interesting to study. Consequently, we explore the cosmological implications of this local limit of NLG. Spatially homogeneous and isotropic (FLRW) cosmological models have been treated in [32,33] in connection with H_0 tension. Therefore, we concentrate here on a class of spatially homogeneous but anisotropic spacetimes.

3. Anisotropic Models

Let us consider a Bianchi type I model with a metric of the form

$$ds^2 = -dt^2 + X^2 dx^2 + Y^2 dy^2 + Z^2 dz^2, \tag{26}$$

where X, Y, and Z are functions of time t. This spacetime is spatially homogeneous, with three spacelike commuting Killing vector fields ∂_x, ∂_y, and ∂_z. A detailed discussion of such spacetimes is contained in Chapter 13 of [39]; for a recent discussion within the context of teleparallelism, see [40].

Einstein's gravitational field equations are

$$ {}^0 G_{\mu\nu} + \Lambda\, g_{\mu\nu} = 8\pi G\, T_{\mu\nu}, \qquad {}^0 G_{\mu\nu} = {}^0 R_{\mu\nu} - \frac{1}{2} g_{\mu\nu}\, {}^0 R, \tag{27}$$

where $T_{\mu\nu}$ is assumed to be due to the presence of a comoving perfect fluid with density $\rho(t)$ and pressure $P(t)$,

$$T_{\mu\nu} = (\rho + P)U_\mu U_\nu + P g_{\mu\nu}, \tag{28}$$

and Λ is the cosmological constant. With respect to the system of coordinates $x^\mu = (t, x, y, z)$, the perfect fluid is comoving with $U^\mu = \delta^\mu{}_0$; hence,

$$T_{\mu\nu} = \mathrm{diag}(\rho, PX^2, PY^2, PZ^2). \tag{29}$$

Moreover, the Einstein tensor ${}^0 G_{\mu\nu}$ is diagonal as well with components

$$ {}^0 G_{00} = \frac{\dot{X}\dot{Y}}{XY} + \frac{\dot{Y}\dot{Z}}{YZ} + \frac{\dot{Z}\dot{X}}{ZX}, \tag{30}$$

$$ {}^0 G_{11} = -X^2\left(\frac{\ddot{Y}}{Y} + \frac{\ddot{Z}}{Z} + \frac{\dot{Y}\dot{Z}}{YZ}\right), \tag{31}$$

$$ {}^0 G_{22} = -Y^2\left(\frac{\ddot{Z}}{Z} + \frac{\ddot{X}}{X} + \frac{\dot{Z}\dot{X}}{ZX}\right), \tag{32}$$

$$ {}^0 G_{33} = -Z^2\left(\frac{\ddot{X}}{X} + \frac{\ddot{Y}}{Y} + \frac{\dot{X}\dot{Y}}{XY}\right). \tag{33}$$

It is interesting to work out the Kretschmann scalar $\mathcal{K}$,

$$\mathcal{K} = {}^0 R_{\mu\nu\rho\sigma}\, {}^0 R^{\mu\nu\rho\sigma}, \tag{34}$$

for metric (26). The result is

$$\frac{1}{4}\mathcal{K} = \left(\frac{\ddot{X}}{X}\right)^2 + \left(\frac{\ddot{Y}}{Y}\right)^2 + \left(\frac{\ddot{Z}}{Z}\right)^2 + \left(\frac{\dot{X}\dot{Y}}{XY}\right)^2 + \left(\frac{\dot{Y}\dot{Z}}{YZ}\right)^2 + \left(\frac{\dot{Z}\dot{X}}{ZX}\right)^2. \tag{35}$$

Detailed discussions of the GR solutions of these models with $\Lambda = 0$ for dust ($P = 0$) can be found, for instance, in [41,42], Section 5.4 of ref. [43], and Section 12.15 of ref. [44]. We give a brief description of these solutions in Section 4 in connection with cosmic deceleration.

We are interested in the extended GR framework. Therefore, consider the class of observers that are spatially at rest with adapted tetrad $e^\mu{}_{\hat{\alpha}}$ field given by

$$e^\mu{}_{\hat{0}} = (1,0,0,0), \quad e^\mu{}_{\hat{1}} = (0, \tfrac{1}{X},0,0), \quad e^\mu{}_{\hat{2}} = (0,0,\tfrac{1}{Y},0), \quad e^\mu{}_{\hat{3}} = (0,0,0,\tfrac{1}{Z}), \tag{36}$$

where the spatial axes point along the Cartesian coordinate directions. We have

$$e_\mu{}^{\hat{0}} = (1,0,0,0) , \quad e_\mu{}^{\hat{1}} = (0, X, 0, 0) , \quad e_\mu{}^{\hat{2}} = (0, 0, Y, 0) , \quad e_\mu{}^{\hat{3}} = (0, 0, 0, Z) . \tag{37}$$

We compute the Weitzenböck torsion tensor (4) in this case, and we find $C_{\mu\nu}{}^0 = 0$, $C_{ij}{}^k = 0$, and the only nonzero components can be obtained from

$$C_{01}{}^1 = \frac{\dot{X}}{X} , \qquad C_{02}{}^2 = \frac{\dot{Y}}{Y} , \qquad C_{03}{}^3 = \frac{\dot{Z}}{Z} . \tag{38}$$

Similarly, we have $C_{\mu\nu 0} = 0$, $C_{ijk} = 0$, and the only nonzero components of $C_{\mu\nu\rho}$ can be obtained from

$$C_{011} = \dot{X} X , \qquad C_{022} = \dot{Y} Y , \qquad C_{033} = \dot{Z} Z . \tag{39}$$

It follows from these results that the torsion vector is given by

$$C_0 = -\left(\frac{\dot{X}}{X} + \frac{\dot{Y}}{Y} + \frac{\dot{Z}}{Z} \right) , \qquad C_i = 0 , \tag{40}$$

while the torsion pseudovector $\check{C}_\mu = 0$ in this case.

The calculations of contorsion (7) and the auxiliary torsion (9) tensors produce similar results. That is, $K_{0\mu\nu} = 0$, $K_{ijk} = 0$, and the only nonzero components of $K_{\mu\nu\rho}$ can be obtained from

$$K_{101} = \dot{X} X , \qquad K_{202} = \dot{Y} Y , \qquad K_{303} = \dot{Z} Z . \tag{41}$$

Moreover, $\mathfrak{C}_{\mu\nu 0} = 0$, $\mathfrak{C}_{ijk} = 0$, and the only nonzero components of $\mathfrak{C}_{\mu\nu\rho}$ can be obtained from

$$\mathfrak{C}_{101} = X^2 \left(\frac{\dot{Y}}{Y} + \frac{\dot{Z}}{Z} \right) , \quad \mathfrak{C}_{202} = Y^2 \left(\frac{\dot{X}}{X} + \frac{\dot{Z}}{Z} \right) , \quad \mathfrak{C}_{303} = Z^2 \left(\frac{\dot{X}}{X} + \frac{\dot{Y}}{Y} \right) . \tag{42}$$

In the local limit of NLG, the constitutive relation of modified TEGR is given by $N_{\mu\nu\rho}(x) = S(x)\mathfrak{C}_{\mu\nu\rho}(x)$, where the gravitational susceptibility S is a property of the background spacetime. In the case of the homogeneous time-dependent background (26), we assume that S is a function of time t. Therefore, $N_{\mu\nu 0} = 0$, $N_{ijk} = 0$, and the only nonzero components of $N_{\mu\nu\rho}$ can be obtained from

$$N_{101} = S(t) X^2 \left(\frac{\dot{Y}}{Y} + \frac{\dot{Z}}{Z} \right) , \quad N_{202} = S(t) Y^2 \left(\frac{\dot{X}}{X} + \frac{\dot{Z}}{Z} \right) , \quad N_{303} = S(t) Z^2 \left(\frac{\dot{X}}{X} + \frac{\dot{Y}}{Y} \right) . \tag{43}$$

We can now compute $Q_{\mu\nu}$ given in Equation (14) and $\mathcal{N}_{\mu\nu}$ given in Equation (18). The results are that these quantities are diagonal with elements

$$Q_{00} = -S(t) \left(\frac{\dot{X}\dot{Y}}{XY} + \frac{\dot{Y}\dot{Z}}{YZ} + \frac{\dot{Z}\dot{X}}{ZX} \right) , \tag{44}$$

$$Q_{11} = -S(t) X^2 \frac{\dot{Y}\dot{Z}}{YZ} , \qquad Q_{22} = -S(t) Y^2 \frac{\dot{Z}\dot{X}}{ZX} , \qquad Q_{33} = -S(t) Z^2 \frac{\dot{X}\dot{Y}}{XY} . \tag{45}$$

For $\mathcal{N}_{\mu\nu}$, however, we find $\mathcal{N}_{00} = 0$ and

$$\mathcal{N}_{11} = -\frac{X^2}{YZ}\frac{d}{dt}[S(Y\dot{Z} + \dot{Y}Z)],\tag{46}$$

$$\mathcal{N}_{22} = -\frac{Y^2}{XZ}\frac{d}{dt}[S(X\dot{Z} + \dot{X}Z)],\tag{47}$$

$$\mathcal{N}_{33} = -\frac{Z^2}{XY}\frac{d}{dt}[S(X\dot{Y} + \dot{X}Y)].\tag{48}$$

Collecting everything, the modified GR field Equation (18) can be expressed as

$$(1+S)\left(\frac{\dot{X}\dot{Y}}{XY} + \frac{\dot{Y}\dot{Z}}{YZ} + \frac{\dot{Z}\dot{X}}{ZX}\right) = \Lambda + 8\pi G\rho\tag{49}$$

and

$$(1+S)\left(\frac{\ddot{Y}}{Y} + \frac{\ddot{Z}}{Z} + \frac{\dot{Y}\dot{Z}}{YZ}\right) = \Lambda - 8\pi GP - \frac{dS}{dt}\left(\frac{\dot{Y}}{Y} + \frac{\dot{Z}}{Z}\right),\tag{50}$$

$$(1+S)\left(\frac{\ddot{X}}{X} + \frac{\ddot{Z}}{Z} + \frac{\dot{X}\dot{Z}}{XZ}\right) = \Lambda - 8\pi GP - \frac{dS}{dt}\left(\frac{\dot{X}}{X} + \frac{\dot{Z}}{Z}\right),\tag{51}$$

$$(1+S)\left(\frac{\ddot{X}}{X} + \frac{\ddot{Y}}{Y} + \frac{\dot{X}\dot{Y}}{XY}\right) = \Lambda - 8\pi GP - \frac{dS}{dt}\left(\frac{\dot{X}}{X} + \frac{\dot{Y}}{Y}\right).\tag{52}$$

3.1. Field Equations

To express the gravitational field equations for the anisotropic models under consideration in a more tractable form, it is useful to consider

$$V(t) = XYZ, \qquad W(t) := \frac{\dot{X}\dot{Y}}{XY} + \frac{\dot{Y}\dot{Z}}{YZ} + \frac{\dot{Z}\dot{X}}{ZX},\tag{53}$$

where $|V(t)| = \sqrt{-g}$ and note that

$$\frac{\dot{V}}{V} = \frac{\dot{X}}{X} + \frac{\dot{Y}}{Y} + \frac{\dot{Z}}{Z}, \qquad \frac{\ddot{V}}{V} = \frac{\ddot{X}}{X} + \frac{\ddot{Y}}{Y} + \frac{\ddot{Z}}{Z} + 2W.\tag{54}$$

Let us add Equations (50)–(52) to obtain

$$(1+S)\left(2\frac{\ddot{V}}{V} - 3W\right) = 3(\Lambda - 8\pi GP) - 2\frac{dS}{dt}\frac{\dot{V}}{V}.\tag{55}$$

Using Equation (49), we can write

$$(1+S)\frac{\ddot{V}}{V} = 3[\Lambda + 4\pi G(\rho - P)] - \frac{dS}{dt}\frac{\dot{V}}{V}\tag{56}$$

or

$$\frac{d}{dt}[(1+S)\dot{V}] = 3V[\Lambda + 4\pi G(\rho - P)].\tag{57}$$

Another interesting result is obtained by writing Equation (49) as $VW = (\Lambda + 8\pi G\rho)V/(1+S)$ and taking the time derivative of both sides. From the relation

$$\frac{1}{V}\frac{d(VW)}{dt} = \frac{\dot{X}}{X}\left(\frac{\ddot{Y}}{Y} + \frac{\ddot{Z}}{Z} + \frac{\dot{Y}\dot{Z}}{YZ}\right) + \frac{\dot{Y}}{Y}\left(\frac{\ddot{X}}{X} + \frac{\ddot{Z}}{Z} + \frac{\dot{X}\dot{Z}}{XZ}\right) + \frac{\dot{Z}}{Z}\left(\frac{\ddot{X}}{X} + \frac{\ddot{Y}}{Y} + \frac{\dot{X}\dot{Y}}{XY}\right),\tag{58}$$

we find

$$(1+S)\frac{d(VW)}{dt} = (\Lambda - 8\pi GP)\dot{V} - 2\frac{dS}{dt}VW. \tag{59}$$

Using Equation (49), we finally obtain

$$\frac{d\rho}{dt} = -(\rho + P)\frac{\dot{V}}{V} - \frac{\frac{dS}{dt}}{(1+S)}\left(\rho + \frac{\Lambda}{8\pi G}\right). \tag{60}$$

Equations (57) and (60) are important consequences of the modified field equations. Furthermore, let us define a new temporal variable τ by

$$\tau := \int_0^t \frac{dt'}{1+S(t')}, \qquad dt = (1+S)d\tau; \tag{61}$$

hence, the spacetime metric in (τ, x, y, z) coordinates is

$$ds^2 = -(1+S)^2 d\tau^2 + X^2(\tau)dx^2 + Y^2(\tau)dy^2 + Z^2(\tau)dz^2, \tag{62}$$

where S is now considered, by an abuse of notation, a function of τ. For instance, let us suppose $S(t) = t$; then, $\tau = \ln(1+t)$, and in the above metric we have in this case $S = -1 + e^\tau$. For metric (62), the Kretschmann scalar $\mathcal{K}$ given by Equation (35) can be expressed in terms of the new temporal variable τ using

$$\frac{\dot{X}}{X} = (1+S)^{-1}\frac{1}{X}\frac{dX}{d\tau} \tag{63}$$

and

$$\frac{\ddot{X}}{X} = (1+S)^{-2}\left(\frac{1}{X}\frac{d^2X}{d\tau^2} - \frac{\frac{dS}{d\tau}}{1+S}\frac{1}{X}\frac{dX}{d\tau}\right). \tag{64}$$

The gravitational field equations can now be written in terms of the temporal variable τ as

$$\frac{1}{XY}\frac{dX}{d\tau}\frac{dY}{d\tau} + \frac{1}{YZ}\frac{dY}{d\tau}\frac{dZ}{d\tau} + \frac{1}{ZX}\frac{dZ}{d\tau}\frac{dX}{d\tau} = (1+S)(\Lambda + 8\pi G\rho), \tag{65}$$

$$\frac{1}{Y}\frac{d^2Y}{d\tau^2} + \frac{1}{Z}\frac{d^2Z}{d\tau^2} + \frac{1}{YZ}\frac{dY}{d\tau}\frac{dZ}{d\tau} = (1+S)(\Lambda - 8\pi GP), \tag{66}$$

$$\frac{1}{X}\frac{d^2X}{d\tau^2} + \frac{1}{Z}\frac{d^2Z}{d\tau^2} + \frac{1}{XZ}\frac{dX}{d\tau}\frac{dZ}{d\tau} = (1+S)(\Lambda - 8\pi GP), \tag{67}$$

$$\frac{1}{X}\frac{d^2X}{d\tau^2} + \frac{1}{Y}\frac{d^2Y}{d\tau^2} + \frac{1}{XY}\frac{dX}{d\tau}\frac{dY}{d\tau} = (1+S)(\Lambda - 8\pi GP). \tag{68}$$

To solve Equations (66)–(68), let us subtract, for instance, Equation (66) from Equation (67) to obtain

$$\frac{1}{X}\frac{d^2X}{d\tau^2} - \frac{1}{Y}\frac{d^2Y}{d\tau^2} + \frac{1}{Z}\frac{dZ}{d\tau}\left(\frac{1}{X}\frac{dX}{d\tau} - \frac{1}{Y}\frac{dY}{d\tau}\right) = 0, \tag{69}$$

which with $V = XYZ$ can be written as

$$\frac{d}{d\tau}\left(\frac{1}{X}\frac{dX}{d\tau} - \frac{1}{Y}\frac{dY}{d\tau}\right) + \frac{1}{V}\frac{dV}{d\tau}\left(\frac{1}{X}\frac{dX}{d\tau} - \frac{1}{Y}\frac{dY}{d\tau}\right) = 0. \tag{70}$$

Hence,

$$V\left(\frac{1}{X}\frac{dX}{d\tau} - \frac{1}{Y}\frac{dY}{d\tau}\right) = \mathcal{C}_{12}, \tag{71}$$

where $\mathcal{C}_{12}$ is a constant of integration and similar results hold for the other metric functions. Finally, in terms of temporal variable τ, Equation (57) can be written as

$$\frac{1}{V}\frac{d^2V}{d\tau^2} = 3(1 + S)[\Lambda + 4\pi G(\rho - P)]. \tag{72}$$

3.2. Special Solutions

We now explore some cases of particular interest.

3.2.1. de Sitter

Let us first consider de Sitter's solution with

$$\rho = P = 0, \qquad X = Y = Z = e^{\lambda t}, \tag{73}$$

where λ is a nonzero constant. The field equations imply

$$3\lambda^2(1 + S) = \Lambda, \qquad 3\lambda^2(1 + S) = \Lambda - 2\lambda\frac{dS}{dt}. \tag{74}$$

Therefore, $dS/dt = 0$ and S must be constant. It follows that de Sitter spacetime is not a solution of the modified TEGR since the susceptibility function is independent of time while the background spacetime is dynamic. This is in agreement with the fact that de Sitter spacetime is not a solution of NLG [25].

3.2.2. Kasner

Let us next consider the standard Kasner metric [45,46]

$$ds^2 = -dt^2 + t^{2p_1}dx^2 + t^{2p_2}dy^2 + t^{2p_3}dz^2, \tag{75}$$

$$p_1 + p_2 + p_3 = p_1^2 + p_2^2 + p_3^2 = 1. \tag{76}$$

In GR, this empty universe model is a solution of the field equations with $\Lambda = 0$ and $\rho = P = 0$. Note that with $p_1 = 1$, say, and hence $p_2 = p_3 = 0$, we recover flat spacetime. Therefore, we can assume $p_1 < p_2 < p_3$; that is,

$$-\frac{1}{3} \le p_1 \le 0, \qquad 0 \le p_2 \le \frac{2}{3}, \qquad \frac{2}{3} \le p_3 \le 1. \tag{77}$$

It follows from field Equations (49) and (50) that

$$\frac{(1 + S)}{t^2}(p_1p_2 + p_2p_3 + p_1p_3) = \Lambda + 8\pi G\rho, \tag{78}$$

$$\frac{(1 + S)}{t^2}(p_2^2 - p_2 + p_3^2 - p_3 + p_2p_3) = \Lambda - 8\pi GP - \frac{dS}{dt}(p_2 + p_3), \tag{79}$$

etc. Because $p_1 p_2 + p_2 p_3 + p_1 p_3 = 0$ and $p_2^2 - p_2 + p_3^2 - p_3 + p_2 p_3 = 0$ together with two other cyclically related terms that vanish, we find that ρ, P, and S must be constants such that

$$-8\pi G\rho = \Lambda \le 0, \qquad \rho + P = 0, \qquad S = \text{constant}. \tag{80}$$

Therefore, Kasner's spacetime with constant S is not a solution of the modified TEGR.

3.2.3. Flat FLRW Model

We now consider $X = Y = Z = a(t)$. Then, field Equations (49) and (50) imply

$$3(1 + S)\left(\frac{\dot{a}}{a}\right)^2 = \Lambda + 8\pi G\rho, \tag{81}$$

$$(1 + S)\left[2\frac{\ddot{a}}{a} + \left(\frac{\dot{a}}{a}\right)^2\right] = \Lambda - 8\pi GP - 2\frac{dS}{dt}\frac{\dot{a}}{a}. \tag{82}$$

A thorough treatment of these equations is contained in a recent paper [32], where they were employed with $\Lambda = 0$ in a detailed discussion of the implications of the modified Cartesian flat cosmology in connection with H_0 tension.

It is important to emphasize that in these time-dependent solutions considered thus far, S must be dependent upon time as well; otherwise, we do not have a physically meaningful solution of the theory.

4. Solution for Dust with $\Lambda = 0$

4.1. $S = 0$

When $S = 0$, we obtain the gravitational field equations in GR. A well-known class of anisotropic solutions corresponds to $\Lambda = 0$ and $P = 0$ [41–44]. In this case, Equations (57) and (60) imply

$$\rho V = \frac{\ell_0}{6\pi G}, \qquad V = \ell_0 t^2 (1 + \Sigma/t), \tag{83}$$

where $\ell_0 > 0$ and $\Sigma > 0$ are integration constants with dimensions of length (or time, since $c = 1$) and we have assumed that $V = 0$ at $t = 0$. Using

$$\int \frac{dt}{V} = -\frac{1}{\ell_0 \Sigma} \ln(1 + \Sigma/t), \tag{84}$$

Equation (71) and similar ones can be integrated with the result that ratios such as X/Y, etc., up to constant coefficients are given by $(1 + \Sigma/t)$ to some constant powers. In the absence of anisotropy ($\Sigma = 0$), we must recover the standard Einstein–de Sitter solution; that is, Σ is the anisotropy parameter for finite $t > 0$. Therefore, we look for the solutions of the field equations such that the metric coefficients are given by

$$X = \ell_0^{q_1} t^{2/3} (1 + \Sigma/t)^{q_1}, \quad Y = \ell_0^{q_2} t^{2/3} (1 + \Sigma/t)^{q_2}, \quad Z = \ell_0^{q_3} t^{2/3} (1 + \Sigma/t)^{q_3}, \tag{85}$$

where q_i, $i = 1, 2, 3$, are constants that must add up to unity in order to satisfy Equation (83). A detailed examination reveals that the field equations are all satisfied in the case of dust with $\Lambda = 0$ provided

$$q_1 + q_2 + q_3 = 1, \qquad q_1^2 + q_2^2 + q_3^2 = 1. \tag{86}$$

It is possible to choose $q_1 < q_2 < q_3$; that is, $-\frac{1}{3} \le q_1 \le 0$, $0 \le q_2 \le \frac{2}{3}$, and $\frac{2}{3} \le q_3 \le 1$. Then, a convenient representation of these constants is given by

$$q_i = \frac{1}{3} - \frac{2}{3}\sin[\theta - 2(i-1)\pi/3], \qquad i = 1, 2, 3, \qquad \frac{\pi}{6} \le \theta \le \frac{\pi}{2}. \tag{87}$$

It is important to note that at late times, $t \to \infty$, we recover the standard Einstein–de Sitter solution.

4.2. $S \neq 0$

Let us note that a constant $S \neq 0$ is equivalent to a constant rescaling of the time coordinate in GR, namely, $t \to (1+S)^{1/2}t$. Therefore, new results are obtained only when S depends upon time, which is necessary for a proper physical interpretation of the susceptibility function S.

We now consider the case $dS/dt \neq 0$. For $P = 0$ and $\Lambda = 0$, it follows from Equations (60) and (72) that

$$(1+S)\rho\,V = \frac{\ell_0}{6\pi G}\,, \qquad V = \ell_0\,\tau^2(1+\Sigma/\tau)\,. \tag{88}$$

Moreover, the metric coefficients are given by

$$X = \ell_0{}^{q_1}\,\tau^{2/3}(1+\Sigma/\tau)^{q_1}\,, \quad Y = \ell_0{}^{q_2}\,\tau^{2/3}(1+\Sigma/\tau)^{q_2}\,, \quad Z = \ell_0{}^{q_3}\,\tau^{2/3}(1+\Sigma/\tau)^{q_3}\,, \tag{89}$$

where $q_1 + q_2 + q_3 = 1$ and $q_1^2 + q_2^2 + q_3^2 = 1$. For $\Sigma = 0$ or $\tau \to \infty$, we recover the isotropic solution corresponding to the modified Einstein–de Sitter model.

It is interesting to explore the implications of these solutions for the dimensionless deceleration parameter defined for the x direction, say, by $Q_x = -(\ddot{X}/X)/(\dot{X}/X)^2$, etc.

Let us first recall that $\Sigma > 0$ is the anisotropy parameter; in fact, for $\Sigma = 0$, we obtain the standard Einstein–de Sitter result that $Q = 1/2$ in every direction. Let anisotropic expansion occur in the z direction; then, Equation (86) allows two possibilities, namely, $(q_1, q_2, q_3) = (0, 0, 1)$ and $(q_1, q_2, q_3) = (2/3, 2/3, -1/3)$. We choose the former case for the sake of simplicity. We are interested in the nature of deceleration parameter Q_z in the z direction. For $S = 0$, we have $Q_x = Q_y = 1/2$ and

$$Q_z = 2\frac{(t+\Sigma)(t-2\Sigma)}{(2t-\Sigma)^2} < \frac{1}{2}\,. \tag{90}$$

Indeed, for $t < 2\Sigma$ we have $Q_z < 0$ and hence acceleration in the z direction that later turns into deceleration for $t > 2\Sigma$ with magnitude $< 1/2$. As $t \to \infty$, $Q_z \to 1/2$; that is, isotropy is recovered at late times.

How do these results change in the presence of $S(t)$? In terms of temporal parameter τ, we find for the x direction, say,

$$Q_x = -X\frac{d^2X}{d\tau^2}\left(\frac{dX}{d\tau}\right)^{-2} + \frac{dS}{dt}X\left(\frac{dX}{d\tau}\right)^{-1}\,. \tag{91}$$

Therefore, with

$$X = Y = \tau^{2/3}\,, \qquad Z = \ell_0\,\tau^{2/3}(1+\Sigma/\tau)\,, \tag{92}$$

the results are

$$Q_x = Q_y = \frac{1}{2} + \frac{3}{2}\tau\,\frac{dS}{dt} \tag{93}$$

and

$$Q_z = 2\frac{(\tau+\Sigma)(\tau-2\Sigma)}{(2\tau-\Sigma)^2} + 3\left(\frac{\tau+\Sigma}{2\tau-\Sigma}\right)\tau\,\frac{dS}{dt}\,. \tag{94}$$

Let us suppose, for instance, that $dS/dt > 0$. Then, the deceleration increases in the x and y directions. The same is true in the z direction for $\tau > 2\Sigma$; however, for $\tau < \Sigma/2$, the presence of $dS/dt > 0$ causes extra acceleration in the z direction. For $2\Sigma > \tau > \Sigma/2$, the sign of Q_z depends upon the magnitude of dS/dt. We note that isotropy is restored at late times $\tau \gg \Sigma$.

5. Solution for Dynamic Dark Energy with $\Lambda = 0$

Let us imagine a universe that in the absence of the cosmological constant ($\Lambda = 0$) is dominated by dynamic dark energy with $P_{de} + \rho_{de} = 0$. Therefore, Equation (60) implies

$$\rho_{de} = \rho_{de}(t_0) \frac{1 + S_0}{1 + S}, \qquad \rho_{de}(t_0) > 0, \quad S_0 = S(t_0), \tag{95}$$

where t_0 refers to some fiducial epoch in the expansion of the universe. The pressure of dark energy is always negative, $P_{de} = -\rho_{de}$, and we assume on the basis of Equation (95) that

$$6\pi G\rho_{de}(1 + S) = 6\pi G\rho_{de}(t_0)\,(1 + S_0) := \frac{1}{\tau_0^2}, \qquad \eta := \frac{\tau}{\tau_0}, \tag{96}$$

where $\tau_0 > 0$ is a constant with the dimensions of time, and henceforth we employ η as the new dimensionless temporal variable. To explore the anisotropic acceleration of this dark energy universe model, we assume for the sake of simplicity that $X = Y$. Thus, we have a cylindrical model with the z direction as the main direction of anisotropy. Let us note here that for $S = 0$ our dynamic dark energy source in effect reduces to a cosmological constant.

With $X = Y$, Equation (68) reduces to

$$\frac{2}{X}\frac{d^2X}{d\eta^2} + \frac{1}{X^2}\left(\frac{dX}{d\eta}\right)^2 = \frac{4}{3}. \tag{97}$$

This nonlinear equation can be easily solved with

$$X := \alpha^{2/3}, \qquad \frac{d^2\alpha}{d\eta^2} - \alpha = 0, \qquad \alpha = C_+\,e^{\eta} + C_-\,e^{-\eta}, \tag{98}$$

where $C_{\pm}$ are integration constants. Next, Equation (65) implies

$$\frac{1}{X^2}\left(\frac{dX}{d\eta}\right)^2 + \frac{2}{Z}\frac{dZ}{d\eta}\frac{1}{X}\frac{dX}{d\eta} = \frac{4}{3}, \tag{99}$$

or

$$\frac{1}{Z}\frac{dZ}{d\eta} = \frac{\alpha}{\beta} - \frac{1}{3}\frac{\beta}{\alpha}, \tag{100}$$

where $\alpha = d\beta/d\eta$ and

$$\beta = \frac{d\alpha}{d\eta} = C_+\,e^{\eta} - C_-\,e^{-\eta}. \tag{101}$$

Hence, $Z = C_0\,\alpha^{-1/3}\,\beta$, where C_0 is an integration constant. Therefore,

$$X = Y = \alpha^{2/3}, \quad Z = C_0\alpha^{-1/3}\beta, \quad V = X^2 Z = C_0\,\alpha\,\beta = C_0(C_+^2\,e^{2\eta} - C_-^2\,e^{-2\eta}). \tag{102}$$

One can check that this is the general solution of the field equations in the present case. For these solutions, the Kretschmann scalar $\mathcal{K}$ can be written as

$$\frac{1}{4}(1 + S)^4\,\tau_0^4\,\mathcal{K} = \frac{16}{9}K_1 - \frac{8}{3}\frac{\beta}{\alpha}\left(1 - \frac{\beta^2}{3\alpha^2}\right)\tau_0\frac{dS}{dt} + K_2\frac{\alpha^2}{\beta^2}\left(\tau_0\frac{dS}{dt}\right)^2, \tag{103}$$

where

$$K_1 = 1 - \frac{2}{3}\frac{\beta^2}{\alpha^2} + \frac{1}{3}\frac{\beta^4}{\alpha^4}, \qquad K_2 = 1 - \frac{2}{3}\frac{\beta^2}{\alpha^2} + \frac{\beta^4}{\alpha^4}. \tag{104}$$

We are interested in the behavior of the acceleration of these anisotropic solutions. For the deceleration parameters of this model, we find

$$Q_x = Q_y = \frac{1}{2}\left(1 - 3\frac{\alpha^2}{\beta^2}\right) + \frac{3}{2}\tau_0\frac{dS}{dt}\frac{\alpha}{\beta}, \tag{105}$$

$$Q_z = -4\left(1 - 3\frac{\alpha^2}{\beta^2}\right)^{-2} - 3\tau_0\frac{dS}{dt}\frac{\alpha}{\beta}\left(1 - 3\frac{\alpha^2}{\beta^2}\right)^{-1}. \tag{106}$$

At late times $\eta \to \infty$, $\alpha/\beta \to 1$, and these expressions reduce to $Q_x = Q_y = Q_z = -1 + (3\tau_0/2)\, dS/dt$. However, for finite η there could be anomalous behavior when $\beta = 0$ or $3\alpha^2 = \beta^2$. If $\beta = 0$, then $Q_x = Q_y = -\infty$, while $Q_z = 0$. On the other hand, if $3\alpha^2 = \beta^2$, then $Q_x = Q_y = \pm(\sqrt{3}\tau_0/2)dS/dt$, while $Q_z = -\infty$. To give an example of the former situation, consider, for instance, the case where $C_+ = C_- = C > 0$ with $\alpha = 2C\cosh\eta$ and $\beta = 2C\sinh\eta$. In this case, $3\alpha^2 > \beta^2$. Then, $X = Y = (2C)^{2/3}\cosh^{2/3}\eta$ and $Z = C_0(2C)^{2/3}\cosh^{-1/3}\eta\,\sinh\eta$. The universe starts from a singular pancake state at $\eta = 0$ and expands with infinite acceleration in the x and y directions but with zero acceleration in the z direction. Indeed, inspection of the Kretschmann scalar $\mathcal{K}$ given by Equation (103) reveals that $\mathcal{K}$ diverges at $\tau = 0$ provided $dS/d\tau \neq 0$.

For an example of the case where $3\alpha^2 = \beta^2$, let us consider, for instance, $C_+ = 1$ and $C_- = -(2 + \sqrt{3})$. Then, the universe starts from a cigar state at $\eta = 0$ with infinite acceleration in the z direction but with finite acceleration $-(\sqrt{3}\tau_0/2)dS/dt$ in the x and y directions assuming that $dS/dt > 0$. The Kretschmann scalar (103) remains finite in this case.

These results could be interesting in connection with recent observational evidence in favor of anomalous anisotropic acceleration of bulk flow in the local universe [34–38].

6. Discussion

The cosmic acceleration in the standard cosmological model is assumed to be isotropic. Large-scale anisotropy in the Hubble flow has been the subject of numerous studies. We have explored anisotropy in cosmic acceleration within the theoretical framework of a recent teleparallel extension of general relativity that corresponds to the local limit of nonlocal gravity. This modified theory involves a function $S(x)$ such that GR is recovered for $S = 0$. We are interested in the cosmological significance of the extra function $S(x)$. In particular, in a dynamic Bianchi type I model that is consistent with the modified GR field equations, we theoretically investigate anisotropy in cosmic acceleration and determine the contribution of $S(t)$ to local anisotropic acceleration. Our results in Sections 4 and 5 for possible local anisotropic cosmic acceleration depend explicitly upon $dS(t)/dt$.

The function $S(t)$ may also possibly contribute to the resolution of an anomaly in the quadrupole anisotropy of CMB temperature. It has been reported that the CMB temperature angular power spectrum suffers from a deficit in its quadrupole moment [47–50]. The anomaly has to do with the low amplitude of the quadrupole anisotropy compared to the prediction of the standard ΛCDM model. On the other hand, in an anisotropic universe described by metric (26), one would in general expect a change in the quadrupole moment of the CMB temperature due to different amounts of redshift suffered by photons traveling after recombination along different directions toward the observer [51–55]. However, Big Bang nucleosynthesis (BBN) puts a tight constraint on the anisotropy of the expansion rate [56]; hence, it is rather difficult for the anisotropic expansion to justify the CMB quadrupole deficit [51]. A suitable susceptibility function $S(t)$ of the modified gravity theory may be able to ameliorate the situation. That is, $S(t)$ may allow the quadrupole deficit anomaly to be alleviated while respecting the BBN constraint. A more complete discussion is beyond the scope of the present paper.

Author Contributions: The authors have contributed equally to this work. All authors have read and agreed to the published version of the manuscript.

Funding: The work of AB has been supported financially by Iran Science Elites Federation.

Data Availability Statement: Not applicable.

Conflicts of Interest: The authors declare no conflict of interest.

References

1. Einstein, A. *The Meaning of Relativity*; Princeton University Press: Princeton, NJ, USA, 1955.
2. Hehl, F.W.; Mashhoon, B. Nonlocal Gravity Simulates Dark Matter. *Phys. Lett. B* **2009**, *673*, 279–282. [CrossRef]
3. Hehl, F.W.; Mashhoon, B. Formal framework for a nonlocal generalization of Einstein's theory of gravitation. *Phys. Rev. D* **2009**, *79*, 064028. [CrossRef]
4. Mashhoon, B. *Nonlocal Gravity*; Oxford University Press: Oxford, UK, 2017.
5. Hopkinson, J. Residual Charge of the Leyden Jar.—Dielectric Properties of different Glasses. *Phil. Trans. R. Soc. Lond.* **1877**, *167*, 599–626.
6. Poisson, S. Mémoire sur la théorie du magnétisme en mouvement. *Mém. Acad. Sci. France* **1823**, *6*, 441–570.
7. Jackson, J.D. *Classical Electrodynamics*, 3rd ed.; Wiley: Hoboken, NJ, USA, 1999.
8. Landau, L.D.; Lifshitz, E.M. *Electrodynamics of Continuous Media*; Pergamon: Oxford, UK, 1960.
9. Hehl, F.W.; Obukhov, Y.N. *Foundations of Classical Electrodynamics: Charge, Flux, and Metric*; Birkhäuser: Boston, MA, USA, 2003.
10. van den Hoogen, R.J. Towards a covariant smoothing procedure for gravitational theories. *J. Math. Phys.* **2017**, *58*, 122501. [CrossRef]
11. Maggiore, M.; Mancarella, M. Nonlocal gravity and dark energy. *Phys. Rev. D* **2014**, *90*, 023005. [CrossRef]
12. Capozziello, S.; Bajardi, F. Non-Local Gravity Cosmology: An Overview. *Int. J. Mod. Phys. D* **2022**, *31*, 2230009. [CrossRef]
13. Kumar, K.S.; Maheshwari, S.; Mazumdar, A.; Peng, J. An anisotropic bouncing universe in non-local gravity. *J. Cosmol. Astropart. Phys.* **2021**, *07*, 025. [CrossRef]
14. Dimitrijevic, I.; Dragovich, B.; Koshelev, A.S.; Rakic, Z.; Stankovic, J. Cosmological Solutions of a Nonlocal Square Root Gravity. *Phys. Lett. B* **2019**, *797*, 134848. [CrossRef]
15. Koshelev, A.S.; Kumar, K.S.; Starobinsky, A.A. Cosmology in nonlocal gravity. *arXiv* **2023**, arXiv:2305.18716.
16. Deser, S.; Woodard, R.P. Nonlocal Cosmology II—Cosmic acceleration without fine tuning or dark energy. *J. Cosmol. Astropart. Phys.* **2019**, *06*, 034. [CrossRef]
17. Chen, C.Y.; Chen, P.; Park, S. Primordial bouncing cosmology in the Deser-Woodard nonlocal gravity. *Phys. Lett. B* **2019**, *796*, 112–116. [CrossRef]
18. Weitzenböck, R. *Invariantentheorie*; Noordhoff: Groningen, The Netherlands, 1923.
19. Maluf, J.W. The teleparallel equivalent of general relativity. *Ann. Phys. (Berlin)* **2013**, *525*, 339–357. [CrossRef]
20. Aldrovandi, R.; Pereira, J.G. *Teleparallel Gravity: An Introduction*; Springer: New York, NY, USA, 2013.
21. Blagojević, M.; Hehl, F.W. (Eds.) *Gauge Theories of Gravitation*; Imperial College Press: London, UK, 2013.
22. Itin, Y.; Obukhov, Y.N.; Boos, J.; Hehl, F.W. Premetric teleparallel theory of gravity and its local and linear constitutive law. *Eur. Phys. J. C* **2018**, *78*, 907. [CrossRef]
23. Cho, Y.M. Einstein Lagrangian as the translational Yang-Mills Lagrangian. *Phys. Rev. D* **1976**, *14*, 2521–2525. [CrossRef]
24. Puetzfeld, D.; Obukhov, Y.N.; Hehl, F.W. Constitutive law of nonlocal gravity. *Phys. Rev. D* **2019**, *99*, 104013. [CrossRef]
25. Mashhoon, B. Nonlocal Gravity: Fundamental Tetrads and Constitutive Relations. *Symmetry* **2022**, *14*, 2116. [CrossRef]
26. Bini, D.; Mashhoon, B. Nonlocal gravity: Conformally flat spacetimes. *Int. J. Geom. Meth. Mod. Phys.* **2016**, *13*, 1650081. [CrossRef]
27. Rahvar, S.; Mashhoon, B. Observational Tests of Nonlocal Gravity: Galaxy Rotation Curves and Clusters of Galaxies. *Phys. Rev. D* **2014**, *89*, 104011. [CrossRef]
28. Chicone, C.; Mashhoon, B. Nonlocal Gravity in the Solar System. *Class. Quantum Gravity* **2016**, *33*, 075005. [CrossRef]
29. Roshan, M.; Mashhoon, B. Dynamical Friction in Nonlocal Gravity. *Astrophys. J.* **2021**, *922*, 9. [CrossRef]
30. Roshan, M.; Mashhoon, B. Characteristics of Effective Dark Matter in Nonlocal Gravity. *Astrophys. J.* **2022**, *934*, 9. [CrossRef]
31. Roshan, M.; Mashhoon, B. Nonlocal Gravity: Modification of Newtonian Gravitational Force in the Solar System. *Universe* **2022**, *8*, 470. [CrossRef]
32. Tabatabaei, J.; Baghram, S.; Mashhoon, B. Local Limit of Nonlocal Gravity: A Teleparallel Extension of General Relativity. *arXiv* **2023**, arXiv:2212.05536.
33. Tabatabaei, J.; Banihashemi, A.; Baghram, S.; Mashhoon, B. Dynamic Dark Energy from the Local Limit of Nonlocal Gravity. *Int. J. Mod. Phys. D* **2023**, *in press*. [CrossRef]

34. Colin, J.; Mohayaee, R.; Rameez, M.; Sarkar, S. Evidence for anisotropy of cosmic acceleration. *Astron. Astrophys.* **2019**, *631*, L13. [CrossRef]
35. Secrest, N.J.; von Hausegger, S.; Rameez, M.; Mohayaee, R.; Sarkar, S.; Colin, J. A Test of the Cosmological Principle with Quasars. *Astrophys. J. Lett.* **2021**, *908*, L51. [CrossRef]
36. Secrest, N.J.; von Hausegger, S.; Rameez, M.; Mohayaee, R.; Sarkar, S. A Challenge to the Standard Cosmological Model. *Astrophys. J. Lett.* **2022**, *937*, L31. [CrossRef]
37. Solanki, R.; Patel, B.; Jaybhaye, L.V.; Sahoo, P.K. Cosmic acceleration with bulk viscosity in an anisotropic $f(R, L_m)$ background. *Commun. Theor. Phys.* **2023**, *75*, 075401. [CrossRef]
38. Perivolaropoulos, L. On the isotropy of SnIa absolute magnitudes in the Pantheon+ and SH0ES samples. *arXiv* **2023**, arXiv:2305.12819.
39. Stephani, H.; Kramer, D.; MacCallum, M.A.H.; Hoenselaers, C.; Herlt, E. *Exact Solutions of Einstein's Field Equations*, 2nd ed.; Cambridge University Press: Cambridge, UK, 2003.
40. Coley, A.A.; van den Hoogen, R.J. Spatially Homogeneous Teleparallel Gravity: Bianchi I. *arXiv* **2023**, arXiv:2305.12168.
41. Heckmann, O.; Schücking, E. World models. In *La Structure et l'Evolution de l'Univers*; Editions Stoops: Brussels, Belgium, 1958; pp. 149–162.
42. Heckmann, O.; Schücking, E. Relativistic Cosmology. In *Gravitation: An Introduction to Current Research*; Witten, L., Ed.; Wiley: New York, NY, USA, 1962; pp. 438–469.
43. Hawking, S.W.; Ellis, G.F.R. *The Large Scale Structure of Space-Time*; Cambridge University Press: Cambridge, UK, 1973.
44. Plebanski, J.; Krasinski, A. *An Introduction to General Relativity and Cosmology*; Cambridge University Press: Cambridge, UK, 2006.
45. Kasner, E. Geometrical Theorems on Einstein's Cosmological Equations. *Am. J. Math.* **1921**, *43*, 217–221. [CrossRef]
46. Chicone, C.; Mashhoon, B.; Rosquist, K. Double-Kasner Spacetime: Peculiar Velocities and Cosmic Jets. *Phys. Rev. D* **2011**, *83*, 124029. [CrossRef]
47. Akrami, Y. et al. [Planck Collaboration]. Planck 2018 results. VII. Isotropy and Statistics of the CMB. *Astron. Astrophys.* **2020**, *641*, A7.
48. Aghanim, N. et al. [Planck Collaboration]. Planck 2018 results. VI. Cosmological parameters. *Astron. Astrophys.* **2020**, *641*, A6. Erratum in *Astron. Astrophys.* **2021**, *652*, C4.
49. Bennett, C.L.; Hill, R.S.; Hinshaw, G.; Larson, D.; Smith, K.M.; Dunkley, J.; Gold, B.; Halpern, M.; Jarosik, N.; Kogut, A.; et al. Seven-Year Wilkinson Microwave Anisotropy Probe (WMAP) Observations: Are There Cosmic Microwave Background Anomalies? *Astrophys. J. Suppl.* **2011**, *192*, 17. [CrossRef]
50. Bennett, C.L.; Larson, D.; Weiland, J.L.; Jarosik, N.; Hinshaw, G.; Odegard, N.; Smith, K.M.; Hill, R.S.; Gold, B.; Halpern, M.; et al. Nine-Year Wilkinson Microwave Anisotropy Probe (WMAP) Observations: Final Maps and Results. *Astrophys. J. Suppl.* **2013**, *208*, 20. [CrossRef]
51. Akarsu, Ö.; Kumar, S.; Sharma, S.; Tedesco, L. Constraints on a Bianchi type I spacetime extension of the standard ΛCDM model. *Phys. Rev. D* **2019**, *100*, 023532. [CrossRef]
52. Campanelli, L.; Cea, P.; Tedesco, L. Ellipsoidal Universe Can Solve The CMB Quadrupole Problem. *Phys. Rev. Lett.* **2006**, *97*, 131302; Erratum in *Phys. Rev. Lett.* **2006**, *97*, 209903. [CrossRef] [PubMed]
53. Campanelli, L.; Cea, P.; Tedesco, L. Cosmic Microwave Background Quadrupole and Ellipsoidal Universe. *Phys. Rev. D* **2007**, *76*, 063007. [CrossRef]
54. Buniy, R.V.; Berera, A.; Kephart, T.W. Asymmetric inflation: Exact solutions. *Phys. Rev. D* **2006**, *73*, 063529. [CrossRef]
55. Sachs, R.K.; Wolfe, A.M. Perturbations of a cosmological model and angular variations of the microwave background. *Astrophys. J.* **1967**, *147*, 73–90. [CrossRef]
56. Barrow, J. Light elements and the isotropy of the Universe. *Mon. Not. R. Astron. Soc.* **1976**, *175*, 359–370. [CrossRef]

Article

Weak Coupling Regime in Dilatonic $f(R,T)$ Cosmology

Francisco A. Brito [1,2,*], Carlos H. A. B. Borges [1,3], José A. V. Campos [2] and Francisco G. Costa [3]

1   Departamento de Física, Universidade Federal da Paraíba, Caixa Postal 5008,
    João Pessoa 58051-970, Paraíba, Brazil; carlos.borges@ifpb.edu.br
2   Departamento de Física, Universidade Federal de Campina Grande, Caixa Postal 10071,
    Campina Grande 58109-970, Paraíba, Brazil; a.campos@uaf.ufcg.edu.br
3   Instituto Federal de Educação Ciência e Tecnologia da Paraíba (IFPB), Campus Campina Grande—Rua
    Tranquilino Coelho Lemos, 671, Jardim Dinamérica I, Campina Grande 58432-300, Paraíba, Brazil;
    francisco.geraldo@ifpb.edu.br
*   Correspondence: fabrito@df.ufcg.edu.br

Abstract: We consider $f(R,T)$ modified theories of gravity in the context of string-theory-inspired dilaton gravity. We deal with a specific model that under certain conditions describes the late time Universe in accord with observational data in modern cosmology and addresses the H_0 tension. This is done by exploring the space of parameters made out of those coming from the modified gravity and dilatonic charge sectors. We employ numerical methods to obtain several important observable quantities.

Keywords: $f(R,T)$ gravity; dilatonic gravity; H_0 tension

check for
updates

Citation: Brito, F.A.; Borges,
C.H.A.B.; Campos, J.A.V.; Costa, F.G.
Weak Coupling Regime in Dilatonic
$f(R,T)$ Cosmology. *Universe* 2024, 10,
134. https://doi.org/10.3390/
universe10030134

Academic Editor: Kazuharu Bamba

Received: 18 December 2023
Revised: 4 March 2024
Accepted: 5 March 2024
Published: 11 March 2024

1. Introduction

Modern observations in cosmology, performed from Type Ia Supernova (SNIa) [1,2], Large Scale Structure (LSS) [3,4], Wilkinson Microwave Anisotropy Probe (WMAP) [5–7] data, Cosmic Microwave Background (CMB) [8,9] and Baryonic Acoustic Oscillations (BAO) [10,11], indicate that the expansion of the Universe has entered an accelerated phase. Furthermore, the same observational data show that 95% of the matter and energy content of the Universe (when described in terms of a fluid effectively entering Einstein's equations) is in the form of unknown species called Dark Matter (DM) and Dark Energy (DE). General Relativity has always been consistent with observational data and one of the most recent examples is the detection of gravitational waves through LIGO [12]. General Relativity analyzes the acceleration of the Universe based on dark energy, according to observations of type Ia supernovae, which counterbalances gravitational attraction. The Dark Energy [13–20] component (ρ_x) is characterized by a negative effective pressure, $p_x < -\rho_x/3$. The simplest candidate for such a dark energy is a positive cosmological constant Λ, but such an identification raises some difficult questions, such as why Λ is so small (in particle physics) [21–23]. This is called the fine-tuning problem for Λ. Another important problem is why $\Lambda = \rho_{x0}$ in present epoch (where ρ_{x0} is the present value of the dark energy density of the Universe, in Planck units). This problem is known as cosmic coincidence [24]. A promising way to solve the above problems is to introduce a single scalar field, dubbed the quintessence [25], whose potential goes asymptotically to zero. The potential associated with this scalar field tends to zero as the field goes to infinity after an infinite (very long) time. On the other hand, we can try to reconcile the observational data with the acceleration of the Universe through modifications of the theory of gravity. One way to do this is to start with a modification of the Einstein–Hilbert action formulation using an arbitrary function of the Ricci scalar, $f(R)$ [26]. In this context, we assume that at large scales the Einstein gravity model breaks down. Some specific types of $f(R)$ models have been proposed in the literature (see [27–29] and related references for a recent review). These theories acquired a lot of interest following the work by Starobinsky on cosmic inflation [30]. The late-time cosmic acceleration of the Universe, in this context, was first explained by a natural modification, adding terms to the action that are proportional to R^n [31,32]. Quintessence issues by taking into account generic models with actions containing $f(R)$ terms were addressed in Ref. [33]. Quantum effects may lead to a generalization of $f(R)$ theories of gravity [34–36]. In Ref. [37], an unusual coupling between matter and geometry was developed by Harko et al., where the gravitational sector is given by an arbitrary function of the Ricci scalar and the trace T of the energy-momentum tensor. This type of theory is known as $f(R, T)$ gravity. Cosmological and astrophysical consequences of $f(R, T)$ models have been explored in the last few years. For instance, in Ref. [38], cosmological solutions for a perfect fluid in a spatially flat Friedmann–Lemàitre–Robertson–Walker (FLRW) metric is investigated. Other studies on cosmological applications of $f(R, T)$ gravity, including the inflationary scenario, can be found in the Refs. [39–45]. An essential component of all superstring models, and consequently, of the cosmological scenarios based on the string effective action, is the dilaton field. This field controls the effective strength of all gauge couplings in the context of "grand-unified" models of all fundamental interactions [46]. This coupling strengths may drive the Universe towards a phase of strong coupling, which can possibly precede the standard decelerated evolution. The dilaton can also be geometrically interpreted as the effective "radius" of the eleventh dimension [47] in the M-theory context. The dilaton may control the inflationary dynamics and play a role in the generation of the primordial spectra of quantum fluctuations amplified by inflation. String theory dilaton may provide a natural implementation of the coupled quintessence scenario, provided the cosmological running of the dilaton does not stop after entering the weak coupling regime $e^\phi \ll 1$, as $\phi \to -\infty$. We shall consider a particular scenario in which the dilaton approaches to zero as $t \to \infty$. This is possible with an exponentially suppressed (non-perturbative) potential. As we shall see, because of the loop corrections, the fields inside the matter action are in general non-minimally and non-universally coupled to the dilaton. This will render dilatonic charge densities that are fundamental to form the new space of parameters of the model.

In Section 2, we will present the formalism of a theory of gravity modified by $f(R, T)$, addressing the variation of the modified action, while in Section 3, we will review a scenario of string theory at low energy $\phi \to -\infty$ with a dilatonic field subject to an effective potential $\tilde{V}$. In Section 4, we will analyze a dilatonic cosmological theory in a scenario of modified gravity of type $f(R, T)$ for a homogeneous and isotropic FLRW universe. Section 5 is dedicated to detail the evolution of cosmological relevant quantities through numerical methods. Our discussions and conclusions are, respectively, present in Sections 6 and 7.

2. Gravitational Field Equations of $f(R, T)$ Gravity

Let us first take the action given in [37]:

$$S = \frac{1}{2\kappa} \int d^4x f(R, T) \sqrt{-g} + \int d^4x \mathcal{L}_m \sqrt{-g}, \tag{1}$$

where $f(R, T)$ is an arbitrary function of the Ricci scalar curvature $R = g^{\mu\nu} R_{\mu\nu}$, $T = g^{\mu\nu} T_{\mu\nu}$ is the trace of the energy-momentum tensor and $\mathcal{L}_m$ is the Lagrangian density of matter and $\kappa = 8\pi G$.

Admitting the definition of the energy-momentum tensor, we can express it in such a way that the Lagrangian density of matter depends only on $g_{\mu\nu}$, that is:

$$T_{\mu\nu} = g_{\mu\nu} \mathcal{L}_m - 2 \frac{\partial \mathcal{L}_m}{\partial g^{\mu\nu}}. \tag{2}$$

By varying the action given in Equation (1) in relation to $g^{\mu\nu}$, we have the field equations given by:

$$
\begin{aligned}
f_R(R,T)R_{\mu\nu} \; &+\; g_{\mu\nu}\Box f_R(R,T) - \nabla_\mu\nabla_\nu f_R(R,T) \\
&+\; f_T(R,T)(T_{\mu\nu} + \Theta_{\mu\nu}) - \frac{1}{2}f(R,T)g_{\mu\nu} - 8\pi T_{\mu\nu} = 0,
\end{aligned}
\tag{3}
$$

so that $(T_{\mu\nu} + \Theta_{\mu\nu})$ corresponds to the variation of the trace with respect to the metric tensor, with $\Theta_{\mu\nu} \equiv g^{\alpha\beta}\dfrac{\delta T_{\alpha\beta}}{\delta g^{\mu\nu}}$ set in [37]. We will denote $f_R(R,T)$ and $f_T(R,T)$ as the derivatives of $f(R,T)$ with respect to the Ricci scalar curvature and the trace of the energy-momentum tensor, respectively.

From the definition of $\Theta_{\mu\nu}$ and using Equation (2), we have:

$$
\Theta_{\mu\nu} = -2T_{\mu\nu} + g_{\mu\nu}\mathcal{L}_m - 2g^{\alpha\beta}\frac{\partial^2 \mathcal{L}_m}{\partial g^{\mu\nu}\partial g^{\alpha\beta}}.
\tag{4}
$$

In other words, $\Theta_{\mu\nu}$ will depend on the Lagrangian of matter, which can refer to the case of the electromagnetic field, the massless scalar field and the case of the perfect fluid, among others.

3. Stringy Cosmology

Let us now consider cosmological scenarios related to the effective action that comes from low-energy string theory in which the dilaton field exerts influence in the dynamics of the Universe. We shall focus on the sector of the effective action, coming from a low-energy string theory, given by the tensor field (the metric $\tilde{g}_{\mu\nu}$) and a scalar field (the dilaton ϕ), where the tilde indicates that we are working on the *string frame*, and where the dilaton couples to the Ricci scalar and dilatonic dynamics in an explicit form—see below. Our starting point is the string-frame, low-energy, gravidilaton effective action, to the lowest order in the α' expansion, but including the dilaton-dependent loop and nonperturbative corrections, encoded in a few "form factors", due to the loop corrections $\psi(\phi)$ and $Z(\phi)$ [48]. $V(\phi)$ is the effective dilaton potential. The model action is [49,50]:

$$
S = -\frac{M_P^2}{2}\int d^4x\sqrt{-\tilde{g}}\left[e^{-\psi(\phi)}\tilde{R} + \tilde{Z}(\phi)\left(\tilde{\nabla}\phi\right)^2 + \frac{2}{M_P^2}\tilde{V}(\phi)\right] + \tilde{S}_m(\phi,\tilde{g},\text{matter}).
\tag{5}
$$

We can discuss the phenomenology of the relic dilaton background by taking into account two possibilities. First, massive dilaton is gravitationally more *strongly coupled* to macroscopic matter. In strong coupling limit $\phi \to \infty$, we assume that it is possible to make an asymptotic Taylor expansion in inverse powers of the coupling constant $g_s^2 = \exp(\phi)$, similar to the context of "induced gravity". In these models, the gravitational and gauge couplings saturate at small, but finite, values because of the very large number N of fundamental gauge bosons presents in the loop corrections [49,50]. By this assumption, we can write $\exp(-\psi(\phi)) = c_1^2 + b_1\exp(-\phi) + \mathcal{O}(\exp(-2\phi))$, $Z(\phi) = -c_2^2 + b_2\exp(-\phi) + \mathcal{O}(\exp(-2\phi))$ and $\tilde{V} = V_0\exp(-\phi) + \mathcal{O}(\exp(-2\phi))$, where c_1^2 and c_2^2 are dimensionless numbers. To be consistent with the tree-level relation $\lambda_P/\lambda_S = M_S/M_P = \exp(\phi/2)$ (for $d = 3$), in which $M_P \simeq 10M_S$, as required by a consistent string unification in the context of gravitational and gauge interactions [51], and we have $c_1^2 \sim c_2^2 \sim 10^2$. On the other hand, very light (or massless) dilaton is weakly coupled to matter. In the rest of this paper we will focus our attention on this second possibility, where we will consider that the dilaton is weakly coupled. In this regime, we will admit that $\phi \to -\infty$, while $\exp(-\psi(\phi)) = Z(\phi) = \exp(-\phi)$.

We can now characterize the dynamical evolution of the Universe with a metric minimally coupled to the dilaton. In this frame, the string effective action is also minimally coupled to perfect fluid sources. Considering the lowest order α', we have [49,50]:

$$S = -\frac{M_P^2}{2} \int d^4x \sqrt{-\tilde{g}}\, e^{-\phi} \left[\tilde{R} - (\tilde{\nabla}\phi)^2 + \frac{2}{M_P^2} \tilde{V}(\phi) \right] + \tilde{S}_m(\phi, \tilde{g}, \text{matter}). \tag{6}$$

We can use a more convenient coordinate system, the so-called *Einstein frame*, in terms of the metric $g_{\mu\nu}$ that is defined by a conformal transformation $\tilde{g}_{\mu\nu} = e^{\phi} g_{\mu\nu}$. In this frame, the action can be written as:

$$S = -\frac{M_P^2}{2} \int d^4x \sqrt{-g} \left[R - \frac{1}{2}(\nabla\phi)^2 + \frac{2}{M_P^2} \hat{V}(\phi) \right] + S_m(\phi, e^{\phi} g_{\mu\nu}, \text{matter}), \tag{7}$$

where we have defined:

$$\hat{V} = e^{\phi} \tilde{V}. \tag{8}$$

Because of the loop corrections, the fields appearing in the action S_m are generally non-minimally and non-universally coupled to the dilaton [52]. The gravitational and dilatonic "charge densities", $T_{\mu\nu}$ and σ, are defined as:

$$\frac{\delta S_m}{\delta g^{\mu\nu}} = \frac{1}{2} \sqrt{-g}\, T_{\mu\nu}, \qquad \frac{\delta S_m}{\delta \phi} = \frac{1}{2} \sqrt{-g}\, \sigma. \tag{9}$$

When $\sigma \neq 0$, the effective gravidilaton theory is very different from a typical scalar-tensor gravity model of the Jordan–Brans–Dicke type.

4. Cosmology in Dilatonic $f(R, T)$ Gravity

Let us now assume that the action in Equation (6) depends not only on R, but on a function $f(R, T)$, so that R is the Ricci scalar curvature and T is the trace of the energy-momentum tensor of the dilatonic field. Since we are interested in investigating the dynamics of the dilaton field in the presence of other sources, we can rewrite this gravidilaton action as:

$$S = -\frac{M_P^2}{2} \int d^4x \sqrt{-g}\, f(R, T) + \frac{M_P^2}{2} \int d^4x \sqrt{-g} \left[\frac{1}{2} g^{\mu\nu} \nabla_\mu \phi \nabla_\nu \phi - \frac{2}{M_P^2} \hat{V}(\phi) \right] + S_m(\phi, e^{\phi} g_{\mu\nu}, \text{matter}). \tag{10}$$

Let us now assume a homogeneous and isotropic Universe described by the Friedmann–Lemaître–Robertson–Walker (FLRW) metric, whose line element is written as:

$$ds^2 = dt^2 - a^2(t) \left(dr^2 + r^2 d\theta^2 + r^2 \sin^2\theta\, d\phi^2 \right). \tag{11}$$

In what follows, we will also consider $M_P^2 \equiv 1/8\pi G = 2$, except otherwise indicated. For this model, the dilatonic Lagrangian density $\mathcal{L}^\phi$ is given by:

$$\mathcal{L}^\phi = \frac{1}{2} g^{\mu\nu} \nabla_\mu \phi \nabla_\nu \phi - \hat{V}(\phi), \tag{12}$$

where $\hat{V}(\phi) = e^{\phi} \tilde{V}(\phi)$ is the effective dilaton potential as previously defined. The scalar dynamics of the model is governed by the energy-momentum tensor of the dilaton field and is given by:

$$T_{\mu\nu}^\phi = \nabla_\mu \phi \nabla_\nu \phi - g_{\mu\nu} \left[\frac{1}{2} g^{\alpha\beta} \nabla_\alpha \phi \nabla_\beta \phi - \hat{V}(\phi) \right], \tag{13}$$

from which we can write:

$$T_{00}^\phi = \frac{1}{2} \dot{\phi}^2 + \hat{V}(\phi), \tag{14}$$

$$T_{ii}^{\phi} = \frac{a^2}{2}\dot{\phi}^2 - a^2\hat{V}(\phi). \tag{15}$$

By tracing the energy-momentum tensor of the dilatonic field, we are left with:

$$T^{\phi} = -\dot{\phi}^2 + 4\hat{V}(\phi). \tag{16}$$

The dilaton equation of motion is given by:

$$\ddot{\phi} + 3H\dot{\phi} + \frac{d\hat{V}}{d\phi} + \frac{1}{2}\sigma = 0. \tag{17}$$

Following (9), σ is the charge associated with the coupling between dilatonic field and the fluid that makes up the Universe. In the usual way, $H = \dot{a}/a$ is the Hubble parameter and a dot denotes differentiation with respect to the Einstein cosmic time. For this model:

$$\rho_{\phi} = \frac{1}{2}\dot{\phi}^2 + \hat{V}(\phi), \qquad p_{\phi} = \frac{1}{2}\dot{\phi}^2 - \hat{V}(\phi), \tag{18}$$

are the energy density and pressure of the dilaton field.

Varying Equation (10) with respect to $g_{\mu\nu}$, we will obtain:

$$f_R(R,T)R_{\mu\nu} - \frac{1}{2}f(R,T)g_{\mu\nu} + (g_{\mu\nu}\Box - \nabla_\mu\nabla_\nu)f_R(R,T) = \frac{1}{2}T_{\mu\nu} - f_T(R,T)T_{\mu\nu} - f_T(R,T)\Theta_{\mu\nu}, \tag{19}$$

where $T_{\mu\nu} = T^f + T^{\phi}$ is total energy-momentum tensor of the perfect fluid (f) that fills the Universe (baryonic matter, radiation and dark matter) plus the dilaton field. Using Equation (4), we can write:

$$\Theta_{\mu\nu} = -2\left(T_{\mu\nu}^{\phi} + T_{\mu\nu}^{f}\right) + g_{\mu\nu}\left(\mathcal{L}^{\phi} + \mathcal{L}^{f}\right), \tag{20}$$

with $\mathcal{L}^{\phi}$ and $T_{\mu\nu}^{\phi}$ given by (12) and (18), respectively. We assume that the energy-momentum tensor of the matter and energy is given by $T_{\mu\nu} = (\rho + p)u_\mu u_{\mu\nu} - pg_{\mu\nu}$, with conditions $u_\mu u^\mu = 1$ and $u^\mu \nabla_\nu u_\mu = 0$ satisfied by the four-velocity u^μ. Therefore, the Lagrangian of the perfect fluid becomes $\mathcal{L}^f = -p$. In this way, we can write the $(0-0)$ and $(i-i)$ components of $\Theta_{\mu\nu}$ as:

$$\Theta_{00} = -\frac{1}{2}\dot{\phi}^2 - 3\hat{V} - 2\rho - p, \quad \Theta_{ii} = -\frac{3}{2}a^2\dot{\phi}^2 + 3a^2\hat{V} - a^2p. \tag{21}$$

In the following, we will consider a simple example of $f(R,T)$ theories in order to study the late time evolution of the Universe with the presence of a single dilatonic field.

4.1. Model $f(R,T) = R + \alpha T^{\phi}$

This model was first studied by Harko et al. in Ref. [37]. These authors reproduce, in the context of $f(R,T)$, a relativistically covariant model of interacting dark energy based on a specific action [53]. Another interesting aspect of this choice is that the gravitational coupling becomes an "effective time dependent coupling", depending on the derivative of $f(T)$ with respect to the trace [37]. In order to reconcile our model with the the evidences which support General Relativity, we will assume a modified gravity model by $f(R,T) = R + \alpha T$. Using the expression (19), we have:

$$R_{\mu\nu} - \frac{1}{2}g_{\mu\nu}R = \left(\frac{1}{2} - \alpha\right)T_{\mu\nu} - \alpha\Theta_{\mu\nu} + \frac{1}{2}g_{\mu\nu}\alpha T. \tag{22}$$

Thus, using the definitions (14)–(16) and (20) in (22), one can obtain $(0,0)$ and $(1,1)$ components of field equations in a FLRW Universe as being:

$$3H^2 = \left(\frac{1}{2} - \alpha\right)\frac{1}{2}\dot{\phi}^2 + \left(\frac{1}{2} + 4\alpha\right)\hat{V}(\phi) + \frac{1}{2}(1 + 3\alpha)\rho - \frac{1}{2}\alpha p, \tag{23}$$

$$2\dot{H} + 3H^2 = -\left(\frac{1}{2} + 3\alpha\right)\frac{1}{2}\dot{\phi}^2 + \left(\frac{1}{2} + 4\alpha\right)\hat{V}(\phi) - \frac{1}{2}(1 + 3\alpha)p + \frac{1}{2}\alpha\rho. \tag{24}$$

The combination of (23), (24) and (17) leads to the coupled conservation equations for the matter (baryonic and dark), radiation and dilaton energy density, respectively:

$$(1 + 3\alpha)\dot{\rho}_b + 3H(1 + 2\alpha)\rho_b + \alpha\left(12H\dot{\phi}^2 + 10\frac{d\hat{V}}{d\phi}\dot{\phi}\right) = 0, \tag{25}$$

$$(1 + 3\alpha)\dot{\rho}_d + \left[3(1 + 2\alpha)H - \frac{1}{2}(1 - 2\alpha)Q\dot{\phi}\right]\rho_d + \alpha\left(12H\dot{\phi}^2 + 10\frac{d\hat{V}}{d\phi}\dot{\phi}\right) = 0, \tag{26}$$

$$\left(1 + \frac{8}{3}\alpha\right)\dot{\rho}_r + 4(1 + 2\alpha)H\rho_r + \alpha\left(12H\dot{\phi}^2 + 10\frac{d\hat{V}}{d\phi}\dot{\phi}\right) = 0, \tag{27}$$

$$\dot{\rho}_\phi + 6H[\rho_\phi - (1 + 8\alpha)\hat{V}] - 10\alpha\frac{d\hat{V}}{d\phi}\dot{\phi} + \frac{1}{2}(1 - 2\alpha)\sigma\dot{\phi} = 0. \tag{28}$$

In the set of equations shown above, we have separate the radiation, baryonic and non-baryonic (dark) matter components of the cosmological fluid by setting:

$$\rho = \rho_m + \rho_r = \rho_b + \rho_d + \rho_r, \quad p_b = p_d = 0, \quad p_r = \frac{1}{3}\rho_r. \tag{29}$$

We assume that ordinary matter and radiation have nearly metric couplings, i.e., that σ_b and σ_r vanish as $\phi \to -\infty$. This agrees, for instance, with the precision tests of Newtonian gravity [54]. In the dark matter sector, we shall assume a specific model to Lagrangian density. For "cold dark matter", one has a dilatonic charge σ_d which gives us the relationship [49,50,55]:

$$Q(\phi) = \frac{\sigma_d}{\rho_d} = Q_0\frac{e^{Q_0\phi}}{c^2 + e^{Q_0\phi}}. \tag{30}$$

For large enough values of the constant c^2, $Q(\phi)$ approaches to finite (non-zero) values in the limit of weak coupling regime, that is, when $\phi \to -\infty$.

Finally, we shall specify the form of the effective (Einstein frame) dilaton potential by choosing the string frame potential as $\tilde{V}(\phi) = V_0$. This allows us to write, quite generically, the simplest potential:

$$\hat{V}(\phi) = V_0 e^\phi, \tag{31}$$

where V_0 is a constant. This potential is in agreement with the assumption of exponential suppression at weak coupling regime.

It is convenient to parameterize the temporal evolution of all variables in terms of the logarithm of the scale factor, $\chi = \ln(a/a_i)$. In this relationship, a_i corresponds to the initial scale factor. Thus, the Einstein Equation (23) and the dilaton Equation (17) can be written, respectively, as:

$$H^2 = \frac{(1 + 3\alpha)(\rho_b + \rho_d) + (1 + \frac{8}{3}\alpha)\rho_r + (1 + 8\alpha)\hat{V}}{6 - \left(\frac{1}{2} - \alpha\right)\left(\frac{d\phi}{d\chi}\right)^2}, \tag{32}$$

$$2H^2\frac{d^2\phi}{d\chi^2} + \left[(1 + 4\alpha)\left(\frac{1}{2}\rho_b + \frac{1}{2}\rho_d + \frac{1}{3}\rho_r\right) + (1 + 8\alpha)\hat{V} - 2\alpha H^2\left(\frac{d\phi}{d\chi}\right)^2\right]\frac{d\phi}{d\chi} + 2\frac{d\hat{V}}{d\phi} + Q\rho_d = 0. \tag{33}$$

The matter and radiation evolution equations become:

$$(1 + 3\alpha)\frac{d\rho_b}{d\chi} + 3(1 + 2\alpha)\rho_b + \alpha\left[12H^2\left(\frac{d\phi}{d\chi}\right)^2 + 10\frac{d\hat{V}}{d\phi}\frac{d\phi}{d\chi}\right] = 0,\tag{34}$$

$$(1 + 3\alpha)\frac{d\rho_d}{d\chi} + 3(1 + 2\alpha)\rho_d - \frac{1}{2}(1 - 2\alpha)Q\rho_d\frac{d\phi}{d\chi} + \alpha\left[12H^2\left(\frac{d\phi}{d\chi}\right)^2 + 10\frac{d\hat{V}}{d\phi}\frac{d\phi}{d\chi}\right] = 0,\tag{35}$$

$$\left(1 + \frac{8}{3}\alpha\right)\frac{d\rho_r}{d\chi} + 4(1 + 2\alpha)\rho_r + \alpha\left[12H^2\left(\frac{d\phi}{d\chi}\right)^2 + 10\frac{d\hat{V}}{d\phi}\frac{d\phi}{d\chi}\right] = 0.\tag{36}$$

Finally, the dilaton conservation equation can be written as:

$$\frac{d\rho_\phi}{d\chi} + 6[\rho_\phi - (1 + 8\alpha)\hat{V}] - 10\alpha\frac{d\hat{V}}{d\phi}\frac{d\phi}{d\chi} + \frac{1}{2}(1 - 2\alpha)\sigma\frac{d\phi}{d\chi} = 0,\tag{37}$$

which is equivalent to Equation (33). Finally, we defined the useful density parameter $\Omega = \rho/\rho_c$, where ρ_c is the critical energy density. Thus, Equation (32) can be rewritten in terms of different density parameters for each component of the Universe:

$$1 = \Omega_r + \Omega_d + \Omega_b + \Omega_\phi,\tag{38}$$

where:

$$\Omega_r = \frac{\rho_r}{3H^2}, \qquad \Omega_d = \frac{\rho_d}{3H^2}, \qquad \Omega_b = \frac{\rho_b}{3H^2}, \qquad \Omega_\phi = \frac{\rho_\phi}{3H^2},\tag{39}$$

are the density parameters of radiation, dark matter, baryonic matter and dilaton field, respectively.

4.2. Numerics

We can solve the set of Equations (34)–(37) by numerical methods with Equation (32) as a constraint on the initial dataset. Here, we also recover the $8\pi G$ factor, which can be given in terms of the Planck mass $M_p \equiv 1/\sqrt{G} = 1.22 \times 10^{19}$ GeV. We shall use a potential of the form (31), by assuming $V_0 = 2.65 \times 10^{-123} M_p^4$, as well as the $Q(\phi)$ Formula (30) with $c^2 = 10$ for different values of the charge Q_0.

As we shall see shortly, (α, Q_0) is the space of parameters that we will explore to address acceptable cosmological scenarios. To ensure that at the current time $\chi = 0$, the values of the energy densities are in agreement with the data from current cosmological measurements, we adjust the initial values of the energy densities of radiation, dark matter, baryonic matter and dark energy, respectively, as $\rho_{ri}(\chi_i) = 8.58 \times 10^{-93} M_p^4$, $\rho_{di}(\chi_i) = 5.28 \times 10^{-98} M_p^4$, $\rho_{bi}(\chi_i) = 9.30 \times 10^{-99} M_p^4$ and $\rho_{\phi i}(\chi_i) = 1.67 \times 10^{-105} M_p^4$, starting the integration at $\chi_i = -20$—this corresponds to big bang nucleosynthesis (BBN) redshift $z_{\mathrm{BBN}} \sim 10^9$. Furthermore, the initial value for the dilaton is $\phi_i = 7 \times 10^{-8} M_p$.

5. Results

5.1. The Density Parameters Ω_r, Ω_d, Ω_b and Ω_ϕ

The density parameters (39) are depicted in Figures 1 and 2. The graph in Figure 1 expresses the behavior of the density parameter Ω of each component as a function of χ, by fixing $\alpha = 2 \times 10^{-2} m_p^{-2}$ and adjusting scenarios with $Q_0 = 2$ and $Q_0 = 20$. Considering the radiation component Ω_r, we note that it presents a degenerate evolution that starts constantly between the intervals $\chi \approx [-20, -13]$, decays quickly between $\chi \approx [-15, -4]$ and becomes null outside both intervals. In other words, Q_0 does not interfere with the evolution of Ω_r. Regarding the dark matter component Ω_d, we have a degenerate behavior in the interval $\chi \approx [-20, -2]$ and a minimal influence of Q_0 on $\chi \approx [-2, 1]$, in order to present a slightly more accentuated decay with the decrease of Q_0. In this curve, the evolution begins with $\Omega_d = 0$ in the intervals $\chi \approx [-20, -13]$ and a sharp increase in $\chi \approx [-13, -2]$ that precede the decay; there is also a degenerate maximum point with $\Omega_d(\chi = -2) \approx 0.9$. For the baryonic matter component Ω_b, we have an evolution with a degeneracy in the interval $\chi \approx [-20, -1]$ and a sufficiently small influence of Q_0 outside this interval. The behavior of Ω_b starts with a null value at $\chi \approx [-20, -13]$ followed by a gentle increase, remaining constant at its maximum point with $\Omega_b = 0.15$ at $\chi \approx [-5, -2]$ and ends with a decay outside these ranges. Regarding the dilaton component Ω_ϕ, we have a beginning with $\Omega_\phi = 0$ that is independent of Q_0 in the interval $\chi \approx [-20, -2]$ and a significant elevation that grows smoothly with the increase in Q_0. Relating the components Ω_r, Ω_d and Ω_b, we note that at $\chi \approx -13$, the first decreases while the second and third increase. On the other hand, at $\chi \approx -2$, Ω_r remains null and Ω_d and Ω_b decrease while Ω_ϕ grows significantly, thus presenting an intersection with Ω_d at $\chi \approx -0.28$ and $\chi \approx -0.41$ for $Q_0 = 2$ and $Q_0 = 20$, respectively. We also have another point of intersection relating the components Ω_d and Ω_r given in $\chi \approx -8.01$ and $\chi \approx -7.97$ for $Q_0 = 2$ and $Q_0 = 20$, respectively.

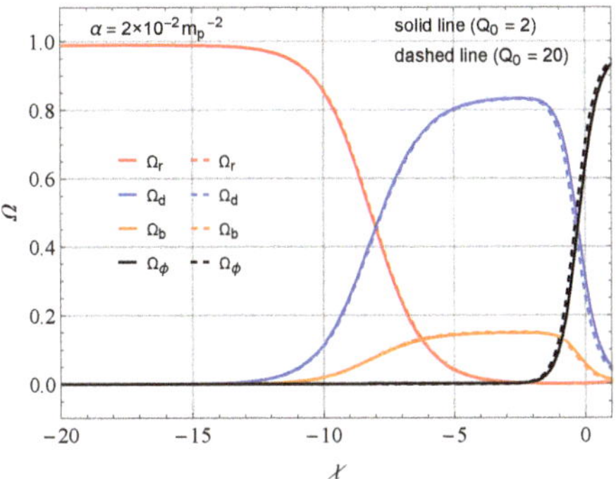

Figure 1. The density Ω as a function of χ for fixed $\alpha = 2 \times 10^{-2} m_p^{-2}$ and $Q_0 = 2$ and for $Q_0 = 20$.

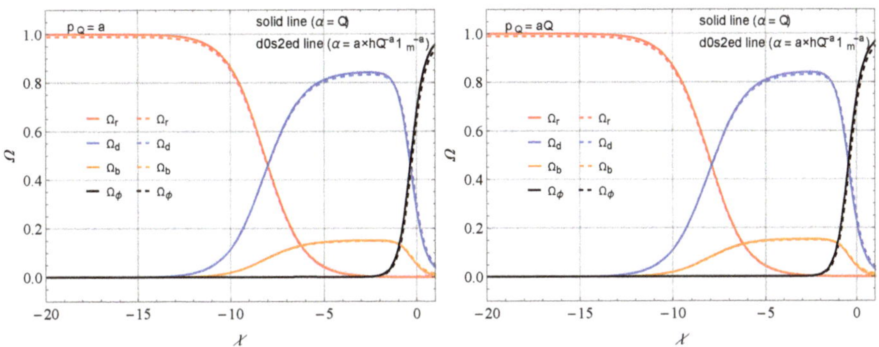

Figure 2. (**left panel**) The density Ω as a function of χ for fixed $Q_0 = 2$ and (**right panel**) $Q_0 = 20$ for $\alpha = 0$ and $\alpha = 2 \times 10^{-2} m_p^{-2}$.

The graphs in Figure 2 (left panel) and Figure 2 (right panel) are similar to each other; however, in Figure 2 (right panel), we set $Q_0 = 20$ and adjust α for two values as in Figure 2 (left panel). Comparing these graphs with Figure 1, we can observe that the increase in Q_0 has minimal influence on the radiation Ω_r, dark matter Ω_d and baryonic Ω_b components. However, for the dilaton component Ω_ϕ, we note that there is a more pronounced rise in the respective evolution curve.

5.2. The Hubble Parameter $H(z)$

The graph in Figure 3 expresses the behavior of the Hubble parameter as a function of redshift z, where we fixed $Q_0 = 2$ and explored different values for α. In this case, we can observe that in all values admitted for α we have a decay followed by an increase. Note that the minimum point of each curve has a redshift z that decreases with the increase in α, that is, for $\alpha = 0$, $\alpha = 5 \times 10^{-3} m_p^{-2}$, $\alpha = 1 \times 10^{-2} m_p^{-2}$ and $\alpha = 2 \times 10^{-2} m_p^{-2}$, we have $z \approx 0.64$, $z \approx 0.61$, $z \approx 0.59$ and $z \approx 0.54$, respectively. On the other hand, the values of the Hubble parameter at $z = 0$, i.e., H_0 increases with the growth of α, assuming values between $H_0 \approx 67$ km s^{-1} Mpc^{-1} and $H_0 \approx 72$ km s^{-1} Mpc^{-1}.

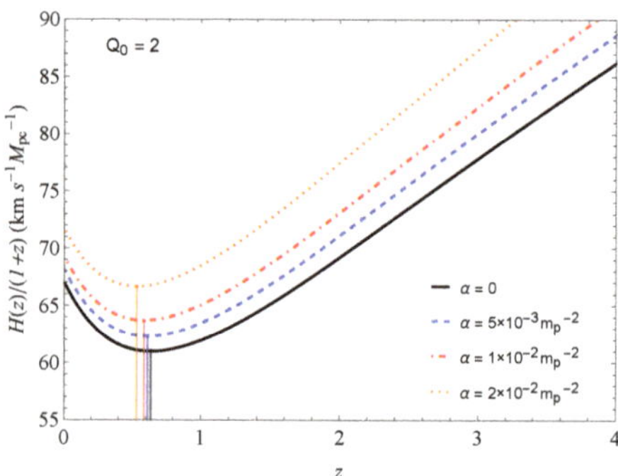

Figure 3. The Hubble parameter $H(z)$ as a function of the redshift z for fixed $Q_0 = 2$ and several values of α.

The graphs in Figure 4 show opposite behavior of the graph shown in Figure 3 with respect to the parameters, where they have interchanged their roles. More specifically, Figure 4 (left panel) shows the behavior of the Hubble parameter as a function of redshift z for different values of Q_0, for $\alpha = 0$ fixed. Notice that in all curves, we have a decay of H followed by an increase in which there is a degeneracy starting at $z \approx 2.6$. For $Q_0 = 2$, $Q_0 = 6$, $Q_0 = 10$ and $Q_0 = 20$, we have a minimum point at $z \approx 0.64$, $z \approx 0.69$, $z \approx 0.76$ and $z \approx 0.85$, respectively. We also note that the values of z, referring to each point of minimum, increase with the increase of Q_0. It is also worth noting that the decay of H to the minimum point becomes more pronounced with the increase of Q_0; additionally, for smaller values of Q_0, we have slower decays. In this case, the values of the Hubble parameter at $z = 0$, i.e., H_0 assume values between $H_0 \approx 67$ km s^{-1} Mpc^{-1} and $H_0 \approx 74$ km s^{-1} Mpc^{-1}. Finally, with respect to the graph in Figure 4 (right panel), we find that it is similar to Figure 4 (left panel), but with $\alpha = 5 \times 10^{-3} m_p^{-2}$ fixed. In this case, we again have a decay followed by an increase that becomes degenerate at $z \approx 2.4$. The curves assume minimum points whose redshift values are $z \approx 0.61$, $z \approx 0.67$, $z \approx 0.74$ and $z \approx 0.82$ corresponding to $Q_0 = 2$, $Q_0 = 6$, $Q_0 = 10$ and $Q_0 = 20$, respectively. We can see that H increases with the increase in α, and consequently, its minimum points take on greater values for H; however, the corresponding redshift values z decrease. Thus, in this case, the values of the Hubble parameter at $z = 0$, i.e., H_0 assume values between $H_0 \approx 68$ km s^{-1} Mpc^{-1} and $H_0 \approx 75.5$ km s^{-1} Mpc^{-1}.

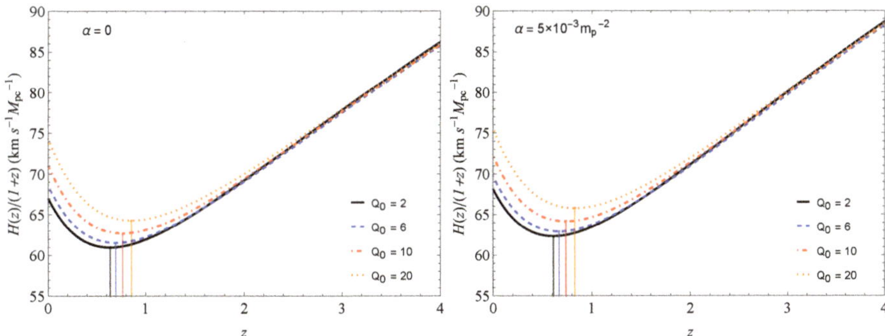

Figure 4. (**left panel**) The Hubble parameter $H(z)$ as a function of the redshift z for fixed $\alpha = 0$ and (**right panel**) $\alpha = 5 \times 10^{-3} m_p^{-2}$ and several values of Q_0.

5.3. The Dilaton Field

The graph in Figure 5 details the behavior of ϕ as a function of χ assuming $\alpha = 2 \times 10^{-2} m_p^{-2}$ being fixed and exploring different values for Q_0. In this case, there is a convergence of $\phi(\chi) = 0$ for each Q_0 in the interval $\chi \approx [-20, -13]$, followed by a decay that ends in $\phi(\chi \approx 1) = -10$. Notice that, in the range $\chi \approx [-10, -1]$, $\phi(\chi)$ presents a decay that becomes more pronounced with the decrease of Q_0, that is, the growth of Q_0 reflects a slower decay to $\phi(\chi)$. This can be observed more clearly by analyzing the decay in which $Q_0 = 20$, where $\phi(\chi \approx -7.5) \approx -1.5$ and $\phi(\chi \approx -2) \approx -2$, while for $Q_0 = 6$, we have $\phi(\chi \approx -8) \approx -1.5$ and $\phi(\chi \approx -7.3) \approx -2$. We also observe that from $\chi \approx -1$, we have an expressive decay that is independent of Q_0.

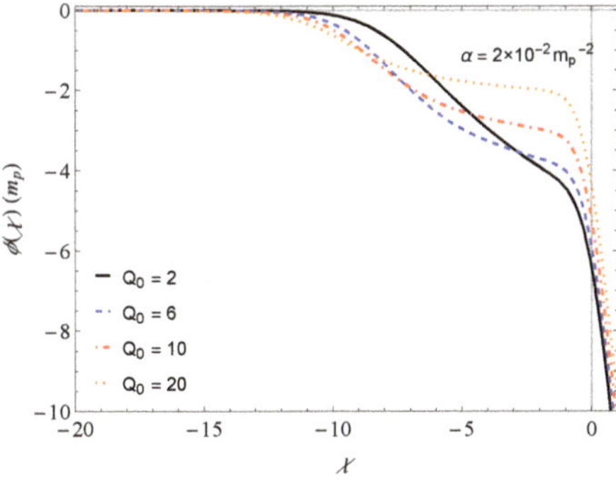

Figure 5. The dilaton field ϕ as a function of χ for fixed $\alpha = 2 \times 10^{-2} m_p^{-2}$ and several values of Q_0.

The graphs in Figure 6 are quite similar. In the first case (left panel), we fix $Q_0 = 2$ and explore three values for α. As we can see, the variation of α presents a sufficiently small change in the behavior of $\phi(\chi)$ that decays in a degenerate way in the interval $\chi \approx [-20, -1]$, while for the range $\chi \approx [-1, 1]$, we have a decay that becomes smoothly slower with increasing α. In the second case (right panel), we fix $Q_0 = 20$ and explore three values for α. Furthermore, we note that the increase in Q_0 contributes to a slower decay in the interval $\chi \approx [-20, -1]$.

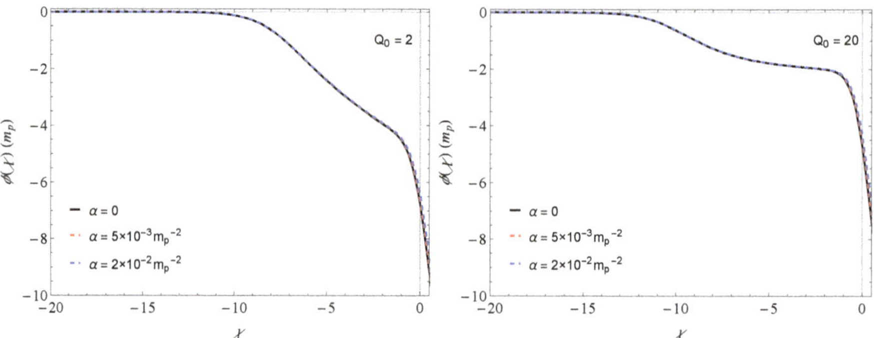

Figure 6. (**left panel**) The dilaton field ϕ as a function of χ for fixed $Q_0 = 2$ and (**right panel**) $Q_0 = 20$ and several values of α.

5.4. The Running of $Q(\phi)$

The graphic in Figure 7 represents the behavior of $Q(\phi)$ as a function of χ by considering different values for Q_0 and fixing $\alpha = 2 \times 10^{-2} m_p^{-2}$. Notice that for $Q_0 = 2$, we have a very slow decay in the interval $\chi \approx [-20, 0]$; however, it is more pronounced outside this range. Concerning the other curves, we have a decay that becomes more accentuated with the increase of Q_0 admitting $\chi \approx [-20, -1]$. Outside this range, we have an even more significant decay.

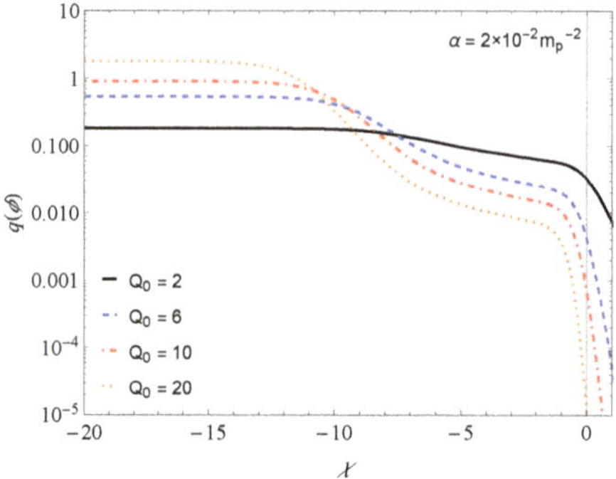

Figure 7. The evolution of $Q(\phi)$ as a function of χ for fixed $\alpha = 2 \times 10^{-2} m_p^{-2}$ and several values of Q_0.

The graphs in Figure 8 develops similar behavior. In the first case (left panel), we keep $Q_0 = 2$ fixed and explore some values of α. We note that $Q(\phi)$ decays degenerately in the interval $\chi \approx [-20, -1]$, that is, without any influence of α. Outside this range, $Q(\phi)$ decays more slowly with increasing α. In the second case (right panel), we keep $Q_0 = 20$ fixed and also explore some values of α and we have a degenerate decay that becomes slower with the increase of Q_0 and without the influence of α, in the range $\chi \approx [-20, -1]$; however, outside this range, we have a sharp decay that increases smoothly with the decrease in α.

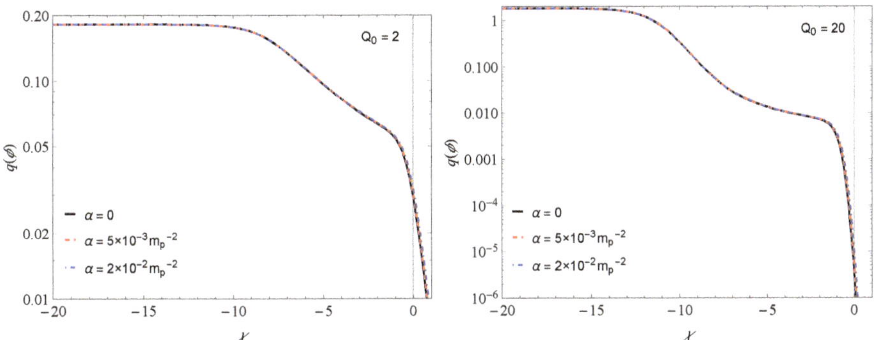

Figure 8. (**left panel**) The evolution of $Q(\phi)$ as a function of χ for fixed $Q_0 = 2$ and (**right panel**) $Q_0 = 20$ and several values of α.

5.5. The Energy Densities ρ_r, ρ_d, ρ_b and ρ_ϕ

The graphs in Figure 9 show the evolution of energy density as a function of χ for each component. In the first case (left panel), we fix $Q_0 = 2$ and consider $\alpha = 0$ and $\alpha = 2 \times 10^{-2} m_p^{-2}$. Notice that for $\alpha = 0$, the components of radiation ρ_r and dark matter ρ_d and baryonic ρ_b present a linear decay, so that for ρ_r, we have a more pronounced behavior in relation to ρ_d and ρ_b. On the other hand, assuming $\alpha = 2 \times 10^{-2} m_p^{-2}$, we will have degenerate models at $\alpha = 0$ in the intervals $\chi \approx [-16, -1]$, $\chi \approx [-20, 1]$ and $\chi \approx [-20, 0.5]$, for ρ_r, ρ_d and ρ_b, respectively. Outside these ranges, each component decays more slowly, so that ρ_r, ρ_d and ρ_b converge their energies after $\chi \approx 3$. For the dilaton component, ρ_ϕ, we have a degenerate decay and non-linear scenarios, considering both α. Such decay becomes slower after $\chi \approx -3$, thus presenting points of intersection with each of the other components, that is, at $\chi \approx -2.5$, $\chi \approx -1$ and $\chi \approx -0.5$ we have the points of intersection between the decays of ρ_ϕ with ρ_r, ρ_b and ρ_d, respectively. Similar behavior can be found for the second case (right panel) for $Q_0 = 20$.

Furthermore, in relation to the dilaton component ρ_ϕ (right panel), it presents a more pronounced decay when compared to the graph in (left panel), that is, ρ_ϕ has a more significant decay with the increase Q_0. For the points of intersection between ρ_ϕ and the other components, we will have the same values corresponding to χ; however, such points occur at a lower energy density when compared to the graph in (left panel).

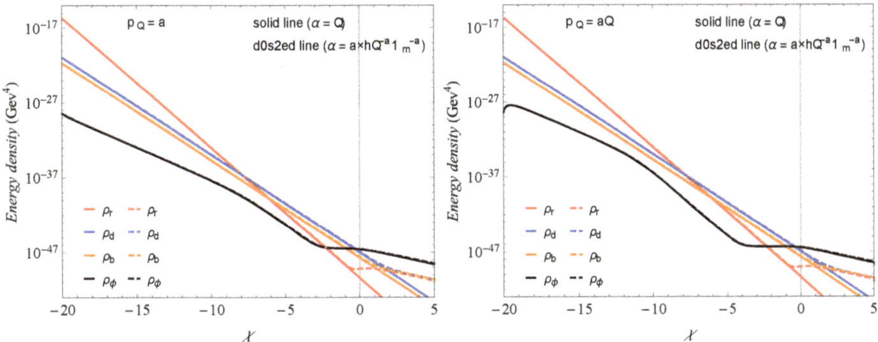

Figure 9. (**left panel**) The evolution of the energy densities for different components as a function of χ for fixed $Q_0 = 2$ and (**right panel**) $Q_0 = 20$ for two values of α.

6. Discussions

We shall first address the issue of H_0 tension. This problem now well known as the "Hubble tension" is related to the divergence of the measurements of the Hubble constant H_0 with respect to different applied techniques. In other words, the measurements of regions in the recent Universe, such as observations from the Hubble Space Telescope of Cepheid variables—see Riess et al. [56,57]—present considerable different results for H_0 as compared with its measurements made in the early Universe by the Planck spacecraft—for the 2018 Planck release, see [58]. The main difference between these two techniques is that in the latter case, the data from Planck CMB observations are processed under the base-ΛCDM model. This has raised some questions about this model and then some extended models have been put forward in the literature in order to solve the "Hubble tension", but according to the analysis performed in [58], none of the extended models solves this tension in a satisfactory way. In our present study, we also offer an alternative model to address this problem. We have considered three region of parameters to accomplish both techniques. We denominate these sets of parameters as scenarios I, II and III—see Table 1 and Figures 3 and 4.

Table 1. Comparison of the ranges of H_0 values obtained by Riess et al. 2019 [56] and Planck 2018 [58] with Scenarios I, II and III obtained by numerical methods through the parameters $\alpha = 0$, $\alpha = 2 \times 10^{-2}m_p^{-2}$, $\alpha = 5 \times 10^{-2}m_p^{-2}$, $Q_0 = 2$ and $Q_0 = 20$. The entire range of values of H_0 in Riess et al. 2019 is included in the range of values set out in Scenario III and partially included in the range of values in Scenario II. As for the range of H_0 values for Planck, these partially comprise Scenarios I, II and III.

Experiments	H_0	α	Q_0
Riess et al. 2019 [56]	74.03 ± 1.42 km s^{-1} Mpc^{-1}	–	–
Planck 2018 [58]	67.3 ± 1.20 km s^{-1} Mpc^{-1}	–	–
Scenarios	H_0	α	Q_0
I	$67 - 72$ km s^{-1} Mpc^{-1}	$0 - 2 \times 10^{-2}m_p^{-2}$	2
II	$67 - 74$ km s^{-1} Mpc^{-1}	0	2–20
III	$68 - 76$ km s^{-1} Mpc^{-1}	$5 \times 10^{-3}m_p^{-2}$	2–20

Scenarios I and II, following the range of parameters properly chosen, can simulate the results from Riess et al. [56,57] and Planck-based ΛCDM model. Scenario III simulates Riess et al. [56,57]. Although this seems to be a reasonable test of viability of our model, a statistical analysis to find the best fit of α and Q_0 according to Planck data should be addressed elsewhere. A similar behavior in easing the Hubble constant tension has recently appeared in the context of W_3 algebras [59].

Concerning the densities Ω_r, Ω_d, Ω_b and Ω_ϕ, depicted in Figures 1 and 2 and Table 2, notice that for the regimes of parameters considered in the Scenario I, II and III, there are no significant changes in relation to the Planck-based ΛCDM model, although they yield substantial changes for the Hubble constant H_0.

Table 2. The table shows the values of density parameters for radiation, dark matter, baryonic matter and dilaton field (dark energy) at $\chi = 0$ for different regions in the space of parameters (α, Q_0).

Ω_r	Ω_d	Ω_b	Ω_ϕ	α	Q_0
9.4×10^{-5}	0.271	0.049	0.679	0	2
7.7×10^{-5}	0.219	0.040	0.740	0	20
16.2×10^{-5}	0.302	0.056	0.636	$2 \times 10^{-2}m_p^{-2}$	2
19.7×10^{-5}	0.266	0.047	0.701	$2 \times 10^{-2}m_p^{-2}$	20

The dilaton field is running to achieve the limit $\phi \to -\infty$ in all scenarios discussed above, as depicted in Figures 5 and 6, which guarantees the weak field field regime as we have previously assumed. Thus, the dilatonic potential (31) approaches zero at this limit, as expected in quintessence scenarios.

The initial values of the energy densities adopted above leads to acceptable current energy densities. For instance, for fixed parameters $\alpha = 0, Q_0 = 2$, at $\chi = 0$, we find $\rho_r = 3.44 \times 10^{-51} \text{GeV}^4$, $\rho_d = 9.87 \times 10^{-48} \text{GeV}^4$, $\rho_b = 1.79 \times 10^{-48} \text{GeV}^4$ and $\rho_\phi = 2.47 \times 10^{-47} \text{GeV}^4$. The matter–radiation equality occurs in the redshift z_{eq}. See in Table 3 different values for such redshift, the Hubble constant H_0 and a_{eq} for some region of parameters.

Table 3. The table shows the acceptable values of redshift of matter–radiation equality $z_{eq} \sim 3000$ and corresponding a_{eq}, and the Hubble constant H_0 for different regions in the space of parameters (α, Q_0). For $\alpha \neq 0$, one finds unacceptable values $z_{eq} \ll 3000$. This may be a sign that a statistical analysis can reveal that the best fit for α is very small or identically zero.

z_{eq}	H_0	a_{eq}	α	Q_0
3391	67.0	2.9×10^{-4}	0	2
3346.6	74.4	3.0×10^{-4}	0	20
604.1	68.1	16.5×10^{-4}	5×10^{-3}	2
425.6	75.5	23.4×10^{-4}	5×10^{-3}	20

7. Conclusions

We explore a model of string-theory-inspired dilaton gravity in realm of modified $f(R, T)$ gravity. The numerical analyses were made in the model and revealed several cosmological quantities to describe dark energy for the late-time Universe. The model displays linear contributions in T (trace of the energy-momentum tensor) and in this preliminary approach, it seems to cover the well-accepted behavior of ΛCDM model for low redshifts, which is in accord with the Planck 2018 data. In this perspective, the model also mimics extensions of the ΛCDM model due to suitable adjusted space of parameters that allows to deal with the Hubble constant H_0 tension. We have shown in three scenarios considered in the present study that by varying appropriately some parameters, such as α and Q_0, one can obtain values of H_0 in a wide range spanning from the Planck results to the SHOES results and beyond. The analysis on the redshift of matter–radiation equality shows better results for $\alpha = 0$. This may be a sign that a statistical analysis can reveal that the best fit for α is very small or identically zero. Although this seems to be a reasonable test of viability of our model, a statistical analysis to find the best fit of α and Q_0 according to Planck data should be addressed elsewhere. As a perspective, we shall address these and other issues in the future. Further studies with different models should be addressed, as for instance, one could apply several investigations on the inflationary regime, such as constraining the parameters through contour plot in the plane made out of scalar spectral indices and tensor-to-scalar ratio [60].

Author Contributions: F.A.B., C.H.A.B.B. and F.G.C. developed the conceptual idea; J.A.V.C. developed the numerics. All authors have read and agreed to the published version of the manuscript.

Funding: This research was funded by CNPq grant number 309092/2022-1, CNPq/PRONEX/FAPESQ-PB grant number 165/2018 and FAPESQ-PB grant number 77/2022. C.H.A.B. Borges also acknowledges PIQIFPB for financial support, PIQIFPB grant number 23325.003350.2022-84

Data Availability Statement: Data are contained within the article.

Conflicts of Interest: The authors declare no conflicts of interest.

References

1. Riess, A.G.; Filippenko, A.V.; Challis, P.; Clocchiatti, A.; Diercks, A.; Garnavich, P.M.; Gilliland, R.L.; Hogan, C.J.; Jha, S.; Kirshner, R.P.; et al. Observational evidence from supernovae for an accelerating universe and a cosmological constant. *Astron. J.* **1998**, *116*, 1009. [CrossRef]
2. Perlmutter, S.; Aldering1, G.; Goldhaber, G.; Knop, R.A.; Nugent, P.; Castro, P.G.; Deustua1, S.; Fabbro, S.; Goobar, A.; Groom, D.E.; et al. Measurements of Ω and λ from 42 High-Redshift Supernovae. *Astrophys. J.* **1999**, *517*, 565. [CrossRef]
3. Koivisto, T.; Mota, D.F. Dark energy anisotropic stress and large scale structure formation. *Phys. Rev. D* **2006**, *73*, 083502. [CrossRef]
4. Daniel, S.F.; Caldwell, R.R.; Cooray, A.; Melchiorri, A. Large scale structure as a probe of gravitational slip. *Phys. Rev. D* **2008**, *77*, 103513. [CrossRef]
5. Komatsu, E.; Kogut, A.; Nolta, M.R.; Bennett, C.L.; Halpern, M.; Hinshaw, G.; Jarosik, N.; Limon, M.; Meyer, S.S.; Page, L.; et al. First-year Wilkinson microwave anisotropy probe (wmap)* observations: Tests of Gaussianity. *Astrophys. J. Suppl.* **2003**, *148*, 119–134. [CrossRef]
6. Spergel, D.N.; et al. [WMAP Collaboration]. First-Year Wilkinson Microwave Anisotropy Probe (WMAP)* Observations: Determination of Cosmological Parameters. *Astrophys. J. Suppl.* **2003**, *148*, 175. [CrossRef]
7. Hinshaw, G.; Larson, D.; Komatsu, E.; Spergel, D.N.; Bennett, C.; Dunkley, J.; Nolta, M.R.; Halpern, M.; Hill, R.S.; Odegard, N.; et al. Nine-year Wilkinson Microwave Anisotropy Probe (WMAP) observations: Cosmological parameter results. *Astrophys. J. Suppl.* **2013**, *208*, 19. [CrossRef]
8. Caldwell, R.R.; Doran, M. Cosmic microwave background and supernova constraints on quintessence: Concordance regions and target models. *Phys. Rev. D* **2004**, *69*, 103517. [CrossRef]
9. Huang, Z.Y.; Wang, B.; Abdalla, E.; Su, R.K. Holographic explanation of wide-angle power correlation suppression in the cosmic microwave background radiation. *J. Cosmol. Astropart. Phys.* **2006**, *0605*, 013. [CrossRef]
10. Eisenstein, D.J.; Zehavi, I.; Hogg, D.W.; Scoccimarro, R.; Blanton, M.R.; Nichol, R.C.; Scranton, R.; Seo, H.J.; Tegmark, M.; Zheng, Z.; et al. Detection of the baryon acoustic peak in the large-scale correlation function of SDSS luminous red galaxies. *Astrophys. J.* **2005**, *633*, 560. [CrossRef]
11. Percival, W.J.; Reid, B.A.; Eisenstein, D.J.; Bahcall, N.A.; Budavari, T.; Frieman, J.A.; Fukugita, M.; Gunn, J.E.; Ivezić, Ž.; Knapp, G.R.; et al. Baryon acoustic oscillations in the Sloan Digital Sky Survey data release 7 galaxy sample. *Mon. Not. R. Astron. Soc.* **2010**, *401*, 2148. [CrossRef]
12. Abbott, B.P.; et al. [LIGO Scientific Collaboration and Virgo Collaboration]. Observation of Gravitational Waves from a Binary Black Hole Merger. *Phys. Rev. Lett.* **2016**, *116*, 061102. [CrossRef]
13. Padmanabhan, T. Dark energy and gravity. *Gen. Relativ. Gravit.* **2008**, *40*, 529–564. [CrossRef]
14. Frieman, J.; Turner, M.; Huterer, D. Dark Energy and the Accelerating Universe. *Ann. Rev. Astron. Astrophys.* **2008**, *46*, 385–432. [CrossRef]
15. Martin, J. Quintessence: A mini-review. *Mod. Phys. Lett. A* **2008**, *23*, 1252–1265. [CrossRef]
16. Caldwell, R.R.; Kamionkowski, M. The Physics of Cosmic Acceleration. *Ann. Rev. Nucl. Part. Sci.* **2009**, *59*, 397–429. [CrossRef]
17. Silvestri, A.; Trodden, M. Approaches to Understanding Cosmic Acceleration. *Rep. Prog. Phys.* **2009**, *72*, 096901. [CrossRef]
18. Bamba, K.; Capozziello, S.; Nojiri, S.; Odintsov, S.D. Dark energy cosmology: The equivalent description via different theoretical models and cosmography tests. *Astrophys. Space Sci.* **2012**, *342*, 155–228. [CrossRef]
19. Li, M.; Li, X.-D.; Wang, S.; Wang, Y. Dark Energy: A Brief Review. *Front. Phys.* **2013**, *8*, 828–846. [CrossRef]
20. Sami, M.; Myrzakulov, R. Late time cosmic acceleration: ABCD of dark energy and modified theories of gravity. *Int. J. Mod. Phys. D* **2016**, *25*, 1630031. [CrossRef]
21. Peebles, P.J.E.; Ratra, B. The Cosmological Constant and Dark Energy. *Rev. Mod. Phys.* **2003**, *75*, 559–606. [CrossRef]
22. Padmanabhan, T. Cosmological constant: The Weight of the vacuum. *Phys. Rep.* **2003**, *380*, 235–320. [CrossRef]
23. Sahni, V. The Cosmological constant problem and quintessence. *Class. Quant. Grav.* **2002**, *19*, 3435–3448. [CrossRef]
24. Velten, H.E.; Vom Marttens, R.F.; Zimdahl, W. Aspects of the cosmological "coincidence problem". *Eur. Phys. J. C* **2014**, *74*, 3160. [CrossRef]
25. Tsujikawa, S. Quintessence: A Review. *Class. Quant. Grav.* **2013**, *30*, 214003. [CrossRef]
26. Buchdahl, H.A. Non-linear Lagrangians and cosmological theory. *Mon. Not. R. Astron. Soc.* **1970**, *150*, 1–8. [CrossRef]
27. Sotiriou, T.P.; Faraoni, V. $f(r)$ theories of gravity. *Rev. Mod. Phys.* **2010**, *82*, 451. [CrossRef]
28. Nojiri, S.; Odintsov, S.D. Unified cosmic history in modified gravity: From F(R) theory to Lorentz non-invariant models. *Phys. Rept.* **2011**, *505*, 59–144. [CrossRef]
29. Nojiri, S.; Odintsov, S.D.; Oikonomou, V.K. Modified Gravity Theories on a Nutshell: Inflation, Bounce and Late-time Evolution. *Phys. Rept.* **2017**, *692*, 1–104. [CrossRef]
30. Starobinsky, A.A. A new type of isotropic cosmological models without singularity. *Phys. Lett. B* **1980**, *91*, 99–102. [CrossRef]
31. Carroll, S.M.; Duvvuri, V.; Trodden, M.; Turner, M.S. Is cosmic speed-up due to new gravitational physics? *Phys. Rev. D* **2004**, *70*, 043528. [CrossRef]
32. Nojiri, S.; Odintsov, S.D. Modified gravity with negative and positive powers of the curvature: Unification of the inflation and of the cosmic acceleration. *Phys. Rev. D* **2003**, *68*, 123512. [CrossRef]

33. Capozziello, S.; Cardone, V.F.; Carloni, S.; Troisi, A. Curvature quintessence matched with observational data. *Int. J. Mod. Phys. D* **2003**, *12*, 1969–1982. [CrossRef]
34. Dzhunushaliev, V.; Folomeev, V.; Kleihaus, B.; Kunz, J. Modified gravity from the quantum part of the metric. *Eur. Phys. J. C* **2014**, *74*, 2743. [CrossRef]
35. Yang, R. Effects of quantum fluctuations of metric on the universe. *Phys. Dark Universe* **2016**, *13*, 87. [CrossRef]
36. Liu, X.; Harko, T.; Liang, S.D. Cosmological implications of modified gravity induced by quantum metric fluctuations. *Eur. Phys. J. C* **2016**, *76*, 420. [CrossRef]
37. Harko, T.; Lobo, F.S.; Nojiri, S.I.; Odintsov, S.D. $f(R,T)$ gravity. *Phys. Rev. D* **2011**, *84*, 024020. [CrossRef]
38. Shabani, H.; Farhoudi, M. $f(R,T)$ cosmological models in phase space. *Phys. Rev. D* **2013**, *88*, 044048. [CrossRef]
39. Xu, M.X.; Harko, T.; Liang, S.D. Quantum Cosmology of $f(R,T)$ gravity. *Eur. Phys. J. C* **2016**, *76*, 449. [CrossRef]
40. Moraes, P.H.R.S.; Sahoo, P.K. The simplest non-minimal matter-geometry coupling in the $f(R,T)$ cosmology. *Eur. Phys. J. C* **2017**, *77*, 480. [CrossRef]
41. Shabani, H.; Ziaie, A.H. Bouncing cosmological solutions from $f(R,T)$ gravity. *Eur. Phys. J. C* **2018**, *78*, 397. [CrossRef]
42. Debnath, P.S. Bulk viscous cosmological model in $f(R,T)$ theory of gravity. *Int. J.Geom. Meth. Mod. Phys.* **2019**, *16*, 1950005. [CrossRef]
43. Bhattacharjee, S.; Sahoo, P. Comprehensive analysis of a non-singular bounce in $f(R,T)$ gravitation. *Phys. Dark Universe* **2020**, *28*, 100537. [CrossRef]
44. Bhattacharjee, S.; Santos, J.R.L.; Moraes, P.H.R.S.; Sahoo, P.K. Inflation in $f(R,T)$ gravity. *Eur. Phys. J. Plus* **2020**, *135*, 576. [CrossRef]
45. Gamonal, M. Slow-roll inflation in $f(R,T)$ gravity and a modified Starobinsky-like inflationary model. *Phys. Dark Universe* **2021**, *31*, 100768. [CrossRef]
46. Witten, E. Some properties of O (32) superstrings. *Phys. Lett. B* **1984**, *149*, 351–356. [CrossRef]
47. Witten, E. String theory dynamics in various dimensions. *Nucl. Phys. B* **1995**, *443*, 85–126. [CrossRef]
48. Damour, T.; Polyakov, A.M. The String dilaton and a least coupling principle. *Nucl. Phys. B* **1994**, *423*, 532. [CrossRef]
49. Gasperini, M. Dilatonic interpretation of quintessence? *Phys. Rev. D* **2001**, *64*, 043510. [CrossRef]
50. Gasperini, M.; Piazza, F.; Veneziano, G. Quintessence as a runaway dilaton. *Phys. Rev. D* **2001**, *65*, 023508. [CrossRef]
51. Veneziano, G. Large N bounds on, and compositeness limit of, gauge and gravitational interactions. *J. High Energy Phys.* **2002**, *6*, 051. [CrossRef]
52. Taylor, T.R.; Veneziano, G. Dilaton couplings at large distances. *Phys. Lett. B* **1988**, *213*, 459. [CrossRef]
53. Poplawski, N.J. A Lagrangian description of interacting dark energy. *arXiv* **2006**, arXiv:gr-qc/0608031.
54. Fischbach, E.; Talmadge, C. Six years of the fifth force. *Nature* **1992**, *356*, 207–215. [CrossRef]
55. Gasperini, M. On the response of gravitational antennas to dilatonic waves. *Phys. Lett. B* **1999**, *470*, 67–72. [CrossRef]
56. Riess, A.G.; Casertano, S.; Yuan, W.; Macri, L.M.; Scolnic, D. Large Magellanic Cloud Cepheid Standards Provide a 1% Foundation for the Determination of the Hubble Constant and Stronger Evidence for Physics beyond ΛCDM. *Astrophys. J.* **2019**, *876*, 85. [CrossRef]
57. Riess, A.G.; Yuan, W.; Macri, L.M.; Scolnic, D.; Brout, D.; Casertano, S.; Jones, D.O.; Murakami, Y.; Anand, G.S.; Breuval, L.; et al. A Comprehensive Measurement of the Local Value of the Hubble Constant with 1 km s^{-1} Mpc^{-1} Uncertainty from the Hubble Space Telescope and the SH0ES Team. *Astrophys. J. Lett.* **2022**, *934*, L7. [CrossRef]
58. Aghanim, N.; et al. [Planck]. Planck 2018 results. VI. Cosmological parameters. *Astron. Astrophys.* **2020**, *641*, A6; Erratum in *Astron. Astrophys.* **2021**, *652*, C4. [CrossRef]
59. Ambjorn, J.; Watabiki, Y. Easing the Hubble constant tension. *Mod. Phys. Lett. A* **2022**, *37*, 2250041. [CrossRef]
60. Santos, J.R.L.; da Costa, S.S.; Santos, R.S. Cosmological models for f(R,T)−Λ(ϕ) gravity. *Phys. Dark Univ.* **2023**, *42*, 101356. [CrossRef]

Article

Bayesian Implications for the Primordial Black Holes from NANOGrav's Pulsar-Timing Data Using the Scalar-Induced Gravitational Waves

Zhi-Chao Zhao [1,†] and Sai Wang [2,3,*,†]

[1]   Department of Applied Physics, College of Science, China Agricultural University, Qinghua East Road, Beijing 100083, China
[2]   Theoretical Physics Division, Institute of High Energy Physics, Chinese Academy of Sciences, Beijing 100049, China
[3]   School of Physical Sciences, University of Chinese Academy of Sciences, Beijing 100049, China
*   Correspondence: wangsai@ihep.ac.cn
†   These authors contributed equally to this work.

Abstract: Assuming that the common-spectrum process in the NANOGrav 12.5-year dataset has an origin of scalar-induced gravitational waves, we study the enhancement of primordial curvature perturbations and the mass function of primordial black holes, by performing the Bayesian parameter inference for the first time. We obtain lower limits on the spectral amplitude, i.e., $\mathcal{A} \gtrsim 10^{-2}$ at 95% confidence level, when assuming the power spectrum of primordial curvature perturbations to follow a log-normal distribution function with width σ. In the case of $\sigma \to 0$, we find that the primordial black holes with $2 \times 10^{-4} - 10^{-2}$ solar mass are allowed to compose at least a fraction 10^{-6} of dark matter. Such a mass range is shifted to more massive regimes for larger values of σ, e.g., to a regime of $4 \times 10^{-3} - 0.2$ solar mass in the case of $\sigma = 1$. We expect the planned gravitational-wave experiments to have their best sensitivity to $\mathcal{A}$ in the range of 10^{-4} to 10^{-7}, depending on the experimental setups. With this level of sensitivity, we can search for primordial black holes throughout the entire parameter space, especially in the mass range of 10^{-16} to 10^{-11} solar masses, where they could account for all dark matter. In addition, the importance of multi-band detector networks is emphasized to accomplish our theoretical expectation.

Keywords: primordial black hole; scalar-induced gravitational waves; Bayesian parameter inference; pulsar-timing data analysis

Citation: Zhao, Z.-C.; Wang, S. Bayesian Implications for the Primordial Black Holes from NANOGrav's Pulsar-Timing Data Using the Scalar-Induced Gravitational Waves. *Universe* **2023**, *9*, 157. https://doi.org/10.3390/universe9040157

Academic Editor: Kazuharu Bamba

Received: 18 February 2023
Revised: 20 March 2023
Accepted: 22 March 2023
Published: 24 March 2023

1. Introduction

It is well-known that the primordial black holes (PBHs) could be produced by gravitational collapses of the enhanced primordial curvature perturbations on small scales [1]. The mass function of PBHs is determined by the power spectrum of primordial curvature perturbations [2–4]. To produce a sizeable quantity of PBHs, the scalar spectral amplitude should conquer a critical magnitude ∼0.1 [5], which is about eight orders of magnitudes larger than the spectral amplitude on large scales. On the other hand, the dynamics of cosmic inflation [6–8] is one of the greatest puzzles of cosmology to be uncovered. In particular, the small-scale primordial curvature perturbations were generated by quantum fluctuations at the late-time stage of inflation. Their power spectrum encodes characteristic information about the late-time dynamics of inflation, which could be different from that of the early-time one measured precisely with experiments on cosmic microwave background (CMB) [9,10] and large-scale structure (LSS) [11,12]. Therefore, we can gain more information about cosmic inflation by investigating the PBHs.

The scenario of PBHs can also be a reasonable candidate of dark matter, or compose a part of the latter [13,14]. The mass function of PBHs is defined as a ratio between the energy density fractions of PBHs and dark matter. Extensive studies have shown observational constraints on the mass function of PBHs (e.g., see reviews in Refs. [15,16]). In addition, the scenario of PBHs was suggested to interpret the formation mechanism of supermassive black holes (SMBHs) [17] and the events of compact binary coalescences [18–25], which emitted the gravitational waves (GWs) observed by the Advanced LIGO, Virgo and KAGRA detectors [26–28].

Accompanied by the formation process of PBHs, the scalar induced gravitational waves (SIGWs) [29–32] were inevitably produced by the enhanced primordial curvature perturbations. Specifically, they were produced through nonlinear couplings of tensor-scalar-scalar modes in the very early cosmos. Given the power spectrum of primordial curvature perturbations, a semi-analytic formula was provided to compute the energy density fraction spectrum of SIGWs [33,34]. It has been widely used to investigate the mass function of PBHs in the literature.

In this work, we will investigate the mass function of PBHs, as well as the power spectrum of primordial curvature perturbations, by analyzing a 12.5-year dataset from North American Nanohertz Observatory for Gravitational Waves (NANOGrav)[1] [35]. Strong evidence of a stochastic common-spectrum process was reported by NANOGrav Collaboration. If this signal has an origin of SIGWs, we can straightforwardly obtain constraints on the power spectrum of primordial curvature perturbations. Since the latter can collapse to form the PBHs due to gravity, we therefore obtain corresponding constraints on the mass function of PBHs. In fact, the scenario of PBHs with different masses has been suggested to interpret such a signal (see, e.g., Refs. [36–41]). However, there is still lack of systematic Bayesian parameter inferences, which will be performed for the first time in this paper.

The rest of the paper is arranged as follows. We briefly review the theory of SIGWs in Section 2. By performing Bayesian analysis, we obtain the NANOGrav constraints on the enhancement of primordial curvature perturbations in Section 3. Correspondingly, we demonstrate the physical implications for the mass function of PBHs in Section 4. We forecast the sensitivity curves of ongoing and planned gravitational-wave detectors in Section 5. The conclusion and discussion are shown in Section 6.

2. Theory of Scalar-Induced Gravitational Waves

As previously shown [33,34], there is a lengthy and tedious derivation process to obtain a semi-analytic formula for the energy density fraction spectrum of SIGWs in the early cosmos. Therefore, we summarize some theoretical results that would be essential to our data analysis. The energy density fraction spectrum of GWs is defined as $\Omega_{GW}(k) = \rho_{GW}(k)/\rho_c$, where k denotes the GW wavenumber, ρ_c is the critical energy density of the cosmos, and $\int \rho_{GW}(k) d \ln k$ denotes the total energy density of GWs. The SIGWs in the subhorizon have the energy density spectrum as $\rho_{GW} \sim \langle \overline{h_{ij,l} h_{ij,l}} \rangle$ [42], where the overbar stands for the oscillation average and the angle brackets defines the power spectrum. The strain of SIGWs, as the second-order tensor perturbations produced by the linear scalar perturbations, can be roughly expressed as $h \sim \phi^2$, where ϕ denotes the linear scalar perturbation. If considering the Gaussian scalar perturbations, we obtain $\rho_{GW} \sim \langle \phi^4 \rangle \sim \langle \phi^2 \rangle \langle \phi^2 \rangle$, where $\langle \phi^2 \rangle$ is determined by the initial value and transfer function of ϕ. Therefore, following Equation (14) of Ref. [33], the energy density fraction spectrum of SIGWs in the radiation-dominated cosmos is given as

$$\Omega_{GW}(k) = \int_0^\infty dv \int_{|1-v|}^{|1+v|} du \, F(v,u) \mathcal{P}_{\mathcal{R}}(vk) \mathcal{P}_{\mathcal{R}}(uk) \,, \tag{1}$$

where $\mathcal{P}_{\mathcal{R}}(uk)$ (and $\mathcal{P}_{\mathcal{R}}(vk)$) is power spectrum of primordial curvature perturbations at wavenumber uk (and vk) with u (and v) being a dimensionless variable, and $F(v,u)$ is a functional of the scalar transfer function describing the post-inflationary evolution of ϕ and can be simplified as

$$F(v, u) = \frac{3}{2^{10} v^8 u^8} \left[4v^2 - (v^2 - u^2 + 1)^2 \right]^2 \left(v^2 + u^2 - 3 \right)^2$$
$$\times \left\{ \left[\left(v^2 + u^2 - 3 \right) \ln \left(\left| \frac{3 - (v+u)^2}{3 - (v-u)^2} \right| \right) - 4vu \right]^2 \right.$$
$$\left. + \pi^2 \left(v^2 + u^2 - 3 \right)^2 \Theta \left(v + u - \sqrt{3} \right) \right\}. \tag{2}$$

Due to a lack of direct measurements, the nature of primordial curvature perturbations on small scales remains unknown, compared with that on the largest scales measured by the cosmic microwave background. For computational simplicity, we take two concrete expressions of $\mathcal{P}_\mathcal{R}(k)$, which are frequently adopted in the literature (e.g., Refs. [33,43–46]), as follows

$$\mathcal{P}_\mathcal{R}(k) = \mathcal{A}\, \delta \left(\ln \frac{k}{k_*} \right), \tag{3}$$

$$\mathcal{P}_\mathcal{R}(k) = \frac{\mathcal{A}}{\sqrt{2\pi}\sigma} \exp \left(-\frac{\ln^2(k/k_*)}{2\sigma^2} \right), \tag{4}$$

where $\mathcal{A}$ is an amplitude, σ is a standard variance, and k_* is a pivot wavenumber. We can relate the wavenumber with the frequency, i.e., $k = 2\pi f$ and $k_* = 2\pi f_*$. During the process of parameter inferences, both $\mathcal{A}$ and f_* are independent parameters. However, we will not let σ be independent, even though different values of σ would alter our results. In fact, it is challenging to explore the full parameter space of σ based on the current dataset. Therefore, we let $\sigma = 1$ for simplicity in this work. We can test the robustness of our conclusions by comparing the results obtained when taking two different values of σ. In particular, we will compare the results from Equations (3) and (4). We leave this point to be demonstrated further in the following sections.

In Figure 1, we show the energy density fraction spectrum of SIGWs, which is normalized with $\mathcal{A}^2$. For comparison, we show the results when taking different versions of σ. Equation (3) can be viewed as a limit of Equation (4) when $\sigma \to 0$. In the regime of $f \ll f_*$, the choice of $\sigma \to 0$ displays the highest spectral amplitude than non-vanishing choices of σ. This implies that in such a case, we can obtain the best sensitivity on measurements of $\mathcal{A}$ for a given GW experiment. Correspondingly, we can obtain the strictest constraints on the mass function of PBHs. We will demonstrate these points in the following sections.

Once the early-universe quantity $\Omega_{\mathrm{GW}}(k)$ is given in Equation (1), we have the energy density fraction spectrum of SIGWs in the present cosmos to be [43]

$$\Omega_{\mathrm{GW},0}(f) = \Omega_{\mathrm{r},0} \left[\frac{g_{*,\rho}(T)}{g_{*,\rho}(T_{\mathrm{eq}})} \right] \left[\frac{g_{*,s}(T_{\mathrm{eq}})}{g_{*,s}(T)} \right]^{\frac{4}{3}} \Omega_{\mathrm{GW}}(k), \tag{5}$$

where $\Omega_{\mathrm{r},0}$ is the present energy density fraction of radiations, and the subscript $_{\mathrm{eq}}$ stands for cosmological quantities at the epoch of matter-radiation equality. Here, we have taken into account contributions from the effective relativistic degrees of the cosmos, i.e., $g_{*,\rho}$ and $g_{*,s}$, as functions of f by interpolating their tabulated data in terms of cosmic temperature T in Ref. [47][2] as well as by considering a relation between T and f in Wang et al. [43], i.e.,

$$\frac{f}{\mathrm{nHz}} = 26.5 \left(\frac{T}{\mathrm{GeV}} \right) \left[\frac{g_{*,\rho}(T)}{106.75} \right]^{\frac{1}{2}} \left[\frac{g_{*,s}(T)}{106.75} \right]^{-\frac{1}{3}}. \tag{6}$$

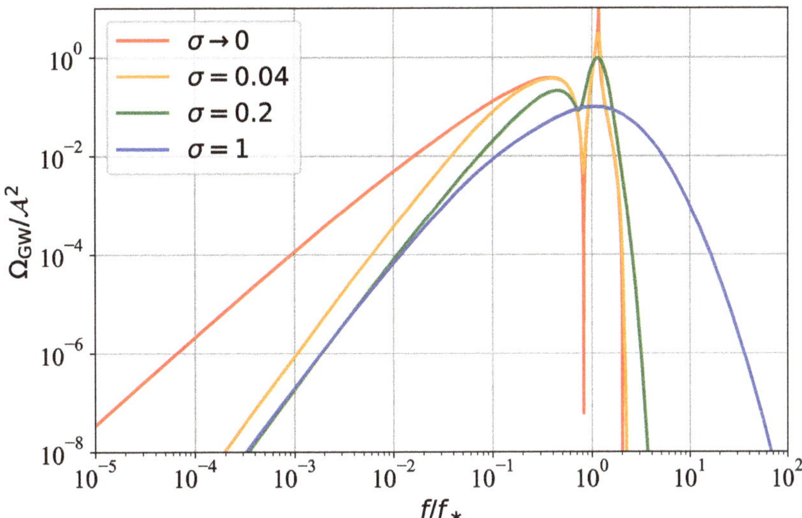

Figure 1. Energy density fraction spectrum of SIGWs normalized with $\mathcal{A}^2$. We adopt $\sigma \to 0$ to label the power spectrum of primordial curvature perturbations in Equation (3). Different choices of σ in Equation (4) are plotted for comparison.

Throughout this paper, we use the measured value of cosmological parameters in the data release 2018 of *Planck* satellite [10]. The publicly available `Astropy` [48–50] software [3] is adopted to evaluate all cosmological quantities. Please note that Equation (5) is one of the leading formulas that will be used during Bayesian analysis in the next section.

3. NANOGrav Constraints on Primordial Curvature Perturbations

When a pulsar timing array (PTA) experiment measures the stochastic gravitaitonal wave background, the timing residual cross-power spectral density is given by $S_{ab}(f) = \Gamma_{ab} h_c^2 / (12\pi^2 f^3)$, which can be shown by combining Equations (1) and (2) in Arzoumanian et al. [51]. Here, Γ_{ab} stands for the overlap reduction function (ORF) that describes the correlation between two pulsars a and b as a function of line-of-sight separation angle between them. Following Equation (17) in Maggiore [52], $h_c(f)$ is the characteristic strain defined as $h_c^2 = 3H_0^2 \Omega_{\text{GW},0} / (2\pi^2 f^2)$, where $H_0 = 100 h_0$ km s^{-1}Mpc^{-1} is the Hubble constant with h_0 being the reduced Hubble constant. Therefore, the timing residual cross-power spectral density becomes [51]

$$S_{ab} = \Gamma_{ab} \frac{H_0^2 f_{\text{yr}}^{-5}}{8\pi^4} \left(\frac{f}{f_{\text{yr}}} \right)^{-5} \Omega_{\text{GW},0}(f) \,, \tag{7}$$

where f_{yr} is a pivot frequency corresponding to a duration time of 1 year, and the formula for $\Omega_{\text{GW},0}(f)$ is shown in Equation (5). For an isotropic background of GWs, e.g., the SIGWs considered in this work, we take ORF to be the Hellings and Downs coefficients [53].

The timing residual data for a single pulsar is decomposed in its individual constituents, i.e., [54]

$$\delta t = M\epsilon + Fa + Uj + n \,. \tag{8}$$

The term $M\epsilon$ stands for the inaccuracies in the subtraction of timing model, where M denotes the timing model design matrix, and ϵ is a vector describing small offsets for the timing model parameters. The matrix M is computed through `libstempo` [55][4], which is a python interface for `TEMPO2` [56,57][5] timing package. We use the latest Jet Propulsion Laboratory (JPL) solar system ephemeris (SSE) model, DE438 [58], in the timing model fits.

The term Fa accounts for all low-frequency signals, including the pulsar-intrinsic red noise. The Fourier design matrix F has alternating sine and cosine functions, and a is a vector comprised of Fourier coefficients at the integer $(1, 2, \ldots, N_{\mathrm{mode}})$ multiples of the harmonic base frequency $1/T_s$, where T_s denotes the span between the minimum and maximum time of arrivals (TOAs) in the array [59]. Described by a per-epoch variance (ECORR) for each receiver and backend system [54], the term Uj denotes the white noise which is fully correlated for simultaneous observations at different frequencies, but fully uncorrelated in time. The matrix U maps all TOAs observed simultaneously at different frequencies to a total TOA, and j is the per-epoch white noise which is fully correlated across all observing frequencies. The term n is the timing residual introduced by Gaussian white noise, which is described by the parameters of the TOA uncertainties (EFAC) and an additive variance (EQUAD) for each receiver and backend system [54].

To estimate the allowed parameter space, we perform Bayesian parameter inferences by analyzing a dataset of 45 pulsars in the 12.5-year data release of NANOGrav Collaboration [35][6]. We list all independent parameters as well as their priors in Table 1. We adopt $N_{\mathrm{mode}} = 30$ frequency bands to the power-law spectrum of pulsar-intrinsic red noise and the common-spectrum process. In practice, we will use the publicly available enterprise [60][7] to compute the likelihoods and PTMCMCSampler [61][8] to perform Markov-Chain Monte-Carlo sampling.

Table 1. Priors used in all analyses performed in this paper.

Parameter	Description	Prior	Comment
	White Noise		
E_k	EFAC per backend/receiver system	Uniform $[0, 10]$	single-pulsar analysis only
Q_k [s]	EQUAD per backend/receiver system	log-Uniform $[-8.5, -5]$	single-pulsar analysis only
J_k [s]	ECORR per backend/receiver system	log-Uniform $[-8.5, -5]$	single-pulsar analysis only
	Red Noise		
A_{red}	power-law spectral amplitude	log-Uniform $[-20, -11]$	one parameter per pulsar
γ_{red}	power-law spectral index	Uniform $[0, 7]$	one parameter per pulsar
	Primordial curvature perturbations		
$\log \mathcal{A}$	spectral amplitude	Uniform $[-3, 0]$	one parameter for PTA
$\log f_*$ [Hz]	spectral pivot frequency	Uniform $[-10, 0]$	one parameter for PTA

Our results are shown in Figure 2, which displays the 95% CL contour plots of $\log \mathcal{A}$ and $\log f_*$ for the power spectra of primordial curvature perturbations with $\sigma \to 0$ (red solid curve) and $\sigma = 1$ (blue solid curve), respectively. We find that $\mathcal{A}$ has lower limits around 10^{-2} for the two different choices of σ. In contrast, the NANOGrav dataset prefers different frequency bands. For a given value of $\mathcal{A}$, the frequency range in the case of $\sigma \to 0$ is almost always larger than that in the case of $\sigma = 1$ by a few times. This result is consistent with the expectations of Figure 1. However, we find strong positive correlations between $\log \mathcal{A}$ and $\log f_*$ in either choice of σ. Furthermore, in Figure 2, we also label the spectral amplitudes that produce the PBHs with total abundance $\bar{f}_{\mathrm{pbh}} = 1$ (dashed curves) and $\bar{f}_{\mathrm{pbh}} = 10^{-10}$ (dotted curves) that will be interpreted in the next section.

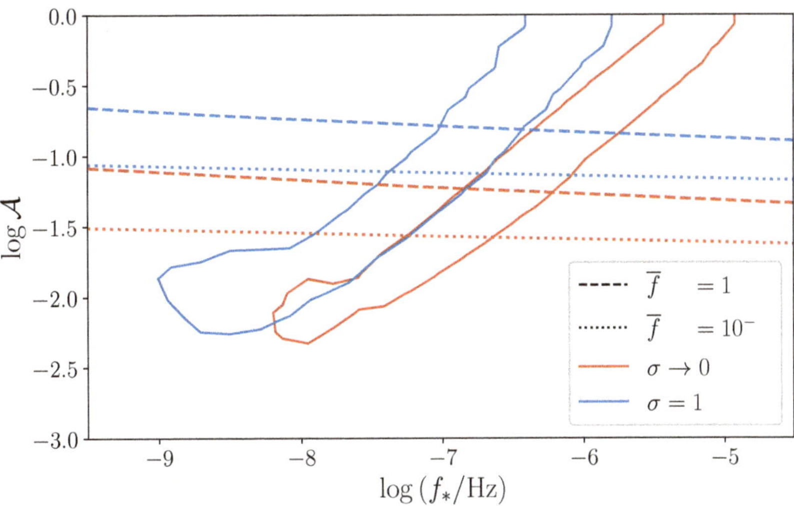

Figure 2. The 95% CL contour plot of $\mathcal{A}$ and f_* inferred from the NANOGrav 12.5yr dataset. Solid curves stand for $\sigma \to 0$ and $\sigma = 1$, while other curves denote $\bar{f}_{\mathrm{pbh}} = 1$ and $\bar{f}_{\mathrm{pbh}} = 10^{-10}$.

4. Implications for Primordial Black Holes

We define the mass function of PBHs as $f_{\mathrm{pbh}}(m) = \Omega_{\mathrm{dm}}^{-1} d\Omega_{\mathrm{pbh}}/d\ln m$, in which Ω_{dm} and Ω_{pbh} stand for the present energy density fractions of dark matter and PBHs with mass m, respectively. We adopt the concept of critical collapse [62,63] and Press-Schechter formalism [64]. Therefore, following Equation (9) in Wang et al. [43], we have

$$f_{\mathrm{pbh}}(m) = \frac{\Omega_{\mathrm{m}}}{\Omega_{\mathrm{dm}}} \int g(T)\tilde{\beta}(m, m_H)d\ln m_H \,, \tag{9}$$

where Ω_{m} is the present energy density fraction of non-relativistic matter. Here, we denote

$$g(T) = \frac{g_{*,\rho}(T)}{g_{*,\rho}(T_{\mathrm{eq}})} \frac{g_{*,s}(T_{\mathrm{eq}})}{g_{*,s}(T)} \frac{T}{T_{\mathrm{eq}}} \,, \tag{10}$$

$$\tilde{\beta}(m, m_H) = \frac{\kappa \mu^{\gamma+1}}{\sqrt{2\pi}\gamma\Delta(k)}\exp\left[-\frac{(\delta_c + \mu)^2}{2\Delta^2(k)}\right] \,, \tag{11}$$

where we have $\mu = [m/(\kappa m_H)]^{1/\gamma}$ with numerical constants $\kappa = 3.3$ and $\gamma = 0.36$, and the critical overdensity for gravitational collapse is $\delta_c = 0.45$. Here, we disregard corrections to δ_c from, e.g., QCD equation of state [65]. The quantity $\tilde{\beta}(m, m_H)$ is a mass distribution function of PBHs, which were produced when the horizon mass was m_H. The subscript $_{\mathrm{eq}}$ stands for cosmological quantities at the epoch of matter-radiation equality. Cosmic temperature T is related to m_H enclosed by the Hubble horizon, i.e., [43]

$$\frac{m_H}{M_\odot} = 4.76 \times 10^{-2} \left(\frac{T}{\mathrm{GeV}}\right)^{-2} \left[\frac{g_{*,\rho}(T)}{106.75}\right]^{-\frac{1}{2}} \,, \tag{12}$$

where $M_\odot$ denotes the solar mass. We can relate f with m_H by combining Equation (6) with Equation (12) and by reducing T from both equations. The coarse-grained fluctuations in the radiation-dominated cosmos are given by [66,67]

$$\Delta^2(k) = \frac{16}{81} \int d\ln q \left[w\left(\frac{q}{k}\right)\right]^2 \left(\frac{q}{k}\right)^4 \mathcal{T}^2\left(q, \frac{1}{k}\right)\mathcal{P}_\mathcal{R}(q) \,, \tag{13}$$

where $w(y) = \exp(-y^2/2)$ is a Gaussian window function and $\mathcal{T}(q, \tau) = 3(\sin x - x \cos x)/x^3$ with $x = q\tau/\sqrt{3}$ is a scalar transfer function. We further define the total abundance of PBHs in dark matter to be $\bar{f}_{\mathrm{pbh}} = \int f_{\mathrm{pbh}}(m) d\ln m$, and define the average mass of PBHs to be $\bar{m} = \int f_{\mathrm{pbh}}(m) dm / \bar{f}_{\mathrm{pbh}}$. The latter roughly displays $\bar{m} \simeq m_*$, where m_* is corresponded to k_*.

Once the constraints on the power spectrum of primordial curvature perturbations are obtained, as shown in Figure 2, we can recast them into constraints on the mass function of PBHs, or more precisely, on the average mass and total abundance of PBHs. We show the results in Figure 3. The shaded regions are allowed by the NANOGrav 12.5-year dataset for $\sigma \to 0$ (red region) and $\sigma = 1$ (blue region). When at least one fraction (e.g., $f_{\mathrm{pbh}} = 10^{-10}$) of dark matter is composed of PBHs, the mass range $2 \times 10^{-4} - 10^{-2} M_\odot$ $(4 \times 10^{-3} - 0.2 M_\odot)$ for $\sigma \to 0$ ($\sigma = 1$) is preferred by the current dataset. Based on Figure 3 in Wang et al. [43], we find that these mass ranges can be cross-checked with high-precision by observing the GWs emitted from inspiraling stage of PBH binaries. In addition, they might be further tested by measuring the anisotropies in stochastic gravitational-wave background (SGWB) [68]. For comparison, we also depict the existing upper limit (cyan curve) on the mass function of PBHs, as reviewed in Carr et al. [15]. We find the SIGW probe to be more powerful than electromagnetic probes, implying that a larger parameter space can be explored with the SIGW probe. This can also be understood by revisiting Figure 2, in which we depicted the curves labeling $\bar{f}_{\mathrm{pbh}} = 10^{-10}$. The latter is corresponded to $\mathcal{A} \simeq \mathrm{few} \times 10^{-2}$. In contrast, the NANOGrav experiment has reached $\mathcal{A} \simeq 10^{-2}$ in the most sensitive frequency band $\sim (10^{-9} - 10^{-8})$ Hz.

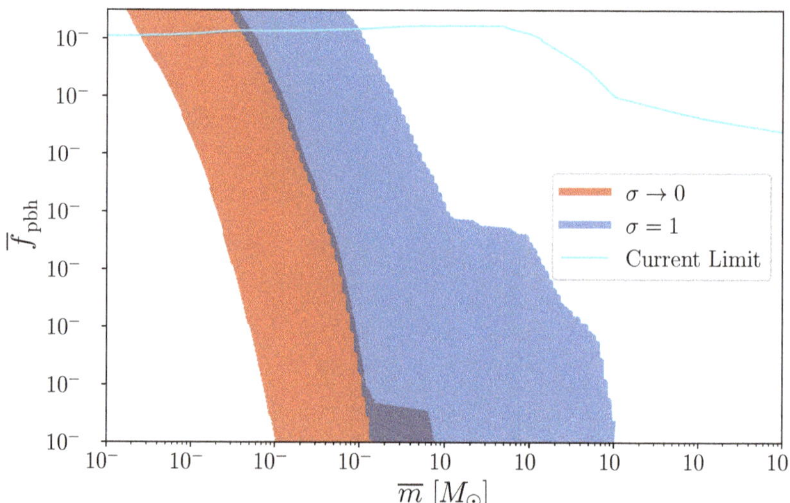

Figure 3. The NANOGrav constraints on the averaged mass and total abundance of PBHs. Shaded regions are allowed by the NANOGrav dataset for $\sigma \to 0$ and $\sigma = 1$. Cyan curve denotes the existing upper limit on the mass function of PBHs [15].

5. Constraints from Ongoing and Planned Gravitational-Wave Detectors

In the future, the power spectrum of primordial curvature perturbations and the mass function of PBHs, which remain unknown until now, can be further explored with ongoing and planned GW experiments, such as Square Kilometer Array (SKA) [69–71], μAres [72], Laser Interferometer Space Antenna (LISA) [73,74], Big Bang Observer (BBO) [75,76], Decihertz Interferometer Gravitational wave Observatory (DECIGO) [77,78], Einstein Telescope (ET) [79] and Advanced LIGO and Virgo [26–28]. Such multi-band observations could explore all possible parameter spaces of primordial curvature perturbations and PBHs.

Other experiments are not considered in this work, but our method can be generalized straightforwardly to study them, if needed.

To estimate the sensitivity curve of a given GW experiment consisting of n_{det} detectors, we define an optimal signal-to-noise ratio (SNR) denoted with ρ as follows [80]

$$\rho^2 = n_{\text{det}} T_{\text{obs}} \int_{f_{\min}}^{f_{\max}} \left[\frac{S_{\text{GW}}(f)}{S_n^{\text{eff}}(f)} \right]^2 df \,, \tag{14}$$

where T_{obs} is a duration time of observing run, the power spectral density (PSD) of SIGWs is defined as

$$S_{\text{GW}}(f) = \left(\frac{3H_0^2}{2\pi^2 f^3} \right) \Omega_{\text{GW},0}(f) \,, \tag{15}$$

and the effective noise PSD of the detector network is denoted as $S_n^{\text{eff}}(f)$ that is a function of f. The concrete setups of aforementioned experiments are summarized in Table 2 of Ref. [81] and references therein. We consider a single detector for LISA, two independent detectors for μAres, BBO and DECIGO, three detectors for ET, and 200 pulsars for SKA. For comparison, we consider one detector for Advanced LIGO with an observing duration of four years and 100% duty circle.

Given a desired value of SNR, which is unity in this work, we obtain the minimal detectable $\mathcal{A}_{\min}$ for the given experiment by resolving Equation (14). Since $\Omega_{\text{GW},0}(f)$ is uniquely determined by $\mathcal{A}$ and f_*, we depict the theoretical expectation of $\mathcal{A}_{\min}$ in the $\mathcal{A} - f_*$ plane in Figures 4 and 5 for the choices of $\sigma \to 0$ and $\sigma = 1$, respectively. For comparison, we plot the exclusion region on $\mathcal{A}$ from the Advanced LIGO and Virgo first three observing runs (red shaded) [82].

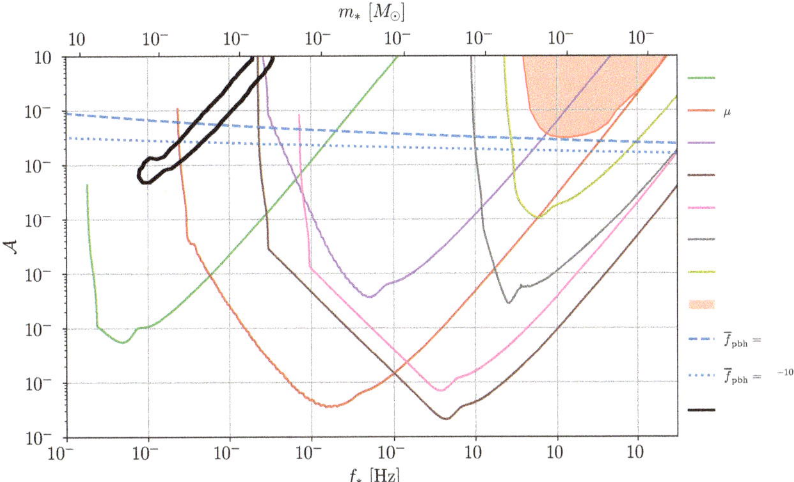

Figure 4. Sensitivities of ongoing and planned GW experiments on measurements of the power spectrum of primordial curvature perturbations with $\sigma \to 0$. The excluded region by Advanced LIGO–Virgo first three observing runs [82] is shown for comparison. We show the allowed region from NANOGrav 12.5-year dataset, as shown in Figure 2. We also depict critical curves corresponded to $\bar{f}_{\text{pbh}} = 1$ and $\bar{f}_{\text{pbh}} = 10^{-10}$.

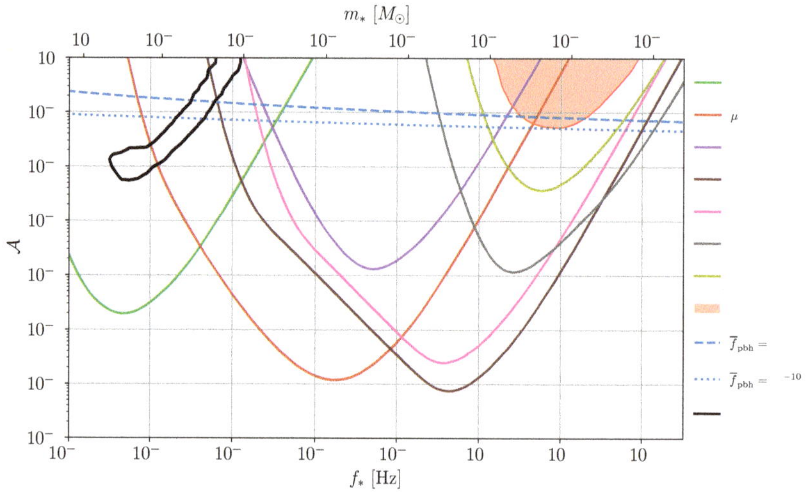

Figure 5. The same as Figure 4 but $\sigma = 1$.

Based on Figures 4 and 5, we find that the regions allowed by the current dataset of NANOGrav (enclosed by black curves) can be thoroughly tested with SKA and μAres. Smaller values of $\mathcal{A}$, i.e., $\sim(10^{-4} - 10^{-7})$, are also expected to be reached by the planned GW experiments. The sensitivity of Advanced LIGO to measure the SIGWs is expected to be improved by one order of magnitude in upcoming observing runs. If other values of SNR are desired, we can obtain revised $\mathcal{A}_{\min}$ by rescaling the above results by multiplying a factor of $(\mathrm{SNR})^{1/2}$. After recasting the expected constraints on $\mathcal{A}$ into constraints on the mass function of PBHs, we further find that the full parameter space of PBHs, that account for at least a fraction of dark matter, e.g., $f_{\mathrm{pbh}} \sim 10^{-10}$, can be thoroughly explored with multi-band GW measurements, e.g., a detector network composed of SKA, μAres and ET, or other detector networks. Therefore, we expect that the scenario of PBHs as a candidate of dark matter can be supported or vetoed by future GW observations.

6. Conclusions and Discussion

In this work, we obtained new constraints on the power spectrum of primordial curvature perturbations and the mass function of PBHs by searching for the energy density fraction spectrum of SIGWs in the NANOGrav 12.5-year dataset. We found the lower limits on $\mathcal{A}$, namely, $\mathcal{A} \gtrsim 10^{-2}$ and showed the 95% CL contours of $\mathcal{A}$ and f_* (see Figure 2). Recasting these contours into the PBH mass-abundance plane, we showed the parameter space of PBHs allowed by the NANOGrav (see Figure 3). We found that at least a fraction of dark matter can be interpreted with the scenario of PBHs in the mass range $\sim(10^{-4} - 10^{-1})M_\odot$. We also studied dependence of the above results on the value of σ. Furthermore, this mass range is expected to be cross-checked with high-precision by observing the GWs from PBH mergers (see Figure 3 in Wang et al. [43]). We also found that the SIGW probe is much more powerful than other probes to search for the PBHs.

We further forecasted the sensitivity curves of ongoing and planned GW experiments on detection of PBHs and the primordial curvature perturbations by searching for SIGWs. We found that the primordial curvature perturbations with spectral amplitude larger than $\sim(10^{-4} - 10^{-7})$ can be measured with planned GW detection programs. The sensitivity of Advanced LIGO-Virgo to detect the SIGWs was expected to be improved by one order of magnitude in the near future. This prediction may promote extensive investigations of cosmic inflation at late-time stages. Meanwhile, the scenario of PBHs within almost whole mass range can be thoroughly tested, since the critical spectral amplitude to form the PBHs is much larger than 10^{-4} (see Figures 4 and 5). In particular, we can search for the PBHs within mass range $(10^{-16} - 10^{-11})M_\odot$, which can compose all of the dark matter and otherwise are beyond capabilities of other probes. In addition, we emphasized the importance of multi-band GW detector networks for accomplishing the above theoretical expectations.

In this paper, we made several assumptions to simplify our computations. First, we disregarded possible contributions of primordial non-Gaussianity to the formation of PBHs [46,83–86] and to the production of SIGWs [45]. It is an interesting topic to study the primordial non-Gaussianity, deserving an independent work. Second, we took into account the median value of the effective relativistic degrees of freedom of the early Universe, but disregarded their uncertainties [47]. In fact, changing the above two assumptions could alter the mass function of PBHs and the energy density fraction spectrum of SIGWs. We would leave a possible study of this question to future works. Third, we disregarded contributions of nonlinear cosmological perturbations to the energy density fraction spectrum of SIGWs, since there is not a complete theory of SIGWs at the third order [87–91]. We might revisit this assumption in future, once the theory is complete.

Author Contributions: Conceptualization, S.W.; methodology, Z.-C.Z. and S.W.; software, Z.-C.Z.; validation, Z.-C.Z. and S.W.; formal analysis, Z.-C.Z. and S.W.; investigation, Z.-C.Z. and S.W.; resources, S.W.; data curation, Z.-C.Z.; writing—original draft preparation, S.W.; writing—review and editing, Z.-C.Z. and S.W.; visualization, Z.-C.Z. and S.W.; supervision, S.W.; project administration, S.W.; funding acquisition, Z.-C.Z. and S.W. All authors have read and agreed to the published version of the manuscript.

Funding: This research was funded by the National Natural Science Foundation of China grant number 12175243 and grant number 12005016, the Key Research Program of the Chinese Academy of Sciences grant number XDPB15, and the science research grants from the China Manned Space Project grant number CMS-CSST-2021-B01.

Data Availability Statement: Publicly available datasets were analyzed in this study. This data can be found here: https://data.nanograv.org (accessed on 21 March 2023).

Acknowledgments: We acknowledge Zu-Cheng Chen, Jun-Peng Li and Ke Wang for helpful discussions.

Conflicts of Interest: The authors declare no conflict of interest.

Notes

1. http://data.nanograv.org, version v3, accessed on 21 March 2023.
2. https://member.ipmu.jp/satoshi.shirai/EOS2018.php, accessed on 1 January 2019.
3. https://www.astropy.org/, version 5.1, accessed on 21 March 2023.
4. https://vallis.github.io/libstempo/, version 2.4.5, accessed on 21 March 2023.
5. https://bitbucket.org/psrsoft/tempo2.git, version 2021.07.1, accessed on 21 March 2023.
6. https://github.com/nanograv/12p5yr_stochastic_analysis, accessed on 1 January 2022.
7. https://zenodo.org/record/4059815, version 3.2.3, accessed on 21 March 2023.
8. https://zenodo.org/record/1037579, version 2.0.0, accessed on 21 March 2023.

References

1. Hawking, S. Gravitationally collapsed objects of very low mass. *Mon. Not. R. Astron. Soc.* **1971**, *152*, 75. [CrossRef]
2. Khlopov, M.Y. Primordial Black Holes. *Res. Astron. Astrophys.* **2010**, *10*, 495–528. [CrossRef]
3. Belotsky, K.M.; Dmitriev, A.D.; Esipova, E.A.; Gani, V.A.; Grobov, A.V.; Khlopov, M.Y.; Kirillov, A.A.; Rubin, S.G.; Svadkovsky, I.V. Signatures of primordial black hole dark matter. *Mod. Phys. Lett. A* **2014**, *29*, 1440005. [CrossRef]
4. Franciolini, G.; Musco, I.; Pani, P.; Urbano, A. From inflation to black hole mergers and back again: Gravitational-wave data-driven constraints on inflationary scenarios with a first-principle model of primordial black holes across the QCD epoch. *arXiv* **2022**, arXiv:2209.05959.
5. Musco, I.; De Luca, V.; Franciolini, G.; Riotto, A. Threshold for primordial black holes. II. A simple analytic prescription. *Phys. Rev. D* **2021**, *103*, 063538. [CrossRef]
6. Guth, A.H. The Inflationary Universe: A Possible Solution to the Horizon and Flatness Problems. *Phys. Rev.* **1981**, *D23*, 347–356. [CrossRef]

7. Linde, A.D. A New Inflationary Universe Scenario: A Possible Solution of the Horizon, Flatness, Homogeneity, Isotropy and Primordial Monopole Problems. *Phys. Lett.* **1982**, *108B*, 389–393. [CrossRef]

8. Starobinsky, A.A. A New Type of Isotropic Cosmological Models Without Singularity. *Phys. Lett.* **1980**, *B91*, 99–102. [CrossRef]

9. Hinshaw, G.; Larson, D.; Komatsu, E.; Spergel, D.N.; Bennett, C.L.; Dunkley, J.; Nolta, M.R.; Halpern, M.; Hill, R.S.; Odegard, N. Nine-Year Wilkinson Microwave Anisotropy Probe (WMAP) Observations: Cosmological Parameter Results. *Astrophys. J. Suppl.* **2013**, *208*, 19. [CrossRef]

10. Aghanim, N.; Larson, D.; Komatsu, E.; Sperge, D.N.; Bennett, C.L.; Dunkley, J.; Nolta, M.R.; Halpern, M.; Hill, R.S.; Odegard, N. Planck 2018 results. VI. Cosmological parameters. *Astron. Astrophys.* **2020**, *641*, A6; Erratum in *Astron. Astrophys.* **2021**, *652*, C4. [CrossRef]

11. Alam, S.; Ata, M.; Bailey, S.; Beutler, F.; Bizyaev, D.; Blazek, J.A.; Bolton, A.S.; Brownstein, J.R.; Burden, A.; Chuang, C.-H. The clustering of galaxies in the completed SDSS-III Baryon Oscillation Spectroscopic Survey: Cosmological analysis of the DR12 galaxy sample. *Mon. Not. R. Astron. Soc.* **2017**, *470*, 2617–2652. [CrossRef]

12. Abbott, T.M.C. et al. [DES Collaboration] Dark Energy Survey Year 3 results: Cosmological constraints from galaxy clustering and weak lensing. *Phys. Rev. D* **2022**, *105*, 023520. [CrossRef]

13. Carr, B.J.; Hawking, S.W. Black holes in the early Universe. *Mon. Not. R. Astron. Soc.* **1974**, *168*, 399–415. [CrossRef]

14. Chapline, G.F. Cosmological effects of primordial black holes. *Nature* **1975**, *253*, 251–252. [CrossRef]

15. Carr, B.; Kohri, K.; Sendouda, Y.; Yokoyama, J. Constraints on primordial black holes. *Rept. Prog. Phys.* **2021**, *84*, 116902. [CrossRef] [PubMed]

16. Escrivà, A.; Kuhnel, F.; Tada, Y. Primordial Black Holes. *arXiv* **2022**, arXiv:2211.05767.

17. Kohri, K.; Sekiguchi, T.; Wang, S. Cosmological 21-cm line observations to test scenarios of super-Eddington accretion on to black holes being seeds of high-redshifted supermassive black holes. *Phys. Rev. D* **2022**, *106*, 043539. [CrossRef]

18. Sasaki, M.; Suyama, T.; Tanaka, T.; Yokoyama, S. Primordial Black Hole Scenario for the Gravitational-Wave Event GW150914. *Phys. Rev. Lett.* **2016**, *117*, 061101. [CrossRef]

19. Raidal, M.; Vaskonen, V.; Veermäe, H. Gravitational Waves from Primordial Black Hole Mergers. *JCAP* **2017**, *09*, 037. [CrossRef]

20. Ali-Haïmoud, Y.; Kovetz, E.D.; Kamionkowski, M. Merger rate of primordial black-hole binaries. *Phys. Rev. D* **2017**, *96*, 123523. [CrossRef]

21. Chen, Z.C.; Huang, Q.G. Merger Rate Distribution of Primordial-Black-Hole Binaries. *Astrophys. J.* **2018**, *864*, 61. [CrossRef]

22. Nishikawa, H.; Kovetz, E.D.; Kamionkowski, M.; Silk, J. Primordial-black-hole mergers in dark-matter spikes. *Phys. Rev.* **2019**, *D99*, 043533. [CrossRef]

23. Bird, S.; Cholis, I.; Munoz, J.B.; Ali-Haimoud, Y.; Kamionkowski, M.; Kovetz, E.D.; Raccanelli, A.; Riess, A.G. Did LIGO detect dark matter? *Phys. Rev. Lett.* **2016**, *116*, 201301. [CrossRef] [PubMed]

24. Wang, S.; Zhao, Z.C. GW200105 and GW200115 are compatible with a scenario of primordial black hole binary coalescences. *Eur. Phys. J. C* **2022**, *82*, 9. [CrossRef]

25. Belotsky, K.M.; Dokuchaev, V.I.; Eroshenko, Y.N.; Esipova, E.A.; Khlopov, M.Y.; Khromykh, L.A.; Kirillov, A.A.; Nikulin, V.V.; Rubin, S.G.; Svadkovsky, I.V. Clusters of primordial black holes. *Eur. Phys. J. C* **2019**, *79*, 246. [CrossRef]

26. Harry, G.M. Advanced LIGO: The next generation of gravitational wave detectors. *Class. Quant. Grav.* **2010**, *27*, 084006. [CrossRef]

27. Acernese, F.; Agathos, M.; Agatsuma, K.; Aisa, D.; Allemandou, N.; Allocca, A.; Amarni, J.; Astone, P.; Balestri, G.; Ballardin, G. Advanced Virgo: A second-generation interferometric gravitational wave detector. *Class. Quant. Grav.* **2015**, *32*, 024001. [CrossRef]

28. Somiya, K. Detector configuration of KAGRA: The Japanese cryogenic gravitational-wave detector. *Class. Quant. Grav.* **2012**, *29*, 124007. [CrossRef]

29. Mollerach, S.; Harari, D.; Matarrese, S. CMB polarization from secondary vector and tensor modes. *Phys. Rev.* **2004**, *D69*, 063002. [CrossRef]

30. Ananda, K.N.; Clarkson, C.; Wands, D. The Cosmological gravitational wave background from primordial density perturbations. *Phys. Rev.* **2007**, *D75*, 123518. [CrossRef]

31. Baumann, D.; Steinhardt, P.J.; Takahashi, K.; Ichiki, K. Gravitational Wave Spectrum Induced by Primordial Scalar Perturbations. *Phys. Rev.* **2007**, *D76*, 084019. [CrossRef]

32. Assadullahi, H.; Wands, D. Constraints on primordial density perturbations from induced gravitational waves. *Phys. Rev.* **2010**, *D81*, 023527. [CrossRef]

33. Kohri, K.; Terada, T. Semianalytic calculation of gravitational wave spectrum nonlinearly induced from primordial curvature perturbations. *Phys. Rev.* **2018**, *D97*, 123532. [CrossRef]

34. Espinosa, J.R.; Racco, D.; Riotto, A. A Cosmological Signature of the SM Higgs Instability: Gravitational Waves. *JCAP* **2018**, *09*, 012. [CrossRef]

35. Arzoumanian, Z.; Baker, P.T.; Blumer, H.; Bécsy, B.; Brazier, A.; Brook, P.R.; Burke-Spolaor, S.; Chatterjee, S.; Chen, S.; Cordes, J.M. The NANOGrav 12.5 yr Data Set: Search for an Isotropic Stochastic Gravitational-wave Background. *Astrophys. J. Lett.* **2020**, *905*, L34. [CrossRef]

36. De Luca, V.; Franciolini, G.; Riotto, A. NANOGrav Data Hints at Primordial Black Holes as Dark Matter. *Phys. Rev. Lett.* **2021**, *126*, 041303. [CrossRef]

37. Vaskonen, V.; Veermäe, H. Did NANOGrav see a signal from primordial black hole formation? *Phys. Rev. Lett.* **2021**, *126*, 051303. [CrossRef]
38. Kohri, K.; Terada, T. Solar-Mass Primordial Black Holes Explain NANOGrav Hint of Gravitational Waves. *Phys. Lett. B* **2021**, *813*, 136040. [CrossRef]
39. Domènech, G.; Pi, S. NANOGrav hints on planet-mass primordial black holes. *Sci. China Phys. Mech. Astron.* **2022**, *65*, 230411. [CrossRef]
40. Atal, V.; Sanglas, A.; Triantafyllou, N. NANOGrav signal as mergers of Stupendously Large Primordial Black Holes. *JCAP* **2021**, *06*, 022. [CrossRef]
41. Yi, Z.; Fei, Q. Constraints on primordial curvature spectrum from primordial black holes and scalar-induced gravitational waves. *arXiv* **2022**, arXiv:2210.03641.
42. Maggiore, M. *Gravitational Waves. Volume 1: Theory and Experiments*; Oxford Master Series in Physics; Oxford University Press: Oxford, UK, 2007.
43. Wang, S.; Terada, T.; Kohri, K. Prospective constraints on the primordial black hole abundance from the stochastic gravitational-wave backgrounds produced by coalescing events and curvature perturbations. *Phys. Rev. D* **2019**, *99*, 103531. [CrossRef]
44. Bugaev, E.; Klimai, P. Induced gravitational wave background and primordial black holes. *Phys. Rev. D* **2010**, *81*, 023517. [CrossRef]
45. Adshead, P.; Lozanov, K.D.; Weiner, Z.J. Non–Gaussianity and the induced gravitational wave background. *JCAP* **2021**, *10*, 080. [CrossRef]
46. Ferrante, G.; Franciolini, G.; Iovino, A.J. Primordial non-gaussianity up to all orders: Theoretical aspects and implications for primordial black hole models. *arXiv* **2022**, arXiv:2211.01728.
47. Saikawa, K.; Shirai, S. Primordial gravitational waves, precisely: The role of thermodynamics in the Standard Model. *JCAP* **2018**, *1805*, 035. [CrossRef]
48. Robitaille, T.P. et al. [The Astropy Collaboration] Astropy: A Community Python Package for Astronomy. *Astron. Astrophys.* **2013**, *558*, A33. [CrossRef]
49. Price-Whelan, A.M. et al. [The Astropy Collaboration] The Astropy Project: Building an Open-science Project and Status of the v2.0 Core Package. *Astron. J.* **2018**, *156*, 123. [CrossRef]
50. Price-Whelan, A.M. et al. [The Astropy Collaboration] The Astropy Project: Sustaining and Growing a Community-oriented Open-source Project and the Latest Major Release (v5.0) of the Core Package*. *Astrophys. J.* **2022**, *935*, 167. [CrossRef]
51. Arzoumanian, Z.; Baker, P.T.; Brazier, A.; Burke-Spolaor, S.; Chamberlin, S.J.; Chatterjee, S.; Christy, B.; Cordes, J.M.; Cornish, N.J.; Crawford, F.; et al. The NANOGrav 11-year Data Set: Pulsar-timing Constraints on the Stochastic Gravitational-wave Background. *Astrophys. J.* **2018**, *859*, 47. [CrossRef]
52. Maggiore, M. Gravitational wave experiments and early universe cosmology. *Phys. Rept.* **2000**, *331*, 283–367. [CrossRef]
53. Hellings, R.W.; Downs, G.S. Upper limits on the isotropic gravitational radiation background from pulsar timing analysis. *Astrophys. J. Lett.* **1983**, *265*, L39–L42. [CrossRef]
54. Arzoumanian, Z.; Brazier, A.; Burke-Spolaor, S.; Chamberlin, S.; Chatterjee, S.; Christy, B.; Cordes, J.; Cornish, N.; Demorest, P.; Deng, X.; et al. The NANOGrav Nine-year Data Set: Limits on the Isotropic Stochastic Gravitational Wave Background. *Astrophys. J.* **2016**, *821*, 13. [CrossRef]
55. Michele, V. Libstempo: Python Wrapper for Tempo2. Astrophysics Source Code Library, Record ascl:2002.017. 2020. Available online: https://bitbucket.org/psrsoft/tempo2.git (accessed on 21 March 2023).
56. Hobbs, G.; Edwards, R.T.; Manchester, R.N. Tempo2, a new pulsar timing package. 1. overview. *Mon. Not. R. Astron. Soc.* **2006**, *369*, 655–672. [CrossRef]
57. Edwards, R.T.; Hobbs, G.B.; Manchester, R.N. Tempo2, a new pulsar timing package. 2. The timing model and precision estimates. *Mon. Not. R. Astron. Soc.* **2006**, *372*, 1549–1574. [CrossRef]
58. Folkner, W.M.; Park, R.S. *Planetary Ephemeris DE438 for Juno*; Technical Report IOM 392R-18-004; CAJet Propulsion Laboratory: Pasadena, CA, USA, 2018.
59. van Haasteren, R.; Vallisneri, M. New advances in the Gaussian-process approach to pulsar-timing data analysis. *Phys. Rev. D* **2014**, *90*, 104012. [CrossRef]
60. Ellis, J.A.; Vallisneri, M.; Taylor, S.R.; Baker, P.T. *ENTERPRISE: Enhanced Numerical Toolbox Enabling a Robust PulsaR Inference SuitE*; Zenodo: Genève, Switzerland, 2020. [CrossRef]
61. Ellis, J.; van Haasteren, R. *jellis18/PTMCMCSampler: Official Release*; Zenodo: Genève, Switzerland, 2017. [CrossRef]
62. Carr, B.; Kuhnel, F.; Sandstad, M. Primordial Black Holes as Dark Matter. *Phys. Rev.* **2016**, *D94*, 083504. [CrossRef]
63. Yokoyama, J. Cosmological constraints on primordial black holes produced in the near critical gravitational collapse. *Phys. Rev.* **1998**, *D58*, 107502. [CrossRef]
64. Press, W.H.; Schechter, P. Formation of galaxies and clusters of galaxies by selfsimilar gravitational condensation. *Astrophys. J.* **1974**, *187*, 425–438. [CrossRef]
65. Byrnes, C.T.; Hindmarsh, M.; Young, S.; Hawkins, M.R.S. Primordial black holes with an accurate QCD equation of state. *JCAP* **2018**, *08*, 041. [CrossRef]
66. Young, S.; Byrnes, C.T.; Sasaki, M. Calculating the mass fraction of primordial black holes. *JCAP* **2014**, *1407*, 045. [CrossRef]

67. Ando, K.; Inomata, K.; Kawasaki, M. Primordial black holes and uncertainties in the choice of the window function. *Phys. Rev.* **2018**, *D97*, 103528. [CrossRef]
68. Wang, S.; Vardanyan, V.; Kohri, K. Probing primordial black holes with anisotropies in stochastic gravitational-wave background. *arXiv* **2021**, arXiv:2107.01935.
69. Dewdney, P.E.; Hall, P.J.; Schilizzi, R.T.; Lazio, T.J.L. The square kilometre array. *Proc. IEEE* **2009**, *97*, 1482–1496. [CrossRef]
70. Weltman, A.; Bull, P.; Camera, S.; Kelley, K.; Padmanabhan, H.; Pritchard, J.; Raccanelli, A.; Riemer-Sørensen, S.; Shao, L.; Andrianomena, S.; et al. Fundamental physics with the Square Kilometre Array. *Publ. Astron. Soc. Austral.* **2020**, *37*, e002. [CrossRef]
71. Moore, C.J.; Cole, R.H.; Berry, C.P.L. Gravitational-wave sensitivity curves. *Class. Quant. Grav.* **2015**, *32*, 015014. [CrossRef]
72. Sesana, A.; Korsakova, N.; Sedda, M.A.; Baibhav, V.; Barausse, E.; Barke, S.; Berti, E.; Bonetti, M.; Capelo, P.R.; Caprini, C.; et al. Unveiling the gravitational universe at μ-Hz frequencies. *Exper. Astron.* **2021**, *51*, 1333–1383. [CrossRef]
73. Amaro-Seoane, P.; Audley, H.; Babak, S.; Baker, J.; Barausse, E.; Bender, P.; Berti, E.; Binetruy, P.; Born, M.; Bortoluzzi, D.; et al. Laser Interferometer Space Antenna. *arXiv* **2017**, arXiv:1702.00786.
74. Robson, T.; Cornish, N.J.; Liu, C. The construction and use of LISA sensitivity curves. *Class. Quant. Grav.* **2019**, *36*, 105011. [CrossRef]
75. Crowder, J.; Cornish, N.J. Beyond LISA: Exploring future gravitational wave missions. *Phys. Rev. D* **2005**, *72*, 083005. [CrossRef]
76. Harry, G.M.; Fritschel, P.; Shaddock, D.A.; Folkner, W.; Phinney, E.S. Laser interferometry for the big bang observer. *Class. Quant. Grav.* **2006**, *23*, 4887–4894. [CrossRef]
77. Sato, S.; Seiji, K.; Masaki, A.; Takashi, N.; Kimio, T.; Akito, A.; Ikkoh, F.; Kunihito, I.; Nobuyuki, K.; Shigenori, M.; et al. The status of DECIGO. *J. Phys. Conf. Ser.* **2017**, *840*, 012010. [CrossRef]
78. Kawamura, S.; Ando, M.; Seto, N.; Sato, S.; Musha, M.; Kawano, I.; Yokoyama, J.; Tanaka, T.; Ioka, K.; Akutsu, T.; et al. Current status of space gravitational wave antenna DECIGO and B-DECIGO. *PTEP* **2021**, *2021*, 05A105. [CrossRef]
79. Punturo, M.; Abernathy, M.; Acernese, F.; Allen, B.; Andersson, N.; Arun, K.; Barone, F.; Barr, B.; Barsuglia, M.; Beker, M. The Einstein Telescope: A third-generation gravitational wave observatory. *Class. Quant. Grav.* **2010**, *27*, 194002. [CrossRef]
80. Schmitz, K. New Sensitivity Curves for Gravitational-Wave Signals from Cosmological Phase Transitions. *JHEP* **2021**, *01*, 097. [CrossRef]
81. Campeti, P.; Komatsu, E.; Poletti, D.; Baccigalupi, C. Measuring the spectrum of primordial gravitational waves with CMB, PTA and Laser Interferometers. *JCAP* **2021**, *01*, 012. [CrossRef]
82. Romero-Rodriguez, A.; Martinez, M.; Pujolàs, O.; Sakellariadou, M.; Vaskonen, V. Search for a Scalar Induced Stochastic Gravitational Wave Background in the Third LIGO-Virgo Observing Run. *Phys. Rev. Lett.* **2022**, *128*, 051301. [CrossRef]
83. Franciolini, G.; Kehagias, A.; Matarrese, S.; Riotto, A. Primordial Black Holes from Inflation and non-Gaussianity. *JCAP* **2018**, *03*, 016. [CrossRef]
84. Gow, A.D.; Assadullahi, H.; Jackson, J.H.P.; Koyama, K.; Vennin, V.; Wands, D. Non-perturbative non-Gaussianity and primordial black holes. *arXiv* **2022**, arXiv:2211.08348.
85. Cai, Y.F.; Ma, X.H.; Sasaki, M.; Wang, D.G.; Zhou, Z. Highly non-Gaussian tails and primordial black holes from single-field inflation. *arXiv* **2022**, arXiv:2207.11910.
86. Kitajima, N.; Tada, Y.; Yokoyama, S.; Yoo, C.M. Primordial black holes in peak theory with a non-Gaussian tail. *JCAP* **2021**, *10*, 053. [CrossRef]
87. Yuan, C.; Chen, Z.C.; Huang, Q.G. Probing primordial–black-hole dark matter with scalar induced gravitational waves. *Phys. Rev. D* **2019**, *100*, 081301. [CrossRef]
88. Zhou, J.Z.; Zhang, X.; Zhu, Q.H.; Chang, Z. The third order scalar induced gravitational waves. *JCAP* **2022**, *05*, 013. [CrossRef]
89. Chang, Z.; Zhang, X.; Zhou, J.Z. Primordial black holes and third order scalar induced gravitational waves. *arXiv* **2022**, arXiv:2209.12404.
90. Chen, C.; Ota, A.; Zhu, H.Y.; Zhu, Y. Missing one-loop contributions in secondary gravitational waves. *arXiv* **2022**, arXiv:2210.17176.
91. Meng, D.S.; Yuan, C.; Huang, Q.G. One-loop correction to the enhanced curvature perturbation with local-type non-Gaussianity for the formation of primordial black holes. *Phys. Rev. D* **2022**, *106*, 063508. [CrossRef]

Article

Do White Holes Exist?

Enrique Gaztanaga [1,2,3]

[1]   Institute of Space Sciences (ICE, CSIC), 08193 Barcelona, Spain; gaztanaga@darkcosmos.com
[2]   Institut d Estudis Espacials de Catalunya (IEEC), 08034 Barcelona, Spain
[3]   Institute of Cosmology & Gravitation, University of Portsmouth, Dennis Sciama Building, Burnaby Road, Portsmouth PO1 3FX, UK

Abstract: In a paper published in 1939, Albert Einstein argued that Black Holes (BHs) did not exist "in the real world". However, recent astronomical observations indicate otherwise. Does this mean that we should also expect White Holes (WHs) to exist in the real world? In classical General Relativity (GR), a WH refers to the time reversed version of a collapsing BH solution that allows the crossing of the BH event horizon inside out. Such solution has been disputed as not possible because escaping an event horizon violates causality. Despite such objections, the Big Bang model is often understood as a WH (the reverse of a BH collapse). Does this mean that the Big Bang breaks causality? Recent measurements of cosmic acceleration indicate that our Big Bang solution is not really a WH, but a BH. Events decelerate when the expansion accelerates and this prevents the crossing of the event horizon from inside out. We present a general explanation of why this happens; the explanation resolves the above causality puzzle and indicates that such apparent WH solutions have a regular Schwarzschild BH exterior.

Keywords: cosmology; dark energy; general relativity; black holes

Citation: Gaztanaga, E. Do White Holes Exist? *Universe* **2023**, *9*, 194. https://doi.org/10.3390/universe9040194

Academic Editors: Gonzalo J. Olmo and Kazuharu Bamba

Received: 9 March 2023
Revised: 9 April 2023
Accepted: 17 April 2023
Published: 19 April 2023

1. Introduction

Decades after Einstein [1] concluded that BHs did not exist, observations have shown that they are real astronomical objects [2–4]. A Schwarzschild Black Hole (BH) solution:

$$ds^2 = g_{\mu\nu}dx^\mu dx^\nu = -[1 - r_S/r]\,dt^2 + \frac{dr^2}{[1 - r_S/r]} + r^2 d\Omega^2, \tag{1}$$

represents a singular point source of mass M. The gravitational radius $r_S \equiv 2GM$ corresponds to an event horizon and prevents us from seeing inside r_S. The Schwarszchild solution applies to the exterior of any BH, no matter what the interior solution is, as long as we can approximate the outer region as empty space. The Hawking–Penrose's theorems [5,6] tell us that nothing can come out of r_S. This has created the BH information lost paradox [7,8]. One possible way around this is to introduce the concept of maximally extended Schwarszchild solution using the Kruskal–Szekeres coordinates $T = T(t,r)$ and $X = X(t,r)$ (see Figure 1), where the future BH event horizon becomes the past White Hole (WH) horizon. Information can escape r_S in a WH. There are two disconnected exterior spaces which could be connected inside with an Einstein–Rosen bridge or Schwarszchild wormhole [9].

If we throw a particle into a BH, the WH solution corresponds to the traveling of that particle back in time to us (from our past), before the particle was sent. Such trajectory might be formally possible (because there is no arrow of time at the fundamental level), but it violates causality, so it makes no physical sense as a classical solution (quantum mechanics effects might provide some way around this [10]). This is related to the example of retarded and advanced potentials in classical electrodynamics: both are mathematical solutions of the wave equations, but only one of them connects cause and effect. The mirror image of the top quadrant in Figure 1 has the arrow pointing downward and not upward. This shows that the time reversed solution (the mirror image) is still a BH (where the particle falls into the gravitational radius r_S) and not a WH (as indicated in the figure).

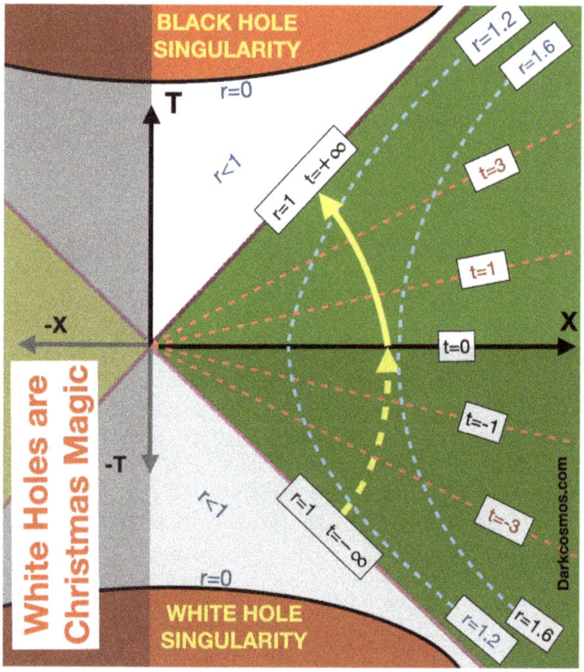

Figure 1. Relation between the Schwarszchild coordinates (t, r) (in units of $r_S = 1$) and the Kruskal–Szekeres coordinates (T, X). The top right quadrant ($T > 0$, $X > 0$) is the regular BH solution (the green region is external to r_S). Lines of constant r are the cyan hyperbolic dashed lines. Lines of constant time t are orange dashed straight lines ($T = 0$ is also $t = 0$). If we radially throw a test particle at $t = 0$, it will follow the continuous yellow arrow. This solution can be formally extended into the bottom right quadrant, which corresponds to a WH, and is the time reversed (horizontal flip) of the BH solution. A particle can escape the event horizon of the WH (dashed yellow arrow), but only before it is thrown! This violates causality and is therefore not a physical solution. With a vertical flip, the solution could be maximally extended to a negative radius $X < 0$. But this generates a disconnected external space (yellow region), which is also not part of the original solution.

Here we will study the more realistic case of classical Lemaitre–Tolman–Bondi (LTB) solutions, which include the FLRW metric, the Oppenheimer–Snyder BH [11] and the thin shell BH [12] as particular cases. For some reason, i.e., the difficulty of a black-to-white hole bounce (see [13,14]), these solutions are usually investigated only as BH collapsing solutions. As we will show, the same solutions also exist, in principle, as WH solutions. The most famous of this is the expanding Big Bang model originally proposed by Friedman in 1922 [15] and Lemaitre in 1927 [16]. However, at closer inspection, these solutions need to be modified to include a surface term. After that correction, we show that the WH correspond, in fact, to BH expanding solutions. Our argument is supported by considering surface terms in the Einstein–Hilbert action of classical GR and also by the recent observation that our cosmic expansion is accelerating.

2. LTB Solutions

The most general metric with spherical symmetry in spherical coordinates $dx^\mu = (dt, dr, d\theta, d\phi)$ can be written as [17]:

$$ds^2 = -A(t,r)\ dt^2 + B(t,r)\ dr^2 + R^2(t,r)d\Omega^2. \tag{2}$$

Other common notation is: $A \equiv e^\nu$, $B \equiv e^\lambda$ and $R^2 \equiv e^\mu$ [11,18,19]. An alternative to this uses proper time $dx^\mu = (d\tau, d\chi, d\theta, d\phi)$:

$$ds^2 = -d\tau^2 + e^{\lambda(\tau,\chi)}d\chi^2 + r^2(\tau,\chi)d\Omega^2, \tag{3}$$

where the radial coordinate χ can be comoving or not (because its evolution can be encoded in the λ and r functions). This last metric is sometimes called the *Lemaitre–Tolman* metric [16,18] or the *Lemaitre–Tolman–Bondi* or *LTB metric*. A metric such as this one, expressed with $g_{00} = 1$ and $g_{0\mu} = 0$ is called *synchronous* (or in a *synchronous frame*) because time lines are geodesics. Either way, it is possible to express the spherical symmetric metric with two functions. The best form in each case depends on the energy content and the observer's frame. In all cases, this is a *local metric* around a reference central point in space which we have set to be the origin ($\vec{r} = 0$).

The advantage of using the proper time and an observer moving with a perfect fluid is that the stress tensor becomes diagonal: $T^\nu_\mu = diag[-\rho, p, p, p]$, where $\rho = \rho(\tau, \chi)$ is the energy density and $p = p(\tau, \chi)$ is the pressure. We will focus here in the matter-dominated case $p = 0$ for simplicity, but we expect similar results to apply to more general situations (see [20]). The solution to the field equation $8\pi G T^1_0 = G^1_0 = 0$ is $\dot{\lambda} r' = 2\dot{r}'$, where dots and primes correspond to time τ and radial χ partial derivatives. This equation can be solved as $e^\lambda = C r'^2$, where $C = C(\chi)$ is an arbitrary function of χ. The choice $C = 1$ corresponds to the particular flat geometry case:

$$ds^2 = -d\tau^2 + [\partial_\chi r]^2 d\chi^2 + r(\tau,\chi)^2 d\Omega^2, \tag{4}$$

The solution for r in this case is easily found:

$$H^2 \ \equiv\ r_H^{-2} \equiv \left(\frac{\dot{r}}{r}\right)^2 = \frac{2GM}{r^3} \tag{5}$$

$$M \ \equiv\ 4\pi \int_0^\chi \rho(\tau,\chi) r' r^2 d\chi = M(\chi) \tag{6}$$

The above expression reproduces the Newtonian energy conservation in free fall: $\frac{1}{2}\dot{r}^2 = GM/r$ [21] and corresponds to an expanding or collapsing relativistic spherical ball. When $\rho = \rho(\tau)$ is uniform, we find $r = a(\tau)\chi$ so that Equation (4) reproduces the flat FLRW metric:

$$ds^2 = -d\tau^2 + a^2(\tau)\left[d\chi^2 + \chi^2 d\Omega^2\right], \tag{7}$$

and Equation (5) reproduces the corresponding solution $3H^2 = 8\pi G\rho$. The next simplest solution to Equation (5) is that of the FLRW uniform cloud with a fixed total mass M_T:

$$M_T \equiv \int_0^\infty \rho\, 4\pi r^2 (\partial_\chi r) d\chi \tag{8}$$

The solution is $r = a\chi$ as in the standard FLRW metric but with a boundary at $R(\tau) \equiv a(\tau)\chi_*$ above which ($\chi > \chi_*$) we have empty space: $\rho = 0$.

This is a consequence of *Birkhoff's theorem* [22] (or *Gauss' law* in non relativistic mechanics), since a sphere cut out of an infinite uniform distribution conserves the same spherical symmetry and the solutions are independent of what is outside. If the outer region is empty space, we just recover the static Schwarzschild solution outside and the FLRW metric inside. Thus, the FLRW metric is both a solution to a global homogeneous (i.e., $M_T = \infty$) uniform background and also to the inside of a local (finite M_T) uniform sphere centered around one particular point. The local solution is called the FLRW cloud (FLRW*) [20]. As we will show next, the LTB solution could in principle be viewed as a BH or a WH, depending on whether the FLRW metric is expanding or collapsing.

A timelike radial geodesic ($d\chi = 0$) in the FLRW cloud has a mass-energy M inside χ which is independent of τ. Such fixed comoving coordinate $\chi = \chi_*$ corresponds to a system with a fixed mass M_T inside (see also [23]), which expands or collapses following the Hubble–Lemaitre law of Equation (5).

From Equation (5), we have $H = H_S(a/a_S)^{-3/2} = \pm 1/\tau$, where H_S is just the value at some arbitrary time ($a = a_S$), when R intersects r_S, so that $r_S = a_S\chi_* = 1/H_S = \pm 2GM_T$. This solution is time reversible and the evolution can cross r_S in both directions. This is a well known solution which includes the Oppenheimer–Snyder BH collapse [11] and the thin shell BH [12]. However, note that when $R < r_S$, we have $R > r_H$ (or $\dot{R} > 1$), which creates a region between $R > r > r_H$ which is acausal during expansion (this is the well known horizon problem in the standard Big Bang cosmology). We can also reproduce the same LTB (or FLRW*) solution using junction conditions to verify that the exterior of r_S is indeed a classical (Schwarszchild) BH despite the looks of Equation (4). The original derivation [20] is reproduced here in Appendix A (with some typos corrected) for reference.

To show that this solution actually crosses the gravitational radius r_S, we can estimate the event horizon (EH), R_{EH}, of the FLRW* metric. This is the maximum distance that a photon emitted at time τ can travel following an outgoing radial null geodesic [24]:

$$R_{EH} = a(\tau) \int_{\tau}^{\infty} \frac{d\tau}{a(\tau)} = a \int_{a}^{\infty} \frac{da}{H(a)a^2} \tag{9}$$

For $H \sim a^{-3/2}$, we have $R_{EH} \sim a^{3/2}$, which grows unbounded with a and therefore crosses r_S, as shown by the dashed red line in Figure 2.

The case $H_S < 0$ corresponds to a collapsing solution, and therefore, a BH. This collapsing solution is protected by the Equivalence principle, as a free fall test particle at $r > R$ is equivalent to a particle moving in empty space and can therefore cross r_S. The case $H_S > 0$ is expanding and is what we have labeled as a WH solution. It just corresponds to a fluid expanding inside r_S. However, what is strange about this solution is that information can actually escape from the interior to the exterior of r_S, which is contrary to all that we have learned about BHs and causality. How is that possible?

The standard objection to this paradox is that this expanding configuration can never be achieved. This is reflected in the fact that $R > r_H$ is not causally connected to its past (the so called horizon problem), which is a similar objection to the one for WH interpretation of the Schwarzschild solution, as discussed in the introduction. Note that this expanding solution corresponds to the matter-dominated Big Bang solution, which is very close to current observations [1]. This is why it is often said that the Big Bang is a WH [2].

Here, we argue that this expanding WH solution is not correct. This is not because it cannot be achieved (as illustrated by the existence of our own observed universe). But because the gravitational radius r_S should be interpreted as a boundary that separates the interior from the exterior manifold. This is strictly the case if the exterior is empty space (as we have assumed here) [3]. Such boundary requires that we change the GR field equations. Appendix B reproduces the original calculation in [20,25] that shows that the Gibbons–Hawking–York (GHY) boundary in the action corresponds to an effective Λ term: $\Lambda = 3/r_S^2$. We will show next how this boundary term transforms the WH solution into a BH solution.

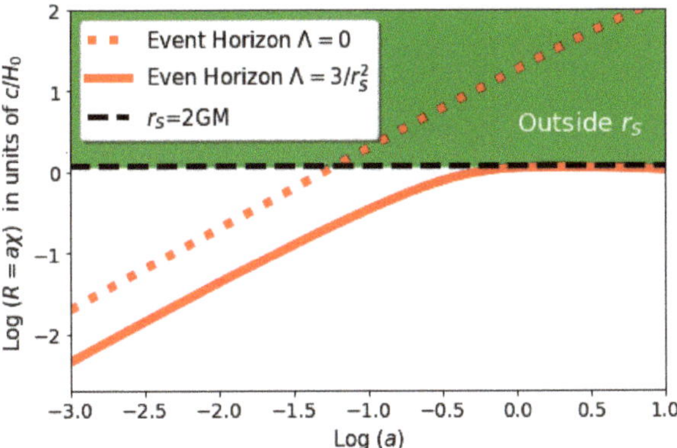

Figure 2. Event horizon in Equation (9) as a function of cosmic time (given by the scale factor *a*) for a matter-dominated FLRW metric ($\Omega_m = 1$, $\Omega_\Lambda = 0$, dashed red line) and for one which also has a $\Lambda = 3/r_S$ term ($\Omega_m = 0.25$, $\Omega_\Lambda = 0.75$, continuous red line).

How a WH Turns into a BH

Let us next consider how the derivation presented in Appendix A, which assumed $\Lambda = 0$, changes when including an effective Λ term: $\Lambda = 3/r_S^2$ inside r_S (as suggested by the GHY boundary argument given above). Such Λ term does not change the form of the FLRW metric itself, but (as it is well known) it changes the field equations and therefore the solution to expansion rate $3H^2 = 8\pi G\rho + \Lambda$. But the Λ term does change the form of the Schwarszchild solution and metric inside. The solution now is the deSitter—Schwarzschild metric: $F = 1 - r_S/R - R^2/r_S^2$. Thus, to find the new junction, we just need to replace F in the definition of β in Equation (A4). The new second junction condition then becomes:

$$R = \left[\frac{r_H^2 r_S^3}{r_S^2 - r_H^2}\right]^{1/3} \quad \text{or} \quad H^2 = \frac{r_S}{R^3} + \frac{1}{r_S^2} \tag{10}$$

which is exactly the new Hubble law with $\Lambda = 3/r_S^2$ and a constant mass $M = 4/3\pi\rho R^3$ in Equation (6). This shows that the LTB (or FLRW*) expanding metric is also a solution to the new field equations with the r_S boundary. However, this solution is no longer a WH, but has become a BH. We can check this by estimating the new EH in Equation (9), now including the effective Λ term in H. The new estimation for R_{EH} is displayed as a red continuous line in Figure 2. As can be seen, the EH is trapped inside r_S, which indicates that no information can escape. The WH solution has now turn into a BH.

3. Conclusions

We have shown that classical WH solutions in GR can be turned into an expanding BH solutions once we account for the fact that the gravitational radius r_S corresponds to a boundary condition in the action of GR.

The matter-dominated case study here is a very good approximation for our universe, because in the later stages of its evolution, it is totally dominated by mater and the effective $\Lambda = 3/r_S^2$. This could also be in general a good approximation for stellar or supermassive BHs with uniform density and pressure because as $a \to \infty$ inside, matter and Λ always dominate. The characteristic gravitational time is quite short:

$$\tau \sim GM \simeq 1.1 \times 10^{-13} \frac{M}{M_\odot} \text{yr}, \tag{11}$$

so, even for a super massive BH ($M \sim 10^9 M_\odot$), time is measured in seconds or hours. In astronomical time-scales, the evolution is quickly dominated by the effective $\Lambda = 1/r_S^2$ term inside. This, by the way, explains the coincidence problem in our Universe [26].

If we think of experimental cosmology before the year 2003 (i.e., ignore cosmic acceleration for a minute), the LTB expanding WH solution in Equation (5) (with $H > 0$) agrees very well with all the observations at that time, which favored a matter-dominated universe (the so called EdS universe with $\Omega_m = 1$). This is why some people still say that the Big Bang is a WH. However, today, we know that the universe has an effective Λ term, and this indicates that we are inside a BH [20,27]. Here, we interpret the observed Λ to be an effective term that corresponds to the gravitational radius $r_S = \sqrt{3/\Lambda} = 2GM_T$ of our local universe. Such BH Universe (BHU) could be within a larger background that may or may not be totally empty. In the later case, r_S will increase because of accretion from the outside. This case needs to be studied in more detail, but it could result in an effective Λ term that slowly decreases with time.

In terms of the proper coordinate radius r in Equation (5), the universe seems to enter a phase of cosmic acceleration because of the effective $\Lambda = 3/r_S^2$ term. However, this description is coordinate (or gauge) dependent. In terms of the more physical R_{EH} radius in Equation (9), the effect of r_S (or Λ) is, in fact, to decelerate events and bring cosmic expansion to a halt or frozen state.

This is illustrated in Figure 2. The case with a dashed line ($\Lambda = 0$) represents the always accelerating solution ($\ddot{R}_{EH} > 0$), whereas the case of the continuous line (with $\Lambda \neq 0$) becomes a decelerating solution that asymptotically stops ($\ddot{R}_{EH} < 0$) at $R_{EH} = r_S$. Thus, events in the universe decelerate (and not accelerate) because of Λ. It is therefore more appropriate to say that our physical universe is decelerating and it is described by an expanding metric inside a BH and not by a WH solution. This same conclusion also applies to the most general BH (or time reverse WH) solutions described by the spherically symmetric metric of Equation (4) with finite mass in Equation (6): the solutions are always trapped BH and not WH solutions.

Funding: This work was partially supported by grants from Spain Plan Nacional (PGC2018-102021-B-100) and Maria de Maeztu (CEX2020-001058-M) and from European Union LACEGAL 734374 and EWC 776247. IEEC is funded by Generalitat de Catalunya.

Data Availability Statement: No new data are presented.

Conflicts of Interest: The author declares no conflict of interest.

Appendix A. Timelike Junction

To match the FLRW metric in Equation (7) to the Schwarszchild metric in Equation (1), we chose a timelike 3D surface Σ that is fixed in comoving coordinates at $\chi = \chi_*$. Thus, Σ corresponds to a spherical shell with radius $R = a\chi_*$ that follows a radial geodesic trajectory. The interior of such shell corresponds to a FLRW cloud of fixed mass M_T that is expanding or contracting. We use 3D indexes (α, β) to label the matching shell. We can then take the 3D subset of FLRW coordinates $dy^\alpha = (d\tau, d\delta, d\theta)$ as cordinates in the shell, so that the induced metric, $h^-_{\alpha\beta}$, is:

$$ds^2_{\Sigma^-} = h^-_{\alpha\beta} dy^\alpha dy^\beta = -d\tau^2 + a^2(\tau)\chi_*^2 d\Omega^2 \tag{A1}$$

Since we have spherical symmetry, we take the solid angle $d\Omega$ and the angular coordinates to be the same in the FLRW and the Schwarszchild metrics. Thus, the only remaining variable is the FLRW comoving time τ. For the outside Schwarszchild frame with coordinates $dx^\mu = (dt, dr, d\delta, d\theta)$ in Equation (1), the same junction Σ^+ is described by some unknown functions $r = R(\tau)$ and $t = T(\tau)$, where t and r are the time and radial coordinates in the physical Schwarszchild frame of Equation (1). We then have:

$$dr = \dot{R}d\tau \;\; ; \;\; dt = \dot{T}d\tau, \tag{A2}$$

where the dot here refers to τ time derivatives. The induced metric h^+ estimated from the outside BH Schwarszchild metric (in Equation (1)) becomes:

$$
\begin{aligned}
ds^2_{\Sigma+} &= h^+_{\alpha\beta}dy^\alpha dy^\beta = -Fdt^2 + \frac{dr^2}{F} + r^2 d\Omega^2 \\
&= -(F\dot{T}^2 - \dot{R}^2/F)d\tau^2 + R^2 d\Omega^2
\end{aligned}
\tag{A3}
$$

where we have used $F \equiv 1 - r_S/R$ to simplify the following equations. Comparing Equation (A1) with Equation (A3), we can see that the matching condition $h^- = h^+$ results in

$$
R(\tau) = a(\tau)\chi_* \; ; \; F\dot{T} = \sqrt{\dot{R}^2 + F} \equiv \beta,
\tag{A4}
$$

so that given $a(\tau)$ and χ_* in the FLRW metric, we can simply find both $R(\tau)$ and $\beta(\tau)$ from the above equations. The second matching condition is that the derivative of the two metrics must also be continuous at Σ. This requires that the extrinsic curvature $K^\pm$ normal to Σ is the same from each side of the hypersurface ($\Sigma^\pm$) as

$$
K_{\alpha\beta} = -[\partial_a n_b - n_c \Gamma^c_{ab}]e^a_\alpha e^b_\beta
\tag{A5}
$$

where n_a is the 4D vector normal to Σ and $e^a_\alpha = \partial x^a/\partial y^\alpha$. Thus, $u^a = e^a_\tau = (1,0,0,0)$ is the outward 4D velocity and $n^- = (0,a,0,0)$ is the normal to Σ^- on the inside. On the outside, $u^a = (\dot{T},\dot{R},0,0)$ and $n^+ = (-\dot{R},\dot{T},0,0)$. We thus have $n_a u^a = 0$ and $n_a n^a = +1$, as expected for a timelike surface, from both from the inside n^- and from the outside n^+. The extrinsic curvature, K^-, estimated for the inside FLRW metric is then:

$$
\begin{aligned}
K^-_{\tau\tau} &= -(\partial_\tau n^-_\tau - a\Gamma^\chi_{\tau\tau})e^\tau_\tau e^\tau_\tau = 0 \\
K^-_{\theta\theta} &= a\Gamma^\chi_{\theta\theta}e^\theta_\theta e^\theta_\theta = -a\chi_* = R
\end{aligned}
\tag{A6}
$$

where we have used the Christoffel symbols for the FLRW:

$$
\begin{aligned}
\Gamma^\tau_{\tau\tau} = \Gamma^\tau_{\tau\chi} = \Gamma^\chi_{\tau\tau} = \Gamma^\chi_{\chi\chi} = 0 \; &; \; \Gamma^\tau_{\theta\theta} = -a^2\chi_*^2 H \\
\Gamma^\chi_{\tau\chi} = \Gamma^\tau_{\chi\chi}a^{-2} = -H \; &; \; \Gamma^\chi_{\theta\theta} = \chi_*
\end{aligned}
\tag{A7}
$$

The extrinsic curvature K^+ estimated for the outside Schwarszchild metric:

$$
\begin{aligned}
K^+_{\tau\tau} &= \ddot{R}\dot{T} - \dot{R}\ddot{T} + \frac{\dot{T}r_S}{2R^2F}(\dot{T}^2F^2 - 3\dot{R}^2) = \frac{\dot{\beta}}{\dot{R}} \\
K^+_{\theta\theta} &= \dot{T}\Gamma^r_{\theta\theta} = \dot{T}FR = \beta R
\end{aligned}
\tag{A8}
$$

where we have used the definition of β in Equation (A4) and the Christoffel symbols for the Schwarszchild metric:

$$
\begin{aligned}
\Gamma^t_{tt} = \Gamma^r_{tr} = \Gamma^t_{\theta\theta} = 0 \; &; \; \Gamma^r_{\theta\theta} = FR; \\
\Gamma^t_{tr} = -\Gamma^r_{rr} &= \Gamma^r_{tt}F^{-2} = \frac{r_S}{2FR^2}
\end{aligned}
\tag{A9}
$$

Note how $K_{\delta\delta} = \sin^2\theta K_{\theta\theta}$, so that $K^-_{\delta\delta} = K^+_{\delta\delta}$ follows from $K^-_{\theta\theta} = K^+_{\theta\theta}$. Comparing Equation (A6) with Equation (A8), the second matching condition: $K^-_{\alpha\beta} = K^+_{\alpha\beta}$ requires $\beta = 1$, which using Equation (A4), results in

$$
R = \left[r_H^2 r_S\right]^{1/3}
\tag{A10}
$$

This just reproduces the LTB (or FLRW*) in Equation (5) with $M = M_T$ inside R in Equation (6). The time equation is $\dot{T} = (1 - R^2H^2)^{-1}$.

Appendix B. The GHY Boundary Term

The Einstein–Hilbert action is given by [17,28–30]:

$$S = S_4 \equiv \int_{V_4} dV_4 \left[\frac{R - 2\Lambda}{16\pi G} + \mathcal{L} \right], \tag{A11}$$

where V_4 is the volume of the 4D spacetime manifold, $dV_4 = \sqrt{-g}\, d^4x$ is the invariant volume element, $R = g^{\mu\nu} R_{\mu\nu}$ is the Ricci scalar and $\mathcal{L}$ is the energy-matter Lagrangian. Einstein's field equations can be obtained for the metric field $g_{\mu\nu}$ by requiring S to be stationary ($\delta S = 0$) under arbitrary variations of the metric $\delta g^{\mu\nu}$. The solution is well known [17,30,31]:

$$G_{\mu\nu} + \Lambda g_{\mu\nu} = 8\pi G\, T_{\mu\nu} \equiv -\frac{16\pi G}{\sqrt{-g}} \frac{\delta(\sqrt{-g}\mathcal{L})}{\delta g^{\mu\nu}},$$

where $G_{\mu\nu} \equiv R_{\mu\nu} - \frac{1}{2} g_{\mu\nu} R$. This solution requires that boundary terms vanish (e.g., see [17,19,32]). Otherwise, we need to add a Gibbons–Hawking–York (GHY) boundary term [33–35] to the action $S = S_4 + S_{GHY}$, where:

$$S_{GHY} = \frac{1}{8\pi G} \oint_{\partial V_4} d^3y \sqrt{-h}\, K. \tag{A12}$$

where h is the induced metric and K is the trace of the extrinsic curvature at the boundary ∂V_4. Since the expansion inside an isolated BH is bounded by the event horizon $r < r_S$, we need to add this GHY boundary term S_{GHY} to the action. The integral is over the induced metric at ∂V_4, i.e., Equation (A1) with $\partial V_4 = \Sigma^-$ at $R = r_S$:

$$ds^2_{\partial V_4} = h_{\alpha\beta} dy^\alpha dy^\beta = -d\tau^2 + r_S^2 d\Omega^2 \tag{A13}$$

We use Equation (A6) to estimate K:

$$K = K^\alpha_\alpha = \frac{K_{\theta\theta}}{R^2} + \frac{K_{\delta\delta}}{R^2 \sin^2\theta} = -\frac{2}{R} = -\frac{2}{r_S}. \tag{A14}$$

We can now estimate S_{GHY}:

$$S_{GHY} = \frac{1}{8\pi G} \int d\tau\, 4\pi r_S^2\, K = -\frac{r_S}{G}\tau \tag{A15}$$

The contribution of Λ to the action in Equation (A11) is

$$S_\Lambda = -\frac{\Lambda}{8\pi G} V_4 = -\frac{r_S^3 \Lambda}{3G}\tau \tag{A16}$$

where we estimated the 4D volume V_4 as that bounded by ∂V_4 inside $r < r_S$: $V_4 = 2V_3\tau$. The factor of two here accounts for the fact that $V_3 = 4\pi r_S^3/3$ can be covered twice (both during collapse and during expansion).

We can then see that we need $\Lambda = 3r_S^{-2}$, or equivalently, $r_\Lambda = r_S$, to cancel the boundary term. In other words: a Λ term in the field equations can be induced by the evolution inside a BH event horizon.

Notes

[1]   This is the case only if we ignore cosmic acceleration or if we consider an observer in a galaxy far away (say at $z = 2$), when matter domination was an excellent approximation.

[2]   Note that both a WH and a BH require a finite total mass. If M_T is infinitely large, then $r_S = \infty$ and there is no WH or BH. This is in fact the standard Big Bang assumption. But this assumption is impossible to implement, even with Inflation: using local laws of gravity we have to create a uniform space of infinite extend within a finite amount of time.

[3] Even if the exterior is not totally empty and there is some small accretion from the outside, the value of r_S will slowly increase as the BH mass increases. But the r_S boundary still needs to be taken into account to evaluate the action inside.

References

1. Einstein, A. On a Stationary System With Spherical Symmetry Consisting of Many Gravitating Masses. *Ann. Math.* **1939**, *40*, 922–936. [CrossRef]
2. Ghez, A.M.; Klein, B.L.; Morris, M.; Becklin, E.E. High Proper-Motion Stars in the Vicinity of Sagittarius A*: Evidence for a Supermassive Black Hole at the Center of Our Galaxy. *Astrophys. J.* **1998**, *509*, 678–686. [CrossRef]
3. Abbott, R. et al. [LIGO Scientific Collaboration and Virgo Collaboration]. GWTC-2: Compact Binary Coalescences Observed by LIGO and Virgo during the First Half of the Third Observing Run. *Phys. Rev. X* **2021**, *11*, 021053.
4. Akiyama, K. et al. [Event Horizon Telescope Collaboration]. First Sagittarius A* Event Horizon Telescope Results. I. The Shadow of the Supermassive Black Hole in the Center of the Milky Way. *Astrophys. J. Lett.* **2022**, *930*, L12.
5. Penrose, R. Gravitational Collapse and Space-Time Singularities. *Phys. Rev. Lett.* **1965**, *14*, 57–59. [CrossRef]
6. Dadhich, N. Singularity: Raychaudhuri equation once again. *Pramana* **2007**, *69*, 23. [CrossRef]
7. Hawking, S.W. Black holes and thermodynamics. *Phys. Rev. D* **1976**, *13*, 191–197. [CrossRef]
8. Mathur, S.D. The information paradox: A pedagogical introduction. *Class. Quantum Gravity* **2009**, *26*, 224001. [CrossRef]
9. Maldacena, J.; Susskind, L. Cool horizons for entangled black holes. *Fortschritte Phys.* **2013**, *61*, 781–811. [CrossRef]
10. Haggard, H.M.; Rovelli, C. Quantum-gravity effects outside the horizon spark black to white hole tunneling. *Phys. Rev. D* **2015**, *92*, 104020. [CrossRef]
11. Oppenheimer, J.R.; Snyder, H. On Continued Gravitational Contraction. *Phys. Rev.* **1939**, *56*, 455–459. [CrossRef]
12. Israel, W. Singular hypersurfaces and thin shells in general relativity. *Nuovo Cimento B Ser.* **1966**, *44*, 1–14. [CrossRef]
13. Barceló, C.; Carballo-Rubio, R.; Garay, L.J. Mutiny at the white-hole district. *Int. J. Mod. Phys. D* **2014**, *23*, 1442022. [CrossRef]
14. Ben Achour, J.; Brahma, S.; Mukohyama, S.; Uzan, J.P. Towards consistent black-to-white hole bounces from matter collapse. *J. Cosmol. Astropart. Phys.* **2020**, *2020*, 020. [CrossRef]
15. Friedmann, A. On the Curvature of Space. *Gen. Relativ. Gravit.* **1999**, *31*, 1991. [CrossRef]
16. Lemaître, G. Un Univers homogène de masse constante et de rayon croissant rendant compte de la vitesse radiale des nébuleuses extra-galactiques. *Ann. S.S. Brux.* **1927**, *47*, 49–59.
17. Padmanabhan, T. *Gravitation*; Cambridge University Press: Cambridge, UK, 2010.
18. Tolman, R.C. Effect of Inhomogeneity on Cosmological Models. *Proc. Natl. Acad. Sci. USA* **1934**, *20*, 169–176. [CrossRef]
19. Landau, L.D.; Lifshitz, E.M. *The Classical Theory of Fields*; Elsevier: Amsterdam, The Netherlands, 1971.
20. Gaztañaga, E. The Black Hole Universe, part I. *Symmetry* **2022**, *14*, 1849. [CrossRef]
21. Faraoni, V.; Atieh, F. Turning a Newtonian analogy for FLRW cosmology into a relativistic problem. *Phys. Rev. D* **2020**, *102*, 044020. [CrossRef]
22. Johansen, N.; Ravndal, F. On the discovery of Birkhoff's theorem. *Gen. Relativ. Gravit.* **2006**, *38*, 537–540. [CrossRef]
23. Gaztañaga, E. The mass of our observable Universe. *Mon. Not. R. Astron. Soc.* **2023**, *521*, L59–L63. [CrossRef]
24. Ellis, G.F.R.; Rothman, T. Lost horizons. *Am. J. Phys.* **1993**, *61*, 883–893. [CrossRef]
25. Gaztañaga, E. The Cosmological Constant as Event Horizon. *Symmetry* **2022**, *14*, 300. [CrossRef]
26. Gaztañaga, E. The Black Hole Universe, part II. *Symmetry* **2022**, *14*, 1984. [CrossRef]
27. Gaztañaga, E.; Camacho-Quevedo, B. What moves the heavens above? *Phys. Lett. B* **2022**, *835*, 137468. [CrossRef]
28. Hilbert, D. Die Grundlage der Physick. *Konigl. Gesell. D. Wiss. GöTtingen Math.-Phys. K* **1915**, *3*, 395–407.
29. Weinberg, S. *Gravitation and Cosmology*; John Wiley & Sons: New York, NY, USA, 1972.
30. Weinberg, S. *Cosmology*; Oxford University Press: Oxford, UK, 2008.
31. Einstein, A. Die Grundlage der allgemeinen Relativitätstheorie. *Ann. Phys.* **1916**, *354*, 769–822. [CrossRef]
32. Carroll, S.M. *Spacetime and Geometry*; Addison-Wesley: Boston, MA, USA, 2004.
33. York, J.W. Role of Conformal Three-Geometry in the Dynamics of Gravitation. *Phys. Rev. Lett.* **1972**, *28*, 1082–1085. [CrossRef]
34. Gibbons, G.W.; Hawking, S.W. Cosmological event horizons, thermodynamics, and particle creation. *Phys. Rev. D* **1977**, *15*, 2738–2751. [CrossRef]
35. Hawking, S.W.; Horowitz, G.T. The gravitational Hamiltonian, action, entropy and surface terms. *Class. Quantum Gravity* **1996**, *13*, 1487. [CrossRef]

Article

Raychaudhuri Equations, Tidal Forces, and the Weak-Field Limit in Schwarzshild–Finsler–Randers Spacetime

Alkiviadis Triantafyllopoulos [1,*], Emmanuel Kapsabelis [1] and Panayiotis C. Stavrinos [2]

1   Section of Astrophysics, Astronomy and Mechanics, Department of Physics, National and Kapodistrian University of Athens, Panepistimiopolis, 15784 Athens, Greece; manoliskapsabelis@yahoo.gr
2   Department of Mathematics, National and Kapodistrian University of Athens, Panepistimiopolis, 15784 Athens, Greece; pstavrin@math.uoa.gr
*   Correspondence: alktrian@phys.uoa.gr

Abstract: In this article, we study the form of the deviation of geodesics (tidal forces) and the Raychaudhuri equation in a Schwarzschild–Finsler–Randers (SFR) spacetime which has been investigated in previous papers. This model is obtained by considering the structure of a Lorentz tangent bundle of spacetime and, in particular, the kind of the curvatures in generalized metric spaces where there is more than one curvature tensor, such as Finsler-like spacetimes. In these cases, the concept of the Raychaudhuri equation is extended with extra terms and degrees of freedom from the dependence on internal variables such as the velocity or an anisotropic vector field. Additionally, we investigate some consequences of the weak-field limit on the spacetime under consideration and study the Newtonian limit equations which include a generalization of the Poisson equation.

Keywords: Finsler geometry; modified theories of gravity; Raychaudhuri equation; geodesics deviation; cosmology; Weak field; tangent bundle

Citation: Triantafyllopoulos, A.; Kapsabelis, E.; Stavrinos, P.C. Raychaudhuri Equations, Tidal Forces, and the Weak-Field Limit in Schwarzshild–Finsler–Randers Spacetime. *Universe* **2024**, *10*, 26. https://doi.org/10.3390/universe10010026

Academic Editor: Kazuharu Bamba

Received: 30 November 2023
Revised: 29 December 2023
Accepted: 7 January 2024
Published: 9 January 2024

1. Introduction

The evolution equation of the quantities that characterize the (gravitational) flow in a given background spacetime is the Raychaudhuri equation [1,2]. The flows are integral curves, geodesics, or they are generated by a vector field. The Raychaudhuri equation is of significant importance since it describes the dynamical evolution of a gravitational fluid, and it is produced by the structure of deviation of nearby geodesics which are dominated by the curvature of space. It was originated by A. Raychaudhuri [1]. When the metric structure of spacetime changes, the equation is modified. The deviation of geodesics and the tidal forces play a fundamental role in general relativity, gravitation, and cosmology because of the interaction of the curvature of spacetime with matter [3]. The profound role of the equation of geodesic deviation (EDG) on Riemannian spacetime has been recognized in general relativity for a long time. The observable deviation of two neighboring geodesics (time-like or null) brings to life an appearance of the curvature of spacetime, namely, the detection of curvature, expresses a property of the matter sector of spacetime that is given by means of EDG, and it is connected with the polarization of gravitational waves and their detection [4]. How does the small-length deviation vector between corresponding points of two nearby geodesics vary as they move along the geodesics? This is the problem of geodesics deviation, the solution of which provides a good insight into the nature and behavior of space. In cosmology, this problem can be connected with the tidal forces and the scale factor $a(t)$ during the expansion of the universe between geodesic motions of two nearby galaxies (see [5]). The form of geodesics and their deviation depend on the spacetime metric, the connection, and the curvature. Raychaudhuri, in his articles [1,6–8], assumes that the universe can be represented by a time-dependent geometry but does not assume homogeneity or isotropy at early times. One of his aims is to see whether non-zero rotation (spin), anisotropy (shear), and/or a cosmological constant can succeed in circumventing the initial singularity [2]. Deviation of geodesics and the Raychaudhuri equation can be studied in a more general geometric framework than the Riemannian one. Finsler geometry consists of a natural metric generalization of Riemannian geometry. During the last years, rapid progress in the field of Finsler geometry and its applications to gravity and cosmology have extended the research in on corresponding topics; some recent works include [9–26]. In generalized metric spaces such as Finsler or Finsler-like spacetime, where the motion/velocity/direction are incorporated in the spacetime structure, internal anisotropy inherent in the EDG [27–29] and Raychaudhuri equations are attributed on the framework of a tangent bundle of spacetime manifold, thus extending the concept of volume θ, shear σ, and vorticity ω [30,31]. In our theory, the concept of volume Θ expresses the total volume on the tangent bundle which includes the standard form of volume θ and the internal anisotropic bulk that is caused by the geometrical structure and its coupling to the standard volume θ due to the additional degrees of freedom (Section 4.3). Additionally, the form of EDG is modified with extra terms that are originated by the connections, torsions, and anisotropic curvature tensors. In this geometric framework, anisotropic tidal phenomena arise from the internal and external structure of spacetime which are imprinted in the generalized EDG and Raychaudhuri equations. The appearance of extra terms in these equations plays the role of additional force fields or self-gravitating actions over spacetime which arise from the richer geometrical structure. The concept of a nonlinear connection in Finsler or Finsler-like spacetime can be interpreted as the interaction between the external and internal structures of spacetime. In a Finslerian gravitational theory on the tangent bundle of spacetime, curvature effects can be considered as total tidal forces which are produced by the external (horizontal) and internal (vertical) curvature tensors. Different considerations for EDG and Raychaudhuri equations on Finsler and Finsler-like spacetimes have been studied by the one of the authors in [30,31]. Einstein–Finsler-like gravitational field equations that govern the motion of matter have been derived in a generalized form of metric spaces on the Lorentz tangent bundle with Finsler-like geometrical structure [18,32,33]. These equations have also been given in a different form in [5,34–40]. In this article, as an additional motivation, we investigate the form of the equation of geodesics deviation, the Raychaudhuri equation in a Schwarzshild–Finsler–Randers (SFR) space adapted on the Lorentz tangent bundle of spacetime, thus extending the investigation on the SFR framework we have given in previous works [17,41,42]. Some physical consequences are also given in this article.

This work is organized as follows. In Section 2 we present the basic elements of the geometrical structure of the model. In Section 3, we study and derive the form of the deviation of geodesics and paths in completely generalized forms, we apply them to the SFR model, and we give some additional information for the deviation equation because of the extra degrees of freedom and the new geometrical concepts. The resulting anisotropic tidal acceleration is of great significance to the investigation of black hole phenomena. Additionally, we give the form of the weak-field limit of the deviation equations for the SFR model. In Section 4, we study the generalized Raychaudhuri equations in a general and special form for the model under consideration. We analyze the derived equations in the horizontal and vertical parts of the Lorentz tangent bundle and we give some interpretations to these equations. Finally, in the conclusion (Section 5) we discuss and summarize our results.

2. Geometrical Structure of the Model

In this section, we present some basic elements of the underlying geometry of the SFR gravitational model as well as the field equations that determine the relation between geometry and matter. A thorough study of this model can be found in [17,32].

2.1. The Lorentz Tangent Bundle

A Lorentz tangent bundle TM over a spacetime four-dimensional manifold M is a fibered eight-dimensional manifold with local coordinates $\{x^\mu, y^a\}$ where the indices of the spacetime variables x are $\kappa, \lambda, \mu, \nu, \ldots = 0, \ldots, 3$ and the indices of the fiber variables y are $a, b, \ldots, f = 0, \ldots, 3$. An extended Lorentzian structure on TM can be provided if the background manifold is equipped with a Lorentz metric tensor of signature $(-1, \ldots, 1)$.

Below, we present some basic geometrical structures of the model.

2.1.1. The Adapted Basis

In order to take a horizontal and vertical basis on a tangent bundle TM, we need to define a nonlinear connection $\mathbf{N}$ to unequivocally divide the bundle into a horizontal and vertical sub-bundles. The nonlinear connection defines a split of the total space TTM into a horizontal subspace $T_H TM$ and a vertical subspace $T_V TM$. The total space is the Whitney sum:

$$TTM = T_H TM \oplus T_V TM \tag{1}$$

We consider a vector field $X = X^\mu \frac{\partial}{\partial x^\mu}$ on a base Riemannian manifold M along a curve $x^\mu(t)$. If the vector field X^μ coincides with tangent vector of the curve $X^\mu = \dot{x}^\mu(t)$ then the geodesics equation on M is the standard equation using the Levi–Civita connection. It is written as:

$$\frac{d^2 x^\mu}{ds^2} + \Gamma^\mu_{\nu\kappa} \dot{x}^\nu \dot{x}^\kappa = 0 \tag{2}$$

The parallel transport of X is given by the equation of geodesics (2):

$$\frac{dX^\mu}{ds} + \Gamma^\mu_{\nu\kappa} X^\nu X^\kappa = 0 \tag{3}$$

where $\Gamma^\mu_{\nu\kappa}$ is the metrical connection.

We can extend the vector field X to the tangent bundle TM of M as

$$\overline{X} = X^\mu \frac{\partial}{\partial x^\mu} + \frac{dX^a}{dt} \frac{\partial}{\partial y^a} \tag{4}$$

where the coefficients X^a are one-to-one equal to the coefficients X^μ. Thus, X^μ and $\frac{dX^a}{dt}$ are the coefficients of the extended vector $\overline{X}$ on TM, the basis for the horizontal and vertical subspace of the tangent bundle.

According to the considerations of proposal (4.2) of p. 28 of [43], we can substitute Equation (3) to Equation (4) and find:

$$\overline{X} = X^\mu \frac{\partial}{\partial x^\mu} - (\Gamma^a_{\nu\kappa} X^\nu X^\kappa) \frac{\partial}{\partial y^a} \tag{5}$$

By assuming a Cartan-type connection $\Gamma^a_{\mu b} y^b = N^a_\mu$ [44] in a Finsler connection, where N^a_μ are the coefficients of a nonlinear connection, we find:

$$\overline{X} = X^\mu \left(\frac{\partial}{\partial x^\mu} - N^a_\mu \frac{\partial}{\partial y^a} \right) = X^\mu \delta_\mu \tag{6}$$

where we define:

$$\delta_\mu(x, y) = \frac{\partial}{\partial x^\mu} - N^a_\mu \frac{\partial}{\partial y^a} \tag{7}$$

to be an adapted basis for the tangent bundle. Therefore, the nonlinear connection induces the basis $\{E_A\} = \{\delta_\mu, \dot{\partial}_a\}$ on the total space, with

$$\delta_\mu = \frac{\delta}{\delta x^\mu} = \frac{\partial}{\partial x^\mu} - N^a_\mu(x, y) \frac{\partial}{\partial y^a} \tag{8}$$

and

$$\dot{\partial}_a = \frac{\partial}{\partial y^a} \tag{9}$$

2.1.2. Metric Structure on TM

A Sasaki-type metric $\mathcal{G}$ on TM is:

$$\mathcal{G} = g_{\mu\nu}(x,y)\,dx^{\mu} \otimes dx^{\nu} + v_{ab}(x,y)\,\delta y^{a} \otimes \delta y^{b} \tag{10}$$

A pseudo-Finslerian metric $f_{ab}(x,y)$ is defined as one that has a Lorentzian signature of $(-,+,+,+)$ and that also obeys the following form:

$$f_{ab}(x,y) = \pm\frac{1}{2}\frac{\partial^{2}F^{2}}{\partial y^{a}\partial y^{b}} \tag{11}$$

where the function F satisfies the following conditions [43]:

1. F is continuous on TM and smooth on $\widetilde{TM} \equiv TM \setminus \{0\}$, i.e., the tangent bundle minus the null set $\{(x,y) \in TM | F(x,y) = 0\}$.
2. F is positively homogeneous to the first degree on its second argument:

$$F(x^{\mu}, ky^{a}) = kF(x^{\mu}, y^{a}), \qquad k > 0 \tag{12}$$

3. The form

$$f_{ab}(x,y) = \pm\frac{1}{2}\frac{\partial^{2}F^{2}}{\partial y^{a}\partial y^{b}} \tag{13}$$

defines a non-degenerate matrix:

$$\det\left[f_{ab}\right] \neq 0 \tag{14}$$

where the plus–minus sign in (11) is chosen such that the metric has the correct signature.

2.1.3. Connection

In this work, we consider a distinguished connection (d-connection) D on TM. This is a linear connection with coefficients $\{\Gamma^{A}_{BC}\} = \{L^{\mu}_{\nu\kappa}, L^{a}_{b\kappa}, C^{\mu}_{\nu c}, C^{a}_{bc}\}$ that preserves the horizontal and vertical distributions using parallelism :

$$D_{\delta_{\kappa}}\delta_{\nu} = L^{\mu}_{\nu\kappa}(x,y)\delta_{\mu}, \quad D_{\dot{\partial}_{c}}\delta_{\nu} = C^{\mu}_{\nu c}(x,y)\delta_{\mu} \tag{15}$$

$$D_{\delta_{\kappa}}\dot{\partial}_{b} = L^{a}_{b\kappa}(x,y)\dot{\partial}_{a}, \quad D_{\dot{\partial}_{c}}\dot{\partial}_{b} = C^{a}_{bc}(x,y)\dot{\partial}_{a} \tag{16}$$

From these, the definitions for partial covariant differentiation follow as usual, e.g., for $X \in TTM$, we have the definitions for covariant h-derivative

$$X^{A}_{|\nu} \equiv D_{\nu}X^{A} \equiv \delta_{\nu}X^{A} + L^{A}_{B\nu}X^{B} \tag{17}$$

and covariant v-derivative

$$X^{A}|_{b} \equiv D_{b}X^{A} \equiv \dot{\partial}_{b}X^{A} + C^{A}_{Bb}X^{B} \tag{18}$$

In our consideration, the d-connection is metric-compatible:

$$D_{\kappa}\,g_{\mu\nu} = 0, \quad D_{\kappa}\,v_{ab} = 0, \quad D_{c}\,g_{\mu\nu} = 0, \quad D_{c}\,v_{ab} = 0 \tag{19}$$

The d-connection coefficients of our model have the following form:

$$L^{\mu}_{\nu\kappa} = \frac{1}{2}g^{\mu\rho}\left(\delta_k g_{\rho\nu} + \delta_\nu g_{\rho\kappa} - \delta_\rho g_{\nu\kappa}\right) \tag{20}$$

$$L^{a}_{b\kappa} = \dot{\partial}_b N^a_\kappa + \frac{1}{2}v^{ac}\left(\delta_\kappa v_{bc} - v_{dc}\,\dot{\partial}_b N^d_\kappa - v_{bd}\,\dot{\partial}_c N^d_\kappa\right) \tag{21}$$

$$C^{\mu}_{\nu c} = \frac{1}{2}g^{\mu\rho}\dot{\partial}_c g_{\rho\nu} \tag{22}$$

$$C^{a}_{bc} = \frac{1}{2}v^{ad}\left(\dot{\partial}_c v_{db} + \dot{\partial}_b v_{dc} - \dot{\partial}_d v_{bc}\right) \tag{23}$$

2.1.4. Curvature and Torsion

Curvatures and torsions on TM are defined by the multi-linear maps:

$$\mathcal{R}(X,Y)Z = [D_X, D_Y]Z - D_{[X,Y]}Z \tag{24}$$

and

$$\mathcal{T}(X,Y) = D_X Y - D_Y X - [X,Y] \tag{25}$$

where $X, Y, Z \in TTM$. We use the following definitions for the curvature components [43,45]:

$$\mathcal{R}(\delta_\lambda, \delta_\kappa)\delta_\nu = R^{\mu}_{\nu\kappa\lambda}\delta_\mu \tag{26}$$

$$\mathcal{R}(\delta_\lambda, \delta_\kappa)\dot{\partial}_b = R^{a}_{b\kappa\lambda}\dot{\partial}_a \tag{27}$$

$$\mathcal{R}(\dot{\partial}_c, \delta_\kappa)\delta_\nu = P^{\mu}_{\nu\kappa c}\delta_\mu \tag{28}$$

$$\mathcal{R}(\dot{\partial}_c, \delta_\kappa)\dot{\partial}_b = P^{a}_{b\kappa c}\dot{\partial}_a \tag{29}$$

$$\mathcal{R}(\dot{\partial}_\delta, \dot{\partial}_c)\delta_\nu = S^{\mu}_{\nu c\delta}\delta_\mu \tag{30}$$

$$\mathcal{R}(\dot{\partial}_\delta, \dot{\partial}_c)\dot{\partial}_b = S^{a}_{bc\delta}\dot{\partial}_a \tag{31}$$

In addition, we use the following definitions for the torsion components:

$$\mathcal{T}(\delta_\kappa, \delta_\nu) = \mathcal{T}^{\mu}_{\nu\kappa}\delta_\mu + \mathcal{T}^{a}_{\nu\kappa}\dot{\partial}_a \tag{32}$$

$$\mathcal{T}(\dot{\partial}_b, \delta_\nu) = \mathcal{T}^{\mu}_{\nu b}\delta_\mu + \mathcal{T}^{a}_{\nu b}\dot{\partial}_a \tag{33}$$

$$\mathcal{T}(\dot{\partial}_c, \dot{\partial}_b) = \mathcal{T}^{\mu}_{bc}\delta_\mu + \mathcal{T}^{a}_{bc}\dot{\partial}_a \tag{34}$$

The h-curvature tensor of the d-connection in the adapted basis and the corresponding h-Ricci tensor have, respectively, the components given from (26):

$$R^{\mu}_{\nu\kappa\lambda} = \delta_\lambda L^{\mu}_{\nu\kappa} - \delta_\kappa L^{\mu}_{\nu\lambda} + L^{\rho}_{\nu\kappa}L^{\mu}_{\rho\lambda} - L^{\rho}_{\nu\lambda}L^{\mu}_{\rho\kappa} + C^{\mu}_{\nu a}R^{a}_{\kappa\lambda} \tag{35}$$

$$R_{\mu\nu} = R^{\kappa}_{\mu\nu\kappa} = \delta_\kappa L^{\kappa}_{\mu\nu} - \delta_\nu L^{\kappa}_{\mu\kappa} + L^{\rho}_{\mu\nu}L^{\kappa}_{\rho\kappa} - L^{\rho}_{\mu\kappa}L^{\kappa}_{\rho\nu} + C^{\kappa}_{\mu a}R^{a}_{\nu\kappa} \tag{36}$$

where

$$R^{a}_{\nu\kappa} = \frac{\delta N^a_\nu}{\delta x^\kappa} - \frac{\delta N^a_\kappa}{\delta x^\nu} \tag{37}$$

are the non-holonomy coefficients, also known as the curvature of the nonlinear connection.

The v-curvature tensor of the d-connection in the adapted basis and the corresponding v-Ricci tensor have, respectively, the components (31):

$$S^{a}_{bcd} = \dot{\partial}_d C^{a}_{bc} - \dot{\partial}_c C^{a}_{bd} + C^{e}_{bc}C^{a}_{ed} - C^{e}_{bd}C^{a}_{ec} \tag{38}$$

$$S_{ab} = S^{c}_{abc} = \dot{\partial}_c C^{c}_{ab} - \dot{\partial}_b C^{c}_{ac} + C^{e}_{ab}C^{c}_{ec} - C^{e}_{ac}C^{c}_{eb} \tag{39}$$

The curvature tensor mixed coefficients are:

$$R^a_{b\kappa\lambda} = \delta_\lambda L^a_{b\kappa} - \delta_\kappa L^a_{b\lambda} + L^c_{b\kappa} L^a_{c\lambda} - L^c_{b\lambda} L^a_{c\kappa} + C^a_{bc} R^c_{\kappa\lambda} \tag{40}$$

$$P^\mu_{\nu\kappa c} = \dot\partial_c L^\mu_{\nu\kappa} - D_\kappa C^\mu_{\nu c} + C^\mu_{\nu b} T^b_{\kappa c} \tag{41}$$

$$P^a_{b\kappa c} = \dot\partial_c L^a_{b\kappa} - D_\kappa C^a_{bc} + C^a_{bd} T^d_{\kappa c} \tag{42}$$

$$S^\mu_{\nu cd} = \dot\partial_d C^\mu_{\nu c} - \dot\partial_c C^\mu_{\nu d} + C^\kappa_{\nu c} C^\mu_{\kappa d} - C^\kappa_{\nu d} C^\mu_{\kappa c} \tag{43}$$

The generalized Ricci scalar curvature in the adapted basis is defined as

$$\mathcal{R} = g^{\mu\nu} R_{\mu\nu} + v^{ab} S_{ab} = R + S \tag{44}$$

where

$$R = g^{\mu\nu} R_{\mu\nu} \quad , \quad S = v^{ab} S_{ab} \tag{45}$$

2.1.5. Hilbert-like Action

A Hilbert-like action on TM can be defined as

$$K = \int_N d^8\mathcal{U} \sqrt{|\mathcal{G}|}\, \mathcal{R} + 2\kappa \int_N d^8\mathcal{U} \sqrt{|\mathcal{G}|}\, \mathcal{L}_M \tag{46}$$

for some closed subspace $N \subset TM$, where $|\mathcal{G}|$ is the absolute value of the metric determinant, $\mathcal{L}_M$ is the Lagrangian of the matter fields, κ is a constant, and

$$d^8\mathcal{U} = dx^0 \wedge \ldots \wedge dx^3 \wedge dy^0 \wedge \ldots \wedge dy^3 \tag{47}$$

where the eight-parallelepiped $d^8\mathcal{U}$ is considered an oriented compact element of volume.

2.2. The SFR Model

In the SFR model, the metric $g_{\mu\nu}$ is the classic Schwarzschild one:

$$g_{\mu\nu} dx^\mu dx^\nu = -f dt^2 + \frac{dr^2}{f} + r^2 d\theta^2 + r^2 \sin^2\theta\, d\phi^2 \tag{48}$$

with $f = 1 - \frac{R_s}{r}$ and $R_s = 2GM$ the Schwarzschild radius (we assume units where $c = 1$).
The metric v_{ab} is derived from a metric function F_v of the a-Randers type:

$$F_v = \sqrt{-g_{ab}(x) y^a y^b} + A_c(x) y^c \tag{49}$$

where $g_{ab} = g_{\mu\nu} \delta^\mu_a \delta^\nu_b$ is the Schwarzschild metric, and $A_c(x)$ is a covector that expresses a deviation from general relativity, with $|A_c(x)| \ll 1$. The nonlinear connection will take the form:

$$N^a_\mu = \frac{1}{2} y^b g^{ac} \partial_\mu g_{bc} \tag{50}$$

The metric tensor v_{ab} of (49) is derived from (11) after omitting the higher-order terms $O(A^2)$:

$$v_{ab}(x,y) = g_{ab}(x) + w_{ab}(x,y) \tag{51}$$

where

$$w_{ab} = \frac{1}{\tilde{a}}(A_b g_{ac} y^c + A_c g_{ab} y^c + A_a g_{bc} y^c) + \frac{1}{\tilde{a}^3} A_c g_{ae} g_{bd} y^c y^d y^e \tag{52}$$

with $\tilde{a} = \sqrt{-g_{ab} y^a y^b}$. The total metric defined from the steps above is called the Schwarzschild–Finsler–Randers (SFR) metric and the corresponding spacetime is called an SFR spacetime.

We remark that the term w_{ab} represents a deviation from the pseudo-Riemannian space. The above-mentioned term can be useful for studying gravitational waves in a locally anisotropic framework of an SFR spacetime.

Variation in the action (46) with respect to $g_{\mu\nu}$, v_{ab} and N_κ^a leads to the following field equations:

$$\overline{R}_{\mu\nu} - \frac{1}{2}(R + S)\, g_{\mu\nu} + \left(\delta_\nu^{(\lambda}\delta_\mu^{\kappa)} - g^{\kappa\lambda}g_{\mu\nu}\right)\left(D_\kappa \mathcal{T}_{\lambda b}^b - \mathcal{T}_{\kappa c}^c \mathcal{T}_{\lambda b}^b\right) = \kappa T_{\mu\nu} \tag{53}$$

$$S_{ab} - \frac{1}{2}(R + S)\, v_{ab} + \left(v^{cd}v_{ab} - \delta_a^{(c}\delta_b^{d)}\right)\left(D_c C_{\mu d}^\mu - C_{vc}^v C_{\mu d}^\mu\right) = \kappa Y_{ab} \tag{54}$$

$$g^{\mu[\kappa}\partial_a L_{\mu\nu}^{v]} + 2\mathcal{T}_{\mu b}^b g^{\mu[\kappa}C_{\lambda a}^{\lambda]} = \frac{\kappa}{2}\mathcal{Z}_a^\kappa \tag{55}$$

with

$$T_{\mu\nu} \equiv -\frac{2}{\sqrt{|\mathcal{G}|}}\frac{\delta\left(\sqrt{|\mathcal{G}|}\,\mathcal{L}_M\right)}{\delta g^{\mu\nu}} = -\frac{2}{\sqrt{-g}}\frac{\delta\left(\sqrt{-g}\,\mathcal{L}_M\right)}{\delta g^{\mu\nu}} \tag{56}$$

$$Y_{ab} \equiv -\frac{2}{\sqrt{|\mathcal{G}|}}\frac{\delta\left(\sqrt{|\mathcal{G}|}\,\mathcal{L}_M\right)}{\delta v^{ab}} = -\frac{2}{\sqrt{-v}}\frac{\delta\left(\sqrt{-v}\,\mathcal{L}_M\right)}{\delta v^{ab}} \tag{57}$$

$$\mathcal{Z}_a^\kappa \equiv -\frac{2}{\sqrt{|\mathcal{G}|}}\frac{\delta\left(\sqrt{|\mathcal{G}|}\,\mathcal{L}_M\right)}{\delta N_\kappa^a} = -2\frac{\delta\mathcal{L}_M}{\delta N_\kappa^a} \tag{58}$$

and $\overline{R}_{\mu\nu} \equiv R_{\mu\nu} - C_{\mu a}^\kappa R_{v\kappa}^a$, where $\mathcal{L}_M$ is the Lagrangian of the matter fields, δ_v^μ and δ_b^a are the Kronecker symbols, $|\mathcal{G}|$ is the absolute value of the determinant of the total metric (10), and

$$\mathcal{T}_{vb}^a = \dot{\partial}_b N_v^a - L_{bv}^a \tag{59}$$

are torsion components, where L_{bv}^a is defined in (21). From the form of (10) it follows that $\sqrt{|\mathcal{G}|} = \sqrt{-g}\sqrt{-v}$, with g, v the determinants of the metrics $g_{\mu\nu}, v_{ab}$, respectively.

The local anisotropy can contribute to the energy–momentum tensors of the horizontal and vertical space $T_{\mu\nu}$ and Y_{ab}. As a result, the energy–momentum tensor $T_{\mu\nu}$ contains the additional information of local anisotropy of matter fields. Y_{ab}, on the other hand, is an extra concept with no equivalent in Riemannian gravity. It contains more information about local anisotropy which is produced from the metric v_{ab}, which includes additional internal structure of spacetime and can be connected to dark energy [33]. Finally, the energy–momentum tensor $\mathcal{Z}_a^\kappa$ reflects the dependence of matter fields with respect to the nonlinear connection N_μ^a, a structure which induces an interaction between internal and external spaces. This is a different form than of that $T_{\mu\nu}$ and Y_{ab}, which depend on just the external or internal structure, respectively.

Solving the field Equations (53)–(55) to the first order in $A_c(x)$ in a vacuum $(T_{\mu\nu} = Y_{ab} = \mathcal{Z}_a^\kappa = 0)$, we obtain [17]:

$$A_c(x) = \left[\tilde{A}_0 \left|1 - \frac{R_S}{r}\right|^{1/2}, 0, 0, 0\right] \tag{60}$$

where R_S is the Schwarzschild radius, r is the radial coordinate, and $\tilde{A}_0$ is a constant, where $|\tilde{A}_0| \ll 1$.

In the SFR model, the horizontal curvature Ricci tensor $R_{\mu\nu}$ is zero, but the internal vertical curvature v-Ricci tensor S_{ab} and the v-scalar S are different from zero. The components of the v-Ricci tensor are:

$$S_{00} = \frac{\tilde{A}_0^2 f \left(\tilde{a}^2 - f y_t{}^2\right)^2}{\tilde{a}^6} \tag{61}$$

$$S_{11} = \frac{\tilde{A}_0^2 \left[Rs^2 y_\theta^2 \left(3\tilde{a}^2 - 4f y_t^2\right) + Rs^2 y_\phi^2 \sin^2\theta \left(3\tilde{a}^2 - 4f y_t^2\right) + 4(f-1)^2 f y_t^2 \left(f y_t^2 - \tilde{a}^2\right) \right]}{4\tilde{a}^6 (f-1)^2 f} \tag{62}$$

$$S_{22} = \frac{\tilde{A}_0^2 Rs^2 \left[Rs^2 y_\theta^2 \left(4f y_t^2 - 3\tilde{a}^2\right) - 3\tilde{a}^2 (f-1)^2 \left(\tilde{a}^2 - f y_t^2\right) \right]}{4\tilde{a}^6 (f-1)^4} \tag{63}$$

$$S_{33} = -\frac{\tilde{A}_0^2 Rs^2 \sin^2\theta \left[Rs^2 y_\phi^2 \sin^2\theta \left(3\tilde{a}^2 - 4f y_t^2\right) + 3\tilde{a}^2 (f-1)^2 \left(\tilde{a}^2 - f y_t^2\right) \right]}{4\tilde{a}^6 (f-1)^4} \tag{64}$$

$$S_{01} = -\frac{\tilde{A}_0^2 y_t \left(f y_t^2 - \tilde{a}^2\right) \sqrt{f \left[f y_t^2 - \tilde{a}^2 - \frac{Rs^2}{(f-1)^2} \left(y_\theta^2 + y_\phi^2 \sin^2\theta\right) \right]}}{\tilde{a}^6} \tag{65}$$

$$S_{02} = \frac{\tilde{A}_0^2 f Rs^2 y_\theta y_t^2 \left(\tilde{a}^2 - f y_t^2\right)}{\tilde{a}^6 (f-1)^2} \tag{66}$$

$$S_{03} = \frac{\tilde{A}_0^2 f Rs^2 y_t y_\phi \sin^2\theta \left(\tilde{a}^2 - f y_t^2\right)}{\tilde{a}^6 (f-1)^2} \tag{67}$$

$$S_{12} = \frac{\tilde{A}_0^2 Rs^2 y_\theta y_r \left[(f-1)^2 \left(\tilde{a}^2 f + 4 y_r^2\right) + 4f Rs^2 y_\theta^2 + 4f Rs^2 y_\phi^2 \sin^2\theta \right]}{4\tilde{a}^6 (f-1)^4 f^2} \tag{68}$$

$$S_{13} = \frac{\tilde{A}_0^2 Rs^2 y_r y_\phi \sin^2\theta \left[(f-1)^2 \left(\tilde{a}^2 f + 4 y_r^2\right) + 4f Rs^2 y_\theta^2 + 4f Rs^2 y_\phi^2 \sin^2\theta \right]}{4\tilde{a}^6 (f-1)^4 f^2} \tag{69}$$

$$S_{23} = \frac{\tilde{A}_0^2 Rs^4 y_\theta y_\phi \sin^2\theta \left(4f y_t^2 - 3\tilde{a}^2\right)}{4\tilde{a}^6 (f-1)^4} \tag{70}$$

and the scalar v-Ricci curvature is:

$$S = \frac{5\tilde{A}_0^2 \left[\tilde{a}^2 - f y_t^2\right]}{2\tilde{a}^4} \tag{71}$$

where $\tilde{a} = \sqrt{-g_{ab} y^a y^b}$, $f \equiv 1 - \frac{Rs}{r}$, and we have set $y^0 \equiv y_t, y^1 \equiv y_r, y^2 \equiv y_\theta, y^3 \equiv y_\phi$.

The internal curvature S_{bcd}^a can give a physical meaning to the anisotropy (dependence on direction) of gravitational waves on the SFR model.

We have derived [41] the nontrivial Kretschmann-like invariants of the metrics $g_{\mu\nu}$ and v_{ab} to the lowest non-vanishing order:

$$K_H \equiv R_{\kappa\lambda\mu\nu} R^{\kappa\lambda\mu\nu} = \frac{12 R_S^2}{r^6} \tag{72}$$

$$K_V \equiv S_{abcd} S^{abcd} = \left(\frac{3S}{5}\right)^2 \tag{73}$$

Finally, the mixed curvature coefficients are all zero in this model.

2.3. The Newtonial Limit

In this section, we will investigate the Newtonian limit of a Finsler-like metric space on TM. The metric on TM will take the form

$$G = \left(\eta_{\mu\nu} + h_{\mu\nu}(x)\right)\mathrm{d}x^\mu \otimes \mathrm{d}x^\nu + \left(\eta_{ab} + w_{ab}(x,y)\right)\delta y^a \otimes \delta y^\beta \tag{74}$$

where $h_{\mu\nu}(x)$ and $w_{ab}(x,y)$ are small perturbations over the flat Minkowski metrics $\eta_{\mu\nu}$ and η_{ab} on the horizontal and vertical space, respectively.

The field Equations (53) and (54) for the metric are written at first order as:

$$R_{\mu\nu} - \frac{1}{2}(R + S) = \kappa T_{\mu\nu} \tag{75}$$

$$S_{ab} - \frac{1}{2}(R + S) = \kappa Y_{ab} \tag{76}$$

or equivalently

$$\frac{1}{2}\left(\partial_\mu \partial_\kappa h^\kappa_\nu + \partial_\nu \partial_\kappa h^\kappa_\mu - \partial^\kappa \partial_\kappa h_{\mu\nu} - \partial_\mu \partial_\nu h\right) - \frac{1}{2}\eta_{\mu\nu}\left(\partial_\kappa \partial_\lambda h^{\kappa\lambda} - \partial^\kappa \partial_\kappa h + \dot\partial_a \dot\partial_b w^{ab} - \dot\partial^a \dot\partial_a w\right) = \kappa T_{\mu\nu} \tag{77}$$

$$\frac{1}{2}\left(\dot\partial_c \dot\partial_a w^c_b + \dot\partial_c \dot\partial_b w^c_a - \dot\partial^c \dot\partial_c w_{ab} - \dot\partial_a \dot\partial_b\right) - \frac{1}{2}\eta_{ab}\left(\partial_\kappa \partial_\lambda h^{\kappa\lambda} - \partial^\kappa \partial_\kappa h + \dot\partial_a \dot\partial_b w^{ab} - \dot\partial^a \dot\partial_a w\right) = \kappa Y_{ab} \tag{78}$$

where $h = h^\mu_\mu$ and $w = w^a_a$. The third field Equation (55) gives:

$$\dot\partial_a\left(\partial^\kappa h - \partial_\mu h^{\mu\kappa}\right) = \mathcal{Z}^\kappa_a \tag{79}$$

which in our case gives $\mathcal{Z}^\kappa_a = 0$. This means that the matter fields in our space do not directly depend on the nonlinear connection. An analytical approach on a weak-field metric over a Lorentz tangent bundle can be found in [33].

The horizontal metric is effectively a Riemannian one, so the usual symmetries apply to it. One can decompose this metric on a scalar part Φ, a vector a_j, a traceless spatial tensor b_{ij}, and the trace Ψ of the spatial part, where all these parts transform independently under spatial rotations. The metric then takes the form:

$$\begin{aligned}G = {}&\left\{-(1 + 2\Phi)\mathrm{d}t^2 + a_i(\mathrm{d}t\mathrm{d}x^i + \mathrm{d}x^i\mathrm{d}t) + \left[(1 - 2\Psi)\delta_{ij} + 2b_{ij}\right]\mathrm{d}x^i\mathrm{d}x^j\right\}\mathrm{d}x^\mu \otimes \mathrm{d}x^\nu \\ &+ \left(\eta_{ab} + w_{ab}(x,y)\right)\delta y^a \otimes \delta y^\beta\end{aligned} \tag{80}$$

Additionally, we can take advantage of the gauge degrees of freedom of the Riemannian metric to set $\partial_i b^{ij} = 0$ and $\partial_i a^i = 0$. Finally, in our setting, the matter content of spacetime is assumed to be dust in its rest frame:

$$T_{\mu\nu} = \rho u_\mu u_\nu \tag{81}$$

where u^μ is the four-velocity field of the matter fluid. We consider a static spacetime, so all the time derivatives will vanish.

Under these assumptions, the field equations are written as:

$$\nabla^2 \Psi + \frac{1}{4}S = \frac{\kappa}{2}\rho \tag{82}$$

$$\nabla^2 a_i = 0 \tag{83}$$

$$(\delta_{ij}\nabla^2 - \partial_i \partial_j)(\Phi - \Psi) - \nabla^2 b_{ij} - \frac{1}{2}S\eta_{ij} = 0 \tag{84}$$

where ∇ is the three-dimensional spatial grad operator. Taking the trace of (84) yields:

$$\nabla^2(\Phi - \Psi) = \frac{3}{4}S \tag{85}$$

Substituting (85) to (82) gives:

$$\nabla^2 \Phi = \frac{\kappa}{2}\rho + \frac{S}{2} \tag{86}$$

This equation is a direct generalization of the Poisson equation of Newtonian physics. It has been shown in [33] that S can describe the effect of a vacuum energy density, so it can be considered to be a dark energy candidate.

Substituting (86) to (84) gives:

$$\frac{1}{4}S\delta_{ij} - \partial_i\partial_j(\Phi - \Psi) - \nabla^2 b_{ij} = 0 \tag{87}$$

Finally, (83) for a well-behaved field gives:

$$a_i = 0 \tag{88}$$

Equations (86)–(88), together with (76) (or (78)), determine the metric (80) up to the boundary conditions. The vertical energy–momentum tensor can be approximated by its GR limit value, i.e.,

$$Y_{ab} = \frac{1}{2}T^\mu_\mu \eta_{ab} \tag{89}$$

(see also [17]).

We remark that in the Newtonian limit of GR, only the scalar Φ is nonzero, while in our case, more degrees of freedom survive, such as the trace Ψ, the traceless tensor b_{ij}, and the vertical curvature S.

3. Generalized Deviation of Geodesics and Paths

In this section, we derive a completely generalized deviation equation using all forms of torsions for the geodesics and the paths. When there are forces acting on particles, they are not moving on geodesics and are accelerating. In our framework, we present some applications to the SFR model. The deviation equation of geodesics for different types of generalized locally anisotropic spacetime has been studied for a long time [27,29]. Here, we also present the weak deviation equation for this space.

3.1. General Equations

We assume that these geodesics and paths take the general form:

$$\frac{dy^a}{d\lambda} + 2G^a = 0, \quad y^a = \delta^a_\mu \frac{dx^\mu}{d\lambda} \tag{90}$$

Geodesics for the SFR model have been derived in a previous work [42]:

$$\ddot{x}^\lambda + \Gamma^\lambda_{\mu\nu}\dot{x}^\mu\dot{x}^\nu + g^{\kappa\lambda}\Phi_{\kappa\mu}\dot{x}^\mu = 0 \tag{91}$$

where $\Gamma^\lambda_{\mu\nu}$ are the Christoffel symbols of Riemann geometry, $\dot{x}^\mu = \frac{dx^\mu}{d\tau}$, $\Phi_{\kappa\mu} = \partial_\kappa A_\mu - \partial_\mu A_\kappa$, and A_μ is the solution Equation (60). We notice that from the definition of $\Phi_{\kappa\mu}$, we obtain a rotation form of geodesics. If A_μ is a gradient of a scalar field, $A_\mu = \dfrac{\partial\Phi}{\partial x^\mu}$, then $\Phi_{\kappa\mu} = 0$, and the geodesics of our model are identified with the Riemannian ones.

These geodesics are a specific case of (90) for

$$G^\lambda = \frac{1}{2}\left(\Gamma^\lambda_{\mu\nu}\dot{x}^\mu\dot{x}^\nu + g^{\kappa\lambda}\Phi_{\kappa\mu}\dot{x}^\mu\right) \tag{92}$$

The geodesics can be explicitly written in the form:

$$\ddot{t} + \frac{1-f}{rf}\dot{r}\dot{t} = -\tilde{A}_0\dot{r}\frac{f^{-3/2}(1-f)}{2r} \tag{93}$$

$$\ddot{r} + \frac{f(1-f)}{2r}\dot{t}^2 - \frac{1-f}{2rf}\dot{r}^2 - rf\left(\dot{\theta}^2 + \sin^2\theta\,\dot{\phi}^2\right) = -\tilde{A}_0\dot{t}\frac{f^{1/2}(1-f)}{2r} \tag{94}$$

$$\ddot{\theta} + \frac{2}{r}\dot{\theta}\dot{r} - \frac{1}{2}\sin 2\theta\,\dot{\phi}^2 = 0 \tag{95}$$

$$\ddot{\phi} + \frac{2}{r}\dot{\phi}\dot{r} + 2\cot\theta\,\dot{\theta}\dot{\phi} = 0 \tag{96}$$

The deflection angle of the SFR model has been studied in a previous work [42] and has been calculated for the model in hand,

$$\delta\phi_{SFR} \approx \left(1 + \frac{a^2}{2}\right)\frac{4GM}{b} \tag{97}$$

with $a = \tilde{A}_0 b/J$, where $b = J/ER$ is a composite constant formed by the ratio of the angular momentum J divided by the energy ER of the particle moving along the geodesic. The deflection angle $\delta\phi$ of GR is $\delta\phi_{SFR} = \delta\phi_{GR}$ when $\lim_{\tilde{A}_0 \to 0}$. Connecting the geometrical concept of the curvature $\kappa_\phi = \frac{d\phi}{d\tau}$ with the quantity $\kappa_\phi = \frac{\delta\phi_{SFR}}{d\tau}$, we obtain the deflection curvature of the SFR model. Comparing this result with that of the deflection angle parameter in Shapiro et al. [46], we find the value

$$2\left(1 + \frac{a^2}{2}\right) \simeq 1 + \gamma_{Shap} \tag{98}$$

or $a \sim 0.0141421$.

The small difference in the deflection angle of the SFR model in comparison to the GR deflection angle can be attributed to the Lorentz violations [9] or on the small amount of energy which is added to the gravitational potential of SFR model.

Subsequently, we will calculate the eight-velocity tangent vector to the geodesics (90) and define a deviation vector of the geodesics. We obtain:

$$U = y^\mu\partial_\mu - 2G^a\dot{\partial}_a = y^\mu\delta_\mu + \left(y^\nu N_\nu^a - 2G^a\right)\dot{\partial}_a \tag{99}$$

We write the decomposition of U to its horizontal and vertical components as $\{U^A\} = \{u^\mu, v^a\}$.

We define the deviation vector J such that, together with U, they form a local coordinate basis:

$$[U, J] = 0 \tag{100}$$

After some calculations, the above commutator relation gives

$$U^C D_C J^B = J^C D_C U^B + U^A J^C \mathcal{T}_{AC}^B \tag{101}$$

The second covariant derivative of the deviation vector is

$$\begin{aligned}
D_U^2 J^I &= U^A D_A(U^B D_B J^I)\\
&= U^A D_A\left(J^B D_B U^I + U^B J^C \mathcal{T}_{BC}^I\right)\\
&= \left(J^A D_A U^B + U^A J^C \mathcal{T}_{AC}^B\right)D_B U^I + U^A J^B\left([D_A, D_B] + D_B D_A\right)U^I + U^A D_A(U^B J^C \mathcal{T}_{BC}^I)
\end{aligned} \tag{102}$$

We can write

$$U^A D_A u^\mu = \frac{du^\mu}{d\lambda} + u^\lambda u^\nu L^\mu_{\nu\lambda} + v^a u^\nu C^\mu_{\nu a}$$
$$= u^\lambda u^\nu L^\mu_{\nu\lambda} + v^a u^\nu C^\mu_{\nu a} - 2G^\mu \tag{103}$$

where we used (90). We define

$$\Delta^\mu \equiv u^\nu u^\lambda L^\mu_{\nu\lambda} + u^\nu v^a C^\mu_{\nu a} - 2G^\mu \tag{104}$$

so we obtain

$$U^A D_A u^\mu = \Delta^\mu \tag{105}$$

We can consider Δ^μ to be a vector expressing the failure of u^μ being parallel transported along the geodesic.

Similarly, we find the covariant derivative of the vertical part of U:

$$U^A D_A v^d = \frac{dv^d}{d\lambda} + u^\lambda v^a L^d_{a\lambda} + v^a v^b C^d_{ab}$$
$$= u^\lambda v^a L^d_{a\lambda} + v^a v^b C^d_{ab} + \frac{du^\nu}{d\lambda} N^a_\nu + u^\nu \frac{dN^a_\nu}{d\lambda} - 2\frac{dG^a}{d\lambda}$$
$$\equiv \Delta^d \tag{106}$$

The commutator of the covariant derivatives is given by the following known relation:

$$[D_A, D_B] U^I = \mathcal{R}^I_{CBA} U^C U^A J^B + \mathcal{T}^C_{BA} D_C U^I \tag{107}$$

where $\mathcal{R}^I_{CBA}$ is the generalized curvature tensor of the connection D on the tangent bundle.

Taking the spacetime part of (102), and after some straightforward calculations, we obtain the result:

$$D^2_U J^\mu = \left(\mathcal{R}^\mu_{BCA} + D_A \mathcal{T}^\mu_{BC} + \mathcal{T}^\mu_{BD} \mathcal{T}^D_{AC}\right) U^A U^B J^C + 2U^{[A} J^{C]} \mathcal{T}^\mu_{BC} D_A U^B + J^B D_B \Delta^\mu \tag{108}$$

where we used the relations (104) and (107).

Similarly, the fiber part of (102) gives:

$$D^2_U J^d = \left(\mathcal{R}^d_{BCA} + D_A \mathcal{T}^d_{BC} + \mathcal{T}^d_{BD} \mathcal{T}^D_{AC}\right) U^A U^B J^C + 2U^{[A} J^{C]} \mathcal{T}^d_{BC} D_A U^B + J^B D_B \Delta^d \tag{109}$$

Equations (108) and (109) denote the deviation equation of horizontal and vertical paths between two nearby time-like paths on the Lorentzian tangent bundle. If the vectors Δ^μ and Δ^d are equal to zero, the trajectories of nearby observers are geodesics. Equation (108) is reduced to the standard geodesic deviation equation of general relativity when all the torsion components and the vector Δ^μ are equal to zero, and in that case, the curvature tensor coincides with the Riemannian one of the Levi–Civita connection. The torsion terms in (108) come from the geometry of our space; they play the role of a perturbation for the geodesic deviation of GR. From a physical point of view, perturbations of geodesics are affected by extra terms in their equations, e.g., because of additional mass, gas, or dark matter which interact gravitationally during their motion [25]. Tidal acceleration phenomena and anisotropic tidal-field perturbations can appear to be coming from different sources of spacetime.

3.2. Application of the Weak-Field Limit

We investigate first-order perturbations of the deviation equation in a weak Finslerian framework on the tangent bundle of the Riemannian space for an SFR space. A weak-field metric takes the form

$$G = \left[g_{\mu\nu}(x) + h_{\mu\nu}(x,y)\right] dx^\mu \otimes dx^\nu + \left[g_{ab}(x) + w_{ab}(x,y)\right] \delta y^a \otimes \delta y^b \tag{110}$$

with $|h_{\mu\nu}| \ll 1$ and $|w_{ab}| \ll 1$. From relations (40)–(43), it is straightforward to see that the mixed term curvatures $R^a_{b\kappa\lambda}$, $P^\mu_{\nu\kappa c}$, $P^a_{b\kappa c}$, and S^μ_{vcd} as well as the vertical curvature S^a_{bcd} are first-order on the perturbations $h_{\mu\nu}$ and w_{ab}.

At first order on $h_{\mu\nu}$ and w_{ab}, the horizontal deviation equation is:

$$D^2_U J^\mu = R^\mu_{\kappa\lambda\nu} u^\kappa u^\nu J^\lambda + \tilde{H}^\mu \tag{111}$$

where $\tilde{H}^\mu$ is a weak correction on the deviation equation, linear on $h_{\mu\nu}$ and w_{ab}:

$$\tilde{H}^\mu = \mathcal{T}^\mu_{vb} \mathcal{T}^b_{\kappa\lambda} u^\kappa u^\nu J^\lambda + P^\mu_{v\kappa c} u^\nu v^c J^\kappa + \left(S^\mu_{vbc} + D_c \mathcal{T}^\mu_{vb} \right) u^\nu v^c J^b + D_\lambda \mathcal{T}^\mu_{vb} u^\lambda u^\nu J^b$$
$$+ \mathcal{T}^\mu_{vb} \left[\left(u^\lambda J^b - v^b J^\lambda \right) D_\lambda u^\nu + \left(v^c J^b - v^b J^c \right) D_c u^\nu \right] + J^B D_B \Delta^\mu \tag{112}$$

The first-order vertical deviation equation is:

$$D^2_U J^a = D_\lambda \mathcal{T}^a_{v\kappa} u^\nu u^\lambda J^\kappa + \mathcal{T}^a_{v\kappa} \left[\left(u^\lambda J^\kappa - u^\kappa J^\lambda \right) D_\lambda u^\nu + (v^c J^\kappa - u^\kappa J^c) D_c u^\nu \right] + J^B D_B \Delta^a + \tilde{V}^a \tag{113}$$

with

$$\tilde{V}^a = S^a_{bcd} v^b v^d J^c + P^a_{b\kappa c} v^b v^c J^\kappa + \left(R^a_{b\kappa\lambda} + D_b \mathcal{T}^a_{\lambda\kappa} \right) v^b u^\lambda J^\kappa + D_c \mathcal{T}^a_{\kappa b} u^\kappa v^c J^b$$
$$+ \left(D_\lambda \mathcal{T}^a_{vb} + \mathcal{T}^a_{v\kappa} \mathcal{T}^\kappa_{\lambda b} \right) u^\nu u^\lambda J^b + \mathcal{T}^a_{vb} \mathcal{T}^b_{\lambda\kappa} u^\nu u^\lambda J^\kappa \tag{114}$$

the perturbation on the vertical deviation equation.

We apply the above equations for the SFR model and the geodesics (91) and we obtain:

$$\nabla^2_U J^\mu = R^\mu_{v\kappa\lambda} u^\kappa u^\nu J^\lambda + 2 u^\lambda u^\nu \nabla_\lambda \left(N^a_v \dot{\partial}_a J^\mu \right) + J^\lambda g^{\kappa\mu} \nabla_\lambda (\Phi_{v\kappa} u^\nu) \tag{115}$$

where ∇ is the Levi–Civita connection on the base manifold, $R^\mu_{v\kappa\lambda}$ is the Riemann curvature tensor of the classic Schwarzschild spacetime, and we have assumed that the y-dependence of J^μ on y is weak, i.e., $|\dot{\partial}_a J^\mu| \ll 1$. Equation (115) is the first-order generalization of the deviation equation on the SFR model.

We remark that a variation in $\Phi_{v\kappa} u^\nu$ along the deviation vector J^μ can induce an anisotropy of J^μ which varies along the geodesics. In the GR limit, the two last terms in (115) vanish and we obtain the classical deviation equation.

The vertical deviation equation in the SFR spacetime is:

$$D^2_U J^a = J^B D_B \Delta^a \tag{116}$$

Relations (115) and (116) show the rate of change of anisotropic deviation equation (tidal fields) at first order.

3.3. The Schwarzschild–Finsler–Randers spacetime

In SFR spacetime, the h-Ricci curvature tensor $R_{\mu\nu}$ and the h-Ricci curvature scalar R, defined in (36) and (45), respectively, are both zero. Consequently, the non-zero components of the h-Riemann curvature tensor, defined in (35), are equal to the ones of [3]:

$$R^t_{rrt} = 2R^\theta_{r\theta r} = 2R^\phi_{r\phi r} = \frac{R_S}{r^2(R_S - r)} \tag{117}$$

$$2R^t_{\theta\theta t} = 2R^r_{\theta\theta r} = R^\phi_{\theta\phi\theta} = \frac{R_S}{r} \tag{118}$$

$$2R^t_{\phi\phi t} = 2R^r_{\phi\phi r} = -R^\theta_{\phi\phi\theta} = \frac{R_S \sin^2 \theta}{r} \tag{119}$$

$$R^r_{\ trt} = -2R^\theta_{\ t\theta t} = -2R^\phi_{\ t\phi t} = c^2 \frac{R_S(R_S - r)}{r^4} \tag{120}$$

The non-zero components of the h-Riemann curvature tensor are the following:

$$R^{r'}_{\ t'r't'} = -R^{\theta'}_{\ \phi'\theta'\phi'} = -\frac{R_S}{r^3} \tag{121}$$

$$R^{\theta'}_{\ t'\theta't'} = R^{\phi'}_{\ t'\phi't'} = -R^{r'}_{\ \theta'r'\theta'} = -R^{r'}_{\ \phi'r'\phi'} = \frac{R_S}{2r^3}. \tag{122}$$

It is fundamental that the geodesic deviation equation shows the tidal acceleration between two observers who are separated by J^μ. It takes the form $D^2 J^{\mu'}/D\tau^2 = -R^{\mu'}_{\ t'\nu't'}J^{\nu'}$. The observable acceleration $(R_S/r^3)c^2 L$ of a body of length L in radial direction extends it and shrinks it in the lateral direction by $-(R_S/(2r^3))c^2 L$. Moving a body to the direction of a black hole causes spaghettification on the sizes of the body. Additionally, the anisotropic curvature S contributes to an anisotropic deformation of the body in our space.

The vertical curvature depends on the position x and the direction y; it is related to the intrinsic mechanism of spacetime where the gravitational field is extended on the total space of the SFR bundle. If we accept that the Schwarzschild spacetime takes an anisotropic structure with a force field (one form) on its metric, the additional energy originates from the internal (vertical) curvature and increases the form of tidal field around of a black hole. Consequently, we consider that both the horizontal and vertical (anisotropic) curvatures may affect the radial and lateral motion of an observer.

4. Generalized Raychaudhuri Equations

In this section, we investigate the generalized Raychaudhuri equations originated by our model and we give the equations for the horizontal and vertical parts of the Lorentz tangent bundle.

4.1. Horizontal Equations

We define the divergence tensor B^μ_ν as:

$$B^\mu_\nu = D_\nu u^\mu \tag{123}$$

where D_ν denotes the horizontal covariant derivative, and u^μ is a horizontal vector tangent to the geodesic congruence. We calculate the acceleration vector along the direction of u^μ as:

$$\frac{Du^\mu}{d\tau} = u^\nu D_\nu u^\mu = u^\nu B^\mu_\nu \tag{124}$$

In order to find the deviation of B^μ_ν we can use Equation (124):

$$\frac{DB^\mu_\nu}{d\tau} = u^\sigma D_\sigma B^\mu_\nu = u^\sigma D_\sigma (D_\nu u^\mu) \tag{125}$$

We can use the commutator of D as:

$$[D_\sigma, D_\nu]u^\mu = R^\mu_{\ \lambda\sigma\nu}u^\lambda - \mathcal{T}^\lambda_{\sigma\nu}D_\lambda u^\mu - R^a_{\sigma\nu}D_a u^\mu \tag{126}$$

where $R^\mu_{\ \lambda\sigma\nu}$ is the horizontal curvature tensor, $\mathcal{T}^\lambda_{\sigma\nu}$ is the torsion tensor, and $R^a_{\sigma\nu}$ is the curvature of the nonlinear connection. If we use Equations (125) and (126), we have:

$$\frac{DB_\nu^\mu}{d\tau} = u^\sigma [D_\sigma, D_\nu] u^\mu + u^\sigma D_\nu D_\sigma u^\mu \Rightarrow$$

$$\frac{DB_\nu^\mu}{d\tau} = u^\sigma \left[R_{\lambda\sigma\nu}^\mu u^\lambda - \mathcal{T}_{\sigma\nu}^\lambda D_\lambda u^\mu - R_{\sigma\nu}^a D_a u^\mu \right] + \left[D_\nu(u^\sigma D_\sigma u^\mu) - (D_\nu u^\sigma)(D_\sigma u^\mu) \right] \Rightarrow$$

$$\frac{DB_\nu^\mu}{d\tau} = R_{\lambda\sigma\nu}^\mu u^\lambda u^\sigma - \mathcal{T}_{\sigma\nu}^\lambda u^\sigma B_\lambda^\mu - R_{\sigma\nu}^a u^\sigma B_a^\mu - B_\nu^\sigma B_\sigma^\mu \tag{127}$$

where we have used Equation (123) and $u^\sigma D_\sigma u^\mu = 0$ since u^μ is tangent to the geodesics.

We define the horizontal projection tensor as:

$$P_\nu^\mu = \delta_\nu^\mu + u^\mu u_\nu \tag{128}$$

The divergence tensor B_ν^μ can be separated in two parts: a symmetric part and an anti-symmetric part. The symmetric part can also be separated into a part with a trace and a traceless part. We can write the decomposition by using the projection tensor as follows:

$$B_\nu^\mu = \frac{1}{3}\theta P_\nu^\mu + \sigma_\nu^\mu + \omega_\nu^\mu \tag{129}$$

Using the projection tensor in Equation (127) we obtain:

$$P_\mu^\nu \frac{DB_\nu^\mu}{d\tau} = -R_{\mu\nu} u^\mu u^\nu - \mathcal{T}_{\sigma\mu}^\lambda u^\sigma B_\lambda^\mu - R_{\sigma\mu}^a u^\sigma B_a^\mu - B_\mu^\sigma B_\sigma^\mu \tag{130}$$

and if we use the decomposition from Equation (129) we find:

$$\frac{d\theta}{d\tau} = -R_{\mu\nu} u^\mu u^\nu - \mathcal{T}_{\sigma\mu}^\lambda B_\lambda^\mu u^\sigma - R_{\sigma\mu}^a B_a^\mu u^\sigma - \frac{1}{3}\theta^2 - \sigma^{\mu\nu}\sigma_{\mu\nu} + \omega^{\mu\nu}\omega_{\mu\nu} \tag{131}$$

The Equation (131) is the generalized Raychaudhuri equation for the horizontal space.

As we can see, Equation (131) disturbs the rate of the volume because of the presence of the nonlinear connection N_μ^a, and the torsion functions affect the evolution of the gravitational fluid for possible singularities/conjugate points in the universe. In the framework of a given congruence of time-like geodesics, the expansion Θ and shear $\sigma_{\mu\nu}$ are described in a generalized form, provided that the generalized type of Raychaudhuri Equation (131) gives additional information on the kinematics. This is possible due to the perturbation of the deviation equation of nearby geodesics or trajectories, which was described in Section 3.

4.2. Vertical Equations

As with the horizontal space, we define the divergence tensor as:

$$B_b^a = D_b v^a \tag{132}$$

where D_b denotes the vertical covariant derivative, and v^a is a vertical vector tangent to the geodesic congruence. In order to find the deviation of B_b^a, we have:

$$\frac{DB_b^a}{d\tau} = v^c D_c B_b^a = v^c D_c(D_b v^a) \tag{133}$$

We can use the commutator of D as:

$$[D_c, D_b] v^a = S_{dcb}^a v^d - S_{cb}^d D_d v^a \tag{134}$$

where S^a_{dcb} is the vertical curvature tensor, and S^d_{cb} is the commutator of the connection coefficients. If we use Equations (133) and (134) we have:

$$\frac{DB^a_b}{d\tau} = u^c[D_c, D_b]v^a + v^c D_b D_c v^a \Rightarrow$$

$$\frac{DB^a_b}{d\tau} = v^c\left[S^a_{dcb}v^d - S^d_{cb}D_d v^a\right] + [D_b(u^c D_c v^a) - (D_b v^c)(D_c v^a)] \Rightarrow$$

$$\frac{DB^a_b}{d\tau} = S^a_{dcb}v^c v^d - S^d_{cb}B^a_d v^c - B^c_b B^a_c \tag{135}$$

where we have used Equation (132) and $v^c D_c v^a = 0$ since v^a is tangent to the geodesics.

As with the horizontal part, we decompose the divergence tensor as follows:

$$B^a_b = \frac{1}{3}\tilde{\theta}P^a_b + \tilde{\sigma}^a_b + \tilde{\omega}^a_b \tag{136}$$

where the projection tensor is written as:

$$P^a_b = \tilde{\delta}^a_b + v^a v_b \tag{137}$$

If we use the projection tensor in Equation (135) we obtain:

$$P^b_a \frac{DB^a_b}{d\tau} = -S_{ab}v^a v^b - S^d_{ca}B^a_d v^c - B^c_a B^a_c \tag{138}$$

and by using the separation from Equation (136), we find:

$$\frac{d\tilde{\theta}}{d\tau} = -S_{ab}v^a v^b - S^d_{ca}B^a_d v^c - \frac{1}{3}\tilde{\theta}^2 - \tilde{\sigma}^{ab}\tilde{\sigma}_{ab} + \tilde{\omega}^{ab}\tilde{\omega}_{ab} \tag{139}$$

Relation (139) is the vertical Raychaudhuri equation.

4.3. Application to the SFR Model

The non-holonomy coefficients of the nonlinear connection $R^a_{\nu\kappa}$ is given by:

$$R^a_{\nu\kappa} = \delta_\kappa N^a_\nu - \delta_\nu N^a_\kappa \tag{140}$$

If we use Equations (8) and (50) we find:

$$\delta_\kappa N^a_\nu = \partial_\kappa N^a_\nu - N^e_\kappa \dot{\partial}_e N^a_\nu$$

$$\delta_\kappa N^a_\nu = \frac{1}{2}y^b \partial_\kappa(g^{ac}\partial_\nu g_{bc}) - \frac{1}{2}y^f g^{ed}\partial_\kappa g_{df}\dot{\partial}_e(\frac{1}{2}y^b g^{ac}\partial_\nu g_{bc})$$

$$\delta_\kappa N^a_\nu = \frac{1}{2}y^b\left(\partial_\kappa g^{ac}\partial_\nu g_{bc} + g^{ac}\partial_\kappa\partial_\nu g_{bc} + \frac{1}{4}\partial_\kappa g_{bc}\partial_\nu g^{ac}\right) \tag{141}$$

So, the non-holonomy coefficients of the nonlinear connection from (140) become:

$$R^a_{\nu\kappa} = \frac{1}{4}y^b\left(\partial_\kappa g^{ac}\partial_\nu g_{bc} - \partial_\nu g^{ac}\partial_\kappa g_{bc}\right) \tag{142}$$

The non holonomy coefficients of the vertical connection C^a_{bc} are:

$$S^a_{bc} = C^a_{bc} - C^a_{cb} = 0 \tag{143}$$

because from Equation (23), C^a_{bc} is symmetric.

The horizontal component of the torsion tensor is given by:

$$\mathcal{T}^{\lambda}_{\sigma\mu} = L^{\lambda}_{\sigma\mu} - L^{\sigma}_{\mu\sigma} = 0 \tag{144}$$

because the from Equation (20) $L^{\lambda}_{\sigma\mu}$ is symmetric.

For simplicity, we will take σ^{a}_{b} and ω^{a}_{b} to be zero in both the horizontal and the vertical Raychaudhuri equations. So, Equations (131) and (139) can be written as:

$$\frac{d\theta}{d\tau} = -R_{\mu\nu}u^{\mu}u^{\nu} - R^{a}_{\sigma\mu}B^{\mu}_{a}u^{\sigma} - \frac{1}{3}\theta^{2} \tag{145}$$

$$\frac{d\tilde{\theta}}{d\tau} = -S_{ab}v^{a}v^{b} - \frac{1}{3}\tilde{\theta}^{2} \tag{146}$$

where $R_{\mu\nu} = 0$ because it is identified with the GR case in first-order approximation. Furthermore, if we use Equation (142) and take the vertical vector $v^{a} = (-1, 0, 0, 0)$ we find:

$$\frac{d\theta}{d\tau} = -\frac{1}{4}y^{b}\left(\partial_{\mu}g^{ac}\partial_{\sigma}g_{bc} - \partial_{\sigma}g^{ac}\partial_{\kappa}g_{bc}\right)B^{\mu}_{a}u^{\sigma} - \frac{1}{3}\theta^{2} \tag{147}$$

$$\frac{d\tilde{\theta}}{d\tau} = -\frac{2\tilde{A}^{2}_{0}}{\tilde{a}^{2}}f\left(1 - f\frac{y^{2}_{t}}{\tilde{a}^{2}}\right)^{2} - \frac{1}{3}\tilde{\theta}^{2} \tag{148}$$

Here, we have used that the time component of the vertical curvature is given by (61), which can be interpreted as the evolution of anisotropic expansion $\tilde{\theta}$. By adding Equations (145) and (146) we find:

$$\frac{d\theta}{d\tau} + \frac{d\tilde{\theta}}{d\tau} + \frac{1}{3}\theta^{2} + \frac{1}{3}\tilde{\theta}^{2} = -R_{\mu\nu}u^{\mu}u^{\nu} - R^{a}_{\sigma\mu}B^{\mu}_{a}u^{\sigma} - S_{ab}v^{a}v^{b} \Rightarrow$$

$$\frac{d}{d\tau}(\theta + \tilde{\theta}) + \frac{1}{3}(\theta + \tilde{\theta})^{2} - \frac{2}{3}\theta\tilde{\theta} = -R_{\mu\nu}u^{\mu}u^{\nu} - R^{a}_{\sigma\mu}B^{\mu}_{a}u^{\sigma} - S_{ab}v^{a}v^{b} \tag{149}$$

The above-mentioned Equation (149) represents the volumes and their changes, and θ and $\tilde{\theta}$ denote the standard volume from the horizontal background part of the tangent bundle and the internal anisotropic bulk which is caused by the anisotropic structure. Likewise, $\theta\tilde{\theta}$ can be considered to be the coupling of the background volume with its anisotropic bulk during the evolution of world lines, and the quantity $\theta + \tilde{\theta}$ expresses the total volume.

5. Discussion and Conclusions

The fully developed equations that characterize the flow in a given background spacetime are the Raychaudhuri equations, which are fundamental since they describe the dynamical evolution of the gravitational fluid. They are produced by the structure of deviation of nearby geodesics, which is dominated by the curvature of space. In general, the correspondence between fluids and gravity can be represented in a realistic way to understanding current theoretical and observational problems.

In this article, we examine and derive the deviation equation of geodesics and paths as well as the form of a Raychaudhuri equation in a completely generalized framework and we apply them to a Schwarzschild–Finsler–Randers model in which we showed that the extra terms in (108), (109), (131), and (139) anisotropically affect the acceleration tidal vector field and the variation in the volume (expansion) during the evolution of fluid lines (geodesics and paths). From a physical point of view, the generalized deviation geodesics equation is influenced by extra degrees of freedom, e.g, because of additional mass, gas, dark matter, etc. [18,25], which interact gravitationally during their motion. Tidal acceleration phenomena and anisotropic tidal-field perturbations can appear from different sources of spacetime. It is remarkable to mention here that acceleration geometrical concepts constitute intrinsic properties on a tangent bundle as, e.g, a vector field. The extended geometrical structure of the SFR model includes the Schwarzschild spacetime and gives us additional information on the kinematics because of the extra degrees of freedom reflected in additional terms of torsion, nonlinear connection and S-Kretschmann-like curvature invariants that are imprinted on the corresponding equations.

Moreover, we investigate the weak-field limit of a Finslerian perturbation on a Riemannian spacetime in the cases of a deviation equation. We also study the Newtonian limit of the model and derive a generalized Poisson equation. Additionally, we presented an interesting application for the deviation angle on our generalized framework and compared the result with that of GR. We also study the Raychaudhuri equation, which is extended in the horizontal and vertical parts of the SFR spacetime. In particular, the concept of nonlinear connection in Finsler or Finsler-like spacetime can be geometrically interpreted as an interaction between external and internal structures on the Lorentz tangent bundle spacetime. In a more specific case, a physical interpretation of nonlinear connection can relate internal scalar fields with the matter sector of spacetime, as, for instance, in [18,25]. It can be understood from all the derived equations of the SFR model that they reduce to standard GR when all the extra terms of generalized geometrical structure are omitted. The anisotropic S-curvature is significant in our cosmological model since it can provide addition information for the evolution of gravitational flow lines and the expansion of the universe as well as singularities (focusing/defocusing) of spacetime. This curvature expresses a very small source of anisotropy as is evident from the Equations (60)–(71) in which all the terms are multiplied by a constant $\tilde{A}_0 \ll 1$ (Equation (60)); this means that all the values of S are very small. Consequently, from a physical point of view, it is possible that in a very early period of the universe, the anisotropies of CMB influenced the geometry during cosmological evolution.

It is of special interest that one investigates the weak-field limit in more detail and connect it to the anisotropic polarization of gravitational waves. This research will be the motivation for a future work.

Author Contributions: Conceptualization, P.C.S.; Methodology, A.T., E.K. and P.C.S.; Writing—original draft, A.T. and E.K.; Writing—review and editing, A.T., E.K. and P.C.S.; Supervision, P.C.S. All authors have read and agreed to the published version of the manuscript

Funding: This research received no external funding.

Data Availability Statement: No new data were created or analyzed in this study. Date sharing is not applicable to this article.

Conflicts of Interest: The authors declare no conflict of interest.

References

1. Raychaudhuri, A. Relativistic Cosmology. I. *Phys. Rev.* **1955**, *98*, 1123–1126. [CrossRef]
2. Kar, S.; SenGupta, S. The Raychaudhuri equations: A Brief review. *Pramana* **2007**, *69*, 49–76. [CrossRef]
3. Misner, C.W.; Thorne, K.S.; Wheeler, J.A. *Gravitation*; Princeton University Press: Princeton, NJ, USA, 2017; ISBN 978-0-691-17779-3.
4. Hou, S.; Gong, Y. Strong Equivalence Principle and Gravitational Wave Polarizations in Horndeski Theory. *Eur. Phys. J. C* **2019**, *79*, 197. [CrossRef]
5. Hawking, S.W.; Ellis, G.F.R. *The Large Scale Structure of Space-Time*; Cambridge University Press: Cambridge, UK, 2023; ISBN 978-0-521-09906-6.
6. Raychaudhuri, A. Condensations in Expanding Cosmologic Models. *Phys. Rev.* **1952**, *86*, 90. [CrossRef]
7. Raychaudhuri, A. Arbitrary Concentrations of Matter and the Schwarzschild Singularity. *Phys. Rev.* **1953**, *89*, 417. [CrossRef]
8. Raychaudhuri, A. Relativistic and Newtonian cosmology. *Z. Astrophys.* **1957**, *43*, 161.
9. Kostelecky, A. Riemann-Finsler geometry and Lorentz-violating kinematics. *Phys. Lett. B* **2011**, *701*, 137–143. [CrossRef]
10. Caponio, E.; Stancarone, G. On Finsler spacetimes with a timelike Killing vector field. *Class. Quant. Grav.* **2018**, *35*, 085007. [CrossRef]
11. Bubuianu, L.; Vacaru, S.I. Black holes with MDRs and Bekenstein–Hawking and Perelman entropies for Finsler–Lagrange–Hamilton Spaces. *Ann. Phys.* **2019**, *404*, 10–38. [CrossRef]

12. Pfeifer, C. Finsler spacetime geometry in Physics. *Int. J. Geom. Meth. Mod. Phys.* **2019**, *16* (Suppl. S2), 1941004. [CrossRef]
13. Javaloyes, M.Á.; Sánchez, M. On the definition and examples of cones and Finsler spacetimes. *Rev. Real Acad. Cienc. Exactas Fis. Nat. Ser. Mat.* **2020**, *114*, 30. [CrossRef]
14. Javaloyes, M.Á. Curvature Computations in Finsler Geometry Using a Distinguished Class of Anisotropic Connections. *Mediterr. J. Math.* **2020**, *17*, 123. [CrossRef]
15. Hohmann, M.; Pfeifer, C.; Voicu, N. Cosmological Finsler Spacetimes. *Universe* **2020**, *6*, 65. [CrossRef]
16. Caponio, E.; Masiello, A. On the analyticity of static solutions of a field equation in Finsler gravity. *Universe* **2020**, *6*, 59. [CrossRef]
17. Triantafyllopoulos, A.; Basilakos, S.; Kapsabelis, E.; Stavrinos, P.C. Schwarzschild-like solutions in Finsler–Randers gravity. *Eur. Phys. J. C* **2020**, *80*, 1200. [CrossRef]
18. Konitopoulos, S.; Saridakis, E.N.; Stavrinos, P.C.; Triantafyllopoulos, A. Dark gravitational sectors on a generalized scalar-tensor vector bundle model and cosmological applications. *Phys. Rev. D* **2021**, *104*, 064018. [CrossRef]
19. Stavrinos, P.; Vacaru, S.I. Broken Scale Invariance, Gravity Mass, and Dark Energy inModified Einstein Gravity with Two Measure Finsler Like Variables. *Universe* **2021**, *7*, 89. [CrossRef]
20. Hohmann, M.; Pfeifer, C.; Voicu, N. Mathematical foundations for field theories on Finsler spacetimes. *J. Math. Phys.* **2022**, *63*, 032503. [CrossRef]
21. Javaloyes, M.Á.; Sánchez, M.; Villaseñor, F.F. On the Significance of the Stress–Energy Tensor in Finsler Spacetimes. *Universe* **2022**, *8*, 93. [CrossRef]
22. Heefer, S.; Pfeifer, C.; van Voorthuizen, J.; Fuster, A. On the metrizability of m-Kropina spaces with closed null one-form. *J. Math. Phys.* **2023**, *64*, 022502. [CrossRef]
23. Bubuianu, L.; Singleton, D.; Vacaru, S.I. Nonassociative black holes in R-flux deformed phase spaces and relativistic models of Perelman thermodynamics. *J. High Energy Phys.* **2023**, *5*, 57. [CrossRef]
24. Hama, R.; Harko, T.; Sabau, S.V. Dark energy and accelerating cosmological evolution from osculating Barthel–Kropina geometry. *Eur. Phys. J. C* **2022**, *82*, 385. [CrossRef]
25. Savvopoulos, C.; Stavrinos, P.C. Anisotropic conformal dark gravity on the Lorentz tangent bundle spacetime. *Phys. Rev. D* **2023**, *108*, 044048. [CrossRef]
26. Hama, R.; Harko, T.; Sabau, S.V. Conformal gravitational theories in Barthel–Kropina-type Finslerian geometry, and their cosmological implications. *Eur. Phys. J. C* **2023**, *83*, 1030. [CrossRef]
27. Asanov, G.S.; Stavrinos, P.C. Finslerian deviations of Geodesics over tangent bundle. *Rep. Math. Phys.* **1991**, *30*, 63–69. [CrossRef]
28. Stavrinos, P.C.; Kawaguchi, H. Deviation of Geodesics in the Gravitational Field of Finslerian Space-Time. *Meml. Shonan Inst. Technol.* **1993**, *27*, 35–40.
29. Balan, V.; Stavrinos, P.C. Weak gravitational fields in generalized metric spaces. In Proceedings of the International Conference of Geometry and Its Applications, Thessaloniki, Greece, 23–26 June 1999; pp. 27–37.
30. Stavrinos, P.C.; Alexiou, M. Raychaudhuri equation in the Finsler–Randers space-time and generalized scalar-tensor theories. *Int. J. Geom. Meth. Mod. Phys.* **2017**, *15*, 1850039. [CrossRef]
31. Stavrinos, P. Weak Gravitational Field in Finsler-Randers Space and Raychaudhuri Equation. *Gen. Rel. Grav.* **2012**, *44*, 3029–3045. [CrossRef]
32. Triantafyllopoulos, A.; Kapsabelis, E.; Stavrinos, P.C. Gravitational Field on the Lorentz Tangent Bundle: Generalized Paths and Field Equations. *Eur. Phys. J. Plus* **2020**, *135*, 557. [CrossRef]
33. Triantafyllopoulos, A.; Stavrinos, P.C. Weak field equations and generalized FRW cosmology on the tangent Lorentz bundle. *Class. Quant. Grav.* **2018**, *35*, 085011. [CrossRef]
34. Penrose, R. Gravitational Collapse and Space-Time Singularities. *Phys. Rev. Lett.* **1965**, *14*, 57. [CrossRef]
35. Hawking, S.W. Occurrence of Singularities in Open Universes. *Phys. Rev. Lett.* **1965**, *15*, 689. [CrossRef]
36. Hawking, S.W. Singularities in the Universe. *Phys. Rev. Lett.* **1966**, *17*, 444. [CrossRef]
37. Wald, R. *General Relativity*; Chicago University Press: Chicago, IL, USA, 1984.
38. Yang, J.Z.; Shahidi, S.; Harko, T.; Liang, S.D. Geodesic deviation, Raychaudhuri equation, Newtonian limit, and tidal forces in Weyl-type $f(Q,T)$ gravity. *Eur. Phys. J. C* **2021**, *81*, 111. [CrossRef]
39. Harko, T.; Lobo, F.S.N. Geodesic deviation, Raychaudhuri equation, and tidal forces in modified gravity with an arbitrary curvature-matter coupling. *Phys. Rev. D* **2012**, *86*, 124034. [CrossRef]
40. Mohajan, H.K. Scope of Raychaudhuri equation in cosmological gravitational, focusing and space-time singularities. *Peak J. Phys. Environ. Sci. Res.* **2013**, *1*, 106–114.
41. Kapsabelis, E.; Triantafyllopoulos, A.; Basilakos, S.; Stavrinos, P.C. Applications of the Schwarzschild–Finsler–Randers model. *Eur. Phys. J. C* **2021**, *81*, 990. [CrossRef]
42. Kapsabelis, E.; Kevrekidis, P.G.; Stavrinos, P.C.; Triantafyllopoulos, A. Schwarzschild–Finsler–Randers spacetime: Geodesics, dynamical analysis and deflection angle. *Eur. Phys. J. C* **2022**, *82*, 1098. [CrossRef]
43. Miron, R.; Anastasiei, M. The Geometry of Lagrange Spaces: Theory and Applications. In *Fundamental Theories of Physics*; Springer: Heidelberg, The Netherlands, 1994. [CrossRef]
44. Miron, R.; Watanabe, S.; Ikeda, S. Some Connections on Tangent Bundle and Their Applications to General Relativity. *Tensor New Ser.* **1987**, *46*, 8–22.

45. Vacaru, S.; Stavrinos, P.C.; Gaburov, E.; Gonta, D. *Clifford and Riemann-Finsler Structures in Geometric Mechanics and Gravity*; Differential Geometry—Dynamical Systems, Monograph 7; Geometry Balkan Press: Bucharest, Romania, 2006.
46. Shapiro, S.S.; Davis, J.L.; Lebach, D.E.; Gregory, J.S. Measurements of the solar gravitational deflection of radio waves using geodetic very-long-baseline interferometry data, 1979–1999. *Phys. Rev. Lett.* **2004**, *92*, 121101. [CrossRef]

Review

The Unsettled Number: Hubble's Tension

Jorge L. Cervantes-Cota [1,*], Salvador Galindo-Uribarri [1,†] and George F. Smoot [2,3,4,5]

1 Department of Physics, National Institute for Nuclear Research, Km 36.5 Carretera Mexico-Toluca, Ocoyoacac C.P. 52750, Mexico State, Mexico
2 Donostia International Physics Center (DIPC), Basque Country, E-48080 San Sebastian, Spain; gfsmoot@lbl.gov
3 Emeritus, Institute for Advanced Study, Hong Kong University of Science and Technology, Clear Water Bay, Kowloon, Hong Kong 999077, China
4 Emeritus, Université Sorbonne Paris Cité, Laboratoire APC-PCCP, Université Paris Diderot, 10 rue Alice Domon et Leonie Duquet, CEDEX 13, 75205 Paris, France
5 Emeritus, Department of Physics and LBNL, University of California, MS Bldg 50-5505 LBNL, 1 Cyclotron Road, Berkeley, CA 94720, USA
* Correspondence: jorge.cervantes@inin.gob.mx
† Salvador Galindo-Uribarri passed away at the later stages of the manuscript preparation.

Abstract: One of main sources of uncertainty in modern cosmology is the present rate of the universe's expansion, H_0, called the Hubble constant. Once again, different observational techniques bring about different results, causing new "Hubble tension". In the present work, we review the historical roots of the Hubble constant from the beginning of the twentieth century, when modern cosmology originated, to the present. We develop the arguments that gave rise to the importance of measuring the expansion of the Universe and its discovery, and we describe the different pioneering works attempting to measure it. There has been a long dispute on this matter, even in the present epoch, which is marked by high-tech instrumentation and, therefore, in smaller uncertainties in the relevant parameters. It is, again, currently necessary to conduct a careful and critical revision of the different methods before one invokes new physics to solve the so-called Hubble tension.

Keywords: Hubble constant; Hubble tension; observational cosmology; history of physics

Citation: Cervantes-Cota, J.L.; Galindo-Uribarri, S.; Smoot, G.F. The Unsettled Number: Hubble's Tension. *Universe* **2023**, *9*, 501. https://doi.org/10.3390/universe9120501

Academic Editor: Kazuharu Bamba

Received: 30 October 2023
Revised: 17 November 2023
Accepted: 20 November 2023
Published: 29 November 2023

1. Introduction

General relativity brought the idea of the expansion of space, not into something containing space, but of space itself. In the standard cosmological picture, the expansion rate of the universe is constantly changing as the cosmos evolves. The present-day expansion rate of the universe is given by a numerical value, the Hubble constant (H_0). This is a fixed number used as a unit of measurement to describe this expansion. The importance of knowing this value with good certainty lies in the fact that this knowledge provides a measurement scale of the present universe through the Hubble Sphere $r_{HS} = c/H_0$ and its time scale through Hubble time $t_H = 1/H_0 \sim 14$ Gyrs. It also serves to estimate the critical density of the universe, $\rho_{critical} = 3\,H_0{}^2/8\,\pi\,G$, which, in general relativity, creates flat space geometry. Since H_0 is our least well-determined cosmological parameter, things are often expressed in terms of $h = H_0/(100\ \mathrm{km/s/Mpc})$, or h^2, in this case, so that the result can be scaled properly when accounting for final errors and results. Other cosmological parameters are intimately related to these quantities. Thus, the precise determination of H_0 could reveal missing pieces in our current understanding of physics. The present review provides a historical description of the origin, measurement, and present status of the Hubble constant.

During the past two decades, estimates of the constant have repeatedly been made using two different approaches. The first involved measuring features in the relative recent universe. The second approach used light left over from shortly after the big bang, the cosmic microwave background (CMB). These two routes are known, respectively, as the "late" and "early" routes to the Hubble constant, as the first involved measuring aspects of the late evolution stages of the universe, while the second focuses on the very early phase.

In recent years, a discrepancy regarding the estimation of H_0 has emerged, involving measurements from the two routes of its assessment. In defiance of all expectations, estimates of H_0 from early route assessments do not agree with those of the late route. This discrepancy is known as the Hubble tension. This tension might be nothing more than a measurement error. However, it has stubbornly prevailed in spite of the fact that all the recent measurements have increased their precision.

At present there is debate on whether experimental glitches in either set of estimates cause the discrepancy, but no one is sure what those glitches would be. However, there have been opinions that the Hubble tension points to something missing from our understanding of the cosmos that might require an adjustment of the standard cosmological model, also known as "ΛCDM": cold dark matter plus constant dark energy Λ.

The scientific importance of resolving the tension has aroused the interest of scientists who are outsiders to the fields of astronomy and cosmology. This work is aimed at those who have interest in the Hubble tension, but work in fields of physical sciences other than those involved in the problem. It is hoped that this work will also appeal to field insiders, in particular, young postdoc researchers who are well-acquainted with the problem but would like to learn some historical details that are frequently missed or not mentioned in the current specialized literature. This present review might serve these individuals as a reference.

To achieve the purposes of this work, we have organized its narrative as follows:

We begin in Section 2, giving an account a debate held on April 1920 at the Smithsonian Museum of Natural History between Harlow Shapley and Heber Curtis on the size of the universe. Shapley maintained that the Milky Way was the entirety of the universe, while Curtis affirmed that the Milky Way was only one of multiple galaxies.

At the root of the debate was the problem of determining the distances at which celestial objects are situated. Thus, in Section 3, our story goes back to 1908, a few years before the debate, when Henrietta Leavitt discovered the relation between the luminosity and the period of Cepheid variables. Leavitt's discovery provided astronomers with the first "standard candle" with which to measure the distances to faraway galaxies.

Section 4 centers its attention on Vesto Melvin Slipher. He was, in 1912, the first to discover that distant galaxies are redshifted, thus providing the first empirical basis for the expansion of the universe. Later, in the 1920s, Edwin Hubble showed that Andromeda was far outside the Milky Way by measuring Cepheid variable stars, proving that Curtis was correct. After Hubble found that the spiral nebulae were actually galaxies like ours, Slipher's findings attracted the attention of some astronomers. On the one hand, most nebulae (now known to be galaxies) were rapidly moving away from Earth; on the other, a few were approaching it. Why did this imbalance exist? What was causing it?

Section 5 deals with De Sitter's solution, found in 1917, to Einstein's field equations for an empty universe. The solution implied an exponentially expanding universe. If De Sitter's model were true, the redshifts observed by Slipher were indeed reflecting the predicted expansion of the universe. This suggested that a relation between redshifts and the distance of those light-emitting galaxies should exist.

Section 6 describes what Hubble found in 1929. What he found was a simple linear relation between the distances to galaxies and their recession speeds. He and his colleague Humason found the value of the constant of the linear relation to be about 500 km/s/Mpc. This was the first determination of what we know as the Hubble constant.

The next sections, Sections 7–10, provide an account of the emergence of cosmological relativity and the contributions of De Sitter, Friedmann, and Lemaître to information regarding the expanding universe. In 1926, Lemaître calculated the value H_0 to be about 625 km/s/Mpc.

Then, in Section 11, we come across "The earliest H_0-Tension". This was a problem involving the age of the Earth based on early-20th radioactive dating as compared to the age of the universe, inferred from the value of the Hubble constant. It turned out that the Earth was found to be older than the Universe. Today, we know that the problem had its origin in the then-accepted value for H_0, which was far from today's estimates.

The problem of correctly determining the value of the Hubble constant depends on the accurate assessment of distances to receding objects. Distances were once measured (and are still measured) by using Cepheids, which are high-luminosity variable stars. This method requires good calibration of the period P of the pulsating variable star to its maximum luminosity L (a P-L calibration). This is discussed in Section 12. Due to an odd coincidence, the P-L calibration used in the 1920s and 1930s was incorrect, consequently producing erroneous values for the Hubble constant.

A better calibration was found independently in the 1940s by Baade and Thackeray. This is presented in Section 13. As result, new attempts to measure the Hubble constant were triggered (Section 14). New values for the Hubble constant appeared, and during the period of 1940–1955, and the value of the constant reduced dramatically. A 1955 measurement by Humason, Mayall, and Sandage produced a value of 180 km/s/Mpc. With a stroke, this erased the issue of the supposed young age of the universe—it was older.

Section 15 mentions the need to measure H_0 at very large distances, since gravitational interactions among our neighboring galaxies may be causing some of them to move much more quickly or slowly than the rest of the universe.

This need led to the development of methods other than the use of Cepheids to reach further distances. However, most of them are, in some way, dependent on comparisons with Cepheid-determined distances.

Part II is devoted to providing brief accounts of several different methods:

Section 16, The Planetary Nebulae Luminosity Function (PNLF);

Section 17, The Tully–Fisher Relation (TFR);

Section 18, The Faber–Jackson Relation (F-J);

Section 19, Fundamental plane, the Dn-σ relation;

Section 20, Surface Brightness Fluctuation Method (SBF);

Section 21, Tip of the red giant branch (TRGB);

Section 22, Global Cluster Luminosity Function GCL.

Then, we take a short break (Section 23) to describe a discussion had by a group of experts at the end of the twentieth century on the merits of the distance measurement methods available at that time.

In Section 23, we describe the Cepheid calibration developments in the twentieth century, followed by the important method involving the use of Type Ia Supernovae (SN Ia) as standard candles in Section 24. This method led to the discovery that, contrary to the supposition at that time (end of twentieth century), the universe's expansion was not slowing down, but rather accelerating. There was a prevailing feeling that a cosmological constant was at work again in the field equations. Section 25 accounts for an alternative candle to measure the Hubble constant using Mira variable stars (Miras).

Section 26 briefly discusses the concept of a deceleration parameter (q_0) whose measurement provides values for the Hubble constant and the average density of matter in the universe. This method requires an accurate determination of the apparent luminosity of an object as a function of its redshift z. Then, we move on to the use of gravitational lensing to measure the Hubble constant.

To end with the description of the "late" route measurements for H_0, Sections 27–29 describe the lensing, masers, and the Sunyaev–Zeldovich Effect, respectively. Then, we move to an "early" method: The cosmic microwave background radiation (CMB) method, which is closely related to baryon acoustic oscillations (BAO) H_0 determination (Section 30). In Section 31, we briefly mention the standard siren method, which is a promising and different technique with which to determine the Hubble constant. In Section 32, we explain how shadows of supermassive black holes can be used to determine H_0, and in Section 33, we explain how fast radio bursts can also be used to constrain the Hubble constant.

In Section 34, we describe the current situation of the H_0 value. The results of "late" route measurements for H_0 all converge to a value within a very narrow range of uncertainty. The same can be said about the "early" route, but its values converge to a different H_0. Thus, the results of both ("early" and "late") sets of measurements do not coincide with each other within the error bars. We conclude this work by presenting some of the views and outlooks on this crisis, which have arisen because of the Hubble tension.

Part I. Early historical roots.

In the first section, we describe the historical events and personalities that played roles in the effort to determine astronomical distances in order to measure the Hubble constant.

2. The Great Debate of the 1920s

A little over a hundred years ago, on 26 April 1920, a debate was hosted by the US National Academy of Sciences. It took place in the Baird auditorium in the nation's capital. The place was an elegant, classically inspired room that featured a domed ceiling of Guastavino tiles. The audience was there to attend a day-long event that comprised an interesting debate. The debate was entitled "The distance scale of the Universe", with Harlow Shapley of Mt. Wilson Solar Observatory and Heber Doust Curtis of Lick Observatory as the debaters [1]. The first debater, Shapley, was a young astronomer who had previously practiced journalism, and the second, Curtis, a veteran astronomer better known for his comprehensive work on spiral nebulae for over a decade.

In addition to discussing the size and extent of the universe, which was the main reason for the debate, an attempt was also made to answer the query "Is the Milky Way an island Universe or is it just one of such galaxies"? Arguments were made in favor of the island universe by Shapley. For his part, Curtis championed the position that other galaxies exist alongside the Milky Way. The event began in the morning, with each of the astronomers' presentations addressing their own technical theses, and the actual debate took place later in the evening.

Shapley believed that the Milky Way contained all of the known and unknown celestial objects. In simple words, for him, the Milky Way was the whole universe. Shapley claimed that the evidence favored the island universe hypothesis, arguing that spiral nebulae (today identified as galaxies) were part of the Milky Way. He had estimated, using his own measurements, that the Milky Way was large (100 kpc). Shapley further argued that novae, which had been observed in spiral nebulae such as the Andromeda nebula, showed the same apparent brightness as those seen in the middle of the Milky Way [2]. Looking into former studies, he calculated that if Andromeda were not in the Milky Way, then it would be as large as 30,000 kpc., an implausibly-sized object for the audience present in the Baird auditorium. Consequently, according to Shapley, Andromeda was part of the island universe.

Perhaps the best argument presented by Shapley during the debate, and certainly one of the most forceful, was Adriaan van Maanen's measurements of the alleged rotation of the Pinwheel galaxy [3]. Van Maanen's measurements (now known to be incorrect) showed that the Pinwheel galaxy rotated on a time scale of years [4]. Shapley correctly argued that if Pinwheel were outside the confines of the Milky Way, then its spin rate would be greater than the speed of light, so it had to be concluded that such a "nebula" was also within the Milky Way.

In the course of the debate, an enigmatic issue concerning an astronomical observation made years earlier by Vesto Melvin Slipher received an easy answer from Shapley. We must recall that, some years before the debate, Vesto Melvin Slipher had measured the first wavelength shifts of spiral nebulae, meaning that they were recessing from us [5]. Regarding this, Shapley commented that Slipher's measurements meant that these objects were somehow repulsed away from the Milky Way's center by some unknown mechanism. For his part, Curtis did not comment on this matter.

In contrast, Heber Curtis was convinced that the Milky Way and Andromeda were independent galaxies, just two of many such bodies. On the subject of the size of the Milky Way, Curtis was of the opinion that, based on measurements of star counts in different regions of the sky, the Milky Way's diameter was only around 10,000 kpc. As for the Pinwheel galaxy argument, Curtis agreed that if the results of van Maanen were correct, Shapley was right. But Curtis rejected Adriaan van Maanen's results on the grounds that he considered van Maanen's results accurately unrealistic. Later astronomers have re-examined the measurements of Van Maanen, and have concluded that he made a serious error.

During the debate, Curtis noted that many novae on Andromeda were dim because their brightness was diminished by their distance from us. If one takes that distance into account, then the brightness of the observed novae approximately agrees with those that are closest to us in our Milky Way. Like Shapley, Curtis also had no explanation for the nebula recession observed by Slipher. Curtis pointed to evidence that the Milky Way had a spiral structure like any other nebula. Both Shapley's and Curtis' accounts were published a year later, in 1921 [6].

The debate ended a few years later when, as we will see below, Edwin Powell Hubble's work on Cepheid variables in several "nebulae" (today recognized as galaxies within the local group) settled the issue of the existence of external galaxies.

3. The First Candle

In 1908, Miss Henrietta Swan Leavitt was working at the Harvard College Observatory as a female "computer", as observatory assistants were familiarly called at that time. Formally, her position had the pompous name of "Curator of Astronomical Photographs". Leavitt had been hired some years prior by the observatory director, Edward Charles Pickering, to measure and register the brightness of stars. Her task was to catalog stars' luminosities by examining photographic plates from the observatory's collection. Her specific assignment was to identify variable stars. She utilized an instrument called a blink comparator that is no longer used today. By using that instrument, she would compare pairs of plates of the same star field taken a few days or weeks apart. This time-consuming operation involved manually flipping a screen back and forth quickly to suppress one image at a time, with the images in question on a pair of plates. In this way, a variable star would show up as a flashing spot. After analyzing plates of the Small Magellanic Cloud (SMC), taken in the lapse between the years 1893 and 1906, she compiled a catalog containing 1777 variable stars [7].

At some point, Leavitt conjectured that there was a relationship between the periodic luminosity changes observed in variable stars and their maximum brightnesses. She restricted her search to Cepheid variables that resided in the SMC. The reason for her opportune choice was her informed assumption that these stars must be at the same distance from Earth and, therefore, the comparison between their luminosities was well-founded. By 1912, Henrietta Swan Leavitt found that 25 Cepheid stars in the SMC would brighten and dim periodically [8]. Figure 1 shows the clear relationship that she found between the maximum (and minimum) brightness of each star and the length of its period, as she suspected.

The discovery made by Miss Henrietta Leavitt led to the first standard-candle method for estimating distances to galaxies. In principle, the road to determining the actual distances to sky systems containing Cepheids was seeded. All that was needed was to find the distance to just one nearby Cepheid variable to calibrate Leavitt's period versus the luminosity relation (P-L relation).

Strangely, in her paper, Leavitt did not explicitly write out a mathematical relation. In modern notation, Leavitt Law is of the form $<M_v> = a + b \log_{10} P$, where P is the period of the luminous oscillation in days and a and b are constants to be determined. This pair of constants is univocally determined by a single point that astronomers call the zero point in their jargon. The zero point is defined as the absolute magnitude of a hypothetical Cepheid with P = 1 day.

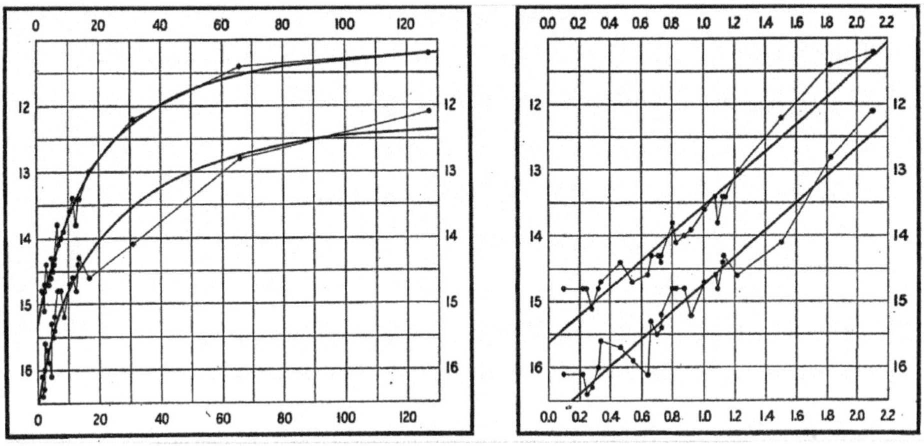

Figure 1. Period–luminosity curves and best fits. **Left**: Abscissas in days are equal to periods, ordinates corresponding to star magnitudes at their maxima and minima. **Right**: abscissas are equal to logarithms of the periods. Figure taken from ref. [8].

Some people regret that Leavitt did not continue her research in determining stellar distances, but it must be remembered that her work at the observatory was limited to the study of stellar luminosities. In spite of this, at the end of her paper, Leavitt anticipated "It is to be hoped, also, that the parallaxes of some variables of this type may be measured".

The importance of performing these future measurements was not overlooked. A short time later, Ejnar Hertzsprung, by incorporating Leavitt's data, established the first period–luminosity calibration: <Mv> = $-0.6 - 2.1 \log_{10}$ P. He also carried out a statistical parallax analysis on 13 Cepheids previously reported by Leavitt for which proper motions were available. Surprisingly, his paper had a misprint, as he reported a distance to SMC Cepheids of only 3000 light-years [9]. This result was well below the value of the actual distance, and perhaps this was the reason why his publication did not receive much attention. Some years later, Harold Shapley and Henry Norris Russell realized the error that had appeared in Hertzsprung's article and corrected the misprinted distance to 30,000 light years, see footnote 2 p. 434 in [10]. An interesting fact in Hertzsprung's work is that he ignored the interstellar absorption effects that cause dimming of light during its passage through the interstellar medium and alter the magnitudes of real stars.

Almost simultaneously to Hertzsprung's work, a very short note was published by Henry Norris Russell in 1913. In it, he estimates "the mean distance and real brightness... by the method of parallactic motion" of several variable stars, including maximum and minimum brightness estimates for the group of Cepheids previously reported by Leavitt (without citing Leavitt's publication) [11]. Some authors mistakenly indicate that Russell, like Hertzsprung, also presented a calibration of Leavitt's Law in this publication; however, this does not seem to be the case, as no such calibration appears in Russell's original paper. The point here is that Russell was among the first to perform an absolute magnitude determination of Cepheids. Yet, Russell did not, similarly to Hertzsprung, consider the effect of interstellar absorption.

However, it is interesting to note that in one of Russell's subsequent publications, this time in co-authorship with Harlow Shapley, absorption was considered [10]. The paper was concerned with the galactic distribution of eclipsing variables and Cepheids, and it came to the conclusion that there exists interstellar absorption of about two visual magnitudes per kpc, as was previously suggested by Edward S. King of the Harvard College observatory.

In effect, the need to consider interstellar absorption in luminosity measurements of stellar objects was implicitly suggested in 1912 by Edward S. King of the Harvard College observatory [12]. Since 1897, King had been involved in investigating how to obtain reliable luminosity measurements of stars in photographic plates [13]. For this purpose, he devised a photometric apparatus [14]. After applying his photometric methods to the moon and planets, he focused on determining the luminosity of stars following a protocol he carefully elaborated [15]. In connection with this task, King discussed the role of the absorbent medium in space and deduced evidence for its existence [16]. Then, he proposed an interstellar absorption of about 2 mag per Kpc [17]. A little later, in 1916, when he had already gathered more observations, he concluded "All indications point to the presence of an absorbing medium in space, or some factor which produces effects similar to absorption, by making the more distant stars redder" [18]. It is important to notice that this redness should not be misinterpreted as being produced by Doppler's effect. It is simply a light scattering effect caused by dust particles in interstellar space. A decade later, in 1927, King presented the hypothesis that a local cloud of absorbing matter, extending from the Sun to at least 100 light years, envelops our local star cluster [19].

Over the next two decades, there was a tendency for many astronomers to ignore the effects of interstellar dust absorption on stellar luminosities, which would ultimately influence the correct determination of Hubble's constant value using candle-based methods for several years. This may have been due to Shapley's change of mind from considering absorption to ignoring its effects. After the 1914 joint publication with Russel, Shapley was just beginning to determine distances using standard-candle methods. In 1915, Shapley published an article titled "Studies of Magnitudes in Clusters, 1. On the Absorption of Light in Space", where one of his conclusions read as follows: "It seems to be necessary to conclude that the selective extinction of light in space is entirely inappreciable, at least in the direction of the Hercules cluster" [20]. In this study, Shapley observed stars of all colors, so he correctly assumed that there was no absorption or redness of light on its path to Earth. In this case, there is indeed little absorption, since the studied clusters are far from the dust band that covers the plane of our galaxy. Therefore, the effects of absorption were imperceptible for Shapley. The problem came from extrapolating particular conclusions. It took him and many others several decades to abandon the trend of ignoring possible effects of interstellar absorption. This would have had a vast effect on the subsequent history of Hubble's constant.

4. The End of the Great Debate

In 1901, the wealthy heir Percival Lowell had spent part of his fortune establishing an astronomical observatory in Flagstaff, Arizona. Lowell was a highly controversial character, as he claimed to have observed a system of canals on Mars (the imaginary Schiaparelli's canals) that he believed criss-crossed the planet's surface, distributing water from the poles all over the red planet [21].

Around 1901, Lowell acquired an expensive state-of-the-art spectrograph for his observatory. In 1906, Lowell asked Vesto Melvin Slipher, one of the observatory's staff members, to survey spiral nebulae. The reason was that Lowell believed that the nebulae may have been solar systems in the process of formation. For this task, Slipher modified the spectrograph for nebular spectroscopy. After modifying the instrument, Slipher focused on the Andromeda Nebula. Not only did he record absorption lines, but he also saw that the lines were shifted toward the blue [22]. Interpreting these shifts as Doppler shifts, Slipher calculated that the Andromeda Nebula was hurtling at about -300 km s^{-1} toward the Earth [23]. The fact that Slipher's discovery came from the premises of Lowell's observatory caused a stir of doubt regarding its soundness. Nevertheless, by 1914, Slipher had already collected data from 15 nebulae, of which 13 were receding and 2 were moving towards Earth [24].

In 1919, Edwin Powell Hubble joined the Mount Wilson observatory staff, where he had access to the 60-inch reflector as well as the just-completed 100-inch Hooker telescope.

This instrument was by far the most powerful telescope in the world. In 1923, Hubble's research program became focused on locating novae in Andromeda using the Hooker telescope. By October 1923, Hubble had discovered what he took to be a nova in a nebula's outer edges. But as Hubble examined previously obtained plates of the same region, he noticed that his newly discovered "nova" regularly exhibited brightness changes, so upon constructing a light curve for its varying brightness, he realized that what he had found was a Cepheid variable, not a nova.

Hubble measured the period and apparent brightness of the Cepheid. Then, by employing Leavitt's Law, he arrived at a distance of around 900,000 light years (estimated to be approximately 2.5 million light years presently), placing Andromeda Nebula well outside the limits of Shapley's estimate of the Milky Way's size. In Hubble's best self-promoting style, his finding was first released by the press [25]. Hubble's formal announcement, entitled "Extragalactic Nature of Spiral Nebulae," was delivered in absentia by Henry Norris Russell to a joint meeting of the American Astronomical Society and the American Association for the Advancement of Science which was held at the end of December 1924 [26]. Thus, in the words of Shapley's opponent in the Great Debate, Heber D. Curtis, it was not until 1924 that "... all doubts as to the island Universe character of the spirals were finally swept away by Hubble's discovery of a Cepheid" [27]. Hubble's results for Andromeda were not formally published in a peer-reviewed scientific journal until 1929 [28], but "The Great Debate" was over in 1924.

However, a delicate conflict had endured at Mount Wilson between Hubble and his observatory colleague Adriaan Van Maanen. Recall that it was Van Maanen's measurements that Shapley used to argue that the spiral nebulae were located within the limits of the Milky Way. This conflict between these two characters (this time, Hubble and Van Maanen), turned into a public dispute, which ended in 1935 with the publication of a brief note by each of them [29,30]. In Van Maanen's note, he conceded that his measurements should be taken with caution.

5. Velocity–Distance Early Searches

After Hubble found that the spiral nebulae were actually galaxies like ours, Slipher's findings attracted the attention of some astronomers. On the one hand, most nebulae (now known to be galaxies) were rapidly moving away from Earth; on the other, a few were approaching it. Why did this imbalance exist? What was causing it?

In 1917, Willem de Sitter, a Dutch mathematician from Leiden University, had discovered a solution to Einstein's field equations of general relativity [31]. De Sitter's solution for an empty universe had very important cosmological consequences: specifically, an exponentially expanding universe. This signified that, if the implications of De Sitter's model were true, the redshifts observed by Slipher were indeed reflecting the predicted expansion of the universe. Whether it was the result of De Sitter or Slipher's findings, the fact is that some astronomers undertook the challenge to search for some relationship between redshifts and the distances of those light-emitting objects. General relativity theory, as interpreted by De Sitter, suggested that this relation should exist.

One of the first astronomers to accept the challenge was Carl Wilhelm Wirtz. His observations produced correlations between radial velocities and nebular distance indicators [32]. As distance indicators, he used comparisons between the diameters of galaxies. He was the first to attempt the search for a relation between radial velocity and distance. As a result, he found V (km/sec) = $2200 - 1200 \log (D_m)$, where D_m represents the angular diameter (scales like $1/r$) in arc minutes of the observed nebula. Clearly, this was incorrect, but his relation followed the right tendency for V as it increases with distance r.

Another scientist undertaking the challenge of finding a velocity–distance relation, albeit for different reasons, was Ludwik Silberstein (Einstein's antagonist [33]). Silberstein attempt to establish a velocity–distance relation was intended to determine the curvature of the universe, which he considered to be fixed. He used De Sitter's results to derive a formula for the shift of the spectral lines emitted by stars [34]. For distant stars, Silberstein found that, with the limit of small velocities, the Doppler shift is $\Delta\lambda/\lambda = \pm\, r/R$, where R is the radius of curvature of the universe and $\Delta\lambda$ is the shift in the wavelength λ of the line. At this point, it is convenient to remember that today, a redshift value "z" is reported ($z = (\lambda_{obs}/\lambda_{em}) - 1$, where λ_{obs} is the observed and λ_{em} is the emitted wavelength). Silberstein applied his formula to a list of stellar clusters as well as to the small and large Magellan clouds. The application of this law to this mix of objects (some approaching and others receding) gave him a value for the radius of curvature of the universe of R of about 10^8 lyr. As expected, it did not take long for Silberstein's work to be criticized by various astronomers, including Arthur Eddington [35].

One earlier attempt to obtain a velocity–distance relationship was that of the Swedish Knut Emil Lundmark. In 1924, he plotted the radial velocity of 44 galaxies against their estimated distances [36]. He assumed that Andromeda was 200,000 pc away. Then, he made rough determinations of the distances to other galaxies by comparing their angular sizes and brightnesses to that of Andromeda. Figure 2 shows Lundmark's plot. He concluded that there may be a relationship between galactic redshifts and distances, but "not a very definite one".

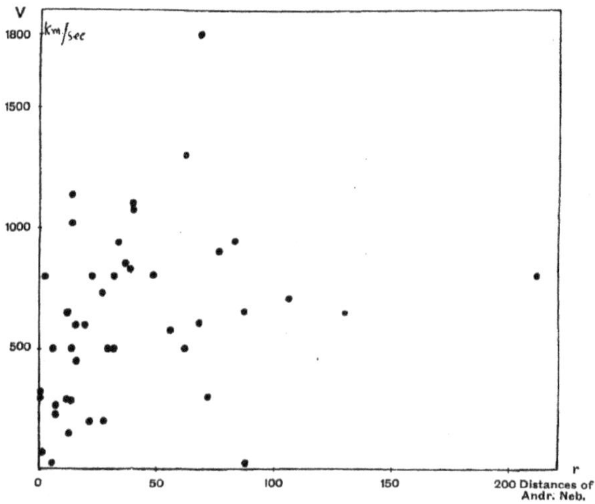

Figure 2. Knut Lundmark's 1924 [36] relation between relative distances and observed radial velocities of spiral nebulae (The scale unit is the distance to the Andromeda "nebula").

By the mid-1920s, there were no convincing studies on a possible relationship between recession velocity and distance. The collected evidence was far from being able to test De Sitter's model. The chief issue involved the imprecise estimates of distances to galaxies. It was then that Hubble decided to involve himself in the problem.

6. Hubble's Entrance

Since entering Mount Wilson, Hubble had made good use of the 100-inch Hooker telescope. By the early 1920s, he had already detected 11 Cepheid variables in Barnard's Galaxy (NGC 6822) and derived an estimate for its distance [37]. Hubble's detection was a milestone in astronomy, as it was the first system beyond the Magellanic clouds to have its distance determined.

However, the obtained value of the distance was well below the currently accepted value, because Hubble used what was then the most recent calibration of Leavitt's Law, made by Shapley in 1925. Today it is known that the usage of Shapley's calibration underestimated distances. Subsequently, by using the same Cepheid method and comparing the mean luminosity between galaxies (a kind of "embryonic ladder"), he continued with the determination of distances to other galaxies, such as Andromeda and the Triangulum Galaxies. Step by step, Hubble and colleagues piled up estimates on galactic distances.

By 1929, the spectral shifts for 46 galaxies had been already measured, the great majority by Vesto Slipher. In that same year, Hubble plotted the recession velocity (which, for nearby objects, is given by v = c·z) versus nebular distances that he considered to be fairly reliable, i.e., most his own estimates plus two by Shapley and four by Humason. Of the 22 redshift values plotted by Hubble, 18 had previously been measured by Slipher. Hubble did not reference Slipher in his publication [38].

Hubble asserted that his plot of redshift versus distance was, at least to the first approximation, best represented by a linear relation. Curiously, instead of expressing the relation into the form that would soon become typical (v = c · z = H · r, with H as a constant and following what would later be termed the Hubble–Lemaitre law), Hubble wrote it in terms of the solar motion equations. Nevertheless, by interpreting redshifts as Doppler shifts, what Hubble found was a simple linear relation between distances to galaxies and their recession speeds (as determined by their redshifted spectral lines). Finally, in his paper, Hubble announced his intention to expand his research and made a concluding remark:

"In order to investigate the matter on a much larger scale, Mr. Humason at Mount Wilson has initiated a program of determining velocities of the most distant nebulae that can be observed with confidence. The outstanding feature, however, is the possibility that the velocity distance relation may represent the De Sitter effect and hence that numerical data may be introduced into discussions of the general curvature of space".

Indeed, as he stated in his 1929 article, in order to extend his research, Hubble enlisted the help of Milton La Salle Humason, a colleague from Mount Wilson who had toured the entire hierarchy of the observatory, from mule skinner in the early days of the establishment through janitor until his curiosity led him to learn from other astronomers and, eventually, become one of them.

Humason focused his instruments on the very faint and presumably farthest galaxies. Thanks to his ability, exposure times of plates taken by him were adequate to allow him to identify ionized Calcium H and K spectral lines, and short enough to prevent entire spectra from becoming continuous. For each of the observed galaxies, Humason calculated the z values of their shift and distances. In March 1931, they presented their extended results [39] (see Figure 3). The value for the H constant which they found for Hubble's empirical relation was about 500 km/s/Mpc.

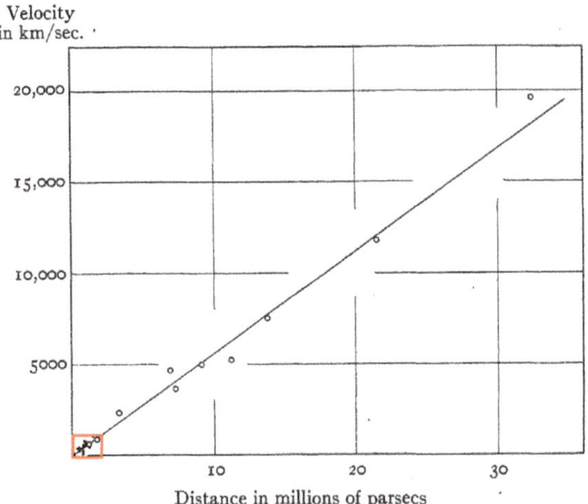

Figure 3. Galactic redshift vs. distance, plotted by Hubble and Humason (1931) [39]; the rectangle in the lower left corner encloses data points plotted in 1929.

7. The Emergence of Cosmological Relativity

Reactions to the appearance of general relativity by the broad spectrum of scientists showed widespread indifference among the less informed. Take, for example, a fragment of the correspondence (1919) between the distinguished astronomer George Ellery Hale and the then-Assistant Secretary at the Smithsonian Institution, Charles Greely Abbot. This epistolary exchange took place a year before the so-called "great debate", about which we have already written in a previous section. In a first letter, Hale had suggested two topics for Abbot to choose from for the keynote debate at the National Academy of Science (NAS) meeting the next following year (April 1920). The suggested topics were general relativity and the size of the Milky Way. Hale, in fact, favored relativity, but Abbot had the last word on this choice. Abbot's response illustrates our point:

> *"As to relativity, I must confess that I would rather have a subject in which there would be a half dozen members of the Academy competent enough to understand at least a few words of what the speakers were saying if we had a symposium upon it. I pray to God that the progress of science will send relativity to some region of space beyond the fourth dimension, from whence it may never return to plague us".* [40]

In fact, the plague had started earlier and was there to stay.

After completing his general theory of relativity in 1915, one of the first things Einstein attempted was to apply it to model the universe at large scale. Einstein's idea of the universe was that of a static entity, an idea shared by the majority of scientists of that time, and there was no known reason at that time for Einstein to doubt it [1].

To ensure the stability of his static universe, Einstein introduced a change in his field equations by adding what he called a "cosmical" term proportional to the metric tensor that he thought would guarantee the large-scale immobility of his model universe. This additional term gave rise, even in empty space, to a repulsive force which would allow his model universe to remain static, counterbalancing the gravitational attraction of the matter within it.

In 1917, Einstein published his renewed version of the field equations, adding the mentioned "cosmical" term to his original field equations. He solved his field equations considering an isotropic and homogeneous cylindrical universe, that is: space dimensions corresponded to a sphere, but the time dimension was uncurved. His paper was titled "Cosmological considerations in the General Theory of Relativity". Thus, his renewed field equations written in standard form were as follows ("λ" is today; "Λ" is the cosmological constant [2]),

$$R_{\mu\nu} - \frac{1}{2}g_{\mu\nu}R - \lambda g_{\mu\nu} = -\kappa T_{\mu\nu},$$

This calculation shows that λ is proportional to ρ, the mean mass density of the universe, and inversely proportional to R, its radius of curvature, i.e., $\lambda = 1/R^2 = \frac{1}{2}\kappa\rho c^2$. Regarding this result, Eddington readily pointed out an issue with the marginal stability of Einstein's universe model when he wrote "the question at once arises, by what mechanism can the value of λ, be adjusted to correspond with M"? In Eddington's book, M is the total mass of the universe [41]. With this comment, Eddington hinted that Einstein's model universe was static, but unstable. In addition, Einstein's world model offered no explanation of the observed redshifts. Nonetheless, it was the first step that marked the beginning of modern cosmology as the scientific study of the origin and structure of the universe. The second step was made by a Dutch mathematician and astronomer named Willem De Sitter.

Beginning in 1911, De Sitter maintained regular correspondence with Arthur Eddington on matters of astronomical interest. De Sitter held the chair of astronomy at Leiden University in the Netherlands, where he also discussed issues related to the theory of relativity and its consequences with his university colleagues, Hendrik A. Lorentz and Paul Ehrenfest. He had also met sporadically with Einstein for the same reasons. De Sitter had contributed to spreading relativity in the English-speaking world, since the Great War that prevailed in those days prevented the free diffusion of Einstein's ideas from Germany. In this regard, in 1916, he published two articles entitled "On Einstein's theory of gravitation and its astronomical consequences, first part and second part", which contributed to the diffusion of relativity [42,43].

In 1916, Paul Erhenfest suggested to De Sitter that some of the difficult problems associated with the idea of an infinite universe could be avoided by assuming a closed universe. It was in 1917, upon learning of Einstein's publication, that De sitter produced three solutions for the field equations, retaining the "cosmical" term. He stipulated that his model should be isotropic and static. He constrained the spatial part of space–time to constant curvature. One of the models was Einstein's own. It contained matter related to the "cosmical" term, but zero pressure. Another contained zero pressure, zero matter, and no "cosmical" term. The last is what we know now as De Sitter's model [44]. Density and pressure were both zero, but he retained the "cosmical" term. As this term was introduced by Einstein to counterbalance the gravitational contraction, and since De Sitter did not include mass, his universe expanded. If some light emitter was found within this universe, what would happen is that if the universe were to expand, a redshift of its light would be observed. This phenomenon was labeled the De Sitter effect in its time.

Despite the fact that the model of the universe proposed by De Sitter was quite far from what would be considered a real universe, the model was very relevant because it, at least, predicted the presence of redshifts.

De Sitter's solutions to Einstein's field equations caught the attention of several theorists among them—Ludwick Silberstein, Hermann Weyl, and Richard Tolman—who began to explore the physical aspects of the given solutions. It was around this time that two individuals unknown to experts in the field independently made great strides in solving the field equations. The first was a young Russian man named Aleksandr Aleksandrovich Friedmann, and the second was a Belgian Catholic priest, Georges Henry Joseph Édouard Lemaître.

8. Friedmann

In 1922, a short note by Einstein of only eleven lines appeared in the Zeitschrift für Physik journal [45]. It reads, *"The results contained in the cited work regarding a non-stationary world seemed suspicious to me. In fact, it turns out that this given solution is not compatible with the field equations"*. In this belittling note, Einstein was referring to a paper by A. Friedmann that had appeared that same year [46].

At that time, Alexander Friedmann held an academic position at the University of Saint Petersburg (formerly Petrograd and renamed Leningrad). Friedman's situation at the end of the Russian civil war must have been difficult, as he was academically isolated from the rest of the world. News about astronomical discoveries like those of the redshift should not have flowed easily at this time, when there were still skirmishes between the red and the overpowered white armies.

In the paper in question, which was belittled by Einstein, Friedmann gave several valid solutions to Einstein's equations with the "cosmic" λ-term, assuming, as Einstein did, a homogeneous, isotropic, and positively curved universe. The main difference that Friedmann introduced, in contrast to Einstein's and De Sitter's models, was allowing the radius of curvature R to vary with time. Friedmann wrote the line element $ds^2 = g_{ik}\,dx^i dx^k$ as:

$$ds^2 = \mathrm{R}^2 \left(dx_1^2 + sin^2 x_1 dx_2^2 + sin^2 x_1 sin^2 x_2 dx_3^2 \right) + M^2 dx_4^2$$

where $M = 2\,\pi^2 \rho R^3$ stands for the mass of the universe, ρ its density, and R a distance scale proportional to the curvature of space.

In this 1922 paper, Friedmann begins by dealing with static cases where R is not a function of time (x_4). Then, he reduces his model to the static Einstein and Sitter universes by setting the function R as equal to R^2/c, where the value R, i.e., the radius of curvature, is constant. To recover Einstein's and De Sitter's models, he set $M = 1$ for Einstein's and

$M = \cos(x_1)$ for De Sitter's, respectively. He demonstrated that these are the only two solutions for static universes. Then, he moved on to the non-static cases.

Using Einstein's fundamental equations and retaining λ, Friedmann then derived equations for the determination of $R = R(x_4)$ and the matter density $\rho = \rho(x_4)$, both of which were allowed to vary with time according to Einstein's equations:

$$R_{ik} - \frac{1}{2} g_{ik} R + \lambda \, g_{ik} = \kappa \, T_{ik}$$

Neglecting pressure terms, the time-dependent $R(x_4)$ is determined by λ and ρ. The energy–momentum tensor reduces to

$$T_{44} = c^2 \rho g_{44} \quad T_{4i} = 0 \quad T_{ik} = 0 \quad for \; i,k = 1,2,3$$

Setting $i = k = 1, 2, 3$, Friedmann obtained his equation for R:

$$\frac{\dot{R}^2}{R^2} + \frac{2R\ddot{R}}{R^2} + \frac{c^2}{R^2} - \lambda = 0,$$

where a dot stands for the time derivative. Setting $i = k = 4$, he obtained the following relationship:

$$\frac{3\dot{R}^2}{R^2} + \frac{3c^2}{R^2} - \lambda = \kappa c^2 \rho,$$

which is known as Friedmann equation today. Since the value of λ was unknown, he considered different possibilities and found that, depending on the ratio between the attraction of the gravitational force due to the total mass of the universe and the "cosmic" term, one can obtain universes that expand or contract or vary periodically as a function of time. In a second paper in Zeitschrift fuer Physik, and in a book published in 1923, he explored the open-geometry structure of the universe [47,48].

Friedmann speculated on the age, mass, and size of some of his universes, but did not make comparisons with astronomical data known at the time because he thought that the uncertainty of these values did not warrant it.

He also did not incorporate the redshifts measured by Slipher, perhaps because he ignored their existence. Friedmann died of typhoid fever in 1925 at age 37. His work did not attract attention and was erased from memory until it was revived years later by Eddington.

As we have already mentioned, at the time Friedmann sent his paper for publication, he held an academic position at the University of Saint Petersburg (Petrograd). One of his colleagues at the university, Ivan Kruthoff, had attended a meeting at Leiden in May 1923 where Einstein was also staying (at Paul Ehrenfest's home). Einstein frequently visited Leiden, as he had the appointment of special professor ("Bijzonder Hoogleraar"), a position promoted by Ehrenfest [49]. Friedmann saw Kruthoff's visit to Leiden as an opportunity to discuss his previously delivered letter to Einstein, in which he presumably provided more details about his solutions. In this way, he ensured that Einstein would read his letter. We must mention that Friedmann and Ehrenfest maintained correspondence, since the former had been an outstanding student of Ehrenfest when he was a lecturer at Saint Petersburg in 1912.

When Einstein realized that the solutions found by Friedmann were mathematically "correct and clarifying", he wrote a short note rectifying the error of appreciation he had made in his previous note, but warning that they were non-static solutions. "They show that in addition to the static solutions to the field equations there are time varying solutions" [50].

9. Lemaître

On Friday, 10 January 1930, at Burlington House, Piccadilly, an important meeting of the Royal Astronomical Society was held. Willem De Sitter gave a lecture on the discovery Hubble had just recently announced: the probable existence of a linear relationship between the recession velocity of spiral nebulae and their distances [51]. De Sitter explained in detail to the audience how the distances to the spirals were determined. He also admitted that it was difficult to reconcile this result with one of the two universe theories, the one proposed by him and the other by Einstein.

During the discussion that followed the talk, Eddington wondered why there were only two static solutions. One of which was Einstein's, and in the other (De Sitter's), the static universe expanded as soon as matter was introduced into it. To Eddington's concern, De Sitter commented that it would be desirable to investigate how to include matter into his model, but he admitted that the difficulty that would arise would be keeping the universe static. However, Eddington conceded that, perhaps, it would not hurt to investigate non-static intermediary solutions. From the discussion that followed the talk, the idea of a static universe prevailed, as it was ingrained in the minds of academics despite Hubble's results. The résumé of the discussion that took place at the meeting was published in the February issue of *The Observatory* [52].

A few weeks after the meeting at Burlington House, Eddington received a letter from the Belgian Jesuit Georges Lemaître which stunned him. But before reproducing part of the letter, it is timely to recall some earlier undertakings of Lemaître.

In the past, from 1923 to 1924, the Belgian Jesuit had spent a season of research on general relativity in the U.K., at Cambridge. There, he took lectures by Eddington on "Relativity Theory of Electrons and Protons" and "Fundamental Theory", giving him the opportunity to advance his knowledge on the subject [53]. As a visiting student, Lemaître produced a paper in which he generalized the definition of simultaneity [54].

After his stay at the English university, Lemaître and Eddington crossed the Atlantic to participate in the 94th annual meeting of the British Association for the Advancement of Science in Toronto (6–13 August 1924), where Eddington spoke on relativity and the bending of starlight. Then, in September 1924, Lemaître arrived at Harvard College Observatory to work on Cepheids under Harlow Shapley's supervision. Lemaître spent 9 months there, where he wrote a paper criticizing De Sitter's universe. The Belgian priest spotted that the geometrical coordinates chosen by De Sitter in his model introduced a spurious inhomogeneity. To be precise, what Lemaître found was that De Sitter's model produced a privileged point with properties different from the others, a fact which violated the principle of spatial homogeneity. De Sitter chose his line element as:

$$ds^2 = R^2\left[-d\chi^2 - sin^2\chi\left(d\theta^2 + sin^2\theta d\phi^2\right) + cos^2\chi d\tau^2\right],$$

where R is the constant radius of the universe with coordinates: χ, θ, ϕ, τ. The spatial part of the line element is constant and time-independent, so it produces a static universe whereas the temporal part, $R^2 cos^2\chi d\tau^2$, depends on the spatial coordinate χ, except when $\chi = 0$, thus distinguishing this point from others. Lemaître noted that *"It is clear that such an introduction of an apparent center in a Universe which, by definition, has none is objectionable for a study of the properties of this Universe"*. That is, De Sitter's model contravened his initial premise of spatial homogeneity. Lemaître published a paper on the subject, proposing a new choice coordinates that avoided the aforementioned problem [55,56].

In late 1924, the Belgian priest had the opportunity to attend the 33rd meeting of the American Astronomical Society in Washington, D.C., where he heard Henry N. Russell announcing Hubble's theory on the distance to Andromeda using the Cepheid observations, a presentation that, as we have already pointed out, was of celebrated consequence as it ended the "Great Debate". Russell also read a paper on "Stellar Evolution" by Eddington, who had already returned to England.

Afterwards, on 18 June 1925, for the purpose of learning more about extragalactic distances, Lemaître traveled to California to meet Hubble in person at Caltech. On his way back to the East Coast, Lemaître stopped at the Lowell Observatory in Arizona to visit Vesto Slipher in order find out more about the method of measuring redshifts. All of these visits allowed him, a theoretician, to become familiar with astronomical observations.

On June 1925, Father Lemaître paid a visit to the Old Continent to attend the second triennial meeting of the International Astronomical Union (14 June to 22 July) at Cambridge, England. During the meeting, Slipher showed his latest redshift measurements, Hubble discussed his proposal on galaxy classification based on morphology, and De Sitter received an honorary degree of Sc. D. from Cambridge [57].

During the meeting, a delicate question was scheduled for discussion. This was the admission of Germany and the Central powers to the International Astronomical Union. At that time, the wounds caused by the First World War were just healing. However, the decision to admit Germany to the Club was deferred until after participants had had the opportunity to discuss the matter informally. Having previously foreseen this situation, the organizers of the event had planned a large series of social activities (visits, banquets, and garden parties) that would allow for the exchange of opinions among the attendees on the aforementioned matter and, of course, on scientific matters [58]. At those events, Lemaître had had the opportunity to renew acquaintances and perhaps to assess the *"air-du-temps"* on a non-static model of the universe.

Back in Belgium in 1927, he learned by mail that MIT had conferred him a doctoral degree (Ph.D.) for a thesis on general relativity [59], and he was exempt from *"viva voce"* (oral defense). Upon returning to Louvain, he became part of the academic staff at the Catholic University. There, in 1927, he finished writing his paper on a non-static model of the universe, which he published in the Annals of the Scientific Society of Brussels in French.

In autumn 1927, Einstein was in Brussels on the occasion of the Fifth Solvay International Conference. The congress met then to discuss the newly formulated quantum theory. Obviously, in those moments, Einstein's thoughts were not centered on cosmology. The new formulation of quantum mechanics made him feel uneasy.

For Father Lemaître, this was an opportunity to exchange ideas directly with the "Pope of Relativity". He took a train from Louvain to Brussels. He himself later described this meeting [60]: "While walking in the alleys of Leopold Park, [Einstein] told me about an article, little noticed, that I had written the previous year on the expansion of the Universe and that a friend [Auguste Piccard who was also present during the stroll] had made him read. After some favorable technical remarks, he concludes by saying that from the physical point of view it seemed to him completely abominable (*tout à fait abominable*)". During that encounter, Lemaître also learned from Einstein of Friedmann's previous work published in 1922. Lemaître did not know German, which may explain why he had ignored the existence of Friedmann's paper. Later that day, during a ride in a taxi with Einstein and Picard, Lemaître developed the impression that "Einstein was hardly aware of the astronomical facts" [61]. Einstein's comment (*"magister dixit"*) must have surprised Lemaître, and in spite of his Einstein's *"anathema"*, Lemaître was tenacious with his ideas.

Before meeting Einstein, Father Lemaître had already sent his manuscript to Eddington soon after receiving reprints of his paper, but received no response; his former mentor classified the manuscript without really reading it. In July 1928, Lemaître traveled to Leiden, where De Sitter chaired the third assembly of the International Astronomical Union. Unfortunately, Lemaître did not have a chance to discuss this matter with him.

As mentioned above, the report from the January 10th meeting at Burlington House reproducing the discussion between Eddington and De Sitter regarding the possibility of finding non-static intermediate solutions to Einstein's field equations was published in 1930 in *The Observatory*. As soon as an issue reached Lemaître, he immediately wrote a letter to Eddington attaching copies of his 1927 paper and asking Eddington to send a copy to De Sitter [62] (A similar solution had previously been given by A. Friedman [63]):

"Dear Professor Eddington, I just read the February No., of the Observatory and your suggestion of investigating of non-statical intermediary solutions between those of Einstein and de Sitter. I made these investigations two years ago. I consider a Universe of curvature constant in space but increasing in time. And I emphasize the existence of a solution in which the motion of the nebulae is always a receding one from time minus infinity to plus infinity".

From reading the full content of Lemaître's missive, Eddington also learned that Alexander Friedmann had produced (years before, in 1922) a non-stationary solution to Einstein's field equations. Friedmann's solution was like the independently rediscovered one, which was attached to the letter sent by Lemaître in 1927. The letter also indicated that, two years prior, the Jesuit had sent De Sitter a copy and lamented that he likely did not read it [64].

Upon learning of this, Eddington warned De Sitter. The latter was, at that moment, writing a manuscript on the same topic, focusing on the astronomical consequences of relativity. De Sitter, realizing the great importance of the solution, modified his article to include and comment on Friedmann's and Lemaître's accomplishments [65].

In the first part of his paper, De Sitter argues why "static solutions to field equations must be rejected and that the true solution represented in nature must be a dynamical solution". Then, he continues: "A dynamical solution [to the field equations] ... is given by Dr. G. Lemaitre which had unfortunately escaped my notice until my attention was called to it by Professor Eddington a few weeks ago in a paper published in 1927". Next, he goes over the solution for English-speaking readers. At the end of his publication, De Sitter praises Lemaître's achievement.

For his part, Eddington published an article in which he discussed the instability of Einstein's model and also applauded the Belgian priest: "...we learnt of a paper by Abbé G. Lemaître which gives a remarkably complete solution of the various questions connected with the Einstein and De Sitter cosmogonies" "...my original hope of contributing some definitely new result has been forestalled by Lemaître's brilliant solution" [66].

Eddington embraced Lemaître's model with great enthusiasm, distributing it to colleagues and arranging for it to be translated into English and republished in the Monthly Notices of the Royal Astronomical Society. We shall comment on a mathematical formula that disappeared in the translation, but first, we will describe the contents of Lemaître's celebrated paper.

10. Lemaître's Expanding Universe

Lemaître's paper begins by considering an Einstein universe, where the radius is allowed to vary in an arbitrary way. Then, he restricts himself to those solutions in which the three-dimensional space has complete spherical symmetry [3] and is filled with matter comparable to a rarified gas, that is, uniformly and homogeneously distributed through space where its molecules are the extragalactic nebulae. Hence, the corresponding total density of energy in matter $\delta(t)$ depends on time ($\delta(t)$ is the trace of the energy momentum). For his line element, he considers:

$$ds^2 = -\mathrm{R}(t)^2\, d\sigma^2 + dt^2,$$

where $\mathrm{R}(t)$ is the radio of curvature of the 3D space and σ is the spatial volume element.

Then, he uses some lines to explain why he considers the matter's contribution to pressure to be negligible, but not p, the radiation pressure. Thus, he denotes the total energy density ρ as $\rho = \delta + 3p$. Also, he assumes a closed universe, so energy is conserved. Under these assumptions, the expression for the conservation of energy turns out to be:

$$\frac{d\rho}{dt} + \frac{3\dot{\mathrm{R}}}{\mathrm{R}}(\rho + p) = 0$$

The volume of space is $V = (4\pi/3)\,R^3$, so energy conservation becomes $d(V\rho) + pdV = 0$. This means that the work done by radiation pressure plus the variation in total energy add up to zero. Then, he explains why pressure coming from the mass loss of stars that is converted or transformed into energy does not contribute to the pressure. Also, he recovers Einstein's model by setting the condition of the constancy of the universe's radius while setting $\rho = 0$, as well as retrieving De Sitter's solution.

$$\frac{3\dot{R}^2}{R^2} + \frac{3c^2}{R^2} - \lambda = \kappa c^2 \rho$$

Then, Lemaître turns his attention to Doppler's effect. Here, Lemaître explains that the cosmological nature of spectral shifts is not due to relative motion between the observed object and the observer, but to the variation of the radius of the universe. He notes that the large recession velocities of extragalactic nebulae are a cosmological effect due to the fast expansion of the universe, perhaps due to the pressure of radiation. Next, he directs his interest to the lightshift.

For a light ray emitted at space coordinates σ_1 and observed at σ_2, and since it follows a geodesic ($ds = 0$) from Lemaître's line element, the equation for a light ray is:

$$\sigma_2 - \sigma_1 = \int_{t_1}^{t_2} \frac{dt}{R}$$

If the light ray emission in σ_1 occurs at time $t_1 + dt_1$, where dt_1 is the period of the emitted light and is observed in σ_2 at $t_2 + dt_2$, where dt_2 is the period of the observed light, then the above integral implies:

$$\frac{dt_2}{R_2} - \frac{dt_1}{R_1} = 0, \qquad \frac{dt_2}{dt_1} - 1 = \frac{R_2}{R_1} - 1.$$

where R_1 and R_2 the values of R at t_1 and t_2. Since wave length λ is related to wave period dt by $\lambda = cdt$, then light emitted with wavelength λ_1 at σ_1 will arrive at σ_2 with a wavelength

$$\lambda_2 = \lambda_1 \frac{R_2}{R_1}$$

λ_1 is subtracted from both sides, and the equation is rearranged:

$$\frac{\lambda_2 - \lambda_1}{\lambda_1} = \left(\frac{R_2}{R_1} - 1\right) = \frac{\Delta\lambda}{\lambda_1}$$

If the redshift $\Delta\lambda/\lambda$ is interpreted as being due to Doppler's effect due to velocity v, for $v \ll c$, $\Delta\lambda/\lambda = v/c$. Thus, we can assume that $\lambda_2 - \lambda_1 = \Delta\lambda \ll \lambda_1$.

$$\frac{\Delta\lambda}{\lambda_1} = \frac{v}{c} = \left(\frac{R_2}{R_1} - 1\right) = \frac{R_2 - R_1}{R_1} = \frac{dR}{R}$$

Lemaître assumed that extragalactic nebulae existed at a distance r, very close to us, as compared to the radius of the universe, $r \ll R$. For this case, the spatial part $R(t)d\sigma$ of the line element $ds^2 = -R(t)^2 d\sigma^2 + dt^2$ can be replaced by r, so $ds^2 = -r^2 + dt^2$. Furthermore, in this situation, it is valid to set $ds = 0$, since light travels on null geodesics. Therefore, $r = dt$ expresses the time in units of r/c. Hence, Lemaître obtained the following approximate formulae:

$$\frac{v}{c} = \frac{dR}{R} = \frac{\dot{R}}{R}dt = \frac{\dot{R}}{R}r$$

Thus, he obtained the Hubble relation $v = \frac{\dot{R}}{R} c \cdot r$ (Equation (24) in Lemaître's paper), which is written as follows in modern terms:

$$v = H_0\, r$$

Afterward, using the radial velocities of spirals determined by Strömberg [67] in 1925 and their apparent magnitudes measured by Hubble [68] in 1926, Lemaître calculated the value H_0 (known as the Hubble constant today) to be about 625 km/s/Mpc.

In the English translation of Lemaître's paper, produced upon request from Eddington and published in 1931, the brief analysis of the linear velocity–distance relation as well as equation No. 24 (Hubble law), published in the original French version, were omitted in the translated paper [69]. For a long time, the reason for this remained a mystery. Many commentators formulated hypotheses on this matter, ranging from conspiracy to carelessness. But it turned out that it was Lemaître himself who decided not to incorporate them. For Lemaître's, motives an ample explanation is given by Mario Livio's account and the references therein [70].

11. The Earliest H_0-Tension

Towards the beginning of the 1930s, the main objection to the interpretation of the Hubble constant as the indicator of the universe's expansion was that its measured value implied a universe younger than Earth's accepted age.

Before the 1930s, scientific estimates of the age of the Earth dated from the mid-19th century, from Helmholtz–Kelvin gravitational contraction ages [71] to those resting on geology (i.e., the amount of salt in the oceans, sediments). Finally, after a fierce debate between geologists and physicists [72], an estimate of age based on early-20th radioactive dating was determined [73]. This produced an estimate of around 2 to 3 billion years by the 1930s [74].

On the other hand, the cosmological age of the universe was estimated by taking the reciprocal of the value of the Hubble constant (Hubble's time). Hubble's first evaluation of his eponymous constant resulted in a value of around 500 km per second per megaparsec. This implied an age of the universe of about two billion years, which was in a tense contradiction with the estimated age of the Earth—about three billion years. As a consequence, this mismatch created room for doubt. Some critics questioned that the observed nebulae redshifts were, in fact, a manifestation of Doppler's effect. Such was the paradigm that reigned in the early 1930s.

This incongruity raised two possible explanations: on the one hand, measurements were wrong, or on the other hand, Doppler's shift needed a novel interpretation, perhaps through "new" physics.

Regarding the latter case, in 1929, Fritz Zwicky proposed the concept of "tired light" as an alternative explanation for the redshift–distance relationship [75]. This was a hypothetical redshift mechanism where photons lost energy over time via collisions with other particles in their trajectories through a static universe. Thus, the more distant objects would appear redder than more nearby ones.

Zwicky's idea was not taken lightly by the scientific community, in view of the tension between the apparent age of Earth and the predicted age of the universe. In fact, in 1935, Edwin Hubble and Richard Tolman wrote [76]:

> "... both incline to the opinion, however, that if the red-shift is not due to recessional motion, its explanation will probably involve some quite new physical principles [... and] use of a static Einstein model of the Universe, combined with the assumption that the photons emitted by a nebula lose energy on their journey to the observer by some unknown effect, which is linear with distance, and which leads to a decrease in frequency, without appreciable transverse deflection".

Let us now turn our attention to the possibility of an erroneous Hubble constant measurement in the 1930s. This, as we all know, requires only the measurement of the distance and velocity of a sky object belonging to the galaxy in question. The measurement of recession velocity was straightforward even in those days. Spectral recordings had already achieved good accuracy. However, distances were measured, as we shall see next, by observing standard candles.

12. Cepheids

We must now travel back some years to narrate how early distances to nearby galaxies were obtained. Their assessment was necessary in order to achieve a reliable figure for the Hubble constant. It was known from the very beginning that parallax was of little use for faraway objects. Hence, making a comparison between a standard candle and the stellar object whose distance was in question was the next reasonable step to determine its proximity. Cepheid stars were the natural choice.

Classical Cepheids are high-luminosity, radially-pulsating, variable stars. Their intrinsic brightness periodically fluctuates within a range from $-2 > M > -7$ mag at visual wavelengths, making them, in principle, ideal standard candles for distance indicators on galactic and extragalactic scales. The periodic variation of one of these stars was first discovered in 1784 by John Goodricke. He discerned the period of the star δ Cepheid, the prototypical example of a Cepheid [77]. Then, in 1913, as we have already mentioned, the discovery by Leavitt of a P-L (period–luminosity) relation of classical Cepheids residing in the SMC led Hertzsprung to perform a preliminary P-L calibration, but in his work, he ignored interstellar absorption. Afterwards, as we have pointed out, in a subsequent publication following the suggestion made by Edward Skinner King, Russell and Shapley estimated that interstellar absorption diminished stellar magnitudes by about 2 units per kpc.

By 1918, Harlow Shapley had applied a more sophisticated color correction method to the conversion of Leavitt's photographic magnitudes into visual ones, producing what seemed at that time to be a consistent calibration of the P-L Cepheids. However, today, we know that Shapley's calibration was wrong from its beginning. It had a zero point error of around 1.5 in magnitude. The main two reasons for this gross imprecision were that Shapley did not take into account interstellar absorption in spite of his previous work with Russell, and that he unknowingly included the two types of "Cepheids" that we know to exist into today a single set of variable stars. This latter fact was unknown to Shapley.

Let us describe this in more detail. There are two classes of Cepheids: Type I, also known as classical Cepheids, and Type II, sometimes rarely called W Virginis stars after their prototype, W Virginis. Figure 4 shows a modern sketch of regions of type I and type II populations, and the variables RR Lyrae are expected to occur in a P-L graph. With regard to the RR Lyrae shown in Figure 4, these stars pulse in a manner similar to Cepheid variables. They are commonly found in globular clusters, but the nature of these stars is rather different. It is pertinent to point out that RR Lyraes are also used as standard candles.

Shapley's way of obtaining a $P-L$ relation was to first inspect data on type I Cepheid stars located in the SMC to determine the zero point of the calibration. For the already-exposed reasons involving absorption, the observed luminosities of these type I stars were dimmer by approximately 1.5 magnitude. Coincidentally, this magnitude difference is what roughly separates type I from type II Cepheid curves (see Figure 4). If the observed type II stars happen to be close to us (i.e., in Milky Way globular clusters), their luminosities are comparatively unaffected by absorption. This means that a type I region apparently overlaps with a type II region (see Figure 5). Another remarkable coincidence is that the slopes of types I and II are very similar to each other.

In constructing a P-L relation, Shapley not only unknowingly included Cepheids of both types I and II from seven different stellar systems into a single P-L graph, but went even further and also added in a set of RR Lyrae variables. Figure 6 shows the P-L graph that Shapley obtained. There, one can observe that the linear PL relation broke down at the bottom right corner of the figure due to the inclusion of what Shapley called "cluster type Cepheids", now termed RR Lyrae variables. He explicitly made the distinction. The reason for this bending in the P-L linear relation was that, unlike Cepheid variables, RR Lyrae variables do not follow a strict period–luminosity linear relationship at visual wavelengths, although they do in the infrared K band [78].

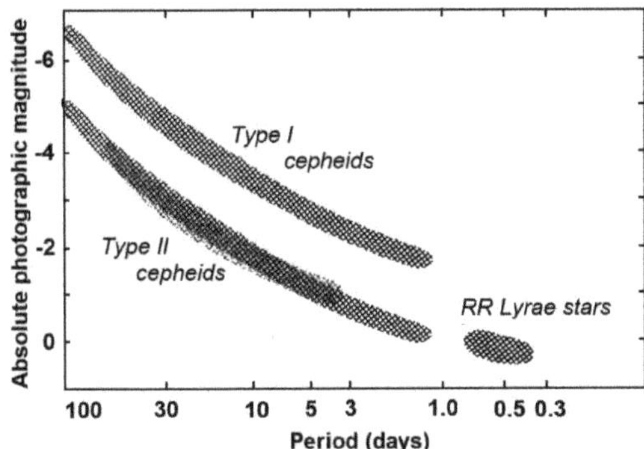

Figure 4. Approximate regions of type I, type II, and variable RR Lyrae populations that are expected to appear.

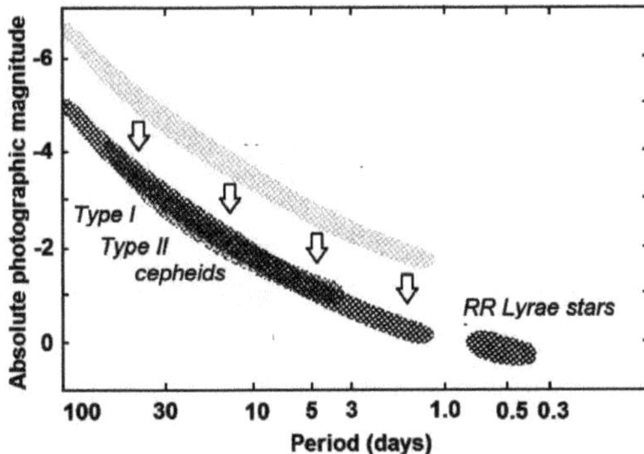

Figure 5. Apparent overlap between regions of type I and type II Cepheids. Arrows indicate that type I were misidentified as type II Cepheids. Also shown are RR Lyrae.

By the mid-1920s, the Shapley calibration of the P-L Cepheids with its later adjustments was considered the most appropriate and reliable method for determining stellar distances [79,80]. In spite of several justified criticisms (on this matter, see Fernie's review paper [81]), generalized acceptance of Shapley's P-L calibration prevailed for several decades. It is interesting to recall that, as late as 1935, Hubble, in his Silliman Memorial Lectures at Yale, declared regarding Shapley's calibration: "Further revision is expected to be of minor importance" [82].

Nevertheless, the extraordinary coincidences that led to Shapley's calibration affected future estimates of Hubble's constant. These coincidences were not spotted for a while, until Walter Baade and Thackeray discovered Shapley's error.

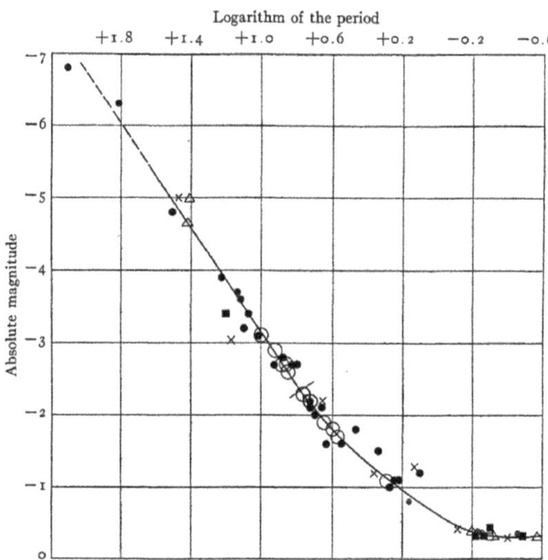

Figure 6. Luminosity–period curve of Cepheid variation. The various symbols designate variables from seven different systems. Credit: figure taken from Shapley's 1918 paper [79].

13. Baade and Thackeray, a Good Fresh Breeze

In the late 1940s, on cloudy winter nights that hampered observation at Mount Wilson, Edwin Hubble and Walter Baade filled the time by chatting. One of those nights, Hubble mentioned to Baade a concern that had haunted him since 1931, when he investigated the globular clusters of the Andromeda nebula. He noted that the upper limit of luminosity of the globular clusters was about 1.5 magnitudes fainter than the upper limit of those of our own galaxy. Baade argued that this discrepancy was difficult to understand unless *"the discrepancy had entered through some loophole in one of the distance determinations"* [83]. Hubble, in turn, argued that there might be a real difference between the clusters of Andromeda and those of our own galaxy. To Baade's surprise, Hubble added that he found that the brightest globular clusters in M33 were still fainter than those of Andromeda. According to Baade, the discussion ended when they realized that none had a convincing explanation for the discrepancy. This conversation may have seeded doubt in Baade regarding Shapley's calibration, but it was only after the recognition that there were two types of Cepheids that *"the first serious doubts arose concerning the accepted form of the period-luminosity relation"* [83]. The question was, then, how to disprove Shapley's P-L relation.

Fortunately for Baade, the 200-inch Hale telescope at Mount Palomar Observatory was close to completion. This offered him the ideal opportunity to select a nearby galaxy which contained stellar populations of variables so their luminosities could be compared *"side by side"*. For Baade, there was no doubt that the Andromeda was the most suitable object for such an investigation and that the 200-inch telescope could answer his questions. However, although the telescope was dedicated in 1948, small optical defects were found, so it took the entire next year of 1949 to correct them. Thus, the telescope was not operational until 1950. Naturally, Baade was very eager to settle these disturbing questions.

In 1949, Baade received a letter dated February 16th from Andrew David Thackeray, based at the Radcliffe Observatory in South Africa, asking for advice. Thackeray explained *"...I have examined several* [globular clusters] *in the Mag.* [Magellanic] *Clouds but I fancy such work will overlap a good deal with Harvard* [Shapley's adscription]*..."*. On March 29th, Baade replied, celebrating that the Radcliffe telescope was in operation and saying that he was sure it would settle many questions that could not be answered from the northern hemisphere. He suggested to Thackeray that, among other topics, it would be highly desirable to investigate whether there were "truly" globular clusters in the LMC. Then, he clarified why he made this odd remark. He explained that Shapley had always been talking about globular clusters in the LMC, but had not yet found variables of the type Baade called globular variables, which were supposed to be there. This seemed very odd to Baade. He further explained that these stars must have existed there, unless the very unlikely fact was true that globular clusters in the Magellanic clouds were of a different nature to those in our galaxy. Recall that this query arose between Hubble and Baade during their chats on cloudy nights at the Mount Wilson Observatory.

Baade ended his letter to Thackeray, not without showing a hint of personal rivalry between him and Harlow Shapley:

"...Whatever the final outcome we would know where we stand in the view of a most vexing question. Both Hubble and I hope that Shapley's tendency to consider the Magellanic Clouds as his personal property will not deter you from attacking this problem. He has monopolized the Clouds all too long and it is high time that the barbed wire fences and the warning signs "Keep out. This means you!" are taken down. Monopolies in science are intolerable and should never be respected. Moreover lately Shapley has worked his gold mine only if he needed money for booze (some stuff for publication). The whole situation has become intolerable and a good fresh breeze is most desirable...". [84]

For his part, the research program undertaken by Baade using the powerful and brand-new Palomar telescope was intended to identify variable stars and compare their luminosities, since he suspected that, in reality, the origin of the problem with Shapley's calibration was not the existence of different globular clusters, but of two different populations of Cepheids. Three years later, in 1953, at the IAU General Assembly in Rome, Walter Baade reported his findings [85].

Baade reported that RR Lyrae variables could not be detected in the Andromeda Nebula, even when using the 200-inch telescope. The reason for this negative result, according to Baade, was that Shapley's calibration underestimated distances. His reasoning was simple and straightforward: Baade assumed that if the Shapley calibration was correct, the distance to Andromeda would be around 275 to 300 kpc, and in that case, it would be simple and straightforward to compare the variables' luminosities with the use of the 200-inch telescope. Instead, he could only observe the very brightest Population II stars on limiting exposures, but not a single RR Lyrae star. Baade concluded that Shapley's calibration underestimated the distance to M31, and that RR Lyrae stars were not visible because they exist much further away. The argument made by Baade seemed convincing, but strictly speaking, it was not an incontrovertible reason to assert the failure of Shapley's relation, only indicating a very strong possibility.

However, the incontrovertible evidence came immediately after Baade had spoken, when Thackeray announced that he and A.J. Wasselink had discovered the first Lyrae variables in the SMC globular cluster (NCG 121). Recall that the SMC was the galaxy in which Leavitt originally established the first P-L relation using classical Cepheids.

This discovery meant that the absolute magnitudes of RR Lyrae and Cepheids (type i) could be directly compared. According to Shapley's calibration, these RR Lyrae should have appeared at a magnitude of 17.5, but it should be of no surprise to the reader that they were actually fainter by 1.5 mag. This proved Shapley's mistake.

As result, the distance to M31 was twice as far as originally calculated. The perceived universe doubled in size; the Hubble constant reduced its value by half; and the age of the universe doubled, partially easing the paradox of Earth being older than the universe.

In retrospect, Baade commented on this years later: *"...there were good reasons to suspect that unknowingly Shapley had made a fatal step when he linked the cluster-type variables to the type I Cepheids through the type II Cepheids in globular clusters and that in reality we were dealing with two different period-luminosity relations, the one valid for the type I Cepheids, the other for the type II Cepheids"*.

14. The Changing Value of H_0

In this section, we do not attempt to list all measurements made during the period from Lemaître's first estimate of H_0 in 1927 to the mid-twentieth century measurements. This has already been carried out many times, often by the original researchers or their close collaborators (see, e.g., Fernie 1969 [81]). Our intention here is to show the downward trend experienced over time during the second quarter of the last century for the values of the Hubble constant. This trend was not due to an arbitrary value-adjustment of H_0 from the observers, but to efforts to improve observations, "revise" bias and confusions made by others on observed objects, and to consider the sources of errors that were previously overlooked or ignored (shown in Table 1). As we have already seen in the previous section, by the mid-twentieth century, Shapley's calibration had been found to be erroneous and was substituted by Baade and Thackeray's calibration.

Table 1. Trend of values for H_0 in the period of 1930–1955.

Year	Author	Value (km/s/Mpc)	Method	Ref.
1927	Lemaître	600	Used data of Hubble and Strömberg	[86]
1929	Hubble	500	Cepheids; Shapley's calibration P−L	[38]
1931	Hubble and Humason	$526 \pm 10\%$	Cepheids; Shapley's calibration P−L	[39]
1946	Mineur	320	Interstellar absorption corrections making a major recalibration to Shapley's scale. Used zero point $M_0 = -1.54$, see Baade (1956)	[87,88] [83]
1951	Behr	240	He heuristically pre-discovered the Scott effect and made the corresponding corrections to observations by J. Stebbins and A. E. Whitford (Astrophys J. 108 413 (1948), Mt Wilson Contr. 753, using Shapley's calibration	[89]
1952	Baade and Thackeray	280 ± 30	RR Lyrae stars; corrected Shapley's calibration (see next section)	[84,90]
1956	Humason, Mayall, and Sandage	180 ± 20	Comparison of photographic magnitudes for 576 galaxies to their redshifts; their calibration set the brightest galaxies in clusters equal to M31 luminosity	[91]

Inherently, the observed trending decrease in the value of the Hubble constant increased the Hubble time. This, in turn, reduced the tension that existed between the supposed age of the universe and its estimates obtained by various means, namely, from estimates made according to the ages of some stars. Figure 7 shows some selected Hubble constant values in relation to the decreasing trend that H_0 estimates experienced during the second quarter of the twentieth century.

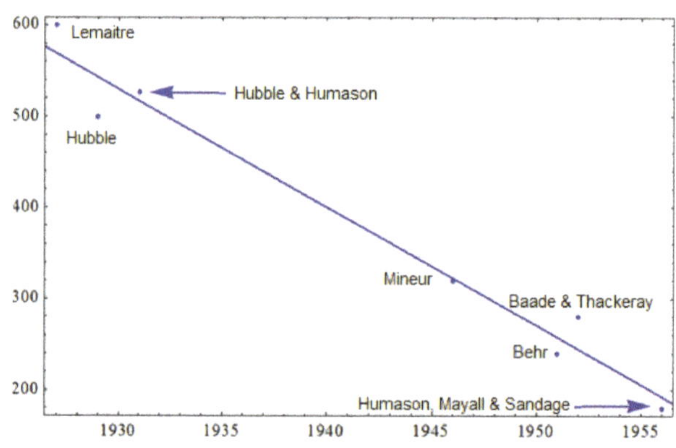

Figure 7. Some values determined for H_0 in the period of 1930–1955, showing the trend.

15. The Need to See Further

When Walter Baade and David Thackeray revealed that Hubble had made a miscalculation of the distances to galaxies by using a wrong Cepheid gauge calibration, they erased, at a stroke, the issue on the supposed young age of the universe—it was confirmed to be older.

However, a hurdle would soon arise on the use of Cepheid stars to determine the value of H_0. This obstacle was that, even though the Cepheid stars were bright (on average, about 10^4 brighter than the sun) their luminosity was not sufficient for them to be observed them at that time beyond a few dozen parsecs. On this scale, this meant that the universe could not be assumed to be isotropic or homogeneous.

In addition, it had been known since Vesto Slipher's early measurements that not all nearby galaxies were receding; on the contrary, a pair of them was actually approaching us. This meant that, although nearby galaxies participate in the expansion of the universe, gravitational interactions among neighboring galaxies may cause some of them to move much faster or slower than the rest of the universe. In fact, it was later discovered that our local group of galaxies experiences a "Virgocentric flow". This is a preferential movement directed towards the Virgo Cluster, so corrections must be applied to obtain the Hubble flow (see Figure 8).

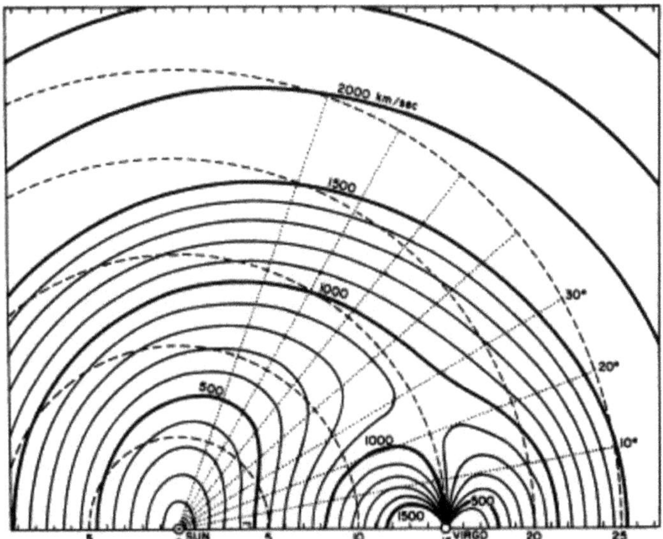

Figure 8. One of two 2D grids with Earth and the Virgo cluster on the *x*-axis. Redshift contours are plotted for a Virgocentric flow. Note that a pure Hubble flow would be concentric (from Tonry and Davis 1981) [92].

The fact that the genuine Hubble flow is affected by local gravitational interactions forced astronomers to accurately determine the Hubble constant at remote distances. The need to measure further away was clear, as the large-scale structure of the universe makes it isotropic. This task is extremely difficult.

Part II. Building and adding rungs to the cosmic ladder

During the second half of the last century, several independent groups of astronomers developed methods for measuring the distances to galaxies, leading to new estimates of the Hubble constant and generating debates on the merits of each technique. We shall next provide a brief description of the most relevant methods that appeared in the second half of the 20th century.

In 1970, Sidney Van den Bergh [93] from the David Dunlap Observatory in Toronto reviewed nine methods used at that time to determine the Hubble constant. These were: diameters of HII regions (obtaining 91^{+19}_{-15} km/s/Mpc), luminosity classification of galaxies (98^{+19}_{-17} km/s/Mpc), brightest globular clusters in galaxies (72^{+15}_{-12} km/s/Mpc), mass-to-light ratios (105^{+33}_{-26} km/s/Mpc), third-brightest cluster galaxy (126^{+48}_{-35} km/s/Mpc), supernovae and the extra-galactic distance scale (123^{+47}_{-34} km/s/Mpc), surface brightness and diameter of galaxies (89^{+46}_{-30} km/s/Mpc), brightest stars in galaxies (95^{+15}_{-12} km/s/Mpc), and regional variations of the Hubble constant (not specified). Van den Bergh computed their mean value, obtaining 95^{+15}_{-12} km/s/Mpc.

These methods, as originally thought, are no longer employed today. New methods were to arise in the following years. We account for them in the following sections.

16. Planetary-Nebula Luminosity Functions

In simple terms, planetary nebulae (PNe) are expanding shells of bright, luminous gas expelled by dying stars of low-to-intermediate mass (~1 to 8 $M_\odot$) towards the end of their evolution. Planetary nebulae is an inaccurate term for these objects because they are unrelated to planets or exoplanets. PNe are important tools used to estimate distances, despite the brief lifespans of their terminal phases (~30,000–50,000 y).

In 1963, Karl G. Henize and Bengt E. Westerlund suggested that PNe in the Milky Way and in the SMC might all have the same maximum brightness [94]. Their hypothesis opened the way to using bright PNe as extragalactic standard candles. But it was not until 1978 that Holland Cole Ford and David Charles Jenner estimated the distance to the large spiral M81 by making observations on eight PNe in the exterior of this Local Group galaxy [95]. They noted that these PNe appeared to be around 20 times fainter than those found in M31. Hence, assuming that the suggestion made by Henize and Westerlund of maximum brightness equality was valid, they estimated that M81 was about 3 Mpc away. This distance agreed quite well with other independent assessments made at that time.

The advent of charged coupled devices (CCDs) in the early 1980s, combined with superior telescopes, allowed Ford, together with George Howard Jacoby and Robin Bruce Ciardullo, to develop an alternative and superior method by which to use PNe as distance indicators [96–99]. They employed the concept of luminosity function, which characterizes the distributions of PNe by counting the total number of PNe in each luminosity or absolute magnitude interval in a given volume. Jacoby and Ciardullo used a narrow band filter to register the [O III] emission-line ($\lambda5007$) of PNe [100]. These objects glow predominantly at this wavelength, so this facilitated the subtraction of sky backgrounds. In this way, Jacoby and Ciardullo found and recorded several hundred extragalactic PNe.

During their investigation, Jacoby and Ciardullo noted that, when they studied a galaxy in this way, they detected many faint PNe, but few bright PNe. By counting the number of PNe of each level of brightness, they defined the planetary nebulae luminosity function (PNLF) [101]. The important finding was that the shape of the PNLF appeared to be the same for every galaxy.

The number of PNe decreased smoothly with the increasing luminosity, and there was a limiting luminosity beyond which they could not detect more PNe. The cut-off was sharp. Jacoby and colleagues proposed that the PNLF was of the form suggested by Ciardullo et al. (1989) [98]:

$$N(M) \propto e^{0.307M}\left(1 - e^{3(M^* - M)}\right),$$

where $N(M)$ is the number of PNe with absolute magnitude M, and M* is the absolute magnitude of the most luminous PNe (i.e., the bright cut-off magnitude), which is currently accepted to be [102] M* = 4.47 $\pm$ 0.05. This characteristic cut-off in the PNLF provides an excellent standard candle. A comparison of the PNLF in a remote galaxy with that in M31 gives the distance in terms of the M31 distance. Figure 9 shows the PNLF for M31.

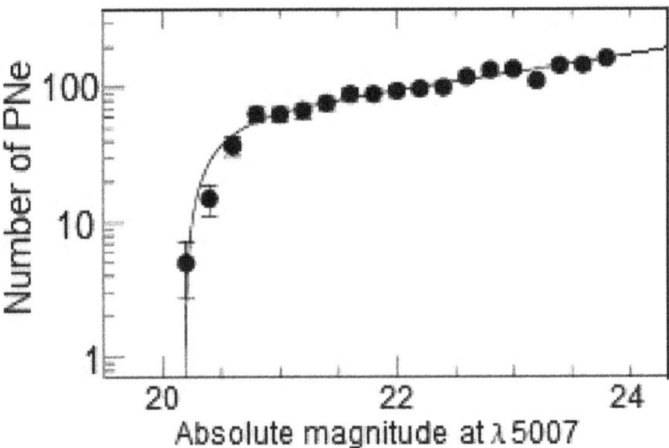

Figure 9. The planetary luminosity function for M31 from Merrett et al. [103]. The sharp cut-off at the bright end of the [O III] PNLF is characteristic of all galaxies. Note the good fit of the standard PNLF.

Before satellite-based telescopes, ground-based observations using the PNLF were limited to about 20 Mpc. Beyond that distance, dimmer PNe become too faint to observe. Therefore, it was a useful complement to calibrations using Cepheids. Today, the cut-off of PNLF constitutes a robust standard candle. It is now used to calibrate Type Ia supernovae (e.g., Jacoby et al., 2006 [104]).

One the questions concerning the [OIII] PNLF technique is why there is such observational invariance in the bright cut-off magnitude along the morphological sequences of galaxies, i.e., from spirals undergoing active star formation to the old populations in elliptical and lenticular galaxies. This is a current point at issue, and the answer has not yet been firmly established. However, the PNLF helps to tie together Population I and Population II strands in the distance ladder [105].

17. The Tully–Fisher Relation (TFR)

In 1977, Richard Brent Tully and James Richard Fisher discovered an empirical power law relation between the luminosity of a spiral galaxy and its rotational velocity [106]. Their proposal was to measure the HI 21 cm line width to find the rotational velocity of the galaxy; use this to calculate the luminosity; and compare it with the apparent magnitude of the galaxy to deduce its distance.

To check their proposal, they observed spiral galaxies in two nearby clusters, namely, the Virgo cluster and the Ursa Major cluster. Then, they made a comparison between galaxies in the same cluster. This assured them that, in principle, galaxies are situated essentially at the same distance. Finally, by applying their idea to nearby, Cepheid-calibrated galaxies, Tully and Fisher judged their method, and it agreed with previous distance measurements. The numerical value they obtained for H_0 using TRF in 1977 was 80 km/s/Mpc.

Subsequent work by Marc Aaronson, John Huchra, and Jeremy Mould [107] found that the power law relationship favored a power law exponent to the fourth, $L \approx V^4$, if the spirals possessed some properties. In 1983, Aaronson suggested an improvement to the TFR method by measuring luminosities in the infrared (K band) radiation rather than the optical band to reduce scattering effects [108].

Since then, the TFR started serving as a rung on the cosmic distance ladder. TFR was calibrated by means of the Cepheid gauge. By the end of the 20th century, TRF had established itself as standard technique, extending the ladder to a greater distance.

By the turn of the century, it was clear that the TFR fit better when luminosity was replaced by the galaxy's total baryonic mass [109]. This latter variation is known as the baryonic Tully–Fisher relation (BTFR). It states that baryonic mass is proportional to velocity, again, to the power of roughly 4 [110].

18. The Faber–Jackson Relation (F-J)

During the 1970s, efforts were carried out to find possible correlations between various parameters of elliptical galaxies, such as their luminosities, sizes, central velocity dispersions, abundances of heavy elements, etc.

In 1976, the astronomers Sandra Moore Faber and Robert Earl Jackson discovered a relationship connecting the luminosity of elliptical galaxies to the velocity dispersion (σ) of stars near their respective centers [111]. They found that the luminosity of an elliptical galaxy increases with the velocity dispersion of the stars to a power of n ($L = \sigma^n$). Originally, Faber and Jackson found that n $\approx$ 4. But later, it was found that the value of n depends on the range of galaxy luminosity. For low-luminosity elliptical galaxies, the F-J relation is fitted with a value of n $\approx$ 2. This value was found by a team led by Roger Davies [112]. A value of n $\approx$ 5 for luminous elliptical galaxies was reported by Paul L. Schechter [113]. These findings have been confirmed observationally by many authors. For a comprehensive reference list in the literature, see, for instance, Markovic and Guzman [114].

Like the TFR, the F-J relation has been useful in providing another means of estimating distances, in this case to elliptical galaxies. One possible advantage of the F-J method is that it uses the more easily observable properties of elliptical galaxies by measuring the central stellar velocity dispersion, which can be achieved relatively easily by using spectroscopy to measure the Doppler shift of light emitted by the stars. Afterwards, an estimate of the true luminosity of the galaxy via the F-J relation can be obtained. Finally, a comparison can be made to the apparent magnitude of the galaxy. This comparison provides an estimate of the distance modulus and, thus, the distance to the elliptical galaxy. However, there is still a major disadvantage. As we have already pointed out, elliptical galaxies do not faithfully follow the F-J power law, since their coefficient n varies from 2 to 5 depending on their luminosity.

In 1991, Donald H. Gudehus suggested combining a galaxy's central velocity dispersion with measurements of its central surface brightness and radius parameter [115]. By using this combination, or "reduced galaxian radius parameter", Gudehus asserted that it was possible to improve the estimate of the galaxy's distance, yielding results free of systematic bias.

19. Fundamental Plane, the D_n-σ Relation

During the decade that followed the discovery of the F-J relation, extensive research was carried out attempting to find out other possible correlations between various parameters in elliptical galaxies. But it was not until 1987 that the American astronomers Marc Davis and Stanislav George Djorgovski [116] replaced the luminosity of an elliptical galaxy in the F-J relation with two novel parameters: the effective radius of the elliptical galaxy (R_e) and the average surface brightness (I_e) within that radius. They found a relationship connecting the three parameters: $R_e = k \, \sigma^{1.36} \, I_e^{-0.85}$, where k is a constant.

Any one of the three parameters (average surface brightness, velocity dispersion, and effective radius) may be estimated from the other two; together, they describe a plane which Davis and Djorgovski called "the fundamental plane" for ellipticals. In the Davis and Djorgovski method, I_e and σ are measured to obtain an effective radius R_e. This effective radius is used as a standard rod. Once this rod's length has been obtained, by directly measuring its angular size, it becomes easy to determine the distance from the observer to the galaxy through small-angle approximation.

The Davis and Djorgovski method stands out against the F-J relation, where σ is measured and luminosity is used as a standard candle. Incidentally, the F-J relation can be viewed as a line projected on the fundamental planes of elliptical galaxies.

An alternative variant correlation in the fundamental plane was proposed in 1987 by the group of the "Seven Samurais", as they were known (1987) [117]. The group was composed of American, Argentinian, and British astronomers: David Burstein, Roger Davies, Alan Dressler, Sandra Faber, Donald Lynden-Bell, Roberto Terlevich, and Gary A. Wegner. The Seven Samurais found that there is an excellent correlation between σ and a quantity they called D_n, which represents the diameter of a central circular region of an elliptical galaxy within which the total average surface brightness is some particular value (20.75 magnitude per square second of arc). The correlation can be expressed as:

$$\frac{D_n}{kpc} = \left(\frac{\sigma}{100\text{km/s}}\right)^{1.33}$$

This relationship has a scatter of 15% between galaxies, as it represents a slightly oblique projection of the fundamental plane. The "Seven Samurai" published a series of papers providing the results of an all-sky survey of an elliptical, building a large store of data on elliptic galaxies.

20. Surface Brightness Fluctuation Method (SBF)

The surface brightness fluctuation method (SBF) was proposed in 1988 by the American astronomers John Landis Tonry and Donald P. Schneider [118], and later was further developed by Tonry [119].

This method aims to estimate the distance from an observer to a galaxy by measuring the degree of brightness fluctuations at its surface. The most common situation for a telescope is not being able to resolve a population of bright stars in distant galaxies. However, the discrete nature of stars causes fluctuations in the surface brightness of the galaxy. The further away the galaxy is located, the more "diffuse" and less "dotted" it will appear.

The SBF method measures the variance in a galaxy's light distribution, which arises from fluctuations in the amounts and luminosities of individual stars for each resolution element.

The galaxy surface brightness is independent of the distance from the galaxy to the observer, but the variance (measured in the Fourier space) is not. This variance changes as the inverse of the square of the distance "d" in which the galaxy is located (d^{-2}).

In practice, the CCD image of the galaxy under observation must be excised from external background and foreground sources. Then, an isophotal model is fitted to the image. Next, the model is subtracted from the galaxy image. The remaining image is Fourier-transformed, and the SBF pattern is measured according to the power spectrum of the residual image that is left behind. The amplitude of the spectrum gives the luminosity of the galaxy. Pertinent corrections for interstellar dust absorption must also be accounted for. Tonry and colleagues measured the Hubble constant for over a decade, obtaining a range of values from 88 to 77 km/s/Mpc [120,121].

21. Tip of the Red Giant Branch (TRGB)

The foundation of this method is similar to that of the previous one, which is based on the constancy of luminosities of stars at the "tip" of the red giant branch in the Hertzsprung–Russell diagram. This happens when the degenerate helium nuclei in low-mass stars reach their critical mass and start burning carbon. Thus, the star is in the horizontal branch afterwards. The TRGB luminosity is typically measured in the I-band, where the luminosity has little dependency on stellar age or stellar metallicity, so it is a well-established standard candle. With the advent of CCDs, this method was consolidate [122–125].

By the mid-1990s, the most reliable primary distance indicator for nearby galaxies was the Cepheid P-L relation, which served as a basis for other secondary distance indicators of other, more distant targets. During this time, the TRGB was shown to be an independent distance calibrator which, in fact, was applicable to a wider type of galaxies, not exclusively to the late-type galaxies used for Cepheid calibration. Both methods were then competing in terms of their precision in determining distances [126]. At that time, in a Space Telescope Science Institute meeting on the extragalactic distance scale, Allan Sandage reported a value of 55^{+5}_{-5} km/s/Mpc, while the TRGB group reported 73^{+6}_{-6} (statistical) $^{+8}_{-8}$ (systematic) km/s/Mpc [127]. As Wendy Freedman put it in a short review in 1998 [128], the results seemed to reach an uncertainty of 10% for the Hubble constant when it was measured using different methods, leaving in the past a factor of two that astronomers had used throughout the previous two decades.

The Carnegie–Chicago Hubble Program (CCHP), put forward later, in the 2010s, was designed to provide an alternative route for the calibration of SNe Ia and, thereby, to yield an independent determination of H_0 via measurement of the TRGB in nearby galaxies. Their results will be outlined below. The most recent TRGB result was 72.94 ± 1.98 km/s/Mpc, with an additional uncertainty due to algorithm choices of 0.83 km/s/Mpc [129].

22. Global Cluster Luminosity Function (GCLF)

Globular clusters (GC) are compact and spherical sets of stars found in the halos of large galaxies. They are among the most luminous objects in galaxies. Due to their brightness, as early as 1955, they were recognized as potentially important distance indicators. It was in this year that William Alvin Baum tried to use a GC to estimate the distance to a particular galaxy. Baum's strategy involved a simple comparison between the brightest GC in M87 (in the Virgo cluster) to that in M31, assuming that the luminosities were identical. Then, he calculated the distance to M87 with the inverse square law, using M31 as a gauge [130]. Incidentally, in his publication, he deduced from his distance measurement a value of 150 km/s/Mpc for H_0.

Later, it was observed by Gérard de Vaucouleurs, and independently by other astronomers, that the brightest star clusters of a galaxy are correlated to the luminosity of the galaxy to which they belong [131]. Thus, the assumption made by Baum in 1955 of a fixed luminosity for the brightest clusters in a galaxy was incorrect.

In the late 1970s, René Racine suggested that the globular cluster luminosity function (GCLF) could be a "potentially important distance indicator in extragalactic studies" [132]. His working hypothesis was that the magnitude distribution of old globular clusters exhibited a universal shape. This distribution was characterized by its mean value and standard deviation. In turn, the mean value was used as the standard candle to evaluate distances. For unknown reasons, the shape of the GCLF appeared to be lognormal in shape, with a universal mean and standard deviation (nonetheless, with some dependences on other factors, such as metallicity).

As first introduced by David Hanes, astronomers usually fit a Gaussian curve to $\phi(m)$, the relative number of globular clusters, as a function of their magnitude, m [133]:

$$\varphi(m) = A \, e^{-(m-m_0)/2\sigma^2},$$

where m_0 is the peak or turnover point TO (i.e., the magnitude at which most GC are found), σ is the width of the distribution, and A is the normalization constant. The distribution is then characterized by only two parameters, σ and TO. The latter is the standard candle, and the distance measurement is relative to the Milky Way GCLF TO value or to that of M31. The GCLF is a secondary distance indicator, since the absolute distances to either our galaxy or M31 globular clusters must be known. This method has been used to estimate distances and, hence, the Hubble constant as well.

23. Cepheid Calibration Development in the Twentieth Century

In the early 1990s, a group of experts on cosmic distance determination wrote an article discussing which were the most reliable methods for that purpose [102]. These were according to their opinion: planetary nebula luminosity functions (PNLF) (cf., Section 16), Tully–Fisher relations (cf., Section 17), fundamental-plane relationships for elliptical galaxies (cf., Section 19), surface brightness fluctuations (SBF) (cf., Section 20), Globular-Cluster Luminosity Functions (GCLF) (cf., Section 22), and Type-Ia supernovae (SN Ia) (see Section 24). We have already described these in the present work, except for the SN Ia method, which we shall describe in the next section.

Before we report on the SN Ia method, we must note that the methods already described so far are all secondary distance indicators; that is, they need to be calibrated, usually by finding a Cepheid distance to a galaxy which in turn requires knowledge of an accurate period–luminosity relation. Cepheid variable stars are still primary calibrators for secondary standard candles. Because of their central role in establishing the distance scale, we must mention the evolution of Cepheid's period–luminosity relation values during the twentieth century.

As we have already stated, it took some years after the revelation of Baade and Thackeray that the P-L relation was wrong for the astronomical community to initiate the search for a reliable and accurate Cepheid P-L calibration. Over the next decades of the second half of the last century, several independent groups of astronomers made great efforts to improve the Cepheid period–luminosity relation. Table 2 shows some of the most influential calibrations made during the 20th century.

Table 2. Leading calibrations of the Cepheid period in the twentieth century. Luminosity relation (<Mv> = −a − b log10 P, P (days)).

Year	a	b	<Mv>	Author
1913	0.60	2.10	−2.70	Hertzsprung [134]
1918	0.72	2.10	−2.82	Shapley [135]
1961	1.67	2.54	−4.21	Kraft [136]
1968	1.43	2.80	−4.23	Sandage and Tammann [137]
1987	1.35	2.78	−4.13	Feast and Walker [138]
1991	1.40	2.76	−4.16	Madore and Freedman [139]
1997	1.38	2.77	−4.15	Tanvir [140]
1997	1.43	2.81	−4.24	Feast and Catchpole [141]

It is noteworthy that, in the last decade of the twentieth century, the parameters of the period–luminosity relationship began to concur with each other.

In the late 1960s, Wisniewski and Johnson [142] presented high-precision photometry of Milky Way Cepheids that led to the understanding of two properties of light curves: (1) the decrease in amplitude moving from the ultraviolet to the near-infrared, and (2) the shift in the relative phase of maximum light, also with the increasing wavelength. These represented major advances in our understanding of Cepheids [143].

24. Type Ia Supernovae (SN Ia)

Since the initial surveys of intergalactic distances, Edwin Hubble had a feeling that novae, as they were generically called (novae and supernovae were not differentiated at that time), might serve as criteria for determining intergalactic distances. His belief was based on the fact that their great luminosity made them detectable at great distances and that, despite the scant available data, it seemed that their maximum brightnesses reached similar standard values. However, he lamented that the sporadic frequency of their appearances in the sky would prevent them from contributing much to the problem of obtaining distances [144]. Walter Baade and Fritz Zwicky coined the term supernovae in the abstract of their joint paper "Supernovae and Cosmic Rays", presented orally by Zwicky at an American Physical Society meeting at Stanford in December of 1933. The term supernova was created to distinguish them from ordinary novae, which are far less luminous [145]. It is interesting to mention that, at the time of the publication of Hubble's "Realm of Nebulae" (1936), only 115 events (novae and supernovae all together) had been recorded [144].

Beginning in 1937, Rudolph Leo Bernard Minkowski, while working at Mount Wilson, made spectroscopic observations of supernovae. Within a few years, he and Baade realized, based on their spectra, that there were two different types of supernovae: Type I and Type II. The first type shows no hydrogen spectroscopic lines, whereas Type II supernovae have hydrogen. By 1986, S. E. Woosley and T. A. Weaver had evaluated accumulating evidence revealing that Type I SN can be separated in into three subclasses [146]: Type Ia class is silicon-rich, Type Ib is helium-rich, and the objects which have neither silicon nor helium in abundance are classified as Type Ic. In turn, the Type II class is divided into II-P, which have ~100-day "plateaus" in their light curves; II-L, which have "linear" declines in their light curves; and II-n, which have narrow lines in their spectra [147].

Type Ia supernovae (SN Ia) are the most luminous and homogeneous kind. An SN Ia supernova event is the blast that occurs when a dead star becomes a natural thermonuclear bomb. They originate from old, low-mass stars in binary systems. To be precise, a white dwarf star near the Chandrasekhar limit ($1.4\ M_\odot$) accretes mass from its companion star until it becomes denser, and a thermonuclear blast occurs. The explosion blows the dwarf star completely apart, spewing out material at remarkably high speeds (10^4 km/s). The glow of this expanding fireball takes a few weeks to reach its peak brightness, and afterwards, its luminosity declines over a period of months in a normal way. Figure 10 shows a typical light curve of an SN Ia.

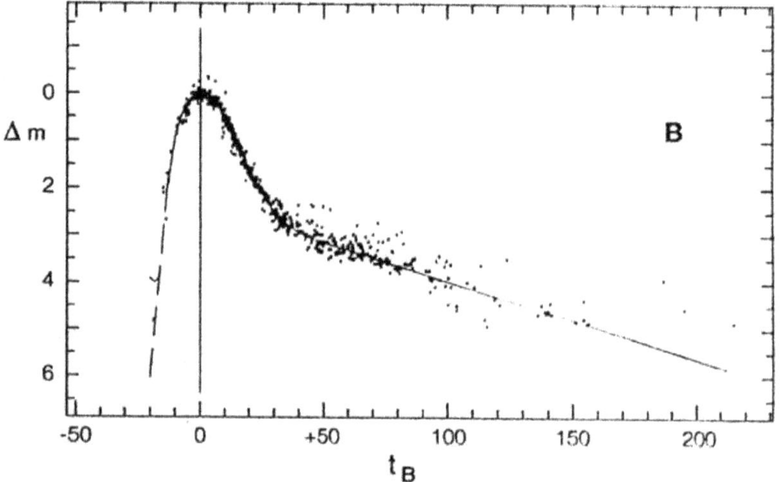

Figure 10. Typical B light curve of an SN Ia (t_B = days). Credit: Figure taken from the review cited in [148], based on data of 22 supernovae, by R. Cadonau and B. Leibundgut [149].

In 1968, Charles T. Kowal of Mount Wilson and Palomar published a redshift–magnitude graph relation for 19 type I supernovae, assuming a value for the Hubble constant (100 km/s/Mpc) [150]. Figure 11 shows little dispersion of his plotted points, concluding that "There is therefore considerable hope that the magnitudes of type I supernovae, [...], can be used as reliable distance indicators, [...], and visible at very great distances".

In his 1968 paper, Kowal also predicted the future use of supernovae to determine cosmic acceleration: "It may even be possible to determine the second order term in the redshift-magnitude relation when light curves become available for very distant supernovae". The "second-order term" would be the one that indicated cosmic acceleration or deceleration". This was proven to be the case almost three decades later. It is pertinent to mention here that the original idea of tracking supernovae as deceleration indicators goes back to the 1930s, when Walter Baade and Fritz Zwicky thought it could be possible to learn how much the universe was decelerating by observing supernovae [151].

On 22 June 1990, E. Thouvenot, at the Observatory of the Côte d'Azur, discovered a supernova SN 1990N significantly before its maximum [152]. This timely discovery (14 days before peak luminosity) allowed astronomers at the Cerro Tololo Inter-American Observatory (CTIO) to opportunely record its light curves [153]. Five years later, Allan Sandage asked the Hubble Space Telescope (HST Proposal, 5981) to allocate time for searching Cepheids in the SN 1990N parent galaxy [154]. As result, 20 Cepheids were located. Sandage and collaborators obtained the distance to the parent galaxy by measuring the periods and magnitudes of 20 Cepheid variable stars with HST. With this assessment, they obtained a value of 58 ± 4 km/s/Mpc for the Hubble constant based on a Cepheid–supernova distance scale [155].

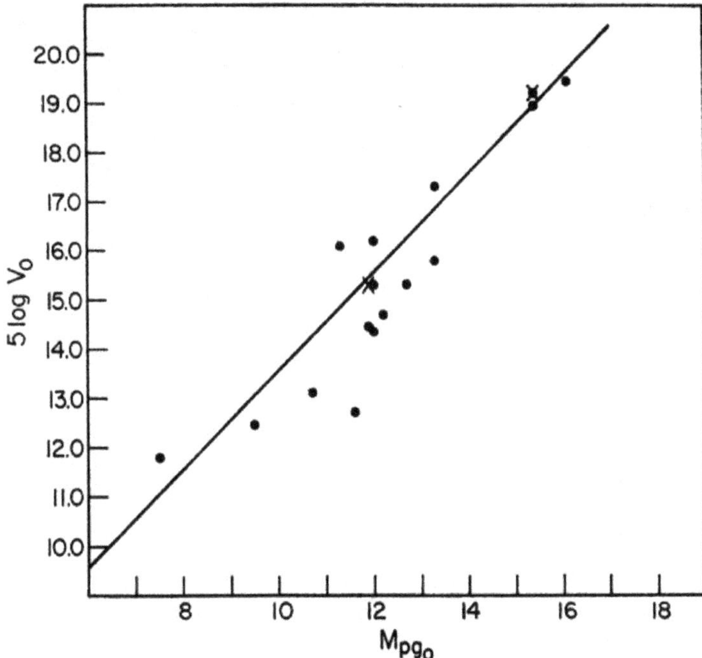

Figure 11. Kowal's redshift–magnitude relation. The crosses represent the average velocity and average magnitude of supernovae in the Virgo and Coma clusters (Kowal 1968 [150]).

Throughout all these years of sporadic observations of supernovae, it became clear that the use of Type Ia supernovae as standard candles had potential to become one of the best methods of determining distances. This is due to their fantastic luminosities and the small dispersion in their inherent peak brightness. However, it was also known that there is a drawback: their visible occurrence is infrequent, and they do not last long at their peak brightness.

This motivated several astronomers to start collaborative ventures using CCD coupled with large telescopes located in different countries. However, this was not an easy task. Before the launch of the HST, a group of Danish astronomers were among the first to use CCDs to search for SN from an Earth-grounded telescope [156]. After a two-year search, in 1985, they managed to find a single SN Ia [157]. It was clear that the job of discovering new SN required a larger collaborative effort, and that they needed to wait for the launch of the HST.

Two years later, in 1987, Saul Perlmutter and Carl Pennypacker, both researchers in Rich Muller's group at UC Berkeley were encouraged by the introduction of new sensitive CCD sensors into astronomy, as well as the new computers that were just then becoming fast enough to analyze significant amounts of data. They decided to measure the deceleration of the universe's expansion using supernovae. The scheme of a universe slowing down its expansion, possibly coming to a halt and, furthermore, perhaps collapsing, was still prevalent at that time.

Based on the Danish group's unfortunately meager results (as we have mentioned above), the approach the Berkeley group took was to develop the capability to gaze at more than a small cluster of galaxies at a time (as the Danish program had achieved). Thus, the Berkeley group developed a novel optical system capable of observing thousands of galaxies at a time. This novel instrument was used with a large enough telescope, the Anglo-Australian 4 m telescope. In addition, Perlmutter advanced a new search strategy for supernovae. It consisted of collecting wide-field images just after the new moon, then subsequently collecting a second set of images just before the next new moon (and then subtracting the second set from the first set). The short time scale between the two sets of images ensured that the discovered supernovae would not have enough time to reach maximum brightness or start fading away. This gave time for large telescopes to record SN light curves. Over the years, the Berkeley group incorporated more astronomers and observatories from around the world to form a larger project known as the Supernova Cosmology Project (SCP), led by Perlmutter.

Almost simultaneously to the progress of the SCP project, a similar effort was developed. In 1989, motivated by the suggestion of Allan Sandage and by discussions at the UC Santa Cruz meeting on supernovae [158], Mario Hamuy, Nicholas B. Suntzeff, Mark Phillips, and Jose Maza joined efforts to begin a search for supernovae as a collaboration between the University of Chile and the Cerro Tololo Inter-American Observatory (CTIO) [159]. The survey, called the Calán/Tololo survey, aimed to rigorously test SN Ia as standard candles, using redshift as an accurate proxy for relative distance

The Calán/Tololo survey used the CTIO Curtis Schmidt telescope to take plates, each covering a field of 25 sq-deg on the sky. The plates were developed and dispatched the next morning to the University of Chile's Department of Astronomy, where they were examined. Objects that could be supernovae were observed the next day using a 0.9 m telescope equipped with a CCD camera, which had been exclusively set aside for that task. Between 1990 and 1993, 50 supernovae were identified, of which 32 were Type Ia [159].

In addition, the Calán/Tololo survey provided a uniform photometric and spectroscopic dataset of supernovae which led to the discovery of a method which used Type Ia supernovae as reliable standard candles: the "Phillips relationship" [160]. This method is based on the discovery of a tight correlation between the rate of decline from maximum light and peak luminosity of SN Ia light curves. Its application reduces dispersion in measurements of SN Ia peak luminosities.

In 1994, Brian P Schmidt, then a post-doctoral research associate at Harvard University, and Nick Suntzeff, a staff astronomer at CTIO, submitted a proposal to the CTIO called "A Pilot Project to Search for Distant Type Ia Supernova". This program (in Schmidt and Suntzeff's words) was "the next step in the Calán/Tololo SN survey". Old colleagues from the Calán/Tololo survey joined the program, as did astronomers from other parts of the world. The group, later named the High-Z Team (HZT), used the Keck telescopes in Hawaii and those of the European Southern Observatory (ESO) in Chile, in addition to having access to Hubble data. The HZT elected Brian P. Schmidt, then a postdoctoral fellow at Mt. Stromlo Observatory in Australia, to manage the team.

Under the leadership of Schmidt, the HZT developed a series of scripts to automatically subtract the massive amounts of imaging data. The procedure, in simple words, consisted of making the two images of two different epochs as identical as possible by aligning them, and then matching and scaling the image point spread functions between the two epochs. These two images were subtracted, and then the difference image was searched for new objects.

As a result of the HZT program, by 1998, the team had studied 48 SN Ia, of which 14 were high redshift. Similar results were achieved by the competing team, the SCP, by whom 42 high and 18 low redshift supernovae were studied. Measuring q_0, the deceleration parameter, had always been part of the aims of both the HZT and SCP programs. Both projects found in 1998 that high-redshift SN appeared to be 10% to 15% more distant than expected, which meant that the deceleration parameter q_0 was negative ($q_{1998} \sim -0.75$[18]). In simple words, the universe's expansion was not slowing down, but rather accelerating, and there was a prevailing feeling that a cosmological constant was at work again in the field equations. The interpretation of this as a new acceleration was actually "decided" by vote at a scientific congress held in May 1998, the month during which Alan Riess of the HZT headed the first publication on the subject [161], and a short while later, a similar publication was authored by the SCP group [162]. In 2011, Saul Perlmutter, Brian Schmidt, and Adam Riess received the Nobel Prize in Physics for this discovery.

25. Mira Variables

In more recent times, other astrophysical *stars* have entered into the cosmological set. Mira variable stars have also been used as independent, intermediate-range distance (5–50 Mpc) indicators. Miras have absolute magnitudes comparable to those of Cepheids, and follow useful period–luminosity relations in the near infrared. As they are not in the optical range, measurements represent a challenge, as does the long luminosity periods of few hundred days. But Miras are ubiquitous and are present in all types of galaxy morphologies, so they are potentially interesting calibrators in galaxies where SN Ia are also present. Recently, Caroline D. Huang et al. [163] presented the first observations of Miras in an SN Ia host, yielding the first Mira-based calibration of the luminosity of SN Ia. Using this technique, they found an H_0 of around 73 ± 4 km/s/Mpc. The intention is to find Mira variables in a wide range of local SN Ia hosts in order to better determine the uncertainties.

26. The Deceleration Parameter q_0

In the second half of the twentieth century, the accepted models of the universe were based on general relativity under a pair of assumptions. The first was that the universe is homogeneous and isotropic on large scales, and the second assumption was that it contains normal matter, i.e., matter whose density falls directly in proportion to the volume of space which it occupies. Within this framework, in 1961, Allan Sandage, an astronomer close to Baade and Hubble, proposed observational tests to discriminate between selected world models (steady-state models, exploding cosmologies, etc.) [164]. These tests were often known as classical tests of cosmology.

Through the 20th century, the late universe was assumed to be dominated by a single component of matter with a density ρ_i compared to a critical density, ρ_{crit}. The ratio of the average density of matter compared to the critical density is called the density parameter, $\Omega_M = \rho_i / \rho_{crit}$. This critical density is the value where the gravitational attraction of matter in the universe causes space to become geometrically flat. As an alternative, a universe could have open hyperbolic geometry if its density were below this critical value, and if it were above, the universe would be characterized by closed spherical geometry.

What Sandage's proposed tests tried to discern was the geometry of space. For this purpose, it was necessary to measure a deceleration parameter q_0 that was equivalently related to Ω_M through solutions to the Friedmann equation, assuming a universe consisting solely of normal matter. This relation turned out to be:

$$q_0 = \Omega_M / 2$$

Astronomers needed observables to measure, and one of Sandage's tests included measuring the luminosity of an object as a function of its redshift z [165].

$$D_L = \frac{c}{H_o q_0^2} \left[q_0 z + (q_0 - 1)\left(\sqrt{1 + 2q_0 z} - 1 \right) \right]$$

where D_L is the luminosity distance, defined as the inverse square law of an object of luminosity, L, and observed flux, f:

$$D_L \equiv \sqrt{L/4\pi f}$$

These measurements in turn provided values for Hubble's constant and the average density of matter in the universe. Additional work by Sandage and others extended these types of measurements to further objects.

27. Gravitational Lenses

Until now, the techniques we have described to establish the H_0 value have been based on using the cosmic ladder to obtain a final estimate of absolute distance to the object being observed. Each rung of the ladder requires gauging a particular distance indicator, and each is subject to inaccuracies that contribute, in the end, to increasing the final uncertainty. Direct and more precise methods to estimate absolute distances, such as orbital parallax, are limited to nearby objects, and are difficult to apply to objects outside our own galaxy.

The ultimate solution would be finding primary indicators that are reliable and bright enough to directly calculate extragalactic distances, thus avoiding many rungs on the cosmic ladder. One of these techniques involves the use of gravitational lensing.

The use of lensing to measure the Hubble constant was first suggested by Sjur Refsdal as part of his PhD thesis. In 1964, Sjur Refsdal wrote a pair of fundamental papers, which were both communicated to the Royal Astronomical Society by Hermann Bondi. In the first, he described the properties of a point-mass gravitational lens, simplifying calculations previous made by Gavil A. Tikhov in 1937, arguing that geometrical optics could be used for gravitational lensing [166]. In the second paper, he studied the effect of a gravitational lens on a supernova lying far behind a distant galaxy, close to its line of sight. In this case, the gravitational lens was able to split the hypothetical supernova image into two images. The two different paths that light followed from the supernova reached the observer at different moments. Refsdal showed that this time delay, Dt, between both images can be used to determine the distance to the source d_S and that to the lensing galaxy d_L. Refsdal showed that the delay is given by [167]:

$$\Delta t = k \left(\frac{d_L d_S}{d_S - d_L} \right)$$

The factor k depends on the angular separation between images and the mass distribution of the lens. The angular separation is observable, while for the mass distribution, certain assumptions have to be taken. An additional correction to Dt (not mentioned by Refsdal in that paper) must be applied due to the Shapiro time delay [168]. The Hubble parameter follows from the redshift ratio of the lens and the source. The difficulty of the method lies in that the mass distribution of the lensing galaxy or galaxies must be assumed to model the lens, and the angular separation of the lensed images might be difficult to measure.

In the same publication, Refsdal called attention to the potential importance of quasars in distance measurements using gravitational lenses. Here, it is pertinent to recall that some years before (1963) the first quasar, a "quasi-stellar", compact, very luminous, and distant source was identified by Maarten Schmidt [169].

Two more articles were published by Refsdal in 1968, where he suggested the use of lensing for testing cosmological theories such as the still-popular steady-state theory, and theories based on general relativity [170]. The second paper dealt with the conditions needed to determine the mass and distance of a star which acts as a gravitational lens, if the lens effect can be observed from the Earth and from at least one distant space observatory [171].

By that time, it had already been realized that light propagation in a real universe whose mass distribution is not homogeneous should differ from a homogeneous universe, because there are regions of space with greater mass density than others and, consequently, with different lensing effects. This fact must be considered. A third paper was published by Refsdal at the beginning of 1970 on the influence of lensing on the apparent luminosity of very distant light sources in static and flat universes with inhomogeneous mass distributions [172]. In this article, he studied the validity and limitations of this model, as well as possible extensions to expanding and curved universes.

In 1979, the first lensed quasar was discovered by Dennis Walsh, Robert F. Carswell, and Ray J. Weymann [173]. They reported the double quasar Q0957+561 as two quasar images with the same color, redshift, and spectra, separated by only 6.1 arc-seconds, produced by a gravitational lens. However, for a time, some radio astronomers disputed this interpretation, but further investigations confirmed it.

In this case (quasar Q0957+561), Refsdal's method could not be applied immediately due to two obstacles. The first was modeling the mass distribution of the lens, which turned out to be difficult as the lens consisted of at least two adjacent galaxies. The second difficulty was the measurement of Dt, which required long observation periods [174,175]. These were out of reach for many astronomers due to limited telescope allocation times and the usual problems of ground-based optical astronomy, such as the annual occultation by the Sun causing periodic gaps of 4 to 5 months in data acquisition. In addition, the flux variations in both images were small, and such measurements required extreme care and were difficult to carry out [176]. In addition to these difficulties, the works of Falco, Gorenstein, and Shapiro [177,178] and their coworkers [179] were key to realizing that the lens modeling was ambiguous and that some degeneracies existed in the parameter estimation. In particular, they pointed out that the Mass-Sheet Degeneracy (at that time known as Magnification Transformation) generates an ambivalence to estimate H_0 from the time delays between gravitationally lensed images. They also pointed out that this ambivalence can be addressed with a measurement of the lensing galaxy's velocity dispersion. The velocity dispersion constrains the lensing galaxy's mass, which breaks the degeneracy and leads to a point value for H_0 [180].

During 1991, Ramesh Narayan and other independent groups were the first to put Refsdal's idea into practice [181]. Narayan combined a model of mass distribution in the gravitational lens (Q0957+561) system, with measurements of the time delay, to obtain a value for H_0 of 37 ± 14 km/s/Mpc. Another independent estimate was made of the same quasar by Roberts et al. [182], and a value of 46 ± 14 (42 ± 14) km/s/Mpc was given for H_0. A third estimate was made by G. Rhee [183] that yielded 50 ± 17 km/s/Mpc.

All of these values had significant uncertainties. Part of the uncertainty was in modeling the mass distribution of the lensing system. The conspicuous, bright galaxy that apparently acted as the lens was part of a cluster of lensing galaxies. The time delay was also doubtful due to limited durations of the telescope observations.

In 1994, a team lead by Edwin Lewis Turner of Princeton gained access to the 3.5 m Apache Point Observatory in NM. From late 1994 through May 1995, every night, weather permitting, they observed the double quasar 0957+561 A B. They found the time delay of the trailing image B to be 415 days, with respect to the leading image A [184]. Their predicted time delay was confirmed in 1996, when the variability pattern between both images was 417 days. The result was published in 1997, together with their estimate of $H_0 = 64 \pm 13$ km/s/Mpc. They argued that their estimate was of comparable quality to those based on more conventional techniques at that time. In the same paper, the authors pointed out the advantages of the gravitational lens method. Briefly, it is a geometrical method based on the well-understood physics of general relativity in the weak-field limit. It yields a direct, single-step method for measuring H_0 and thus avoids error propagation along the "distance ladder", which is no more secure than its weakest rung. It measures distances to cosmologically faraway distant objects, thus precluding the possibility of confusing a local with a global expansion rate. It gives an independent measurement of H_0 in two or more lensed systems with different sources and lens redshifts. This provides an internal consistency check of the obtained H_0 value, i.e., if a small number of time delay measurements all give the same H_0, this value can be regarded as correct with considerable confidence.

Further advances toward a model independent characterization of strong lenses have been developed in recent years [185], as have studies on the intrinsic degeneracies of the gravitational lensing formalism [186]. With these works, one was finally able to derive the most general class of lensing degeneracies, make physical sense out of the mathematical parameters, and put them into the overall context of related cosmological work. These results gave a more definitive basis on which to understand strong lensing in cosmology. As Jenny Wagner put it, "...after this major breakthrough, the constraints on H_0 from lensing that had been accused of being a black-box, can now be put on solid grounds" [187].

The gravitational lens method increased in importance as more gravitationally lensed quasars were discovered for which the mass distribution of the lens was simpler than that of Q0957+561, but that was to occur in the twenty-first century.

28. Megamasers

Another distance indicator that is independent of the cosmic ladder is the mega-maser technique, i.e., a maser powerful enough for us to see at cosmic distances. This powerful emitter turned out to be a mega H_2O maser produced inside vast clouds of water molecules located in the vicinities of galaxy centers and pumped by infrared radiation. These masers, as we shall see, facilitated a more precise distance assessment of the NGC4258 galaxy.

In 1995, a group of American and Japanese astronomers found compelling evidence for the existence of a supermassive compact object at the center of the NGC4258 galaxy [188]. This evidence was gathered using the very long baseline array (VLBA). The VLBA is a is a system of ten 25-meter radio telescopes distributed along the USA's territory and operated remotely from Socorro, New Mexico. This set of antennas works together as a very extended baseline interferometer. Its baseline is about 8600 km, thus producing superb angular resolution.

The team measured the intensity of the narrow spectral line emissions (1.35 cm) of water maser sources. During this surveillance, they registered spikes of H_2O maser emissions, with each spike corresponding to a lump of H_2O masering material. These spikes came from masers with radial velocities equal to the average velocity of the galaxy, but they also found spikes from masers with higher and lower radial velocities. The simplest explanation is that there is a disk of H_2O gas orbiting a central mass in NGC4258. The high and low velocity spikes originated from masers at the edges of the disk. From the dynamical characteristics of the Keplerian motion of the disk, the group deduced that there was a black hole (BH) at the center of rotation, with a mass of at least $3.6 \times 10^7 \, M_\odot$. The discovery made the news. Incidentally, at nearly that time, there were independent investigations by Reinhard Genzel and Andrea Ghez regarding the existence of a BH at the center of our Milky Way that, years later, resulted in Nobel prizes for this latter pair of scientists.

It was not until 1999 that further investigations conducted by members of the same original team of astronomers introduced a novel geometrical method to infer the extra-galactic distance from their direct measurement of orbital motions of H_2O masers in the disk of masing clouds surrounding the nucleus of the NGC4258 (M106) galaxy [189].

This time, astronomers reported that the disc had an estimated a radius of 0.13 pc, and from its angular radius (4.1 mas), they deduced its distance: 6.4 ± 0.9 Mpc. The novelty of this geometric method was that required no assumptions or calibrations, just precise measurements of angles and velocities. The method turned out to be much more accurate than other previous distance determinations of NGC4258.

In subsequent observations of the mega H_2O masers in NGC 4258, measurements continued to be refined over the years. Table 3 shows the values of the distance to the NGC 4258 galaxy, with uncertainty estimates. The reader will notice that, currently, the margin of error is small. This, as we shall later comment, has also resulted in a dramatic reduction in uncertainty surrounding the Hubble constant.

Over the 25 years after the first report of the distance to NGC 4258 in 1995, the number of VLBI observations and geometric distance estimates increased. These new estimates were based on modeling the Keplerian orbits of H_2O masers orbiting around the BH using larger data sets. Table 3 shows these newer estimates of distance to NGC 4258, with the last three reports showing only very small changes in the estimated distance, but with successive improvements in the uncertainty.

Table 3. Estimates of the distance to NGC 4258.

Reference	Distance (Mpc)
Miyoshi et al. (1995) [188]	64 $\pm$ 0.9
Herrnstein et al. (1999) [189]	7.2 $\pm$ 0.5
Humphreys et al. (2013) [190]	7.596 $\pm$ 0.228
Riess et al. (2016) [191]	7.540 $\pm$ 0.197
Reid et al. (2019) [192]	7.576 $\pm$ 0.112

Also using this technique, the Megamaser project (MCP) [193] aims to determine the Hubble constant. They have been working for over a decade, and recently, they found a value of $H_0 = 73.9 \pm 3.0$ km/s/Mpc [194] based on measurements of five megamaser galaxies in the Hubble flow. Their measurements are consistent with the high end of H_0.

29. The Sunyaev–Zeldovich Effect

Another way to estimate the Hubble constant is through looking at distortions in the CMB radiation caused by the scatter of hot gas in galaxy clusters. The CMB light travels from the last scattering surface to us, and in between, it finds a series of clusters that possess a hot gas of electrons that, in turn, scatters the CMB light, provoking a CMB brightness decrement in the direction of the cluster. This effect was proposed in 1970 by R. A. Sunyaev and Ya. B. Zel'dovich [195,196], and it is named to honor these two scientists—the SZ effect, for short. There are two effects: the thermal, which is caused by energetic electrons in the cluster, and the kinematic, caused by the relative motion of the cluster with respect to the Hubble flow.

In the late 1970s, the idea was put forward that a comparison of the thermal SZ effect with the X-ray properties of a cluster, also caused by high-energy electrons, could help to constrain cosmological parameters such as the Hubble constant, among others. These two effects have different degeneracies of their amplitudes, so its comparison could help to break parameter dependencies. The method is also independent of a standard candle or ruler, relying only on the properties of highly ionized plasma, with temperatures around 10 keV. Using the same density of electrons in the cluster, one can obtain the angular diameter distance, and hence infer the Hubble constant. For a summary of the physics of this technique and related works up to the early 2000s see the website cited in [197].

The first works to consider this idea were Cavaliere, Danese, and de Zotti [198] in 1977; Cowie and Perrenod in 1978 [199]; Gunn et al. in 1979 [200]; Silk and White in 1978 [201]; and Cavaliere et al. in 1979 [202]. Early estimations were made by Birkinshaw in 1979 [203]. Later, in the 1990s, M. Birkinshaw and J.P. Hughes [204] measured it in a finer way, though the uncertainties at that time were still big (± 7 km/s/Mpc). A decade later, other measurements were carried out, for example by Reese et al. in 2000 [205] and 2002 [206]; Patel et al. in 2000 [207]; Mason et al. in 2001 [208]; Sereno in 2003 [209]; Udomprasert et al. in 2004 [210]; Schmidt et al. in 2004 [211]; and Jones et al., who used ground-based radio interferometers, in 2005 [212]. In that epoch, the value of the Hubble constant was still around 50–100 km/s/Mpc, not only because of the uncertainties in the measurements, but also because of the large spread of the central values obtained via different methods.

The SZ technique has been further refined with the advent of CMB probes such as WMAP and, later, SPT (South Pole Telescope); ACT (Atacama Cosmology Project); and Planck. These are presently obtaining much smaller uncertainty values (± 3 km/s/Mpc) [213].

30. H_0 Measurements with CMB Probes and BAO

There are two other modern techniques used to measure the Hubble constant, both of them related to the physics of the early universe's acoustic oscillations: CMB and BAO. CMB measurements are now an important basis of modern cosmology. The CMB is the fingerprint of the universe, providing us information about not only the time it formed, but also the initial conditions of the universe long before the first second of life. As if that were not enough, since this light comes from such remote distances and from moments so far away, on its journey to us, it collects information about the properties of the universe on its path. The general features of this very important fingerprint were theoretically calculated prior to its observation using the Big Bang model. The light from the CMB comes from the farthest region and, therefore, from the earliest time in our universe that we will ever be able to directly see, no matter how powerful future telescopes or satellites become. This is because all the light that had previously interacted with the electrons and protons in a very efficient way has ceased to exist, although new light is generated. As the universe cooled, the temperature dropped and the light stopped interacting; the free electrons combined with protons to form neutral hydrogen atoms (one proton and one electron). When this happened, the universe was about 380,000 years old. The last scattered photons formed the so-called last scattering surface, located 13,400 millions of years in our past. The light stemming from it traveled to us in an almost unperturbed way. The CMB light was measured in 1964 by Arno Penzias and Robert Wilson, who deserved the 1978 Nobel Prize in Physics. They observed a noise in their microwave antenna, and after examining possible sources of noise, even down to the pigeon droppings on the antenna, they deduced that it was an observed noise coming from a celestial source of extragalactic origin. The correct interpretation of cosmic background radiation was made by scientists in the Robert H. Dicke group, which included P. J. E. Peebles, P. G. Roll, and D.T. Wilkinson, who had been seeking to measure it at that time, in addition to the group of Russian scientists from the Yaakov B. Zel'dovich group from Moscow, which included A. G. Doroshkevich and I. D. Novikov. The measured temperature was approximately $-270\ ^\circ$C, about 3K above absolute zero. Time passed, and towards the end of the 1980s, a modern version of Penzias and Wilson's experiment was carried out. This experiment was commanded by one of the authors of this article (G.F.S.) and his collaborators, and was carried out with the COsmic Background Explorer (COBE) satellite. Thus, this began a new experimental era of high precision in cosmology. The COBE team, in the early 1990s, revealed with great accuracy that the universe is homogeneous and isotropic, but not entirely. Tiny anisotropies discovered by COBE, one part in a hundred thousand, were responsible for the subsequent formation of stars, galaxies, galaxy clusters, and all the large-scale structures of our universe. Furthermore, COBE found a blackbody spectrum from the CMB indicating that the early universe was in thermodynamic equilibrium, with (almost) the same temperature everywhere and in a homogeneous state. These were two of the most important discoveries in 20th-century astronomy, and they are why George F. Smoot and John Mather were awarded the 2006 Nobel Prize in Physics.

The origin of the minuscule anisotropies was most likely due to the primordial fluctuations of a quantum field which was present in the very early universe, during the inflationary era. Modern quantum field theories, along with cosmological models, help to understand how small fluctuations evolved at cosmic scales, and then how COBE is able to detect them tens of billions of years later. This early period of inflation was a stage of accelerated growth of our universe, and not only gave rise to the formation of primordial fluctuations, but also, although it seems contradictory, explains why our observable universe is so homogeneous and isotropic. Inflation was postulated in the early 1980s independently by Alex Starobinsky and Alan Guth, and was further developed by Andrei Linde and many others. It is an essential part of the Big Bang model in the first instances of the existence of the universe.

The COBE satellite and other cosmological probes, such as the BOOMERANG and MAXIMA hot air balloons, launched in Antarctica in the late 1990s; the more recent WMAP

(Wilkinson Microwave Anisotropy Probe) satellite launched in 2001; and the Planck satellite in 2012, have confirmed these peculiar characteristics of the origin of our universe. The CMB not only provides information on the fundamental fluctuations, but also on the behavior, of the universe at that time. Regions of the space that subtend an angle of less than one degree from us to the last scattering surface have undergone oscillations in their growth due to the interactions between protons, electrons, and light just before the hydrogen atoms were formed. The competition between the force of gravity, which attempted to bring all the particles together, and the pressure force of light, which fought against it, generated these oscillations to the primordial plasma, imprinting a unique fingerprint on light and matter. At present, when we measure the acoustic pattern of the CMB light, we are able to measure different cosmological parameters that are fitted to CMB maps in a delicate manner, given the precision of Planck. An efficient way to compare maps with model parameters is through the use of the Monte Carlo Markov Chain (MCMC) technique for the model predictions. The theoretical prediction is performed by solving the Friedmann equation for the background evolution together with perturbed Boltzmann equations for each component of the universe: "baryons" (nuclei and electrons), light, neutrinos, dark matter, and dark energy (or the constant Λ). There are different Boltzmann solvers, starting from the pioneering code of Uros Seljak and Matias Zaldarriaga, called CMB-Fast, in the mid 1990s, as well as many others that came later. Today, the most used ones are CAMB and CLASS, which are coupled to MCMC solvers. There are six cosmological parameters to solve (the present amount of baryons and dark matter; the angle of the CMB acoustic scale; the reionization parameter; and two inflation signatures: the amplitude of primordial perturbations and the spectral index). This may depend on additional parameters for extended models, depending on, e.g., the universe's curvature, the number of relativistic species, the mass of neutrinos, etc. The Hubble constant is tightly constrained in combination with density parameters, so this technique permits us to indirectly determine it. As already mentioned, the preferred value for the flat ΛCDM model is $H_0 = 67.4 \pm 0.5$ km/s/Mpc [214].

Once light decouples from plasma, baryonic matter begins to fall to potential wells of DM, attracting DM to it while keeping the oscillatory patterns formed by the plasma. This imprints on the matter clustering a specific feature at a large scale that serves as a standard ruler and tells us about the size of the horizon at last scattering (or drag time, to be more specific). These baryon acoustic oscillations (BAOs) represent an amazing pattern of cosmological dimension in which over-densities in the matter field are found, as a rule, at a comoving size every ca. 150 megaparsecs. These acoustic oscillations were predicted long before they were measured. BAOs were estimated in the mid-1990s, and their measurement was reported in 2005 for first time by two independent groups, the Sloan Digital Sky Survey (SDSS) collaboration led by Daniel Eisenstein [215] and the 2dF collaboration led by Shaun Cole [216]. This was another amazing and very beautiful success in science. By localizing galaxies in the 3D universe through the BAO technique, one is able to compute the distances to the BAO feature at different times in the past. In the 2000s, the 2dF, 6dF, WiggleZ, and SDSS (BOSS) collaborations measured the BAO feature and computed the distance to it at different redshifts. Again, the Hubble constant was indirectly measured, given a (ΛCDM) model. The most refined measurements to date were obtained from the eBOSS collaboration, in which different tracers were used (see, e.g., [217]), and their results were in accordance with those of Planck.

31. Gravity Waves Standard Sirens

Astrophysical events at cosmological distances may also be useful in determining the Hubble constant. We will see examples of these in the following sections.

A new, interesting, and different method to determine H_0 is through gravity waves emitted from the coalescence of neutron stars at the final stages of binary systems. In principle, coalescence of black holes emits gravity waves as well, but does not provide redshifts nor precise distances. An electromagnetic counterpart is needed, and that is why one employs neutron stars. This method is considered to be applicable to every neutron coalescence. The name "standard siren" was picked to distinguish it from the known standard candles [218]. This method was proposed long ago by Bernard Schutz [219], but its realization materialized just few years ago. The method relies on the physics behind the collapse of the binary, and it is free of the systematics of local distance scales. This technique provides luminosity distances, and because they are well modeled during its coalescence phase, their uncertainties could be potentially small. On 17 August 2017, LIGO/Virgo collaboration detected a pulse of gravitational waves, called GW170817, associated with the merging of two neutron stars in NGC 4993, an elliptical galaxy in the constellation Hydra located 43 Mpc from us. The collaboration paper [220] reported a Hubble constant of 70^{+12}_{-8} km/s/Mpc. With this, we are entering into a new era of gravitational wave multi-messenger astronomy. The associated uncertainties will be greatly reduced in the coming years as more events of this type are measured.

One can also make a pure gravitational wave estimation of Hubble's constant using neutron star–black hole mergers [221]. H_0 can be derived purely from the gravitational waves of neutron star–black hole mergers. This new method provides an estimate of H_0 spanning the redshift range of $z < 0.25$, with the sensitivity of current gravity waves and without the need for any afterglow detection. The authors employed the inherently tight neutron star mass function together with the merge's waveform amplitude and frequency to estimate the distance and redshift, respectively, thereby obtaining H_0 statistically. Their first estimate is $H_0 = 86^{+55}_{-46}$ km/s/Mpc for the secure neutron star–black hole events GW190426 and GW200115. One expects that with ten more such events, one may reach a precision of $\delta H_0 / H_0 \lesssim 20\%$.

32. Black Hole Shadows

Recently, another alternative method to determine the cosmic distance scale has been proposed to measure the Hubble constant [222]. It is based on the measurement of the shadows of supermassive black holes in order to use them as standard rulers; a historical account of black hole shadows can be found in this reference [223]. With a known mass of a black hole, its shadow is unique within GR and has a known physical size. By measuring its angular size, its angular diameter distance can be determined, which in turn depends on H_0. Technically, however, the problem is intricate, since light paths must be computed under the influence of the local action of gravity and, at the same time, in an expanding background. This technique is applicable to the late universe for small redshifts ($z \lesssim 0.1$), for large black hole masses ($\geq 10^9$ solar masses), and potentially also for much larger redshifts in the coming years. Even though the angular diameter distance decreases for higher redshifts, the angular size of a shadow is expected to increase for high redshifts [224]. This scope will allow us to study angular diameter distances as functions of time. The current precision of H0 is on the order of 10% according to data such as those from the Einstein Horizon Telescope [225], or a small (few) percentage when considering probes in the near future [226–228]. A note of caution has been raised on this technique regarding the difficulty of obtaining reliable estimates of a black hole's mass; achieving the required angular resolution; and having sufficient knowledge of high redshift accretion dynamics, especially challenging high redshift measurements [229]. Other possible effects may come from alternative models to ΛCDM in which pressure singularities may exist, which change the shadows of black holes and, therefore H_0. This could eventually resolve the Hubble tension [230,231].

33. Fast Radio Bursts

Fast radio bursts (FRBs) are millisecond-duration pulses of radio emission observed at frequencies from hundreds of MHz to a few GHz, now recognized to have a cosmological origin. Their physical source is, however, still under debate, and there have been many progenitor models proposed. FRBs have the potential to act as cosmological probes, and in particular to determine the Hubble constant at redshifts $z < 1$, among other applications.

FRB pulses scatter as they travel through the ionized intergalactic medium, with the inferred total dispersion measurement being a powerful probe of the column density of ionized electrons along the line of sight. The dispersion measure and the signal-to-noise ratio of the pulse, together with the redshift associated with the host galaxy of the burst, provide a direct constraint on H_0. One approach to computing H_0 is to assume that the FRB energetics do not depend on redshift, being a kind of standard candle, and the sensitivity to H_0 is given through the signal-to-noise ratio of the pulse. It appears unlikely that they are standard candles unless there is a correction factor, since repeater pulses are far from identical from one pulse to the next. The other approach considers the cosmic contribution to the FRB dispersion measure, whose average depends directly on both the Hubble constant and the baryon density parameter; thus, synergies with CMB and nucleosynthesis results could help to constrain H_0. In the study of C.W. James et al. [232], a detailed methodology and bias estimates are provided. Using a sample of 9 FRBs, Steffen Hagstotz et al. [233] found a Hubble constant of 62.3 ± 9.1 kms/Mpc/s, whereas Qin Wu et al. [234] used 18 localized FRBs and found an H_0 with a smaller uncertainty of $68.81^{+4.99}_{-4.33}$ kms/Mpc/s. Finally, using a sample of 16 localized and 60 unlocalized FRBs, C.W. James et al. [235] found a best fit of 73^{+12}_{-8} kms/Mpc/s. This new approach is certainly developing, and will surely provide more stringent constraints on H_0 in the near future from the CHIME catalogs [236], among others.

34. The Current Situation and Final Remarks

By the early 1990s, many estimates of the Hubble constant had already been made using several methods. As a result of these measurements, a dichotomy of results surged. Values for H_0 were grouped into one of two separate intervals. In the first set of the two, dubbed the "long" timescale, the value of H_0 was situated in the interval between 40 to 60 km/s/Mpc, while in the second set, labeled the "short" timescale, H_0 was located between 80 to 100 km/s/Mpc (i.e., a large value of the Hubble constant and a "short" timescale for the cosmological expansion). This indicated a "factor-of-two" difference!

Recognized astronomers such as Allan Sandage, and Gustav A. Tammann, and some supernova theorists have argued that the value of the Hubble constant is in the range of the long timescale, while Gérard de Vaucouleurs and others have championed for the short-scale value. Debates between defenders of both positions lasted several years and were inconclusive.

On the one hand, short-timescale critics brought up the "Malmquist bias" as one of the most important sources of their opponents' miscalculations. This bias, in a brightness-limited survey of luminous objects in the sky, is an unintentional type of censorship where luminous objects below a certain apparent observational brightness threshold are excluded. They cannot be observed. This unintentional exclusion is the source of the bias. On the other hand, "long"-scale critics have mentioned accumulative systematic errors made by their opponents in the processes of measurements and treatment of data originally used to gauge the ladder scale.

Over the years, this controversy created an unsatisfactory and uncomfortable situation. Astronomers could not reach a consensus on how they could adjust their data to account for various effects that might lead to bias in their observations. Differences in the choice of secondary distance indicator methods were at the root of almost all the current debate at that time concerning the value of the Hubble constant.

Later on, with the advent of the HST, uncertainties diminished. Using different techniques, Wendy Freedman et al. reported the final results from the measured a value of H_0 of around 72 to ca. 10% accuracy [237]. This value was in the middle of the short and long timescales.

Alternatives arose to end this stagnation. They were proposed by independent groups, and were not based on climbing the rugs of the cosmic ladder. They included, namely, gravitational lensing, the S-Z method, and CMB and BAO analysis.

In recent years, Cepheids techniques have made progress in different directions in different projects; for an account of this, see reference [143]. It turned out that methods anchored to distance ladder techniques (Cepheids, supernovae, Mira variables) have been finding Hubble constant values in the range of 72–74 km/s/Mpc using the standard flat ΛCDM model. Recent works of the late-universe teams, such as the SH0ES collaboration, which used the distances to supernovae calibrated with Cepheids, reported an H_0 of ~73 ± 1 km/s/Mpc; these results were confirmed using the latest James Webb Space Telescope (JWST) measurements by Adam Riess et al. [238]. Other methods, such as strong lensing techniques, also obtain a Hubble constant in this interval. For instance, the H0LICOW collaboration reported $H_0 = 73.3\,^{+1.7}_{-1.8}$ km/s/Mpc at 2.4% precision using the light from six multiply-imaged quasar systems [239]; although different results have been obtained since the modeling of the lens, mass distribution is an important systematic. These methods use late-time physics. On the contrary, early-universe physics at last scattering are anchored to the sound horizon, yielding CMB and BAO results that indicate smaller values of H_0, around 67 km/s/Mpc. The uncertainties reported in these papers show a discrepancy of around four sigma, or larger in some cases. In Figure 12, we show the curves of the ΛCDM model that predict SH0ES data together with clustering data from the BOSS collaboration, which measured the Hubble expansion rate as a function of redshift. The differences are clear.

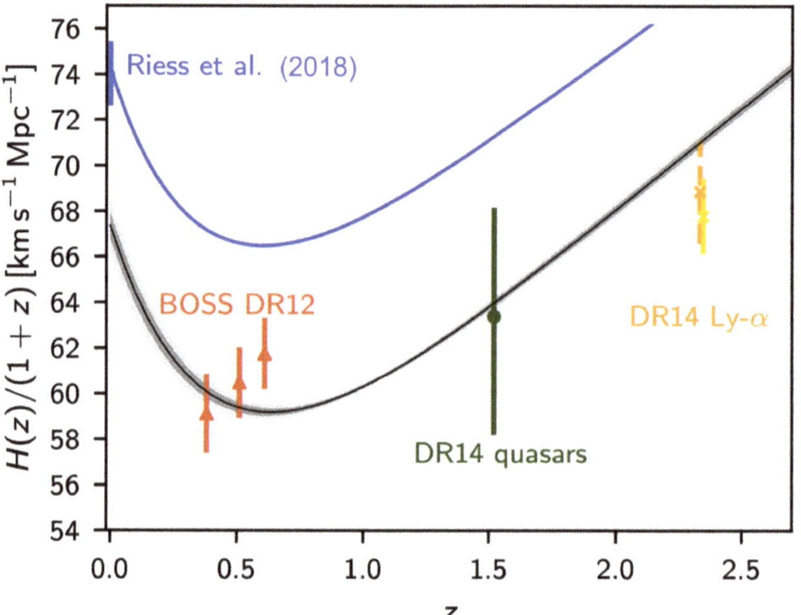

Figure 12. Hubble expansion rate, as observed in the present (2023) and earlier (vs. redshift), along with the best-fitted cosmological model predictions from CMB observations (the lower curve). The upper curve is the model used for fitting the "nearby" supernova data, which extended out to z ~1. The actual data fit fairly tightly on this curve. Note that by quoting as the Hubble Constant H_0 one suppresses the model dependence of the nearby observations. Doing so allows us to see that the model would need to have "new physics" to bring these into agreement. If we assume some form of continuity in the expansion of the universe, then "new physics" has to jump between the light-based standard candles and the size-based standard candles (cosmic rulers). Credit: Figure was modified by us after a plot from reference [214]. The blue point corresponds to the measurement in in ref. [240].

On the other hand, based on TRGB the CCHP collaboration recent results by Freedman et al. [241], lies somewhat in the middle of the tension, with an H_0 around 70 ± 2, as Figure 13 shows. These results ameliorate the discrepancy, and eventually may point to a solution within the ΛCDM model. A likely way to resolve the apparent discrepancy is for at least two of the methods to have additional systematic errors that are underestimated. The big question is which are those two or three out of three?

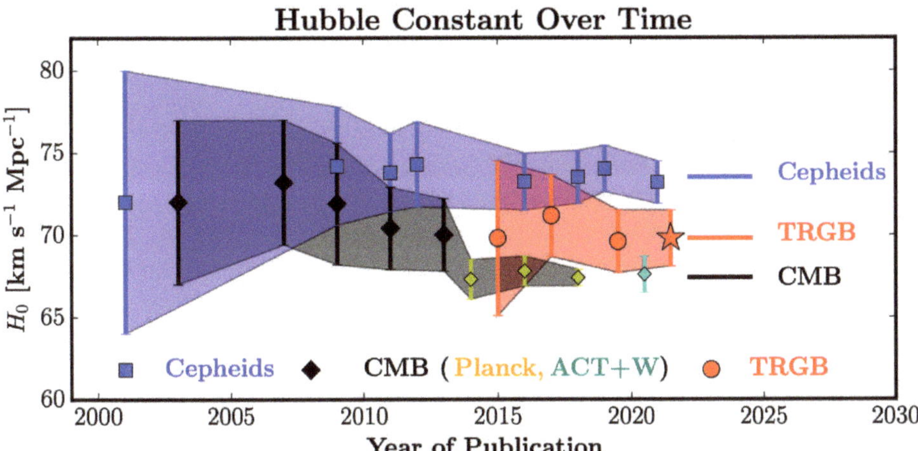

Figure 13. Determination of Hubble constant using different techniques in the last two decades, with shadows showing the associated uncertainties. Credit: Wendy Freedman, reference [241].

In the meantime, this controversy has inspired many theoretical papers to attempt to resolve this discrepancy. Recent papers have appeared to explain the discrepancy of the Hubble constant determination using late- and early-universe physics. They are based on different types of physics, anchored in different scales, but both are apparently correct. A compilation of recent H_0 measurements is shown in Figure 14. It is clear that early physics determinations were smaller than those of late physics, except for those measurements from the TRGB obtained by CCHP. See further details in reference [242], and this reference [243] added during galley proofing corrections that accounts for the different modern techniques to measure H_0.

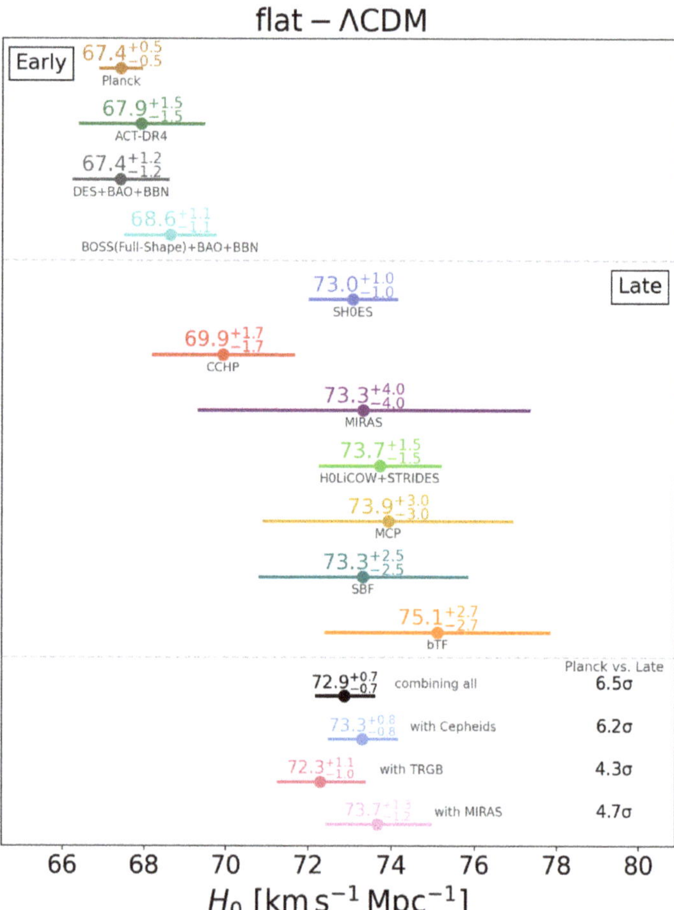

$H_0 \ [\mathrm{km\,s^{-1}\,Mpc^{-1}}]$

Figure 14. The spread of the Hubble constant measurements performed in recent years, showing small values for CMB and clustering physics (**upper** panel) and large values for different rugs used in late measurements (**middle** panel). Also, different combinations of methods are shown, as well as their discrepancies with the early-physics results (**bottom** panel). Credit: Vivien Bonvin and Martin Millon [244].

In the near future, one hope lies in the emerging data that the JWST is certainly collecting. The JWST is making new infrared and more precise optical observations available. As stated by Freedman and Madore [143], the JWST has a resolution three times that of the HST, with nearly ten times the sensitivity. Thus, it will reduce the Cepheids, TRGB, and Miras systematics, among others, in order to better determine the Hubble constant. At the same time, new CMB experiments are nearly ready, such as The Simons Observatory and plans such as CMB-S4. In addition, new BAO probes such as DESI, EUCLID, and LSST are underway; these will improve our precision in calculating the Hubble constant. Will these coming measurements relax the Hubble tension or increase it? A more precise answer on the expansion rate of the universe and the Hubble constant will make it clear whether the ΛCDM model will persist or whether new physics will need to be added to it, and new theoretical horizons will surge. In fact, there are already plenty of possible explanations to the Hubble tension; see, for instance, a review of solutions [245,246]. Also, several hints have been put forward [247] suggesting that modifications to both the early and late universe are necessary in order to solve the problem of Hubble tension.

In conclusion, we will remark that there is a prejudice introduced when one labels a parameter that changes with time as a constant. In fact, the Hubble constant is the value the parameter H(z) has at the present time, but it will have a very different value if measured 10 billion years before or after the present time. H_0 is not a fundamental physical constant. It provides a scale for the present-day universe, as it is a reference. There is even some evidence that H_0 might have a decreasing trend when computed using data at higher redshifts (see reference [239] and others [248,249]). Now, its different estimations resulting from diverse standards based on different anchors may be indicative of some new physics, although this also might simply be a difference due to different systematic errors in the various techniques. We believe that time, as well as additional and improved observations, will settle the argument, just as occurred with the Great Shapley–Curtis debate.

Author Contributions: All authors contributed in an equal manner. All authors have read and agreed to the published version of the manuscript.

Funding: J.L.C.-C. was funded by CONAHCYT grant number 283151.

Data Availability Statement: No new data were created or analyzed in this study. Data sharing is not applicable to this article.

Acknowledgments: This research made use of NASA's Astrophysics Data System Bibliographic Services.

Conflicts of Interest: The authors declare no conflict of interest.

Notes

[1] For thousands of years people had referred to the "fixed stars". In astronomy, the fixed stars (Latin: *stellae fixae*) are the luminary points, mainly stars, that appear not to move relative to one another against the darkness of the night sky in the background. This is in contrast to those lights visible to naked eye, namely planets and comets, that appear to move slowly among those "fixed" stars. So, we had thousands of years prejudice for a static Universe and a human lifetime natural time scale while the stars move on time scales of millions of years or more.

[2] What Einstein did not imagine is that the cosmological constant that he would later regret having introduced, and called his "biggest blunder" would later represent the dark energy that is an essential ingredient of the modern ΛCDM model.

[3] All initial cosmological models assumed spherical (actually homogeneous and isotropic) symmetry in order to make cosmology a tractable issue. In the first half century this was an assumption but later radio surveys and most importantly the isotropy of the Cosmic Background Radiation have justified and improved this assumption. The CMB limits are uniformity to the part in 100,000 level or better. This is sufficient to find the first order solutions and treat the rest with perturbation theory.

References

1. Hoskin, M.A. The 'great debate': What really happened. *J. Hist. Astron.* **1976**, *7*, 169–182. [CrossRef]
2. Shapley, H. Note on the Magnitudes of Novae in Spiral Nebulae. *Publ. Astron. Soc. Pac.* **1917**, *29*, 213–217. [CrossRef]
3. Van Maanen, A. Preliminary evidence of internal motion in the spiral nebula Messier 101. *Proc. Natl. Acad. Sci. USA* **1916**, *2*, 386. [CrossRef]
4. Berendzen, R.; Hart, R. Adriaan Van Maanen's Influence on the Island Universe Theory: Part 2. *J. Hist. Astron.* **1973**, *4*, 73–98. [CrossRef]
5. Slipher, V.M. Nebulae. *Proc. Am. Philos. Soc.* **1917**, *56*, 403–409.
6. Shapley, H.; Curtis, H.D. The Scale of the Universe. *Bull. Natl. Res. Counc.* **1921**, *2*, 171–217.
7. Leavitt, H.S. 1777 variables in the Magellanic Clouds. *Ann. Harvard Coll. Obs.* **1908**, *60*, 87.
8. Leavitt, H.S.; Pickering, E.C. Periods of 25 Variable Stars in the Small Magellanic Cloud. *Harv. Coll. Obs. Circ.* **1912**, *173*, 1–3.
9. Hertzsprung, E. On the Distribution of Variables of the delta Cephei Type. *Astron. Nachr.* **1913**, *196*, 201–210.
10. Russell, H.N.; Shapley, H. On the distribution of eclipsing variable stars in space. *Astrophys. J.* **1914**, *40*, 434. [CrossRef]
11. Russell, H.N. Notes on the Real Brightness of Variable Stars. *Science* **1913**, *37*, 651.
12. King, E.S. Standard tests of photographic plates. *Ann. Harv. Coll. Obs.* **1912**, *59*, 1–32.
13. Bailey, S.I. The history and work of Harvard observatory, 1839 to 1927; an outline of the origin, development, and research of the Astronomical observatory of Harvard college together with brief biographies of its leading members. *Harv. Obs. Monogr.* **1931**, *4*, 137.
14. King, E.S. Photographic photometry on a uniform scale. *Ann. Harv. Coll. Obs.* **1912**, *59*, 33–62.
15. King, E.S. Photographic magnitudes of bright stars. *Ann. Harv. Coll. Obs.* **1912**, *59*, 95–126.
16. King, E.S. Photographic magnitudes of 76 stars. *Ann. Harv. Coll. Obs.* **1912**, *59*, 127–155.

17. King, E.S. Photographic magnitudes of 153 stars. *Ann. Harv. Coll. Obs.* **1912**, *59*, 157–186.
18. King, E.S. Absorbing medium in space. *Ann. Harv. Coll. Obs.* **1916**, *76*, 1–10.
19. King, E.S. Possible Local Cloud of Absorbing Matter. *Harv. Coll. Obs. Circ.* **1927**, *299*, 1–13.
20. Shapley, H. Studies of Magnitudes in Star Clusters: I. On the Absorption of Light in Space. *Proc. Natl. Acad. Sci. USA* **1916**, *2*, 12. [CrossRef]
21. Lowell, P. *Mars and its Canals*; MacMillan & Co.: New York, NY, USA, 1906.
22. Slipher, V.M. The radial velocity of the Andromeda Nebula. *Lowell Obs. Bull.* **1913**, *2*, 56–57.
23. Smith, R.W. The road to radial velocities: V. M. Slipher and mastery of the spectrograph. *Astron. Soc. Pac. Conf. Ser.* **2013**, *471*, 143–164.
24. Slipher, V.M. Spectrographic observations of nebulae. *Pop. Astron.* **1915**, *23*, 21–24.
25. The New York Times Archives November 23, 1924, Page 6 "Finds Spiral Nebulae are Stellar Systems; Dr. Hubbell [sic] Confirms View that They Are 'Island Universes' Similar to Our Own". Available online: https://www.nytimes.com/1924/11/23/archives/finds-spiral-nebulae-are-stellar-systems-dr-hubbell-confirms-view.html (accessed on 10 August 2022).
26. Hubble, E. Paper read at the 33rd Meeting of the American Astronomical Society. *Pop. Astron* **1924**, *252*, 1925.
27. Curtis, H.D. *Handbuch der Astrophysik*; Springer: Berlin, Germany, 1933; p. 837.
28. Hubble, E.P. A spiral nebula as a stellar system, Messier 31. *Astrophys. J.* **1929**, *69*, 103. [CrossRef]
29. Hubble, E.P. Angular rotations of spiral nebulae. *Astrophys. J.* **1935**, *81*, 334–335. [CrossRef]
30. Van Maanen, A. Internal motion in spiral nebulae 1935. *Astrophys. J.* **1935**, *81*, 336–337. [CrossRef]
31. de Sitter, W. On the relativity of inertia. Remarks concerning Einstein's latest hypothesis. *K. Ned. Akad. van Wet. Proc.* **1917**, *19*, 1217–1225.
32. Wirtz, C. de Sitters Kosmologie und die Radialbewegungen der Spiralnebel. *Astron. Nachrichten* **1924**, *222*, 21. [CrossRef]
33. Flin, P.; Duerbeck, H.W.; Bajan, K. Ludwik Silberstein and General Relativity. *Astron. Astrophys. Trans.* **2017**, *30*, 63–70.
34. Silberstein, L. The curvature of de Sitter's space-time derived from globular clusters. *Mon. Not. R. Astron. Soc.* **1924**, *84*, 363. [CrossRef]
35. Eddington, A.S. Radial velocities and the curvature of space-time. *Nature* **1924**, *113*, 746–747. [CrossRef]
36. Lundmark, K. The determination of the curvature of space-time in de Sitter's world. *Mon. Not. R. Astron. Soc.* **1924**, *84*, 747–770. [CrossRef]
37. Hubble, E.P. NGC 6822, a remote stellar system. *Astrophys. J.* **1925**, *62*, 409–433. [CrossRef]
38. Hubble, E.P. A Relation between Distance and Radial Velocity among Extra-Galactic Nebulae. *Proc. Natl. Acad. Sci. USA* **1929**, *15*, 168–173. [CrossRef]
39. Hubble, E.; Humason, M.L. The velocity-distance relation among extra-galactic nebulae. *Astrophys. J.* **1931**, *74*, 43. [CrossRef]
40. Letter of Abbot to Hale, 20 January 1920 (Hale Microfilm). Quoted in Reference [1]. Available online: https://apod.nasa.gov/diamond_jubilee/1920/cs_real.html (accessed on 1 July 2022).
41. Eddington, A.S. *The Mathematical Theory of Relativity*; The Cambridge University Press: Cambridge, UK, 1923; p. 280.
42. De Sitter, W. On Einstein's theory of gravitation and its astronomical consequences. First paper. *Mon. Not. R. Astron. Soc.* **1916**, *77* (Suppl. 9), 155. [CrossRef]
43. De Sitter, W. On Einstein's theory of gravitation and its astronomical consequences. Second paper. *Mon. Not. R. Astron. Soc.* **1916**, *77*, 155–184. [CrossRef]
44. De Sitter, W. Einstein's theory of gravitation and its astronomical consequences. Third paper. *Mon. Not. R. Astron. Soc.* **1917**, *78*, 3–28. [CrossRef]
45. Einstein, A. Bemerkung zu der Arbeit von A. Friedmann „Über die Krümmung des Raumes". *Z. Phys.* **1922**, *11*, 326. [CrossRef]
46. Friedman, A. Über die Krümmung des Raumes. *Z. Phys.* **1922**, *10*, 377–386.
47. Friedman, A. Über die Möglichkeit einer Welt mit konstanter negativer Krümmung des Raumes. *Z. Phys.* **1924**, *21*, 326–332. [CrossRef]
48. Friedmann, A. Die Welt als Raum und Zeit. *Phys. J.* **2003**, *2*, 60–62.
49. Einstein in Leiden. Available online: https://www.lorentz.leidenuniv.nl/history/einstein/einstein.html (accessed on 1 July 2022).
50. Einstein, A. Notiz zu der Arbeit von A. Friedmann „Über die Krümmung des Raumes". *Z. Phys.* **1923**, *16*, 228. [CrossRef]
51. De Sitter, W. Proceedings of the RAS. *Observatory* **1930**, *53*, 37–39.
52. Eddington, A.S. January 10 meeting of the Royal Astronomical Society. *Observatory* **1930**, *53*, 33–44.
53. Archives de Georges Lemaître 2.9.4.1. Cours suivis par Georges Lemaître N. 589 p 68. Available online: https://uclouvain.be/archives (accessed on 1 August 2022).
54. Lemaitre, G. XIX. The motion of a rigid solid according to the relativity principle. *Lond. Edinb. Dublin Philos. Mag. J. Sci.* **1924**, *48*, 164–176. [CrossRef]
55. Lemaître, G. Note on de Sitter's Universe. *J. Math. Phys.* **1925**, *4*, 188–192. [CrossRef]
56. Lemaître, G. Note on de Sitter's Universe. *Publ. Lab. D'astronomie Geod. L'universite Louvain* **1926**, *2*, 37–41. [CrossRef]
57. van Maanen, A. The Cambridge Meeting of the International Astronomical Union, July 14-22, 1925. *Publ. Astron. Soc. Pac.* **1925**, *37*, 244–255. [CrossRef]
58. Green, W.K. The Meeting of the International Astronomical Union. *Science* **1925**, *62*, 555–557. [CrossRef] [PubMed]

59. Lemaître, G.H. The Gravitational Field in a Fluid Sphere of Uniform Invariant Density According, to the Theory of Relativity. Ph.D. Thesis, Massachusetts Institute of Technology, Cambridge, MA, USA, 1927.
60. Lemaitre, G. Rencontres avec Einstein. *Rev. Quest. Sci.* **1958**, *129*, 129–132.
61. Godart, O.; Heller, M. *Cosmology of Lemaitre*; History of Astronomy Series; Pachart Publishing House: Tucson, Arizona, 1985; Volume 3, p. 57, ISBN 978-0881262834.
62. Lemaître, G. Un Univers homogene de masse constante et de rayon croissant, rendant compte de la vitesse radiate des nebuleuses extra-galactiques. *Ann. Soc. Sci. Brux.* **1927**, *47*, 49.
63. Friedman, A. On the curvature of space. *Gen. Relativ. Gravit.* **1999**, *31*, 1991–2000. [CrossRef]
64. Nussbaumer, H.; Bieri, L. *Discovering the Expanding Universe*; Cambridge University Press: Cambridge, UK, 2009; pp. 122–123.
65. De Sitter, W. On the distances and radial velocities of extra-galactic nebulae, and the explanation of the latter by the relativity theory of inertia. *Proc. Natl. Acad. Sci. USA* **1930**, *16*, 474. [CrossRef]
66. Eddington, A.S. On the instability of Einstein's spherical world. *Mon. Not. R. Astron. Soc.* **1930**, *90*, 668–678. [CrossRef]
67. Stromberg, G. Analysis of radial velocities of globular clusters and non-galactic nebulae. *Astrophys. J.* **1925**, *61*, 353–362. [CrossRef]
68. Hubble, E.P. Extragalactic nebulae. *Astrophys. J.* **1926**, *64*, 321–369. [CrossRef]
69. Lemaître, G. Expansion of the Universe, A homogeneous Universe of constant mass and increasing radius accounting for the radial velocity of extra-galactic nebulae. *Mon. Not. R. Astron. Soc.* **1931**, *91*, 483–490. [CrossRef]
70. Livio, M. Lost in translation: Mystery of the missing text solved. *Nature* **2011**, *479*, 171–173. [CrossRef]
71. Burchfield, J.D. *Lord Kelvin and the Age of the Earth*; University of Chicago Press: Chicago, IL, USA, 1990.
72. Hellman, H. *Great Feuds in Science: Ten of the Liveliest Disputes ever*; Wiley: New York, NY, USA, 1998.
73. Lewis, C. *The Dating Game: One Man's Search for the Age of the Earth*; Cambridge University Press: Cambridge, UK, 2002.
74. Badash, L. Rutherford, Boltwood, and the age of the earth: The origin of radioactive dating techniques. *Proc. Am. Philos. Soc.* **1968**, *112*, 157–169.
75. Zwicky, F. On the Redshift of Spectral Lines Through Interstellar Space. *Proc. Natl. Acad. Sci. USA* **1929**, *15*, 773–779. [CrossRef]
76. Hubble, E.; Tolman, R.C. Two methods of investigating the nature of the nebular redshift. *Astrophys. J.* **1935**, *82*, 302. [CrossRef]
77. Hoskin, M. Goodricke, Pigott and the Quest for Variable Stars. *J. Hist. Astron.* **1979**, *10*, 23–41. [CrossRef]
78. Catelan, M.; Pritzl, B.J.; Smith, H.A. The RR Lyrae Period-Luminosity Relation. I. Theoretical Calibration. *Astrophys. J. Suppl. Ser.* **2004**, *154*, 633. [CrossRef]
79. Shapley, H. Studies based on the colors and magnitudes in stellar clusters. VI. On the determination of the distances of globular clusters. *Astrophys. J.* **1918**, *48*, 154–181. [CrossRef]
80. Shapley, H. Studies based on the colors and magnitudes in stellar clusters. IX. Three notes on Cepheid variation. *Astrophys. J.* **1919**, *49*, 24–41. [CrossRef]
81. Fernie, J.D. The period-luminosity relation: A historical review. *Publ. Astron. Soc. Pac.* **1969**, *81*, 707–731. [CrossRef]
82. Hubble, E. *The Realm of the Nebulae 1936*; The 1935 Silliman Memorial Lectures Series; Reprint Edition; Yale University Press: New Haven, CT, USA, 1982; Volume 25, p. 16.
83. Baade, W. The period-luminosity relation of the Cepheids. *Publ. Astron. Soc. Pac.* **1956**, *68*, 5–16. [CrossRef]
84. Feast, M. Stellar populations and the distance scale: The Baade-Thackeray correspondence. *J. Hist. Astron.* **2000**, *31*, 29–36. [CrossRef]
85. Douglas, A.V. Eight General Assembly of the International Astronomical Union. *J. R. Astron. Soc. Can.* **1952**, *46*, 217.
86. Lemaître, G. Bruxelles. *Ann. Soc. Sei.* **1927**, *47A*, 49.
87. Mineur, H. Zéro de la relation période-luminosité et absorption de la lumière dans l'espace interstellaire. *Ann. d'Astrophys.* **1944**, *7*, 160.
88. Berthod-Zaborowski, H. Recherche d'une deuxième correction au zéro de la relation période luminosité et à la constante de l'absorption dans l'espace interstellaire. *Ann. D'astrophysique* **1946**, *9*, 123.
89. Behr, A. Zur Entfernungsskala der extragalaktischen Nebel. *Astron. Nachrichten* **1951**, *279*, 97–104. [CrossRef]
90. Francis Thackeray, J. Doubling the age and size of the Universe at the IAU in Rome in 1952: Contributions by David Thackeray, Walter Baade and Harlow Shapley. *S. Afr. J. Sci.* **2020**, *116*, 1–2.
91. Humason, M.L.; Mayall, N.U.; Sandage, A.R. Redshifts and magnitudes of extragalactic nebulae. *Astron. J.* **1956**, *61*, 97–162. [CrossRef]
92. Tonry, J.L.; Davis, M. Velocity Dispersions of Elliptical and s0 Galaxies-Part Two-Infall of the Local Group to Virgo. *Astrophys. J.* **1981**, *246*, 680. [CrossRef]
93. Van den Bergh, S. Extra-galactic Distance Scale. *Nature* **1970**, *225*, 503–505. [CrossRef]
94. Henize, K.G.; Westerlund, B.E. Dimensions of Diffuse and Planetary Nebulae in the Small Magellanic Cloud. *Astrophys. J.* **1963**, *137*, 747. [CrossRef]
95. Ford, H.C.; Jenner, D.C. Planetary Nebulae in the nuclear bulge of M81: A new distance determination. *Bull. Am. Astron. Soc.* **1978**, *10*, 665.
96. Ciardullo, R.; Jacoby, G.H.; Ford, H.C.; Neill, J.D. Planetary nebulae as standard candles. II-The calibration in M31 and its companions. *Astrophys. J.* **1989**, *339*, 53–69. [CrossRef]
97. Jacoby, G.H.; Ciardullo, R.; Ford, H.C.; Booth, J. Planetary nebulae as standard candles. III-The distance to M81. *Astrophys. J.* **1989**, *344*, 704–714. [CrossRef]

98. Ciardullo, R.; Jacoby, G.H.; Ford, H.C. Planetary nebulae as standard candles. IV-A test in the Leo I group. *Astrophys. J.* **1989**, *344*, 715–725. [CrossRef]

99. Jacoby, G.H.; Ciardullo, R.; Ford, H.C. Planetary nebulae as standard candles. V-The distance to the Virgo Cluster. *Astrophys. J.* **1990**, *356*, 332–349. [CrossRef]

100. Ford, H.C.; Jacoby, G.H. Planetary nebulae in local group galaxies. V-The Andromeda Galaxy. *Astrophys. J.* **1978**, *219*, 437–444. [CrossRef]

101. Jacoby, G.H. Planetary nebulae as standard candles. I-Evolutionary models. *Astrophys. J.* **1989**, *339*, 39–52. [CrossRef]

102. Jacoby, G.H.; Branch, D.; Ciardullo, R.; Davies, R.L.; Harris, W.E.; Pierce, M.J.; Pritchet, C.J.; Tonry, J.L.; Welch, D.L. A critical review of selected techniques for measuring extragalactic distances. *Publ. Astron. Soc. Pac.* **1992**, *104*, 599. [CrossRef]

103. Merrett, H.R.; Merrifield, M.R.; Douglas, N.G.; Kuijken, K.; Romanowsky, A.J.; Napolitano, N.R.; Arnaboldi, M.; Capaccioli, M.; Freeman, K.C.; Gerhard, O.; et al. A deep kinematic survey of planetary nebulae in the Andromeda galaxy using the Planetary Nebula Spectrograph. *Mon. Not. R. Astron. Soc.* **2006**, *369*, 120–142. [CrossRef]

104. Jacoby, G.H.; Phillips, M.M.; Feldmeier, J.J. Calibrating Type Ia SNe Using the Planetary Nebula Luminosity Function. *Proc. Int. Astron. Union* **2006**, *2* (Suppl. 234), 435–436. [CrossRef]

105. Ciardullo Ciardullo, R.; Feldmeier, J.J.; Jacoby, G.H.; De Naray, R.K.; Laychak, M.B.; Durrell, P.R. Planetary nebulae as standard candles. XII. Connecting the population I and population II distance scales. *Astrophys. J.* **2002**, *577*, 31. [CrossRef]

106. Tully, R.B.; Fisher, J.R. A new method of determining distances to galaxies. *Astron. Astrophys.* **1977**, *54*, 661–673.

107. Aaronson, M.; Huchra, J.P.; Mould, J.R. The infrared luminosity/H I velocity-width relation and its application to the distance scale. *Astrophys. J.* **1979**, *229*, 1–13. [CrossRef]

108. Aaronson, M.; Mould, J. A distance scale from the infrared magnitude/HI velocity-width relation. IV-The morphological type dependence and scatter in the relation; the distances to nearby groups. *Astrophys. J.* **1983**, *265*, 1–17. [CrossRef]

109. McGaugh, S.S.; Schombert, J.M.; Bothun, G.D.; de Blok, W.J.G. The Baryonic Tully-Fisher Relation. *Astrophys. J. Lett.* **2000**, *533*, L99. [CrossRef] [PubMed]

110. Torres-Flores, S.; Epinat, B.; Amram, P.; Plana, H.; Mendes de Oliveira, C. GHASP: An Hα kinematic survey of spiral and irregular galaxies–IX. The near-infrared, stellar and baryonic Tully–Fisher relations. *Mon. Not. R. Astron. Soc.* **2011**, *416*, 1936–1948. [CrossRef]

111. Faber, S.M.; Jackson, R.E. Velocity dispersions and mass-to-light ratios for elliptical galaxies. *Astrophys. J.* **1976**, *204*, 668–683. [CrossRef]

112. Davies, R.L.; Efstathiou, G.; Fall, S.M.; Illingworth, G.; Schechter, P.L. The kinematic properties of faint elliptical galaxies. *Astrophys. J.* **1983**, *266*, 41–57. [CrossRef]

113. Schechter, P.L. Mass-to-light ratios for elliptical galaxies. *Astron. J.* **1980**, *85*, 801–811. [CrossRef]

114. Matkovic, A.; Guzman, R. Kinematic properties and stellar populations of faint early-type galaxies–I. Velocity dispersion measurements of central Coma galaxies. *Mon. Not. R. Astron. Soc.* **2005**, *362*, 289–300. [CrossRef]

115. Gudehus, D.H. Systematic bias in cluster galaxy data, affecting galaxy distances and evolutionary history. *Astrophys. J.* **1991**, *382*, 1–18. [CrossRef]

116. Djorgovski, S.G.; Davis, M. Fundamental properties of elliptical galaxies. *Astrophys. J.* **1987**, *313*, 59–68. [CrossRef]

117. Dressler, A.; Lynden-Bell, D.; Burstein, D.; Davies, R.L.; Faber, S.M.; Terlevich, R.; Wegner, G. Spectroscopy and photometry of elliptical galaxies. I-A new distance estimator. *Astrophys. J.* **1987**, *313*, 42–58. [CrossRef]

118. Tonry, J.; Schneider, D.P. A new technique for measuring extragalactic distances. *Astron. J.* **1988**, *96*, 807–815. [CrossRef]

119. Tonry, J.L.; Dressler, A.; Blakeslee, J.P.; Ajhar, E.A.; Fletcher, A.B.; Luppino, G.A.; Metzger, M.R.; Moore, C.B. The SBF Survey of Galaxy Distances. IV. SBF Magnitudes, Colors, and Distances. *Astrophys. J.* **2001**, *546*, 681–693. [CrossRef]

120. Tonry, J.L.; Ajhar, E.A.; Luppino, G.A. Surface Brightness Fluctuations and the Distance to the Virgo Cluster. *Astrophys. J. Lett.* **1989**, *346*, L57. [CrossRef]

121. Tonry, J.L.; Blakeslee, J.P.; Ajhar, E.A.; Dressler, A. The SBF Survey of Galaxy Distances. II. Local and Large-Scale Flows. *Astrophys. J.* **2000**, *530*, 625–651. [CrossRef]

122. Freedman, W.L. Stellar content of nearby galaxies. I–BVRI CCD photometry for IC 1613. *Astron. J.* **1988**, *96*, 1248–1306. [CrossRef]

123. Mould, J.; Kristian, J. The Stellar Population in the Halos of M31 and M3. *Astrophys. J.* **1986**, *305*, 591. [CrossRef]

124. Da Costa, G.S.; Armandroff, T.E. Standard Globular Cluster Giant Branches in the (M(I), (V–I)o) Plane. *Astron. J.* **1990**, *100*, 162. [CrossRef]

125. Lee, M.G.; Freedman, W.L.; Madore, B.F. The Tip of the Red Giant Branch as a Distance Indicator for Resolved Galaxies. *Astrophys. J.* **1993**, *417*, 553. [CrossRef]

126. Sakai, S.; Madore, B.F.; Freedman, W.L. Tip of the Red Giant Branch Distances to Galaxies. III. The Dwarf Galaxy Sextans A. *Astrophys. J.* **1996**, *461*, 713. [CrossRef]

127. Donahue, M.; Livio, M. (Eds.) *The Extragalactic Distance Scale*; Cambridge University Press: Cambridge, UK, 1997.

128. Freedman, W.L. Measuring the Hubble constant. *Phys. Rep.* **1998**, *307*, 45–51. [CrossRef]

129. Scolnic, D.; Riess, A.G.; Wu, J.; Li, S.; Anand, G.S.; Beaton, R.; Casertano, S.; Anderson, R.I.; Dhawan, S.; Ke, X. CATS: The Hubble Constant from Standardized TRGB and Type Ia Supernova Measurements. *Astrophys. J. Lett.* **2023**, *954*, L31. [CrossRef]

130. Baum, W.A. The distribution of luminosity in elliptical galaxies. *Publ. Astron. Soc. Pac.* **1955**, *67*, 328. [CrossRef]

131. De Vaucouleurs, G. The brightest star clusters in galaxies as distance indicators. *Astrophys. J.* **1970**, *159*, 435. [CrossRef]

132. Harris, W.E.; Racine, R. Globular clusters in galaxies. *Annu. Rev. Astron. Astrophys.* **1979**, *17*, 241–274. [CrossRef]
133. Hanes, D.A. Globular clusters and the Virgo cluster distance modulus. *Mon. Not. R. Astron. Soc.* **1977**, *180*, 309–321. [CrossRef]
134. Hertzsprung, E. Über die räumliche Verteilung der Veränderlichen vom delta Cephei-Typus. *Astron. Nachrichten* **1913**, *196*, 201.
135. Shapley, H. Studies based on the colors and magnitudes in stellar clusters. VII. The distances distribution in space, and dimensions of 69 globular clusters. *Astrophys. J.* **1918**, *48*, 154–181. [CrossRef]
136. Kraft, R.P. Color Excesses for Supergiants and Classical Cepheids. V. The Period-Color and Period-Luminosity Relations: A Revision. *Astrophys. J.* **1961**, *134*, 616. [CrossRef]
137. Sandage, A.; Tammann, G.A. A composite period-luminosity relation for Cepheids at mean and maximum light. *Astrophys. J.* **1968**, *151*, 531. [CrossRef]
138. Feast, M.W.; Walker, A.R. Cepheids as distance indicators. *Annu. Rev. Astron. Astrophys.* **1987**, *25*, 345–375. [CrossRef]
139. Madore, B.F.; Freedman, W.L. The Cepheid distance scale. *Publ. Astron. Soc. Pac.* **1991**, *103*, 933. [CrossRef]
140. Tanvir, N.R. Cepheids as distance indicators. The Extragalactic Distance Scale. In Proceedings of the ST ScI May Symposium, Baltimore, MD, USA, 7–10 May 1996; Livio, M., Donahue, M., Panagia, N., Eds.; Cambridge University Press: Cambridge, UK, 1997; p. 91.
141. Feast, M.W.; Catchpole, R.M. The Cepheid period-luminosity zero-point from HIPPARCOS trigonometrical parallaxes. *Mon. Not. R. Astron. Soc.* **1997**, *286*, L1–L5. [CrossRef]
142. Wisniewski, W.Z.; Johnson, H.L. UBVRIJKL Light Curves of Classical Cepheids. *Commun. Lunar Planet Lab.* **1968**, *7*, 57.
143. Freedman, W.L.; Madore, B.F. The Cepheid Extragalactic Distance Scale: Past, Present and Future. Invited Review for IAU Symposium 376. *arXiv* **2023**, arXiv:2308.02474.
144. Hubble, E.P. *Realm of the Nebulae*; Yale University Press: New Haven, CT, USA, 1936; pp. 154–155.
145. Osterbrock, D.E. Who Really Coined the Word Supernova? Who First Predicted Neutron Stars? *Am. Astron. Soc. Meet. Abstr.* **2001**, *199*, 15.
146. Woosley, S.E.; Weaver, T.A. The physics of supernova explosions. *Annu. Rev. Astron. Astrophys.* **1986**, *24*, 205–253. [CrossRef]
147. Filippenko, A.V. Optical spectra of supernovae. *Annu. Rev. Astron. Astrophys.* **1997**, *35*, 309–355. [CrossRef]
148. Branch, D.; Tammann, G.A. Type ia supernovae as standard candles. *Annu. Rev. Astron. Astrophys.* **1992**, *30*, 359–389. [CrossRef]
149. Cadonau, R.; Leibundgut, B. Supernova studies. I. A catalogue of magnitude observations of supernovae I. *Astron. Astrophys. Suppl. Ser.* **1990**, *82*, 145–178.
150. Kowal, C.T. Absolute magnitudes of supernovae. *Astron. J.* **1968**, *73*, 1021–1024. [CrossRef]
151. Baade, W. The Absolute Photographic Magnitude of Supernovae. *Astrophys. J.* **1938**, *88*, 285. [CrossRef]
152. Maury, A.; Thouvenot, E.; Buil, C.; Brunetto, L.; Albanese, D.; Prat, G.; Cappellaro, E.; Benetti, S.; Turatto, M.; Kirshner, R.; et al. Supernova 1990N in NGC 4639. *Int. Astron. Union Circ.* **1990**, *5039*, 1.
153. Leibundgut, B.; Kirshner, R.P.; Filippenko, A.V.; Shields, J.C.; Foltz, C.B.; Phillips, M.M.; Sonneborn, G. Premaximum observations of the Type Ia SN 1990N. *Astrophys. J.* **1991**, *371*, L23–L26. [CrossRef]
154. Sandage, A. Calibration of Nearby Type IA Supernovae as Standard Candles: NGC 4639. *HST Propos.* **1995**, *Cycle 5, ID*, 5981. Available online: https://archive.stsci.edu/proposal_search.php?id=5981&mission=hst (accessed on 15 September 2023).
155. Sandage, A.; Saha, A.; Tammann, G.A.; Labhardt, L.; Panagia, N.; Macchetto, F.D. Cepheid calibration of the peak brightness of Type Ia Supernovae: Calibration of SN 1990N in NGC 4639 averaged with six earlier Type Ia Supernova calibrations to give H0 directly. *Astrophys. J. Lett.* **1996**, *460*, L15. [CrossRef]
156. Hansen, L.; Nørgaard-Nielsen, H.U.; Jørgensen, H.E. Search for supernovae in distant clusters of galaxies. *Messenger* **1987**, *47*, 46–49.
157. Nørgaard-Nielsen, H.U.; Hansen, L.; Jørgensen, H.E.; Salamanca, A.A.; Ellis, R.S.; Couch, W.J. The discovery of a type IA supernova at a redshift of 0.31. *Nature* **1989**, *339*, 523–525. [CrossRef]
158. Leibundgut, B. Supernovae Type-Ia as Standard Candles, Supernovae. In Proceedings of the Tenth Santa Cruz Workshop in Astronomy and Astrophysics, Lick Observatory, Mt Hamilton, CA, USA, 9–21 July 1989; Woosley, S.E., Ed.; Springer-Verlag: New York, NY, USA, 1991; p. 1751.
159. Hamuy, M.; Maza, J.; Phillips, M.M.; Suntzeff, N.B.; Wischnjewsky, M.; Smith, R.C.; Antezana, R.; Wells, L.A.; Gonzalez, L.E.; Gigoux, P.; et al. The 1990 calán/tololo supernova search. *Astron. J.* **1993**, *106*, 2392–2407. [CrossRef]
160. Phillips, M.M. The absolute magnitudes of Type IA supernovae. *Astrophys. J.* **1993**, *413*, L105–L108. [CrossRef]
161. Riess, A.G.; Filippenko, A.V.; Challis, P.; Clocchiatti, A.; Diercks, A.; Garnavich, P.M.; Gilliland, R.L.; Hogan, C.J.; Jha, S.; Kirshner, R.P.; et al. Observational Evidence from Supernovae for an Accelerating Universe and a Cosmological Constant. *Astron. J.* **1998**, *116*, 1009–1038. [CrossRef]
162. Perlmutter, S.; Aldering, G.; Goldhaber, G.; Knop, R.A.; Nugent, P.; Castro, P.G.; Deustua, S.; Fabbro, S.; Goobar, A.; Groom, D.E.; et al. Measurements of Ω and Λ from 42 High-Redshift Supernovae. *Astrophys. J.* **1999**, *517*, 565–586. [CrossRef]
163. Huang, C.D.; Riess, A.G.; Yuan, W.; Macri, L.M.; Zakamska, N.L.; Casertano, S.; Whitelock, P.A.; Hoffmann, S.L.; Filippenko, A.V.; Scolnic, D. Hubble Space Telescope Observations of Mira Variables in the SN Ia Host NGC 1559: An Alternative Candle to Measure the Hubble Constant. *Astrophys. J.* **2020**, *889*, 5. [CrossRef]
164. Sandage, A. The Ability of the 200-INCH Telescope to Discriminate between Selected World Models. *Astrophys. J.* **1961**, *133*, 355. [CrossRef]

165. Mattig, W. Über den Zusammenhang zwischen Rotverschiebung und scheinbarer Helligkeit. *Astron. Nachrichten* **1958**, *284*, 109. [CrossRef]
166. Refsdal, S.; Bondi, H. The gravitational lens effect. *Mon. Not. R. Astron. Soc.* **1964**, *128*, 295–306. [CrossRef]
167. Refsdal, S. On the possibility of determining Hubble's parameter and the masses of galaxies from the gravitational lens effect. *Mon. Not. R. Astron. Soc.* **1964**, *128*, 307–310. [CrossRef]
168. Shapiro, I.I. Fourth test of general relativity. *Phys. Rev. Lett.* **1964**, *13*, 789. [CrossRef]
169. Schmidt, M. 3C273: A star-like object with large red-shift. *Nature* **1963**, *197*, 1040. [CrossRef]
170. Refsdal, S. On the possibility of testing cosmological theories from the gravitational lens effect. *Mon. Not. R. Astron. Soc.* **1966**, *132*, 101–111. [CrossRef]
171. Refsdal, S.; Rosseland, S. On the possibility of determining the distances and masses of stars from the gravitational lens effect. *Mon. Not. R. Astron. Soc.* **1966**, *134*, 315–319. [CrossRef]
172. Refsdal, S. On the Propagation of light in Universes with inhomogeneous mass distribution. *Astrophys. J.* **1970**, *159*, 35. [CrossRef]
173. Walsh, D.; Carswell, R.F.; Weymann, R.J. 0957+ 561 A, B: Twin quasistellar objects or gravitational lens? *Nature* **1979**, *279*, 381. [CrossRef] [PubMed]
174. Vanderriest, C.; Schneider, J.; Herpe, G.; Chevreton, M.; Moles, M.; Wlerick, G. The value of the time delay Delta t (A, B) for the'double'quasar 0957+ 561 from optical photometric monitoring. *Astron. Astrophys.* **1989**, *215*, 1–13.
175. Schild, R.E.; Cholfin, B. CCD camera brightness monitoring of Q0957+ 561 A, B. *Astrophys. J.* **1986**, *300*, 209–215. [CrossRef]
176. Falco, E.E.; Wambsganss, J.; Schneider, P. The role of microlensing in estimates of the relative time delay for the gravitational images of Q0957+ 561. *Mon. Not. R. Astron. Soc.* **1991**, *251*, 698–706. [CrossRef]
177. Falco, E.E.; Gorenstein, M.V.; Shapiro, I.I. On model-dependent bounds on h0 from gravitational images: Application to Q0957+561A,B. *Astrophys. J.* **1985**, *289*, L1–L4. [CrossRef]
178. Gorenstein, M.V.; Falco, E.E.; Shapiro, I.I. Degeneracies in parameter estimates for models of gravitational lens systems. *Astrophys. J.* **1988**, *327*, 693–711. [CrossRef]
179. Gorenstein, M.V.; Cohen, N.L.; Shapiro, I.I.; Rogers, A.E.; Bonometti, R.J.; Falco, E.E.; Bartel, N.; Marcaide, J.M. VLBI observations of the gravitational lens system 0957 + 561: Structure and relative magnification of the A and B images. *Astrophys. J.* **1988**, *334*, 42–58. [CrossRef]
180. Falco, E.E.; Gorenstein, M.V.; Shapiro, I.I. New model for the 0957+561 gravitational lens system: Bounds on masses of a possible black hole and dark matter and prospects for estimation of H0. *Astrophys. J.* **1991**, *372*, 364–379. [CrossRef]
181. Narayan, R. Gravitational lensing, time delay, and angular diameter distance. *Astrophys. J.* **1991**, *378*, L5–L9. [CrossRef]
182. Roberts, D.; Lehár, J.; Hewrtt, J.; Burke, B.F. The Hubble constant from VLA measurement of the time delay in the double quasar 0957+561. *Nature* **1991**, *352*, 43–45. [CrossRef]
183. Rhee, G. An estimate of the Hubble constant from the gravitational lensing of quasar Q0957+561. *Nature* **1991**, *350*, 211–212. [CrossRef]
184. Kundic, T.; Colley, W.N.; Gott, J.R.; Malhotra, S.; Pen, U.L.; Rhoads, J.E.; Stanek, K.Z.; Turner, E.L. An event in the light curve of 0957+561A and prediction of the 1996 image B light curve. *Astrophys. J.* **1995**, *455*, L5–L8. [CrossRef]
185. Wagner, J. Generalised model-independent characterisation of strong gravitational lenses IV: Formalism-intrinsic degeneracies. *Astron. Astrophys.* **2018**, *620*, A86. [CrossRef]
186. Wagner, J. Generalised model-independent characterisation of strong gravitational lenses VI: The origin of the formalism intrinsic degeneracies and their influence on H0. *Mon. Not. R. Astron. Soc.* **2019**, *487*, 4492–4503. [CrossRef]
187. Wagner, J.; (Bahamas Advanced Study Institute & Conferences, Stella Maris, Long Island, The Bahamas). Private communication, 2023.
188. Miyoshi, M.; Moran, J.; Herrnstein, J.; Greenhill, L.; Nakai, N.; Diamond, P.; Inoue, M. Evidence for a black hole from high rotation velocities in a sub-parsec region of NGC4258. *Nature* **1995**, *373*, 127–129. [CrossRef]
189. Herrnstein, J.R.; Moran, J.M.; Greenhill, L.J.; Diamond, P.J.; Inoue, M.; Nakai, N.; Miyoshi, M.; Henkel, C.; Riess, A. A geometric distance to the galaxy NGC4258 from orbital motions in a nuclear gas disk. *Nature* **1999**, *400*, 539–541. [CrossRef]
190. Humphreys, E.M.L.; Reid, M.J.; Moran, J.M.; Greenhill, L.J.; Argon, A.L. Toward a new geometric distance to the active galaxy NGC 4258. III. Final results and the Hubble constant. *Astrophys. J.* **2013**, *775*, 13. [CrossRef]
191. Riess, A.G.; Macri, L.M.; Hoffmann, S.L.; Scolnic, D.; Casertano, S.; Filippenko, A.V.; Tucker, B.E.; Reid, M.J.; Jones, D.O.; Silverman, J.M.; et al. A 2.4% determination of the local value of the Hubble constant. *Astrophys. J.* **2016**, *826*, 56. [CrossRef]
192. Reid, M.J.; Pesce, D.W.; Riess, A.G. An improved distance to NGC 4258 and its implications for the Hubble constant. *Astrophys. J. Lett.* **2019**, *886*, L27. [CrossRef]
193. The Megamaser Cosmology Project (MCP). Available online: https://safe.nrao.edu/wiki/bin/view/Main/MegamaserCosmol ogyProject (accessed on 15 July 2023).
194. Pesce, D.W.; Braatz, J.A.; Reid, M.J.; Riess, A.G.; Scolnic, D.; Condon, J.J.; Gao, F.; Henkel, C.; Impellizzeri, C.M.V.; Kuo, C.Y.; et al. The Megamaser Cosmology Project. XIII. Combined Hubble Constant Constraints. *Astrophys. J. Lett.* **2020**, *891*, L1. [CrossRef]
195. Sunyaev, R.A.; Zel'dovich, Y.B. The Spectrum of Primordial Radiation, its Distortions and their Significance. *Comments Astrophys. Space Phys.* **1970**, *2*, 66.
196. Sunyaev, R.A.; Zel'dovich, Y.B. The Observations of Relic Radiation as a Test of the Nature of X-Ray Radiation from the Clusters of Galaxies. *Comments Astrophys. Space Phys.* **1972**, *4*, p–173.

197. Measuring the Hubble Constant with the Sunyaev-Zel'dovich Effect. Available online: https://ned.ipac.caltech.edu/level5/March04/Reese/frames.html (accessed on 1 July 2023).
198. Cavaliere, A.; Danese, L.; de Zotti, G. Unborn clusters. *Astrophys. J.* **1977**, *217*, 6–15. [CrossRef]
199. Cowie, L.L.; Perrenod, S.C. The origin and distribution of gas within rich clusters of galaxies - The evolution of cluster X-ray sources over cosmological time scales. *Astrophys. J.* **1978**, *219*, 354. [CrossRef]
200. Gunn, J.E.; Longair, M.S.; Rees, M.J.; Abell, G.O. Observational Cosmology. *Phys. Today* **1979**, *32*, 58–60.
201. Silk, J.; White, S.D.M. The determination of q0 using X-ray and microwave observation of galaxy clusters. *Astrophys. J.* **1978**, *226*, L103.
202. Cavaliere, A.; Danese, L.; de Zotti, G. Cosmic distances from X-ray and microwave observations of clusters of galaxies. *Astron. Astrophys.* **1979**, *75*, 322.
203. Birkinshaw, M. Limits to the value of the Hubble constant deduced from observations of clusters of galaxies. *Mon. Not. R. Astron. Soc.* **1979**, *187*, 847–862. [CrossRef]
204. Birkinshaw, M.; Hughes, J.P. A Measurement of the Hubble Constant from the X-ray Properties and the Sunyaev-Zel'dovich effect of Abell 2218. *Astrophys. J.* **1994**, *420*, 33. [CrossRef]
205. Reese, E.D.; Mohr, J.J.; Carlstrom, J.E.; Joy, M.; Grego, L.; Holder, G.P.; Holzapfel, W.L.; Hughes, J.P.; Patel, S.K.; Donahue, M. Sunyaev-Zeldovich Effect-derived Distances to the High-Redshift Clusters MS 0451.6–0305 and Cl 0016+16. *Astrophys. J.* **2000**, *533*, 38. [CrossRef]
206. Reese, E.D.; Carlstrom, J.E.; Joy, M.; Mohr, J.J.; Grego, L.; Holzapfel, W.L. Determining the Cosmic Distance Scale from Interferometric Measurements of the sunyaev-zeldovich effect. *Astrophys. J.* **2002**, *581*, 53–85.
207. Patel, S.K.; Joy, M.; Carlstrom, J.E.; Holder, G.P.; Reese, E.D.; Gomez, P.L.; Hughes, J.P.; Grego, L.; Holzapfel, W.L. The Distance and Mass of the Galaxy Cluster Abell 1995 Derived from Sunyaev-Zeldovich Effect and X-Ray Measurements. *Astrophys. J.* **2000**, *541*, 37.
208. Mason, B.S.; Myers, S.T.; Readhead, A.C.S. A measurement of h0 from the Sunyaev-Zeldovich effect. *Astrophys. J.* **2001**, *555*, L11.
209. Sereno, M. Simultaneous determination of ΩM0 and H_0 from joint Sunyaev-Zeldovich effect and X-ray observations with median statistics. *Astron. Astrophys.* **2003**, *412*, 341.
210. Udomprasert, P.S.; Mason, B.S.; Readhead, A.C.S.; Pearson, T.J. An Unbiased Measurement of H0 through Cosmic Background Imager Observations of the Sunyaev-Zel'dovich Effect in Nearby Galaxy Clusters. *Astrophys. J.* **2004**, *615*, 63–81.
211. Schmidt, R.W.; Allen, S.W.; Fabian, A.C. An improved approach to measuring H0 using X-ray and SZ observations of galaxy clusters. *Mon. Not. R. Astron. Soc.* **2004**, *352*, 1413.
212. Jones, M.E.; Edge, A.C.; Grainge, K.; Grainger, W.F.; Kneissl, R.; Pooley, G.G.; Saunders, R.; Miyoshi, S.J.; Tsuruta, T.; Yamashita, K.; et al. H_0 from an orientation-unbiased sample of Sunyaev–Zel'dovich and X-ray clusters, *Mon. Not. R. Astron. Soc.* **2005**, *357*, 518. [CrossRef]
213. Kozmanyan, A.; Bourdin, H.; Mazzotta, P.; Rasia, E.; Sereno, M. Deriving the Hubble constant using Planck and XMM-Newton observations of galaxy clusters. *Astron. Astrophys.* **2019**, *621*, A34. [CrossRef]
214. Aghanim, N.; Akrami, Y.; Ashdown, M.; Aumont, J.; Baccigalupi, C.; Ballardini, M.; Banday, A.J.; Barreiro, R.B.; Bartolo, N.; Basak, S.; et al. Planck 2018 results VI. Cosmological parameters. *Astron. Astrophys.* **2020**, *641*, A6. [CrossRef]
215. Eisenstein, D.J.; Zehavi, I.; Hogg, D.W.; Scoccimarro, R.; Blanton, M.R.; Nichol, R.C.; Scranton, R.; Seo, H.; Tegmark, M.; Zheng, Z.; et al. Detection of the Baryon Acoustic Peak in the Large-Scale Correlation Function of SDSS Luminous Red Galaxies. *Astrophys. J.* **2005**, *633*, 560–574. [CrossRef]
216. Cole, S.; Percival, W.J.; Peacock, J.A.; Norberg, P.; Baugh, C.M.; Frenk, C.S.; Baldry, I.; Bland-Hawthorn, J.; Bridges, T.; Cannon, R.; et al. The 2dF Galaxy Redshift Survey: Power-spectrum analysis of the final data set and cosmological implications. *Mon. Not. R. Astron. Soc.* **2005**, *362*, 505–534. [CrossRef]
217. Wang, Y.; Zhao, G.-B.; Zhao, C.; E Philcox, O.H.; Alam, S.; Tamone, A.; de Mattia, A.; Ross, A.J.; Raichoor, A.; Burtin, E.; et al. The clustering of the SDSS-IV extended baryon oscillation spectroscopic survey DR16 luminous red galaxy and emission-line galaxy samples: Cosmic distance and structure growth measurements using multiple tracers in configuration space. *Mon. Not. R. Astron. Soc.* **2020**, *498*, 3470–3483. [CrossRef]
218. Holz, D.E.; Hughes, S.A. Using Gravitational-Wave Standard Sirens. *Astrophys. J.* **2005**, *629*, 15. [CrossRef]
219. Schutz, B.F. Determining the Hubble constant from gravitational wave observations. *Nature* **1986**, *323*, 310–311. [CrossRef]
220. The LIGO Scientific Collaboration and The Virgo Collaboration; Abbott, B.P.; Abbott, R.; Acernese, F.; Ackley, K.; Adams, C.; Adams, T.; Addesso, P.; Adhikari, R.X.; Adya, V.B.; et al. A gravitational-wave standard siren measurement of the Hubble constant. *Nature* **2017**, *551*, 85–88.
221. Fung, L.W.H.; Broadhurst, T.; Smoot, G.F. Pure Gravitational Wave Estimation of Hubble's Constant using Neutron Star-Black Hole Mergers. *arXiv* **2023**, arXiv:2308.02440v1.
222. Tsupko, O.Y.; Fan, Z.; Bisnovatyi-Kogan, G.S. Black hole shadow as a standard ruler in cosmology. *Class. Quantum Grav.* **2020**, *37*, 065016. [CrossRef]
223. Cervantes-Cota, J.L.; Galindo-Uribarri, S.; Smoot, G.F. The Legacy of Einstein's Eclipse, Gravitational Lensing. *Universe* **2020**, *6*, 9. [CrossRef]
224. Bisnovatyi-Kogan, G.S.; Tsupko, O.Y. Shadow of a black hole at cosmological distances. *Phys. Rev. D* **2018**, *98*, 084020. [CrossRef]
225. Event Horizon Telescope. Available online: https://eventhorizontelescope.org (accessed on 1 July 2022).
226. Qi, J.-Z.; Zhang, X. A new cosmological probe using super-massive black hole shadows. *Chin. Phys. C* **2020**, *44*, 055101. [CrossRef]

227. Renzi, F.; Martinelli, M. Climbing out of the shadows: Building the distance ladder with black hole images. *Phys. Dark Univ.* **2022**, *37*, 101104. [CrossRef]

228. Escamilla-Rivera, C.; Castillejos, R.T. H0 Tension on the Light of Supermassive Black Hole Shadows Data. *Universe* **2023**, *9*, 14. [CrossRef]

229. Vagnozzi, S.; Bambi, C.; Visinelli, L. Concerns regarding the use of black hole shadows as standard rulers. *Class. Quant. Grav.* **2020**, *37*, 087001. [CrossRef]

230. Odintsov, S.D.; Oikonomou, V.K. Did the Universe experience a pressure non-crushing type cosmological singularity in the recent past? *EPL* **2022**, *137*, 39001. [CrossRef]

231. Odintsov, S.D.; Oikonomou, V.K. Dissimilar donuts in the sky? Effects of a pressure singularity on the circular photon orbits and shadow of a cosmological black hole. *EPL* **2022**, *139*, 59003. [CrossRef]

232. James, C.W.; Prochaska, J.X.; Macquart, J.P.; North-Hickey, F.O.; Bannister, K.W.; Dunning, A. The z–DM distribution of fast radio bursts. *Mon. Not. R. Astron. Soc.* **2022**, *509*, 4775–4802. [CrossRef]

233. Hagstotz, S.; Reischke, R.; Lilow, R. A new measurement of the Hubble constant using fast radio bursts. *Mon. Not. R. Astron. Soc.* **2022**, *511*, 662–667. [CrossRef]

234. Wu, Q.; Zhang, G.-Q.; Wang, F.-Y. An 8 per cent determination of the Hubble constant from localized fast radio bursts. *Mon. Not. R. Astron. Soc. Lett.* **2022**, *515*, L1–L5. [CrossRef]

235. James, C.W.; Ghosh, E.M.; Prochaska, J.X.; Bannister, K.W.; Bhandari, S.; Day, C.K.; Deller, A.T.; Glowacki, M.; Gordon, A.C.; E Heintz, K.; et al. A measurement of Hubble's Constant using Fast Radio Bursts. *Mon. Not. R. Astron. Soc.* **2022**, *516*, 4862–4881. [CrossRef]

236. CHIME/FRB. Available online: https://www.chime-frb.ca/catalog (accessed on 1 September 2023).

237. Freedman, W.L.; Madore, B.F.; Gibson, B.K.; Ferrarese, L.; Kelson, D.D.; Sakai, S.; Mould, J.R.; Kennicutt, J.R.C.; Ford, H.C.; Graham, J.A.; et al. Final Results from the Hubble Space Telescope Key Project to Measure the Hubble Constant. *Astrophys. J.* **2001**, *553*, 47–72. [CrossRef]

238. Riess, A.G.; Anand, G.S.; Yuan, W.; Casertano, S.; Dolphin, A.; Macri, L.M.; Breuval, L.; Scolnic, D.; Perrin, M.; Anderson, R.I. Crowded No More: The Accuracy of the Hubble Constant Tested with High Resolution Observations of Cepheids by JWST. *arXiv* **2023**, arXiv:2307.15806. [CrossRef]

239. Wong, K.C.; Suyu, S.H.; Chen, G.C.-F.; E Rusu, C.; Millon, M.; Sluse, D.; Bonvin, V.; Fassnacht, C.D.; Taubenberger, S.; Auger, M.W.; et al. H0LiCOW—XIII. A 2.4 per cent measurement of H0 from lensed quasars: 5.3σ tension between early- and late-Universe probes. *Mon. Not. R. Astron. Soc.* **2020**, *498*, 1420–1439. [CrossRef]

240. Riess, A.G.; Casertano, S.; Yuan, W.; Macri, L.; Anderson, J.; MacKenty, J.W.; Bowers, J.B.; Clubb, K.I.; Filippenko, A.V.; Jones, D.O.; et al. New Parallaxes of Galactic Cepheids from Spatially Scanning the Hubble Space Telescope: Implications for the Hubble Constant. *Astrophys. J.* **2018**, *855*, 136. [CrossRef]

241. Freedman, W. Measurements of the Hubble Constant: Tensions in Perspective. *Astrophys. J.* **2021**, *919*, 16. [CrossRef]

242. Verde, L.; Treu, T.; Riess, A.G. Tensions between the Early and the Late Universe. *Nat. Astron.* **2019**, *3*, 891–895. [CrossRef]

243. Verde, L.; Schöneberg, N.; Gil-Marín, H. A tale of many H0. *arXiv* **2023**, arXiv:2311.13305.

244. H0LiCOW-public/H0_tension_plots. Available online: https://github.com/shsuyu/H0LiCOW-public/tree/master/H0_tension_plots (accessed on 15 September 2023).

245. Di Valentino, E.; Mena, O.; Pan, S.; Visinelli, L.; Yang, W.; Melchiorri, A.; Mota, D.F.; Riess, A.G.; Silk, J. In the Realm of the Hubble tension—A Review of Solutions. *Class. Quant. Grav.* **2021**, *38*, 153001. [CrossRef]

246. Abdalla, E.; Abellán, G.F.; Aboubrahim, A.; Agnello, A.; Akarsu, Ö.; Akrami, Y.; Alestas, G.; Aloni, D.; Amendola, L.; Anchordoqui, L.A.; et al. Cosmology Intertwined: A Review of the Particle Physics, Astrophysics, and Cosmology Associated with the Cosmological Tensions and Anomalies. *J. High Energy Astrophys.* **2022**, *34*, 49–211. [CrossRef]

247. Vagnozzi, S. Seven Hints That Early-Time New Physics Alone Is Not Sufficient to Solve the Hubble Tension. *Universe* **2023**, *9*, 393. [CrossRef]

248. Krishnan, C.; Colgáin, E.; Ruchika, A.A.S.; Sheikh-Jabbari, M.M.; Yang, T. Is there an early Universe solution to Hubble tension? *Phys. Rev. D* **2020**, *102*, 103525. [CrossRef]

249. Colgáin, E.Ó.; Sheikh-Jabbari, M.M.; Solomon, R.; Dainotti, M.G.; Stojkovic, D. Putting Flat ΛCDM In The (Redshift) Bin. *arXiv* **2022**, arXiv:2206.11447.

MDPI AG

Grosspeteranlage 5

4052 Basel

Switzerland

Tel.: +41 61 683 77 34

Universe Editorial Office

E-mail: universe@mdpi.com

www.mdpi.com/journal/universe